DIFFERENTIAL EQUATIONS FOR MATHEMATICS, SCIENCE, AND ENGINEERING

DIFFERENTIAL EQUATIONS FOR MATHEMATICS, SCIENCE, AND ENGINEERING

PAUL W. DAVIS

Worcester Polytechnic Institute

Prentice Hall, *Englewood Cliffs, New Jersey 07632*

Library of Congress Cataloging-in-Publication Data

Davis, Paul W., (date)
  Differential equations for mathematics, science,
and engineering / by Paul W. Davis
    p.   cm.
  Includes bibliographical references and index.
  ISBN 0-13-211236-1
  1. Differential equations.   I. Title.
QA372.D29   1991   91-25604
515'.35–dc20   CIP

QA
372
.D34
1992

Acquisition Editor: Steve Conmy
Editorial/Production Supervision and Interior Design: Judi Wisotsky
Editorial Assistant: Joan Dello Stritto
Copy Editor: Linda Thompson
Cover Design: Butler/Udell
Prepress Buyer: Paula Massenaro
Manufacturing Buyer: Lori Bulwin

Credits for chapter opening quotes:
Chapter 1: YA GOT TROUBLE from ''The Music Man''
by Meredith Willson
© 1957, 1958, 1966 FRANK MUSIC CORP. and MEREDITH WILLSON MUSIC
© Renewed 1985, 1986 FRANK MUSIC CORP. and MEREDITH WILLSON MUSIC
International Copyright Secured   All Rights Reserved   Used By Permission
Chapters 3 and 10 are reprinted by permission from *Peter's Quotations: Ideas
for Our Times* by Laurence J. Peter, William Morrow and Co., Inc., 1977.
Chapter 5: *Whole Lotta Shakin' Goin' On* published by Whole Lotta Shakin'
Music.
Chapter 8: from *In the Outlaw Area* by Calvin Tomkins, *The New Yorker,*
January 8, 1966, p. 54.

Printed in the United States of America
10 9 8 7 6 5 4 3 2 1

ISBN 0-13-211236-1

Prentice-Hall International (UK) Limited, *London*
Prentice-Hall of Australia Pty. Limited, *Sydney*
Prentice-Hall Canada Inc., *Toronto*
Prentice-Hall Hispanoamericana, S.A., *Mexico*
Prentice-Hall of India Private Limited, *New Delhi*
Prentice-Hall of Japan, Inc., *Tokyo*
Simon & Schuster Asia Pte. Ltd., *Singapore*
Editora Prentice-Hall do Brasil, Ltda., *Rio de Janeiro*

24141839

To Sharon

CONTENTS

LIST OF FIGURES

LIST OF TABLES

PREFACE

TO THE READER

This text offers a bridge between an important branch of mathematics and your experiences in other areas of science and engineering. Constructing this bridge demands an understanding of mathematical ideas and techniques as well as a mastery of a variety of physical concepts.

Few of the problems you encounter in professional practice will be neatly labeled "Use mathematics on this one" or "Physical intuition will tell you all you need here". Instead, you must make your own judgments about the characteristics of a satisfactory solution and the means to obtain it.

To provide such experiences here, many exercises ask for your opinion or for an explanation. You may be more accustomed to a straight-forward request for the mechanics of some mathematical manipulation. Even if you feel uncomfortable with such vague questions, do not let that discomfort prevent your attacking the problem.

Begin by forming an idea of what might constitute a satisfactory answer. What information might lead toward that answer? Where might you find such information?

If physical intuition suggests an answer, try to confirm it mathematically. Interpret mathematical answers in light of the original physical problem to be sure they are sensible. Ask yourself whether unreasonable mathematical answers are consequences of faulty mathematics or signs of poor physical assumptions in the construction of the problem.

The art of blending physical and mathematical reasoning is not acquired quickly. It is certainly much more subtle than the mechanical manipulations you may have mastered so quickly in other areas of study. To paraphrase one experienced applied mathematician, your ability to think for yourself is your best guarantee of success.

Important terms are shown in **bold face** when they are first defined. Particularly important ideas are given formal definitions. Use the explanatory material that accompanies these definitions to form an intuitive basis for each concept.

Figures, tables, examples, and referenced equations are numbered consecutively within each chapter. For example, equation (6.4) is the fourth numbered equation of chapter 6. Citations to the bibliography at the end of the text are enclosed in square brackets; e.g., [4].

The exercises for each section contain an exercise guide, a table listing the exercises which use particular ideas developed in that section. Use it to select exercises to practice the topics that trouble you most. Review the topics listed in the exercise guide to check your mastery of the concepts of that section.

Summaries appear at key points throughout the text. Read them carefully to be sure they agree with your understanding of the material. Consult them when you need help separating the forest from the trees.

Notes in smaller type are scattered throughout the text to provide additional information. Skip them on your first reading, but return to them later.

Do not try to read for complete understanding in one sitting. Reread the material repeatedly, looking for deeper understanding each time. Pass over details at first. Take assertions for granted until the ideas fall into place. Keep a pencil handy to make notes of your own ideas.

PATTERNS

Part of the power and beauty of mathematics is its ability to capture common patterns among apparently disparate ideas. Identifying those patterns is part of productive mathematical thinking. The more similarities you see, the less you need to remember. To guide you, marginal notes highlight these patterns.

Some of the patterns that appear in this text are

- Models are derived by coupling a physical law (perhaps Newton's law $F = ma$ or a conservation law) with experimental observations, perhaps a constitutive law like Hooke's law $F = kz$. Identify those key components in each model you study.
- Definitions of basic terms like *solution, homogeneous, linear,* etc. are presented successively for first-order equations, second-order equations, and systems. Find the similarities among these situations to see the pattern of generalization.
- Most solution method are introduced in a simple first-order setting, then generalized to higher-order equations and systems of first-order equations. Look for the underlying pattern so that you can use what you already know in the new situation.

Exercises and chapter projects ask you to extend these patterns to new settings, giving you the opportunity to develop your skills of problem solving and generalization.

ACKNOWLEDGMENTS

I own an enormous debt of gratitude to my family, friends, colleagues, and students for their ideas and encouragement throughout the writing of this book. At the risk of omitting some, I must especially thank John M. Boyd, Vincent F. Connolly, Mayer Humi, Roger Y.-M. Lui, William B. Miller, Janos Turi, Samuel M. Rankin, and David A. Torrey. Pat Brown and Paul Christodoulides prepared the solutions to the exercises. Bob Sickles, Steve Conmy, Judi Wisotsky and their associates at Prentice Hall have played key roles.

I am grateful to the reviewers who generously shared their ideas and suggestions: Merle D. Roach (University of Alabama, Huntsville), Robert Broschat (South Dakota State University), Frank M. Cholewinski (Clemson University), Allan M. Krall (Pennsylvania State University), J. M. Cushing (University of Arizona), Harold T. Jones (Andrews University), Thomas G. Kudzma (University of Lowell), and Terry L. Herdman (Virginia Polytechnic Institute and State University). Their assistance has been significant.

Most important, Sharon, Aaron, and Alicia have been good sports and my inspiration all along.

Paul Davis

DIFFERENTIAL EQUATIONS FOR MATHEMATICS, SCIENCE, AND ENGINEERING

1

PROLOGUE

It takes judgment, brains, and maturity . . .
Professor Harold Hill in Meredith Willson's *The Music Man*

This prologue introduces the amalgam of mathematical and physical ideas in which differential equations are embedded. The themes and ideas introduced here will be elaborated upon in subsequent chapters to help your judgment, mathematical maturity, and problem solving skills grow.

1.1 GOALS

Differential equations constitute a language that scientists and engineers use to make careful mathematical statements about the problems they confront. This text is an introduction to that language which emphasizes understanding rather than the formalities of grammar.

We discuss three major issues:

1. *Formulating a model* using differential equations;
2. *Analyzing the model*, both by solving the differential equation and by extracting qualitative information about the solution from the differential equation;
3. *Interpreting the analysis* in light of the physical setting we modeled in step 1.

This text aims to develop skill with solution techniques as well as judgment and sensitivity in using differential equations to understand complex physical phenomena.

The balance of the quotation at the head of this chapter is "... to line up a three-rail billiard shot." The logic of this shot uses a simple model, and judgment corrects for spin. Your judgment and maturity in mathematics will grow with experience, too.

1.2   A MODELING EXAMPLE

A question to which modeling may contribute some answers is:

> What parameters affect the acceleration, velocity, and displacement of a rock thrown vertically upward?

This simple question may not have a simple answer. Does the weight of the rock matter? Gravity is obviously important, but what about wind and air resistance? Does the shape of the rock matter? The speed with which the rock leaves our hand ought to affect how high it goes. Could a long wind-up or a sharp whack with a bat force it higher, even though its starting speed is the same?

A mathematical model allows us to make a list of just which effects we will consider, and it permits us to state precisely the ways in which they act. We can simplify the problem by omitting the effects we do not understand, can not describe clearly, or think do not matter much. The model itself will be a set of mathematical relations involving quantities we know and quantities we do not know, such as the velocity of the rock as a function of time.

Intuition suggests that gravity should be the predominant factor affecting the rock's motion, so we will try to include it and omit everything else, such as wind, the shape of the rock, etc. If we are wrong, the error should appear in one of two ways: The model will not be complete without adding more information, or the model will predict something that counters our physical experience.

To derive a model in this setting, we have two tools at our disposal, a physical law and an experimental fact.

> **Newton's Law** The total force F_t on a body of mass m experiencing acceleration a is the product of mass and acceleration: $F_t = ma$.
>
> **Experimental Fact** The force of gravity on a body is proportional to its mass.

>> Both *law* and *fact* are misused. A *law* is usually a postulate that has repeatedly led to mathematical models predicting physically reasonable behavior. A *fact* is a summary of physical observations that are sufficiently accurate for the purpose at hand.

Newton's Law is given in an unambiguous mathematical statement, but the experimental fact is not. To restate the latter in mathematical terms, we need notation and agreement on a coordinate system to determine the signs of forces and the like. Borrowing the notation of Newton's law, taking up as the positive direction, and letting g be the (positive) constant of proportionality, the experimental fact can be written

$$F_g = -mg \,,$$

where F_g is the force due to gravity acting on a body of mass m. The minus sign enters because of the up-is-positive orientation we have chosen; see figure 1.1.

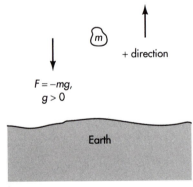

Figure 1.1: A coordinate direction for the rock model. This orientation means positive velocities are upwards, negative are downwards.

Neglecting air resistance and other troublesome effects means that the force of gravity is the only force acting on the rock. Hence, the total force on the rock is just the force of gravity; i.e., $F_t = F_g$ or, when m is canceled,

$$a = -g. \tag{1.1}$$

Equation (1.1) is a model for the motion of a rock thrown vertically upwards subject only to the influence of gravity.

The mass m of the rock appears nowhere in (1.1). Therefore, our model predicts that the motion of the rock is independent of its mass.

Does our physical experience suggest that the motion of the rock is independent of its mass? Can't a marble be thrown higher than a bowling ball? Is it the mass of the rock that matters, then, or the speed with which you release it? Or is it the size of the rock, a factor that the model (1.1) has neglected?

That wave of questions certainly raises doubts about the validity of the model, but there is a clue that mass may not matter in Galileo's legendary experiments in Pisa: Mass did not affect the speed of an object's fall. So it may not be unreasonable to suggest that the motion of the rock, at least in its fall back to earth, is independent of its mass.

> Galileo actually rolled brass spheres down an inclined plane, timing them with a water clock.

Let's recast this model in search of more information about the motion of the rock. Since acceleration is the derivative of velocity with respect to time, the equation $a = -g$ can also be written

$$\frac{dv}{dt} = -g, \tag{1.2}$$

a model that now incorporates velocity.

Analysis of a model seeks to understand the qualitative relations among the variables appearing in it. Those relations are often best expressed by simple graphs that capture the principal features of the behavior that concerns us. We will use that approach now and repeatedly in what follows.

Since g is a positive number, the differential equation (trumpets sound!) $dv/dt = -g$ asserts that $dv/dt < 0$; i.e., the slope of the velocity-versus-time curve is always negative. The velocity of the rock is always decreasing.

Moreover, the model $dv/dt = -g$ tells us that the slope dv/dt of the velocity-versus-time curve is actually constant. The velocity-versus-time curve must look like one of the straight lines shown in figure 1.2.

If velocity is decreasing with time, then it must eventually become negative, as the graphs of figure 1.2 illustrate. Recalling our agreement to take the upwards direction as positive, we conclude that negative velocity corresponds to the rock falling.

Knowing that the rock eventually will fall hardly earns us gasps of admiration, but it does point out a flaw in the model. Both graphs of figure 1.2 (and any other lines with negative slopes you care to draw) become more and more negative as time increases. That is, the rock continues to fall faster and faster, forever.

The flaw, of course, is simple. Our model omitted the ground, which stops the fall of the rock. However, our interpretation has identified a clear limitation: The model $dv/dt = -g$ is not valid for all time.

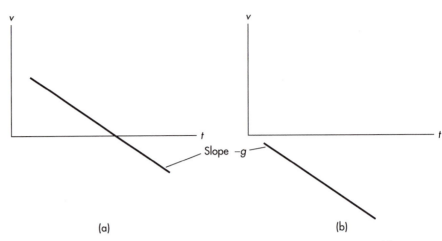

Figure 1.2: Some possible velocity-versus-time curves compatible with model 2.

Some additional thought indicates that figure 1.2(b) cannot be a reasonable graph of velocity versus time, for velocity is never positive there. We need a graph such as that of figure 1.2(a), which includes a period of positive velocity corresponding to throwing the rock up into the air.

But we cannot draw the graph more precisely without more information. We know that the line has slope $-g$. If we had the intercept with either the t-axis or the v-axis, then we could draw the unique line that is the graph of velocity versus time.

These intercepts are physically significant. The t-intercept (horizontal axis intercept) is a point where velocity is momentarily zero, as the rock's velocity changes from positive to negative. Plainly, this transition occurs at the highest point of the rock's trajectory, and the t-intercept must be the time from release to peak. On the other hand, the v-intercept (vertical axis intercept) is the velocity at time $t = 0$, the instant the rock was released.

The velocity at $t = 0$, written $v(0)$, is a quantity we can control directly as we release the rock, whereas the time from release to peak is a bit more remote. It appears then that we ought to specify $v(0)$, the *initial velocity*, as part of the model if we want to obtain an unambiguous graph of velocity versus time.

Prompted by this argument, we propose a new model,

$$\frac{dv}{dt} = -g \,, \tag{1.3}$$

$$v(0) = v_i \,, \tag{1.4}$$

where the constant v_i is the specified initial velocity. The extra data given by (1.4), a specification of the value of the unknown function (or one of its derivatives) at the initial time, is called an **initial condition**.

Using the reasoning discussed before, we can draw a unique graph of velocity versus time as shown in figure 1.3. Throwing the rock upward corresponds to $v_i > 0$.

Our ability to draw a unique velocity-versus-time curve suggests that we have a firm grasp on the motion of the rock, at least within the bounds set by the assumptions underlying the model. The model (1.3)–(1.4)

$$\frac{dv}{dt} = -g, \; v(0) = v_i,$$

still does not involve the mass of the rock, but it does indicate that we need an additional piece of data, the initial velocity. Perhaps the motion of a massive rock differs from that of a smaller one because we cannot give the same initial velocity to both. If there is only one velocity-versus-time curve corresponding to the model (1.3)–(1.4), then the motion of a marble and a bowling ball should be identical if we can impart the *same initial velocity* to both.

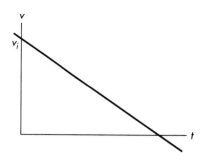

Figure 1.3: The particular velocity-versus-time curve obtained from model 3. The v-intercept is specified by equation (1.4), $v(0) = v_i$.

EXAMPLE 1   *The Guiness Book of World Records claims that Nolan Ryan pitched a ball at 45.1 m/s (100.9 mi/h) while playing for the Houston Astros in 1974. Put Mr. Ryan on his back and implement the model (1.3)–(1.4) in this case.*

His world's record pitch corresponds to an initial velocity of $v_i = 45.1$ m/s. For consistency of units, we must use $g = 9.8$ m/s^2. Hence, the governing model is

$$\frac{dv}{dt} = -9.8\,,$$

$$v(0) = 45.1\,.$$

According to this model, any object that can be given an initial velocity of 45.1 m/s will reach the same height as Nolan Ryan's baseball. Obviously, a bowling ball must be thrown by an arm stronger even than Mr. Ryan's to achieve the same initial velocity.   ■

As you have observed, $dv/dt = -g$ is a differential equation. The preceding analysis suggested ways in which we can extract information from the differential equation without actually writing down its solution. We will learn more of these indirect techniques for extracting qualitative information about solutions, and we will learn techniques for explicitly calculating solutions. We will also devote some attention to questions such as the one suggested in the paragraph preceding the example: How many different solutions, or velocity-versus-time curves, does (1.3)–(1.4) possess? (A sound answer will require a definition of *solution* so that we know what *different* means.)

1.3   DIFFERENTIAL EQUATIONS AND SOLUTIONS

1.3.1   Some Terminology

Both models 2 and 3 involved a *differential equation*, an equation for an unknown function $v(t)$ that involved the derivative of the function. Other modeling processes can lead to differential equations involving higher derivatives of the unknown (see exercise 5) or both the unknown and its derivative (see equation (2.1)).

A **differential equation** is a relation involving one or more derivatives of an unknown function and perhaps the function itself. The unknown function, e.g., $v(t)$ in (1.3),

$$\frac{dv}{dt} = -g,$$

is called the **dependent variable**, and its argument, e.g., t in the last equation, is called the **independent variable**. The order of the highest derivative of the dependent variable appearing in the equation is called the **order** of the differential equation; $v' = -g$ is a *first*-order equation.

EXAMPLE 2   *Identify the dependent variable, the independent variable, and the order of the differential equation*

$$y'' + 2y' + 16 \sin y = -3 \cos \pi t.$$

Since y is differentiated, it is the dependent variable. The order of the differential equation is two because the highest derivative that appears is the second derivative y''.

The independent variable is perhaps less obvious; y'' could denote d^2y/dx^2 or d^2y/ds^2 or d^2y/dr^2 or $\ldots$. But the variable that appears on the right side is t. We conclude that the independent variable must be t.   ■

We call the term g in the differential equation $v' = -g$ a **parameter**. It represents a constant in a given setting (say, 9.8 m/s^2 on the surface of the earth), but it would change if we used different units or moved our rock tossing experiment to another planet.

The initial velocity v_i is also a parameter in the model (1.3)–(1.4),

$$\frac{dv}{dt} = -g, \; v(0) = v_i.$$

It is a constant for a given trial, but we could reasonably ask how the motion of the rock changes as we vary v_i, perhaps by asking someone other than Nolan Ryan to pitch.

Since dv/dt, $v'(t)$, and v' are all symbols for the derivative of v with respect to t, the following are equivalent forms of the same differential equation:

$$\frac{dv}{dt} = -g,$$

$$v'(t) = -g,$$

$$v' = -g.$$

Of course, other letters can represent the same variables. The differential equation

$$y'(x) = k,$$

where k is a constant, is a mathematical problem that is equivalent to those listed in the preceding paragraph. We can identify y with v, x with t, and k with $-g$. (The context will tell us that v' means dv/dt, while y' means dy/dx.)

A **solution of a differential equation** is a sufficiently differentiable function that renders the relation defining the equation an identity when substituted for the unknown. For example, some solutions of $v' = -g$ are

$$v(t) = -gt + 2,$$

$$v(t) = -gt - 6.3,$$

$$v(t) = -gt + C, \ C \text{ any constant,} \tag{1.5}$$

as you can verify by differentiating each expression for $v(t)$ and substituting in the equation $v' = -g$. *Sufficiently differentiable* normally means that the derivative of the solution of the highest order appearing in the equation is piecewise continuous. Unless it is obvious from the context, we must also state the interval of the independent variable for which the solution is valid.

You can *always* verify a solution by substitution.

More generally, the generic first-order differential equation is often written

$$y' = f(x, y);$$

e.g., we could write the rock tossing equation as $v' = f(t, v)$ where f is the constant function $f(t, v) = -g$. The function $u(x)$ is a solution of $y' = f(x, y)$ for $a \leq x \leq b$ if u is differentiable on $[a, b]$ and $u'(x) = f(x, u(x))$, $a \leq x \leq b$.

How did we obtain these solutions? The equation $v' = -g$ just presents an antidifferentiation problem whose solution is (1.5), $v(t) = -gt + C$. (Unfortunately, other differential equations can require more complex methods.)

When can we determine the arbitrary constant C in a solution such as $v = -gt + C$? When we are given additional information such as the initial velocity condition $v(0) = v_i$. Setting $t = 0$ in the solution $v(t) = -gt + C$ yields

$$v(0) = C,$$

while the initial condition requires

$$v(0) = v_i.$$

Hence, $C = v_i$, and a solution of the model $v' = -g$, $v(0) = v_i$, is

$$v(t) = -gt + v_i. \tag{1.6}$$

This solution is just the equation of the straight line velocity-versus-time graph shown in figure 1.3, a graph we produced without actually solving the initial-value problem $v' = -g$, $v(0) = v_i$. Using $v(0) = v_i$ to find C is the same as using that initial condition to determine which of the lines in figure 1.2 is the actual graph of the rock's velocity.

Note the pattern: The solution method gives C; the extra data $v(0) = v_i$ determines C.

Extra information such as $v(0) = v_i$ is called an **initial condition**. A problem such as

$$v' = -g, \ v(0) = v_i,$$

which couples the auxiliary data $v(0) = v_i$ with a differential equation, is called an **initial-value problem** if all the data is given at one value of the independent variable. The reason for the name *initial value* is obvious in this model, where the data are specified at $t = 0$.

1.3 Differential Equations and Solutions

7

Initial conditions specify the value of the dependent variable and perhaps one or more of its derivatives at a fixed value of the independent variable. These values are called **initial values**. The number of initial values is equal to the order of the differential equation, and the highest derivative whose initial value is specified is one less than the order of the differential equation.

A **solution of an initial-value problem** is a solution of the differential equation that satisfies the given initial condition.

EXAMPLE 3 *Solve the initial-value problem of example 1, which models Nolan Ryan's vertical pitch,*

$$v' = -9.8 \text{ m/s}^2, \tag{1.7}$$

$$v(0) = 45.1 \text{ m/s}. \tag{1.8}$$

Verify that the function obtained is indeed a solution of this initial-value problem.

We can obtain a solution of the differential equation (1.7) by antidifferentiation, just as we obtained the solution (1.5):

$$v(t) = -9.8t + C.$$

The initial condition (1.8) leads to

$$v(0) = C = 45.1 \text{ m/s}.$$

Hence, the solution of the initial-value problem (1.7)–(1.8) is

$$v(t) = 9.8t + 45.1.$$

This function is a solution of the initial-value problem (1.7)–(1.8) because it satisfies the differential equation (1.7),

$$\frac{dv}{dt} = \frac{d(-9.8t + 45.1)}{dt} = -9.8,$$

as well as the initial condition (1.8),

$$v(0) = -9.8 \cdot 0 + 45.1 = 45.1. \qquad \blacksquare$$

Finding the solution of a differential equation amounts to finding a formula for the unknown function. The analysis and interpretation stages of the modeling process demand an understanding of the qualitative behavior of the solution. We have seen that much of that qualitative behavior can be deduced without knowing a solution formula. Consequently, finding a solution is only one part of using the language of differential equations productively.

1.3.2 More Examples

EXAMPLE 4 *Solve the initial-value problem*

$$\frac{dy}{dx} = xe^{x^2}, \ y(0) = 2.$$

Since the unknown function does not appear on the right side of the differential equation, we can use antidifferentiation to obtain

$$y(x) = \int xe^{x^2}\,dx + C = \frac{1}{2}e^{x^2} + C.$$

To evaluate the arbitrary constant C, apply the initial condition:

$$y(0) = \frac{1}{2} + C = 2,$$

$$C = \frac{3}{2}.$$

Hence, the solution of the initial-value problem is

$$y(x) = \frac{1}{2}e^{x^2} + \frac{3}{2}. \qquad \blacksquare$$

EXAMPLE 5 *Find the solution of the second-order differential equation*

$$\frac{d^2z}{ds^2} = \sin s.$$

As in the preceding example, we can use antidifferentiation, but two steps are required now because of the second derivative:

$$\frac{dz}{ds} = \int \sin s\,ds + C_1 = -\cos s + C_1,$$

$$z(s) = \int (-\cos s + C_1)\,ds + C_2 = -\sin s + C_1 s + C_2,$$

The problem statement asks for *the* solution. But this differential equation has many solutions, one for each value of C_1 and C_2. We can not speak of *the* solution.

Since no initial conditions are given, we cannot determine the arbitrary constants C_1, C_2. $\qquad \blacksquare$

EXAMPLE 6 *Verify that*

$$z(s) = -\sin s + C_1 s + C_2 \qquad (1.9)$$

is indeed a solution of the differential equation

$$\frac{d^2z}{ds^2} = \sin s. \qquad (1.10)$$

To substitute the solution (1.9) into the differential equation (1.10), we must compute $z''(s)$. From the solution formula (1.9), we calculate

$$z'(s) = -\cos s + C_1,$$

$$z''(s) = \sin s.$$

Hence, we do have a solution of the differential equation (1.10). $\qquad \blacksquare$

EXAMPLE 7 *Verify that for any value of the constant C.*

$$y(t) = Ce^{4t}$$

is a solution for all t of the differential equation

$$\frac{dy}{dt} = 4y.$$

We use the proposed solution $y(t) = Ce^{4t}$ to compute dy/dt, the expression on the left side of the differential equation:

$$\frac{dy}{dt} = \frac{d(Ce^{4t})}{dt} = C4e^{4t}.$$

The right side of the differential equation is just

$$4y = 4(Ce^{4t}).$$

The two expressions are obviously identical. We have shown that if $y = Ce^{4t}$, then $dy/dt = 4y$ for all t. To put it differently, for any value of C, the expression $y = Ce^{4t}$ is a solution for all t of the differential equation $dy/dt = 4y$. ∎

> The differential equation $dy/dt = 4y$ also has a less interesting solution, the function $y(t) \equiv 0$. Obviously, $d0/dt = 4 \cdot 0$. The identically zero solution is called the **trivial solution**.

1.3.3 Separation of Variables—a Quick Look

How did we find the solution to an equation such as $dy/dt = 4y$? Simple antidifferentiation fails:

$$y(t) = \int \frac{dy}{dt} \, dt = \int 4y \, dt.$$

We cannot evaluate the antiderivative $\int 4y \, dt$ because we do not know the formula for y as a function of t. (If we did, we would already know the solution!)

A technique that works for this particular equation and for many other first-order equations is **separation of variables**. To illustrate it quickly here, notice that $dy/dt = 4y$ can be written with all the y variables on one side of the equation and all the t variables on the other,

Separate variables.

$$\frac{dy}{y} = 4 \, dt,$$

We will verify $y \neq 0$ when we have a formula for y.

providing $y \neq 0$. We accomplished this *separation of variables* by replacing the derivative dy/dt by a quotient of the differentials dy and dt.

Proceeding formally (that is, without careful mathematical justification), we find the antiderivative of each side of the separated equation:

Integrate.

$$\int \frac{dy}{y} = \ln|y|$$

$$= \int 4 \, dt = 4t + c,$$

or

$$\ln |y| = 4t + c.$$

The two different arbitrary constants that arose from the two antiderivatives were collected on the right side of the equality and their difference labeled c.

An exponential inverts the natural logarithm to give

Solve for y.

$$|y| = e^{4t+c} = e^c e^{4t} = |C| e^{4t}.$$

We have defined $|C| = e^c$.

Since $|y| = \pm y$, according to whether y is positive or negative, $y = \pm C e^{4t}$. The exponential is always positive, so the sign of $C e^{4t}$ is determined by the sign of C; that is, our solution is

$$y = C e^{4t},$$

with C chosen to be either positive or negative, as appropriate. If $C \neq 0$, then $y \neq 0$, as we required.

To verify that this solution is correct, turn back to example 7.

That is all there is to separation of variables, when it works: Put all terms involving the dependent variable on one side of the equation and all terms involving the independent variable on the other side. Then integrate.

EXAMPLE 8 *Use separation of variables to find a solution of the initial-value problem*

$$\frac{dy}{dt} = 4y - 6, \; y(0) = -3.$$

First we find a function that satisfies the differential equation; then we use the initial condition to determine any arbitrary constant that appears in the solution formula.

Separating variables yields

$$dy = (4y - 6)dt,$$

$$\frac{dy}{4y - 6} = dt,$$

providing $4y - 6 \neq 0$.

Antidifferentiation follows:

$$\int \frac{dy}{4y - 6} = \frac{1}{4} \ln |4y - 6| = \int dt = t + c.$$

Multiplying by 4, using the exponential to invert the natural logarithm, writing $|C| = e^{4c}$, and disposing of the absolute value signs as above gives a solution of the differential equation,

$$y(t) = \frac{3}{2} + C e^{4t}.$$

Using the initial condition $y(0) = 2$ to determine C, we find

$$y(0) = \frac{3}{2} + C = -3\,,$$

$$C = -\frac{9}{2}\,.$$

Check $4y - 6 \neq 0$ yourself.

The solution of the initial-value problem is obtained by substituting this value of C into the solution formula of the preceding paragraph:

$$y(t) = \frac{3}{2} - \frac{9}{2}e^{4t}\,. \qquad \blacksquare$$

EXAMPLE 9 *Use separation of variables to find a solution of*

$$\frac{du}{dt} = u^2 \cos \pi t.$$

For what values of t is the solution you find valid?

We replace du/dt by a quotient of the differentials du and dt and collect terms involving u on the left and those involving t on the right:

$$\frac{du}{u^2} = \cos \pi t \, dt\,,$$

providing $u \neq 0$. Then we antidifferentiate:

$$\int \frac{du}{u^2} = \int \cos \pi t \, dt\,,$$

$$-\frac{1}{u} = \frac{1}{\pi}\sin \pi t + C\,. \qquad (1.11)$$

Solving for the dependent variable u yields

$$u(t) = -\frac{\pi}{\sin \pi t + C}\,.$$

This solution is valid for any value of t for which $\sin \pi t \neq -C$.

There was some algebraic sleight of hand. Inverting (1.11), multiplying through by -1, and multiplying the right side by π/π actually yields

$$u(t) = \frac{-\pi}{\sin \pi t + \pi C}\,.$$

Since C is an arbitrary constant, we have replaced the expression πC by C for the sake of simplicity. $\blacksquare$

We close this chapter with two questions, designed to point out two morals.

1. Does the formula $u = -\pi/(\sin \pi t + C)$ for the solution of the differential equation

$$\frac{du}{dt} = u^2 \cos \pi t$$

easily reveal information about the solution that would help you sketch its graph and understand its behavior?

2. Can you use either simple antidifferentiation or separation of variables to find the solution of

$$\frac{dy}{dt} = 4y + \cos \pi t?$$

To answer the first question, we see that the solution formula does tell us that the solution will blow up at values of t for which $\sin \pi t + C = 0$. But sketching the graph of this curve from the formula $u = -\pi/(\sin \pi t + C)$ may not be easy. The differential equation $u' = u^2 \cos \pi t$ itself might be more useful because it immediately provides us information about the slope of the solution, an essential ingredient for curve sketching.

Moral 1 Solution formulas are useful, but they are often an incomplete answer if your goal is understanding the phenomenon modeled by the differential equation.

The answer to the second question is *no*. You can not solve $y' = 4y + \cos \pi t$ by separation of variables because you cannot separate the variables, try as you might.

Moral 2 We need additional solution techniques. We need ways other than trial and error to tell us which techniques will solve a given equation.

We confront each of these issues in detail in succeeding chapters.

1.4   CHAPTER 1 SUMMARY

In our context, a **model** involves a differential equation whose unknown function is a quantity of interest, such as the velocity of the thrown rock. The differential equation in the model is usually derived from a **physical law** such as Newton's Law using an **experimental fact** to relate the quantities of interest. For example, we used the experimental fact that the force of gravity is proportional to mass, $F_g = -mg$.

A **differential equation** is a relation involving one or more derivatives of an unknown function and perhaps the function itself. A **solution of a differential equation** is a sufficiently smooth function which renders the differential equation an identity when it is substituted for the unknown function. The **order of a differential equation** is the order of the highest derivative appearing in the equation.

The unknown function is called the **dependent variable**. The argument of the unknown function, the variable with respect to which it is differentiated, is the **independent variable**. A **parameter** is a term that is constant for any particular occurrence of the differential equation.

An **initial-value problem** is a differential equation coupled with a specification of the value of the dependent variable and perhaps one or more its derivatives at a fixed value of the independent variable. These values are called **initial values**, and such specifications are called **initial conditions**.

If the differential equation simply gives an expression for a derivative of the unknown in terms of the independent variable, it can be solved by **antidifferentiation**. If all the dependent variables in a first-order differential equation can be collected on one side of the equation and all the independent variables on the other, then the equation can be solved by **separation of variables**.

1.5   HOW TO READ A PROBLEM

Few of the problems you will encounter in your career will come complete with instructions for their solution. Indeed, life may be just an ambiguous word problem. This section outlines some strategies for solving the sorts of technical problems you will encounter in this text and later in your academic and professional career.

PROBLEM   *Complete the following steps to derive a model describing the height above the surface of the earth of a rocket launched vertically upwards. Use Newton's Law ($F = ma$) and the following experimental facts:*

> **Fact 1:**   The force of gravity on the rocket is proportional to its mass and inversely proportional to the square of its distance from the center of the earth.
>
> **Fact 2:**   A known function $T(t)$ gives the thrust force of the rocket at time t measured from the instant of ignition.

1. *Introduce a coordinate system and write a formula for the force F_g of gravity given by the first experimental fact in terms of the height $y(t)$ of the rocket above the surface of the earth at time t.*
2. *Use Newton's law to obtain a differential equation relating the height $y(t)$, one or more of its derivatives, the gravity expression from part 1, and the thrust $T(t)$ of the rocket to one another.*
3. *Complete the model by adding appropriate initial conditions. (Is the rocket moving at the instant its motor is ignited?)*

Begin your attack on the problem by deciding what the problem is about:

*Follow the steps given next to derive a **model** describing the height above the surface of the earth **of a rocket launched vertically upwards**.*

As you study a problem, underline important words and phrases.

The words that describe the heart of the problem are shown here in bold type.

What information have we been given?

... Use Newton's Law ($F = ma$) and the following experimental facts:

Fact 1: The *force of gravity* on the rocket is proportional to its mass and inversely proportional to the square of its distance from the center of the earth.

Fact 2: A known function $T(t)$ gives the **thrust force of the rocket** at time t measured from the instant of ignition.

We see that we are given $F = ma$, information about the force of gravity on the rocket, and the name of a function describing the thrust force of the rocket.

What specifically does the problem ask of us? What constitutes a solution? The first part directs:

1. *Introduce a coordinate system and write a formula for the force F_g of gravity given by the first experimental fact in terms of the height $y(t)$ of the rocket above the surface of the earth at time t.*

We will have completed this part of the problem when we have defined a coordinate system and when we have written a formula relating F_g from fact 1 to $y(t)$. (This notation is defined in the problem statement.)

A natural coordinate system is one that points vertically upward and places its origin at the surface of the earth. Such a vertical coordinate is just the height y, as illustrated in figure 1.4.

To write a formula for F_g, the force of gravity on the rocket, re-read Fact 1. It provides the recipe for the rest of the answer to part 1:

*The force of gravity on the rocket is **proportional to its mass** and **inversely proportional to the square of its distance** from the center of the earth.*

To write this functional relation, we need notation for a proportionality constant, for the mass of the rocket, and for the distance of the rocket from the center of the earth.

Figure 1.4 uses R for the radius of the earth. Then the distance from the rocket to the center is $R + y$. A natural choice for the mass of the rocket is m. How about G for the proportionality constant? We will leave to you the job of using this notation to turn the recipe in fact 1 into a formula for F_g.

Having settled part 1 (if you did your share by writing a formula for F_g), we turn to part 2. It directs:

2. *Use Newton's Law to **obtain a differential equation** relating the height $y(t)$, one or more of its derivatives, the gravity expression from part 1, and the thrust $T(t)$ of the rocket to one another.*

The problem says "... obtain a differential equation ...." But how? Look again:

2. *Use Newton's Law ....*

Newton's law is $F = ma$; F is the total force on the rocket, m is its mass, and a is its acceleration.

What forces sum to give F? The problem statement lists

... the gravity expression from part 1, and the thrust $T(t)$ of the rocket ....

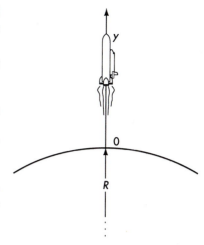

Figure 1.4: A vertical coordinate system whose origin is at the surface of the earth.

You do not know the *rule* for the thrust force, just the name of the function. But you need only the name. If the actual formula appears later, you can always substitute it into your results.

In part 1, you have written a formula for F_g, the gravity force. You are handed the name $T(t)$ of the formula for the thrust force. Hence, $F = F_g + T(t) = ma$.

The acceleration a of the rocket is the second derivative of displacement. That observation relates a to y. Now you can turn $F = ma$ into the required differential equation.

> *... relating the height $y(t)$, one or more of its derivatives, the gravity expression from part 1, and the thrust $T(t)$ of the rocket to one another.*

Part 2 is finished when you have written this differential equation.

Part 3 gives one more instruction, and it asks a parting question.

3. **Complete the model** *by adding appropriate initial conditions.* (**Is the rocket moving** *at the instant its motor is ignited?*)

"Complete the model ...." How? " ... by adding appropriate initial conditions." When you complete the details of part 2, you will find that the differential equation contains the second derivative d^2y/dt^2. Hence, two initial conditions are required. Natural choices are initial position and initial velocity. What is the initial position in the coordinate system of figure 1.4?

"Is the rocket moving at the instant its motor is ignited?" Who knows? But it is natural to suppose that the rocket was fixed on its launch pad and that its initial velocity $y'(0)$ is zero. The question suggests a specific numerical value for the second initial condition.

We have stepped through this problem item by item, emphasizing the need to obtain from the problem statement the information you need to decide what constitutes a solution. We have not completed all the details. They are left to exercise 17.

We close by summarizing this approach. By underlining appropriate phrases, by making a separate list, or by some other means of your choice, identify

- What constitutes a solution of the problem you face,
- What information you are given,
- What directions, if any, you are given for obtaining a solution.

Use these three categories of information to formulate a reasonable solution.

1.6 CHAPTER EXERCISES

EXERCISE GUIDE

To gain experience . . .	Try Exercises
Deriving models	5(a) 8–9(a–b), 17–18
Analyzing and interpreting models	1, 4, 5(b–d), 8–9(c–e), 10, 11, 12
With the basic ideas of these models	2, 3,
Solving differential equations and initial value problems	6, 7, 13, 16, 19(a)
With differential equations concepts	14, 15, 19

1. Use the solution (1.6), $v(t) = -gt + v_i$, to determine the time t_p required for the rock to reach the peak of its trajectory. (*Hint:* What is its velocity at the peak?) Locate t_p on the graph of figure 1.3.

2. Suppose we had decided to formulate model 3,

$$\frac{dv}{dt} = -g, \ v(0) = v_i,$$

by specifying the t-intercept t_p of the velocity-versus-time curve (see figure 1.3) instead of the v-intercept v_i. Write the model in this form. Determine the constant C in (1.5), $v(t) = -gt + C$, in terms of t_p.

3. *A model with too much auxiliary data.* Suppose we try to specify both initial velocity v_i and time-to-peak t_p in the rock-tossing model; that is, suppose we pose the model

$$v'(t) = -g, \tag{1.12}$$

$$v(0) = v_i, \ v(t_p) = 0. \tag{1.13}$$

(The second condition takes advantage of the observation that velocity is zero at the peak of the rock's trajectory.) Show that the two conditions in (1.13) contradict one another unless $t_p = v_i/g$. In other words, show that the model (1.12–1.13) has no solution unless the auxiliary data in (1.13) satisfy a special relation. (A problem of this sort is called *overdetermined* because only part of the data is truly necessary to specify the solution.)

Interpret the observations of this exercise geometrically in terms of figure 1.3. Does figure 1.3 show the relation that must hold between the two pieces of auxiliary data in (1.13)?

4. You have been asked to determine the velocity with which a sling shot releases a stone. Propose an experiment based on the model $v' = -g, \ v(0) = v_i$, to accomplish this task. (*Hint:* See exercises 1 and 10.)

5. **(a)** Model 2, $v' = -g$, was obtained from $a = -g$ by writing acceleration as the derivative of velocity. Continue this logic by writing velocity as the derivative of height y: $v = dy/dt$. (Figure 1.1 will need a y-axis. Be sure to mark its origin.)

   Derive the following model involving height for the motion of the rock:

$$\frac{d^2y}{dt^2} = -g. \tag{1.14}$$

   Solve this equation by anti-differentiation and argue that two pieces of initial data must be pre-scribed to determine all arbitrary constants. Suggest reasonable choices for this data, and state the resulting initial-value problem.

   (b) Provide some analysis and interpretation of this initial-value problem model. For example, what is the time required to fall from the peak of the trajectory back to the point of release? What is the rock's velocity at the instant it returns to the point of release?

   What type of curve has a constant second derivative, as in (1.14)? How much data is needed to specify such a curve? Sketch such a curve and show the data you propose to use on the sketch.

   (c) What is the maximum height of the rock? How does the maximum height depend upon the parameters in the model? (Interpret; do more than recite equations.)

   (d) Solve the initial-value problem from part (a), unless you have already. Compare your analysis with this solution.

6. Use antidifferentiation to find solutions involving the specified number of arbitrary constants for the following differential equations:
   (a) $dy/dt = \cos t$, one constant
   (b) $d^2w/dt^2 = \sqrt{t}$, two constants
   (c) $dy/dx = 4e^x$, one constant

7. Solve the following initial-value problems.
   (a) $dy/dt = \cos t, \ y(0) = 0$
   (b) $d^2w/dt^2 = \sqrt{t}, \ w(1) = 2, \ w'(1) = -2$
   (c) $dy/dx = 4e^x, \ y(2) = \pi$

8. *A model of the rock's motion with air resistance.* To include the effects of air resistance, we can appeal to an additional experimental fact:

   The force due to air resistance is proportional to the velocity, and it acts in the direction opposite to velocity.

   (a) *Formulate a mathematical statement of the experimental fact.* Argue that a correct mathematical expression of this experimental fact is $F_a = -kv$, where k is a positive constant of proportionality. (We expect k to depend upon the shape of the rock but not its mass.)

   (b) *Derive a model.* The total force acting on the rock in this model is $F_g + F_a$. Couple this observation with Newton's law to derive a more sophisticated model of the rock's motion,

$$\frac{dv}{dt} = -g - \frac{k}{m}v, \ v(0) = v_i.$$

(c) *Analyze the model.* How does the v-versus-t curve of this model compare with that of figure 1.3, the corresponding curve for model 3? Do the two curves begin at the same point? How do their slopes compare at $t = 0$? How do their slopes compare when $v = 0$? How does the slope of the graph of this model change as the velocity decreases from v_i to zero?

(d) *Interpret.* Does air resistance appear to modify the rock's motion in a manner consistent with physical intuition? Which term in the differential equation accounts for air resistance? Are the comments in the text about the effects of the rock's mass on its motion still valid?

(e) *Interpret the model in a different setting.* Suppose the rock were dropped from a balloon. Would this model still apply? Would the velocity of the falling rock increase indefinitely?

9. Shotgun pellets and other small spherical metal particles are formed by dropping a bit of molten metal down a long hollow tube, called a shot tower. As the mass falls, air resistance retards its motion with a force proportional to the *square* of its velocity.

    Let m denote the mass of the particle and v its velocity. Let r be the proportionality constant for the force of air resistance. Take *up* as the *positive* direction.

(a) *Formulate a mathematical statement of the experimental facts.* Write an expression for the force of gravity on the pellet. Write an expression for the force of air resistance on the pellet. (Be careful with signs!.)

(b) *Derive a model.* Use Newton's law ($F_{total} = ma$) to derive a model for the velocity of the pellet.

(c) *Analyze the model.* How does the v-versus-t curve for this model compare with that of figure 1.3, the corresponding curve for model 3?

(d) *Interpret.* Will the pellet fall faster and faster or will it attain a constant terminal velocity? If the latter occurs, find the value of the terminal velocity.

10. How long would Nolan Ryan's vertical pitch require to reach the top of its trajectory? (See example 1.)

11. Draw a careful graph of the solution of the model for Nolan Ryan's vertical pitch. (See example 1.) Use a piece of graph paper so that the slope and the intercept with the vertical axis can be plotted exactly.

12. Use the model of exercise 5 to determine the maximum height Nolan Ryan's vertical pitch could reach. (See example 1.)

13. Use separation of variables to solve each of the following differential equations. State the range of values of the independent variable for which the solution is valid.

(a) $dy/dt = -8y$
(b) $dP/dt = 0.0102P$
(c) $dP/dt = kP$, where k is a constant
(d) $dP/dt = 0.015P - 0.209$
(e) $y' = y^3 - y$
(f) $dT/dt = -0.0002(T - 5)$
(g) $dT/dt = -(Ak/cm)(T - T_{out})$, where A, k, c, m, T_{out} are all constants
(h) $dy/dx = e^{2x}$
(i) $dy/dx = e^{2y}$
(j) $du/dy = y^2 - 1$

14. (a) Which of the differential equations in the preceding exercise have the same generic form as the differential equation in part (a); that is, which equations have the form $dy/dx = ax$ for some constant a? In each case, identify the variables that correspond to y and to x and to the parameter a.

(b) Which of the differential equations in the preceding exercise have the same generic form as the differential equation in part (d); that is, which equations have the form $dy/dx = ax + b$ for some constants a, b? In each case, identify the variables that correspond to y and to x and to the parameters a and b.

(c) Which of the differential equations in the preceding exercise have a generic form that is completely distinct from all others in that exercise? In each case, write the form of the equation using y for the dependent variable, x for the independent variable, and $a, b, c, \ldots$ for constants.

15. Which of the differential equations listed in exercise 13 possess the trivial solution? Which do not? Verify the claim in each case by substituting the identically zero function into the equation and showing that equality does or does not hold.

16. Give an example of a differential equation (other than the one in the text) involving an unknown function and only its first derivative that cannot be solved by separation of variables.

17. Follow the steps given below to derive a model describing the height above the surface of the earth of a rocket launched vertically upwards. Use Newton's law ($F = ma$) and the following experimental facts:

Fact 1 The force of gravity on the rocket is proportional to its mass and *inversely* proportional to the square of its distance from the *center* of the earth.

Fact 2 A known function $T(t)$ gives the thrust force of the rocket at time t measured from the instant of ignition.

(a) Introduce a coordinate system and write a formula for the force F_g of gravity given by the first experimental fact in terms of the height $y(t)$ of the rocket above the *surface* of the earth at time t.

(b) Use Newton's law to obtain a differential equation relating the height $y(t)$, one or more of its derivatives, the gravity expression from part (a), and the thrust $T(t)$ of the rocket to one another.

(c) Complete the model by adding appropriate initial conditions. (Is the rocket moving at the instant its motor is ignited?)

18. In his *Dialogues Concerning Two New Sciences* (see [15, p. 344], Galileo asserts the proposition

The spaces described by a body falling from rest with a uniformly accelerated motion are to each other as the squares of the time intervals employed in traversing these distances.

(In a more modern translation of Galileo's Latin, we might replace *spaces described* with *distances traveled*.)

(a) Derive a model that relates the displacement of the body to a uniform acceleration a. (*Hint:* Can you use something like (1.14)?)

(b) Use the solution of the model to prove Galileo's assertion that distance traveled is proportional to the square of the elapsed time.

19. Differential equations of the form $dy/dt = ky$ are often called *equations of growth and decay*. Explain this name as follows.

(a) Solve the equation and argue that y is increasing for all t if $k > 0$ and $y(0) > 0$.

(b) Solve the equation and argue that y is decreasing for all t if $k < 0$ and $y(0) > 0$.

(c) Argue directly from the differential equation that y is increasing if $k > 0$ and $y(0) > 0$, because $y' > 0$.

(d) Argue directly from the differential equation that y is decreasing if $k < 0$ and $y(0) > 0$, because $y' < 0$.

We encounter both forms of this equation in the next chapter.

1.7 CHAPTER PROJECTS

1. **A more sophisticated model of the rock's motion** To include the effects of air resistance in a slightly more accurate manner than in exercise 8, appeal to a different experimental fact:

The force due to air resistance is proportional to the *square* of the velocity, and it acts in the direction opposite to velocity.

(a) Argue that a correct mathematical expression of this experimental fact is $F_a = -kv|v|$, where k is a positive constant of proportionality. Couple this observation with Newton's law to derive a more sophisticated model of the rock's motion,

$$\frac{dv}{dt} = -g - \frac{k}{m}v|v|, \quad v(0) = v_i \,.$$

(b) How does the v-versus-t curve of this model compare with that of figure 1.3, the corresponding curve for model 3? Do the two curves begin at the same point? How do their slopes compare at $t = 0$? How do their slopes compare when $v = 0$? How does the slope of the graph of this model change as the velocity decreases from v_i to zero?

(c) Does air resistance appear to modify the rock's motion in a manner consistent with physical intuition? Which term in the differential equation accounts for air resistance? How does the rock's mass affect its motion? Does the velocity of the falling rock increase indefinitely? Could one speak of a *terminal velocity*?

2

FIRST-ORDER MODELS

Come now let us consider the generations of the Human Race . . .
Aristophanes

The generations of humans and, more generally, populations of animals, insects, chemicals, and even money provide a rich source of models with a common feature, growth. Heat-flow models offer another alternative, the potential for decay. The population and heat-flow models developed in this chapter are a wellspring of mathematical challenges.

2.1 POPULATION MODELS

2.1.1 Data and a Simple Model

Population surveys reveal an apparently innocuous fact: Most countries in the world have populations that increase between 1% and 3% per year; that is, each succeeding year's population is 0.01 to 0.03 greater than the preceding year's. Some typical annual fractional increases are listed in table 2.1. These numbers suggest that the populations of industrialized nations increase at about a one percent annual rate, while some third-world nations run closer to a two or three percent annual rate.

Do these modest differences in annual rates of population increase have any great significance? Do they foreshadow future population imbalances? What population levels might we anticipate? How does varying the annual rate of increase affect population levels a generation or two hence? How will limited resources, immigration, disease, and other external factors influence future population levels?

We shall develop, analyze, and interpret mathematical models that will provide some answer to these questions. We emphasize that we will not find

Table 2.1

Populations and annual growth rates of selected countries and regions for 1978 and 1985. The population data are from [5, pp. 147, 166–169].

Country	1978		1985	
	Population (millions)	Growth Rate	Population (millions)	Growth Rate
World Total	4,258	0.0183	4,837	0.0165
United States	218.72	0.0085	239.28	0.0097
Brazil	115.40	0.0282	135.56	0.0216
Argentina	26.39	0.0129	30.56	0.0154
USSR	261.26	0.0109	277.54	0.0094
PRC	933.03	0.0123	1,059.52	0.0120
Indonesia	145.10	0.0232	163.39	0.0217
India	638.39	0.0197	750.86	0.0203
Japan	114.90	0.0084	120.75	0.0061
Denmark	5.10	0.0039	5.11	0.0020
German Federal Republic	61.31	0.0005	61.02	0.0005
German Democratic Republic	16.76	–0.0007	16.64	–0.0012
Great Britain	49.12	0.0012	49.92	0.0032
Papau New Guinea	2.99	0.0301	3.33	0.0300

Note:

$$1978 \text{ annual growth rate} = \frac{(1979 \text{ population}) - (1978 \text{ population})}{1978 \text{ population}}$$

The 1985 annual growth rate is defined similarly.

the answer to any of these questions; do not let fancy mathematics and strings of numbers persuade you that any question about issues as complex as human population has a unique answer.

This text is written for students of science and engineering. Those who will never encounter population models might well ask, "Why waste time on something I don't need? Let's get to the differential equations I'm going to use in circuit theory or thermodynamics or ...."

Population models are a good setting for studying modeling because little specialized knowledge is required at our introductory level. The talents for modeling, analysis, and interpretation you develop in this setting can easily be transferred to your specialty, but learning will not be complicated by excess technical baggage. For example, predicting the mass of a radioactive isotope involves populations of atoms. Determining the charge on a capacitor involves populations of electrons.

World population is also an important social concern. Predictions of population trends and proposals for solving population problems often are based on a mathematical model. A critical understanding of modeling can make you a more effective world citizen.

The modest spread of annual rates of increase shown in table 2.1 could presage dramatic differences in future populations. For example, pushing a few buttons on a calculator (see exercise 15) reveals the following:

If the initial population is 1,000,000, then the population 10 years later is

- 1,104,622 at an annual rate of increase of 1%, and
- 1,343,916 at an annual rate of increase of 3%.

If these population differences do not seem significant, look ahead 50 years!

Aren't these numbers enough? No! We need models that will illuminate the effects of parameters like the annual growth rate. Numerical calculations can show us trends, but they probably cannot identify the mechanisms by which different parameter values affect population growth.

To begin modeling, we need governing laws and experimental facts, such as Newton's Law and "the force of gravity is proportional to mass" fact used in the rock motion models of chapter 1. The basic law is so obvious in this setting that it hardly seems worth stating:

The $F = ma$ of population models.

Law of Conservation of Population The net change in a population over a given period of time is the number of individuals added less the number of individuals removed.

The experimental facts we need depend upon the causes of population change we choose to consider. Let us restrict our attention to births and "natural" deaths only, neglecting migration, wars, disasters, plagues, etc. The population of the United States over the last few decades corresponds to these conditions. The data of table 2.2 suggest that the net annual growth rate has hovered around 0.01 over that period. An easy simplification is to suppose that the birth rate and the death rate are each constant. This idea leads us to accept the following.

The "force of gravity is proportional to mass" of populations.

Experimental Fact 1 The fraction of individuals in a population who reproduce in a unit time period is constant.

Experimental Fact 2 The fraction of individuals in a population who die in a unit time period is constant.

The constants mentioned in these experimental facts are loosely called the *birth rate* and the *death rate*, respectively. More precisely, they are the number of deaths and births per unit time per individual. They are illustrated in figure 2.1. If time is measured in years, then these constants would have units of per year per person.

To derive a model, let k_b denote the birth rate constant and k_d the death rate constant mentioned in the experimental facts. Then $k_b P(t)$ is the rate at which individuals are being added to the population due to births at time t. Similarly, $k_d P(t)$ is the rate at which individuals are being removed from the population due to deaths.

We multiply rate $k_b P(t)$ by elapsed time Δt. "Rate times time" is only approximate because the rate $k_b P(t)$ is not constant.

The number of individuals added from t to $t + \Delta t$ is approximately $k_b P(t)\Delta t$, while the number lost is approximately $k_d P(t)\Delta t$.

The law gives a relation among the experimental facts.

Since the net change in population over this period of time is $P(t + \Delta t) - P(t)$, we can use the law of conservation of population to write

$$P(t + \Delta t) - P(t) = k_b P(t)\Delta t - k_d P(t)\Delta t \,.$$

TABLE 2.2

United States population data for 1970–88 taken from [3]. The arithmetic mean of the 18 growth rates for 1970–1987 is 0.0102.

Year	Population (millions)	Annual Growth Rate
1970	205.05	0.0127
1971	207.66	0.0108
1972	209.90	0.0096
1973	211.91	0.0092
1974	213.85	0.0099
1975	215.97	0.0096
1976	218.04	0.0101
1977	220.24	0.0107
1978	222.59	0.0111
1979	225.06	0.0120
1980	227.76	0.0104
1981	230.14	0.0103
1982	232.52	0.0098
1983	234.80	0.0094
1984	237.00	0.0096
1985	239.28	0.0097
1986	241.61	0.0096
1987	243.92	0.0090
1988	246.11	

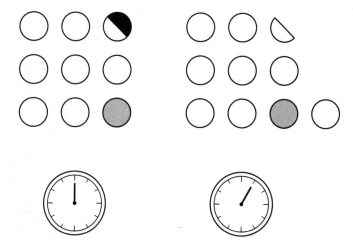

Figure 2.1: A fixed fraction of this population reproduces (light shading) and a fixed fraction (dark shading) dies during a unit time. In this example, $k_b = \frac{1}{9}$ and $k_d = \frac{1}{18}$.

We have used the conservation law in the form "net change equals number added less number lost."

Dividing this last expression by Δt gives an expression dangerously close to a derivative (which is a good thing, considering the title of the book):

$$\frac{P(t + \Delta t) - P(t)}{\Delta t} = (k_b - k_d)P(t).$$

Unfortunately, P is a function that counts people; it takes only integer values. Since it is not even continuous, it can never have a derivative.

We can hedge as follows. If the population is large, then the addition of one person to the population produces only a small relative change in the value of P. So we shall smooth out those integer jumps in P, pretend P is continuous, even differentiable, and ignore fractions that occur because of this smoothing.

Now we can let Δt approach zero, to obtain from the preceding difference quotient

$$\frac{dP}{dt} = (k_b - k_d)P.$$

Intuition and our experience with the velocity model in chapter 1, which also involved a first-order equation, both suggest that we need an initial condition, the size of the starting population. Let us use the population P_i at the beginning of the first year of study and label the first year as t_0. Then the initial condition is $P(t_0) = P_i$.

Hence, our complete model is

MODEL

$$\frac{dP}{dt} = kP, \tag{2.1}$$

$$P(t_0) = P_i, \tag{2.2}$$

where we have written $k = k_b - k_d$. The constant $k = k_b - k_d$ is called the *net growth rate* or the *intrinsic rate of population growth* for the population. This growth rate is "net" because it includes both births and deaths. It is positive if more people were born each year than died.

> In a population of radioactive isotopes, k is negative because atoms "die" as they decay, but none are "born."

The net *annual* growth rate is the change in population over the given year divided by the population at the start of the year. (See the note in table 2.1.) The growth rates illustrated in the tables are *net* growth rates. We can estimate intrinsic growth rates from annual growth rates, but we are making the classic approximation, *equating an average rate of change over a finite time interval to the instantaneous rate $dP/dt = kP$*

The differential equation $dP/dt = kP$ appearing in this model is a consequence of our postulate that birth and death rates are constant. That equation says that the rate of change of population dP/dt is proportional to the current population P. The constant of proportionality is k, the population growth rate parameter.

We can use separation of variables to find a formula for the function $P(t)$ that satisfies the population model initial-value problem (2.1)–(2.2). But first let us learn what we can about the behavior of the population function in a specific case.

EXAMPLE 1   *Implement the population model $dP/dt = kP$, $P(t_0) = P_i$, for the United States in the decade of the 1970s. Comment on the sign of the slope of the graph of P versus t.*

Table 2.2, suggests that we might take an average annual growth rate of $k = 0.0102$, since that value is the mean of the growth rates shown there. For this case, the model (2.1)–(2.2) becomes

$$\frac{dP}{dt} = 0.0102\,P$$

$$P(1970) = 205.05 \quad (P \text{ in millions}).$$

The slope of the population curve at $t = 1970$ is

$$\frac{dP}{dt} = (0.0102)(205.05) = 2.09 \text{ million people per year.}$$

Hence, the population grows initially, and it continues to grow faster and faster because the rate of growth increases with the population. The slope of the population curve at any instant is 0.0102 times the population at that instant. ■

Rather than study this model for just a particular numerical situation, as in the previous example, let us analyze the model $dP/dt = kP$, $P(t_0) = P_i$, for arbitrary (positive) values of the growth rate parameter k.

If the initial population P_i is positive (negative populations are unlikely), then at $t = t_0$ the slope of the P-versus-t curve is positive, for the differential equation $dP/dt = kP$ combines with the initial condition $P(t_0) = P_i$ to yield

If you become confused in a general analysis, replace the letters with specific numbers. Here you can take $k = 0.0102$, $t_0 = 1970$, and $P_i = 205.05$.

ANALYSIS

$$\frac{dP}{dt} = kP_i > 0 \text{ at } t = t_0\,.$$

Hence, the population increases initially.

Indeed, the equation $dP/dt = kP$ forces the P-versus-t curve to have a positive slope as long as P remains positive. Since P begins with a positive value, the graph of P versus t always has a positive slope.

As P increases, the right-hand side of $dP/dt = kP$ becomes larger. Since P increases with time, the slope of the population curve increases with time. A curve with this sort of behavior is sketched in figure 2.2.

The population curve of figure 2.2, whose shape was suggested by the preceding analysis, increases without bound at an ever more rapid rate. It appears that there is no upper limit to the population in this model and that larger populations increase at more rapid rates. If a model of this type is indeed applicable to human populations, then overpopulation is no idle threat!

INTERPRETATION

Have we omitted important features from this model that might limit this ever-increasing population growth? Might the death rate increase as larger populations compete for a fixed supply of food? We will consider such possibilities later.

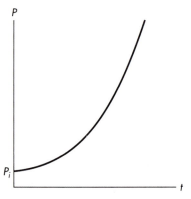

Figure 2.2: A curve with the continuously increasing slope predicted by equation (2.1), $dP/dt = kP$.

To obtain an alternate view of the predictions of this model, let us return to the specific case of the United States during the 1970s. In the next example, we will find a formula for population as a function of time by solving the initial-value problem that constitutes the model.

EXAMPLE 2   *Use separation of variables to find the solution of the population model for the United States in the decade of the 1970s,*

$$\frac{dP}{dt} = 0.0102\,P\,,$$

$$P(1970) = 205.05\,.$$

(See example 1.) Does the solution formula exhibit the same unbounded increase suggested by our analysis of the general equation $dP/dt = kP$?

To solve the differential equation, we can follow *precisely* the steps we used to solve $dy/dt = 4y$ in section 1.3.3, replacing y with P, 4 with k, and so on.

Collecting terms involving the variable P on the left and those involving t on the right yields

Separate variables.

$$\frac{dP}{P} = 0.0102\,dt\,,$$

Integrate.

providing $P \neq 0$. Integrating both sides produces

$$\ln|P| = 0.0102\,t + c\,.$$

Solve for P.

Exponentiation removes the natural logarithm,

$$|P| = e^{0.0102t+c}\,.$$

Finally, we can write $|C| = e^c$ and let the sign of C determine the sign of P:

$$P(t) = Ce^{0.0102\,t}\,.$$

This function solves $dP/dt = 0.0102\,P$ for *any* value of the arbitrary constant C. Clearly, $P \neq 0$ if $C \neq 0$.

Find C.

It remains to choose C to satisfy the initial condition $P(1970) = 205.05$. Substituting $t = 1970$ and $P = 205.05$ in the solution formula of the preceding paragraph leads to

$$205.05 = Ce^{0.0102\cdot1970}\,.$$

We find

$$C = 205.05e^{-0.0102\cdot1970}\,.$$

Substituting this expression for C into the solution formula $P(t) = Ce^{0.0102\,t}$ yields the solution of the initial-value problem $dP/dt = 0.0102P$, $P(1970) = 205.05$

$$P(t) = 205.05e^{0.0102(t-1970)}\,.$$

That is, this function satisfies both the differential equation $dP/dt = 0.0102\,P$ *and* the initial condition $P(1970) = 205.05$. (Exercise 1(a) asks you to verify these claims.) We have a formula for the population $P(t)$ predicted by the model $dP/dt = 0.0102P$, $P(1970) = 205.05$.

To check, verify $dP/dt = 0.0102P$ and $P(1970) = 205.05$.

Does this solution formula exhibit the sort of unbounded growth suggested in figure 2.2? Indeed, this solution shows we are facing *exponential* population growth. ∎

Could we obtain a formula for the solution of the general population model $dP/dt = kP$, $P(t_0) = P_i$? Certainly. We need only mimic the steps in the preceding example. See part (f) of exercise 2.

Compare the two analyses. One used the differential equation directly. The other found a formula for the solution of the initial-value problem.

> Thomas Malthus (1766-1834) is credited with the first observation of the hypothesis underlying the model (2.1)–(2.2): The rate of increase of a population is proportional to its size. This and other population models are discussed in Smith's article [17]. The sobering predictions described in that article may pique your interest in population modeling.
>
> Frauenthal's monograph [6] gives a good view of some of the scientific challenges of demography. Murray [12] provides a comprehensive account of many applications of mathematics to biology. For more information about the role of conservation laws in applied mathematics, see Lin and Segel's pioneering text [11].

2.1.2 Exercises

EXERCISE GUIDE

To gain experience . . .	Try Exercises
Deriving models	8, 10
Analyzing models from differential equations	4, 6, 9(a), 11(a), 14
Analyzing models from a solution	3, 5, 7, 9(b–c), 11(b–c), 13(b–c)
Interpreting models	1(b), 4–5, 7, 14
With concrete examples of models	1–5, 7, 12, 13
Solving problems like these models	1–3, 5, 7, 9(b), 11(b)–13(b)
With the basic ideas of these models	4, 6, 12, 15–17
With the mathematical subtleties of model derivation	18, 19

1. Example 2 used separation of variables to find the solution

$$P(t) = 205.05e^{0.0102(t-1970)}$$

of the initial-value problem $dP/dt = 0.0102\,P$, $P(1970) = 205.05$ modeling the population of the United States during the 1970s. (Population is measured in millions.)

(a) Verify that this function actually solves the initial-value problem by showing that it satisfies both the differential equation and the initial condition.

(b.) Use this solution formula to compute $P(1978)$ and $P(1988)$. How accurately has this model predicted the actual populations shown in table 2.2?

2. Use separation of variables to solve each of the given differential equations. Verify that you have actually obtained a solution in each case by substituting your proposed solution into the original differential equation and verifying that it reduces to an identity. If

a differential equation is accompanied by an initial condition, use it to determine the arbitrary constant C that appears in your solution.

(a) The differential equation given in example 1 as part of the model of the population of the United States in the decade of the 1970s,

$$\frac{dP}{dt} = 0.0102P.$$

(b) The model for the population of the United States in the decade of the 1970s,

$$\frac{dP}{dt} = 0.0102P, \ P(1970) = 205.05.$$

(c) A model for the population of Brazil beginning in 1978,

$$\frac{dP}{dt} = 0.0282P, \ P(1978) = 115.40.$$

(d) The differential equation from a model for the population of Brazil beginning in 1978,

$$\frac{dP}{dt} = 0.0282P.$$

(e) The differential equation for the general population model derived in this section,

$$\frac{dP}{dt} = kP.$$

(*Hint:* In section 1.3.3, we first applied separation of variables to an equation that looks very much like this one.)

(f) The general population model derived in this section,

$$\frac{dP}{dt} = kP, \ P(t_0) = P_i.$$

3. For each of the indicated regions or countries,

i. Write a population model in the form $dP/dt = kP$, $P(t_0) = P_i$, using 1978 as the initial year. Obtain growth rates and initial population data from table 2.1

ii. Use separation of variables to find the solution of the initial-value problem.

iii. Evaluate your solution at $t = 1985$ and compare the population predicted by your model with that given in table 2.1. What might account for the differences you observe?

(a) The world
(b) United States
(c) U.S.S.R.
(d) People's Republic of China (P.R.C.)
(e) German Democratic Republic
(f) Papau, New Guinea

4. Table 2.1 shows that the population of the German Democratic Republic has a negative growth rate in both 1978 and 1985. Is its population growing or decaying? What can you say about the relative values of k_b and k_d for the German Democratic Republic?

5. Use the data of Table 2.1 for 1985 to write a model for the population of the German Democratic Republic. Solve the initial-value problem, and use the solution of your model to determine when the population will have declined to half of its 1985 value. This period is called the *half-life* of this declining population. Repeat the calculation using the data for 1978. Does the half-life appear to be sensitive to the value of the growth rate parameter?

6. What does the model $P' = (k_b - k_d)P$, $P(0) = P_i$, predict if $k_b < k_d$? Is that behavior consistent with the physical significance of the inequality $k_b < k_d$?

7. One way of gauging the speed with which a population is growing is to determine the time required for its size to double. You can use the data of table 2.1 and the solution of the appropriate model to make this calculation.

(a) Using the data for 1985, determine the time required for the population of the world to double.

(b) Using the data for 1985, determine the time required for the population of Indonesia to double.

(c) Using the general population equation $dP/dt = kP$, $k > 0$, find a formula in terms of the growth rate k for the time required for a population to double. Does the initial population matter? Find the time needed to double the slowest and fastest growing 1985 populations in table 2.1. Comment on the difference.

8. Consider a "population" of atoms of a radioactive isotope. No new isotope is created, but a fixed fraction s of the atoms decay to a more stable state ("die") in unit time. State a form of the law of conservation of population that applies to isotopes, state the appropriate experimental fact(s), and derive the model

$$\frac{dN}{dt} = -sN, \ N(0) = N_i,$$

for the number of atoms $N(t)$ of the isotope at time t. Explain the physical significance of N_i. (Actually, physicists and chemists do not use units that directly count the number of isotopes, as we have suggested here. Instead, they use *moles*, multiples of Avogadro's number — 6.023×10^{23} — of isotopes.)

9. (a) Argue directly from the differential equation in the model of the preceding exercise that the number of isotopes N is decreasing with time.
   (b) Use separation of variables to solve the initial-value problem of the preceding exercise. Argue that your solution formula shows that N is in fact decaying to zero.
   (c) For a given value of the decay constant s, find the time required for the amount of isotope present to be reduced to half of its initial value. This time is known as the half-life of the isotope. Show that the half-life is independent of the initial amount of isotope.

10. Consider a "population" of dollars in a bank account. No money is being removed from the account. A fixed fraction i of the amount in the account is added to it each day as the account earns daily interest. (The daily interest rate i corresponds to a daily birth rate k_b for a population.) State a form of the law of conservation of population that applies to money in the bank, state the appropriate experimental fact(s), and derive the model

$$\frac{dN}{dt} = iN, \ N(0) = N_i,$$

for the number of dollars $N(t)$ in the account at time t. Explain the physical significance of N_i.

11. (a) Argue directly from the differential equation in the model of the preceding exercise that the amount of money N in the account is increasing with time.
   (b) Use separation of variables to solve the initial-value problem of the preceding exercise. Argue that your solution formula shows that N is in fact increasing without bound.
   (c) For a value of the daily interest rate i of your choice, use the solution formula to determine the time required for the initial sum in the account to double. Show that the time required to dou-

ble the amount in the account is independent of the initial deposit.

12. The data in table 2.3 gives the population of the United States at 10-year intervals.
   (a) Has the annual growth rate been constant? Estimate values of k, the growth rate parameter in the governing equation $P' = kP$, for the decades 1790–1800, 1850–1860, 1900–1910, and 1960–1970.
   (b) Use the annual growth rate estimated for the decade 1790–1800 to predict the 1970 population using the approach of exercise 3. (Write an initial-value problem model using the data in table 2.3, solve the initial-value problem, and evaluate the solution at 1970.) How does your result compare with the actual population? Repeat using the 1850 annual growth rate.

Table 2.3

United States population from 1790–1980 (from [3, p. 7])

Year	Population (millions)	Year	Population (millions)
1790	3.93	1890	63.06
1800*	5.30	1900	76.09
1810*	7.22	1910	92.41
1820	9.62	1920	106.46
1830	12.90	1930	123.19
1840*	17.12	1940	132.12
1850	23.26	1950	151.68
1860	31.51	1960*	180.67
1870	39.91	1970	204.88
1880	50.26	1980	226.55

* Indicates decades in which the land area included was increased.

13. The data in table 2.3 give the population of the United States at 10-year intervals. It can be used to estimate the net annual growth rate for any of the decades shown there.
   (a) Use the data of table 2.3 to estimate the annual growth rate of the United States for the decade 1790-1800. Using that data, state an initial-value problem of the form $P' = kP$, $P(t_0) = P_i$, which models the population of the United States beginning in 1790. (Give appropriate numerical values for the parameters k, t_0, P_i.)

(b) Use separation of variables to find a solution for the initial-value problem you found in part (a). Compare the value given by your solution formula with the population value in table 2.3 in 1970. What factors might account for any discrepancies you observe?

(c) Repeat the preceding two steps beginning with 1850.

14. The analysis of the population model

$$\frac{dP}{dt} = kP, \ P(t_0) = P_i\,,$$

showed that the slope of the population curve is always positive when P is positive. Further, it suggested that the slope increases as P increases. Confirm this suggestion by showing that the population curve is concave up because $P''(t) > 0$. (Calculate an expression for P'' directly from $P' = kP$.) Is this behavior consistent with the sketch of figure 2.2?

Determine $P''(1970)$ for the United States population model of examples 1 and 2. According to this model, how fast is population growth accelerating?

Can the solution of this initial-value problem exhibit local maxima or minima?

15. Work out the details of the back-of-the-envelope calculation mentioned in the fourth paragraph of this section. That is, show that if the initial population is $1,000,000$ and the annual growth rate is 1%, then the population after 10 years is $1,104,622$, while at a growth rate of 3% the population rises to $1,343,916$. (If the annual growth rate is 3%, then $P_1 = 1.03P_0$, $P_2 = 1.03P_1 = (1.03)^2 P_0$, and so forth, where P_n is the population at the end of year n.)

16. The calculations in the preceding exercise suggest a population model of the form $P(t) = (1 + k)^t P(0)$, where k is the annual growth rate. Derive such a model, and explore its qualitative features. Are there limits to growth in such a model? Is your answer to that question consistent with the hypotheses upon which the model is based?

17. The analysis of the model $P' = kP$, $P(t_0) = P_i$, suggested that the population curve predicted by this model would have the general shape of figure 2.2.

(a) Starting with the population of Brazil in 1978 (see table 2.1), approximate the curve of figure 2.2 by drawing a straight line through the point $(1978, P(1978))$ with slope dP/dt given by $dP/dt = kP$. Use this line to estimate $P(1979)$. Then repeat the line drawing process beginning

at $(1979, P(1979))$. (The differential equation $dP/dt = kP$ will give you the slope of this new line segment using your value for $P(1979)$.) Continue for 10 years. (You are actually approximating the solution of the model $dP/dt = kP, P(t_0) = P_i$ by a numerical technique known as the Euler method. See section 3.5.)

(b) Establish a connection between this approximation technique and the model of exercise 15.

18. In deriving the population models in this section, we have repeatedly used logic that amounts to the following:

The net growth rate at time t is $kP(t)$. If Δt is not too large, then the increase in population is $kP(t)\Delta t$, the product of rate with time.

But no matter how small Δt is, the rate $kP(t)$ will not be constant. Could the approximation we are making here lead to a basic flaw in our population models?

This exercise asks you to carry out a more precise derivation, which shows that the approximation we described in the preceding paragraph is valid; our models are not flawed on this account.

To be definite, limit your attention to a population that has a constant net growth rate and which is free of immigration or emigration

(a) Argue that the increase in population from time t to $t + \Delta t$ is

$$\int_t^{t+\Delta t} kP(s)\, ds\,.$$

(*Hint:* The instantaneous rate of change — the population "velocity" — is $kP(t)$. How do you obtain distance traveled from the velocity function? Note that the variable of integration is s to avoid confusion with the particular time t at which the interval of interest begins.)

(b) Use the preceding result and the law of conservation of population to show that

$$P(t + \Delta t) - P(t) = \int_t^{t+\Delta t} kP(s)\, ds\,.$$

What is the corresponding equation in the derivation in the text?

(c) The next steps in the text's derivation are division by Δt and finding the limit as Δt approaches zero. The left-hand term in the preceding equation obviously reduces to dP/dt. Argue that the right-hand term can be written

$$\lim_{\Delta t \to 0} \frac{f(t + \Delta t) - f(t)}{\Delta t} = f'(t),$$

where $f(t) = \int_0^t kP(s)\, ds$.

(d) Use the fundamental theorem of calculus (see the Appendix) to show that $f'(t) = kP(t)$.

(e) Conclude that this more careful derivation leads to the same model: $dP/dt = kP$, $P(t_0) = P_i$.

19. The preceding exercise used the fundamental theorem of calculus in a careful derivation of the population model equation $dP/dt = kP$. Instead, use the mean-value theorem for integrals to show that

$$\int_t^{t+\Delta t} kP(s)\, ds = kP(\tilde{t})\Delta t$$

for some $\tilde{t}$, $t < \tilde{t} < t + \Delta t$. Then carry out a careful derivation of $dP/dt = kP$ similar to that in the preceding exercise. What becomes of $\tilde{t}$ as Δt approaches zero? (You can find a statement of the mean-value theorem for integrals in your calculus text.)

2.2 EMIGRATION AND COMPETITION

2.2.1 Emigration

In the fall of 1989, a sudden change in the policies of the government of the German Democratic Republic permitted mass emigration that briefly reached levels of approximately $12,000$ people per day, or an effective annual rate of 4.39×10^6 people per year. (If t is measured in years, we must express all rates in the same units.) Similarly, in the spring of 1980, the Cuban government allowed departures from its borders that amounted to about $5,000$ people per day, or 1.83×10^6 per year. (The qualitative features of such emigration rates are shown in figure 2.3.)

Emigration is people leaving, with apologies to those who know their prefixes.

If we retain the experimental facts of constant birth and death rates, how does the addition of emigration affect population levels?

To apply the law of conservation of population, we must account for all ways of increasing the population and all ways of decreasing the population over an arbitrary time interval Δt. There is only one way to increase population, births, and we have argued earlier that the increase in population over the time period Δt due to births is

$$k_b P(t)\Delta t$$

We have two ways to lose population, death and emigration. As before, the number lost to deaths over the time period Δt is

$$k_d P(t)\Delta t$$

Figure 2.3: An emigration function that reflects a sudden change in government policy to temporarily permit exit. The German Democratic Republic began permitting exits in September of 1989; Cuba, in May of 1980.

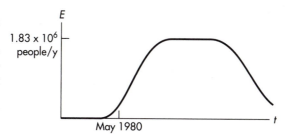

If individuals are leaving the population at the rate $E(t)$, the number lost due to emigration over the time period Δt is

$$E(t)\Delta t .$$

The law of conservation of population asserts that the net change in the population over the time interval from t to $t + \Delta t$ is the number of individuals added due to births less the number lost to deaths *and* to emigration. That is, conservation of population requires

$$P(t + \Delta t) - P(t) = k_b P(t)\Delta t - k_d P(t)\Delta t - E(t)\Delta t .$$

In words, we have just written

$$\text{net change over} \Delta t = \text{number added by births}$$

$$-\text{number lost to deaths}$$

$$-\text{number lost to emigration.}$$

Dividing the conservation of population expression by Δt and finding the limit as Δt approaches zero, just as in the derivation of (2.1), yields

$$\frac{dP}{dt} = kP - E ,$$

where we have again written $k = k_b - k_d$ for the net population growth rate.

Adding the same initial condition as before completes the model:

MODEL

$$\frac{dP}{dt} = kP - E , \tag{2.3}$$

$$P(t_0) = P_i . \tag{2.4}$$

We could try to find a formula for the population function $P(t)$ predicted by this model. Instead, we will learn as much as we can about the behavior of this population *directly* from the differential equation $dP/dt = kP - E$.

ANALYSIS

To begin an analysis of this model, note that we cannot be sure that dP/dt is positive at $t = t_0$. What is the critical initial population level that guarantees $dP/dt > 0$ at $t = t_0$? Can you argue that dP/dt remains positive if it is positive at $t = t_0$? That it remains negative if it is negative at $t = t_0$?

INTERPRETATION

If the questions just posed can be answered yes, then it appears that a sufficiently small initial population could be completely eliminated by emigration while a sufficiently large population could survive. Alternatively, we might use the initial population to determine the maximum rate of emigration a population could sustain.

EXAMPLE 3   *Using the following data, implement the emigration model (2.3)–(2.4) for Ireland during the period 1847-1850, the worst of the horrors of the potato famine.*

An average of about $209,000$ people per year left Ireland during this period. To simplify the situation, let us take a constant emigration rate equal to this average value:

$$E = 0.209 \text{ million people per year.}$$

One might estimate a net annual rate of population increase of 1.5%, or $k = 0.015$. The population of Ireland in 1847 was about 8 million.

The model (2.3)–(2.4) then has the form

$$\frac{dP}{dt} = 0.015P - 0.209, P(1847) = 8.$$

The slope of the population curve at $t = 1847$ is

$$\frac{dP}{dt} = 0.015P(1847) - 0.209 = 0.015 \cdot 8 - 0.209 = -0.089.$$

The population is decreasing! Convince yourself that the slope of this model's population curve will be forever negative. Had emigration continued unchecked, Ireland would have become deserted. ∎

Could we obtain a formula for the solution of the initial value problem that models Ireland during the potato famine? Yes, indeed, and either exercise 1 or 2(b) offers you that opportunity. Could we obtain a formula for the solution of the general emigration model $dP/dt = kP - E$, $P(t_0) = P_i$? Again, yes. See exercise 2(g).

2.2.2 Competition

The models we have examined so far have omitted an important constraint, competition. As population grows, food and space can become scarce, causing death by starvation and disease.

To model a population whose growth is limited by competition for resources, we could suppose that the birth rate is unaffected by competition, while the death rate rises with population to reflect competition. Hence, we postulate:

Tentative Idea 1 A constant fraction a of the population reproduces per unit time.

Similar to experimental fact 1 with a replacing k_b.

(Our ideas are too tentative to warrant a title as grand as "experimental fact.") Thus, the number of individuals added to the population in the time interval from t to $t + \Delta t$ is approximately $aP(t)\Delta t$.

We will suppose that the death rate varies in a way that reflects the number of encounters between individuals, because it is those encounters in the search for food and shelter that can limit population growth. A single individual in a population of size P could encounter $P - 1$ other individuals. Since there are P single individuals in the population, the total number of encounters among all individuals is proportional to $P(P - 1) \approx P^2$. Hence, we postulate:

Tentative Idea 2 The number of individuals who die per unit time is proportional to P^2.

Similar to experimental fact 2.

We denote the constant of proportionality by s; s is the analogue of k_d in experimental fact 2 at the beginning of the previous section.

> In essence, we have postulated that a certain fraction of encounters between competitors leads to the death of one of them. This cold logic conceals considerable anguish over the reality of this scenario in many parts of the world.

With these two tentative "facts" at hand, we can then claim that the increase in the population from t to $t + \Delta t$ is $aP\Delta t$, while the decrease is $sP^2\Delta t$. An application of the now familiar law of conservation of population yields

$$P(t + \Delta t) - P(t) = aP\Delta t - sP^2\Delta t.$$

The usual steps then yield

MODEL

$$\frac{dP}{dt} = aP - sP^2, \tag{2.5}$$

$$P(t_0) = P_i.$$

The differential equation in this limited population growth model is sometimes called the *logistic equation*. We defer the analysis and interpretation of this model to project 2. Likewise, the opportunity to find the solution of this initial value problem awaits in exercise 2.

> P. F. Verhulst (1804-1849) obtained the logistic equation by modifying the Malthusian hypothesis of constant growth rate to permit the net growth rate to decline.

2.2.3 Summary of Population Models

We again coupled a law with experimental facts to derive models. The law we used was conservation of population, rather than Newton's law. Conservation of population says that the change in population over a period of time is the number added less the number lost.

The three models we derived are initial value problems. The differential equation for the simplest growth model is

$$\frac{dP}{dt} = kP.$$

The growth rate constant is $k = k_b - k_d$, the difference of the birth rate and the death rate. The differential equation for a model of the same population subject to emigration at a rate E is

$$\frac{dP}{dt} = kP - E.$$

The logistic equation

$$\frac{dP}{dt} = aP - sP^2$$

arose in a model that supposed the death rate increased with population. Each of the models was completed by imposing the initial condition

$$P(t_0) = P_i.$$

We analyzed the behavior of these models in two ways, by directly obtaining slope information from the differential equation $dP/dt = \cdots$ and by finding the formula for the solution of the initial value problem.

EXERCISE GUIDE

To gain experience . . .	Try Exercises
Deriving models	5(a), 6– 7, 8(b), 9–10, 12, 18
Analyzing models from differential equations	3, 8(a), 11, 14–16, 17, 19
Analyzing models from a solution	4, 5(c), 13
Interpreting models	3–4, 5(b–c,) 11, 13–15
With concrete examples of models	1– 2, 5, 20
Solving problems like these models	1– 2, 4, 5(b), 13
With the basic ideas of these models	5, 8, 11, 14
With the mathematical subtleties of model derivation	21– 22

1. This exercise leads you through the process of finding a solution of the initial-value problem

$$\frac{dP}{dt} = 0.015\,P - 0.209, \ P(1847) = 8,$$

which models the Irish potato famine, and it asks you to use the solution formula you obtain to corroborate the analysis of the text.

(a) Separate variables in the differential equation so that you are left with the antidifferentiation problem

$$\int \frac{dP}{0.015\,P - 0.209} = \int dt.$$

Evaluate each of these antiderivatives.

(b) Exponentiate, bring a constant out of the exponential, and obtain the formula

$$P(t) = Ce^{0.015t} + \frac{0.209}{0.015},$$

where C is an arbitrary constant.

(c) Using the initial condition $P(1847) = 8$ to evaluate the constant C, obtain the solution

$$P(t) = (8 - \frac{0.209}{0.015})e^{0.015(t-1847)} + \frac{0.209}{0.015}$$

of the complete initial-value problem.

(d) Verify by substituting into the differential equation and the initial condition that you do indeed have a function which satisfies both.

(e) From an analysis of the differential equation itself, the text concludes that this model predicts the population of Ireland will decline. Show that the solution formula you have just obtained predicts precisely the same effect. When would the population of Ireland have been depleted? Could you have obtained that information without the solution formula?

2. Use separation of variables to solve each of the differential equations listed below. Verify that you have actually obtained a solution in each case by substituting your proposed solution into the original differential equation and verifying that it reduces to an identity. If a differential equation is accompanied by an initial condition, use it to determine the arbitrary constant C that appears in your solution.

(a) The differential equation given in example 3 as part of the model of Ireland during the potato famine,

$$\frac{dP}{dt} = 0.015P - 0.209.$$

(b) The model given in example 3, of Ireland during the potato famine,

$$\frac{dP}{dt} = 0.015P - 0.209, \ P(1847) = 8.$$

(c) The differential equation from a possible model for the German Democratic Republic during the fall of 1989, assuming emigration continued unchecked,

$$\frac{dP}{dt} = -0.0012P - 4.39.$$

(Population is in millions, time in years. From where in the text did we get these numbers?)

(d) A possible model for the German Democratic Republic during the fall of 1989, assuming emigration continued unchecked,

$$\frac{dP}{dt} = -0.0012P - 4.39, \ P(1989) = P_i \, .$$

(The growth rate value is that for 1985. The initial population appears just as the parameter P_i, since we do not have its exact value. How might you use the data of table 2.1 to make a reasonably accurate estimate of P_i?)

(e) The differential equation from a model of the population of Cuba during the spring of 1980, assuming emigration continued unchecked,

$$\frac{dP}{dt} = kP - 1.825 \, .$$

(Population is in millions, time in years. The growth rate appears just as the parameter k since we do not have its value. Where might you find a value for k?)

(f) The differential equation of the population model with constant emigration rate E,

$$\frac{dP}{dt} = kP - E.$$

(What difficulties might you encounter if E were a function of time rather than a constant?)

(g) The population model with constant emigration rate E,

$$\frac{dP}{dt} = kP - E, \ P(t_0) = P_i \, .$$

(h) The differential equation of the logistic population model,

$$\frac{dP}{dt} = aP - sP^2.$$

(i) The logistic population model,

$$\frac{dP}{dt} = aP - sP^2, \ P(t_0) = P_i \, .$$

3. Suppose the government had decided to limit Irish emigration in 1847 to a level that would keep the pop-

ulation constant. Use the data of example 3 to determine the number of people per year who could have emigrated then.

4. Follow the first four steps of exercise 1 to obtain a formula for the solution of the general emigration model $dP/dt = kP - E$, $P(0) = P_i$. Derive from your formula an expression for the critical initial population level that is required to prevent depletion of the population by a given emigration rate E. Find the time when the population reaches zero if the initial population is less than this critical value.

5. (a) Adapt the data from table 2.1 and the law of conservation of population to derive an emigration model for the German Democratic Republic during the fall of 1989, when approximately 12,000 people per day were emigrating.

(b) Find the solution of this model. If this rate of emigration had continued indefinitely, when would the country have been stripped of its population?

(c) What is the maximum emigration rate the government could have permitted without causing the population to decline to zero? Is emigration the sole cause of population decline in this model?

6. A population of yeast spores increases due to reproduction by a fixed proportion R (i.e., by $100R\%$) each day. The spores are harvested at a rate of H spores per day.

(a) Write an expression for the *rate* per day at which the number of yeast spores is increasing due to reproduction alone.

(b) Write an expression for the *rate* per day at which the number of yeast spores is decreasing due to harvesting alone.

(c) State a law of conservation of population and use it to derive a model for the number of yeast spores in this population.

7. For most of the past two centuries, the number of immigrants entering the United States has exceeded the number of emigrants departing. Derive an *immigration model* for such a situation. Carefully state the growth rate assumptions you use, and precisely define any terms you introduce to characterize immigration. Derive your model from the law of conservation of population.

8. The preceding exercise requests a careful derivation of a model for a population subject to *immigration*. The text derives a model for a population with constant net growth rate subject to *emigration*; the governing differential equation is the emigration equation $P'(t) = kP(t) - E(t)$.

(a) Call the rate of immigration $I(t)$. Comparing this situation with the emigration model suggests that an immigration model equation should have the form $P'(t) = kP \pm I(t)$. Which sign do you think is correct? Explain your reasoning.

(b) Derive the model as requested in the preceding exercise. Did you guess the form of the differential equation correctly? If not, what did you do wrong?

9. A population of amoeba in a particular region of a petri dish is under observation. The birth rate of the amoeba is proportional to the population. The rate at which amoeba die is proportional to the *square* of the population (to account for crowding effects).

In addition, amoeba move into or out of this region is response to an attractant released by other amoeba. The rate of movement of amoeba is proportional to the difference between the population in the region and the population A_{out} outside the region. The amoeba move *toward the higher population* level.

Let $A(t)$ denote the population of amoeba in the region at time t.

(a) Letting b denote the proportionality constant, write an expression for the *number* (not rate) of amoeba added to the population in the region by births over a time period of length Δt.

(b) Letting d denote the proportionality constant, write an expression for the *number* (not rate) of amoeba lost from the population in the region by deaths over a time period of length Δt.

(c) Letting r denote the proportionality constant, write an expression for the *number* (not rate) of amoeba added to the population in the region by the effect of the attractant over a time period of length Δt. (Note that population is added if $A > A_{out}$ and lost if $A < A_{out}$.

(d) State an appropriate conservation law for the amoeba population in the region. Derive a model for the population function $A(t)$ from this law.

10. A population of yeast spores increases by a fixed proportion k each day. Spores are harvested at a rate of E spores per day. Argue that this spore population is governed by the emigration model $P' = kP - E,\ P(t_0) = P_i$. Derive this model from the law of conservation of population.

11. An amount of money m is earning interest at a daily rate of I (or $100I$ percent). Regular deposits are made at a rate of $D(t)$ dollars per day. The account was opened with a starting balance of S dollars. Determine which three of the following initial-value

problems could *not* possibly be a reasonable model of this situation. *Justify* rejecting each unacceptable choice.

(a) $m' + Im = D, m(0) = S$
(b) $m' - Im = -D, m(0) = S$
(c) $m' - Im = D, m(0) = S$
(d) $m' + Im = -D, m(0) = S$

(Interest is being compounded continuously on the funds in this account.)

12. Show that the one initial-value problem you did not reject in the previous exercise is the correct model by deriving it from the "law of conservation of money": The net change in the balance in an account over an interval of time equals the amount added less the amount removed.

13. Assuming that the rate of deposit D is constant, solve the one initial-value problem you did not reject in exercise 11. Show that the solution function $m(t)$ you obtain exhibits behavior that is consistent with the situation being modeled there, a bank account that is earning interest and receiving regular deposits.

14. One of the initial-value problems listed in exercise 11 has exactly the same form as the emigration model $dP/dt = kP - E,\ P(t_0) = P_i$. Identify it. (Recall that exercise 11 considers a model of the balance in a bank account earning interest and receiving regular deposits.) By comparing the physical situations being modeled (population with emigration, bank account with regular deposits), argue that this initial-value problem can not possibly model the bank account.

15. The analysis of the model $P' = kP - E,\ P(t_0) = P_i$, raises several questions. Answer them and pursue the directions suggested by the interpretation of that model:

What is the critical initial population level that guarantees $dP/dt > 0$ at $t = t_0$? Can you argue that dP/dt remains positive if it is positive at $t = t_0$? That it remains negative if it is negative at $t = t_0$?

It appears that a sufficiently small initial population could be completely eliminated by emigration while a sufficiently large population could survive. Instead of determining a critical initial population level, use the initial population to determine the maximum rate of emigration a population could sustain.

16. Use the model $P' = kP - E,\ P(t_0) = P_i$, to determine the maximum emigration rate five countries of your choice from table 2.1, could sustain. (*Hint:* Use

the answer to the challenge posed by the last sentence of exercise 15.)

17. The analysis of the emigration model

$$\frac{dP}{dt} = kP - E, \ P(t_0) = P_i,$$

showed that the slope of the population curve at t_0 could be either positive or negative, depending on the size of the initial population P_i. If the emigration rate E is constant, argue that a population curve that starts with a positive (negative) slope always has a positive (negative) slope. Further, argue that the second derivative of the solution of this initial-value problem has the same sign as the first derivative, so that an increasing (decreasing) population curve is concave up (down). (*Hint:* Calculate P'' directly from $P' = kP - E$.)

Can population ever become negative in this model? Can the population curve exhibit local maxima or minima?

18. The logistic model $dP/dt = aP - sP^2$, $P(0) = P_i$, was derived in the text by assuming that the number of individuals dying per unit time is proportional to the square of the population. This exercise outlines a different approach.

The competition effect is difficult to assess quantitatively because it could affect both births and deaths, but it should act to decrease the net growth rate k as population increases. A linear, decreasing relationship between k and P may be the easiest way to achieve this variation. Mathematically, we can

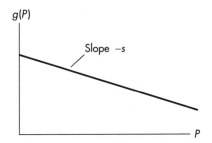

Figure 2.4: A growth rate that reflects competition by decreasing with increasing population.

write such a variable growth rate as

$$g(P) = a - sP$$

for some (positive) constants a, s. The variable growth rate g replaces the constant growth rate k used earlier.

The last equation is our new experimental "fact," such as it is. If s is a small positive constant, then $g(P)$ has the qualitative behavior sketched in figure 2.4. The constants a and s must be chosen to fit specific population data. Derive the logistic equation model by invoking conservation of population with population additions due only to the *net* growth term $g(P)P$ and with no removal term.

19. Repeat exercise 18 of section 2.1 for a population with a constant net growth rate k that is losing individuals to emigration at a rate of $E(t)$ individuals per unit time.

20. Repeat exercise 18 of section 2.1 for a population with a variable net growth rate $g(P) = a - sP$. Assume there is neither emigration nor immigration.

2.3   A HEAT-LOSS MODEL

A power transistor in a critical circuit nearly overheats. Will it cool sufficiently between duty cycles? The rate of cooling of molten metal in a casting determines the microstructure and ultimate strength of the part. How does the thickness of the walls of the mold affect the rate of cooling?

We could ask similar questions about heat loss in a chemical reactor or about a temperature sensitive experiment aboard the space shuttle. A parallel set of questions could inquire how rapidly devices warm up when they are exposed to hotter surroundings.

The mathematical models that provide answers to these questions all have similar form. To introduce such models and to minimize technical jargon, we will seek answers to a question about a very common object:

How does insulation affect the way the temperature drops in an unheated home?

We need two experimental facts.

Experimental Fact 1 The heat energy Q of a body is proportional to its mass and to its absolute temperature T.

Heat energy is exhibited in the vigor of the random motion of the molecules of the body. The constant of proportionality is called the **specific heat** of the body; a given weight of a material with a high specific heat, such as water, holds more heat energy per degree of temperature than an equal mass of material of lower specific heat, such as copper or air. The relationship is

You feel colder in the water at the beach than in the air because water's higher specific heat steals heat from your body faster.

$$Q = cmT, \qquad (2.6)$$

where c denotes the specific heat.

If Q is measured in calories, m in grams, and T in degrees Kelvin (degrees Celsius plus $273°$), then specific heat c has units of calories per gram per degree Kelvin (cal/g·K). Table 2.4 lists specific heats of some common materials.

The Kelvin temperature scale is the Celsius scale with its zero shifted $273°$ from the freezing point of water down to "absolute zero." For example, a comfortable room temperature of $20°$C is 293 K.

Although scientists usually use the calorie as a unit of heat energy, engineers in the United States often use the British thermal unit (Btu). One Btu is about 252 cal.

A calorie is the energy required to raise the temperature of one gram of water from $15°$C to $16°$C. Consequently, the specific heat of water is $c = 1$ cal/g·K.

We will also use the following.

Experimental Fact 2 The rate of flow of heat energy through a unit area of a surface (such as the wall of a house) is proportional to the temperature difference on either side of that surface. The heat flows in the direction of decreasing temperature, from hot to cold.

Table 2.4

Specific heats and thermal conductivities of selected materials. The thermal conductivity values are for a sample 1 cm thick.

material	specific heat c (cal/g • K)	thermal conductivity k (cal/s • K • cm^2)
Water	1.0	—
Glass	0.20	0.0025
Marble	0.21	0.0071
Copper	0.09	0.908
Wood	0.42	0.0003
Air	0.24	—
Red brick	—	0.0015
Flannel	—	0.00023

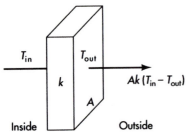

Figure 2.5: Rate of heat flow through a surface of area A. The direction from the inside to the outside is positive.

The proportionality constant, denoted by k, is known as the **thermal conductivity**. It depends on both the material and the thickness of the wall. If we take the direction inside-towards-outside as positive, the *rate of heat flow through a wall of area A* is

$$Ak(T_{\text{in}} - T_{\text{out}})$$

where T_{in} is the temperature of the inside surface and T_{out} is the temperature of the outside surface. See figure 2.5.

Thermal conductivity k for a material of given thickness has units of calories per second (rate of heat energy flow) per degree Kelvin (the temperature difference) per square centimeter (surface area). Table 2.4 lists thermal conductivities of some common materials.

As with population models, the governing law is one of conservation:

change = amount added − amount removed

Law of Conservation of Heat Energy The net change in the heat energy of a body over a given period of time is the amount of heat energy added less the amount of heat energy removed.

The net change of heat energy within a house is reflected in the change in the temperature of its interior as air, floors, walls, furniture, stereos, etc. gain or lose heat. Causes of heat-energy change include loss or gain by conduction through the walls and roof, heat supplied by external sources such as radiant heating by the sun, heat loss or gain from internal sinks or sources such as air conditioning or a furnace, and so on.

For the moment, suppose that our model home is unheated, without air conditioning, and situated in the shade; that is, it can gain or lose heat only by conduction through its walls and roof.

To derive a model for the temperature of the house, we begin with notation. Let $T(t)$ denote the temperature of the interior of the house at time t, and let T_{out} denote the outside temperature. If A denotes the total area of walls and roof combined and if k is the thermal conductivity of these surfaces (assuming roof and walls are equally well insulated), then the second experimental fact shows that the total heat energy lost in time Δt is

Rate times elapsed time Δt.

$$Ak(T(t) - T_{\text{out}})\Delta t \,.$$

This expression is indeed the heat *lost* because it is positive when $T(t) > T_{\text{out}}$. Heat energy flows out of the house when it is warmer than its surroundings.

If $Q(t)$ denotes the total heat energy of the house at time t, then the net change in heat energy over time Δt is just $Q(t + \Delta t) - Q(t)$. The law of conservation of heat energy tells us how to combine this expression for change in heat energy with the heat-loss term $Ak(T - T_{\text{out}})\Delta t$:

Conservation law: change = amount added − amount removed.

$$Q(t + \Delta t) - Q(t) = 0 - Ak(T(t) - T_{\text{out}})\Delta t \,.$$

(The zero following the equal sign represents the amount of heat added.) Just as in the derivation of the population model (2.1), we divide by Δt and take a formal limit as Δt approaches zero to obtain

$$\frac{dQ}{dt} = -Ak(T(t) - T_{\text{out}}) \,.$$

This differential equation contains two unknown functions, the heat energy Q of the house and its temperature T. But the first experimental fact relates these two quantities: $Q = cmT$. Here m is the mass of the entire house and its contents, and c is the (average) specific heat of the material in the house. From $Q = cmT$, we immediately obtain

$$\frac{dQ}{dt} = cm\frac{dT}{dt}.$$

Combining the last two displayed equations yields a differential equation for a single unknown, the temperature T of the interior of the house. Adding the starting temperature of the house T_i as an initial condition completes the model for the temperature of an unheated house,

$$\frac{dT}{dt} = -\frac{Ak}{cm}(T(t) - T_{out}),\qquad(2.7)$$

$$T(0) = T_i.\qquad(2.8)$$

If the house is initially warmer than the outside, that is, if $T_i > T_{out}$, then the model equation $T' = -(Ak/cm)(T - T_{out})$ shows that the temperature T of the house will decrease. As long as $T > T_{out}$, this decrease will continue. If $T = T_{out}$, then $T' = 0$, suggesting that the temperature of the house would remain at this constant value or *steady state* if it ever reached it.

The differential equation $T' = -(Ak/cm)(T - T_{out})$ reveals that the rate of change of temperature T' is proportional to the thermal conductivity k of the walls and roof of the house. All other factors being equal, we can reduce the rate of temperature drop in a house by a factor of two if we can reduce the thermal conductivity of the walls and roof by a factor of two through increased insulation.

Home insulation is rated by an R factor; R denotes *resistance*. It is the reciprocal of conductivity k. Doubling R corresponds to reducing k by a factor of two. A common attic insulation is rated R 25 for an 8-in. thickness. Figure 2.6 shows a newspaper ad for common building insulations.

The building industry's R factor has units $h/°F/ft^2/Btu$. The formula for conversion from thermal conductivity k in cal/s·K·cm^2 is $R = 3.45 \times 10^{-4}/k$, where k is measured for a sample 1 cm thick and R is the thermal resistance per inch of material.

To comment on specific temperature changes produced by variations in wall and roof insulation, we need an expression for the solution of the initial value problem (2.7)–(2.8).

EXAMPLE 4   *For a typical ranch house, the quotient Ak/cm might have the value 0.012 min^{-1}. The interior of the house is at a comfortable $20°$ C when the furnance quits. How cold will the house be one hour later?*

To write the model $T' = -(Ak/cm)(T - T_{out})$, $T(0) = T_i$, for this situation, we can use $Ak/cm = 0.012$ and $T_i = 20 + 273 = 293$K. Time zero is the instant the furnace stopped. But T_{out} is unspecified. The house might even heat up if it's hotter outside than in.

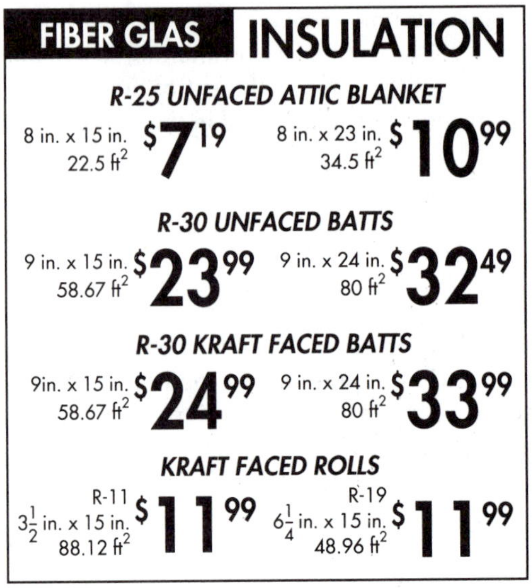

FIBER GLAS INSULATION

R-25 UNFACED ATTIC BLANKET

8 in. x 15 in. 7^{19}
22.5 ft²

8 in. x 23 in. 10^{99}
34.5 ft²

R-30 UNFACED BATTS

9 in. x 15 in. 23^{99}
58.67 ft²

9 in. x 24 in. 32^{49}
80 ft²

R-30 KRAFT FACED BATTS

9in. x 15 in. 24^{99}
58.67 ft²

9 in. x 24 in. 33^{99}
80 ft²

KRAFT FACED ROLLS

R-11
$3\frac{1}{2}$ in. x 15 in. 11^{99}
88.12 ft²

R-19
$6\frac{1}{4}$ in. x 15 in. 11^{99}
48.96 ft²

Figure 2.6: Building insulation with various R values; thermal resistance R is the reciprocal of thermal conductivity k.

The heat-loss model (2.7)–(2.8) takes the form

$$\frac{dT}{dt} = -0.012[T(t) - T_{\text{out}}], \ T(0) = 293 \,.$$

Perhaps we can use separation of variables to write a formula for the temperature of the house, leaving T_{out} as a parameter to be supplied later.

Separate.

If $T - T_{\text{out}} \neq 0$, then the separated differential equation is

$$\frac{dT}{T - T_{\text{out}}} = -0.012 \, dt \,.$$

Integrate.

If T_{out} is constant, we can integrate to obtain

$$\ln|T - T_{\text{out}}| = -0.012t + b \,,$$

where b denotes the constant of integration, to avoid confusion with the specific heat c. Exponentiation, writing $|C| = e^b$, and the usual argument that the sign of C determines the sign of $T - T_{\text{out}}$ yield

Solve for T.

$$T(t) = T_{\text{out}} + Ce^{-0.012t} \,.$$

Find C.

The initial condition $T(0) = 293$ requires $293 = T_{\text{out}} + C$, or $C = 293 - T_{\text{out}}$. The temperature of the house in degrees Kelvin t minutes after the failure of the furnace is

You check $T - T_{\text{out}} \neq 0$.

$$T(t) = T_{\text{out}} + (293 - T_{\text{out}})e^{-0.012t} \,.$$

Since $e^{-.012 \cdot 60} = 0.49$, the temperature 1 h after the furnace failed is $T = 293 \cdot 0.49 + T_{\text{out}} \cdot 0.51 = 143 + 0.51T_{\text{out}}.$ ∎

EXERCISE GUIDE

To gain experience . . .	Try Exercises
Deriving models	4, 5, 11–12(a), 13
Analyzing models from differential equations	4(b), 8, 11(b–c), 12(c)
Analyzing models from a solution	4(c), 5(b–c), 10
Interpreting models	4(b), 8, 10–11(b), 12(d)
With concrete examples of models	4(a), 5(a), 9
Solving problems like these models	1, 4(c), 5(b), 10(a), 11(d), 14(b)
With the basic ideas of these models	5(a), 7–9, 14
With the mathematical form of these models	6, 12(b)

1. Solve each of the following heat-loss model initial-value problems.

   (a) A model for the temperature of a typical ranch house at 20°C when its furnace breaks,

   $$\frac{dT}{dt} = -0.012(T - T_{\text{out}}), \; T(0) = 293 .$$

   (Time is in minutes and temperature in degrees Kelvin. Add the details omitted from the example in the text.)

   (b) The general heat-loss model,

   $$\frac{dT}{dt} = -\frac{Ak}{cm}(T - T_{\text{out}}), \; T(0) = T_i .$$

   (c) A model for the temperature of a house whose roof and walls have different thermal conductivities,

   $$\frac{dT}{dt} = -\frac{A_r k_r + A_w k_w}{cm}(T - T_{\text{out}}), \; T(0) = T_i .$$

2. A power transistor of mass m and surface area A is immersed in an oil cooling bath whose temperature is T_{out}. The thermal conductivity through the wall of the transistor housing to the oil bath is k. The specific heat of silicon is c. Derive a model for the temperature T of the transistor between duty cycles, when it is generating no heat of its own. Comment on any similarities with the house heat loss model $T' = -(Ak/cm)(T - T_{\text{out}}), T(0) = T_i$.

3. Suppose a diligent homeowner has insulated her attic so that its thermal conductivity is less than that of the house walls. Derive a model similar to $T' = -(Ak/cm)(T - T_{\text{out}}), T(0) = T_i$, for her house.

4. An individual of mass 70 kg will be traveling in winter temperatures of 5°C. He can choose to wear a suit of 1 cm thickness of red flannel or 1 cm of copper. (In one case he will be disguised as Santa Claus and in the other, as a droid.)

   (a) Using appropriate parameter values from table 2.4, write a form of the heat-loss model $T' = -(Ak/cm)(T - T_{\text{out}}), T(0) = T_i$, for the body temperature of this explorer while wearing each of these costumes. Neglect the heat generated by his body. (Since the human body is largely water, you can approximate this individual's specific heat as 1 cal/g·K. What should you use for an initial temperature? Can you estimate the area of a human body? Take care with units.)

   (b) Find the ratio of the rate of temperature decrease wearing copper to that wearing flannel. Which costume would you recommend for warmth? (*Neither* would be the answer if style were the main consideration.)

   (c) Find the temperature of his body after a 30-min exposure in each of these costumes, again neglecting the heat his body generates.

   Should you think of his clothing as a reservoir of heat or as insulation? How should you regard his body?

5. The example in the text considers a house for which Ak/cm has the value 0.012.

   (a) What would that value be if the insulation in the walls and roof were doubled, thereby halving the thermal conductivity of those surfaces?

(b) If the house is initially at $20°C$, what is its temperature after one hour without heat?

(c) For two specific values of outside temperature of interest to you, determine how much warmer this better insulated house is than the house in the example after an hour without heat.

6. (a) Is the heat-loss model $T' = -(Ak/cm)(T - T_{out})$, $T(0) = T_i$, similar in mathematical form to any of the models derived in sections 2.1 or 2.2? Identify any such similarities precisely.

(b) If you find a population model that is similar to this heat-loss model, indicate which variables and parameters in the two models are analogous.

(c) Write one differential equation that includes both models as special cases, depending on whether the unknown is interpreted as population or temperature.

7. The equation $T' = -(Ak/cm)(T - T_{out})$ contains the parameter m, the mass of the body losing heat. How might you reasonably compute m when modeling heat loss in a house?

8. How does changing any one of the parameters A, k, c, or m affect the rate of temperature drop in the model $T' = -(Ak/cm)(T - T_{out})$, $T(0) = T_i$? If the surface area were doubled, what could you do to keep the rate of temperature drop the same?

9. The temperature of a house dropped from $25°C$ to $19°C$ in 45 min when the outside temperature was $-2°C$. Estimate Ak/cm for this house.

10. One way of quantifying the relative efficiency of the insulation of a house is determining the time required for the temperature of an unheated house to drop half the way from its initial value of T_i to its final value of T_{out}.

(a) Find a formula for the time required for this temperature drop.

(b) If the total mass of the house could be doubled, perhaps by adding stone flooring or an indoor swimming pool, how would this decay time change? To increase this decay time, should you add materials with a high specific heat or a low specific heat?

11. A more accurate formulation of experimental fact 2 asserts that the rate of heat loss through a surface depends on the *fourth* power of the temperatures: the rate of heat energy flow through a surface of area A and thermal conductivity k is $Ak(T_{in}^4 - T_{out}^4)$.

(a) Using this version of the second experimental

fact, derive the model

$$\frac{dT}{dt} = -\frac{Ak}{cm}(T^4 - T_{out}^4), \quad T(0) = T_i.$$

(b) Does this model still exhibit a temperature decrease when the house is warmer than its surroundings? Is a steady-state temperature of T_{out} still possible?

(c) How do rates of temperature decrease compare between this model and the original model $T' = -(Ak/cm)(T - T_{out})$, $T(0) = T_i$?

(d) What antiderivative must you evaluate to solve this new model by separation of variables? List a set of steps that would evaluate it. Would you anticipate finding a simple solution formula?

12. (a) The model $T' = -(Ak/cm)(T - T_{out})$, $T(0) = T_i$, assumes the house contains no heat source. Derive a similar model for the temperature of a house that contains a furnace by incorporating into the conservation law a term that adds heat energy. Specifically identify the parameter(s) that characterizes the furnace.

(*Hint:* You could follow the lead of heating contractors and specify the heat output rate of the furnace; they usually use units of Btu/h. Alternatively, you could think of the furnace as a small, warm "house" inside the house you are modeling. You could then determine the rate at which it releases heat energy into the house.)

(b) How does the mathematical form of your new model equation compare with $T' = -(Ak/cm)(T - T_{out})$?

(c) Use your model to determine the rate at which the furnace must release heat to maintain the house at a specified constant temperature.

(d) A heating contractor would like a simple formula for calculating the size (heat release rate) of the furnace to be installed in a house. What can you suggest?

13. A power transistor of mass m and surface area A is immersed in an oil cooling bath whose temperature is T_{out}. The thermal conductivity through the wall of the transistor housing to the oil bath is k. The specific heat of silicon is c. When it is operating, it is generating heat internally at a rate of S cal/s. By allowing for this extra source of heat in the conservation law, derive a model for the temperature T of the transistor during a duty cycle. Comment on any similarities with the house heat loss model $T' = -(Ak/cm)(T - T_{out})$, $T(0) = T_i$. Do you see any similarities with the preceding problem, which seeks a model for a house with a furnace?

14. (a) Repeat exercise 17(a) of section 2.1.2 for the heat loss model $T' = -(Ak/cm)(T - T_{out})$, $T(0) = T_i$. The quantity Ak/cm might be about 0.012 min^{-1} for a typical ranch house. Try 5 minute time steps, and estimate the temperature change in one hour for conditions that interest you.

(b) Compare your estimate with the temperature given by the exact solution of the model for these conditions.

2.4 CHAPTER EXERCISES

1. You are given the following:

   Experimental fact A radioactive isotope decays at a rate proportional to its mass.

   Let $k > 0$ be the constant of proportionality and let $y(t)$ be the mass of isotope present at time t.

   (a) Derive the governing differential equation

   $$\frac{dy}{dt} = -ky,$$

   and complete the model by imposing appropriate initial conditions.

   (b) Analyze this model, and interpret your analysis in the light of your intuitive ideas about radioactive decay.

   (c) Solve your model initial value problem and interpret the behavior of the solution in the light of your intuitive ideas about radioactive decay.

   (d) Find the time required for the mass of the isotope to decay to one half its original value. Show that this half life is independent of the original mass.

   (e) Suggest appropriate units for y and k.

2. Repeat parts (a-c) of the preceding problem in the case in which the isotope is bombarded by a neutron beam which creates new unstable isotope at a rate of Q g/s. (Your model will include the differential equation $dy/dt + ky = Q$.)

   Conceivably, the processes of radioactive decay and creation of new isotope by the neutron beam could reach an equilibrium in which losses balance gains and the amount of isotope remains constant. Is this possible in your model? If so, what is the steady-state amount of isotope predicted by the model? Do both the differential equation and the solution of the initial value problem predict the same steady-state amount of isotope?

3. Use the following information to derive a model for the charge q on the capacitor in the RC circuit shown in figure 2.7:

   Kirchhoff's Voltage Law The sum of the voltage changes around a closed loop is zero.

   Experimental Fact 1 The voltage drop v_R across a resistor is proportional to dq/dt with proportionality constant R; R is the *resistance*.

   Experimental Fact 2 The voltage drop v_C across a capacitor is proportional to $q(t)$ with proportionality constant $1/C$; C is the *capacitance*.

   The following steps will lead you to the desired model:

   (a) Write a mathematical statement of each experimental fact.

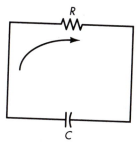

Figure 2.7: A source-free series RC circuit

(b) Write Kirchhoff's voltage law in terms of v_R and v_C.

(c) Combine the results of the preceding two steps to derive the governing differential equation

$$R\frac{dq}{dt} + \frac{1}{C}q = 0.$$

Complete the model for a series RC circuit by imposing an appropriate initial condition.

4. Suppose the RC circuit of the preceding problem now includes a voltage source (imposed emf) E. Derive the following model for this circuit:

$$R\frac{dq}{dt} + \frac{1}{C}q = E,\ q(0) = q_i.$$

See figure 2.8. Note that the source E *increases* voltage when the circuit is traversed in the direction of the arrow, while the resistor R and the capacitor C *decrease* voltage.

5. Analyze the models of the RC circuits in the two preceding problems by answering the following questions for each:

(a) By studying the sign of dq/dt in the differential equation, determine whether a given initial charge q_i on the capacitor will grow or decay (Both R and C are positive constants.)

(b) By studying the solution of each initial value problem, determine whether a given initial charge q_i on the capacitor will grow or decay.

(c) If charge stops flowing in either circuit, what is the steady state charge on the capacitor?

6. The preceding problems deal with RC circuits. Figure 2.9 shows an unusual RC circuit, Ben Franklin's kite flying in a lightning storm. The lightning arises from a voltage difference between the atmosphere and the kite. Current can flow down the wet kite string to the Leyden jar, a kind of capacitor, on the ground. In principle, you could calculate the capacitance of the Leyden jar and the resistance of the kite

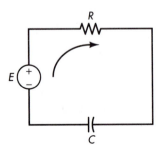

Figure 2.8: A series RC circuit with a voltage source E

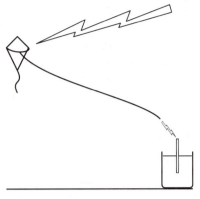

Figure 2.9: A schematic diagram of Ben Franklin's kite flying in a lightning storm; charge accumulates in the Leyden jar at the lower right.

string. Derive a model that could help determine the potential difference that caused the lightning bolt. Describe how you would use the model to estimate this potential difference.

7. Move the setting of the heat-loss model (2.7)–(2.8) from New England to the Southwest: A house whose air conditioner has failed will begin to heat up because the air around it is so hot. Derive a model for this setting, in which $T_{out} > T_i$. Show that you obtain *precisely* the same model: $T' = -(Ak/cm)(T - T_{out})$, $T(0) = T_i$. Perhaps this model would be better called a *heat-flow model*, since it can represent either heat loss or heat gain.

8. A particular chemical reaction produces a chemical at a rate proportional to the concentration of that chemical; the proportionality constant, called the reaction rate coefficient, is k. Diluting the mixture reduces the concentration of the chemical at a rate of d g/cm^3·s; d is positive. Which initial-value problem is an appropriate model of this situation? Justify your rejection of each unacceptable choice.

(a) $c'(t) + kc(t) = d,\ c(0) = c_i$

(b) $c'(t) + kc(t) = -d,\ c(0) = c_i$

(c) $c'(t) - kc(t) = -d,\ c(0) = c_i$

(d) $c'(t) - kc(t) = d,\ c(0) = c_i$

9. Consider the situation described in the previous problem.

(a) Extract from the problem statement the necessary experimental facts, state an appropriate conservation law, and derive a model. If the model you derive is not the same as the initial value problem you choose in the previous problem, find your error and explain it.

(b) Solve the initial value problem you accepted in the previous problem. Show that the behavior of this solution is consistent with the physical situation being modeled.

(c) Solve each of the three initial value problems you rejected in the previous problem. Show that each solution exhibits some physically unacceptable behavior.

10. A thin-walled mold containing molten plastic will be plunged into a cold-water bath to solidify the plastic rapidly. Introduce the necessary notation and derive a model for the temperature of the plastic in the mold. Does this model look like any of the models studied in this chapter?

11. A radioactive isotope is decaying at a rate proportional to its mass; the proportionality constant is $k > 0$. A neutron beam bombarding the sample creates new isotope at a rate of $i(t)$ g/s. If $c(t)$ represents the amount of isotope at time t, which initial value problem is a correct model of this situation? Justify your rejection of each unacceptable choice.
(a) $c'(t) + kc(t) = i(t)$, $c(0) = c_i$
(b) $c'(t) + kc(t) = -i(t)$, $c(0) = c_i$
(c) $c'(t) - kc(t) = i(t)$, $c(0) = c_i$
(d) $c'(t) - kc(t) = -i(t)$, $c(0) = c_i$

12. Consider the situation described in the previous problem.
(a) Extract from the problem statement the necessary experimental facts, state an appropriate conservation law, and derive a model. If the model you derive is not the same as the initial value problem you choose in the previous problem, find your error and explain it.
(b) Solve the initial value problem you accepted in the previous problem. Show that the behavior of this solution is consistent with the physical situation being modeled.
(c) Solve each of the three initial value problems you rejected in the previous problem. Show that each solution exhibits some physically unacceptable behavior.

13. Wastewater dumped into waterways often contains organic matter whose decay consumes oxygen in the water, depriving fish and other aquatic life of the oxygen they need to survive. Environmentalists term such pollution *biological oxygen demand* (BOD); it is measured in units of amount of oxygen per unit volume of water.
(a) Experiments suggest that BOD decays at a rate proportional to the amount present. Derive a model for the BOD level in a closed lake into which BOD is dumped at a given rate.
(b) Either by studying the differential equation in

your model or by solving the initial value problem, determine the fate of the lake. Will it be overwhelmed by BOD and die, or can the natural decay of BOD balance the steady influx of pollutants?
(c) Could the lake cleanse itself of BOD if the pollution were halted? Characterize how quickly the water quality would improve in terms of the parameters you have introduced in your model.

14. An entrepeneur has established a line of credit at a bank. She is charged a daily interest rate of 0.0493% on the total amount she owes the bank at the end of each business day. She borrows at a rate of $b(t)$ dollars per day, and her initial loan was in the amount a_i. If $a(t)$ denotes the amount she *owes* t days after the initial loan, which of the following would be a reasonable model for the amount she owes? Justify your rejection of each unacceptable choice.
(a) $a'(t) + 0.000493a = b(t)$, $a(0) = a_i$
(b) $a'(t) - 0.000493a = b(t)$, $a(0) = a_i$
(c) $a'(t) + 0.000493a = -b(t)$, $a(0) = a_i$
(d) $a'(t) - 0.000493a = -b(t)$, $a(0) = a_i$

15. An insect population (denoted by y) is being destroyed by an insecticide at a constant rate d. It is increasing at a rate proportional to the current population; the constant of proportionality is g. Which of the following initial value problems might be a reasonable model of this situation? Justify rejecting each unacceptable choice and keeping each reasonable choice.
(a) $y' + gy = d$, $y(0) = y_i$
(b) $y' - gy = d$, $y(0) = y_i$
(c) $y' + gy = -d$, $y(0) = y_i$
(d) $y' - gy = -d$, $y(0) = y_i$

16. A possible model for the temperature $T(t)$ of a house heated by a wood stove is

$$T' = -\frac{Ak}{cm}(T - T_{\text{out}}) \pm \frac{S}{cm}, \quad T(0) = T_i,$$

where S is a (positive) parameter representing the constant heat release rate of the stove. Should the last term in the differential equation be $+S/cm$ or $-S/cm$? Justify your choice.

17. Consider the equation $y' = ky + f(t)$.
(a) Examine each of the models derived in this and the previous chapter and in the exercises you have been assigned. Which can be written in this general form? In each case, identify the unknown $y(t)$, the constant k, and the forcing term $f(t)$.

Table 2.5
Quantities added to and removed from populations by various effects

Model name	Mechanisms considered by the model	Number added over Δt	Number lost over Δt
Simple	Natural birth at rate k_b Natural death at rate k_d	$k_b \Delta t$	$k_d \Delta t$
Net change over $\Delta t = k_b \Delta t - k_d \Delta t$			
Emigration	Natural birth at rate k_b Natural death at rate k_d Emigration at rate E	$k_b \Delta t$	$k_d \Delta t$ $E \Delta t$
Net change over $\Delta t = \cdots$			

(b) When $f(t) \equiv 0$, equations of this form are sometimes called *equations of growth or decay*. Why? How does the sign of k determine whether growth or decay occurs?

(c) Why could $f(t)$ reasonably be labeled a "source term"? What must be its sign to make it a genuine source? What is its sign when it is a so-called sink, or loss, term?

18. How might you use the date of table 2.1, page 21, for the G.D.R. and the D.D.R. to estimate the growth rate of a united Germany?

19. A trout farm raises fish in a closed pond. Young fish are added to the pond at a fixed rate A, and mature fish are harvested at a rate R. Derive a model for the population F of the fish in the pond, assuming that a fixed fraction of the trout reproduce per unit time.

20. Table 2.5 shows two population models and the mechanisms for adding or removing population which each considers. The entry for the simple population model is complete.
(a) What additional steps are required to derive the simple population model $P' = kP$, $P(0) = P_i$ from the expression "net change (in the population) over $\Delta t = \cdots$"? What law(s), if any, must you invoke?
(b) Complete the table entry for the emigration model and derive the model $P' = kP - E$, $P(0) = P_i$. Answer the previous questions for this model. Are your answers any different?

(c) Develop a table entry for the logistic model $P' = aP - sP^2$, $P(0) = P_i$.
(d) Develop a table entry for the "population" of isotopes described in exercise 1.
(e) Develop a table entry for the "population" of isotopes described in exercise 2.
(f) Develop a table entry for the "population" of chemical reactant described in exercise 8.
(g) Develop a table entry for the "population" of biological oxygen demand (BOD) described in exercise 13.
(h) Develop a table entry for the "population" of dollars borrowed on the line of credit described in exercise 14.
(i) Develop a table entry for the population of insects described in exercise 15.

21. Construct a table analogous to table 2.5 for each of the following heat-flow models:
(a) the house in exercise 7,
(b) the plastic mold described in exercise 10,
(c) the house heated by a wood stove described in exercise 16.

22. The preceding two problems consider population models (in various guises) and heat-flow models. Comment on the similarities between the tables derived in those two exercises. Can you draw a parallel between the mechanisms for change considered in each case? Can you draw a parallel between the underlying conservation laws?

2.5 CHAPTER PROJECTS

1. **Fitting population data** The simplest population equation is $P' = kP$. Argue that a plot of log P versus t should be a straight line for populations which are accurately modeled by this equation.

(a) Examine the United States population data in table 2.2, page 23, and find some periods of time over which this model is not unreasonable. What is the significance of the slope and the intercept of a line drawn through a plot of the *logarithm* of these data vs. time?

(b) Draw a line by eye through this log data plot. Use that line to estimate the growth rate k. Repeat for every time period for which the model $P' = kP$ seems reasonable.

(c) Using a spreadsheet or other software that will fit a line to data using least squares, estimate k for each of the cases considered in part (b). How do these estimated values compare with the annual growth rates listed in table 2.2?

2. **Analyzing the logistic model** The curve sketching tools developed in calculus can provide a qualitative interpretation of the logistic model

$$\frac{dP}{dt} = aP - sP^2, \ P(t_0) = P_i \, .$$

The idea is to learn as much as possible about the sign of the first and second derivatives of P to aid in sketching the population curve. (Plotting points precisely requires a formula for the solution of this initial value problem, but plotting points doesn't give much more information than this project.)

(a) Write the differential equation as $P' = (a - sP)P$ and argue that the sign of $P'(t_0)$ depends upon the relative values of P_i and the ratio a/s.

(b) Compute P'' directly from $P' = aP - sP^2$ and use information about the sign of P' to find the sign of P'' in the intervals $0 < P < a/2s$, $a/2s < P < a/s$, $a/s < P$. Does the population curve exhibit any inflection points?

(c) Note that $P' = 0$ when $P = a/s$. Argue that populations that are initially larger than a/s decay to the level a/s while those initially smaller than a/s grow to a/s. In either case, the population curve does not actually cross the horizontal line $P = a/s$. (Take this last assertion on faith for the moment; it is a consequence of the uniqueness of the solution of initial value problems. See section 3.9.) What is the growth rate of this population when $P = a/s$?

(d) Can the population curve exhibit any local maxima or minima?

(e) Combine this information into sketches of the population curves for initial populations in each of the three intervals listed in part (b).

(f) The ratio a/s is sometimes called the *carrying capacity* of the population. Explain this term.

(g) What happens to the growth rate of the population at $P = a/2s$?

3. **Growth rate in the logistic model** By writing the logistic equation (2.5), page 34, in the form $dP/dt = (a - sP)P$, we can think of it as arising from the simple population equation $dP/dt = kP$ when the constant growth rate k is replaced by a declining growth rate $g(P) = a - sP$.

Does the declining growth rate function $g(P)$ fit the growth rates obtained for the United States from table 2.3, page 29, over any periods of time? Attempt to estimate the parameters a and s (using regression software in a spreadsheet program if you wish). Use the results of the analysis requested in Project 2 (specifically, that all populations tend toward the value a/s) to predict a maximum United States population.

4. **How hot is the amplifier?** A stereo system amplifier is enclosed in a steel cabinet. Much of the electric power it consumes is released as heat by the components inside the cabinet. If the amplifier were not ventilated, what temperature might it reach?

   A typical amplifier enclosure is about $45 \times 15 \times 30$ cm. Its mass is about 9 kg. The thermal conductivity of the steel cabinet is about 0.05 cal/s·K·cm^2. The specific heat of its components is about 0.08 cal/g·K. If its power consumption in watts is P, then its components are releasing heat at about $4P$ cal/s. (1 watt = 4.186 cal/s; assume the other 0.186 cal/s goes to audible power.)

   (a) Write an expression for the *rate* at which heat energy is lost through the walls of the amplifier cabinet. Write an expression for the *rate* at which heat energy is being released by the components inside the cabinet. Use these expressions and the law of conservation of heat energy to derive a model for the temperature of the amplifier.

   (b) Obtain from your model an expression for the constant operating temperature of the unvented amplifier. What is the operating temperature of a 20-watt amplifier? A 50-watt amplifier? Would you recommend adding ventilation slots to amplifiers to lower their operating temperatures?

3

FIRST-ORDER
SOLUTION METHODS

I can evade questions without help; what I need is answers.
John F. Kennedy

This chapter develops solution techniques for first-order differential equations. It begins with definitions and terminology that classify differential equations in order to determine the structure of solutions and to guide the choice of solution methods.

3.1   BASIC DEFINITIONS

3.1.1   Equations and Solutions

Using examples, we recall the terminology introduced in section 1.3, and we provide more precise definitions of some important concepts.

The population equation

$$\frac{dP}{dt} = kP,$$

is of **first-order**. The unknown function, or dependent variable, is P and the independent variable is t. In the heat-loss equation

$$\frac{dT}{dt} = -\frac{Ak}{cm}(T - T_{\text{out}}),$$

the dependent variable is T, and t is again the independent variable. More generally, in the generic first-order differential equation

$$\frac{dy}{dx} = f(x, y),$$

y is the **dependent variable** and x is the **independent variable**.

The generic first-order **initial-value problem** seeks a function satisfying the two requirements,

$$\frac{dy}{dx} = f(x, y), \ y(x_0) = y_i.$$

The **initial condition** is $y(x_0) = y_i$. A familiar initial-value problem is the simple population model

$$\frac{dP}{dt} = kP, \ P(t_0) = P_i.$$

Since solutions of differential equations and of initial-value problems are a central theme of this chapter, they warrant a formal definition.

Definition 1 The function u is a **solution** on the interval $x_0 \leq x \leq x_1$ of the first-order differential equation $dy/dx = f(x, y)$ if u is continuously differentiable and if

$$\frac{du}{dx} = f(x, u) \text{ for } x_0 \leq x \leq x_1.$$

Further, u is a **solution of the initial-value problem** $dy/dx = f(x, y)$, $y(x_0) = y_i$, on the interval $x_0 \leq x \leq x_1$ if u is continuously differentiable,

$$\frac{du}{dx} = f(x, u) \text{ for } x_0 \leq x \leq x_1, \text{ and } u(x_0) = y_i.$$

Equivalently, u is a solution of an initial-value problem if u is a solution of the differential equation and if u has the correct initial value, $u(x_0) = y_i$. The interval of existence of a solution can be unbounded in either direction.

The same symbol is commonly used for both the unknown in a differential equation and for a specific solution function, rather than distinguishing them by y and u as in the preceding definition. Example 1 is an illustration.

EXAMPLE 1 *Show that $P' = kP$ has the solutions $P(t) \equiv 0$ and $P(t) = e^{kt}$ for all values of t.*

The differential equation is first-order, and both the functions 0 and e^{kt} have continuous first derivatives for all t.

Substituting $P \equiv 0$ into $P' = kP$ obviously yields an identity,

$$\frac{d0}{dt} = 0 = k0,$$

as does substituting $P = e^{kt}$,

$$\frac{dP}{dt} = \frac{de^{kt}}{dt} = ke^{kt} = kP.$$

Both functions are solutions of $P' = kP$. ∎

EXAMPLE 2   *Show that $P(t) = 6e^{kt}$ is a solution of the initial-value problem*

$$P' = kP, P(0) = 6.$$

The proposed solution $6e^{kt}$ obviously solves the differential equation:

$$P' = (6e^{kt})' = 6ke^{kt} = k(6e^{kt}) = kP.$$

To check the initial condition, set $t = 0$ in the proposed solution:

$$P(0) = 6e^{k0} = 6.$$

Hence, the function $P(t) = 6e^{kt}$ solves this initial-value problem because it satisfies both the differential equation and the initial condition.   ∎

An identically zero solution such as $P \equiv 0$ in Example 1 is called the **trivial solution**. *Trivial solution* does not mean *no solution*. The identically zero function is a legitimate, if boring, function.

A differential equation is **homogeneous** if it possesses a trivial solution. Otherwise, the equation is **nonhomogeneous**. Example 1 shows that $P' = kP$ is homogeneous by showing that $P \equiv 0$ is a solution. The heat-loss equation $T' = -(Ak/cm)(T - T_{\text{out}})$ is *non*homogeneous because $T \equiv 0$ is *not* a solution:

$$\frac{d0}{dt} = 0 \neq -\frac{Ak}{cm}(0 - T_{\text{out}}).$$

(Is this equation homogeneous for some value of T_{out}? For some value of k?)

EXAMPLE 3   *For what values of A is the equation*

$$P' - kP = A \sin \omega t$$

homogeneous?

To test homogeneity, substitute the proposed solution $P \equiv 0$:

$$0' - k0 = 0 = A \sin \omega t.$$

Hence, $P \equiv 0$ is a solution and the equation is homogeneous only if $A = 0$. (Although $A \sin \omega t$ is zero for isolated values of t such as $t = 0, \pi, 2\pi, ...$, this relation must hold for an interval of t values if $P \equiv 0$ is to be a solution.)   ∎

Like differential equations, initial conditions are called **homogeneous** if they are satisfied by the trivial solution and **nonhomogeneous** otherwise. The initial condition $P(0) = P_i$ is nonhomogeneous unless $P_i = 0$.

3.1.2 Linear Equations

The first-order differential equation in which $f(x, y)$ is linear in y is important because new solutions can be formed from sums and constant multiples of old solutions. Such equations are called *linear*.

> **Definition 2** A **linear, first-order ordinary differential equation** is a first-order differential equation that can be written in the form
>
> $$a_1(x)y' + a_0(x)y = r(x). \qquad (3.1)$$

Note that the "right-hand side," or **forcing function** $r(x)$ depends *only* on the independent variable x and *not* on the unknown y. Each term on the left-hand side of the prototype linear equation (3.1) involves but one occurrence of the unknown or its derivative. Neither y nor y' appears as an argument of a function. They are not raised to powers other than one, and y and y' do not multiply one another. That is, the unknown appears only in linear factors.

A first-order equation that cannot be written in the form (3.1) is called **nonlinear.**

EXAMPLE 4 *Show that the population equation $P' = kP$ is linear.*

This equation is linear because it can be written in the form (3.1)

$$1P' - kP = 0.$$

(The coefficient 1 was added in front of P' for emphasis.) We make the identification $a_1 = 1$, $a_0 = -k$, $r = 0$, $y = P$, and $x = t$. ∎

EXAMPLE 5 *Show that the logistic equation $P' = aP - sP^2$ is nonlinear.*

Attempting to put the logistic equation in the form (3.1) forces $r = -sP^2$:

$$1P' - aP = -sP^2.$$

The definition of linear first-order equation requires that the term r depend only on the independent variable, whereas this particular choice of r is a function of the unknown P.

We can not transpose the term $-sP^2$ to the left-hand side of the equation because the general form (3.1) permits on the left only terms involving the unknown raised to the first power. Trying to write sP^2 in the form $(sP)P$ to obtain

$$P' + (-a + sP)P = 0,$$

where $a_1 = 1$ and $a_0 = -a + sP$ does not help either, because the coefficient a_0 then depends on the unknown. ∎

When an equation is written in the linear form (3.1) with all unknowns collected on the left, linearity is, roughly speaking, a property only of the left-hand side. Think of "input" to the differential equation as substituting an expression in the left-hand side, and think of "output" as the resulting expression on the right side of the equality. Then linear equations are important because output is proportional to input.

A rubber band appears to be linear. If you pull it (input) twice as hard, it stretches (output) twice as far. But if you pull it too hard, it becomes nonlinear. First it refuses to stretch farther. Then it breaks.

EXAMPLE 6 *Is the equation*

$$P' - kP = A \sin \omega t$$

linear?

This equation is linear for the same reason that $P' - kP = 0$ is linear: We can still make the identification $a_1 = 1$, $a_0 = -k$, $y = P$, and $x = t$, as we did before. The forcing term on the right just changes to $r = A \sin \omega t$. ∎

3.1.3 Superposition

Linear differential equations are important because they obey the principle of superposition.

> **Theorem 3.1 (Principle of Superposition)** Let $y(x)$ be a solution of the general linear first-order equation with forcing term $r(x)$,
>
> $$a_1(x)y' + a_0(x)y = r(x).$$
>
> Let $z(x)$ be a solution of the same differential expression with forcing term $s(x)$:
>
> $$a_1(x)z' + a_0(x)z = s(x).$$
>
> Then the sum $u(x) = C_1 y(x) + C_2 z(x)$ is a solution of the equation with forcing term $C_1 r(x) + C_2 s(x)$,
>
> $$a_1(x)u' + a_0(x)u = C_1 r(x) + C_2 s(x).$$

PROOF The sum u must have the same differentiability properties as the given solutions y and z. We simply need to verify by substitution that it satisfies the appropriate differential equation:

$$
\begin{aligned}
a_1(x)u' + a_0(x)u &= a_1[C_1 y + C_2 z]' + a_0[C_1 y + C_2 z] \\
&= C_1[a_1 y' + a_0 y] + C_2[a_1 z' + a_0 z] \\
&= C_1 r(x) + C_2 s(x)
\end{aligned}
$$

These calculations freely interchanged the derivative operation, addition, and multiplication by a constant; i.e., they made full use of the *linearity* of the first derivative, a property you have used often but may not have called by that name. ∎

A schematic representation of the principle of superposition is

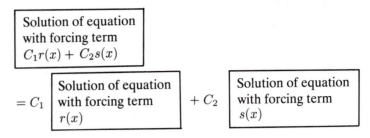

The principle of superposition says that adding two inputs to a linear equation gives as output the sum of their individual outputs, doubling the input to a linear equation doubles the output, and so on.

From the principle of superposition, we immediately deduce

- If y_h is a solution of a linear, homogeneous equation, then Cy_h is again a solution for any constant C.
- If y_p is any particular solution of a linear, *non*homogeneous equation and if y_h is any solution of the corresponding homogeneous equation, then $y_p + Cy_h$ is again a solution of the *non*homogeneous equation.

Exercise 50 asks you to prove these assertions.

The following examples illustrate these two points.

EXAMPLE 7 *The equation $P' = kP$ is homogeneous and linear. It was shown above to have the solution $P_h = e^{kt}$. Show that any multiple of P_h is again a solution.*

Let C be an arbitrary constant. To show that Ce^{kt} is a solution of $P' = kP$, substitute into the equation we hope it satisfies:

$$\frac{d(Ce^{kt})}{dt} = C\frac{d(e^{kt})}{dt} = C(ke^{kt}) = k(Ce^{kt}).$$

Since $(Ce^{kt})' = k(Ce^{kt})$, the expression Ce^{kt} is indeed a solution of $P' = kP$ for *any* value of C. ∎

Use $g = 9.8$ m/s².

EXAMPLE 8 *The rock tossing equation $v' = -9.8$ is linear and nonhomogeneous. One particular solution is $v_p = -9.8t$. Find additional solutions of $v' = -9.8$ by adding to $v_p = -9.8t$ solutions of the corresponding homogeneous equation $v' = 0$.*

Obviously, one solution of the homogeneous equation $v' = 0$ is $v_h = 1$. If C is an arbitrary constant, are expressions of the form $v = -9.8t + Cv_h = -9.8t + C$ solutions of $v' = -9.8$?

Of course, the solutions $v = -9.8t + C$ are just what we obtained by integrating $v' = -9.8$. ∎

EXAMPLE 9 *The heat-loss equation*

$$\frac{dT}{dt} = -\frac{Ak}{cm}(T - T_{\text{out}})$$

3 First-Order Solution Methods

is linear and nonhomogeneous. When T_{out} is constant, one particular solution is $T_p = T_{out}$. Construct new solutions of this equation by adding to T_p constant multiples of a solution of the related homogeneous equation.

Rewritten in the standard linear form, this equation is

$$\frac{dT}{dt} + \frac{Ak}{cm}T = \frac{Ak}{cm}T_{out}. \tag{3.2}$$

Since the derivative of a constant is zero, $T_p = T_{out}$ is obviously a solution, as claimed. (To derive this solution, see exercise 43.)

To obtain the homogeneous form of this equation, replace the forcing term $(Ak/cm)T_{out}$ by zero:

$$\frac{dT}{dt} + \frac{Ak}{cm}T = 0.$$

A solution (obtained by separation of variables) is $T_h = e^{-Akt/cm}$.

Now we must verify that for any constant C,

$$T = T_p + CT_h = T_{out} + Ce^{-Akt/cm}$$

is a solution of the nonhomogeneous heat-loss equation (3.2). Direct substitution of the formulas for T_p and T_h would suffice. Instead, we will use our knowledge of which equation is satisfied by each of the terms in the proposed solution $T_p + CT_h$.

Since T_p solves the nonhomogeneous equation (3.2), we can write

$$\frac{dT_p}{dt} + \frac{Ak}{cm}T_p = \frac{Ak}{cm}T_{out}.$$

Likewise, since T_h solves the homogeneous equation $T' + (Ak/cm)T = 0$, we have

$$\frac{dT_h}{dt} + \frac{Ak}{cm}T_h = 0.$$

Now substitute the proposed solution $T = T_p + CT_h$ into the left side of the heat-loss equation (3.2). From the preceding two equations satisfied by T_p and T_h, we obtain

$$\frac{d(T_p + CT_h)}{dt} + \frac{Ak}{cm}(T_p + CT_h)$$

$$= \frac{dT_p}{dt} + \frac{Ak}{cm}T_p + C\left(\frac{dT_h}{dt} + \frac{Ak}{cm}T_h\right)$$

$$= \frac{Ak}{cm}T_{out} + C \cdot 0$$

$$= \frac{Ak}{cm}T_{out}.$$

Thus, $T_p + CT_h$ is again a solution of the nonhomogeneous heat-loss equation. ∎

In this example, we used our knowledge of which equations T_p and T_h satisfied, not the formulas for these functions. When T_p was the input to the left-hand side,

$$\frac{dT}{dt} + \frac{Ak}{cm} T,$$

the output was the desired right-hand side,

$$\frac{Ak}{cm} T_{\text{out}}.$$

When CT_h was the input to this same left-hand side, the output was zero,

$$C \cdot 0 = 0,$$

because T_h solves the homogeneous equation. Hence, input of the combination $T_p + CT_h$ gave the same output as input of T_p alone.

To state this observation differently, $T_p + CT_h$ and T_p are solutions of the same differential equation because that equation is linear and because T_h solves the homogeneous version of that equation. Adding a multiple of T_h to the solution contributes nothing to the right-hand side of the equation.

What advantage does the solution $T_p + CT_h$ have over the solution T_p? It contains an arbitrary constant C, whose value we can choose to satisfy an initial condition. This property is the basis of a key concept, the *general solution* of a linear equation.

3.1.4 General Solution

First we formalize some terminology. A **particular solution** of a differential equation is just that, any specific solution of the equation. The homogeneous equation corresponding to a given first-order linear equation is that obtained by replacing the forcing term r by zero in the general linear form (3.1). A **homogeneous solution** of a first-order linear equation is any *nontrivial* solution of the corresponding homogeneous equation.

> *Homogeneous solution* is shorter than the more accurate *nontrivial solution of the corresponding homogeneous equation*.

Definition 3 Let y_p be any particular solution of the first-order linear equation (3.1). Let y_h be a homogeneous solution. Let C be an arbitrary constant. Then the combination

$$y_g = y_p + C y_h$$

is a **general solution** of (3.1).

The subscript g in y_g denotes *general*. A general solution contains a whole family of solutions, one for each value of C, the constant of integration. A general solution has two building blocks, a particular solution of the original equation and a nontrivial solution of the corresponding homogeneous equation.

The principle of superposition guarantees that a general solution of a linear equation is *always* a solution. We will see later that *every* solution of a linear differential equation can be obtained from a general solution by choosing C appropriately. Many of the solution techniques we develop will be tailored to finding either a particular solution of the nonhomogeneous equation or a solution of the homogeneous equation.

EXAMPLE 10   *Explain why*

$$v(t) = -9.8t + C$$

is a general solution of the rock-tossing equation $v' = -9.8$.

The differential equation $v' = -9.8$ is certainly first-order and linear; *general solution* is defined. This proposed general solution is just the result of integrating $v' = -9.8$. Clearly, we can identify $v_p = -9.8t$ as a particular solution because $v'_p = (-9.8t)' = -9.8$. Any nonzero constant is a nontrivial solution of the corresponding homogeneous equation $v' = 0$; we could choose the homogeneous solution $v_h = 1$. Then $v = -9.8t + C$ is a general solution because it has the form $v = v_p + Cv_h$, where v_p is a particular solution of the nonhomogeneous equation and v_h is a nontrivial solution of the homogeneous equation. ∎

EXAMPLE 11   *Write a general solution of the heat-loss equation*

$$\frac{dT}{dt} = -\frac{Ak}{cm}(T - T_{\text{out}})$$

for T_{out} constant.

Since this is a first-order linear equation, we can sensibly seek a general solution. From the work in example 9, we have a particular solution $T_p = T_{\text{out}}$ and a homogeneous solution $T_h = e^{-Akt/cm}$. (If you are in doubt, verify by substitution that T_p is a solution of the given equation and that T_h is a solution of the corresponding homogeneous equation $T' + (Ak/cm)T = 0$.)
Then a general solution is

$$T_g = T_p + CT_g = T_{\text{out}} + Ce^{-Akt/cm}.$$ ∎

Now put that constant C in the general solution to use.

EXAMPLE 12   *Find a solution of the initial-value problem modeling heat loss from a house,*

$$\frac{dT}{dt} = -\frac{Ak}{cm}(T - T_{\text{out}}), \ T(0) = T_i,$$

when T_{out} is constant.

Set $t = 0$ in the general solution $T_g = T_{\text{out}} + Ce^{-Akt/cm}$ of the previous example and choose C to satisfy $T(0) = T_i$:

$$T_g(0) = T_{\text{out}} + C = T_i$$

To check, substitute the solution in the equation and verify $T(0) = T_i$.

or $C = T_i - T_{out}$. The solution of the initial-value problem is

$$T(t) = T_{out} + (T_i - T_{out})e^{-Akt/cm}.$$ ∎

By definition, a homogeneous equation always has the trivial solution. Using the trivial solution as a particular solution, we see that a *general solution of a homogeneous equation* is just a constant multiple of any nontrivial solution.

EXAMPLE 13 *Find a general solution of the homogeneous population equation* $P' = kP$.

The obvious particular solution is the trivial solution, $P_p = 0$. A nontrivial solution is $P_h = e^{kt}$. A general solution is $P_g = P_p + CP_h = 0 + Ce^{kt} = Ce^{kt}$, just the solution we obtained earlier by separation of variables.

A particular solution is *any* function that solves the equation.

We also could have chosen $P_p = e^{kt}$. Then we would have written $P_g = P_p + CP_h = (1 + C)e^{kt}$. But $1 + C$ is a single arbitrary constant that we could relabel C. We would find ourselves back at $P_g = Ce^{kt}$.

The homogeneous equation $P' = kP$ has the trivial solution, too. Could we choose $P_h = 0$? No! The homogeneous solution must be nontrivial. ∎

Observing that the general solution of a homogeneous equation has the form Cy_h leads to an alternate statement of the definition of general solution:

A *general solution* of a *non*homogeneous equation is the sum of a particular solution of the *non*homogeneous equation and a general solution of the *homogeneous* equation.

Schematically, that relationship is

General solution of *non*homogeneous equation	=	Particular solution of *non*homogeneous equation	+	General solution of *homogeneous* equation

EXAMPLE 14 *When the emigration rate E is constant, a solution of the nonhomogeneous emigration equation*

$$P' - kP = -E$$

is E/k. Find a general solution of this equation and verify that it satisfies the equation.

We are given a particular solution, $P_p = -E/k$. (You can verify this solution by substitution. To derive it, see exercise 44.)

The corresponding homogeneous equation is $P' - kP = 0$, and we know from Example 7 that its general solution is Ce^{kt}. Hence, a general solution of the nonhomogeneous emigration equation is $P_g = -E/k + Ce^{kt}$.

To verify that P_g is indeed a solution, substitute $P_p(t) + CP_h(t)$ into $P' - kP = -E$:

$$(P_h + P_p)' - k(P_h + P_p) = (P_p' - kP_p) + (P_h' - kP_h)$$

$$= -E + 0 = -E.$$ ∎

Did we need to know the actual formulas for $P_h(t)$ and $P_p(t)$ in the last example? Why not?

3.1.5   Summary

Homogeneous differential equations have the **trivial** (identically zero) solution. **Nonhomogeneous** equations do not.

Linear first-order equations are those that can be written in the special form

$$a_1(x)y' + a_0(x)y = r(x).$$

Every linear equation has a **general solution** of the form

$$y_g = y_p + Cy_h,$$

where y_p is any **particular** solution of the given equation and y_h is a **homogeneous** solution, a nontrivial solution of the corresponding homogeneous equation obtained by setting the original forcing function r to zero. The general solution $y_p + Cy_h$ solves the same equation as does the particular solution y_p because of the **principle of superposition**: Roughly, the sum of the inputs gives the sum of the individual outputs.

Given a general solution of a differential equation, an initial-value problem is easy to solve: Just choose the constant C to satisfy the initial condition.

3.1.6   Exercises

EXERCISE GUIDE

To gain experience . . .	Try Exercises
With linear versus nonlinear	1–26, 29(a–b), 38–39, 42, 48
With homogeneous versus nonhomogeneous	1–28, 29(c–d), 32–33
Checking solutions of equations	30, 41, 43–44(b, d)
With particular solutions	29(e), 30, 38–39, 40–41(a), 43–44(a)
With homogeneous solutions	29(f), 31, 38–39, 40–41(b), 43–44(c), 49
Finding and using general solutions	34–39, 40–41(c), 42, 43–44(e), 45, 47
Solving initial value problems	34–37(b), 46
With the principle of superposition	38–42, 46, 50

For the equations in exercises 1–24:

i. Classify each as linear or nonlinear and as homogeneous or nonhomogeneous, justifying each classification,

ii. Identify the variables corresponding to x and y in the generic form $y' = f(x, y)$ and find the form of f.

1. $v' = -9.8$

2. $v' = -g$

3. $u' = u^2 \cos \pi t$

4. $dy/dt = \cos t$

5. $dy/dx = 4e^x$

6. $v' = -g - (k/m)v$

7. $v' = -g - (k/m)v|v|$

8. $P' = 0.0102P$

9. $dT/dt = -0.0002(T - 5)$

10. $dP/dt = 0.015P - E$

11. $du/dy = y^2 - 1$

12. $P' = kP - E$

13. $P' = kP$

14. $P' = aP - sP^2$

15. $T' = -(Ak/cm)(T - T_{out})$

16. $y' + xy = 0$

17. $y' + 3y = -3x$

18. $y' + xy = -3x^2$

19. $y' + xy = -3y$

20. $y' + xy = -3y^2$

21. $u'(t) - u^2 + u = 0$

22. $u'(t) - t^2 + u = 0$

23. $\sin \pi x = x^3 y'(x)$

24. $\sin \pi y = x^3 y'(x)$

25. Write the heat-loss equation

$$\frac{dT}{dt} = -\frac{Ak}{cm}(T - T_{out})$$

in the standard linear form

$$a_1(x)y' + a_0(x)y = r(x),$$

and verify that the homogeneous form of this equation is

$$\frac{dT}{dt} = -\frac{Ak}{cm}T.$$

26. Write the emigration equation

$$\frac{dP}{dt} = kP - E$$

in the standard linear form

$$a_1(x)y' + a_0(x)y = r(x),$$

and verify that the homogeneous form of this equation is

$$\frac{dP}{dt} = kP.$$

27. For what value(s) of T_{out} is the heat-loss equation

$T' = -(Ak/cm)(T - T_{out})$ homogeneous? For what value(s) of k? A? c? m?

28. For what value(s) of E is the emigration equation $P' = kP - E$ homogeneous? For what value(s) of k?

29. Either explicitly or implicitly, Example 9 makes the following claims about the heat-loss model equation

$$\frac{dT}{dt} = -\frac{Ak}{cm}(T - T_{out}) \qquad (3.3)$$

and its corresponding homogeneous equation

$$\frac{dT}{dt} = -\frac{Ak}{cm}T. \qquad (3.4)$$

Verify each.
   (a) Equation (3.3) is linear. (Explicitly identify a term appearing in (3.3) with each of the symbols $a_1, a_0, r, y,$ and x appearing in the definition (3.1) of linear equation.)
   (b) Equation (3.4) is linear.
   (c) Equation (3.3) is nonhomogeneous. (Are there any values of the parameters in (3.3) for which it is homogeneous? If so, what are the parameters and what are those values?)
   (d) Equation (3.4) is homogeneous. (Does (3.3) reduce to the homogeneous equation (3.4) for the parameter values you found in part (c)?)
   (e) The function $T_p = T_{out}$, T_{out} a constant, is a solution of (3.3).
   (f) The function $T_h = e^{-Akt/cm}$ is a solution of (3.4).

30. You are reading the text and encounter the statement, "Let $y_p(x)$ denote a particular solution of the general linear, first-order equation $a_1(x)y' + a_0(x)y = r(x)$." Use that statement to complete the equation

$$a_1(x)y_p' + a_0(x)y_p = ?$$

31. You are reading the text and encounter the statement, "Let y_h denote any solution of the homogeneous equation $a_1(x)y' + a_0(x)y = 0$." Use that statement to complete the equation

$$a_1(x)y_h' + a_0(x)y_h = ?$$

32. Verify each of the following statements about linear, first-order differential equations.
   (a) If a particular solution is the identically zero function, then the equation is homogeneous.

(b) If the equation is homogeneous, then a particular solution is the identically zero function.

(c) Only homogeneous equations have trivial solutions. (Is this statement related to either of the preceding?)

33. Consider the statement, "If the equation is linear, then it is homogeneous if the forcing term is zero, and it is nonhomogeneous otherwise."

(a) Give examples of a homogeneous linear equation and a nonhomogeneous linear equation which illustrate this statement.

(b) Prove this statement for the general first-order linear equation

$$a_1(x)y' + a_0(x)y = r(x) .$$

34. **(a)** Solve the heat-loss equation $T' = -(Ak/cm)(T - T_{out})$ by separation of variables, assuming T_{out} is constant. Show that you obtain a general solution.

(b) Use this general solution to solve the initial-value problem $T' = -(Ak/cm)(T - T_{out})$, $T(0) = T_i$.

35. **(a)** Solve the population equation $P' = kP$ by separation of variables. Show that you obtain a general solution.

(b) Use this general solution to solve the initial-value problem $P' = kP$, $P(t_0) = P_i$.

36. **(a)** Solve the emigration equation $P' = kP - E$ by separation of variables, assuming E is constant. Show that you obtain a general solution.

(b) Use this general solution to solve the initial-value problem $P' = kP - E$, $P(t_0) = P_i$.

37. **(a)** Solve the rock-tossing equation $v' = -g$ by antidifferentiation. Show that you obtain a general solution.

(b) Use this general solution to solve the initial-value problem $v' = -g$, $v(0) = v_i$.

38. You are given the following information: $T_p(t)$ is a function that solves the equation

$$\frac{dT}{dt} = -\frac{Ak}{cm}(T - T_{out}) ,$$

and $T_h(t)$ solves

$$\frac{dT}{dt} = -\frac{Ak}{cm}T .$$

Show that for any value of the constant C, $T_p + CT_h$ solves the first equation. (You could find formulas for T_p and T_h. Do not. Use only the information given

here. There is an example in the text that will be of great help.)

39. You are given the following information: $P_p(t)$ is a function that solves the equation

$$\frac{dP}{dt} = kP - E ,$$

and $P_h(t)$ solves

$$\frac{dP}{dt} = kP.$$

Show that for any value of the constant C, $P_p + CP_h$ solves the first equation. (You could find formulas for P_p and P_h. Do not. Use only the information given here.)

40. Example 9 shows $T_p + CT_h$ is a solution of the heat-loss equation

$$\frac{dT}{dt} = -\frac{Ak}{cm}(T - T_{out})$$

if T_p is a particular solution of this equation and T_h is a solution of the related homogeneous equation. That example never uses the formulas for either of those solutions,

$$T_p = T_{out}, \quad T_h = e^{-Akt/cm} .$$

(The expression for T_p assumes T_{out} is constant.) *Using these formulas,*

(a) Verify that T_p is a particular solution of $T' = -(Ak/cm)(T - T_{out})$.

(b) Verify that T_h is a solution of the corresponding homogeneous equation. State that equation explicitly.

(c) Verify that $T_p + CT_h$ is a solution of $T' = -(Ak/cm)(T - T_{out})$ for any value of the constant C.

41. According to the principle of superposition, $P_p + CP_h$ should be a solution of the emigration equation

$$\frac{dP}{dt} = kP - E$$

if P_p is a particular solution of this equation and P_h is a solution of the related homogeneous equation. Assuming E is constant, formulas for those solutions are

$$P_p = E/k, \quad P_h = e^{kt} .$$

Using these formulas:

(a) Verify that P_p is a particular solution of $P' = kP - E$.

(b) Verify that P_h is a solution of the corresponding homogeneous equation. State that equation explicitly.

(c) Verify that $P_p + CP_h$ is a solution of $P' = kP - E$ for any value of the constant C.

42. Example 9 argues that $T_p + CT_h$ is a solution of the heat-loss model equation stated there. Mimic those steps to show that $y_p + Cy_h$ is a solution of the general linear equation $a_1(x)y' + a_0(x)y = r(x)$. Instead of a direct verification, could you have appealed to the principle of superposition? Recall that y_p is a particular solution of $a_1(x)y' + a_0(x)y = r(x)$ and that y_h is a solution of the corresponding homogeneous equation.

43. Example 9 claims that a particular solution of the heat-loss model equation

$$\frac{dT}{dt} = -\frac{Ak}{cm}(T - T_{\text{out}})$$

is $T_p = T_{\text{out}}$ (assuming T_{out} is constant) and that a corresponding homogeneous solution is $T_h = e^{-Akt/cm}$.

(a) Obtain this particular solution yourself by

  i. Assuming the solution has the form $T_p = L$, where L is some constant,

  ii. Substituting this guess into the equation and solving for L.

  (Do not worry about making an assumption about a solution you do not yet know. If the assumption is wrong, you will encounter a contradiction of some sort. If it is correct, you will find the solution. Nothing ventured, nothing gained!)

(b) Verify by substitution that $T_p = T_{\text{out}}$ is indeed a solution of the equation.

(c) Obtain the homogeneous solution using separation of variables.

(d) Verify by substitution that the homogeneous solution is correct.

(e) Write a general solution of this differential equation.

44. Example 14 claims that a particular solution of the population model equation with emigration

$$P' - kP = -E$$

is $P_p = E/k$ (assuming E is constant) and that a corresponding homogeneous solution is $P_h = e^{kt}$.

(a) Obtain this particular solution yourself by

  i. Assuming the solution has the form $P_p = c$, where c is some constant,

  ii. Substituting this guess into the equation and solving for c.

(b) Verify by substitution that $P_p = E/k$ is indeed a solution of the equation.

(c) Obtain the homogeneous solution using separation of variables.

(d) Verify by substitution that the homogeneous solution is correct.

(e) Write a general solution of this differential equation.

45. Use the information given in each part to write an expression that is a general solution of the differential equation described there. If the information is insufficient and you can not find what you need or if a general solution is not appropriate, explain.

(a) A nonhomogeneous differential equation has the solution $\sqrt{x}$. The homogeneous version of this equation has the solution e^x.

(b) A linear, first-order, nonhomogeneous differential equation has the solution $\sqrt{x}$. The homogeneous version of this equation has the solution e^x.

(c) The equation $y' + m(x)y = 4x + 1$ has the solution $p(x)$. The equation $y' + m(x)y = 0$ has the solution $h(x)$.

(d) The equation $y' - y = e^x$ has the solution xe^x.

46. If possible, use the given information to find a solution of the indicated initial-value problem. If you can not solve the given initial-value problem, explain why.

(a) Given: y is a solution of $u' - 4u = \tan 2x$, $y(1) = 4$. Solve $u' - 4u = 3\tan 2x$, $u(1) = 12$.

(b) Given: y is a solution of $u' - 4u^2 = \tan 2x$, $y(1) = 4$. Solve $u' - 4u^2 = 3\tan 2x$, $u(1) = 12$.

(c) Given: $P' = \cos \pi x - 4xP$, $H' = -4xH$, $P(0) = H(0) = 2$. Solve $y' + 4xy = \cos \pi x$, $y(0) = 1$.

(d) Given: $P' = \cos \pi x - 4xP$, $H' = -4xH$, $P(0) = -2$, $H(0) = 0$. Solve $y' + 4xy = \cos \pi x$, $y(0) = 1$.

(e) Given: $P' = \cos \pi x - 4xP$, $H' = -4xH$, $P(0) = H(0) = 2$. Solve $y' + 24xy = 6\cos \pi x$, $y(0) = 1$.

47. Verify that

$$y(x) = C\exp\left(-\int_0^x (a_0(s)/a_1(s))\,ds\right)$$

is a general solution of $a_1(x)y' + a_0(x)y = 0$ if $a_1 \neq 0$.

48. Recall the often-repeated calculus slogan, "The derivative of a sum is the sum of the derivatives." Explain why the principle of superposition is a consequence of this statement. (The principle of superposition follows from the linearity of the derivative operation.)

49. *Prove*: The differential equation $y' = f(x, y)$ is homogeneous if and only if $f(x, 0) = 0$ for all x.

50. By making appropriate choices of r, s, C_1, and C_2 in the principle of superposition, verify two claims made in the text:

(a) If y_h is a solution of a linear, homogeneous equation, then Cy_h is again a solution for any constant C.

(b) If y_p is any particular solution of a linear, *non*homogeneous equation and if y_h is any solution of the corresponding homogeneous equation, then $y_p + Cy_h$ is again a solution of the *non*homogeneous equation.

3.2 SEPARATION OF VARIABLES

We review the **separation of variables** solution technique introduced in chapter 1.

Linear, first-order, homogeneous differential equations, among others, are **separable**: All terms involving the unknown (dependent) variable can be placed on one side of the equality, and all terms involving the independent variable can be placed on the other.

Let us apply separation of variables to a general linear, first-order, homogeneous differential equation

$$a_1(x)\frac{dy}{dx} + a_0(x)y = 0.$$

Replacing dy/dx by a quotient of the differentials dy and dx and collecting all terms involving y on the left and all terms involving x on the right lead to

$$\frac{dy}{y} = -\frac{a_0}{a_1}dx, \qquad (3.5)$$

as long as $a_1 \neq 0$ and $y \neq 0$. An antiderivative yields

$$\ln|y(x)| = -\int_0^x \frac{a_0(s)}{a_1(s)}ds + c. \qquad (3.6)$$

We invert the logarithm with the exponential to obtain

$$|y| = \exp\left(-\int_0^x (a_0/a_1)\,ds + c\right) = |C|\exp\left(-\int_0^x (a_0/a_1)\,ds\right),$$

where $|C| = e^c$. But we can remove the absolute value signs around y because the expression

$$y = Ce^{-\int_0^x (a_0/a_1)\,ds} \qquad (3.7)$$

has constant sign and satisfies the differential equation whether C is positive or negative. Furthermore, $y \neq 0$ unless $C = 0$, in which case y is the trivial solution. We have a formula (3.7) for the solution of the linear homogeneous equation $a_1y' + a_0y = 0$.

Do not memorize this formula! It is the first of many, too many to memorize. Understand the process instead. Test your understanding by deriving this formula with the book closed.

The manipulations involving differentials that lead to (3.5) are purely formal. How can we be sure that the expression in (3.7) is really a solution of $a_1 y' + a_0 y = 0$? One approach is to substitute the solution expression (3.7) directly into the differential equation, as exercise 23 requests.

A second approach is to replace the potentially dubious manipulations leading to (3.5) with obviously correct procedures. To that end, write $a_1 y' + a_0 y = 0$ as

$$\frac{1}{y}\frac{dy}{dx} + \frac{a_0(x)}{a_1(x)} = 0,$$

where we assume as before that $a_1 \neq 0$ and $y \neq 0$. Knowing the result we seek, we recognize that the left-hand side is just a derivative,

$$\frac{1}{y}\frac{dy}{dx} + \frac{a_0(x)}{a_1(x)} = \frac{d}{dx}\left(\ln|y| + \int \frac{a_0(x)}{a_1(x)}\,dx\right),$$

providing a_0/a_1 is integrable. Hence,

$$\ln|y| + \int \frac{a_0(x)}{a_1(x)}\,dx = c.$$

We have found our way by more certain means to (3.6), and the derivation of the final formula for y can now continue as before.

Not only has separation of variables given us a solution, it has produced a *general solution* of this first-order, linear, homogeneous equation. The expression (3.7) has the structure of an arbitrary constant multiplying a nontrivial solution of the homogeneous equation. We will see later that *every* solution of the linear equation $a_1 y' + a_0 y = r$ can be obtained from the general solution by choosing an appropriate value for C (providing $a_1 \neq 0$).

EXAMPLE 15   *Solve the population model initial-value problem*

$$\frac{dP}{dt} = 0.02P, \quad P(1960) = 1.2\,.$$

We begin solving any linear initial-value problem by finding a general solution, which separation of variables can provide in this case:

$$\int \frac{1}{P}\,dP = \int 0.02\,dt\,,$$

$$\ln|P| = 0.02t + c\,,$$

$$P(t) = Ce^{0.02t}\,.$$

To reach the last step, replace e^c by $|C|$.

Now use the initial condition $P(1960) = 1.2$ to evaluate the arbitrary constant C. We find $P(1960) = 1.2 = Ce^{(0.02)(1960)} = Ce^{39.2}$ or $C = 1.2e^{-39.2}$. Hence, the solution of this initial-value problem is

$$P(t) = 1.2e^{-39.2}e^{0.02t}.$$

Using $39.2 = 0.02 \cdot 1960$, we can write this solution more naturally as:

$$P(t) = 1.2e^{0.02(t-1960)}.$$

Now the growth-rate parameter 0.02, the starting time 1960, and the initial population level 1.2 are all readily apparent. ∎

Of course, separation of variables is not limited to linear equations.

EXAMPLE 16 *Find a solution of the logistic equation*

$$P' = aP - sP^2.$$

The solution formula (3.7) does not apply to this nonlinear equation.

The appropriate separation is

$$\int \frac{dP}{aP - sP^2} = \int dt.$$

The antiderivative on the left can be evaluated using partial fractions. We obtain

$$\frac{1}{a} \ln \left| \frac{P}{a - sP} \right| = t + c.$$

Multiplying by a, inverting the logarithm with the exponential function, and solving the resulting expression for $P(t)$ finally yields

$$P(t) = \frac{Cae^{at}}{1 + Cse^{at}}, \qquad (3.8)$$

where $|C| = e^{ac}$. These manipulations assume $P, a - sP \neq 0$, and they consider separately the cases $a - sP > 0$ and $a - sP < 0$. ∎

This solution formula (3.8) does not have the form of a general solution; it is not a constant multiplying a homogeneous solution. Of course, the logistic equation is nonlinear, and *general solution* is defined only for linear equations.

EXAMPLE 17 *Find the solution of the logistic population model*

$$P' = aP - sP^2, \qquad P(0) = \frac{a}{s}.$$

To evaluate the constant C in the solution formula (3.8), we set $t = 0$:

$$P(0) = \frac{Ca}{1 + Cs} = \frac{a}{s}.$$

Simplification yields

$$Ca = \frac{a}{s} + Ca,$$

a relation that holds for *no finite value* of C. Is there no solution of this initial-value problem?

There is a solution,

$$P(t) \equiv \frac{a}{s}.$$

Direct substitution shows $P = a/s$ solves $P' = aP - sP^2$. Obviously, $P(0) = a/s$. We simply could not obtain this solution from the separation of variables formula (3.8). What condition does the initial value a/s violate? ∎

We emphasize: *General solution* is defined only for linear equations, and every solution of a linear initial-value problem can be obtained from the general solution by an appropriate choice of the arbitrary constant. But not every solution of a nonlinear equation can be obtained from a solution expression that contains an arbitrary constant.

> Some texts define *general solution* to be any solution that contains an arbitrary constant, whether the differential equation is linear or nonlinear. We have just seen that a "general solution" of a nonlinear equation is not general enough to include all solutions. Hence, we do not define general solution for nonlinear equations.

Having developed this solution method, a natural question is, To which first-order equations can separation of variables be applied? Formula (3.7) shows it can be applied to any homogeneous, first-order linear equation.

Can separation of variables be applied to an arbitrary nonhomogeneous linear first-order equation? In general, the answer is no, unless, for example, the coefficients and the forcing term are all constant.

EXAMPLE 18   *Attempt to solve the differential equation*

$$\frac{dP}{dt} = kP - \alpha e^{-t},$$

which could arise in an emigration model with a decaying emigration term $E = \alpha e^{-t}$.

The forcing term $-\alpha e^{-t}$ gets in the way. Attempting to put P on the left and t on the right always leads to expressions like

$$dP = (kP - \alpha e^{-t})dt.$$

We can not find the antiderivative of the expression on the right; we can not compute $\int P(t)dt$ if we do not know the form for P as a function of t. Attempting to divide through by P leads to a similar quandary. ∎

Note that the same equation with a *constant* emigration rate E can be solved by separation of variables.

Exercise 26 asks you to demonstrate that any differential equation of the form

$$\frac{dy}{dx} = Y(y)X(x)$$

can be solved by separation of variables as long as $Y \neq 0$. An example follows.

EXAMPLE 19 *Solve*

$$\frac{dy}{dx} = y^4 \sin x, \ y(\pi/2) = \frac{1}{2}.$$

Separating variables leads to

$$\frac{dy}{y^4} = \sin x dx.$$

An antiderivative and a little algebra give

$$y(x) = \frac{1}{(3\cos x + C)^{1/3}}.$$

The initial condition forces $1/C^{1/3} = 1/2$ or $C = 8$. Hence, a solution of this initial-value problem is

$$y(x) = \frac{1}{(3\cos x + 8)^{1/3}}.$$ ∎

The differential equation here is an example of the general form $y' = Y(y)X(x)$ with $Y(y) = y^4$ and $X(x) = \sin x$. Does the solution we just obtained fulfill the condition $Y(y) \neq 0$? What happens if the initial condition is changed to $y(\pi/2) = 2$?

3.2.1 Exercises

EXERCISE GUIDE

To gain experience . . .	Try Exercises
Checking solutions	13, 17(a), 23(a)
Using separation of variables to solve equations	1–12, 15–16, 18–19, 21–22(a) 24, 25(b)
With general solutions	3, 5–10, 12, 15, 21–22(a), 23(b)
Solving initial value problems	2–8, 16, 17(c), 18, 20(a), 21–22(b), 23(c)
With conditions for validity of separation of variables	17(b–c), 18, 25(a,c), 26–27
With linear, homogeneous, etc.	14
Analyzing models using solutions	4, 20(b–d), 21(b–c), 22(c)

Solve 1–12. State the conditions under which your solution is valid. If separation of variables can not be applied, explain why.

1. $y' = y^2 \cos x$

2. $y' = y^2 \cos x$, $y(\pi) = \frac{1}{2}$

3. $P' = kP$, $P(t_0) = P_i$

4. $P' = kP^2$, $P(0) = P_i$ (If $k > 0$ and this were a population model, what disaster would it predict?)

5. $T' = -(Ak/cm)(T - T_{out})$, $T(0) = T_i$, T_{out} constant

6. $P' = kP - E$, $P(t_0) = P_i$, E constant

7. $P' = 0.015P - 0.209$, $P(1847) = 8$

8. $P' = 0.015P - 0.209(1 - \sin \pi t)$, $P(1847) = 8$

9. $y' = 2y + 1$

10. $y' = 2y + x$

11. $y' = 2y^2 + 1$

12. $y' = 2x(y + 1)$

13. Check each of the solutions you obtained in exercises 1–12 by substituting into the differential equation and by verifying the initial value, if appropriate.

14. Which of the differential equations given in exercises 1–12 are linear? Homogeneous?

15. Section 3.3.3 will introduce the *integrating factor*, another tool for solving first-order equations. The integrating factor μ is defined by a separable differential equation, such as the following, taken from examples in Section 3.3.3 and exercises in Section 3.3.5. In each case, find μ.
 (a) $\mu' = x\mu$
 (b) $\mu' = \mu \cos x$
 (c) $\mu' = (2/x)\mu$, $x \geq 1$
 (d) $\mu'(t) = (Ak/cm)\mu(t)$
 (e) $\mu' = g(x)\mu$, g continuous for all x

16. Example 19 solves the initial value problem $y' = y^4 \sin x$, $y(\pi/2) = \frac{1}{2}$, but it omits most of the details. Fill them in.

17. Example 19 uses separation of variables to solve the initial value problem $y' = y^4 \sin x$, $y(\pi/2) = \frac{1}{2}$. It obtains the solution

$$y(x) = \frac{1}{(3 \cos x + 8)^{1/3}}.$$

 (a) Verify that this expression is a solution of the initial value problem.
 (b) Verify that the separation of variables process never involves dividing by zero.

 (c) What happens if the initial condition is replaced by $y(\pi/2) = 2$?

18. Example 17 showed that the initial value problem

$$\frac{dP}{dt} = aP - sP^2, \quad P(0) = \frac{a}{s}$$

for the logistic equation could not be solved using the solution formula (3.8) obtained by separation of variables,

$$P(t) = \frac{Cae^{at}}{1 + Cse^{at}}.$$

Study the steps in the derivation of this solution formula, and determine what condition is violated by a solution that assumes the initial value $P(0) = a/s$.

19. Supply the details of the derivation of the solution (3.8) of the logistic equation. In particular, evaluate the antiderivative

$$\int \frac{dP}{aP - sP^2}$$

and carefully solve

$$\frac{1}{a} \ln \left| \frac{P}{a - sP} \right| = t + c,$$

for P.

20. This exercise asks you to analyze the behavior of the logistic population model

$$P' = aP - sP^2, \quad P(0) = P_i,$$

using the solution (3.8) of the differential equation obtained from separation of variables,

$$P(t) = \frac{Cae^{at}}{1 + Cse^{at}}.$$

As you will see, having a solution formula is only a beginning to understanding the behavior predicted by the model.

 (a) Use formula (3.8) for the solution of the logistic equation to write an expression for the solution of the initial value problem. Must you exclude any values of P_i? (*Hint:* See example 17.)

(b) Show that $\lim_{t \to \infty} P(t) = P_\infty$ exists and find its value. How does this behavior for large time differ from that of the simple population model $P' = kP$, $P(0) = P_i$?

(c) Show that if $0 < P_i < P_\infty$, then population is increasing for all time. What happens if $P_\infty < P_i$? If $P_i = P_\infty$?

(d) Based on this analysis, provide an interpretation of the population behavior predicted by the logistic model.

21. Consider the emigration equation

$$\frac{dP}{dt} = kP - E$$

when E is constant.

(a) Find the general solution of this equation.

(b) Find the solution of this equation subject to the initial condition $P(t_0) = P_i$.

(c) Verify that there is a critical initial population level in the sense that P will decay if P_i is too small while it will grow if P_i is sufficiently large.

22. Consider the heat-loss equation

$$\frac{dT}{dt} = -\frac{Ak}{cm}(T - T_{\text{out}})$$

when T_{out} is constant.

(a) Find the general solution of this equation.

(b) Find the solution of this equation subject to the initial condition $T(0) = T_i$.

(c) How do the relative values of T_i and T_{out} influence the behavior of the solution? Is this behavior intuitively reasonable?

23. The text used separation of variables to solve the general homogeneous linear equation

$$a_1(x)y' + a_0(x)y = 0.$$

For $a_1 \neq 0$, it obtained the solution

$$y = C \exp\left(-\int_0^x (a_0/a_1)\, ds\right).$$

(a) Verify by substitution that this expression is indeed a solution.

(b) Verify that this expression is a *general* solution.

(c) Write the solution of this differential equation subject to the initial condition $y(0) = y_i$.

24. Use separation of variables to derive the solution

$$y(x) = y_i \exp\left(-\int_{x_0}^x (a_0/a_1)\, ds\right)$$

of the initial value problem

$$a_1(x)y' + a_0(x)y = 0, \qquad y(x_0) = y_i.$$

Assume $a_1 \neq 0$ and a_1, a_0 continuous.

25. The text contains the following paragraph:

Can separation of variables be applied to an arbitrary nonhomogeneous linear first-order equation? In general, the answer is no, unless, for example, the coefficients and the forcing term are all constant.

(a) Show that if the coefficients a_1, a_0 and the forcing term f in $a_1y' + a_0y = f$ are independent of x, then the equation is separable.

(b) Obtain a formula for the general solution of this equation.

(c) Can you find an example of the equation $a_1y' + a_0y = f$ that is separable without the coefficients and the forcing term being constant?

26. The text observes that we can separate equations of the general form $y' = Y(y)X(x)$. Answer the following questions about such equations and the applicability of separation of variables to them.

(a) Must either X or Y be restricted to be nonzero?

(b) Is such an equation necessarily linear? If your answer is no, find conditions on X and/or Y that guarantee the equation is linear.

(c) Is such an equation necessarily homogeneous? If your answer is no, find conditions on X and/or Y that guarantee the equation is homogeneous.

27. Show that any equation of the form

$$\frac{dy}{dx} = Y(y)X(x)$$

can be solved by separation of variables on an interval in which $Y \neq 0$. Write an expression that defines solutions of this equation. What conditions must be satisfied by X and Y to guarantee that the antiderivatives in this solution actually exist? Is the logistic equation $P' = aP - sP^2$ in this form?

3.3 LINEAR EQUATIONS: SOLUTION METHODS

This section briefly surveys the principal analytic solution methods for the linear first-order equation

$$y' + g(x)y = f(x).$$

Central ideas are highlighted to keep from losing them among the calculations.

3.3.1 Separation and the Homogeneous Equation

We saw in the previous section that separation of variables can quickly solve the homogeneous equation

$$y' + g(x)y = 0$$

if $g(x)$ is continuous. From

$$\frac{dy}{y} = -g(x)\,dx,$$

we obtain

$$\ln|y(x)| = -\int_a^x g(s)\,ds + c.$$

To avoid confusion with the independent variable x, the variable of integration is denoted by s.

Exponentiation and setting $C = \pm e^c$ yield the *general solution of the homogeneous equation* $y' + g(x)y = 0$,

General solution of $y' + g(x)y = 0$

$$y_h(x) = C\exp\left(-\int_a^x g(s)\,ds\right). \tag{3.9}$$

The lower limit of integration a is arbitrary. It is usually chosen to be the value of x at which an initial condition is given.

EXAMPLE 20 *Solve $y' + 6x^2y = 0$, $y(-1) = 2$.*

Separation, integration, and exponentiation yield a general solution:

$$\frac{dy}{y} = -6x^2\,dx,$$

$$\ln|y(x)| = -6\int_{-1}^x s^2\,ds$$

$$y_g(x) = C\exp\left(-6\int_{-1}^x s^2\,ds\right).$$

General solution

The initial condition $y(-1) = 2$ is easy to satisfy because we have taken the initial point $x = -1$ as the lower limit of integration:

$$y_g(-1) = Ce^0 = C = 2.$$

Hence, the solution of the initial-value problem is

$$y(x) = C \exp\left(-6 \int_{-1}^{x} s^2 \, ds\right).$$

Of course, we can evaluate the integral in the exponent to write

Solution of initial-value problem

$$y(x) = 2e^{-2(x^3+1)}.$$ ∎

3.3.2 Variation of Parameters and the Nonhomogeneous Equation

To solve the nonhomogeneous equation

$$y' + g(x)y = f(x),$$

we use a technique known as *variation of parameters*. Although it may seem quite arbitrary at first, closer inspection shows it to be a consequence of linearity.

Suppose we know a nontrivial solution y_h of the homogeneous equation,

$$y_h' + g(x)y_h = 0,$$

for example, from (3.9). **Variation of parameters** begins by assuming that a particular solution of the *non*homogeneous equation can be written in the form

$$y_p(x) = u(x)y_h(x),$$

where u is to be determined.

> This method is also called *variation of constants* because the expression $u(x)y_h(x)$ looks like the homogeneous solution $Cy_h(x)$ with the constant C allowed to vary.

Substituting $y_p = uy_h$ in the nonhomogeneous equation

$$y' + g(x)y = f(x)$$

yields

$$(uy_h)' + g(x)uy_h = f(x),$$
$$u(y_h' + g(x)y_h) + u'y_h = f(x),$$
$$u'(x) = \frac{f(x)}{y_h(x)}.$$

> These calculations raise two questions. What happened to the term $y_h' + g(x)y_h$ in parentheses in the second equation? How can we be sure that $y_h \neq 0$? Exercise 14 asks you for answers.

From the last equation, we find

$$u(x) = \int_a^x \frac{f(s)}{y_h(s)}\,ds + C,$$

and a particular solution of $y' + g(x)y = f(x)$ is

$$y_p(x) = y_h(x) \int_a^x \frac{f(s)}{y_h(s)}\,ds + C y_h(x).$$

Closer inspection reveals that we have a *general* solution of $y' + g(x)y = f(x)$; the first term is a particular solution, and the second is a constant multiple of a nontrivial solution.

From (3.9) with $C = 1$, we can choose $y_h = \exp\left(-\int_a^x g(s)\,ds\right)$. Then a general solution of the nonhomogeneous equation is

General solution of $y' + g(x)y = f(x)$ $\qquad y_g(x) = \int_a^x f(s) \exp\left(-\int_s^x g(r)\,dr\right) ds + C \exp\left(-\int_a^x g(s)\,ds\right).$ (3.10)

> Don't memorize this formula. Understand the process of variation of parameters. The key is $y_p(x) = u(x)y_h(x)$.

EXAMPLE 21 *Solve $y' + 6x^2 y = \cos \pi x$, $y(-1) = -8$.*

From the previous example, we know a solution of the homogeneous equation $y' + 6x^2 y = 0$,

$$y_h(x) = 2e^{-2(x^3+1)}.$$

To use variation of parameters to solve the nonhomogeneous equation, we assume a particular solution has the form $y_p(x) = u(x)y_h(x)$ and substitute. *Without using the formula for y_h,* we find

$$(uy_h)' + 6x^2 uy_h = \cos \pi x,$$

$$u(y_h' + 6x^2 y_h) + u'y_h = \cos \pi x,$$

$$u'(x) = \frac{\cos \pi x}{y_h(x)}.$$

Hence,

$$y_p(x) = y_h(x) \int_{-1}^x \frac{\cos \pi s}{y_h(s)}\,ds + C y_h(x).$$

Since the first term in the last expression solves the nonhomogeneous equation and the second is a general solution of the homogeneous equation, this expression is, in fact, a general solution of $y' + 6x^2 y = \cos \pi x$.

Choosing the lower limit of integration to be the initial point -1 makes evaluation of C easy:

$$y_p(-1) = 0 + C y_h(-1) = C \cdot 2 = -8.$$

Hence, $C = -4$.

Using $y_h(x) = 2e^{-2(x^3+1)}$, we can write the solution of the initial-value problem as

$$y(x) = 2e^{-2(x^3+1)} \int_{-1}^{x} \frac{\cos \pi s}{2e^{-2(s^3+1)}} \, ds - 8e^{-2(x^3+1)}$$

$$= \int_{-1}^{x} e^{-2(x^3-s^3)} \cos \pi s \, ds - 8e^{-2(x^3+1)}. \qquad \blacksquare$$

In the grand scheme of things, the impact of the variation of parameters idea is not as an explicit solution formula such as the one we just obtained. Rather, it is the idea that a solution of the *non*homogeneous equation can be written as the product uy_h, where y_h is a nontrivial solution of the homogeneous equation and u is the solution of a *very* simple differential equation. Much of our general understanding of differential equations is built upon this representation of a solution.

3.3.3 The Integrating Factor

The *integrating factor* method can find a general solution of a linear, nonhomogeneous equation in one operation. It also provides a useful representation of solutions of linear equations.

The nonhomogeneous equation

$$xy' + y = e^{4x} .$$

is easy to solve once we notice that its left side is the derivative of a product:

$$xy' + y = (xy)'.$$

Since the differential equation can be written

$$(xy)' = e^{4x},$$

an antiderivative and simplification yield

$$\int (xy)' \, dx = \int e^{4x} \, dx$$

$$xy = \frac{e^{4x}}{4} + C$$

$$y(x) = \frac{e^{4x}}{4x} + \frac{C}{x}.$$

We have a *general solution* of $xy' + y = e^{4x}$.

But not every equation can be written as a derivative of a product; e.g.,

$$y' + xy = -3x . \qquad (3.11)$$

To obtain such a derivative, we will multiply the entire equation by an **integrating factor** $\mu(x)$

$$\mu(x)y' + x\mu(x)y = -3x\mu(x),$$

and choose μ to make the *left side equal the derivative* $(\mu(x)y)'$.
Since $(\mu(x)y)' = \mu(x)y' + \mu'(x)y$, this condition requires

$$\mu(x)y' + x\mu(x)y = \mu(x)y' + \mu'(x)y.$$

The coefficients of y' are equal. The coefficients of y are equal if

$$\mu' = x\mu.$$

Solving this simple equation for μ by separation of variables yields

$$\frac{d\mu}{\mu} = x\,dx$$

$$\mu(x) = Ce^{x^2/2}.$$

Since we seek *any* nontrivial function that satisfies $\mu' = x\mu$, we can take $C = 1$ for simplicity. We have obtained the *integrating factor*

$$\mu(x) = e^{x^2/2}.$$

Multiplying the original equation $y' + xy = -3x$ by $\mu = e^{x^2/2}$ reduces its left side $y' + xy$ to the derivative of a product:

$$e^{x^2/2}(y' + xy) = \left(ye^{x^2/2}\right)' = -3xe^{x^2/2}.$$

(Verify by computing the derivative on the right.) The transformed equation can then be solved with a single antiderivative:

$$\left(ye^{x^2/2}\right)' = -3xe^{x^2/2}$$

$$ye^{x^2/2} = -3e^{x^2/2} + C$$

$$y = -3 + Ce^{-x^2/2}.$$

The last expression is a general solution of $y' + xy = -3x$.

If we label the unknown in the given differential equation $y(x)$ and assume that the *coefficient of y' is unity*, then the steps in the integrating factor method are as follows:

1. Multiply the given differential equation by the as-yet-undetermined integrating factor $\mu(x)$.
2. Compare the left side of the multiplied equation with the product rule derivative $(\mu y)' = \mu y' + \mu'y$.

3. Force equality between the two expressions from steps 1 and 2 by equating the coefficients of y. Obtain a homogeneous linear equation for $\mu(x)$.

4. Solve for $\mu(x)$.

5. Use $\mu(x)$ to rewrite the left side of the given differential equation as $(\mu y)'$. Solve that transformed equation with a single antiderivative.

As an added benefit, the integrating factor always provides a general solution without having to separately consider the particular solution of the nonhomogeneous equation and the general solution of the homogeneous equation. If the coefficient of y' is not unity, we simply divide it out (as long as it is not zero) before employing the integrating factor process.

We can easily generalize this process. An integrating factor $\mu(x)$ for the equation $y' + g(x)y = \cdots$ is defined by the requirement

$$\mu[y' + g(x)y] = (\mu y)'.$$

Writing $\mu y + \mu g(x)y = (\mu y)' = \mu y' + \mu'y$ and equating coefficients of y lead to a linear, homogeneous equation for μ,

$$\mu' = g(x)\mu.$$

Solving by separation of variables and setting $C = 1$ (compare (3.9), page 72) gives an *integrating factor for $y' + g(x)y$*,

$$\boxed{\mu(x) = \exp\left(\int_a^x g(s)\,ds\right),}$$

where a is an arbitrary constant.

Instead of appealing directly to this formula, the following examples emphasize the process of finding an integrating factor. Mastering that process makes this formula automatic.

EXAMPLE 22   *Find a general solution of*

$$y' + \cos x\, y = \pi \cos x\,.$$

Step 1. Multiply the differential equation by μ:

$$\mu y' + \mu \cos x\, y = \mu\pi \cos x\,.$$

Step 2. Demand μ satisfy

$$(\mu y)' = \mu y' + \mu'y = \mu y' + \mu \cos x\, y.$$

Step 3. Equate coefficients of y:

$$\mu' = \mu \cos x\,.$$

Step 4. Solve for μ (choosing $C = 1$):

$$\mu(x) = e^{\sin x}\,.$$

Step 5. Use $(\mu y)' = \mu \pi \cos x$ to integrate the differential equation and solve for y:

$$y' + \cos x\, y = \pi \cos x$$

$$(ye^{\sin x})' = \pi \cos x e^{\sin x}$$

$$ye^{\sin x} = \pi e^{\sin x} + C$$

$$y(x) = \pi + Ce^{-\sin x}.$$

We have a general solution of the given equation. ∎

EXAMPLE 23 *Find a solution of the initial-value problem*

$$xy' + 2y = 4x^2, \qquad y(1) = 6.$$

The coefficient x of y' is not unity. To permit division by x, restrict consideration to $1 \leq x < \infty$. Divide by x and use an integrating factor to find a general solution of

$$y' + \frac{2}{x}y = 4x, \qquad 1 \leq x < \infty.$$

Step 1. Multiply the differential equation by μ:

$$\mu y' + \mu \frac{2}{x}y = \mu 4x.$$

Step 2. Demand μ satisfy

$$(\mu y)' = \mu y' + \mu' y = \mu y' + \mu \frac{2}{x}y.$$

Step 3. Equate coefficients of y:

$$\mu \frac{2}{x} = \mu'.$$

Step 4. Solve for μ (choosing $C = 1$):

$$\mu(x) = x^2.$$

Step 5. Use $(\mu y)' = \mu 4x$ to integrate the differential equation and solve for y:

$$y' + \frac{2}{x}y = 4x$$

$$(x^2 y)' = x^2 \cdot 4x$$

$$x^2 y = x^4 + C$$

$$y = x^2 + \frac{C}{x^2}.$$

We have a general solution of $xy' + 2y = 4x^2$.

The initial condition $y(1) = 6$ requires that $1 + C = 6$. Hence, $C = 5$, and the solution of the initial-value problem is

$$y(x) = x^2 + \frac{5}{x^2}, \qquad 1 \le x < \infty.$$ ∎

EXAMPLE 24 *Find a general solution of*

$$y' + xy = f(x).$$

where f is an arbitrary continuous function.

The differential expression on the left is the same as that in equation (3.11), $y' + xy = -3x$, for which we found the integrating factor $\mu = e^{x^2/2}$. Given this integrating factor, we can go directly to step 5.

> To demonstrate directly that $\mu = e^{x^2/2}$ is an integrating factor for $y' + xy = f(x)$, show that multiplying by $e^{x^2/2}$ transforms the left side of the equation into the derivative of a product:
>
> $$\left(y e^{x^2/2} \right)' = e^{x^2/2}(y' + xy).$$
>
> This equation defines the integrating factor.

Multiplying $y' + xy = f(x)$ by μ and appealing to this last expression yield

$$\left(y e^{x^2/2} \right)' = e^{x^2/2} f(x).$$

Using the fundamental theorem of calculus, theorem A.2, for the antiderivative involving f, we obtain

$$y e^{x^2/2} = \int_a^x f(s) e^{s^2/2}\, ds + C,$$

where a and C are constants.

To solve for y, divide by the exponential term on the left and bring it inside the definite integral. (It is independent of the variable of integration s.) A general solution of the differential equation is

$$y(x) = \int_a^x f(s) e^{(s^2 - x^2)/2}\, ds + C e^{-x^2/2}.$$

Why did we require that f be continuous? ∎

EXAMPLE 25 *Find a general solution of*

$$y' + g(x)y = \ln x, \qquad x \ge 1,$$

where g is an arbitrary continuous function.

We know that

$$\mu(x) = \exp\left(\int_1^x g(s)\, ds \right)$$

is an integrating factor for $y' + g(x)y$. For convenience, choose $a = 1$, the lower end of the interval where the solution is to be defined.

Proceeding directly to step 5, multiply the differential equation by μ, use $(\mu y)' = \mu y' + \mu g(x)$, and integrate:

$$(\mu y)' = \mu \ln x$$

$$\mu(x)y(x) = \int_b^x \mu(s) \ln s \, ds + C \,,$$

where b is the constant lower limit from the fundamental theorem of calculus, and C is a constant of integration. We choose $b = 1$, the start of the x interval for this equation.

To solve for y, divide by $\mu(x)$ and bring it inside the definite integral on the right side of the last equation. (Note that $\mu(x)$ does not involve the variable of integration s.) Taking care to distinguish dummy variables of integration from limits of integration, we can write the expression $\mu(s)/\mu(x)$ as

$$\frac{\mu(s)}{\mu(x)} = \exp\left(\int_1^s g(r)\,dr - \int_1^x g(r)\,dr\right)$$

$$= \exp\left(-\int_s^x g(r)\,dr\right).$$

Thus, a general solution of $y' + g(x)y = \ln x$ is

$$y(x) = \int_1^x \exp\left(-\int_s^x g(r)\,dr\right) \ln s \, ds + C \exp\left(-\int_1^x g(r)\,dr\right). \qquad \blacksquare$$

The last two examples have used integrating factors to solve an equation with a general coefficient, $y' + g(x)y = \ln x$, and one with a general forcing term, $y' + xy = f(x)$. Exercise 34 asks you to combine those two arguments to show that a general solution of the linear equation

$$y' + g(x)y = f(x)$$

is

$$y(x) = \int_a^x \exp\left(-\int_s^x g(r)\,dr\right) f(s)\, ds + C \exp\left(-\int_a^x g(r)\,dr\right) \,, \quad x > 0\,,$$

where a is arbitrary and f, g are continuous functions. Compare this formula with the general solution (3.10), which combines a homogeneous solution from separation of variables with a particular solution from variation of parameters.

> Don't memorize this formula. Understand the integrating factor process.

3.3.4   Other Solution Methods

The preceding discussion essentially settles the question of finding solution formulas for linear first-order equations. However, there are two special methods for constant-coefficient equations, *characteristic equations* and *undetermined coefficients*, that are so simple they are hard to overlook. These generalize easily to higher-order equations, as does variation of parameters. (Neither separation of variables nor integrating factors generalize.)

Sometimes the integrals in solution formulas are so complex that analytic evaluation is prohibitively difficult or even impossible. Many nonlinear equations have no analytic solution whatsoever. In those situations, we must settle for a table of values of the solution of a specific initial-value problem instead of a general formula for a solution of the differential equation. We then turn to *numerical solution methods* such as the Euler method, which is described later in this chapter.

In a great many cases, only numerical methods can analyze in detail a specific model of a complex physical phenomenon. On the other hand, solution formulas can illuminate the behavior that is characteristic of an entire class of models. Comparing the results of the two approaches separates behavior that is unique to the specific situation from that which is generic to the class of models.

3.3.5 Exercises

EXERCISE GUIDE

To gain experience . . .	Try Exercises
Solving homogeneous equations via separation of variables	1–10, 18
Solving nonhomogeneous equations via variation of parameters	1–10, 13–15
Finding general solutions via an integrating factor	1–10, 23
Deriving integrating factors	21, 24, 25, 28–34
Solving initial value problems	1–10, 16, 26(b), 27(b), 34(a)
Verifying solutions and general solutions	11, 12, 15(b), 20, 22, 26(a), 27(a)
Studying the behavior of solutions	17, 18

In exercises 1–10,

i. Find a general solution of the homogeneous form of the equation by separation of variables.

ii. Find a particular solution of the nonhomogeneous equation using variation of parameters.

iii. Combine the preceding results to write a general solution of the given differential equation.

iv. Find a general solution using an integrating factor. Show that it is equivalent to the general solution obtained in part (iii).

v. Write a solution of the given initial-value problem.

1. $y' - 2xy = x^2$, $y(0) = 3$

2. $T' = -(Ak/cm)(T - T_{\text{out}})$, $T(0) = T_i$, T_{out} constant

3. $v' = -g$, $v(0) = v_i$, g constant

4. $dy/dt + ky = Q$, $y(0) = y_i$, k, Q constant

5. $x' - x \sin \pi t = 4 \sin \pi t$, $x(\pi/4) = 3$

6. $4y' + 12x^3 y = x^3 + 1$, $y(1) = 1$

7. $y - y' = \sqrt{x}$, $y(1) = 4$

8. $R \, dq/dt + q/C = E$, $q(0) = q_i$, R, C, E constant

9. $P' - kP = A \cos 2\pi t$, $P(0) = P_i$, k, A constant

10. $P' - kP = -E$, $P(t_0) = P_i$, k, E constant

11. Using the fundamental theorem of calculus, verify by direct substitution that

$$y_h(x) = C \exp\left(-\int_a^x g(s)\, ds\right)$$

is a solution of $y' + g(x)y = 0$.

12. Using the fundamental theorem of calculus and the

fact that y_h satisfies $y_h' + g(x)y_h = 0$, verify by direct substitution that

$$y(x) = y_h(x) \int_a^x \frac{f(s)}{y_h(s)}\, ds + Cy_h(x).$$

is a solution of $y' + g(x)y = f(x)$.

13. Example 21 uses variation of parameters to solve $y' + 6x^2 = \cos \pi x$. Without using the actual formula

$$y_h(x) = 2e^{-2(x^3+1)}$$

for a solution of $y' + g(x)y = 0$, it substitutes $y_p = uy_h$ into $y' + 6x^2 = \cos \pi x$ and obtains

$$(uy_h)' + 6x^2 uy_h = \cos \pi x,$$

$$u(y_h' + 6x^2 y_h) + u'y_h = \cos \pi x,$$

$$u'(x) = \frac{\cos \pi x}{y_h(x)}.$$

(a) Using the formula for y_h, verify these steps.
(b) Explain why you do not really need a formula for y_h. Why is it enough to know that y_h is a nontrivial solution of $y' + g(x)y = 0$?

14. To apply variation of parameters to the equation

$$y' + g(x)y = f(x),$$

the text substitutes $y = uy_h$ and obtains

$$(uy_h)' + g(x)uy_h = f(x),$$

$$u\, \boxed{(y_h' + g(x)y_h)} + u'y_h = f(x),$$

$$u'(x) = \frac{f(x)}{y_h(x)}.$$

It then asks two questions:
(a) What happened to the term $y_h' + g(x)y_h$ in parentheses in the second equation? (It is boxed here for emphasis. Would this term arise if the equation were nonlinear?)
(b) How can we be sure that $y_h \neq 0$?
Answer each question.

15. The text uses $y_h = \exp\left(-\int_a^x g(s)\, ds\right)$ to derive the general solution

$$y(x) = \int_a^x f(s) \exp\left(-\int_s^x g(r)\, dr\right) ds$$

$$+ C \exp\left(-\int_a^x g(s)\, ds\right)$$

of $y' + g(x)y = f(x)$ from the variation of parameters formula

$$y_p(x) = y_h(x) \int_a^x \frac{f(s)}{y_h(s)}\, ds + Cy_h(x).$$

(a) Verify by substitution that the first expression is a solution of $y' + g(x)y = f(x)$.
(b) Verify that it is a *general* solution.
(c) Carry out the details of deriving the first expression from the second.

16. Use the results of this section to find formulas for the solution of each of the following initial-value problems.
(a) $y' + g(x)y = 0$, $y(a) = y_i$
(b) $y' + g(x)y = f(x)$, $y(a) = y_i$.

17. Use the representation $y_h = C \exp\left(-\int_a^x g(s)\, ds\right)$, equation (3.9), for the solution of

$$y' + g(x)y = 0$$

obtained in this section to prove the following. Assume g is continuous for all x.
(a) The solution of the initial-value problem $y' + g(x)y = 0$, $y(0) = y_i$, is either identically zero or never zero. (*Hint:* What value must y_i have to give this initial-value problem the trivial solution?)
(b) If $g(x) \geq \delta > 0$ for all x and for some δ, then the solution y of $y' + g(x)y = 0$, $y(0) = y_i$, satisfies

$$\lim_{x \to \infty} y(x) = 0.$$

18. Use the representation

$$y_g(x) = \int_a^x f(s) \exp\left(-\int_s^x g(r)\, dr\right) ds$$

$$+ C \exp\left(-\int_a^x g(s)\, ds\right),$$

equation (3.10), for a general solution of

$$y' + g(x)y = f(x)$$

to find a set of conditions on g, f which guarantee that every solution will decay to zero as x grows without bound.

19. Why do we require that the coefficient $g(x)$ in the equation $y' + g(x)y = 0$ be continuous in order to

derive the solution formula

$$y_h(x) = C \exp\left(-\int_a^x g(s)\, ds\right)?$$

Exercises 20–32 ask you to verify results of discussions and examples in the text or to complete details of calculations summarized there.

20. Verify that

$$y = \frac{e^{4x}}{4x} + \frac{C}{x}$$

is a general solution of

$$xy' + y = e^{4x}.$$

(See page 75.) For which values of x is this solution valid?

21. Verify the integrating factor equality

$$e^{x^2/2}(y' + xy) = \left(ye^{x^2/2}\right)'$$

by computing the derivative on the right. (See the work following equation (3.11).)

22. Verify that

$$y = -3 + Ce^{-x^2/2}$$

is a general solution of

$$y' + xy = -3x.$$

(See the work following equation (3.11).)

23. Carry out the details of the calculations required to obtain the solution

$$y = \pi + Ce^{-\sin x}$$

of the equation $y' + \cos x\, y = \pi \cos x$ from the transformed equation

$$\left(ye^{\sin x}\right)' = \pi \cos x e^{\sin x}$$

in example 22. Verify that this solution is indeed a *general* solution of $y' + \cos x\, y = \pi \cos x$.

24. Derive an integrating factor for

$$y' + xy = f(x).$$

Show that it is identical with the integrating factor $e^{x^2/2}$ obtained in the text for $y' + xy = -3x$, to within a constant multiple. Argue from this example that the integrating factor depends upon the terms involving the unknown and not upon the forcing function.

25. Why did example 24 which considered the equation $y' + xy = f(x)$, ask that the forcing function f be continuous? Would any weaker condition have sufficed?

26. **(a)** Verify by direct substitution that the solution

$$y(x) = \int_1^x \exp\left(-\int_s^x g(r)\, dr\right)\ln s\, ds$$

$$+ C \exp\left(-\int_1^x g(r)\, dr\right)$$

of $y' + g(x)y = \ln x$, which was obtained in example 25 is a general solution. Use the fundamental theorem of calculus, theorem A.2.

(b) Solve the initial-value problem $y' + g(x)y = \ln x$, $y(1) = -7$, $x \geq 1$.

27. **(a)** Verify that $y = x^2 + C/x^2$, the solution of $xy' + 2y = 4x^2$ obtained in example 23, is indeed a general solution.

(b) Find a solution of the initial-value problem $xy' + 2y = 4x^2$, $y(1) = 6$.

28. Example 22 finds the integrating factor

$$\mu(x) = e^{\sin x}$$

for $y' + \cos x\, y$ from first principles. Verify that you can obtain the same integrating factor from the general expression

$$\mu(x) = \exp\left(\int_a^x g(s)\, ds\right)$$

with an appropriate choice of a.

29. Example 23 finds the integrating factor

$$\mu(x) = x^2$$

for $y' + 2y/x$ from first principles. Verify that you can obtain the same integrating factor from the general expression

$$\mu(x) = \exp\left(\int_a^x g(s)\, ds\right)$$

with an appropriate choice of a.

30. Example 24 uses the integrating factor

$$\mu(x) = e^{x^2/2}$$

for $y' + xy$. Verify that you can obtain the same integrating factor from the general expression

$$\mu(x) = \exp\left(\int_a^x g(s)\,ds\right)$$

with an appropriate choice of a.

31. Carry out the details of deriving the integrating factor

$$\mu(x) = \exp\left(\int_1^x g(s)\,ds\right)$$

for the general linear equation $y' + g(x)y = \cdots$. In particular, show carefully the steps in the solution of the linear homogeneous equation defining μ,

$$\mu' = \mu g(x).$$

32. Following the procedures in the text, derive an integrating factor for

$$xy' + y = e^{4x},$$

and use it to obtain the solution $y = e^{4x}/4x + C/x$. Does your derivation produce the same integrating factor as the text obtained by inspection?

33. Use an integrating factor to find a general solution of the constant-coefficient, linear equation

$$b_1 y' + b_0 y = f(x),$$

where f is continuous. Find the solution of the initial-value problem $b_1 y' + b_0 y = f(x)$, $y(0) = y_i$.

34. Assume that f and g are both continuous functions.
   (a) Derive an integrating factor for the expression $y' + g(x)y$.
   (b) Use the integrating factor obtained in part (a) to solve the initial-value problem

$$y' + g(x)y = f(x),\, y(0) = y_i.$$

You will find a general solution along the way. Show that it is equivalent to the one given at the end of this section.

3.4 LINEAR EQUATIONS: EXISTENCE AND UNIQUE-NESS

Here we prove that if f and g are continuous, then the first-order linear equation

$$y' + g(x)y = f(x) \tag{3.12}$$

has a solution defined on the entire interval in which f, g are continuous, a result known as an *existence theorem*. We also prove that under the same conditions, the initial-value problem

$$y' + g(x)y = f(x),\ y(x_0) = y_i,$$

has exactly one solution, a *uniqueness theorem*.

The existence theorem tells us that linear, first-order models such as the emigration model or the heat-loss model are not vacuous; e.g., the model equations $P' = kP - E$ and $T' = -(Ak/cm)(T - T_{\text{out}})$ always have solutions. In fact, these particular equations have solutions that are defined for all time. (The physical significance of those solutions is another question.)

The uniqueness theorem says that a model with a prescribed initial condition can describe only one specific response. For example, the emigration model with a specified initial population can describe only one pattern of population variation with time. Changing that initial population will change the resulting population curve. Furthermore, two different population curves starting from different initial values can never cross. Otherwise, their common point would be an initial population value with two distinct solutions emanating from it, in violation of the uniqueness theorem.

Finally, the combination of the existence and uniqueness theorems allows us to assert that *every solution* of the linear equation $y' + g(x)y = f(x)$ can be represented as a *general solution* with an appropriate choice of the arbitrary constant.

Theorem 3.2 (Existence for Linear Equations)   Let f, g be continuous on $x_1 \leq x \leq x_2$. Then there exists a continuously differentiable solution of

$$y' + g(x)y = f(x)$$

defined by

$$y(x) = \int_a^x \exp\left(-\int_s^x g(r)\,dr\right) f(s)\,ds + C \exp\left(-\int_a^x g(r)\,dr\right),$$

$$x_1 \leq x \leq x_2, \tag{3.13}$$

where C is an arbitrary constant and a is a constant in $x_1 \leq x \leq x_2$.

PROOF   Solve $y' + g(x)y = f(x)$ using the integrating factor

$$\mu(x) = \exp\left(\int_a^x g(s)\,ds\right).$$

The solution formula (3.13) results. (Exercise 12 asks you to supply the details.)

Expression (3.13) certainly defines a function y that is continuous and satisfies $y' + g(x)y = f(x)$ for $x_1 \leq x \leq x_2$. Since f, g, and y are all continuous on that interval, $y' = f(x) - g(x)y$ is continuous there as well.   ∎

The constant coefficients and forcing terms in equations such as $P' = kP - E$ and $T' = -(Ak/cm)(T - T_{\text{out}})$ are certainly continuous for all t. Hence, their solutions are defined for all t. But existence of a solution over long intervals of the independent variable can fail if the equation is not linear.

EXAMPLE 26   *The first-order nonlinear equation $y' - y^2 = 0$ has constant (hence, continuous for all x) coefficients, but a solution starting at $x = 0$ with initial value $y(0) = \frac{1}{2}$ is undefined at $x = 2$.*

Use separation of variables to solve $y' - y^2 = 0$:

$$\frac{dy}{y^2} = dx$$

$$\frac{-1}{y} = x + C$$

$$y = \frac{-1}{x + C}.$$

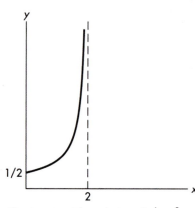

Figure 3.1: The solution of $y' - y^2 = 0$, $y(0) = \frac{1}{2}$, does not exist at $x = 2$ even though the (constant) coefficients of the equation are continuous for all x.

The initial condition $y(0) = \frac{1}{2}$ requires $C = -2$. Hence, a solution of $y' - y^2 = 0$, $y(0) = \frac{1}{2}$, is

$$y = \frac{1}{2 - x}.$$

But this solution is obviously undefined at $x = 2$. See figure 3.1.

Solutions of *non*linear equations may not exist throughout the full interval where their coefficients are continuous. ∎

The proof of the existence theorem is constructive: We have actually exhibited the solution in equation (3.13). Exercise 11 asks you to show that this expression is a *general solution*.

The presence of the arbitrary constant C in the solution expression (3.13) confirms what we already knew: The equation $y' + g(x)y = f(x)$ has an infinite number of solutions, one for each value of C. We cannot speak of uniqueness, of there being only one solution, unless we add an initial condition that fixes the value of the constant C.

Theorem 3.3 (Uniqueness for Linear Initial-Value Problems) *Let f, g be continuous on $x_1 \leq x \leq x_2$. Let $x_1 \leq x_0 \leq x_2$. Then the linear initial-value problem*

$$y' + g(x)y = f(x), \qquad y(x_0) = y_i,$$

has at most one continuously differentiable solution defined for $x_1 \leq x \leq x_2$.

PROOF   Suppose there are two continuously differentiable functions u, v that are solutions of this initial-value problem for $x_1 \leq x \leq x_2$:

$$u' + g(x)u = f(x), \qquad u(x_0) = y_i$$
$$v' + g(x)v = f(x), \qquad v(x_0) = y_i$$

Let $z = u - v$. Then from the principle of superposition, theorem 3.1, page 55, z satisfies the differential equation with forcing term $f(x) - f(x)$; i.e.

$$z' + g(x)z = 0.$$

Now apply the integrating factor $\mu = \exp\left(\int_{x_0}^{x} g(s)\,ds\right)$ to this homogeneous equation,

$$\left[z \exp\left(\int_{x_0}^{x} g(s)\,ds\right)\right]' = 0$$

$$z \exp\left(\int_{x_0}^{x} g(s)\,ds\right) = C,$$

where C is an arbitrary constant and $x_1 \leq x \leq x_2$. To evaluate C, set $x = x_0$. Then $z(x_0) = u(x_0) - v(x_0) = 0$, and we find $C = 0 \cdot e^0 = 0$. Hence, for $x_1 \leq x \leq x_2$,

$$z(x) \exp\left(\int_{x_0}^{x} g(s)\,ds\right) = 0.$$

Since the exponential is never zero, we conclude that

$$z(x) \equiv 0, \ x_1 \leq x \leq x_2.$$

We have shown that any two solutions of this initial-value problem must be identical. Hence, there is at most one solution. ∎

The proof of this theorem is equivalent to showing that the homogeneous initial-value problem

$$y' + g(x)y = 0, \ y(0) = y_i,$$

has only the trivial solution.

An important consequence of the uniqueness theorem is the observation that no two solution curves of a linear differential equation can intersect at a single point. If they did, we could choose that crossing point as an initial condition. But two distinct solutions can not originate from the same initial value without violating theorem 3.3. Figure 3.2 shows this impossible situation. *Two solutions of a linear differential equation that agree at a single point must be identical* throughout their interval of definition.

But this uniqueness result can fail if the differential equation is nonlinear.

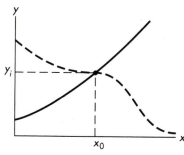

Figure 3.2: Two solution curves of a linear initial-value problem can not cross. Otherwise, there would be two distinct solutions satisfying the one initial condition $y(x_0) = y_i$.

EXAMPLE 27 *Although its (constant) coefficients are continuous, the nonlinear initial-value problem*

$$y' = \frac{3}{2}y^{1/3}, \ y(0) = 0$$

has three distinct solutions.

This initial-value problem clearly has the trivial solution $y \equiv 0$. Separation of variables reveals two others:

$$\frac{2}{3}\frac{dy}{y^{1/3}} = dx$$

$$y^{2/3} = x + C.$$

The initial condition $y(0) = 0$ requires $C = 0$. Two other solutions appear to be $y = \pm x^{3/2}$.

Since $y = \pm x^{3/2}$ are zero at $x = 0$, the separation of variables derivation of these solutions is flawed by division by zero. However, direct substitution, which exercise 19 requests, verifies that $\pm x^{3/2}$ are indeed solutions. Hence, the nonlinear equation $y' = 3y^{1/3}/2$ has *three* solutions satisfying the initial condition $y(0) = 0$; they are $y = 0, \pm x^{3/2}$. ∎

We can combine theorems 3.2 and 3.3 into a single existence and uniqueness theorem for the linear initial-value problem. Exercise 21 asks you for a proof.

Corollary 3.4 (Existence and Uniqueness) *Let f, g be continuous on $x_1 \leq x \leq x_2$. Let $x_1 \leq x_0 \leq x_2$. Then the linear initial-value problem*

$$y' + g(x)y = f(x), \ y(x_0) = y_i,$$

has exactly one *continuously differentiable solution defined on* $x_1 \leq x \leq x_2$. *That solution is*

$$y(x) = \int_{x_0}^{x} \exp\left(-\int_{s}^{x} g(r)\,dr\right) f(s)\,ds + y_i \exp\left(-\int_{x_0}^{x} g(r)\,dr\right),$$
(3.14)

the general solution (3.13) *with* $C = y_i$.

From these results, we can argue that *a general solution includes every possible solution of a linear equation* (3.12) with continuous coefficient and forcing term. By construction, a general solution, if one exists, is always a solution. Theorem 3.2 proves the existence of a general solution (3.13).

To show that every solution of a linear initial-value problem is just the general solution (3.13) with an appropriate choice of C, suppose that $u(x)$ is a solution of $y' + g(x)y = f(x)$ defined on $x_1 \leq x \leq x_2$. From Corollary 3.4, there is exactly one solution of the initial-value problem

$$y' + g(x)y = f(x), \qquad y(x_0) = u(x_0).$$

That solution must be $u(x)$, and it can be written using (3.14) with $y_i = u(x_0)$. But that expression is then the general solution (3.13) with $C = u(x_0)$.

These arguments establish the *power of the general solution*: There are no solutions of the linear differential equation $y' + g(x)y = f(x)$ with f and g continuous that can not be obtained from a general solution.

3.4.1 Exercises

EXERCISE GUIDE

To gain experience . . .	Try Exercises
Applying existence and uniqueness criteria	1–10
With the proof of the existence theorem	11, 21
With the proof of the uniqueness theorem	12, 13, 17, 18, 21
With limitations of the existence theorem	14–16
With limitations of the uniqueness theorem	19, 20
With general solutions	11, 22
With the behavior of solutions	23, 24

Exercises 1–10 list initial-value problems that model a variety of situations considered earlier in the text or in exercises. For which of these models do the results of this section guarantee the existence of a unique solution? For what range of values of the independent variable? If the theorems in this section do not apply, explain why.

1. Rock model: $v' = -g$, $v(0) = v_i$.

2. Rock model with air resistance: $v' = -g - kv$, $v(0) = v_i$.

3. Improved rock model with air resistance: $v' = -g - kv|v|$, $v(0) = v_i$.

4. Simple population model: $P' = kP$, $P(t_0) = P_i$.

5. Emigration model: $P' = kP - E$, $P(t_0) = P_i$.

6. Logistic population model: $P' = aP - sP^2$, $P(t_0) = P_i$.

7. Potato famine model: $P' = 0.015P - 0.209$, $P(1847) = 8$.

8. Potato famine model with time-dependent emigration rate: $P' - 0.015P = -E(t)$, $P(1847) = 8$, where

$$E(t) = \begin{cases} 0.209, & 1847 \leq t < 1851 \\ 0, & 1851 \leq t. \end{cases}$$

(This modification abruptly ends emigration in 1851, when the effects of the famine had begun to wane.)

9. Heat-loss model: $T' = -(Ak/cm)(T - T_{out})$, $T(0) = T_i$.

10. Heat-loss model with furnace: $T' = -(Ak/cm)(T - T_{out}) + F/cm$, $T(0) = T_i$.

11. Show that the solution formula

$$y(x) = \int_a^x \exp\left(-\int_s^x g(r)\, dr\right) f(s)\, ds$$
$$+ C \exp\left(-\int_a^x g(r)\, dr\right)$$

obtained in the existence theorem 3.2 is, in fact, a *general solution* of $y' + g(x)y = f(x)$.

12. Supply the details of the proof of Theorem 3.2. In particular, carry out the integrating factor construction described there. (*Hint:* See example 24.)

13. The solution formula

$$y(x) = \int_a^x \exp\left(-\int_s^x g(r)\, dr\right) f(s)\, ds$$
$$+ C \exp\left(-\int_a^x g(r)\, dr\right)$$

obtained in the existence theorem 3.2 is the same as equation (3.10), which was obtained using separation of variables and variation of parameters. Prove theorem 3.2 using separation of variables and variation of parameters.

14. Find the interval of existence of a solution of the nonlinear differential equation $y' - y^2 = 0$ satisfying the initial condition $y(0) = y_i$. Express the interval of existence in terms of y_i, considering y_i positive, negative, and zero. Does the solution exist for all $x \geq 0$ for any y_i?

15. Example 26 exhibits a nonlinear equation having a solution that does not exist throughout the full interval

in which the (constant) coefficients of the differential equation are continuous. Construct a similar example using $y' - y^3 = 0$.

16. Example 26 exhibits a nonlinear equation having a solution that does not exist throughout the full interval in which the (constant) coefficients of the differential equation are continuous. Using equations of the form $y' - y^n = 0$, $n > 2$, construct a family of examples of this phenomenon.

17. To prove the uniqueness theorem 3.3 the text supposes that there are two functions u, v satisfying the same initial-value problem,

$$u' + g(x)u = f(x), \qquad u(x_0) = y_i,$$
$$v' + g(x)v = f(x), \qquad v(x_0) = y_i.$$

It then appeals to the principle of superposition to show that $z = u - v$ is a solution of the homogeneous differential equation $z' + g(x)z = 0$. Instead, directly verify by substitution that z solves this equation.

18. Verify that $\mu = \exp\left(\int_{x_0}^x g(s)\, ds\right)$ is indeed an integrating factor for the homogeneous equation $z' + g(x)z = 0$ considered in the proof of the uniqueness theorem 3.3.

19. Example 27 exhibits three solutions of the nonlinear initial-value problem

$$y' = \frac{3}{2}y^{1/3}, \qquad y(0) = 0.$$

They are $y = 0, \pm x^{3/2}$. Verify by substitution that each of these functions is indeed a solution of this initial-value problem.

20. Example 27 shows that the nonlinear initial-value problem $y' = 3y^{1/3}/2$, $y(0) = 0$, does not have a unique solution. Construct a family of examples of this phenomenon using $y' = Ay^\alpha$, $y(0) = 0$, with $0 < \alpha < 1$. Choose A to be a convenient value.

21. Prove Corollary 3.4, existence and uniqueness for the linear initial-value problem.

22. Corollary 3.4 displays the solution (3.14)

$$y(x) = \int_{x_0}^x \exp\left(-\int_s^x g(r)\, dr\right) f(s)\, ds$$
$$+ y_i \exp\left(-\int_{x_0}^x g(r)\, dr\right)$$

of the initial-value problem $y' + g(x)y = f(x)$, $y(x_0) = y_i$. Argue that if y_i is interpreted as an arbitrary constant, then this expression is a general solution of $y' + g(x)y = f(x)$.

23. Consider the equation $y' + g(x)y = 0$, where g is continuous and $g(x) \geq \delta > 0$ for $0 \leq x < \infty$.
    (a) Show that the general solution of this equation decays to zero as x grows.

(b) Use the ideas of this section to argue that *all* nontrivial solutions of this equation decay as x grows.

24. Formulate and solve a problem similar to the preceding one in the case when g is *negative* and continuous for $0 \leq x < \infty$.

3.5 NUMERICAL METHODS

The previous sections have developed several approaches to extracting the information an initial-value problem contains. For some equations, separation of variables or variation of parameters or some other analytic method can produce a formula for the solution. Other equations simply do not have closed-form, analytic solutions.

When solution methods fail or become too cumbersome or when we need just a graph of a solution for one particular set of initial condition and parameter values, we can use numerical methods. These processes can be implemented with a computer or a calculator to produce a table of numerical solution values or points on a graph.

The simplest numerical method is based on a natural approximation of the derivative. Since the derivative is defined by

$$y'(t) = \lim_{\Delta t \to 0} \frac{y(t + \Delta t) - y(t)}{\Delta t},$$

an obvious approximation is

$$y'(t) \approx \frac{y(t + \Delta t) - y(t)}{\Delta t},$$

providing Δt is not too large.

Using this idea, we can approximate the generic differential equation $y' = f(t, y(t))$ by the relation

$$\frac{y(t + \Delta t) - y(t)}{\Delta t} \approx f(t, y(t)).$$

If we solve for the "new" value $y(t + \Delta t)$ in terms of both the "old" value $y(t)$ and f evaluated at that old value of y, we have

$$y(t + \Delta t) \approx y(t) + f(t, y(t))\Delta t,$$

the essence of the *Euler method*. The Euler method steps from a solution value $y(t)$ at time t to a new value $y(t + \Delta t)$. Of course, the numerical values calculated with this scheme only approximate the true solution values because of the approximation $y' \approx [y(t + \Delta t) - y(t)]/\Delta t$.

To illustrate these ideas with a specific initial-value problem, consider

$$\frac{dP}{dt} = 0.0282\,P,$$

$$P(1978) = 115.40, \tag{3.15}$$

a model for the population of Brazil using the data from table 2.1. (Time t is measured in years and population P, in millions of people.)

The derivative approximation is

$$P'(t) \approx \frac{P(t + \Delta t) - P(t)}{\Delta t},$$

and the corresponding approximation to the differential equation is

$$P(t + \Delta t) \approx P(t) + 0.0282 P(t) \Delta t.$$

Choosing $\Delta t = 10$ y, for instance, and beginning with the initial time $t_0 = 1978$, we calculate

$$P(1988) \approx P(1978) + 0.0282 P(1978) \cdot 10$$
$$= 115.40 + 0.0282 \cdot 115.40 \cdot 10 = 147.94.$$

Then we can repeat with another step $\Delta t = 10$ from $t = 1988$ using $P(1988) \approx 147.94$,

$$P(1998) \approx P(1988) + 0.0282 P(1988) \cdot 10$$
$$= 147.94 + 0.0282 \cdot 147.94 \cdot 10 = 189.66,$$

and another,

$$P(2008) \approx P(1998) + 0.0282 P(1998) \cdot 10$$
$$= 189.66 + 0.0282 \cdot 189.66 \cdot 10 = 243.15,$$

and so on, leapfrogging to approximations of $P(2018)$, $P(2028)$, ...

To summarize this process, use P_1 to denote the first approximation we computed, $P(1988) \approx 147.94$, use P_2 to denote the second approximation, $P(1998) \approx 189.66$, and so on; that is, P_n is the approximation to the exact solution $P(t_n)$, where $t_n = 1978 + n\Delta t$.

> Writing P with a subscript denotes an element of the sequence $P_0, P_1, P_2,\ldots$ of *approximations* to the exact solution of the initial-value problem; P with parentheses denotes the corresponding values of the *exact* solution function $P(t_0), P(t_1), P(t_2), \ldots$.

Using this notation, let $P_0 = P(t_0) = P(1978)$. Then the preceding computations can be written

$$P(1988) = P(t_1) \approx P_1 \quad \text{where} \quad P_1 = P_0 + 0.0282 P_0 \Delta t,$$
$$P(1998) = P(t_2) \approx P_2 \quad \text{where} \quad P_2 = P_1 + 0.0282 P_1 \Delta t,$$
$$P(2008) = P(t_3) \approx P_3 \quad \text{where} \quad P_3 = P_2 + 0.0282 P_2 \Delta t.$$

Table 3.1 summarizes the numerical values we obtained.

The general step in these computations can be written

$$P(t_{n+1}) \approx P_{n+1} \quad \text{where} \quad P_{n+1} = P_n + 0.0282 P_n \Delta t, \quad n = 0, 1, \quad \ldots,$$

Table 3.1

Values of the approximate solution of the initial value problem $P' = 0.0282\, P$, $P(1978) = 115.40$, computed using the Euler method

n	t_n	P_n
0	1978	115.40
1	1988	147.94
2	1998	189.66
3	2008	243.15

and we begin at $t_0 = 1978$ with $P_0 = P(t_0) = 115.40$. This relation is the *Euler method* applied to the initial-value problem $P' = 0.0282P$, $P(1978) = 115.40$.

The extension to the generic initial-value problem $y' = f(t, y)$, $y(t_0) = y_i$, is straightforward:

Euler method   Given the initial-value problem

$$y'(t) = f(t, y), \ y(t_0) = y_i \,,$$

then the **Euler method** approximation to the solution of this initial-value problem is just the sequence $y_1, y_2, y_3, \ldots$ computed using

EULER METHOD

$$\boxed{y_{n+1} = y_n + f(t_n, y_n)\Delta t, \qquad n = 0, 1, 2, \ldots}$$

beginning at $t = t_0$ with the initial point $y_0 = y_i$.

The value y_n approximates the exact solution $y(t)$ at $t = t_n = t_0 + n\Delta t$.

This general statement of the Euler method has an instructive geometric interpretation. The differential equation $y' = f(t, y)$ is a formula for the slope y' at the point (t, y); e.g., in the population model $P' = 0.0282\, P$, the slope P' of the population curve at P is just $0.0282\, P$. If we let y'_n denote the slope $f(t_n, y_n)$, then the Euler method can be written

$$y_{n+1} = y_n + y'_n \Delta t.$$

In words, the Euler method is

Geometric interpretation of the Euler method

$$\boxed{\text{new value}} = \boxed{\text{old value}} + \boxed{\text{slope}} \times \boxed{\text{step size.}}$$

The Euler method follows the tangent line through the solution curve at the current point (t_n, y_n) to find the next approximate solution point (t_{n+1}, y_{n+1}). Figure 3.3 illustrates this idea.

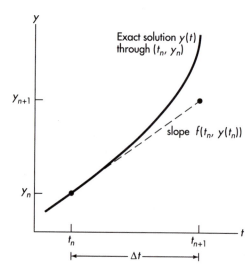

Figure 3.3: The Euler method: The slope $y'_n = f(t_n, y_n)$ of the exact solution through the current point (t_n, y_n) directs us to the next point (t_{n+1}, y_{n+1}).

For example, the first step of the Euler method solution of the population model $P' = 0.0282\,P$, $P(1978) = 115.40$, with $\Delta t = 10$ is

$$P_1 = P_0 + 0.0282\,P_0\Delta t$$
$$= 115.40 + 0.0282 \cdot 115.40\Delta t$$
$$= 115.40 + 3.254 \cdot 10 \approx 147.94\,.$$

We reach P_1 by traveling along the line tangent to the solution curve at $(t_0, P_0) = (1978, 115.40)$. That line has slope $0.0282\,P = 3.254$, and the step length is $\Delta t = 10$ along the horizontal axis, as illustrated in figure 3.4.

Figure 3.5 illustrates the second step of the Euler method for this problem:

$$P_2 = P_1 + 0.0282\,P_1\Delta t$$
$$= 147.94 + 0.0282 \cdot 147.94\Delta t$$
$$= 147.94 + 4.172 \cdot 10 \approx 189.66\,.$$

We reach P_2 by traveling along the line with slope $0.0282\,P = 4.172$ with a step of length $\Delta t = 10$ along the horizontal axis.

> The line we follow is tangent to the exact solution curve through $(t_1, P_1) = (1988, 147.94)$, but that curve is not shown in figure 3.5. Only the solution curve through $(t_0, P_0) = (1978, 115.40)$, the graph of the solution of the initial-value problem, appears there.

Beyond providing a geometric interpretation, figures 3.4 and 3.5 also illustrate a problem with the Euler method. The approximate solution values it generates come from a curve that is actually a connected sequence of straight lines, a polygonal arc, not the smoothly increasing exponential curve we know represents the solution of the initial-value problem $P' = 0.0282\,P$, $P(1978) = 115.40$. The direction followed by a true solution of this initial-value problem must vary continuously, but the Euler method permits changing direction only after a discrete step. For example, even the first step of the numerical solution of $P' = 0.0282\,P$, $P(1978) = 115.40$, can not land at exactly the correct solution value because it is forced to follow the line with slope $0.0282 \cdot 115.40 = 3.254$ for a finite length of time, $\Delta t = 10$, rather than change slope continuously, as the right side of $P' = 0.0282\,P$ demands.

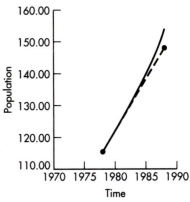

Figure 3.4: The first step away from the initial value using the Euler method for the model $P' = 0.0282\,P$, $P(1978) = 115.40$. The dashed line has slope $0.0282\,P(1978) = 3.254$. The solid line is the exact solution through $(1978, 115.40)$.

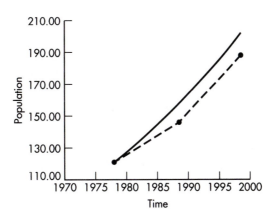

Figure 3.5: The exact solution (smooth curve) and the Euler method approximation after two steps with $\Delta t = 10$ (solid circles) to the population model $P' = 0.0282\,P$, $P(1978) = 115.40$. The dashed line extending from $t = 1988$ to $t = 1998$ has slope $0.0282\,P_1 = 4.172$.

How bad is this error in the Euler method? A numerical experiment provides some estimates.

EXAMPLE 28 *Determine the error in the Euler method approximation using* $\Delta t = 10$ *to the solution of* $P' = 0.0282\,P$, $P(1978) = 115.40$, *at* $t = 2008$.

From table 3.1 the Euler method approximation is $P(2008) = P(t_3) \approx P_3 = 243.15$.

We easily find that the exact solution of the initial-value problem is

$$P(t) = 115.40 e^{0.0282(t-1978)} .$$

The value of the exact solution at $t = 2008$ is

$$P(2008) = 115.40 e^{0.0282 \cdot 3} = 268.92 .$$

The error in the approximation at step 3 is

$$e_3 = P(2008) - P_3 = 268.92 - 243.15 = 25.77$$

an error of almost 10%. The Euler method is clearly subject to significant errors.

This error is illustrated graphically in figure 3.6, which shows both the true solution curve and the approximate solution points that result from taking steps with constant slope. ∎

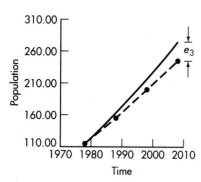

Figure 3.6: The exact solution (smooth curve) and the Euler method approximation after three steps with $\Delta t = 10$ (solid circles) to the population model $P' = 0.0282\,P$, $P(1978) = 115.40$. The error at the end of the third step is e_3.

> The important problem of estimating the accuracy of an approximate solution when we can not check it against the exact solution is a major task of the discipline of *numerical analysis*. The Catch-22 of scientific computing is that we never know the exact answer; if we did, we would not be doing the computing. The next section introduces some of the fundamental ideas of error analysis.

The error in the Euler method seems to occur because we take a step whose direction is fixed to be the slope of the direction field at the beginning of the step; see figure 3.4, for example. If we had an idea of the slope at the point at the *end* of the step, perhaps we could modify the direction of our step to use the knowledge of the direction field both at the *beginning* of the step and at the *end* of the step.

Consider, for example, the situation shown in figure 3.4. The slope at the beginning of the first step is $P'_0 = 0.0282\,P_0 = 3.254$. The slope at the end of that step is $P'_1 = 0.0282\,P_1 = 4.172$.

Back up, remembering the slope information we just obtained. Now take a single step from 1978 to 1988 using the *average* of these two slopes, one from the beginning of the step and the other from the end of a conventional Euler step. This technique is called the *modified Euler method* or the *Heun method*.

To restate the Heun method in the notation of the population model, we begin at the point (t_0, P_0). We first compute the Euler method approximation, which we will denote with an overbar,

$$\bar{P}_1 = P_0 + 0.0282\,P_0 \Delta t.$$

The slope of the direction field at $(t_1, \bar{P}_1)$ is $0.0282 \bar{P}_1$. Now we take a step from the initial point in the direction whose slope is the *average* of the initial and final slopes. That is, we compute

$$P_1 = P_0 + \frac{1}{2}(0.0282\, P_0 + 0.0282\, \bar{P}_0)\Delta t$$

to obtain the approximate solution at time t_1.

More generally, we have the **modified Euler method** or **Heun method** for the initial-value problem

$$y'(t) = f(t, y), \ y(t_0) = y_i,$$

the **Heun method** is the two-step process

$$\bar{y}_{n+1} = y_n + f(t_n, y_n)\Delta t, \qquad\qquad (3.16)$$

$$y_{n+1} = y_n + \frac{1}{2}[f(t_n, y_n) + f(t_{n+1}, \bar{y}_n)]\Delta t, \qquad (3.17)$$

beginning with $y_0 = y_i$.

HEUN METHOD

The preceding statement of the Heun method emphasizes its geometric interpretation, but it wastes effort by evaluating $f(t_n, y_n)$ twice. A more efficient form is

$$\boxed{\begin{aligned} k_1 &= f(t_n, y_n)\Delta t, \\ k_2 &= f(t_{n+1}, y_n + k_1)\Delta t, \\ y_{n+1} &= y_n + (k_1 + k_2)/2, \end{aligned}}$$

EFFICIENT HEUN METHOD

EXAMPLE 29 *Repeat example 28 using the Heun method in place of the Euler method.*

We begin with $P(1978) = 115.40$. Using $\Delta t = 10$ and (for geometric emphasis) the inefficient form (3.16)–(3.17), the pattern of the calculations is:

$$\bar{P}_1 = 115.40 + 0.0282 \cdot 115.40 \cdot 10 = 147.94$$

$$P_1 = 115.40 + \frac{1}{2}0.0282(115.40 + 147.94)10 = 152.53$$

$$\bar{P}_2 = 152.53 + 0.0282 \cdot 152.53 = 195.54$$

$$P_2 = 152.53 + \frac{1}{2}0.0282(152.53 + 195.54)10 = 210.61$$

$$\vdots$$

We find ultimately that $P_3 = 266.48$, the Heun method approximation to $P(2008)$. These values appear in table 3.2.

Table 3.2

Approximate solution of the initial value problem $P' = 0.0282P$, $P(1978) = 115.40$, using the Heun method.

n	t_n	P_n
0	1978	115.40
1	1988	152.53
2	1998	201.61
3	2008	266.48

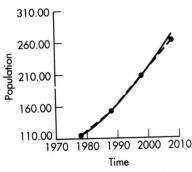

Borrowing the exact solution from example 28 yields the error in the Heun method approximation,

$$P(2008) - P_3 = 268.92 - 266.48 = 2.44 \, ,$$

an error of less than 1%, a significant improvement over the Euler method. ∎

A graph of the Heun method approximation obtained in the last example appears in figure 3.7. The improved accuracy is evident when one compares that figure with figure 3.6, the corresponding plot for the Euler method.

The next section describes even more accurate methods as well as an orderly procedure for obtaining them.

Figure 3.7: The exact solution (smooth curve) and the Heun method approximation with $\Delta t = 10$ (solid circles) to the population model $P' = 0.0282 \, P$, $P(1978) = 115.40$

3.5.1 Exercises

EXERCISE GUIDE

To gain experience . . .	Try Exercises
Applying the Euler method	1(a), 3(a), 5, 7(a), 9(a), 11(a), 14, 16(a), 26(a), 27−29(a)
Applying the Heun method	2(a), 4(a), 6, 8(a), 10(a), 12(a), 13, 15, 16(b), 25(a), 33
With error in the Euler method	1(b), 3(b), 7(b), 9(b), 11(b), 14, 16(c), 24, 26(b−c), 27−29(b)
With error in the Heun method	2(b), 4(b), 8(b), 10(b), 12(b), 15, 16(c), 25(b−c)
Interpreting the Euler method geometrically	30, 32(b)
Interpreting the Heun method geometrically	31−32
Studying solution behavior numerically	17−23, 27−29(c)

1. **(a)** Using five steps of the Euler method, find an approximation to the solution of $y' + 2y = \cos t$, $y(0) = -4$, at $t = 2$.
   (b) Find the exact solution and determine the error in your numerical approximation.

2. **(a)** Using five steps of the Heun method, find an approximation to the solution of $y' + 2y = \cos t$, $y(0) = -4$, at $t = 2$.
   (b) Find the exact solution and determine the error in your numerical approximation.

3. **(a)** Choosing $\Delta t = 0.5$, use the Euler method to find an approximation to the solution of $u' = 1 - u$, $u(0) = 4$, at $t = 2$.
   (b) Find the exact solution and determine the error in your numerical approximation.

4. **(a)** Choosing $\Delta t = 0.5$, use the Heun method to find an approximation to the solution of $u' = 1 - u$, $u(0) = 4$, at $t = 2$.
   (b) Find the exact solution and determine the error in your numerical approximation.

5. Use four steps of the Euler method to approximate the solution of $u' = u^2 - 1$, $u(0) = -2$, at $t = 1$.

6. Use four steps of the Heun method to approximate the solution of $u' = u^2 - 1$, $u(0) = -2$, at $t = 1$.

7. **(a)** Use *one step* of the Euler method to estimate the time required for the population governed by $P' = kP$, $P(0) = P_i$, to double.
   (b) Find an exact formula for the time at which the population has doubled and check the accuracy of your answer.

8. Repeat exercise 7 using the Heun method.

9. (a) Use *one step* of the Euler method to estimate the time required for the difference between the initial temperature and the outside temperature to decrease to one-half its initial size for the heat flow model $T' = -(Ak/cm)(T - T_{out})$, $T(0) = T_i$.

   (b) Find an exact formula for the time at which the temperature difference is at one-half its initial value and check the accuracy of your answer.

10. Repeat exercise 9 using the Heun method.

11. (a) Table 3.1 contains the first few steps of the approximate solution by the Euler method of the initial-value problem $P' = 0.0282\,P$, $P(1978) = 115.40$. Extend that table to the year 2048. What is the corresponding value of n?

    (b) Find the exact solution of this initial-value problem and use it to add a third column to table 3.1, the error $P(t_n) - P_n$. How does the accuracy of the approximate solution vary with n?

12. Repeat the preceding problem for table 3.2, which applies the Heun method to the same initial-value problem. Which method is more accurate for this problem, Euler or Heun?

13. Example 29 approximates the solution to a model for the population of Brazil

$$P' = 0.0282P, \quad P(1978) = 115.40$$

using the inefficient form (3.16–3.17) of the Heun method with $\Delta t = 10$.

   (a) Repeat those calculations using the efficient form. Verify that you obtain the same values of P_n.

   (b) How many multiplication operations do you save at each time step? How many would you save if $0.0282P$ were replaced by a function which required 100 multiplications for its evaluation?

14. For each of the following initial-value problems:

   i. Use 10 steps of the Euler method to find an approximate value of the solution at the indicated value of the independent variable.

   ii. Use available software to confirm the values you just computed.

   iii. Using an analytic solution method of your choice, find the exact value of the solution at the given value of the independent variable.

   iv. Calculate the error in the solution.

   v. Using available software, experimentally determine

the number of steps the Euler method requires to reduce the error you found in part iii by a factor of 10.

   (a) $P' = 0.0282P$, $P(1978) = 115.40$; approximate $P(1990)$. (This initial-value problem is the population model for Brazil mentioned in the text.)

   (b) $T' = -0.72(T - 10)$, $T(0) = 28$; approximate $T(10)$. (This initial-value problem is the heat-loss model (2.6–2.7) with time measured in hours and temperature in degrees Celsius.)

   (c) $P'(t) = 0.015P - 0.209$, $P(1847) = 8$; approximate $P(1850)$. (This initial-value problem is a model for the population of Ireland during the potato famine; see example 3, page 32.)

15. Complete the preceding problem using the Heun (modified Euler) method.

16. Choose a country from the population data of table 2.1 of chapter 2. Determine the values of the parameters k and P_i for an initial-value problem of the form $P' = kP$, $P(1978) = P_i$.

   (a) Use the Euler method to approximate the population in 1985. Compare with the appropriate entry in table 2.1 and comment on the differences you observe.

   (b) Repeat part (a) using the Heun method.

   (c) Compare your numerical results, the data from the table, and the exact solution of the model. Comment on the differences you observe.

17. Using available software and the approximation method of your choice, verify the accuracy of the solution curve sketches you obtained in project 2 of section 2.5 for the population model with competition

$$P'(t) = aP - sP^2, \quad P(t_0) = P_0.$$

Choose appropriate values of the parameters a, s, P_0, and t_0 to duplicate the conditions you considered in that project.

18. Using available software and the approximation method of your choice, estimate the year in which the population of Ireland would have dropped to zero according to the model developed in example 3, page 32.

$$P'(t) = 0.015P - 0.209, \quad P(1847) = 8.$$

(Recall that population is measured in millions and time in years.)

19. Exercise 4 of section 4.1.3, asks you to sketch solution curves for the differential equation $y' = y^3 - y$

for various ranges of initial values of y. Instead, use available software and the approximation method of your choice to graph representative solutions of this equation.

20. Repeat exercise 19 for the initial-value problem $y' = y - y^3$, $y(0) = 0.1$, from exercise 6 of section 4.1.3.

21. Using available software and the approximation method of your choice, select the one graph from among those accompanying exercise 7 of section 4.1.3, which most accurately represents the solution of the initial-value problem $y' = 1 - y^2$, $y(0) = 0$. Justify your choice.

22. Using available software and the approximation method of your choice, select the one graph from among those accompanying exercise 8 of section 4.1.3, which most accurately represents the solution of the initial-value problem $y' = (4 - y)^3$, $y(0) = 0$. Justify your choice.

23. Using available software and the approximation method of your choice, select the one graph from among those accompanying exercise 9 of section 4.1.3, which most accurately represents the solution of the initial-value problem $y' = y^2$, $y(0) = 1$. Justify your choice.

24. Derive the exact solution formula used in example 28 and verify the exact value of $P(2008)$ exhibited there.

25. (a) Using a calculator or available software, verify the Heun method approximation P_3 exhibited in example 29.
   (b) Repeat the calculation with $\Delta t = 5$. For what value of n is P_n an approximation to $P(2008)$? By what factor has the error changed?
   (c) Compare this change in error with that obtained for the Euler method in the corresponding part of the following problem. Which method gives the greater improvement in accuracy in return for doubling the computational effort by halving Δt?

26. (a) Using a calculator, available software, or the formula in the following problem, verify the Euler method approximation P_3 exhibited in example 28.
   (b) Repeat the calculation with $\Delta t = 5$. For what value of n is P_n an approximation to $P(2008)$? By what factor has the error in the approximation to $P(2008)$ changed?
   (c) Compare this change in error with that obtained for the Heun method in the corresponding part of exercise 25. Which method gives the greater

improvement in accuracy in return for doubling the computational effort by halving Δt?

27. (a) Argue that the solution $P(t)$ of the population model (2.1)–(2.2),

$$P'(t) = kP(t), \qquad P(t_0) = P_0,$$

is approximated using the Euler method at time $t_{n+1} = t_0 + (n+1)\Delta t$ by

$$P_{n+1} = (1 + k\Delta t)P_n = (1 + k\Delta t)^2 P_{n-1}$$
$$= \cdots = (1 + k\Delta t)^{n+1} P_0.$$

   (b) How does this expression compare with the exact solution $P(t_{n+1})$?
   (c) Provide a physical interpretation for the approximation formula $P_{n+1} = (1 + k\Delta t)^{n+1} P_0$ obtained in part (a). Is your interpretation approximately consistent with the assumptions used in deriving the differential equation model?

28. Repeat exercise 27 for the heat-loss model (2.6)–(2.7),

$$T'(t) = -\frac{Ak}{cm}(T(t) - T_{\text{out}}), \; T(0) = T_i.$$

29. Try to repeat the preceding two exercises for the population model with competition (2.5),

$$P'(t) = aP - sP^2, \qquad P(t_0) = P_0.$$

What property of this differential equation apparently prevents your writing a closed-form expression analogous to $P_{n+1} = (1 + k)^{n+1} P_0$ for the Euler method approximation to the solution of this problem?

30. Draw a *careful* graph of the exact solution both of the initial-value problem $P' = 0.0282\,P$, $P(1978) = 115.40$, considered in example 28 and of the first few steps of the Euler method approximation using $\Delta t = 20$. (That is, draw an analogue of figure 3.6 for this example.) Explain geometrically why the error in the Euler method, the difference between the exact solution and the approximation, grows with each step.

31. Carry out the preceding problem using the Heun method rather than the Euler method.

32. When the Heun method is applied to the initial-value problem $y' = f(t,y), y(t_0) = y_0$, the second step computes

$$y_{n+1} = y_n + \frac{1}{2}(f(t_n, y_n) + f(t_{n+1}, \bar{y}_{n+1}))\Delta t.$$

(a) Which of the terms in this expression correspond to the slope of the differential equation's direction field at the *beginning* of the step?

(b) Which of the terms in this expression correspond to the slope of the direction field at the *end* of an Euler method step?

(c) What term tells you that you are averaging these two slopes, not weighting one more heavily than another?

33. Write a single expression for the Heun method approximation P_{n+1} to the solution of the simple population model equation $P' = kP$. (Eliminate the $\bar{P}_{n+1}$ term.)

34. There have been a number of suggestions in our earlier analytic studies of the population model with competition,

$$P'(t) = aP - sP^2, \qquad P(t_0) = P_0,$$

that the added term sP^2 acts to provide an upper bound on population, in contrast to the simple population model that predicts continued exponential growth.

Suppose we apply this model to the population of Brazil beginning in 1978. A comparison with the simple model considered in the text,

$$\frac{dP}{dt} = 0.0282\,P, \qquad P(1978) = 115.40,$$

suggests that we might take $a = 0.0282$ with the obvious choices for t_0, P_0.

(a) Using available software and the approximation method of your choice, conduct experiments to determine a value of s that gives a stable population level in about the year 2050. What is the value of the limiting population?

(b) Confirm this result analytically.

3.6 ERROR ANALYSIS AND BETTER NUMERICAL METHODS

3.6.1 Global and Local Error

A comparison of figures 3.6 and 3.7 shows that the Heun method is more accurate than the Euler method, at least for the population model solved there. What precisely is the accuracy advantage of Heun over Euler? What improvement in accuracy can we expect with either method if we reduce the step size? How might we construct methods that are better than either of these? This section develops answers to these questions.

Our primary interest is in the **global error** after N steps with a given method,

$$e_N = y(t_N) - y_N.$$

GLOBAL ERROR

Here $y(t)$ is the *exact* solution of the initial-value problem $y' = f(t,y)$, $y(t_0) = y_i$, and y_N is the last in a sequence of *approximate* solutions $y_0, y_1, \ldots, y_N$ generated with one of these methods. Examples 28 and 29 computed the global error e_3 for Euler and Heun applied to a population model.

We can not expect to find a computable formula for global error; if we had such a formula, then we could find the error and add it to the approximate solution y_N to obtain the exact solution. But we can determine the proportionate change in global error due to a change in Δt.

A key contributor to global error is the **local truncation error**, the error at the end of *one* step of a method, *beginning with an exact solution value,*

$$E_L = y(t_1) - y_1,$$

where $y(t_0) = y_0$. This difference is called *truncation* error because it arises from the approximation $y'(t) \approx [y(t + \Delta t) - y(t)]/\Delta t$, which truncates the limit in the definition of derivative. Figure 3.8 illustrates these two forms of error.

What is the relation between local truncation error and global error? To approximate the solution at some fixed time T using a step size Δt requires $N = T/\Delta t$ steps. The error accumulates as we step from point to point, in part from the local truncation error that is introduced even when the step begins with an exact solution value and in part from the effect of beginning each step after the first with an incorrect solution value. If the latter effect is not too pronounced, intuition suggests that the global error accumulates in the same way as the sum of the local truncation errors. Since there are $N = T/\Delta t$ steps, global error might behave something like the sum of N local error values—that is, like

$$NE_L = T\frac{E_L}{\Delta t}.$$

Indeed, if the solution of the initial-value problem has sufficiently smooth derivatives, than it can be proven that **global error** at time T is proportional to

$$\frac{E_L}{\Delta t}.$$

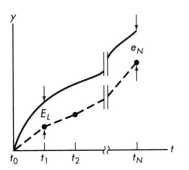

Figure 3.8: Local truncation error E_L and global error e_N in a numerical solution method

3.6.2 Error in the Euler Method

To find an expression for the global error in the Euler method, we must evaluate the local truncation error E_L. Since $y' = f(t, y)$, the Euler method formula $y_1 = y_0 + f(t_0, y_0)\Delta t$ can also be written

One step of the Euler method

$$y_1 = y_0 + y_0'\Delta t,$$

where $y_0' = f(t_0, y_0)$. This expression is dangerously close to a Taylor expansion for the exact solution $y(t_1)$ about the initial point t_0,

Taylor expansion for exact solution

$$y(t_1) = y(t_0) + y'(t_0)(t_1 - t_0) + \frac{y''(t_0)(t_1 - t_0)^2}{2}$$

$$+ \cdots + \frac{y^{(n-1)}(t_0)(t_1 - t_0)^{n-1}}{(n-1)!} + R_n, \tag{3.18}$$

where the remainder term is

$$R_n = \frac{y^{(n)}(\xi)(t_1 - t_0)^n}{n!}$$

for some ξ, $t_0 \leq \xi \leq t_1$. (See Taylor's theorem A.3.) This observation is the key to analyzing local truncation error.

To make a closer connection between the Euler method,

$$y_1 = y_0 + y_0' \Delta t,$$

and the Taylor expansion, write $t_1 - t_0 = \Delta t$ in (3.18), assume y has two continuous derivatives, and stop the expansion at $n = 2$:

$$y(t_1) = y(t_0) + y'(t_0)\Delta t + R_2.$$

Then the local truncation error is

$$
\begin{aligned}
E_L &= y(t_1) - y_1 \\
&= [y(t_0) + y'(t_0)\Delta t + R_2] - (y_0 + y_0'\Delta t) \\
&= y(t_0) - y_0 + (y'(t_0) - y_0')\Delta t + R_2.
\end{aligned}
$$

The definition of local truncation error requires that the Euler method begin with the value of the exact solution; i.e., $y_0 = y(t_0)$. As a consequence, the initial derivative values are identical as well:

$$y_0' = f(t_0, y_0) = f(t_0, y(t_0)) = y'(t_0).$$

Consequently, the local truncation error expression for the Euler method reduces to

$$E_L = R_2 = \frac{y''(\xi)}{2}\Delta t^2,$$

Local error for the Euler method

where $t_0 \le \xi \le t_1$. We have proved the following.

Theorem 3.5 (Local Truncation Error for the Euler Method) Let the solution $y(t)$ of the initial-value problem $y' = f(t, y)$, $y(t_0) = y_i$, have two continuous derivatives in the interval $[t_0, t_1]$. Then within that interval, the local truncation error in the Euler method applied to this problem is $E_L = y''(\xi)\Delta t^2/2$ for some ξ in $[t_0, t_1]$.

Reducing Δt by 2 will decrease the local truncation error by 4, and so on. Since

Local truncation error for the Euler method is proportional to Δt^2

and since global error is proportional to $E_L/\Delta t$, we conclude that

Global error for the Euler method is proportional to Δt

if the exact solution has two continuous derivatives.

Halving Δt will halve the global error (and halving Δt will require twice as much computation to reach the same ending time T). Because its global error is proportional to Δt to the *first* power, the Euler method is called a **first-order method**.

3.6.3 Better Methods

Global error is always proportional to $E_L/\Delta t$. A method with improved global error must exhibit local truncation error E_L proportional to a power of Δt higher than the Δt^2 factor in the local error expression for the Euler method. To improve local truncation error, an improved method needs to match more than the first two terms in the Taylor expansion (3.18) of the exact solution.

Computational experience shows that one improvement on Euler is the Heun method,

Heun method

$$k_1 = f(t_n, y_n)\Delta t,$$

$$k_2 = f(t_{n+1}, y_n + k_1)\Delta t,$$

$$y_{n+1} = y_n + (k_1 + k_2)/2.$$

It averages two slope values to step from y_n to y_{n+1}. Generalizing this idea, we shall seek a method of the form

General method

$$k_1 = f(t_n, y_n)\Delta t,$$

$$k_2 = f(t_n + \alpha\Delta t, y_n + \beta k_1)\Delta t,$$

$$y_{n+1} = y_n + ak_1 + bk_2, \tag{3.19}$$

where the constants $a, b, \alpha,$ and β are to be determined to match as many terms as possible in the Taylor expansion (3.18).

Obviously, the Heun method corresponds to choosing $\alpha = \beta = 1$ and $a = b = \frac{1}{2}$.

To parallel the analysis of local truncation error in the Euler method, we need to write the first step of this method as if it were a Taylor expansion. Letting $y_0' = f(t_0, y_0)$, we have

$$k_1 = y_0'\Delta t,$$

$$k_2 = f(t_0 + \alpha\Delta t, y_0 + \beta k_1)\Delta t,$$

$$y_1 = y_0 + ay_0'\Delta t + bk_2.$$

It remains to write k_2 using a two-variable Taylor expansion about (t_0, y_0) (see section A.3 of the Appendix):

$$
\begin{aligned}
k_2 &= f(t_0 + \alpha\Delta t, y_0 + \beta k_1)\Delta t \\
&= f(t_0, y_0)\Delta t + \frac{\partial f(t_0, y_0)}{\partial t}(t_0 + \alpha\Delta t - t_0)\Delta t \\
&\quad + \frac{\partial f(t_0, y_0)}{\partial y}(y_0 + \beta k_1 - y_0)\Delta t + \mathcal{O}(\Delta t^3) \\
&= y_0'\Delta t + \frac{\partial f(t_0, y_0)}{\partial t}\alpha\Delta t^2 + \frac{\partial f(t_0, y_0)}{\partial y}\beta y_0'\Delta t^2 + \mathcal{O}(\Delta t^3).
\end{aligned}
$$

We have used $y_0' = f(t_0, y_0)$ and $k_1 = y_0'\Delta t$. The term $\mathcal{O}(\Delta t^3)$ ($\mathcal{O}$ for *order of*) denotes terms multiplied by Δt^n, $n \geq 3$; those terms decay to zero at least as fast as Δt^3.

The *big-O* notation $\mathcal{O}(\Delta t^3)$ is a convenient way of lumping together terms that are less significant when Δt is small than those with factors of Δt or Δt^2. More precisely, we write $g(\Delta t) = \mathcal{O}(\Delta t^3)$ for a given function g if

$$\lim_{\Delta t \to 0} \frac{g(\Delta t)}{\Delta t^3} = c$$

for some constant c.

From this expression for k_2, the first step of the general method is

$$y_1 = y_0 + ay_0'\Delta t + bk_2$$

$$= y_0 + ay_0'\Delta t$$

$$+ b\left(y_0'\Delta t + \frac{\partial f(t_0, y_0)}{\partial t}\alpha\Delta t^2 + \frac{\partial f(t_0, y_0)}{\partial y}\beta y_0'\Delta t^2 + \mathcal{O}(\Delta t^3)\right)$$

$$= y_0 + (a + b)y_0'\Delta t + \left(\alpha b\frac{\partial f(t_0, y_0)}{\partial t} + \beta b\frac{\partial f(t_0, y_0)}{\partial y}y_0'\right)\Delta t^2$$

$$+ \mathcal{O}(\Delta t^3). \tag{3.20}$$

y_1 from the method

We wish to compare this expression with the Taylor expansion about t_0 of the exact solution $y(t_1)$,

$$y(t_1) = y(t_0) + y'(t_0)\Delta t + \frac{1}{2}y''(t_0)\Delta t^2 + R_3$$

$$= y_0 + y_0'\Delta t + \frac{1}{2}y''(t_0)\Delta t^2 + R_3.$$

The terms involving Δt^2 in the Taylor expansion multiply y'', while the corresponding terms in the method multiply various partial derivatives of f. To reconcile those terms, we use $y' = f(t, y)$ and the chain rule to write

$$y''(t) = \frac{\partial f(t, y(t))}{\partial t} + \frac{\partial f(t, y(t))}{\partial y}y'(t).$$

Evaluated at $t = t_0$ with $y(t_0) = y_0$ and $y'(t_0) = y_0'$, this expression becomes

$$y''(t_0) = \frac{\partial f(t_0, y_0)}{\partial t} + \frac{\partial f(t_0, y_0)}{\partial y}y_0'.$$

Hence, the Taylor expansion for the exact solution is

$$y(t_1) = y_0 + y_0'\Delta t + \left(\frac{1}{2}\frac{\partial f(t_0, y_0)}{\partial t} + \frac{1}{2}\frac{\partial f(t_0, y_0)}{\partial y}y_0'\right)\Delta t^2 + R_3. \tag{3.21}$$

$y(t_1)$ from the exact solution

Now we can match terms in the method (3.20) with those in the Taylor expansion (3.21) for the exact solution in an attempt to make the order of the local truncation error E_L as high as possible. Comparing coefficients of like powers of Δt in (3.20) and (3.21), we find

$$\Delta t : \quad a + b = 1$$

$$\Delta t^2 : \quad \alpha b = \frac{1}{2} \qquad (3.22)$$

$$\beta b = \frac{1}{2}.$$

(A more detailed analysis would show that we cannot match the terms in Δt^3.) For *any* choice of a, b, α, β satisfying these three equations, the method (3.19) will have local truncation error

$$E_L = y(t_1) - y_1 = R_3 - \mathcal{O}(\Delta t^3).$$

But the Taylor remainder term R_3 is proportional to Δt^3, providing the exact solution $y(t)$ has three continuous derivatives. Consequently, the local truncation error in this method is proportional to Δt^3 and *the global error is proportional to Δt^2.*

Any method of the form (3.19) with a, b, α, β satisfying the conditions (3.22) is called a **second-order Runge-Kutta** method. It is of *second* order because its global error is proportional to Δt to the *second* power. Halving Δt decreases the global error by a factor of four while only doubling the amount of work. Clearly, these Runge-Kutta methods are more efficient than the Euler method.

Of course, one member of this family is the familiar Heun method. As we observed before, it corresponds to the choices $\alpha = \beta = 1$ and $a = b = \frac{1}{2}$, which satisfy the conditions (3.22).

Precisely these same ideas can be used to develop a family of *fourth-order Runge-Kutta methods*. Beginning with a general formulation similar to (3.19), but one that averages four slope values rather than two, we find conditions similar to (3.22) on a set of constants to ensure that the local truncation error is proportional to Δt^5. Hence, these methods are fourth order globally. One choice of those constants leads to the following.

Fourth-order Runge-Kutta method For the initial-value problem

$$y' = f(t, y), \ y(t_0) = y_i$$

the **fourth-order Runge-Kutta method** successively computes the intermediate quantities $k_1, \ldots, k_4$ to find y_{n+1}:

$$
\begin{aligned}
k_1 &= f(t_n, y_n)\Delta t \\
k_2 &= f(t_n + \Delta t/2, y_n + k_1/2)\Delta t \\
k_3 &= f(t_n + \Delta t/2, y_n + k_2/2)\Delta t \\
k_4 &= f(t_n + \Delta t, y_n + k_3)\Delta t \\
y_{n+1} &= y_n + (k_1 + 2k_2 + 2k_3 + k_4)/6.
\end{aligned}
$$

It begins with $y_0 = y_i$.

This method, or one of its fourth-order relatives, is commonly called the *Runge-Kutta method*. The second-order analogues we derived previously are known by other names such as Heun, modified Euler or, for another choice of a, b, α, β, *midpoint* (see exercise 8). There are Runge-Kutta methods of order greater than four, too.

> Ortega and Poole [13] discuss in more detail the analysis of errors in numerical methods for initial-value problems.

3.6.4 Exercises

EXERCISE GUIDE

To gain experience . . .	Try Exercises
With global error in the Euler method	1, 4
With global error in the Heun method	2, 5
With global error in the Runge-Kutta method	3
Computational efficiency	6, 7
With variants of familiar methods	8, 9, 15
Analyzing local truncation error	10, 13, 14
Interpreting the methods geometrically	11, 12, 15(a)

1. Using the Euler method and available software, approximate the solution of $y' + xy = 4x$, $y(0) = 2$, at $x = 4$ with 20, 40, 60, and 80 steps. Determine the global error in your approximations to $y(4)$. Plot error versus Δt to verify that the global error in the Euler method is proportional to Δt.

2. Repeat exercise 1 for the Heun method, plotting global error against the appropriate power of Δt.

3. Repeat exercise 1 for the fourth-order Runge-Kutta method, plotting global error against the appropriate power of Δt.

4. The text claims that the global error in the Euler method is proportional to Δt. Reducing Δt by one-half should reduce the error by roughly the same amount. Test this assertion using available software and an initial-value problem of your choice. Compute an approximation to the solution at a fixed value of the independent variable using the Euler method with 20, 40, and 80 steps. Compare the three approximate solution values you obtain with the exact solution. Do you believe that the error in the Euler method is proportional to Δt?

5. Since the Heun method is a second-order Runge-Kutta method, its global error is proportional to Δt^2. Reducing Δt by one-half should reduce the error by about $\frac{1}{4}$. Test this assertion using available software

and an initial-value problem of your choice. Compute an approximation to the solution at a fixed value of the independent variable using the Heun method with 20, 40, and 80 steps. Compare the three approximate solution values you obtain with the exact solution. Do you believe that the error in the Heun method is proportional to Δt^2?

6. The computational cost of an algorithm is sometimes measured by the number of times it must evaluate a complicated function that appears in it. An appropriate measure for the Euler and Heun algorithms is the number of times each has to evaluate the function f appearing in the differential equation $y' = f(t, y)$. Accuracy is also important; an efficient but terribly inaccurate algorithm is seldom useful.

   Using the global error results in this section, determine how Δt must be changed for each method to reduce the global error at a given time by one-fourth. How will the computational cost of each algorithm change? What recommendations would you make regarding the trade-off between accuracy and computational cost?

7. Referring to the second-order Runge-Kutta methods (3.19), the book says

   > Halving Δt decreases the global error by a factor of four while only doubling the amount of

work. Clearly, these Runge-Kutta methods are more efficient than the Euler method.

Justify these statements.

8. Another acceptable choice of constants for the second-order Runge-Kutta method (3.19) is $a = 0, b = 1, \alpha = \beta = \frac{1}{2}$, leading to the *mid-point method*.
   (a) Use a sketch or other geometric argument to explain the name *mid-point*.
   (b) Verify directly that its local error E_L is proportional to Δt^3 and, hence, that its global error is of second order.
   (c) Using available software, a programmable calculator, or a program of your own, verify as follows that the global error in this method is indeed of second order: Approximate the solution of the initial-value problem $y' = -y$, $y(0) = 1$, at $t = 1$ using this method with 10, 20, and 40 steps; compare the approximate and exact solutions in each case; show that the error is proportional to Δt^2.

9. A variant of the Euler method, known as *backward Euler* or *implicit Euler*, is

$$y_{n+1} = y_n + f(t_{n+1}, y_{n+1})\Delta t.$$

(See the note following exercise 15 to explain the adjective *implicit*.)
   (a) Use a sketch or other geometric argument to explain the name *backward*.
   (b) Verify directly that its local error E_L is proportional to Δt^2 and, hence, that its global error is of first order.
   (c) Using available software, a programmable calculator, or a program of your own, verify as follows that the global error in this method is indeed of first order: Approximate the solution of the initial-value problem $y' = -y$, $y(0) = 1$, at $t = 2$ using this method with 10, 20, and 40 steps; compare the approximate and exact solutions in each case; show that the error is proportional to Δt.

10. Is there a choice of the constants a, b, α, β that reduces the general method (3.19) to the Euler method? Does that choice of constants satisfy the second-order error conditions (3.22)?

11. Explain the geometric significance of the terms $k_1/\Delta t$ and $k_2/\Delta t$ that appear in the general second-order Runge-Kutta method (3.19).

12. The general second-order Runge-Kutta method (3.19) computes $y_{n+1} = y_n + ak_1 + bk_2$. We found that a, b must satisfy $a + b = 1$; see (3.22). Can you interpret this method as stepping from y_n to y_{n+1} by averaging two slopes, as we did with the Heun method?

13. By writing out the terms in (3.20) that involve Δt^3, find the exact expression for the local truncation error in the Heun method. For the Heun method, formulate the analogue of the Euler local error theorem 3.5. Must you demand more continuity of the exact solution than with the Euler method?

14. By writing out the terms in (3.20) that involve Δt^3, find the exact expression for the local truncation error in the general second-order Runge-Kutta method (3.19). What choice(s) of a, b, α, β make the coefficient of this error term a minimum?

15. Suppose you tried to improve the Heun method by eliminating the first Euler step. Then you would be using an approximation scheme that has the form

$$y_{n+1} = y_n + \frac{1}{2}[f(t_n, y_n) + f(t_{n+1}, y_{n+1})]\Delta t.$$

   (a) Give a verbal description of the slopes you would be averaging in this process.
   (b) Calculate a few iterations for the population model $P' = 0.0282 P$, $P(1978) = 115.40$.
   (c) Try to calculate a few iterations for the population model with competition

$$P'(t) = aP - sP^2, \quad P(t_0) = P_0.$$

   Choose values of the parameters a, s, t_0, and P_0 that suit you. What happens?
   (d) What is the difference between the two differential equations? How does that difference affect your ability to implement this method?

This method is known as an *implicit method*. The adjective *implicit* is used because, in general, we cannot solve this iteration scheme to give an explicit expression for the approximation y_{n+1} at the end of the step.

3.7 CHARACTERISTIC EQUATIONS

A **constant-coefficient**, first-order, linear differential equation is one of the form

$$b_1 y' + b_0 y = f(x).$$

where b_0, b_1 are constants. This section and the next develop two methods for solving these special equations, one for homogeneous and the other for nonhomogeneous equations. Although more general methods, such as separation of variables and the integrating factor, can always be applied to constant coefficient equations, these special methods are simple enough to deserve separate study.

We begin with the homogeneous equation $b_1 y' + b_0 y = 0$. A simple example is the population equation

$$P' - kP = 0 \, .$$

Solving this equation amounts to finding a function whose derivative is a constant multiple of itself: $P' = kP$.

What function is its own derivative, give or take a constant multiple? The exponential, of course.

To find precisely which exponential function satisfies $P' - kP = 0$, let this last bit of intuition guide us. *Assume* the solution is an exponential of the form

$$P(t) = e^{rt}$$

for some yet-to-be-determined constant r, and substitute $P = e^{rt}$,

$$re^{rt} - ke^{rt} = 0 \, .$$

Since $e^{rt} \neq 0$ for any finite r and t, this last equality holds only if

$$r - k = 0.$$

We conclude that $r = k$, a constant value consistent with our initial assumption. Hence, a nontrivial solution of $P' = kP$ is $P(t) = e^{kt}$, and a general solution is

$$P_g(t) = Ce^{kt} \, .$$

where C is an arbitrary constant. This is the same general solution as obtained by separation of variables.

The key equation

$$r - k = 0$$

that determined r is called the **characteristic equation** of the differential equation $P' = kP$. It is *characteristic* because it has captured the character of the behavior of the solution by determining the growth or decay factor r.

The steps in the **characteristic equation method** for the first-order, linear, constant-coefficient, homogeneous differential equation

$$b_1 y' + b_0 y = 0 \, .$$

are:

1. Substitute the assumed solution $y = e^{rx}$, r a constant, into the differential equation and simplify to obtain the characteristic equation

$$b_1 r + b_0 = 0 \, .$$

2. Solve the characteristic equation to obtain $r = -b_0/b_1$ and write the general solution

$$y_g = Ce^{-b_0 x/b_1} .$$

Don't memorize this formula. Understand the process.

This method seems to be based on guessing. How can we *assume* anything about an unknown solution of a differential equation? Guessing has no place in mathematics.

Not so. Guessing, or following intuition, is an important part of creative mathematics.

We made an educated guess about the form of the solution. If our guess were wrong, we would encounter some inconsistency; e.g., k would not be constant. We can always verify our final solution by substituting it into the differential equation. The process is foolproof.

EXAMPLE 30 *The data of table 2.1, page 21, suggest that a model for the growth of the world's population is*

$$P' = 0.0183P, \qquad P(1978) = 4258 ,$$

where time is measured in years and population in millions of people. Solve this initial value problem by the characteristic equation method.

Since this differential equation is homogeneous and has constant coefficients, the characteristic equation method is applicable. Instead of using the general solution $P_g(t) = Ce^{kt}$ with $k = 0.0183$, we attack this problem directly.

Assume a solution of the form $P(t) = e^{rt}$, substitute it into $P' = 0.0183P$ to obtain

$$re^{rt} - 0.0183e^{rt} = 0 ,$$

and divide by $e^{rt} \neq 0$ to find the characteristic equation

CHARACTERISTIC EQUATION

$$r - 0.0183 = 0 .$$

Since $r = 0.0183$, a general solution of $P' = 0.0183P$ is

GENERAL SOLUTION

$$P(t) = Ce^{0.0183t} .$$

Find C.

We use the initial condition $P(1978) = 4258$ to find

$$C = 4258e^{-(0.0183)(1978)} .$$

SOLUTION OF INITIAL-VALUE PROBLEM

The solution of the initial-value problem is

$$P(t) = 4258e^{0.0183(t-1978)} . \qquad \blacksquare$$

Could we obtain a characteristic equation for a *non*homogeneous equation like $P' - kP = \alpha e^{-t}$? What would happen when we attempted to divide out the exponential factor?

What would happen if the equation were homogeneous but lacked constant coefficients? Would the resulting "characteristic equation" give a constant value for r?

These questions are explored in the next three examples.

EXAMPLE 31 *Try to solve the nonhomogeneous equation*

$$P' - kP = \alpha e^{-t}$$

by the characteristic equation method.

Assuming $P = e^{rt}$, where r is a constant to be determined, substituting, and dividing by e^{rt} yields

$$r = k - \alpha \frac{e^{-t}}{e^{rt}} = k - \alpha e^{-(1+r)t}.$$

This "characteristic equation" certainly does not give a constant value for r, contradicting our initial assumption. Hence, this differential equation does not have a solution of the form $P = e^{rt}$, where r is a constant. The characteristic equation method has failed. ∎

EXAMPLE 32 *Try to solve the variable-coefficient equation*

$$P' = t^2 P$$

by the characteristic equation method.

Assuming $P = e^{rt}$, substituting, and dividing by e^{rt} yields

$$r = t^2,$$

another "characteristic equation" that fails to have a constant solution consistent with our initial assumption. ∎

EXAMPLE 33 *Try to solve to the nonlinear logistic equation*

$$P' = aP - sP^2$$

by the characteristic equation method.

Assuming $P = e^{rt}$, substituting, and dividing by e^{rt} yields

$$r = a - se^{rt},$$

yet another "characteristic equation" that fails to have a constant solution. Our fundamental assumption that r is constant has been violated again. ∎

To put the moral of these three examples in positive terms, the characteristic equation method finds a general solution of *constant-coefficient, linear, homogeneous* differential equations. Check the form of the equation before you apply the method.

In exercises 1–11, find the characteristic equation and use it to find either a general solution or a solution of the initial-value problem. If the characteristic equation method is not applicable, explain why.

1. $y' - 6y = 0$

2. $xy' - 6y = 0$

3. $2u' + 7u = 0$, $u(-2) = 6$

4. $dw/dx = x$

5. $dw/dx = w$

6. $P' - kP = 0$

7. $dT/dt + (Ak/cm)(T - T_{\text{out}}) = 0$, $T(0) = T_i$

8. $y' - 3y^3 = 0$, $y(1) = 2$

9. $y' - 3y = 0$, $y(2) = 4e^6$

10. $4u' + 2u = 0$

11. $v' = -g$

12. Classify each of the equations in exercises 1–11 as
   (a) Linear or nonlinear,
   (b) Homogeneous or nonhomogeneous,
   (c) Constant coefficient or variable coefficient.

13. Verify, as follows, the two steps of the characteristic equation method given in the text.
   (a) Show that the characteristic equation for the general first-order, linear, homogeneous, constant coefficient differential equation $b_1 y' + b_0 y = 0$ is

$$b_1 r + b_0 = 0.$$

   Begin by assuming a solution of the form $y(x) = e^{rx}$.
   (b) Show that a general solution of $b_1 y' + b_0 y = 0$ is

$$y(x) = Ce^{-b_0 x/b_1}.$$

14. Solve the general constant-coefficient, linear, first-order, homogeneous differential equation $b_1 y' + b_0 y = 0$ by *separation of variables*. Show that you obtain the same general solution as given by the characteristic equation method.

15. Solve the general constant-coefficient, linear, first-order, homogeneous differential equation $b_1 y' + b_0 y = 0$ using an *integrating factor*. Show that you obtain the same general solution as given by the characteristic equation method.

16. Recall the expression e^{rt}, the form we assumed for the solution of $P' - kP = 0$. The quotient $1/r$ is sometimes called the **time constant** for the solution. Explore the meaning of this term by proving the following:
   (a) If $r > 0$, then e^{rt} doubles in a time interval proportional to $1/r$.
   (b) If $r < 0$, then e^{rt} is halved in a time interval proportional to $-1/r$.
   (c) The time constant has units of time.

17. See exercise 16 for the definition of *time constant*.
   (a) What are the time constants for the populations of three countries of your choice from Table 2.1, page 21? If these populations obey the simple population equation $P' = kP$, how much time would each need to double in size? Comment on these results.
   (b) What is the time constant for the heat-loss equation

$$\frac{dT}{dt} = -\frac{Ak}{cm}(T - T_{\text{out}})?$$

   Comment on the relation between the parameters in this problem and the size of the time constant.

18. Find *homogeneous solutions* of the following equations using the methods of this section.
    (a) The emigration equation $P' = kP - E$
    (b) The equation for emigration with decay, $P' = kP - \alpha e^{-t}$
    (c) The heat-loss equation $T' = -(Ak/cm)(T - T_{\text{out}})$
    (d) The equation for Ireland during 1847-1850, the height of the potato famine, $P' = 0.015P - 0.209$
    (e) The rock-tossing equation $v' = -g$

19. Attempt to apply the characteristic equation approach to the constant coefficient, *non*homogeneous differential equation

$$y' - 2y = x^3 .$$

Comment on the results.

20. Attempt to apply the characteristic equation approach to the homogeneous, *non*constant-coefficient differential equation

$$y' - xy = 0 .$$

Comment on the results. How might you find a general solution of this equation?

21. Attempt to apply the characteristic equation approach to the constant coefficient, homogeneous, *non*linear differential equation

$$y' + 4 \sin y = 0 .$$

Comment on the results. Verify that this equation is indeed homogeneous.

3.8 UNDETERMINED COEFFICIENTS

3.8.1 The Main Ideas

We seek particular solutions of *constant-coefficient*, nonhomogeneous equations such as the emigration equation

$$P'(t) - kP(t) = -E,$$

when the emigration rate E is constant. Although techniques such as variation of parameters certainly could be used, we hope to find quicker methods for this restricted class of problems.

The physical situation may help us make an educated guess of the functional form of a particular solution. Under the right conditions, might not growth exactly balance emigration and the population remain constant? If we find that constant population value, we have a simple particular solution.

To pursue that idea, assume that $P' - kP = -E$ has a particular solution P_p which is constant, say

$$P_p \equiv A,$$

for some yet-to-be-determined constant A. (The subscript p denotes *particular solution*.) To find A, we substitute this proposed solution into the differential equation. Since $A' = 0$, we obtain

$$P'_p - kP_p = A' - kA = -kA = -E .$$

Hence, $A = E/k$, and a particular solution of the emigration equation $P' - kP = -E$ is

$$P_p = E/k.$$

Try this educated guessing approach on another problem,

$$y' - 2y = 18e^{-4x} .$$

We must guess a solution of a form that can be combined on the left side of this equation with its derivative to yield the given forcing function e^{-4x} on the right. Since derivatives of exponentials are again exponentials, we guess

Guess a particular solution.

$$y_p(x) = Ae^{-4x},$$

where A is an undetermined constant. We guessed an exponential of the form e^{-4x} because that exponential must appear as the forcing term on the right when our guess is substituted into the left side of the equation $y' - 2y = 18e^{-4x}$.

Substitute.

Substituting $y_p(x) = Ae^{-4x}$ into $y' - 2y = 18e^{-4x}$ yields

$$-4Ae^{-4x} - 2Ae^{-4x} = 18e^{-4x}.$$

Canceling the common factor e^{-4x} gives an equation for the unknown constant A,

Find the undetermined coefficient A.

$$-6A = 18,$$

from which we obtain $A = -3$. Hence, a particular solution of $y' - 2y = 18e^{-4x}$ is

Write the particular solution.

$$y_p(x) = -3e^{-4x}.$$

We have just shown two simple examples of the **method of undetermined coefficients**, a grand name for what might better be called educated guessing. In each of these examples, the *undetermined coefficient* is the unknown constant A.

These examples suggest the following preliminary rules to guide our guessing of the form of a particular solution of constant-coefficient, linear equations:

1. If the forcing function is a constant, guess that a particular solution is a constant.
2. If the forcing function is an exponential, guess that a particular solution is a constant multiple of that exponential.

EXAMPLE 34   *Find a particular solution of*

$$P'(t) - 0.015P(t) = -0.209,$$

the equation from the Irish potato famine model of section 2.2.

Guess.

Rule 1 above suggests that we guess $P_p(t) = A$, where A is the undetermined coefficient. Substituting our guess into the differential equation immediately yields

Substitute.

$$-0.015A = -0.209,$$

Find coefficient.
Write solution.

or $A = 0.209/0.015 = 13.9$. Hence, a particular solution of $P'(t) - 0.015P(t) = -0.209$ is $P_p = 13.9$. ∎

EXAMPLE 35   *Find a particular solution of*

$$P'(t) - kP(t) = -\alpha e^{-t},$$

a population equation with a decaying emigration rate.

Rule 2 suggests that we guess

$$P_p(t) = Ae^{-t}.$$

Here A is the undetermined coefficient. (Of course, in the differential equation k and α are fixed parameters.) Substituting the guess $P_p = Ae^{-t}$ into $P' - kP = -\alpha e^{-t}$ yields

$$-Ae^{-t} - kAe^{-t} = -\alpha e^{-t},$$

or

$$A(1 + k) = \alpha.$$

If $k \neq -1$, we can solve the last expression to find $A = \alpha/(1 + k)$. Hence, a particular solution of $P' - kP = -\alpha e^{-t}$ is

$$P_p(t) = \frac{\alpha}{1 + k}e^{-t}, \qquad k \neq -1. \qquad \blacksquare$$

Of course, the differential equation loses its meaning as a population model when the growth rate k becomes negative.

3.8.2 The Homogeneous Solution Problem

Why did we encounter the restriction $k \neq -1$ in the last example? Certainly, we needed to avoided dividing by zero in $A(1 + k) = \alpha$. To see if there are any deeper reasons, we reconsider this differential equation in the special case $k = -1$,

$$P' + P = -\alpha e^{-t}.$$

Since the forcing term is unchanged, we could again try to guess $P_p = Ae^{-t}$. Substituting that proposed particular solution into $P' + P = -\alpha e^{-t}$ yields

$$P'_p + P_p = -Ae^{-t} + Ae^{-t} = -\alpha e^{-t}$$

or

$$0 = -\alpha e^{-t},$$

an equation that holds only if $\alpha = 0$. What went wrong?

When we substituted the proposed particular solution into the left side of the differential equation, it reduced to zero. Why? Because our guess Ae^{-t} is a solution of the *homogeneous* equation $P' + P = 0$. We will never be able to evaluate the constant A because it disappears when the proposed solution is substituted into the differential equation.

We will encounter the same phenomenon in the next example, but it will guide us toward a solution to this dilemma.

EXAMPLE 36 *Find a particular solution of the rock-tossing equation,*

$$v'(t) = -g,$$

where g is the (constant) acceleration of gravity.

3.8 Undetermined Coefficients

Since the forcing term in this equation is a constant, rule 1 suggests a particular solution of the form

$$v_p(t) = A,$$

where A is an undetermined constant. Substituting into $v' = -g$ yields

$$0 = -g.$$

Our guess for the particular solution has again been reduced to zero when substituted into the differential equation. The form assumed for the particular solution actually satisfies the *homogeneous* equation. The undetermined constant disappears; we will never find it. ∎

We must modify our guessing rules when the proposed particular solution contains a solution of the corresponding homogeneous equation. To learn what that modification should be, we reexamine the last example.

Directly integrating $v' = -g$ yields

$$v(t) = -gt + C,$$

a general solution. The arbitrary constant C is a general solution of the homogeneous equation $v'(t) = 0$, and $-gt$ is a particular solution of the nonhomogeneous equation.

Our first guess for a particular solution was $v_p = A$. The particular solution we just obtained is $v_p(t) = -gt$, which has the form of a constant multiplied by t. That is, the actual particular solution has the form t *times our initial guess*. Let's try again using this idea.

EXAMPLE 37   *Find a particular solution of*

$$v'(t) = -g$$

by assuming a solution of the form t multiplying our first guess, which was $v_p = A$.

Guess.

We are directed to seek a particular solution of the form

$$v_p(t) = At,$$

where A is a constant to be determined. Substituting in the differential equation yields

Substitute.

$$v_p'(t) = (At)' = A = -g.$$

Find coefficient.
Write solution.

The undetermined constant A has been found: $A = -g$. A particular solution of the given differential equation is $v_p(t) = -gt$. ∎

If our first guess for the form of the particular solution satisfies the homogeneous differential equation, it appears that we should multiply that guess by the independent variable to obtain a form that will lead to a solution of the nonhomogeneous equation. Let's try that idea on another problem.

EXAMPLE 38 *Using the "multiply the first guess by t" idea, find a particular solution of*

$$P' + P = \alpha e^{-t}.$$

Our first guess, which turned out to solve the homogeneous equation $P' + P = 0$, was $P_p = Ae^{-t}$. The idea we have just developed suggests we should try instead

Guess.

$$P_p = Ate^{-t}.$$

Substituting this expression into $P' + P = \alpha e^{-t}$ yields

Substitute.

$$Ae^{-t} + At(-e^{-t} + e^{-t}) = -\alpha e^{-t}$$

or

Find coefficient.

$$A = -\alpha.$$

We have found the particular solution $P_p = -\alpha te^{-t}$ of $P' + P = \alpha e^{-t}$.

Write solution.

We conclude that rules 1 and 2 must be supplemented with a third rule:

3. If the proposed particular solution contains a solution of the homogeneous equation, multiply that proposed solution by the independent variable.

EXAMPLE 39 *Find a particular solution of*

$$y' - 3y = 6e^{3x}.$$

Rule 2 first suggests the particular solution $y_p(x) = Ae^{3x}$. To check whether this proposed solution could be a solution of the corresponding homogeneous equation

Make a first guess.

$$y' - 3y = 0,$$

we could substitute directly into that equation. Alternatively, we could find the general solution of this homogeneous equation and determine by inspection whether the proposed particular solution solves the homogeneous equation.

To follow the latter tack, we use the characteristic equation method. Substitute the assumed homogeneous solution $y_h(x) = e^{rx}$ into $y' - 3y = 0$ to obtain the characteristic equation $r - 3 = 0$. Since $r = 3$, the general solution of the homogeneous equation is

Check for homogeneous solution in first guess.

$$y_h(x) = Ce^{3x}.$$

Hence, our proposed particular solution $y_p(x) = Ae^{3x}$ is a solution of the homogeneous equation. Rule 3 now requires that we multiply our first guess by x. The new candidate for the particular solution is

Make a better guess.

$$y_p(x) = Axe^{3x}.$$

Substituting this expression into the nonhomogeneous differential equation $y' - 3y = 6e^{3x}$ yields

$$Ae^{3x} + 3Axe^{3x} - 3Axe^{3x} = 6e^{3x} ,$$

Find the coefficient.

from which we obtain $A = 6$. Hence, a particular solution of $y' - 3y = 6e^{3x}$ is

Write the particular solution.

$$y_p(x) = 6xe^{3x} .$$

With no additional effort, we can combine this particular solution with $y_h(x) = Ce^{3x}$, the general solution of the homogeneous equation found earlier, to obtain a general solution of the nonhomogeneous equation $y' - 3y = 6e^{3x}$,

Combine particular and homogeneous to write general solution.

$$y(x) = Ce^{3x} + 6xe^{3x} .$$

The effort of finding the homogeneous solution wasn't wasted; it enabled us to write a general solution of this equation. ∎

3.8.3 The Whole Story

The basic steps in the **method of undetermined coefficients** are:

1. Assume a particular solution of the appropriate form containing one or more undetermined coefficients.
2. Substitute the assumed solution form into the differential equation and solve for coefficients by equating similar terms.

In the preceding examples, the particular solution is assumed to be the same type function as the forcing term. That approach is also successful with forcing functions that are cosines, sines, or polynomials. The appropriate educated guess for the form of the particular solution in each case is shown in table 3.3, which summarizes and extends the educated guessing rules developed in the previous examples. That table is the guide to step 1 of the undetermined-coefficient solution method.

Recall that the *order*, or *degree*, of a polynomnial is the highest power appearing in it; e.g., $1 - x^4$ is a fourth-order polynomial, and the constant 6 is a polynomial of order zero (because $6 = 6x^0$).

We can restate the contents of table 3.3 as follows:

1. If the forcing function is
   (a) An exponential, the trial particular solution is an exponential,
   (b) A polynomial (including just a constant), the trial particular solution is a polynomial of the same order,
   (c) A sine and/or a cosine, the trial particular solution includes both a sine *and* a cosine.
2. If the trial particular solution solves the related homogeneous equation, multiply the trial solution by the independent variable.

We can carry these rules even further to cover combinations of the functions listed in table 3.3. For example, if the forcing term is a polynomial multiplying an exponential, say $x^3 e^{-2x}$, then the first guess for a particular solution is $A_3 y^3 + \cdots + A_1 x + A_0 e^{-2x}$. We'll neglect this additional complexity until we study higher-order equations.

Note that forcing functions involving *either* a sine or a cosine give rise to particular solution forms involving *both* a sine and a cosine. The following example illustrates this situation.

EXAMPLE 40 *Find a particular solution of*

$$y' - 2y = 13 \cos 3x \,.$$

Guess.

Table 3.3 recommends the form $y_p(x) = A \cos 3x + B \sin 3x$. In anticipation of substituting into the left side of the differential equation, we compute

$$y_p'(x) = -3A \sin 3x + 3B \cos 3x.$$

Note that the $\cos 3x$ term gave rise to $\sin 3x$ in the derivative. If our initial guess had included only the cosine, substituting into the left side of the equation would have given an equation of the form

$$-3A \sin 3x - 2A \cos 3x = \cos 3x \,,$$

which can never be satisfied for any choice of A as long as x is free to vary.

Now we substitute the candidate particular solution and its derivative into $y' - 2y = 13 \cos 3x$ in hopes of determining the constants A and B. After collecting coefficients of like functions, we obtain

Substitute.

$$(-3A - 2B) \sin 3x + (3B - 2A) \cos 3x = 13 \cos 3x \,.$$

Since no $\sin 3x$ term appears on the right, the coefficient of $\sin 3x$ on the left must be zero:

Find coefficients.

$$-3A - 2B = 0 \,.$$

Table 3.3

Particular solutions via undetermined coefficients
for the constant-coefficent, first-order linear equation
$b_1 y' + b_0 y = f(x)$

Forcing Term $f(x)$	Trial Particular Solution
$a_n x^n + \cdots + a_1 x + a_0$	$A_n x^n + \cdots + A_1 x + A_0$
$a e^{qx}$	$A e^{qx}$
$a \cos px + b \sin px$	$A \cos px + B \sin px$

If the assumed form of the particular solution solves the
corresponding homogeneous equation, multiply the
assumed form by x.

Lowercase letters a, a_i, b are constants given in the
forcing function. The coefficients to be determined are
denoted by uppercase letters A, A_i, B.

Similarly, the coefficient of $\cos 3x$ on the left must equal the coefficient of $\cos 3x$ on the right:

$$3B - 2A = 13.$$

Write solution.

Solving this pair of simultaneous linear equations for the unknowns A and B yields $A = -2$, $B = 3$. A particular solution of $y' - 2y = 13 \cos 3x$ is

$$y_p(x) = -2 \cos 3x + 3 \sin 3x.$$ ∎

To solve the simultaneous equations from the last example,

$$-3A - 2B = 0$$
$$-2A + 3B = 13$$

combine the two equations to eliminate one of the unknowns. For example, multiply the first equation by 2, the second by -3, and add to eliminate A, leaving $-13B = -39$, or $B = 3$. Then either equation yields $A = -2$.

3.8.4 More Examples

This subsection offers more examples of particular solutions of constant-coefficient, linear equations. They are constructed using undetermined coefficients. Homogeneous solutions are found using characteristic equations, and these are coupled with the particular solutions to form general solutions and to solve initial-value problems.

EXAMPLE 41 *Find a particular solution of*

$$2y' + 5y = 15x^3 - x.$$

Since the forcing term here is a polynomial of order 3, Table 3.3 recommends assuming a particular solution that is also a polynomial of order 3,

Guess.

$$y_p(x) = A_3 x^3 + A_2 x^2 + A_1 x + A_0.$$

Substitute.

Substituting this form of y_p and collecting coefficients of like powers of x yields

$$5A_3 x^3 + (6A_3 + 5A_2)x^2 + (4A_2 + 5A_1)x + (2A_1 + 5A_0) = 15x^3 - x.$$

Find the coefficients.

We equate coefficients of corresponding powers of x on either side of this equation:

$$x^3 : \qquad 5A_3 = 15$$
$$x^2 : \quad 6A_3 + 5A_2 = 0$$
$$x^1 : \quad 4A_2 + 5A_1 = -1$$
$$x^0 : \quad 2A_1 + A_0 = 0$$

Beginning with the first equation and solving successively gives the values of the undetermined coefficients:

$$A_3 = 3, \qquad A_2 = \frac{-18}{5}, \qquad A_1 = \frac{67}{25}, \qquad A_0 = \frac{-134}{25}.$$

Hence, a particular solution of $2y' + 5y = 15x^3 - x$ is

Write the solution.

$$y_p(x) = 3x^3 - \frac{18x^2}{5} + \frac{67x}{25} - \frac{134}{25}.$$ ∎

We have taken a small shortcut in the last example by not explicitly check- ing whether any part of our proposed particular solution was a solution of the corresponding homogeneous equation. We are using an observation that the exercises ask you to draw from the characteristic equation for the homogeneous differential equation

$$b_1 y' + b_0 y = 0.$$

Its general solution is either an exponential or a constant (if $b_0 = 0$).

If the proposed particular solution contains only a sine and a cosine, no part of it could solve the corresponding homogeneous equation. In the last ex- ample, none of the powers of x except the constant in the proposed particular solution could possibly be solutions of the corresponding homogeneous equa- tion there. However, the homogeneous equation $2y' + 5y = 0$ clearly does not have a constant nontrivial solution.

EXAMPLE 42 *Use table 3.3 to determine the* form *of the particular solution of*

$$y' = 15x^3 - x.$$

Do not evaluate the undetermined coefficients in the proposed particular solution.

The forcing term is the same as that of the last example, but the differential expression on the left is different. Since the forcing terms are identical, our proposed particular solution should be the same as that in the previous example, providing no part of the guess is a solution of the homogeneous version of this equation, $y' = 0$. Hence, we tentatively propose a particular solution of the form

First guess.

$$y_p(x) = A_3 x^3 + A_2 x^2 + A_1 x + A_0.$$

But $y' = 0$ obviously has the general solution $y_h = C$. Since a constant A_0 also appears in the assumed form of the particular solution, table 3.3 dictates that we multiply our entire guess by x. Hence, the correct form for the proposed particular solution of this equation is

Make a better guess.

$$y_p(x) = x(A_3 x^3 + A_2 x^2 + A_1 x + A_0).$$ ∎

To appreciate the need for a particular solution of this form, simply solve the differential equation $y' = 15x^3 - x$ by antidifferentiation:

$$y(x) = \frac{15x^4}{4} - \frac{x^2}{2} + C.$$

The term of highest order in this solution is x^4, as is the term $x(A_3 x^3 + \cdots)$ in the particular solution form of the last example. But the highest-order term would have been only x^3 if the first form of the particular solution had not been multiplied by x to compensate for the presence of a homogeneous solution.

EXAMPLE 43 *Use table 3.3 to determine the* form *of the particular solution of*

$$3y' + 6y = 1 + 2\cos 4x + \sin 4x - 3\sin \pi x - 6e^{-2x} + x^3.$$

Do not evaluate the undetermined coefficients in the proposed particular solution.

This forcing term is a combination of the various forcing terms appearing in table 3.3. Since the differential equation is linear, we can break this problem into a series of smaller tasks. In each subproblem, we seek a particular solution of an equation whose forcing term is just one of the entries in table 3.3. The sum of these individual particular solutions will be a particular solution of the original equation. (The *principle of superposition* justifies this decomposition.)

To identify these subproblems, collect the terms on the right side of the differential equation into groups of similar type:

Make a set of guesses.

Forcing Term Group	Proposed Particular Solution
polynomial: $1 + x^3$	$A_3 x^3 + A_2 x^2 + A_1 x + A_0$
sine, cosine: $2\cos 4x + \sin 4x$	$B_c \cos 4x + B_s \sin 4x$
sine, cosine: $-3\sin \pi x$	$D_c \cos \pi x + D_s \sin \pi x$
exponential: $6e^{-2x}$	Ee^{-2x}

Two groups of sines and cosines appear, one with argument $4x$, the other with argument πx.

Check for homogeneous solution.

To determine if any of these assumed forms should be multiplied by x, we find a general solution of the homogeneous equation $3y' + 6y = 0$. Its characteristic equation is $3r + 6 = 0$, from which we obtain $r = -2$ and

$$y_h(x) = Ce^{-2x}.$$

Since the exponential group appearing last in this list contains a term that is a constant multiple of e^{-2x}, that particular solution term must be multiplied by x to yield the revised form of a particular solution,

Improve one guess.

$$Exe^{-2x}.$$

None of the other proposed solution forms contains e^{-2x}, so no other is changed.

The constants in each group can be determined individually; e.g., E can be chosen so that $y_p(x) = Exe^{-2x}$ is a particular solution of

Substitute in separate equations.

$$3y' + 6y = -6e^{-2x},$$

and B_c and B_s can be chosen so that $y_p(x) = B_c\cos 4x + B_s\sin 4x$ is a particular solution of

$$3y' + 6y = 2\cos 4x + \sin 4x.$$

In accord with the principle of superposition, the sum of the particular solutions for all these individual forcing term groups will be a particular solution of the original equation

$$3y' + 6y = 1 + 2\cos 4x + \sin 4x - 3\sin \pi x - 6e^{-2x} + x^3. \qquad \blacksquare$$

Recall that a general solution of a first-order, linear, nonhomogeneous differential equation is a sum of a general solution of the corresponding homogeneous equation and a particular solution of the original nonhomogeneous equation, $y_g = Cy_h + y_p$.

We now have the tools to find these two component parts for constant-coefficient, linear, first-order differential equations. The characteristic equation method finds the general solution of the homogeneous equation, and the method of undetermined coefficients finds particular solutions of equations whose forcing terms are exponentials, polynomials, or sines and cosines.

EXAMPLE 44 *Find a general solution of*

$$y' - 2y = 13\cos 3x \,.$$

We need two components, a particular solution of the given nonhomogeneous equation and a general solution of the corresponding homogeneous equation. The particular solution component is available from example 40:

PARTICULAR SOLUTION

$$y_p(x) = -2\cos 3x + 3\sin 3x \,.$$

The corresponding homogeneous equation is

$$y' - 2y = 0 \,.$$

Its characteristic equation is $r - 2 = 0$. Hence, a general solution of the homogeneous equation is

HOMOGENEOUS SOLUTION

$$y_h(x) = Ce^{2x} \,.$$

and a general solution of the nonhomogeneous equation $y' - 2y = 13\cos 3x$ is

GENERAL SOLUTION

$$y(x) = Ce^{2x} - 2\cos 3x + 3\sin 3x \,. \qquad \blacksquare$$

EXAMPLE 45 *Find a general solution of*

$$2y' + 5y = 15x^3 - x \,.$$

Again, we need both a particular solution, such as the one provided by example 41:

PARTICULAR SOLUTION

$$y_p(x) = 3x^3 - \frac{18x^2}{5} + \frac{67x}{25} - \frac{134}{25} \,,$$

and a general solution of the corresponding homogeneous equation

HOMOGENEOUS SOLUTION

$$2y' + 5y = 0 \,.$$

Since the characteristic equation for this homogeneous differential equation is $2r + 5 = 0$ (yielding $r = -\frac{5}{2}$), its general solution is $y_h(x) = Ce^{-5x/2}$.

Thus, a general solution of $2y' + 5y = 15x^3 - x$ is

GENERAL SOLUTION

$$y(x) = Ce^{-5x/2} + 3x^3 - \frac{18x^2}{5} + \frac{67x}{25} - \frac{134}{25} \,. \qquad \blacksquare$$

To see that a general solution is the appropriate starting point for solving a linear initial-value problem, think of all the unknown terms in the differential equation grouped on the left and the forcing terms on the right. Substitute the general solution in the left side.

Then the particular solution ensures that the correct forcing term will appear on the right. Although the homogeneous solution contributes zero to the forcing term, it adds an arbitrary constant in the solution itself. It is that constant we choose to satisfy the initial condition.

The next example illustrates the contributions of each component of the general solution to the forcing term, and the succeeding examples use general solutions to solve initial-value problems.

EXAMPLE 46   *Example 44 showed that a general solution of*

$$y' - 2y = 13\cos 3x$$

is

$$y(x) = Ce^{2x} - 2\cos 3x + 3\sin 3x \,.$$

What does each part of this general solution contribute to the right-hand side of $y' - 2y = 13\cos 3x$?

The differential equation is written with all its unknown terms on the left. If we substitute the general solution in the left side, we find

$$y' - 2y = (Ce^{2x})' - 2Ce^{2x} + (-2\cos 3x + 3\sin 3x)' - 2(-2\cos 3x + 3\sin 3x)$$

$$= 2Ce^{2x} - 2Ce^{2x} + 6\sin 3x + 9\cos 3x + 4\cos 3x - 6\sin 3x$$

$$= 0 + 13\cos 3x = 13\cos 3x \,.$$

The term Ce^{2x}, which is the general solution of the homogeneous differential equation, contributes zero to the right-hand side, while the particular solution $-2\cos 3x + 3\sin 3x$ has provided the correct forcing term $13\cos 3x$ on the right.

This result is another example of the *principle of superposition*.   ∎

To solve a linear initial value problem, we find a general solution of the given differential equation and choose the arbitrary constant in it to satisfy the initial condition. The next examples illustrate the process.

EXAMPLE 47   *Find the solution of the initial-value problem*

$$y' - 2y = 13\cos 3x, \qquad y(\pi/2) = 5 \,.$$

GENERAL SOLUTION

Example 44 provides a general solution,

$$y(x) = Ce^{2x} - 2\cos 3x + 3\sin 3x \,.$$

Although the homogeneous solution term Ce^{2x} adds nothing to this expression's ability to satisfy the given nonhomogeneous differential equation, it does provide the arbitrary constant we need to satisfy the initial condition $y(\pi/2) = 5$.

Applying the initial condition to this general solution yields

$$y(\pi/2) = Ce^{\pi} - 2\cos(3\pi/2) + 3\sin(3\pi/2) = Ce^{\pi} - 3.$$

To satisfy $y(\pi/2) = 5$, we choose C so that $Ce^{\pi} - 3 = 5$ or

Find C.

$$C = 8e^{-\pi}.$$

Using this value of C in the general solution $y(x) = Ce^{2x} - 2\cos 3x + 3\sin 3x$ gives the solution of the initial-value problem,

SOLUTION OF INITIAL-VALUE PROBLEM

$$y(x) = 8e^{x-\pi} - 2\cos 3x + 3\sin 3x.$$

To verify that we have correctly solved the initial-value problem, substitute this solution into both the differential equation $y' - 2y = 13\cos 3x$ and the initial condition $y(\pi/2) = 5$. ∎

EXAMPLE 48 *Find the solution of the initial-value problem*

$$2y' + 5y = 15x^3 - x, \qquad y(0) = -7.$$

The pattern is precisely that of the preceding example. The general solution found in Example 45 is

GENERAL SOLUTION

$$y(x) = Ce^{-5x/2} + 3x^3 - \frac{18x^2}{5} + \frac{67x}{25} - \frac{134}{25}.$$

Requiring $y(0) = -7$ yields

$$y(0) = C - \frac{134}{25} = -7$$

or

Find C.

$$C = -\frac{41}{25}.$$

The solution of the initial value problem is

SOLUTION OF INITIAL-VALUE PROBLEM

$$y(x) = \frac{-41e^{-5x/2}}{25} + 3x^3 - \frac{18x^2}{5} + \frac{67x}{25} - \frac{134}{25}.$$

Again, this solution can be checked by showing that it satisfies both the differential equation $2y' + 5y = 15x^3 - x$ and the initial condition $y(0) = -7$. ∎

3.8.5 Exercises

EXERCISE GUIDE

To gain experience . . .	Try Exercises
Using undetermined coefficients	1–18, 21
With particular solutions	1–18, 22(a)
With homogeneous solutions	19(a), 23–24
With general solutions	19(b), 22(b)
Solving initial-value problems	19(c), 22(c),
With the foundation of undetermined coefficients	20–21, 25–27

In exercises 1–18, use undetermined coefficients to find a particular solution of the given equation. If undetermined coefficients are not applicable, explain why.

1. $u' - 4u = 4e^{\pi x}$

2. $u' - 4u = 4\cos \pi x$

3. $u' - 4u = 4\tan \pi x$

4. $u' - 4u = \pi e^{4x}$

5. $u' - 4u = 4x$

6. $u' - 4u = 4x + \pi e^{4x}$

7. $u' - 4xu = \pi e^{-4x}$

8. $u' - 4u = 4u$

9. $u' - 4u = 0$

10. $dy/dx = (6 - x)(6 + x) - 2y$

11. $y' - 3x = 2y$

12. $y' - 3x^2 = 2y$

13. $y' - 3x^2 = 2y^2$

14. $T' + 4T = (2 - t)^2$

15. $y' + \pi y^2 = \cos x$

16. $3w' + 2w = t + 2$

17. $dx/dt = x + 4 - t$

18. $2u' = u - 4(\sin 2t + 5t)$

19. In those of exercises 1–18 in which you found a particular solution,
    (a) find a nontrivial solution of the corresponding homogeneous equation,
    (b) write a general solution of the nonhomogeneous equation,
    (c) find the solution of the equation that has the value 1 when the independent variable is zero.

20. The method of undetermined coefficients finds particular solutions of constant-coefficient, linear, nonhomogeneous equations. To which of the equations in exercises 1–18 is this method *not* applicable because the equation
    (a) has variable coefficients?
    (b) is nonlinear?
    (c) is homogeneous?

21. In each of the following, find the *form* of a particular solution. *Do not evaluate* the coefficients.
    (a) $y' + 3y = 1 + x^4 - e^{3x} + 2\cos 3x$
    (b) $y' + 3y = 1 + x^4 - e^{-3x} + 2\cos 3x$
    (c) $y' + 3x = 1 + x^4 - e^{-3x} + 2\cos 3x$
    (d) $y' + 3y = 2 - \sin 2x + \pi \cos(\pi x/4) + (5 - 3x)^2$
    (e) $y' + 3y = (x + 2)^2 - (x - 3)^3 + 6(2 + \pi e^{-3x})$
    (f) $P' + kP = Ae^{kt} - \sin kt$, k, A constant

22. Assuming that E is constant,
    (a) verify that E/k is a particular solution of
    $$P' - kP = -E,$$
    (b) obtain a general solution of this equation,
    (c) solve the initial-value problem
    $$P' - kP = -E, \qquad P(t_0) = P_0,$$
    the population model with emigration.

23. The text claims that the general linear, constant-coefficient, homogeneous equation
    $$b_1 y' - b_0 y = 0,$$
    has a general solution that is either an exponential or a constant (if $b_0 = 0$). Use the characteristic equation method to verify this assertion.

24. Table 3.3 contains the statement, "If the assumed form of the particular solution solves the corresponding homogeneous equation, multiply the assumed form by x." Does this warning apply to sine or cosine forcing functions? Why? Can a sine or a cosine be a solution of a constant-coefficient, linear, first-order, homogeneous equation?

25. To see that the method of undetermined coefficients is essentially limited to constant-coefficient equations, attempt to solve the variable coefficient equation
    $$xy' + y = e^{4x}$$
    by assuming a particular solution of the form $y_p(x) = Ae^{4x}$, A an undetermined constant. What goes wrong?

26. Here's an idea for solving the logistic equation $P' - aP = sP^2$.

    When the logistic equation is written in this form, the left side is linear with constant coefficients. The right side is a polynomial, so table 3.3 suggests a particular solution that is a polynomial of the same order, $P_p = A_2 P^2 + A_1 P + A_0$. All I have to do is substitute to find the undetermined constants A_2, A_1, A_0.

    What is wrong with this idea? What is a reasonable method for solving this equation?

27. The text suggests that table 3.3 could be extended to provide particular solution forms for forcing terms that are combinations of the functions shown there;

e.g., a forcing term that is a product of a polynomial and an exponential should lead to a particular solution that is a product of a polynomial of the same order with a similar exponential. With that hint,

    i. find the *form* of a particular solution of each of the following.

ii. find a particular solution of each of the following.

    (a) $y' + y = xe^x$
    (b) $y' + y = xe^{-x}$
    (c) $y' + y = x \cos 2x - \sin x$
    (d) $y' + y = e^x \cos 2x$

3.9 UNIQUENESS AND EXISTENCE

3.9.1 Uniqueness

Under what conditions can we guarantee that the general nonlinear initial-value problem

$$y' = f(x, y), \qquad y(0) = y_i, \qquad (3.23)$$

has no more than one solution—i.e., that its solution is *unique*? One answer is to limit consideration to equations of the form $y' = -g(x)y + f(x)$ with f and g continuous, since the uniqueness theorem (theorem 3) for linear equations guarantees then that there is no more than one solution. But we want to treat more than just linear equations.

To state a uniqueness theorem for the general nonlinear equation, we ask, in effect, that $f(x, y)$ be able to be linearized. We can then prove uniqueness by using an integrating factor, just as in the proof of the linear uniqueness theorem.

For the nonlinear initial-value problem as for the linear initial-value problem, we find the following.

    Consequence of Uniqueness: Two solutions of an initial-value problem that agree at one point must be identical throughout their common interval of existence.

In other words, two solution graphs can not cross at an isolated point. Either the two graphs are identical because they are superimposed or they never cross. See figure 3.2, page 87.

    Theorem 3.6 (Uniqueness) Suppose that

    i. $u(x)$, $v(x)$ are both (continuously differentiable) solutions of the initial value problem (3.23) for $x_0 \leq x \leq x_1$ for some x_1, and
    ii. $f(x, y)$ and $\partial f(x, y)/\partial y$ are continuous in x and y for $x_0 \leq x \leq x_1$ and for

$$\min\{u(x), v(x) \mid x_0 \leq x \leq x_1\}$$
$$\leq y \leq \max\{u(x), v(x) \mid x_0 \leq x \leq x_1\}. \qquad (3.24)$$

Then

$$u(x) \equiv v(x), \qquad x_0 \leq x \leq x_1;$$

i.e., the initial-value problem (3.23) has at most one solution.

PROOF   We must show $u(x) \equiv v(x)$, $x_0 \le x \le x_1$. We prove the theorem by contradiction: assume there is some interval $a < x < b$, $x_0 \le a < b \le x_1$, where $u \ne v$ while $u(a) = v(a)$. (One candidate for a is x_0. If there were no such interval $a < x < b$, then u and v would be identical at every point of $x_0 \le x \le x_1$.)

Let $w(x) = u(x) - v(x)$. Since u, v both satisfy the given differential equation, we have $w' = f(x, u) - f(x, v)$. We can apply Taylor's theorem A.3, to the second argument of f to obtain

$$w' = f(x, u) - f(x, v) = \frac{\partial f(x, \xi)}{\partial y} w(x),$$

where ξ lies between $u(x)$ and $v(x)$. But the continuous function $|\partial f(x, \xi)/\partial y|$ is bounded, say by L, for $x_0 \le x \le x_1$, and ξ in the interval (3.24). Hence,

$$w' \le L|w|, \qquad x_0 \le x < x_1.$$

Without loss of generality, we can suppose that

$$w(x) = u(x) - v(x) > 0, \qquad a < x < b,$$

so that $|w(x)| = w(x)$ for $a < x < b$ and

$$w' \le Lw, \qquad a \le x < b.$$

Now multiply this differential *inequality* by the (positive) integrating factor $\mu(x) = e^{-Lx}$. The differential inequality $w' - Lw \le 0$ becomes

$$\left(we^{-Lx}\right)' \le 0.$$

Integrating from $x = a$ yields

$$w(x)e^{-Lx} - w(a)e^{-La} \le 0, \qquad a \le x < b.$$

By hypothesis, $w(a) = u(a) - v(a) = 0$; hence,

$$w(x)e^{-Lx} \le 0,$$

forcing $w(x) \le 0$, $a < x < b$, in violation of the hypothesis $w(x) > 0$, $a < x < b$.

We conclude that $w(x) = u(x) - v(x) \equiv 0$, $x_0 \le x \le x_1$.   ∎

Recall example 27, page 87: The initial-value problem

$$y' = \frac{3}{2}y^{1/3}, \qquad y(0) = 0.$$

has *three* distinct solutions, $y = 0, \pm x^{3/2}$. Has the uniqueness theorem 3.6 failed?

No. The second hypothesis of the uniqueness theorem is violated, for with

$$f(x, y) = \frac{3}{2}y^{1/3},$$

the partial derivative

$$\frac{\partial f}{\partial y} = \frac{1}{2y^{2/3}}$$

(3.25)

is hardly continuous at $y = 0$.

But solutions of

$$y' = \frac{3}{2}y^{1/3}, \qquad y(0) = 1,$$

are unique because the partial derivative (3.25) is certainly continuous in a neighborhood of $y = 1$. (Since $y' > 0$ for $y > 0$, solutions of this initial-value problem increase so that $y(x) \geq 1$.)

3.9.2 EXISTENCE

Under what conditions does the initial-value problem (3.23), $y' = f(x, y), y(0) = y_i$, have a solution? How large is the x interval upon which solutions are defined?

The answer to the first question is provided by an *existence theorem*. The second question asks about the size of the *maximum interval of existence* of the solution of an initial-value problem.

The existence theorem hinges on two key ideas. One is the equivalence between the initial-value problem

$$y' = f(x, y), \qquad y(0) = y_i,$$

and the **integral equation**

$$y(x) = y_i + \int_0^x f(s, y(s)) \, ds.$$

(3.26)

If the function y satisfies the integral equation, then obviously it satisfies the initial condition $y(0) = y_i$. Differentiating with respect to x and using the fundamental theorem of calculus recovers the differential equation itself. To go from the initial-value problem to the integral equation, integrate the differential equation and use $y(0) = y_i$.

The second important idea is *successive approximations*, a technique that leads from the integral equation to a sequence whose limit is a solution of the integral equation and, hence, of the initial-value problem. Successive approximation begins with some guess $y_0(x)$ for the solution of the initial-value problem and computes an improved (we hope) guess $y_1(x)$ using one side of the integral equation (3.26):

$$y_1(x) = y_i + \int_0^x f(s, y_0(s)) \, ds.$$

The process repeats endlessly:

$$y_{n+1}(x) = y_i + \int_0^x f(s, y_n(s)) \, ds, \qquad n = 0, 1, 2, \ldots.$$

(3.27)

The following example should build confidence in successive approximation.

EXAMPLE 49   *Use successive approximation (3.27) to find the solution of*

$$y' = ky, \qquad y(0) = 1.$$

Lacking a better idea, we choose the initial value $y_i = 1$ as the starting approximation: $y_0(x) = 1$. Then with $f(x, y) = ky$, the next approximation is

$$y_1(x) = 1 + \int_0^x k y_0(s) \, ds = 1 + \int_0^x k \, ds = 1 + kx.$$

A second iteration yields

$$y_2(x) = 1 + \int_0^x k \left(1 + ks\right) \, ds = 1 + kx + \frac{(kx)^2}{2},$$

and a third,

$$y_3(x) = 1 + \int_0^x k \left(1 + ks + \frac{(ks)^2}{2}\right) ds = 1 + kx + \frac{(kx)^2}{2} + \frac{(kx)^3}{3!}.$$

Studying the pattern of the iterates reveals that

$$y_n(x) = 1 + kx + \frac{(kx)^2}{2} + \cdots + \frac{(kx)^n}{n!},$$

the Taylor polynomial approximation to the exact solution e^{kx}. Clearly, the limit of the sequence $\{y_n(x)\}$ is the exact solution of this initial-value problem. ∎

Theorem 3.7 (Existence)   *Let $f(x, y)$ and $\partial f(x, y)/\partial y$ be defined and continuous in a domain D that includes the initial point $(0, y_i)$. Then the initial-value problem*

$$y' = f(x, y), \qquad y(0) = y_i,$$

has a solution $y(x)$ defined in a neighborhood of $x = 0$.

Of course, $f(x, y)$ must be defined in a neighborhood of the initial point $(0, y_i)$. The hypothesis that f be continuous allows a limit of the sequence of successive approximations to be brought inside the function. The requirement that $\partial f/\partial y$ be continuous can be relaxed somewhat, but we will not pursue that additional generality here.

PROOF   For some $a, b > 0$, the domain D within which f is defined includes the rectangle

$$R = \{(x, y) \; : \; |x| \le a, \quad |y - y_i| \le b\}.$$

Let M denote the maximum value of $|f(x, y)|$ on R and L the maximum value of $|\partial f(x, y)/\partial y|$ on R.

Mimicking the previous example, we construct a sequence of successive approximations $\{y_n(x)\}$ from (3.27) beginning with $y_0(x) = y_i$. Since f is defined in a neighborhood of $(0, y_i)$, $f(s, y_i)$ is defined providing $|s| \leq a$. Hence,

$$y_1(x) = y_i + \int_0^x f(s, y_i)\, ds$$

is defined for $|x| \leq a$.

Calculating the next iterate requires that $f(s, y_1(s))$ be defined. Hence, we must guarantee $|s| \leq a$ and $|y_1(s) - y_i| \leq b$ to remain in R. Since $|f| \leq M$ on R, (3.27) yields

$$|y_1(x) - y_i| = |\int_0^x f(x, y_i)\, ds| \leq \int_0^x |f(x, y_i)|\, ds \leq Mx.$$

To keep $(x, y_1(x))$ in R, we must further restrict x so that $Mx \leq b$; that is, we now demand $x \leq \min(a, b/M)$.

This argument applies to each of the iterates: $y_n(x)$ is defined as long as $|x| \leq \min(a, b/M)$.

To examine the convergence of the iterates, restrict x still further by requiring $|x| \leq \min(a, b/M, \alpha/L)$ for some number $\alpha < 1$. In addition, if $y(x)$, $z(x)$ are any two functions defined for such values of x, let

$$||y - z|| = \max\{|y(x) - z(x)| \; : \; |x| \leq \min(a, b/M, \alpha/L)\}.$$

We will use $|| \cdots ||$ to compare the maximum difference between iterates of (3.27).

To compare $y_m(x)$ and $y_n(x)$ for arbitrary integers m, n, compute

$$|y_m(x) - y_n(x)| = |\int_0^x [f(s, y_{m-1}(s)) - f(s, y_{n-1}(s))]\, ds$$

$$\leq \int_0^x |f(s, y_{m-1}(s)) - f(s, y_{n-1}(s))|\, ds$$

$$\leq \int_0^x \left|\frac{\partial f}{\partial y}\right| |y_{m-1}(s) - y_{n-1}(s)|\, ds$$

$$\leq L \int_0^x |y_{m-1}(s) - y_{n-1}(s)|\, ds$$

$$\leq Lx ||y_{m-1} - y_{n-1}||.$$

Since $Lx \leq \alpha$, this string of inequalities yields

$$||y_m - y_n|| \leq \alpha ||y_{m-1} - y_{n-1}||. \tag{3.28}$$

Since $\alpha < 1$, we have just shown that the rule or *mapping* (3.27) that takes y_{m-1} into y_m is a *contraction mapping*. The distance between iterates as measured by $|| \cdots ||$ contracts with each iteration.

If $m > n$, then

$$||y_m - y_n|| \leq \alpha ||y_{m-1} - y_{n-1}|| \leq \alpha^2 ||y_{m-2} - y_{n-2}||$$

$$\leq \cdots \leq \alpha^n ||y_{m-n} - y_0||.$$

But the triangle inequality applies to the maximum expression $|| \cdots ||$,

$$||y + z|| \leq ||y|| + ||z||.$$

Using the triangle inequality and the contraction relation (3.28), we obtain

$$\begin{aligned}
||y_{m-n} - y_0|| &= ||y_{m-n} - y_{m-n-1} + \cdots + y_2 - y_1 + y_1 - y_0|| \\
&\leq ||y_{m-n} - y_{m-n-1}|| + \cdots + ||y_2 - y_1|| + ||y_1 - y_0|| \\
&\leq \left(\alpha^{m-n-1} + \cdots + \alpha + 1 \right) ||y_1 - y_0|| \\
&\leq \frac{b}{1-\alpha}.
\end{aligned}$$

This last inequality used $1 + \alpha + \cdots + \alpha^{m-n-1} \leq 1/(1-\alpha)$ from the sum of a geometric series as well as $||y_1 - y_0|| = ||y_1 - y_i|| \leq b$.

Combining the results of these two blocks of inequalities, we have

$$||y_m - y_n|| \leq \frac{\alpha^n b}{1-\alpha}, \qquad m \geq n.$$

That is, since $\alpha < 1$, the maximum difference between arbitrary elements of the sequence $\{y_n(x)\}$ can be made as small as desired by choosing n (and, hence, m) sufficiently large. Such a sequence is known as a *Cauchy sequence*, and it can be shown to *converge uniformly* to a continuous function $y(x)$ for $|x| \leq \min(a, b/M, \alpha/L)$.

We can take limits on both sides of the successive approximation expression (3.27) and use uniform convergence and the continuity of f to bring the limit on the right under the integral and inside f. We find that the limiting function $y(x)$ satisfies (3.26), the integral equation equivalent to the initial value problem (3.23). Hence, $y(x) = \lim_{t \to \infty} y_n(x)$ is a solution of the initial value problem. ∎

Since the requirements for f in the existence theorem are the same as those of the uniqueness theorem, the solution whose existence we have just demonstrated is the *only* solution of the initial-value problem.

> **Corollary 3.8 (Existence and Uniqueness)** *If f and $\partial f/\partial y$ are defined and continuous on some domain that includes the initial point $(0, y_i)$, then the initial-value problem (3.23) has exactly one solution defined for x in some neighborhood of $x = 0$.*

In the proof of the existence theorem, x was restricted to a neighborhood of $x = 0$ by the requirement $x \leq \min(a, b/M, \alpha/L)$. Consequently, this result is a *local existence theorem*.

Could x be allowed to grow to larger values? The primary limitation seems to be keeping the solution point $(x, y(x))$ within the domain D upon which $f(x, y)$ is defined. Indeed, if the solution is *not* defined for all x, then we can show, roughly speaking, that as we approach some finite value of x,

- Either the solution point $(x, y(x))$ leaves the domain D because x is nearing the limit for which $f(x, \cdot)$ is defined.
- Or y is growing without bound.

Because any two solutions that agree within a neighborhood of $x = 0$ must agree throughout their common interval of definition, we can always choose the solution with the larger interval of existence. Consequently, we can speak of a **maximum interval of existence** of an initial-value problem. That interval may be infinite or finite, and the breakdown at a finite endpoint can occur for either of the reasons listed here. The following example exhibits the second phenomena, blow up at finite x.

EXAMPLE 50   *Show that the initial-value problem*

$$y' = y^2, \qquad y(0) = 1,$$

has a finite maximum interval of existence.

Solving by separation of variables, we find

$$y(x) = \frac{1}{1 - x}.$$

Clearly, the interval of existence of this solution is $-\infty < x < 1$. Since $f(x, y) = y^2$ is defined for all y, the interval of existence is limited because the solution of the initial value problem grew without bound, not because f became undefined at a finite value of x.   ∎

Existence and uniqueness are widely discussed. Accessible references include the texts of Hurewicz [9], Sánchez (treating systems) [14], and Simmons [16]. A classic advanced treatment is that of Coddington and Levinson [4].

3.9.3   Exercises

EXERCISE GUIDE

To gain experience . . .	Try Exercises
Verifying the hypotheses of the existence and uniqueness theorems	1
Applying the uniqueness theorem	2–6
Applying the existence theorem	6, 14
With successive approximation	10–20
With integral equations	13
With interval of existence	8–9
With the proof of the uniqueness theorem	7
With the proof of the existence theorem	10–12, 15

1. Write each of the following differential equations in the form $y' = f(x, y)$, compute $\partial f / \partial y$, and give the values of y for which this derivative is continuous for all x.

   (a) $y' - ky + y^3 = 0$
   (b) $y' - \sin(ty) = t^2 - 1$
   (c) $P' = aP - sP^2$ (logistic population equation)
   (d) $P' = kP - E(t)$ (emigration equation)
   (e) $v' = -g - kv|v|/m$ (improved air resistance model)

2. For what values of y_i does the uniqueness theorem 3.6 guarantee unique solutions of the initial-value problem

$$y' = \frac{3}{2}y^{1/3}, \qquad y(0) = y_i?$$

3. For what values of the parameter α does the uniqueness theorem 3.6 guarantee that the initial-value problem

$$y' = y^\alpha, \qquad y(0) = 0,$$

has only the trivial solution?

4. Suppose $f(x,y)$ and $\partial f/\partial y$ are both continuous in x and y and that $f(x,0) = 0$ for all x. Argue that the initial-value problem

$$y' = f(x,y), \qquad y(x_0) = 0,$$

has only the trivial solution.

5. While the logic is ultimately circular, use the uniqueness theorem 3.6 to prove that the linear initial-value problem

$$y' + g(x)y = 0, \qquad y(0) = y_i,$$

has at most one solution if g is continuous. Explicitly display the partial derivative $\partial f/\partial y$ for this equation.

6. Let $p(x,y)$ be a polynomial in y with coefficients that are continuous functions of x:

$$p(x,y) = a_0(x) + a_1(x)y + a_2(x)y^2 + \cdots + a_n(x)y^n.$$

Use the uniqueness theorem 3.6 to argue that the initial-value problem

$$y' = p(x,y), \qquad y(0) = y_i,$$

has at most one solution any value of y_i. Which of the models we have studied can be cast in this form?

7. Why in the proof of theorem 3.6 was it important that the differential inequality $w' \leq Lw$ be multiplied by a *positive* integrating factor $\mu = e^{Lt}$?

8. How does the maximum interval of existence of $y' = y^2$, $y(0) = y_i$, depend upon y_i?

9. Show that $y' = y^{\alpha+1}$, $y(0) = 1$, has a finite interval of existence for $\alpha > 0$.

10. Show that the successive approximation scheme (3.27) leads to the correct solution of the emigration model $P' = kP - E$, $P(0) = P_i$, when E is constant.

11. Show that the successive approximation scheme (3.27) leads to the correct solution of $y' = -2y$, $y(0) = y_i$.

12. Use successive approximation (3.27) to approximate the solution of $y' = y$, $y(0) = y_i$. Show that the sequence you obtain is converging to the correct solution.

13. Write the integral equation (3.26) that is equivalent to the initial-value problem $y' = x^2$, $y(0) = 2$. Carry out the integration and show that you obtain the correct solution of the initial-value problem. Why are you able to carry out the integration?

14. What hypotheses must you impose on $r(x), g(x)$ to ensure that the linear initial-value problem $y' + g(x)y = r(x)$, $y(0) = y_i$, has a unique solution? Are those hypotheses consistent with corollary 3.4, page 87, the existence and uniqueness result for linear problems?

15. How does the proof of the existence theorem change if the initial-condition is $y(x_i) = y_i$ rather than $y(0) = y_i$?

3.10 SOLUTION METHOD SUMMARY

Before you begin solving an equation, take time to diagnose the problem. Carefully decide which solution technique ought to be appropriate. Know what to expect. Be prepared to revise your judgment as you see the results of your work unfold.

Table 3.4 provides a decision tree for choosing an appropriate solution method for first-order equations. Note that *separation of variables* is the only choice for *nonlinear* equations, and it may fail. The *integrating factor*

$$\mu(x) = \exp\left(\int_a^x g(s)\,ds\right)$$

always works for *linear* equations of the form $y' + g(x)y = \cdots$. *Numerical methods* will provide a numerical solution of *any* initial-value problem, linear or nonlinear.

Table 3.4

A decision tree for choosing solution methods for first-order equations

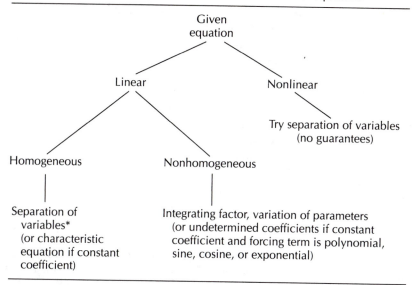

*Linear, homogeneous equations may also be solved by an integrating factor. If all else fails, numerical methods will solve almost any initial-value problem.

3.11 CHAPTER EXERCISES

EXERCISE GUIDE

To gain experience . . .	Try Exercises
Applying basic definitions	1, 24, 26–28
Finding homogeneous solutions	26–27, 39
Finding particular solutions	35–38(a), 39
Finding general solutions	2–25, 26–27(a), 35–38(b)
Solving initial-value problems	2–9, 14, 17, 26–27(b), 28, 29–30(a), 35–38(c)
Using an integrating factor	34
Analyzing solution behavior	29–30(b), 31–33, 35–38(d), 39

1. Write an example of a first-order differential equation
   (a) That is homogeneous but can not be solved by separation of variables.
   (b) That is linear and homogeneous but can not be solved by separation of variables.
   (c) That is linear and has constant coefficients but can not be solved by undetermined coefficients.
   (d) That can not be solved by an integrating factor.
   (e) That is linear and can not be solved by an integrating factor.

   (f) That is linear and has constant coefficients but can not be solved by the characteristic equation method.

   In exercises 2–25, find either a general solution of the given differential equation or the solution of the given initial-value problem, as appropriate. If the solution methods available to you fail or if a general solution is not appropriate, explain.

2. $y' - e^x y = 0$, $y(0) = 1$
3. $y' - e^x y = \cos x$, $y(0) = 1$

4. $y' - e^x y = e^x$, $y(0) = 1$

5. $y' + \cos xy = 0$, $y(\pi) = -1$

6. $y' + \cos xy = \frac{1}{x} e^{-\sin x}$, $y(\pi) = 0$

7. $xy' = 4y$, $y(1) = 3$

8. $xy' = 4x$, $y(1) = 3$

9. $xy' = 4y - \cos x^3$, $y(1) = 3$

10. $dz/dx = Qz$, Q a constant

11. $dz/dx = Qx$, Q a constant

12. $y' + \cos xy = \sin x$

13. $y' + \pi y^2 = 0$

14. $y' + \pi y = e^{-\pi x} + \cos x$, $y(0) = 0$

15. $y' + xy = \cos x$

16. $y' + 2y = 4x^2 - 3e^{2x}$

17. $y' - 2y = 4x^2 - 3e^{2x}$, $y(0) = 4$

18. $y' + 5y = \cos 5x + e^{-5x}$

19. $P' - kP = -E \cos \omega t$, ω constant (a model for a population subject to emigration and immigration)

20. $R \, dq/dt + q/C = 0$, R, C constant

21. $R \, dq/dt + q/C = -E \sin \omega t$, R, C, E, ω constant

22. $P' = aP - sP^2$, a, s constants

23. $P' = aP$, a constant

24. You are given two functions $H(t)$ and $P(t)$ with the following properties:

    i. $H(t)$ is a solution of $y' + a(t)y = 0$,
    ii. $H(0) = 3$,
    iii. $P(t)$ is a solution of $y' + a(t)y = f(t)$,
    iv. $P(0) = -1$.

Express each of the following in terms of $H(t)$ and $P(t)$. If you lack sufficient information, explain what is missing.
(a) The general solution of $y' + a(t)y = f(t)$
(b) The solution of $y' + a(t)y = f(t)$, $y(0) = 5$
(c) The free and forced response of the initial-value problem in part (b)
(d) The steady-state and transient responses of the system in part (b)

25. You are given the following information:

    ■ A solution of $y' + f(x)y = 0$ is $S(x)$.
    ■ A solution of $y' + f(x)y = g(x)$ is $R(x)$.
    ■ $S(0) = 2, R(0) = -3$.

(a) Find a general solution of $y' + f(x)y = g(x)$.
(b) Solve the initial-value problem $y' + f(x)y = g(x)$, $y(0) = 5$.

Express your answers in terms of the functions R and S.

26. Suppose that f is a function that satisfies the differential equation $f' + xf = x^4$ and that $f(1) = 6$. The function $z(x)$ satisfies the equation $z' + xz = 0$. Write a solution of the initial-value problem

$$y' + xy = x^4, \qquad y(1) = 0,$$

in terms of the functions f and z. Explain your reasoning.

27. (a) Solve the heat-loss model initial-value problem

$$\frac{dT}{dt} = -\frac{Ak}{cm}(T - T_{\text{out}}), \qquad T(0) = T_i,$$

when T_{out} is constant.
(b) Show that your solution exhibits the following behavior: If $T_i > T_{\text{out}}$, then the solution decays to a steady state T_{out}, and if $T_i < T_{\text{out}}$, then the solution grows to the same steady state. In this behavior physically reasonable?

28. (a) Solve the emigration model initial-value problem

$$P' = kP - E, \qquad P(0) = P_i,$$

when E is constant.
(b) Show that your solution exhibits the following behavior: If $P_i < E/k$, then the solution decays, and if $P_i > E/k$, then the solution grows. What happens if $P_i = E/k$? Is this behavior physically reasonable?

29. Several of the exercises in chapter 2 offer you a choice of four initial-value problems as possible models of situations described there. Using uniform notation, the four equations are
(a) $c'(t) + kc(t) = d$
(b) $c'(t) + kc(t) = -d$
(c) $c'(t) - kc(t) = d$
(d) $c'(t) - kc(t) = -d$

Assume the constants k, d are positive.
The situations being modeled include the following:

■ *Radioactive decay with bombardment.* A radioactive isotope is decaying at a rate proportional to its mass $c(t)$; the proportionality constant is k. A neutron beam bombarding the sample creates new isotope at a rate of d g/s. (See exercise 2, page 45.)

■ *Chemical reaction.* A particular chemical reaction produces a chemical at a rate proportional to the concentration $c(t)$ of that chemical; the proportionality constant is k. Diluting the mixture reduces the concentration of the chemical at a rate of d g/cm^3- sec. (See exercise 8, page 46.)

- *Money in the bank.* An amount of money $c(t)$ is earning interest, continuously compounded, at a daily rate of k (or $100k$ percent). Regular deposits are made at a rate of d dollars per day. (See exercise 11, page 37.)

Find a general solution of each of these equations. From the behavior of the general solution, identify which equation might be a reasonable model for each situation listed. Use your intuitive ideas about these situations to guide your selection.

30. A population of yeast spores increases due to reproduction by a fixed proportion R each day. The spores are harvested at a constant rate of H spores per day. Exercise 6 of section 2.2.4 asks you to model this situation. One model you might obtain is

$$y' = Ry - H, \qquad y(0) = y_i,$$

where y is the population of spores. Find a solution of this initial-value problem and use it to predict the long-term fate of this population. Do the relative values of the parameters R, H, y_i matter?

31. Use the model for the population of Ireland during the potato famine, $P' = 0.015P - 0.209$, $P(1847) = 8$, to predict the date when the island would have become deserted. (Recall that t is measured in years, P in millions of people.)

32. Use an integrating factor to derive a general solution of the linear, constant-coefficient equation $b_1 y' + b_0 y = 0$. Show that this solution is the same as that given by the characteristic equation method.

33. Exercise 2, page 45, develops a model for the mass y of a radioactive isotope being bombarded by a neutron beam,

$$\frac{dy}{dt} + ky = Q, \qquad y(0) = y_i,$$

where k is the decay constant of the isotope and Q is

the (constant) rate of production of new isotope due to the neutron beam.
(a) Find a particular solution of this equation.
(b) Find a general solution of this equation.
(c) Solve the governing initial-value problem.
(d) What becomes of the mass of the isotope after a long time has passed?

34. Exercise 4, page 46, develops a model for the charge q on the capacitor in an RC circuit subject to an imposed voltage drop E,

$$R\frac{dq}{dt} + \frac{1}{C}q = -E, \qquad q(0) = q_i.$$

The resistance R and capacitance C in the circuit are constant. Assume E is constant as well.
(a) Find a particular solution of this equation.
(b) Find a general solution of this equation.
(c) Solve the governing initial-value problem.
(d) What becomes of the charge on the capacitor after a long time has passed?

35. Repeat the analysis of the RC circuit of exercise 34 when the imposed voltage drop varies sinusoidally, $E = A \sin \omega t$, where the amplitude A and frequency ω are constant. (Do not confuse the amplitude A with an undetermined coefficient!)

36. Repeat the analysis of the RC circuit of exercise 34 when the imposed voltage drop decays exponentially, $E = A e^{-\alpha t}$, where A and $\alpha > 0$ are constant.

37. Solve the heat-loss model,

$$\frac{dT}{dt} = -\frac{Ak}{cm}(T - T_{\text{out}}), \qquad T(0) = T_i,$$

when the outside temperature T_{out} varies periodically between a high L and a low ℓ,

$$T_{\text{out}}(t) = \ell + L\frac{1 + \cos(\pi t/12)}{2}.$$

In addition, verify that this expression for $T_{\text{out}}(t)$ varies as advertised. If t is measured in hours, what is the period of T_{out}?

3.12 CHAPTER PROJECTS

1. **Integrating factors and homogeneous solutions** General solutions obtained by integrating factors always seem to contain a term such as C/μ that appears to be the homogeneous solution part of the general solution $y_p + Cy_h$.

   For example, the text solved the heat-loss model using the integrating factor $\mu = e^{Akt/cm}$. The general solution contained the term

$Ce^{-Akt/cm} = C/\mu$, which certainly satisfies the homogeneous equation $T' + AkT/cm = 0$.

Use one or both of the following approaches to show that this observation is no accident: If μ is an integrating factor for a linear equation, then $1/\mu$ is a solution of the corresponding homogeneous equation.

(a) Find the integrating factor $\mu(x)$ for the general linear equation $y' + g(x)y = f(x)$. (Some examples from the text will help.) Differentiate and use the fundamental theorem of calculus to show that $1/\mu(x)$ is a solution of $y' + g(x)y = 0$. Explicitly state any conditions that must be imposed on f or g.

(b) We find the integrating factor for $y' + g(x)y = f(x)$ by requiring that $\mu(y' + g(x)y) = (\mu y)'$. This equality holds if $\mu' = g(x)\mu$. Use $\mu' = g(x)\mu$ to show that $y_h = 1/\mu(x)$ satisfies $y_h' + g(x)y_h = 0$ and, hence, that $1/\mu$ is a homogeneous solution.

Note that this approach never needs an explicit expression for μ. It only needs to know the differential equation of which μ is a solution. But how do you show $\mu \neq 0$?

2. **Heat loss and emigration with variable forcing** Consider the heat-loss model when the outside temperature T_{out} varies with time,

$$T'(t) = -\frac{Ak}{cm}[T(t) - T_{out}(t)], \qquad T(0) = T_i.$$

Find an expression for the solution of this initial value problem. Use that expression to determine a relation between T_i and $T_{out}(t)$ and other conditions, if necessary, that guarantee that the solution decreases for all time. Are these conditions physically reasonable?

Repeat this process for the emigration model when the emigration rate $E(t)$ varies with time:

$$P' = kP - E(t), \qquad P(t_0) = P_i,$$

but now seek conditions that make the solution *increasing* for all time.

3. **Numerical methods via integration** Formally integrating the generic initial value problem

$$y' = f(t, y), \qquad y(t_i) = y_i$$

leads to

$$y(t) = y_i + \int_{t_i}^{t} f(s, y(s))\, ds.$$

Think of the integrand as a function of s alone, say $g(s) = f(s, y(s))$. Then the integral $\int_{t_i}^{t} g(s)\, ds$ can be approximated in many different ways, including the rectangle rule,

$$\int_{t_i}^{t} g(s)\, ds \approx g(t_i)\Delta t,$$

and the trapezoid rule,

$$\int_{t_i}^{t} g(s)\, ds \approx [g(t_i) + g(t)]\frac{\Delta t}{2},$$

where $\Delta t = t - t_i$. If $t = t_{n+1}$ and $t_i = t_n$, then each approximation to the integral gives a different numerical method for approximating the solution of the initial-value problem.

Identify the two methods given here. What integral approximation corresponds to the Heun method? Describe the *mid-point rule*, the method that results if the rectangle rule is used with g evaluated at the mid-point of the interval rather than at the end. What happens if the evaluation point is the *right* end of the interval?

Determine the local truncation error in the mid-point rule. Conduct some numerical experiments to compare it with Euler and Heun.

4. **The logistic map and chaos** Applying the Euler method to the logistic equation $P' = aP - sP^2$ leads to

$$P_{n+1} = (1 + a\Delta t)P_n - s\Delta t P_n^2, \qquad n = 0, 1, 2, \dots .$$

By defining $b = 1 + a\Delta t$ and $y_n = s\Delta t P_n/b$, those Euler iterations can be written

$$y_{n+1} = b(y_n - y_n^2), \qquad n = 0, 1, 2, \dots .$$

This relation is the *logistic map*. For initial values y_0 in $0 < y_0 < 1$ and b in $0 < b < 4$, it exhibits behavior much richer than anything we might anticipate from the logistic population model.

For some values of b (say, $b = 3.2$), the iterations quickly settle into cycles of period 2, alternating between two fixed values with each iteration, regardless of the initial value. As b increases, say to 3.5, the period suddenly doubles to 4; every fourth iterate is the same. Then the period doubles again to 8 just before $b = 3.6$. This *period doubling* continues as b increases. Interspersed among the periodic iterations are ranges of values of b for which the iterates seem randomly to cover almost every point in an interval, regions of *chaotic behavior*. Furthermore, changing y_0 in such a region leads to a completely different pattern of iterates, almost as if the iterates were chosen at random.

Write a simple program to iterate the logistic map using given values of b and y_0. Skip the first 100 iterates to allow transient effects to disappear. Experiment with b in the neighborhood of 2.8, 3.2, 3.45, 3.5, 3.55, 3.6, and 3.8. If you are more ambitious, provide some graphical output by marking the iterates on the unit interval. (A period two mapping will appear as two points, period four as four points, and so on. Chaotic behavior will begin to fill an interval.)

The chaotic behavior of the logistic map, its apparent randomness for certain values of b, arises not from some external uncertainty but from its nonlinearity. The change in behavior, the period doubling, as b is varied is called *bifurcation*.

Nonlinear dynamics, of which the logistic map is a simple example, offer new insights into such phenomena as variations in the weather and turbulent fluid flow. For an account of many of the people and ideas involved in these discoveries, see Gleick's popular book [7]. The mathematical concepts of nonlinear dynamics are introduced in Baker and Gollub [2].

QUALITATIVE BEHAVIOR OF FIRST-ORDER EQUATIONS

. . . life . . . is only to be understood.
Marie Curie

Understanding the information a differential equation carries can require more than just a formula for its solution. This chapter explains some of the tools and ideas needed for that broader understanding.

4.1 QUALITATIVE BEHAVIOR AND SOLUTION GRAPHS

4.1.1 Solution Graphs

The basic shape of the graph of a solution of a differential equation can tell us a great deal about the behavior of the phenomena that the equation models. That qualitative behavior, a "feel" for the solution, is often more important than the precise details of a point-by-point plot of an exact solution. Consequently, we will give specific attention to the kinds of qualitative information we can deduce about the solution from the differential equation itself.

You learned all the tools you need when you studied curve sketching in calculus. There you plotted a few points and examined the first derivative to find regions where the function was increasing or decreasing. The first derivative also determined the location of critical points, those points that were candidates for local maxima or minima because the first derivative failed to exist or was zero there. The second derivative showed concavity and located inflection points, where the concavity of the curve changed direction.

Those same tools still work, even when we don't have an explicit formula for the solution function. An initial condition gives us one point on the solution curve. A first-order differential equation provides an expression for the first derivative of the unknown, although that expression may well involve the unknown itself. An expression for the second derivative can be calculated from the differential equation, sometimes more easily than from the solution formula itself (assuming we can even find such a formula).

We have hinted at some of these ideas earlier. The following examples bring them together.

EXAMPLE 1   *Sketch graphs of solutions of the rock-tossing model*

$$v'(t) = -g, \; v(0) = v_i \,,$$

for various values of initial velocity v_i; $g > 0$ is the acceleration of gravity.

The differential equation $v' = -g$ tells us that the v curve must have constant, negative slope. The initial condition $v(0) = v_i$ determines a point $(0, v_i)$ of this downward-sloping straight line.

Nonbelievers can compute $v''(t)$ from the differential equation $v' = -g$:

$$v'' = \frac{dv'}{dt} = \frac{d(-g)}{dt} = 0 \,.$$

Since $v'' = 0$, the graph of v must indeed be a line.   ∎

A family of solution curves of $v' = -g$, $v(0) = v_i$, for different choices of the initial velocity v_i is shown in figure 4.1. Do all portions of such solution curves necessarily represent physically valid behavior?

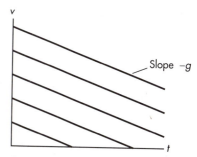

Figure 4.1: A family of solution curves for the initial value problem $v' = -g$, $v(0) = v_i$, the rock-tossing model, for various values of v_i.

EXAMPLE 2   *Sketch a family of solution curves of the simple population model*

$$P' = kP, \; P(0) = P_i \,,$$

for various values of the initial population P_i.

Using logic we have employed before, we see from $P' = kP$ that P' has the same sign as P. If P_i is positive, then $P'(0)$ will be positive, causing P to increase beyond its initial value and thereby remain positive. Hence, P' will always be positive if P_i is positive. Likewise, P' will always be negative if P_i is negative.

The derivative $P' = kP$ is zero only when $P = 0$, and it is defined for all finite P. Therefore, the only critical points occur when $P = 0$. But since $P' = 0$ when $P = 0$, the population remains constant at $P = 0$; a population that starts with nothing never grows.

To examine concavity, compute P'' from $P' = kP$,

$$P'' = \frac{dP'}{dt} = \frac{d(kP)}{dt} = kP' = k^2 P \,.$$

The sign of the second derivative is the same as the sign of P. The graph of a positive population is concave up; that of a negative population, concave down.

> Remember that a curve is concave up because its derivative is increasing and concave down because its derivative is decreasing.

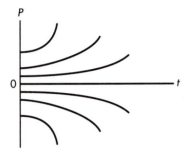

Figure 4.2: Graphs of the solution of the simple population model $P' = kP$, $P(0) = P_i$, for various values of P_i. Solution curves are shown for $P < 0$, but these normally have no physical significance.

Which of these cases occurs — an increasing, concave up population graph or the converse — is determined solely by the initial value P_i. Population curves that begin with a positive value are increasing and concave up. Those that begin with a negative value are decreasing and concave down, as figure 4.2 illustrates. (Of course, negative populations have no direct physical meaning. Only the first quadrant of figure 4.2 can be assigned any physical significance.) ■

Compare the conclusions of this example with the exponential solution $P = Ce^{kt}$ of $P' = kP$.

EXAMPLE 3 *Sketch a family of solution curves of the population model with emigration,*

$$P' - kP = -E, \ P(0) = P_i,$$

for various values of the initial population P_i when the emigration rate E is constant.

We can mimic the techniques of the previous example.

Since $P' = kP - E$, P' is positive when $kP - E$ is positive; i.e., $P' > 0$ if $P > E/k$. Likewise, $P' < 0$ if $P < E/k$. If the initial population P_i exceeds E/k, then $P'(0)$ is positive, and $P(t)$ will remain greater than E/k for all t. If $P_i < E/k$, then $P'(0) < 0$, and $P(t) < E/k$ for all t.

If $P_i = E/k$, then $P'(0) = 0$. In this case, $P' = kP - E$ has the constant solution $P(t) = E/k$, as direct substitution confirms.

Since E is constant, we use the differential equation to compute

$$P'' = \frac{d(kP - E)}{dt} = kP'.$$

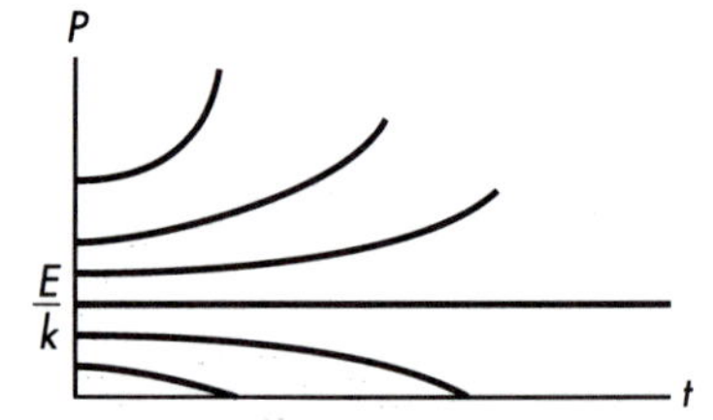

Hence, P'' and P' have the same sign; the population curve is concave up when it is increasing, and it is concave down when it is decreasing.

Evidently, the relative sizes of P_i and E/k determine whether the solution increases or decreases. Initial populations greater than E/k lead to increasing, concave up population curves. Those below E/k lead to decreasing, concave down population curves. If $P_i = E/k$, then the population remains at this constant value. This behavior is illustrated in figure 4.3. ■

The initial population level E/k separates populations that grow at an increasingly more rapid rate from those that decay more and more rapidly. Compare the conclusions of this example with a solution of $P' = kP - E$, $P(0) = P_i$.

Figure 4.3: Graphs of the solution of the emigration model $P' = kP - E$, $P(0) = P_i$, for various values of P_i. Note that $P_i = E/k$ leads to the constant solution $P(t) = E/k$.

EXAMPLE 4 *Describe the slope and concavity characteristics of the graph of the solution of the logistic model*

$$P' = aP - sP^2, \ P(0) = P_i,$$

for various positive values of P_i.

From the differential equation $P' = aP - sP^2$, we have

$$P' = P(a - sP) > 0$$

when both $P > 0$ and $a - sP > 0$. Since we are considering only $P > 0$, we conclude that $P' > 0$ for $a - sP > 0$ or, equivalently, for $P < a/s$. Conversely, $P' < 0$ for $P > a/s$.

We again compute P'' directly from the differential equation:

The chain rule yields $(P^2)' = 2PP'$.

$$P'' = \frac{d(aP - sP^2)}{dt} = aP' - 2sPP'$$
$$= P'(a - 2sP).$$

The sign of P'' is determined by the sign of P' and by the sign of $a - 2sP$. The latter is positive when $P < a/2s$ and negative when $P > a/2s$. In addition, $P'' = 0$ when either $P' = 0$ or $P = a/2s$.

This information is summarized in figure 4.4.

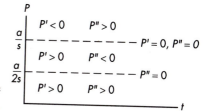

Figure 4.4: A summary of the slope and concavity behavior of the logistic equation $P' = aP - sP^2$.

EXAMPLE 5 *Which of the graphs in figure 4.5 could not be a solution of the differential equation*

$$y' = y^3 - y\,?$$

Writing the differential equation as $y' = y(y^2 - 1)$ reveals that $y' = 0$ when $y = -1, 0, 1$. The sign of y' is determined on the intervals $y < -1$, $-1 < y < 0$, $0 < y < 1$, and $1 < y$ by examining the signs of the two factors y and $y^2 - 1$.

Using the chain rule, as in the preceding example, reveals

$$y'' = \frac{dy'}{dx} = 3y^2 y' - y' = y'(3y^2 - 1).$$

Hence, $y'' = 0$ when $y' = 0$ or when $y = \pm\sqrt{\frac{1}{3}} = \pm 0.577\cdots$. The signs of these two factors determine the sign of y''.

The results of this analysis are summarized in figure 4.6.

Curve a of figure 4.5 is decreasing and concave up over most of the interval $0 < y < 1$. Its negative slope is consistent with the slope behavior summarized in figure 4.6. However, figure 4.6 indicates that a solution of $y' = y^3 - y$ should have an inflection point at $y = \sqrt{\frac{1}{3}}$, a condition violated by curve a in figure 4.5. Hence, curve a of 4.5 is not a graph of a solution of $y' = y^3 - y$.

We reject curve b of figure 4.5 immediately because it is increasing on $0 < y < 1$, where solutions of $y' = y^3 - y$ must be decreasing.

Curve c of figure 4.5 is increasing and concave up for $y > 1$. Both types of behavior are consistent with the properties of figure 4.6. It is a reasonable candidate for a graph of a solution of $y' = y^3 - y$.

A portion of curve d of figure 4.5 is increasing for $y < -1$ while figure 4.6 requires that solutions be decreasing for $y < -1$.

Only curve c figure 4.5 can possibly represent a graph of a solution of $y' = y^3 - y$.

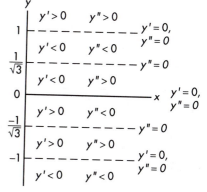

If this equation arose from a model, you could now make some judgment of the model's suitability for the problem at hand. Could you obtain this picture of a solution of $y' = y^3 - y$ as easily from a solution formula? What would a solution formula for this equation look like?

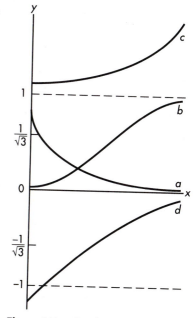

Figure 4.5: Graphs of four proposed solutions of $y' = y^3 - y$.

Figure 4.6: A summary of the slope and concavity behavior of $y' = y^3 - y$.

4.1.2 Direction Fields

In the preceding examples, we captured the qualitative behavior of solutions by sketching graphs. A principal tool was the equation itself, for it allowed us to determine the slope of its solution curve at any point. For example, the simple population model $P' = kP$, $P(0) = P_i$, tells us that the slope of its population curve at $t = 0$ is $P'(0) = kP(0) = kP_i$.

The differential equation is a kind of compass that guides us as we shuffle along the path of a solution curve. The equation specifies the slope of the curve at each point, heading us in the proper direction. How do those compass headings change as we move around the plane?

One way to capture this compass-like information is to plot the slopes determined by the differential equation at a variety of points of the plane. This plane full of slopes is called the **direction field** of the differential equation.

EXAMPLE 6   *Plot the direction field of the simple population equation*

$$P' = kP .$$

Since the right side of this equation is not a function of the independent variable t, the slope of a solution curve depends only on P. In other words, all solution curves have the same slope $P' = kP$ at the same value of P. The arrows in figure 4.7 indicating the directions dictated by the differential equation do not change along horizontal lines.

Since $P' = kP$ and $k > 0$, the slopes must increase as P increases. This behavior is illustrated in figure 4.7 as well.

Do the solution curves for this equation, which we sketched in figure 4.2, follow the directions prescribed in figure 4.7? Is figure 4.7 indeed a set of compass readings that would guide us along the solution curves shown in figure 4.2? ∎

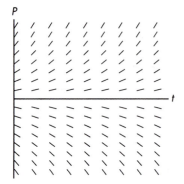

Figure 4.7: The direction field for the simple population equation $P' = kP$.

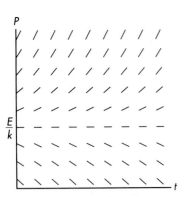

Figure 4.8: The direction field for the emigration equation $P' = kP - E$, E constant.

EXAMPLE 7   *Figure 4.8 shows the direction field for the emigration equation $P' = kP - E$, E constant. What does that figure reveal about the significance of the population level $P = E/k$?*

When $P = E/k$, the direction arrows are horizontal. A population that begins at E/k stays at that equilibrium value. Since the arrows point downward for $P < E/k$, populations initially below this level will decrease. Furthermore, the rate of decrease will become more rapid because the slope of the arrows is steeper as P decreases. The situation reverses when $P > E/k$.

Evidently, $P = E/k$ is a critical initial population level, separating growing populations from decaying populations. Populations that are too small will die out. Larger ones can overcome the loss due to emigration and grow. ∎

The solution $P = E/k + (P_i - E/k)e^{kt}$ of the emigration model $P' = kP - E$, $P(0) = P_i$, contains this same information. But a family of solution curves such as figure 4.3 or a direction field such as figure 4.8 makes immediately evident the presence of the critical population level E/k.

EXAMPLE 8   *Plot the direction field of the logistic equation*

$$P' = aP - sP^2 .$$

Like the simple population model $P' = kP$, the value of the slope P' is independent of t.

Figure 4.4 summarizes our study of this equation from an earlier example. The sign of the derivative is reflected in the slope of the direction arrows in figure 4.9. The concavity information tells us whether slopes will increase or decrease as we move vertically through the direction field toward larger values of P.

To complete our study of this direction field, note that the direction arrows are horizontal at $P = a/s$, where $P' = 0$. Given an initial point, could you use the directions prescribed in figure 4.9 to trace a solution curve? Would its slope and concavity behavior be consistent with figure 4.4? ∎

Study figure 4.9 carefully. As a population curve grows toward $P = a/s$, what becomes of its slope? Can the curve cross $P = a/s$? How do populations larger than a/s behave? Does the logistic equation $P' = aP - sP^2$ predict bounded or unbounded populations? Is this behavior obvious from the solution formula obtained in chapter 3,

$$P(t) = \frac{Cae^{at}}{1 + Cse^{at}} ?$$

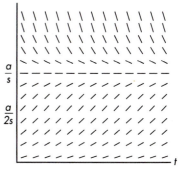

Figure 4.9: The direction field for the logistic equation $P' = aP - sP^2$.

4.1.3 Exercises

EXERCISE GUIDE

To gain experience . . .	Try exercises
Drawing direction fields	1–3(a)
Interpreting direction fields	1(b), 2(c)
Sketching solution graphs	2–3(b), 4–5, 6(a), 7–9, 11
Solving differential equations	2(c), 4(c), 6(b)
Analyzing and interpreting solution behavior	3(c), 5, 6(c), 10, 12

1. (a) Draw the direction field for the rock tossing equation $v' = -g$. Mark your axes carefully, and use the units you prefer. (On the earth, $g = 9.8$ m/s^2 or $g = 32$ ft/s^2.)

   (b) How would your picture of the direction field change if your model were moved to the moon, which has a gravity constant smaller than that of the earth's?

2. (a) Draw the direction field for the differential equation $P' = 0.015P - 0.209$ of the model for the population of Ireland during the potato famine of the last century. (The units for P are millions of people and those for t are years.)

   (b) Using the differential equation, sketch graphs of the solution for various initial populations. What initial population level separates growing from dwindling populations?

   (c) Find a solution of this equation and use it to confirm your qualitative analysis.

3. (a) Figure 4.4 summarizes the slope and concavity behavior of the solution of the logistic equation

   $$P' = aP - sP^2 .$$

   Supply the details needed to complete the construction of this figure from the start given in the text.

   (b) Sketch graphs of the solutions of this equation for various initial populations. Consider initial values in the intervals $(0, a/2s]$ and $(a/2s, a/s]$ as well as values larger than a/s.

   (c) The quantity a/s is sometimes called the **carrying capacity** of the population in this model. Use the curves you drew in part (b) to justify this name.

4. (a) Figure 4.6 summarizes the slope and concavity behavior of the solutions of the differential equation

$$y' = y^3 - y.$$

Supply the details needed to complete the derivation of figure 4.6 from the start given in the text.

(b) Sketch a solution of this equation satisfying the initial condition $y(0) = y_i$ for y_i in each of the indicated intervals:

i. $\sqrt{\frac{1}{3}} < y_i < 1$

ii. $0 < y_i < \sqrt{\frac{1}{3}}$

iii. $-\sqrt{\frac{1}{3}} < y_i < 0$

iv. $-1 < y_i < -\sqrt{\frac{1}{3}}$

v. $y_i < -1$.

(c) Find a solution of the initial-value problem for a value of y_i within each of the intervals in part (b). Is the qualitative behavior you sketched evident from the solution formulas? Do the solution formulas provide any information that is not evident in your graphs?

5. Differential equations of the form

$$u'(t) = u^2 - u$$

arise in the study of materials whose elastic behavior changes dramatically with temperature. The dependent variable u represents the magnitude of a disturbance in the material and t is time. Large disturbances tend to grow, while small ones tend to die out in time.

Show that this equation exhibits such behavior by sketching its solutions for initial values between 0 and 1 and for initial values greater than 1. What happens when $u(0) = 1$?

6. (a) Sketch the best graph you can of the solution of the initial-value problem

$$y' = y - y^3, y(0) = 0.1.$$

Carefully mark regions where the graph is increasing, decreasing, concave up, or concave down. Show inflection points.

(b) Find a solution of this initial-value problem.

(c) Which approach would give the information in your graph more easily, analysis of the differential equation as in part (a) or analysis of the solution you derived in part (b)? Does the solution formula provide some information that isn't apparent from the graph?

7. Which of the graphs in figure 4.10 might be a graph of a solution of $y' = 1 - y^2$, $y(0) = 0$? Justify rejecting each unacceptable choice and keeping each acceptable choice.

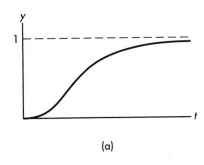

(a)

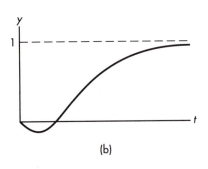

(b)

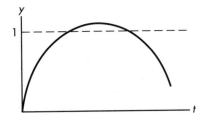

(c)

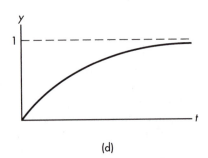

(d)

Figure 4.10: Possible solution graphs for the initial-value problem $y' = 1 - y^2, y(0) = 0$.

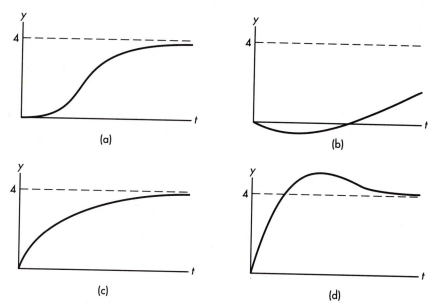

Figure 4.11: Possible solution graphs for the initial-value problem $y' = (4 - y)^3, y(0) = 0$

8. Which of the graphs in figure 4.11 might be a graph of a solution of $y' = (4 - y)^3$, $y(0) = 0$? Justify rejecting each unacceptable choice and keeping each acceptable choice.

9. Which of the graphs in figure 4.12 might be a graph of a solution of $y' = y^2$, $y(0) = 1$? Justify rejecting each unacceptable choice and keeping each acceptable choice.

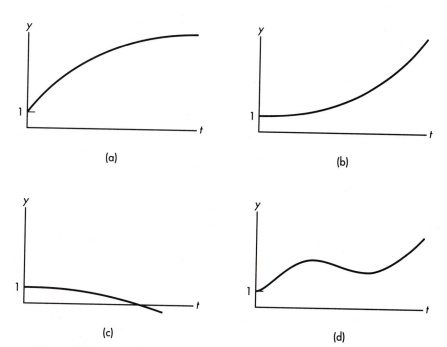

Figure 4.12: Possible solution graphs for the initial-value problem $y' = y^2$, $y(0) = 1$.

10. What range of values of the constant k will ensure that the solution of the initial-value problem $y'+ky = 0$, $y(0) = 1$, decays? Could this problem reasonably model a population of some sort? The mass of a decaying radioactive isotope?

11. Sketch the direction field defined by each of the following differential equations.
    (a) $u' = u^2 - u$
    (b) $y' = y - y^3$
    (c) $y' = 1 - y^2$
    (d) $y' = (4 - y)^3$
    (e) $y' = y^2$

12. Referring to the logistic equation $P' = aP - sP^2$, the closing paragraph of this section states

Study figure 4.9 carefully. As a population curve grows toward $P = a/s$, what becomes of its slope? Can the curve cross $P = a/s$? How do populations larger than a/s behave? Does the logistic equation $P' = aP - sP^2$ predict bounded or unbounded populations? Is this behavior obvious from the solution formula obtained in chapter 3,

$$P(t) = \frac{Cae^{at}}{1 + Cse^{at}} ?$$

Answer these questions.

4.2 EQUILIBRIA, ETC.

We want to understand more fully the solution of initial-value problems such as the heat-loss model

$$\frac{dT}{dt} = -\frac{Ak}{cm}(T - T_{\text{out}}), \ T(0) = T_i .$$

There are several informative perspectives from which to view the solutions of such problems.

4.2.1 Transients, Steady States, and Stability

The first perspective focuses on the behavior over time of the system being modeled. If T_{out} is constant, the solution

$$T(t) = T_{\text{out}} + (T_i - T_{\text{out}})e^{-Akt/cm} ,$$

of the heat-loss model has a component, called the *transient*, that decays exponentially as t grows,

$$T_{\text{trans}}(t) = (T_i - T_{\text{out}})e^{-Akt/cm} .$$

When decay of a transient is governed by an exponential term of the form e^{-rt}, $1/r$ is called the **time constant**. The time constant characterizes the rate of decay of the transient. When $t = 1/r$, then $e^{-rt} = e^{-1} \approx 0.37$, and the transient has decayed to roughly one-third its original size. The time constant in the heat-loss model is cm/Ak.

One of the purposes of mathematical modeling is identifying a parameter group such as the time constant whose value succinctly characterizes the behavior of the system being studied. For example, the time constant of your house tells you whether it is well or poorly insulated. Does added insulation increase or decrease the time constant? Does the heat-loss time constant cm/Ak actually have units of time?

If $T_i = T_{out}$, then we discover another aspect of this solution, the constant solution, or *steady state*, or *equilibrium* $T_{ss} = T_{out}$. We can find such an equilibrium simply by seeking a constant solution of the differential equation. For example, if T_{ss} is to be a constant solution of the heat-loss equation, then $T'_{ss} = 0 = -(Ak/cm)(T_{ss} - T_{out})$. We find $T_{ss} = T_{out}$.

In the heat-loss model, *every* initial state T_i eventually finds its way to this equilibrium,

$$T_{ss} = \lim_{t \to \infty} T(t) = T_{out} .$$

Equilibria that are the limit of solutions which start near them are called *stable*. A steady state we can never reach unless we start exactly on it is called *unstable*.

The steady state T_{out} of the heat-loss equation is *stable*. The emigration equation $P' = kP - E$ has a steady state E/k that is *unstable*. The instability is evident from the direction field diagram of figure 4.8. The flow shown by the arrows is always away from the equilibrium E/k so even a small departure from this state can never return.

The logistic equation $P' = aP - sP^2$ has equilibria at $P = 0, a/s$. Study the direction field diagram of figure 4.9. What do you judge the stability of these steady states to be?

Figure 4.13 illustrates the transient and steady-state parts of the solution of the heat-loss model. The steady state T_{out} is evidently stable.

Not all solutions exhibit steady states or transients, and not all steady states are attained as a transient dies away. For example, the emigration model

$$P' = kP - E, \qquad P(0) = P_i ,$$

with E constant has the solution

$$P(t) = \frac{E}{k} + \left(P_i - \frac{E}{k} \right) e^{kt} .$$

But this solution has no transient because $\lim_{t \to \infty} P(t)$ does not exist. On the other hand, we attain the equilibrium $P_{ss} = E/k$ if (and only if) $P_i = E/k$.

The exponential solution $P = P_i e^{kt}$ of the simple population model exhibits neither a transient nor a steady state.

A **steady state** or **equilibrium** of a differential equation is any constant solution. The **transient** component of a solution is that part (if any) whose limit is zero for large values of the independent variable.

An equilibrium y_{ss} is **stable** if all solutions whose initial values are near y_{ss} approach y_{ss} in the limit. More formally, we have

DEFINITION 1   *The equilibrium y_{ss} is stable if there is some $\delta > 0$ such that every solution having $y_{ss} - \delta \le y(0) \le y_{ss} + \delta$ satisfies $\lim_{t \to \infty} y(t) = y_{ss}$.*

> We could speak equally well of a stable *periodic limiting state*, a periodic particular solution to which nearby solutions decay. See example 11 below.

Perturbing a steady state corresponds to starting the system at an initial value near the equilibrium. If the perturbed problem has a decaying transient, then the equilibrium is stable. For example, if the house in the heat-loss model is at its equilibrium temperature of T_{out}, a perturbation of that steady state is modeled by the heat-loss initial-value problem with $T(0) \approx T_{out}$. Since that

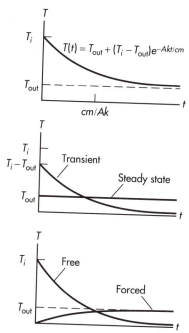

Figure 4.13: The transient and steady-state parts and the free and forced components of the solution of the heat-loss model $T' + (Ak/cm)T = (Ak/cm)T_{out}$, $T(0) = T_i$

problem always has a decaying transient, the equilibrium T_{out} is stable. If the perturbation grows, then the equilibrium is unstable, as is the population level E/k in the emigration model.

> An apple resting in the bottom of a bowl is in a stable equilibrium. A small perturbation, pushing the apple slightly away from that steady state, starts a motion that dies away as the apple settles back to the bottom of the bowl.
> An apple balanced on the edge of a bowl is in an unstable equilibrium. The smallest perturbation, giving it an initial position only slightly different from the steady state, will lead to a large, permanent departure from that tenuous equilibrium on the edge of the bowl. The apple will roll either to the bottom of the bowl or onto the table.

EXAMPLE 9 *Use the definition just given to verify that the equilibrium solution $T_{ss} = T_{\text{out}}$ of the heat-loss equation $T' = -(Ak/cm)(T - T_{\text{out}})$ is stable.*

We must show that $\lim_{t \to \infty} T(t) = T_{ss}$ when $T(t)$ solves

$$\frac{dT}{dt} = -\frac{Ak}{cm}(T - T_{\text{out}}), \ T(0) = T_i$$

for values of T_i near T_{out}.

But we have already seen that the solution

$$T(t) = T_{\text{out}} + (T_i - T_{\text{out}})e^{-Akt/cm}$$

of this initial-value problem approaches T_{out} in the limit,

$$\lim_{t \to \infty} \left(T_{\text{out}} + (T_i - T_{\text{out}})e^{-Akt/cm} \right) = T_{\text{out}} \,,$$

for *every* value of T_i. Hence, T_{out} is a stable equilibrium of the heat-loss equation, and it is stable to arbitrarily large perturbations. ∎

EXAMPLE 10 *Show that the logistic equation $P' = aP - sP^2$ has the steady state $P_{ss} \equiv a/s$ and that this equilibrium is stable.*

Constant solutions of the logistic equation must satisfy

$$0 = aP_{ss} - sP_{ss}^2 = P_{ss}(a - sP_{ss}) \,.$$

Hence, $P_{ss} = 0$ and $P_{ss} = a/s$ are equilibrium solutions.

The direction field diagram of figure 4.9 suggests that $P_{ss} = a/s$ is a stable equilibrium because the flow of the direction field is toward this constant solution. To prove stability, we must show that solutions of

$$P' = aP - sP^2, \qquad P(0) = P_i \,,$$

satisfy

$$\lim_{t \to \infty} P(t) = a/s$$

when P_i is near a/s. That is, we must show that small perturbations of the steady state a/s die out with time.

To study such a perturbation $p(t)$ explicitly, define it by

$$P(t) = \frac{a}{s} + p(t),$$

Solution = steady state + perturbation

where P is the solution of $P' = aP - sP^2$, $P(0) = P_i$. With a little algebra we can find the differential equation of which $p(t)$ is the solution:

$$\left(\frac{a}{s} + p\right)' = a\left(\frac{a}{s} + p\right) - s\left(\frac{a}{s} + p\right)^2$$

$$p' = ap - 2s\frac{a}{s}p - sp^2 + a\frac{a}{s} - s\left(\frac{a}{s}\right)^2$$

$$p' = -ap - sp^2$$

The initial value for p is $p(0) = P_i - a/s$.

Unfortunately, the perturbation is defined by a nonlinear initial-value problem that is no easier to solve than the original logistic equation *unless we use the smallness of the perturbation p to linearize*. Examining solutions that start near the steady state a/s means that the perturbation is small initially; $p(0) = P_i - a/s$ and $P_i \approx a/s$. If p is small, then p^2 is even smaller. The approximation $p^2 \approx 0$ reduces $p' = -ap - sp^2$ to the *linear* equation

$$p' = -ap.$$

This *linearized equation* determines the fate of the perturbation p of the steady state a/s as long as the perturbation is small.

> The linearized equation is an approximation, and the approximation fails if the neglected term sp^2 ever becomes large compared with the remaining term ap.

Since $p' = -ap$ has exponentially decaying solutions, the perturbation dies away to zero, and $\lim_{t\to\infty} P(t) = a/s$. The steady state a/s is stable to perturbations that are near enough a/s to permit the linearizing approximation $p^2 \approx 0$. ∎

The technique we have just illustrated, finding an equation for the perturbation and linearizing it by neglecting small nonlinear terms, is called a *linear stability analysis*.

Compare these last two examples. The heat-loss equation is linear. Its equilibrium solution is stable for every initial condition; starting near the equilibrium is not important. The logistic equation is nonlinear. We had to start near enough the equilibrium to be able to neglect the nonlinear term sp^2 to conduct a linear stability analysis. We can conclude only that the steady state is stable to small perturbations.

A phase plane diagram, a plot of P' versus P as shown in figure 4.14, provides another view of the stability of the steady states $P = 0, a/s$. The parabolic curve in the figure is the graph of P' versus P. When the curve is above the P axis, $P' > 0$ and vice versa. It crosses the horizontal axis at the equilibrium points $P = 0, a/s$ because $P' = 0$ at equilibria.

Perturbations of these steady states correspond to moving to the left or right along the P axis away from $P = 0$ or $P = a/s$. For example, a perturbation

4.2   Equilibria, etc.

149

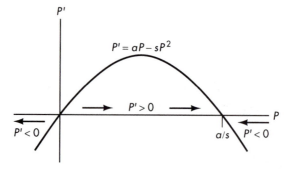

Figure 4.14: A phase plane diagram for $P' = aP - sP^2$. The horizontal arrows on either side of the equilibrium points $P = 0, a/s$ show how the sign of P' near those points determines their stability.

to the right from $P = a/s$ moves to a point on the P axis where $P' < 0$. But $P' < 0$ forces the perturbed P to decrease back towards its equilibrium at a/s, as the left pointing arrow just to the right of $P = a/s$ shows. Similarly, perturbation of the steady state $P = 0$ moves into regions where the sign of P' pulls the perturbation further from the unstable state $P = 0$.

EXAMPLE 11   *Identify the steady-state and transient parts, if any, of the solution of*

$$R\frac{dq}{dt} + \frac{1}{C}q = 120\sin 120\pi t, \; q(0) = q_i \,,$$

a model for the charge q on a capacitor in an RC circuit with an imposed voltage. (See the chapter 2 exercises.)

Using an integrating factor or a combination of undetermined coefficients and a characteristic equation, we find the solution

$$q(t) = B_c \cos 120\pi t + B_s \sin 120\pi t + (q_i - B_c)e^{-t/RC}$$

of this initial-value problem, where

$$B_s = \frac{120C}{1 + (120\pi RC)^2}, \; B_c = -120\pi RC B_s \,.$$

Since resistance R and capacitance C are positive, the transient part of this solution is clearly

$$q_{\text{trans}} = (q_i - B_c)e^{-t/RC} \,.$$

The sine and cosine terms are neither growing nor decaying. The time constant for the decay of the transient is RC.

> If this circuit is meant to store charge to keep a timer module running in a VCR, then we want the transient to decay slowly. If this circuit is meant only to oscillate and the transient will produce an ugly noise in the speakers, then we want it to die out quickly. The model has identified the parameter group RC that determines the rate of decay, helping a designer achieve a desired end.

4   Qualitative Behavior of First-Order Equations

This model has no steady equilibrium. If there were a *constant* steady state q_{ss}, it would have to satisfy

$$R\frac{dq_{ss}}{dt} + \frac{1}{C}q_{ss} = 120\sin 120\pi t\,,$$

or

$$q_{ss} = 120C\sin 120\pi t\,,$$

contradicting the assumption that q_{ss} is constant. In fact, this problem has a *periodic* limiting state, not a *steady* state. The decay of the transient carries us closer to a periodic state. ∎

If the long-term behavior of a system is important, we give little attention to the transient. If we are isolated by a three day blizzard, we know that the temperature of the unheated house modeled by $T' = -(Ak/cm)(T - T_{\text{out}})$, $T(0) = T_i$, will certainly drop to the outside temperature T_{out}. The transient tells us how quickly the house will cool.

The transient would receive most of our attention if we were interested in the speed of approach to the steady state or in the possibility of some unacceptable, extreme behavior occurring before settling into the steady state.

4.2.2 Free and Forced Response

Instead of focusing on the behavior over time, we can take a perspective that breaks the solution into parts representing the response due to being in a particular initial state and the response due to external effects. The appropriate terminology follows.

Nonhomogeneous equations are also called **forced** equations; the **forcing term** in $T' + (Ak/cm)T = (Ak/cm)T_{\text{out}}$ is $(Ak/cm)T_{\text{out}}$. Homogeneous equations are called **unforced**, or **free**.

The solution of the forced equation subject to a homogeneous initial condition is called the **forced response**. The solution of the corresponding unforced equation subject to the original initial condition is the **unforced**, or **free response**.

For example, if we write the heat-loss model so the forcing term appears on the right side of the differential equation, ORIGINAL INITIAL–VALUE PROBLEM

$$\frac{dT}{dt} + \frac{Ak}{cm}T = \frac{Ak}{cm}T_{\text{out}}\,,$$

$$T(0) = T_i\,,$$

then the *forced response* is the solution of FORCED INITIAL VALUE PROBLEM

$$\frac{dT}{dt} + \frac{Ak}{cm}T = \frac{Ak}{cm}T_{\text{out}}\,,$$

$$T(0) = 0\,, \quad \leftarrow \text{zero initial condition}$$

and the *free response* is the solution of

$$\frac{dT}{dt} + \frac{Ak}{cm}T = 0, \quad \leftarrow \text{zero forcing term}$$

$$T(0) = T_i.$$

A general solution is the perfect tool for finding free and forced response. A general solution of the heat-loss equation $T' + (Ak/cm)T = (Ak/cm)T_{\text{out}}$ is

$$T_g = T_p + T_h = T_{\text{out}} + Ce^{-Akt/cm},$$

where

$$T_h = Ce^{-Akt/cm}$$

is the general solution of the corresponding homogeneous equation

The general solution T_g already satisfies the forced differential equation $T' + (Ak/cm)T = (Ak/cm)T_{\text{out}}$. To find the forced response, we have only to choose C so that T_g satisfies the initial condition in the forced problem, $T(0) = 0$. That is, C must satisfy $T_g(0) = T_{\text{out}} + C = 0$ or $C = -T_{\text{out}}$. Hence, the

T_{for} solves the forced initial-value problem.

forced response is

$$T_{\text{for}} = T_{\text{out}} - T_{\text{out}}e^{-Akt/cm}.$$

Similarly, the homogeneous solution T_h already satisfies the homogeneous equation $T' + (Ak/cm)T = 0$. We have only to choose C so that T_h satisfies the initial condition in the free problem, $T(0) = T_i$. That is, C must satisfy $T_h(0) = C = T_i$. Hence, the free response is

T_{free} solves the free initial-value problem.

$$T_{\text{free}} = T_i e^{-Akt/cm}.$$

Note that the sum

$$T_{\text{for}} + T_{\text{free}} = T_{\text{out}} + (T_i - T_{\text{out}})e^{-Akt/cm}$$

solves the original initial-value problem

$$T' + (Ak/cm)T = (Ak/cm)T_{\text{out}}, \qquad T(0) = T_i.$$

Superposition!

Is this an accident? No! The principle of superposition guarantees that the sum $T_{\text{for}} + T_{\text{free}}$ will satisfy the forced differential equation; substituting T_{for} into $T' + (Ak/cm)T$ recovers the forcing term $(Ak/cm)T_{\text{out}}$, while substituting T_{free} yields zero. Linearity also guarantees that the sum satisfies the initial condition $T(0) = T_i$:

$$T_{\text{for}}(0) + T_{\text{free}}(0) = 0 + T_i = T_i.$$

Figure 4.13 illustrates the decomposition of the solution of the complete initial-value problem into its forced and free components.

For the record, we define the forced and free components of the general

linear initial-value problem

$$a_1(x)y' + a_0(x)y = f(x), \; y(0) = y_i.$$

The **forced response** is the solution of the initial-value problem

Forced Initial-Value Problem

$$a_1(x)y' + a_0(x)y = f(x), \ y(0) = \boxed{0}.$$

The **free response** is the solution of the initial-value problem

Free Initial-Value Problem

$$a_1(x)y' + a_0(x)y = \boxed{0}, \qquad y(0) = y_i.$$

The boxed terms emphasize the difference between the forced and free problems and the original initial-value problem.

The decomposition into forced and free components helps us understand what drives a system. The forcing term $(Ak/cm)T_{out}$ in the heat-loss equation $T' + (Ak/cm)T = (Ak/cm)T_{out}$ is an external cause of heat flow; you can see it driving the forced component of the solution up toward T_{out} in the third graph of figure 4.13. The forced response started from zero to avoid the complications of various initial temperatures T_i.

The free response problem $T' + (Ak/cm)T = 0$, $T(0) = T_i$, neglects effects of the outside temperature. It describes the behavior of the system as it evolves from its given initial state T_i, free of external influences.

4.2.3 Exercises

EXERCISE GUIDE

To gain experience . . .	Try exercises
Finding transients and steady states	4-, 5(a), 6(a), 10–12(a), 15, 17(a,d), 18–19, 22, 23
Determining stability of equilibria	4–5(b), 6(b), 9, 10–12(b), 13(a–b), 14(a), 17(b–c), 19
Finding time constants, etc.	2(a–b), 3, 4(c), 18
Finding free and forced response	4(d), 6(c), 10–12(d), 16, 20
Using the principle of superposition	16, 20, 21
Analyzing and interpreting solution behavior	1, 2(c–d), 3, 4(e), 6(a,c), 7, 8, 9, 10–12(c), 13(c), 14(b)

1. Move figure 4.13 from New England to the Southwest. Redraw it for the case $T_i < T_{out}$. (If necessary, convince yourself that the model $T' = -(Ak/cm)(T - T_{out})$, $T(0) = T_i$, remains valid in this case.) Is the description "heat-loss model" still accurate?

2. For a typical ranch house, the parameter group Ak/cm in the heat-loss model has the value 2×10^{-4} s^{-1}.
   (a) What is the corresponding time constant?
   (b) How long is required for the temperature difference $T - T_{out}$ to drop to half its initial value?
   (c) How could you use such information to select a heating (or cooling) plant for the house? Is the information you need for this design decision contained in the transient or the steady-state portion of the solution?
   (d) Would these considerations be fundamentally different if your concern were the temperature of a power transistor or the cargo bay of the space shuttle instead of a house?

3. The text asks whether added insulation increases or decreases the time constant for heat loss from a house. What do you think? Verify that the time constant cm/Ak has units of time.

4. A mass m falling through air experiences a resisting force proportional to its velocity. The proportionality constant k varies with shape. A model for the velocity

v of the mass is

$$v' = -g - \frac{k}{m}v, \quad v(0) = v_i.$$

(See exercise 8 of the chapter 1 exercises.)

(a) Find and sketch the graphs of the transient and steady-state solutions of this problem. Explain their physical significance. (*Hint:* Sky divers speak of a *terminal velocity*.)

(b) Determine the stability of the steady state. Explain the physical significance of your stability conclusion.

(c) What is the time constant for the transient? What does it mean?

(d) Find and sketch the free and forced response of this system.

(e) Interpret the free and forced response in light of the situation being modeled.

5. A more sophisticated model of the situation in the preceding problem supposes that air resistance is proportional to the square of the velocity of the mass. The resulting model is

$$v' = -g - \frac{k}{m}v|v|, \quad v(0) = v_i.$$

(See the chapter 1 project.)

(a) Find the equilibrium solution(s) of this model.

(b) Determine their stability and interpret your results.

6. A population of yeast spores increases due to reproduction by a fixed proportion R each day. The spores are harvested at a rate of H spores per day. A model for the spore population y is the initial-value problem

$$y' - Ry = -H, \quad y(0) = y_i.$$

(See exercise 6, in section 2.2.4.)

(a) Is there a steady-state yeast population? If so, what is its value? Explain the physical significance of this equilibrium.

(b) Is this equilibrium stable? (A direction field diagram might help you decide.) What is the physical significance of your stability conclusions?

(c) Sketch the *free* response of this system. Explain its physical significance.

7. An example in the text uses the solution

$$q(t) = B_c \cos 120\pi t + B_s \sin 120\pi t + (q_i - B_c)e^{-t/RC},$$

where

$$B_s = \frac{120C}{1 + (120\pi RC)^2}, \quad B_c = -120\pi RC B_s,$$

of the RC circuit model

$$R\frac{dq}{dt} + \frac{1}{C}q = 120\sin 120\pi t, \quad q(0) = q_i.$$

Derive this solution.

8. The text finds that perturbations p of the steady state a/s of the logistic equation $P' = aP - sP^2$ are governed by the linearized equation $p' = -ap$. It states, "Since $p' = -ap$ has exponentially decaying solutions ...." Verify this claim.

9. The text claims that the steady state $P \equiv 0$ of the simple population equation $P' = kP$ is unstable. Verify this claim and explain the physical significance of this instability.

10. A particular chemical reaction produces a chemical at a rate proportional to the concentration of that chemical; the proportionality constant, called the reaction rate, is k. Diluting the mixture reduces the concentration of the chemical at a constant rate of d g/cm^3·s. A model for this situation is the initial-value problem

$$c'(t) - kc(t) = -d, \quad c(0) = c_i.$$

(See exercise 8 of the chapter 2 exercises.)

(a) Find the steady state of this equation.

(b) Show this steady state is stable.

(c) Explain the physical significance of this equilibrium.

(d) Find the free and forced response of this system and explain the significance of each.

11. A radioactive isotope is decaying at a rate proportional to its mass; the proportionality constant is $k > 0$. A neutron beam bombarding the sample creates new isotope at a constant rate of i g/s. A model for this situation is the initial-value problem

$$c'(t) + kc(t) = i, \quad c(0) = c_i.$$

(See exercise 11 of the chapter 2 exercises.)

(a) Find the steady state of this equation.

(b) Show this steady state is stable.

(c) Explain the physical significance of this equilibrium.

(d) Find the free and forced response of this system and explain the significance of each.

12. A capacitor, a resistor, and a constant voltage drop E are connected in series. A model for the charge q on the capacitor in this RC circuit is the initial-value problem

$$Rq' + \frac{1}{C}q = E, \qquad q(0) = q_i .$$

(See exercise 4 of the chapter 2 exercises.)
   (a) Find the steady state of this equation.
   (b) Show this steady state is stable.
   (c) Explain the physical significance of this equilibrium.
   (d) Find the free and forced response of this system and explain the significance of each.

13. The text observes that the logistic equation $P' = aP - sP^2$ has the equilibria 0 and a/s.
   (a) Verify this claim.
   (b) Use the direction field diagram of figure 4.9 to determine the stability of each equilibrium.
   (c) Explain the physical significance of your observations.

14. The text shows that the steady state a/s of the logistic equation $P' = aP - sP^2$ is stable.
   (a) Use a linear stability analysis to determine the stability of the other equilibrium, $P \equiv 0$.
   (b) Explain the physical significance of these stability observations.

15. State some conditions which guarantee that the linear equation $y' + g(x)y = f(x)$ has a steady-state solution.

16. **(a)** Find the forced and free responses of the general constant-coefficient, constant forcing term initial-value problem $b_1 y' + b_0 y = h$, $y(0) = y_i$.
   (b) Find the solution y of this initial-value problem and verify that $y_{\text{for}} + y_{\text{free}} = y$.

17. Consider the constant-coefficient equation $b_1 y' + b_0 y = h$ subject to the constant forcing term h.
   (a) Find the steady state(s), if any, of this equation.
   (b) Find conditions which guarantee this equation has a decaying transient.
   (c) Find conditions that guarantee the steady state you found in part (a) is stable.
   (d) Show that the following model equations assume the form $b_1 y' + b_0 y = h$ for appropriate choices of b_1, b_0, h. Verify that their specific steady-state and transient properties match your general conclusions.

   i. The heat-loss equation $T' = -(Ak/cm)(T - T_{\text{out}})$ with outside temperature T_{out} constant

   ii. The emigration equation $P' = kP - E$ with emigration rate E constant
   iii. The RC circuit equation $Rq' + q/C = E$ with imposed voltage E constant (See exercise 4 of the chapter 2 exercises.)
   iv. The rock tossing model with air resistance $v' = -g - kv/m$ (See exercise 8 of the chapter 1 exercises.)

18. Argue that *any* RC circuit modeled by

$$R\frac{dq}{dt} + \frac{1}{C}q = E(t), \qquad q(0) = q_i ,$$

where $E(t)$ is an imposed voltage, has a transient with time constant RC. Does this circuit necessarily have a steady-state charge level q_{ss}? (See exercise 4 of the chapter 2 exercises.)

19. For each of the following equations,

   i. Find all equilibria,
   ii. Find a linear equation governing a perturbation of each steady state,
   iii. Indicate whether the perturbation equation is exact or approximate,
   iv. Determine the stability of each equilibrium,
   v. Sketch a direction field to confirm your stability analysis.

   (a) $y' = 1 - y$
   (b) $y' = 1 - y^2$
   (c) $y' = y - y^3$
   (d) $w' = (2 - w)(4 + w)$
   (e) $P' = aP - sP^3$
   (f) $v' = -g - kv^2/m$
   (g) $v' = -g + kv^2/m$
   (h) $y' = 1 - e^{y-2}$

20. **(a)** Find formulas for the forced and free response of the general linear initial-value problem $y' + g(x)y = f(x)$, $y(0) = y_i$.
   (b) Find a formula for the solution y of this initial-value problem and verify that $y_{\text{for}} + y_{\text{free}} = y$.

21. Suppose we decide to define the following decomposition of the solution of the initial-value problem $y' + g(x)y = f(x)$, $y(0) = y_i$. The *full response* is the solution of $y' + g(x)y = f(x)$, $y(0) = y_i$ and the *flat response* is the solution of $y' + g(x)y = 0$, $y(0) = 0$. Why is such a decomposition a joke? What is y_{flat}?

22. What is the maximum number of steady states the linear equation $y' + g(x)y = f(x)$ can have?

23. The nonlinear equation $y' = y^3 - y$ is one example of a general class of nonlinear, first-order equations of the form $y' + n(y) = 0$; here $n(y) = y - y^3$.

(a) Show that the steady states of $y' + n(y) = 0$ are the solutions of $n(y_{ss}) = 0$. Illustrate this general result for $y' = y^3 - y$.

(b) If $n(y)$ is a quadratic, what is the maximum num-

ber of steady states $y' + n(y) = 0$ can have? Must it have any equilibrium points at all?

(c) Give an example of a nonlinear first-order equation having the equilibrium points ± 2.

(d) What can you say about the possible number of equilibrium points if $n(y)$ is a fourth-order polynomial?

4.3 MISCELLANEOUS EXAMPLES

At this point, we have seen a variety of modeling, analysis, and solution techniques. We will use some of these ideas in different settings to guide you in making your own decisions about which methods are best suited to particular situations.

EXAMPLE 12 *The heat-loss model*

$$\frac{dT}{dt} = -\frac{Ak}{cm}(T - T_{\text{out}}), \qquad T(0) = T_i,$$

was derived for T measured in degrees Kelvin, the absolute temperature scale. How is the model changed when temperature is measured in degrees Celsius?

This question is very typical of the sort of manipulation required when using differential equation models. We must replace the dependent variable T in this initial value problem by one that has units of degrees Celsius.

For the moment, let $T(t)$ denote temperature measured on the Kelvin scale and let $U(t)$ denote temperature measured on the Celsius scale. The difference between the two scales is $273°$; that is,

$$T = U + 273.$$

Since $T' = (U + 273)' = U'$, substituting for T in $T' = -(Ak/cm)(T - T_{\text{out}})$ yields

$$\frac{dU}{dt} = -\frac{Ak}{cm}\left[(U + 273) - (U_{\text{out}} + 273)\right]$$

$$= -\frac{Ak}{cm}(U - U_{\text{out}}).$$

Of course, U_{out} is just the outside temperature measured in degrees Celsius. Similarly, the initial condition $T(0) = T_i$ becomes $U(0) = T_i - 273$. ∎

We see that the model is essentially unchanged. The heat-loss equation is invariant whether written in the T or the U variable. The initial condition is simply the original Kelvin condition $T(0) = T_i$ with the initial temperature converted to degrees Celsius. Henceforth, we can simply let T denote temperature in degrees Celsius without further comment.

EXAMPLE 13 *A home is to be fitted with a furnace capable of maintaining its temperature at a desired level, even when the outside temperature is at the lowest recorded temperature for that area. Formulate a mathematical problem that will answer the question, How big should the furnace be?*

We must translate this question into a mathematical problem by introducing notation and selecting or deriving an appropriate model. Answering the question will require analyzing the model and interpreting the results.

A furnace is able to keep a house warm if it can release heat faster than heat is lost to the cold air around the house. Consequently, the question How big should the furnace be? must be asking for the heat energy output *rate* of the furnace. Denote this heat output rate by H; it might be in units of cal/s or, more typically in the building industry, in Btu/h. (Figure 4.15 shows a furnace advertisement containing its references to Btu/h.)

We must choose H so that the house maintains a desired temperature, say T_d (d for desired) when the outside temperature is at some minimum value, say T_m (m for minimum). If the house is to be maintained at this temperature T_d, then we are apparently asking for a furnace output rate that will lead to a steady-state temperature of T_d while the outside temperature is T_m.

One of the problems in section 2.3.1, asks you to derive a model for a house containing a furnace. One form of that model is

$$\frac{dT}{dt} = -\frac{Ak}{cm}(T - T_{\text{out}}) + \frac{S(t)}{cm}, \qquad T(0) = T_i,$$

where $S(t)$ is the rate of heat output of the furnace at time t (S for source). This model should certainly be relevant here. Furthermore, the first example shows that we can use the more natural temperature scale of degrees Celsius.

> Allowing the heat output rate to vary assumes that the furnace is not running continuously. In most homes, the furnace is controlled not by a clock but by a thermostat. In that case, the heat output rate depends upon temperature, not time.
>
> If we were modeling heat buildup in a power transistor instead of in a house, then $S(t)$ might turn on and off as the transistor cycled in and out of its duty cycle.

To see that this differential equation is part of a potentially plausible model for this situation, note that the first term on the right,

$$-\frac{Ak}{cm}(T - T_{\text{out}}),$$

tends to make T' negative when $T > T_{\text{out}}$. That is, this term tends to cool a warm house. The second term,

$$\frac{S(t)}{cm},$$

represents the heat output rate of the furnace. Whenever it is positive, it tends to increase the temperature of the house. Balancing the two terms could lead to a steady state.

The heat output rate in the model equation varies with time, complicating the analysis. We could simplify the mathematical problem and remain consistent with the physical problem by examining a worst-case situation. We can suppose that the furnace will have to run continuously to maintain the desired tempera-

Figure 4.15: Furnaces are sized by their heat output rates in Btu/h; *per hour* is implied.

ture on the coldest day of the year, when $T_{\text{out}} = T_m$. When the furnace operates continuously, then $S(t) \equiv H$, the constant maximum heat energy output rate of the furnace.

With these choices of parameters, the model equation becomes

$$T' = -\frac{Ak}{cm}(T - T_m) + \frac{H}{cm}.$$

Our problem is thus that of choosing H so that this equation has the steady-state solution T_d. ∎

EXAMPLE 14   *Answer the question of furnace size by solving the mathematical problem posed in the last example: Choose the furnace heat output rate H so that*

$$T' = -\frac{Ak}{cm}(T - T_m) + \frac{H}{cm}$$

has the steady-state temperature T_d we seek for the house.

First we find a steady-state solution T_{ss} of the differential equation. Then we choose H so that the steady temperature is T_d.

Since $T'_{ss} = 0$, the equilibrium temperature is determined by

$$0 = -\frac{Ak}{cm}(T_{ss} - T_m) + \frac{H}{cm}.$$

Hence,

$$T_{ss} = T_m + \frac{H}{Ak}.$$

To choose the size of the furnace, select H so that $T_{ss} = T_d$:

$$H = Ak(T_d - T_m).$$

We should have been able to guess this result! It simply requires that the heat output rate H of the furnace balance the heat-loss rate $Ak(T_d - T_m)$ through the walls and roof when the house is at the desired temperature and the outside temperature is at its minimum.

There is one more point to check. Can this steady-state temperature be realized physically? Is the equilibrium T_d stable? Fortunately, it is, as we now argue.

The governing equation

$$T' = -\frac{Ak}{cm}(T - T_m) + \frac{H}{cm},$$

has the general solution

$$T = T_m + \frac{H}{Ak} + Ce^{-Akt/cm}.$$

The transient is $Ce^{-Akt/cm}$. Hence, every solution approaches the equilibrium

$$T_{ss} = T_m + \frac{H}{Ak}$$

as t grows. But if the furnace is sized properly, $H = Ak(T_d - T_m)$, and the equilibrium we approach is just $T_{ss} = T_d$.

If this equilibrium were not stable, it would be little more than a theoretical possibility. The smallest perturbation could drive the temperature of the house away from T_d. ∎

The situation in this example is not unusual. Modeling has shown us a result that is physically reasonable, at least in hindsight. Since we know precisely the assumptions underlying our model, we can state precisely the conditions under which this reasonable result holds. Without mathematical modeling as a guide, we would have no assurance our intuition is correct.

EXAMPLE 15   *The temperature outside a house often fluctuates in a regular pattern. The nights are a bit colder and the days a bit warmer than some average temperature that is typical of the season. Write a form of the heat-loss model $T' = -(Ak/cm)(T - T_{\text{out}})$, $T(0) = T_i$, that models this situation.*

We must provide a specific functional form for T_{out} that captures this sort of variation. We should look for a form that shows a regular variation about a constant average. The period of the variations should be 24 hours. If t is measured in hours, an appropriate form might be

$$T_{\text{out}}(t) = T_{\text{av}} + V \sin \frac{\pi t}{12}.$$

Here, T_{av} is the average temperature for the season and V is the amplitude (size) of the variations about that mean. The periodic term $\sin(\pi t/12)$ has been chosen to have a period of 24 h. (When $t = 24$ h, $\pi t/12 = 2\pi$, the period of the sine function.) It supposes that $t = 0$ is mid morning, say, about the time the temperature actually assumes its average value. The resulting model is

$$T'(t) = -\frac{Ak}{cm}\left(T(t) - T_{\text{av}} - V\sin\frac{\pi t}{12}\right), \qquad T(0) = T_i. \qquad ∎$$

EXAMPLE 16   *Find the solution of the preceding initial-value problem.*

The differential equation is linear, has constant coefficients, and is forced by a constant term and a sine:

$$T'(t) + \frac{Ak}{cm}T(t) = \frac{AkT_{\text{av}}}{cm} + \frac{AkV}{cm}\sin\frac{\pi t}{12}.$$

We'll use the method of undetermined coefficients to find the particular solution and the characteristic equation method to solve the homogeneous equation.

The homogeneous equation

$$T'(t) + \frac{Ak}{cm}T(t) = 0,$$

has the characteristic equation $r + Ak/cm = 0$, from which we obtain its general solution,

$$T_h(t) = Ce^{-Akt/cm}.$$

The form of the particular solution must contain an undetermined constant to account for the constant AkT_{av}/cm in the forcing term. In addition, it must contain both a sine and a cosine with argument $\pi t/12$ to accommodate the sine term in the forcing function. Hence, we begin finding a particular solution with the guess

$$T_p(t) = D + E\cos\frac{\pi t}{12} + F\sin\frac{\pi t}{12},$$

where D, E, and F are undetermined coefficients. By substituting in the appropriate equation, we see that none of these terms are solutions of the homogeneous equation.

We substitute the assumed form of T_p into the equation,

$$T' = \left(\frac{AkT_{av}}{cm}\right) + \left(\frac{AkV}{cm}\right)\sin\frac{\pi t}{12},$$

and collect like terms to obtain

$$T_p' + \frac{Ak}{cm}T_p = \frac{AkD}{cm} + \left(\frac{\pi F}{12} + \frac{AkE}{cm}\right)\cos\frac{\pi t}{12}$$
$$+ \left(\frac{-\pi E}{12} + \frac{AkF}{cm}\right)\sin\frac{\pi t}{12}$$
$$= \frac{AkT_{av}}{cm} + \frac{AkV}{cm}\sin\pi t/12.$$

Equating constant terms yields

$$\frac{AkD}{cm} = \frac{AkT_{av}}{cm}$$

or $D = T_{av}$.

Equating coefficients of the cosine and sine terms yields a pair of simultaneous equations for E and F,

$$\frac{Ak}{cm}E + \frac{\pi}{12}F = 0$$

$$\frac{-\pi}{12}E + \frac{Ak}{cm}F = \frac{AkV}{cm}.$$

Solving this system gives

$$E = -\frac{\pi AkV}{12cm}\Delta,$$

$$F = \left(\frac{Ak}{cm}\right)^2 V\Delta,$$

where

$$\Delta = \left(\left(\frac{Ak}{cm}\right)^2 + \left(\frac{\pi}{12}\right)^2\right)^{-1}.$$

4   Qualitative Behavior of First-Order Equations

Hence, a particular solution of (13) is

$$T_p(t) = T_{av} + \Delta \frac{AkV}{cm} \left(\frac{Ak}{cm} \sin \frac{\pi t}{12} - \frac{\pi}{12} \cos \frac{\pi t}{12} \right).$$

A general solution of the nonhomogeneous equation is

$$T_g(t) = Ce^{-Akt/cm} + T_p(t), \qquad (4.1)$$

where $T_p(t)$ is defined as before. We need to choose C so that the initial condition $T(0) = T_i$ is satisfied—i.e. so that

$$T_g(0) = C + T_p(0)$$
$$= C + T_{av} - \Delta \frac{AkV}{cm} \frac{\pi}{12} = T_i.$$

We can solve for C and write the solution of the original initial-value problem as

$$T(t) = \left(T_i - T_{av} + \Delta \frac{Ak\pi}{12cm} \right) e^{-Akt/cm}$$
$$+ T_{av} + \Delta \frac{AkV}{cm} \left(\frac{Ak}{cm} \sin \frac{\pi t}{12} - \frac{\pi}{12} \cos \frac{\pi t}{12} \right). \qquad (4.2)$$

What can we learn from this mess? The time constant for the transient is still cm/Ak; that quantity is unaffected by the forcing function. If this house is normally insulated, the transient will have decayed to almost nothing in a few hours and the temperature of the house will begin to follow the swings of the outside temperature. The oscillations of the temperature of the house will have the same period as those outside, 24 h, but the combination of the sine and cosine in the solution shows that the house will be out of *phase* with the outside. That is, its maximum and minimum temperatures will not occur at the same time as those outside. ∎

EXAMPLE 17 *Find the free and forced responses of the heat-loss model with periodically varying outside temperature,*

ORIGINAL INITIAL-VALUE PROBLEM

$$T' = -\frac{Ak}{cm} \left(T - T_{av} - V \sin \frac{\pi t}{12} \right), \qquad T(0) = T_i.$$

We have very little work to do. The general solutions we need were found in the preceding example.

The forced response is defined by the initial-value problem

FORCED INITIAL-VALUE PROBLEM

$$T'_{for} - \frac{Ak}{cm} T_{for} = \frac{Ak}{cm} \left(T_{av} + V \sin \frac{\pi t}{12} \right), \qquad T_{for}(0) = 0;$$

i.e., by the original differential equation subject to a zero initial condition.

Since the differential equation is unchanged, the general solution $T_g = Ce^{-Akt/cm} + T_p$ of the nonhomogeneous equation found in example 16 may be used here as well. We need only choose the arbitrary constant C to satisfy the initial condition $T(0) = 0$.

Hence, C must satisfy

$$T_g(0) = C + T_p(0)$$

$$= C + T_{\text{av}} - \Delta\frac{Ak\pi}{12cm} = 0 .$$

Solving for C and substituting it into the expression (4.1) for the general solution yields the solution of the initial-value problem defining the forced response,

$$T_{\text{for}}(t) = \left(\Delta\frac{Ak\pi}{12cm} - T_{\text{av}}\right)e^{-Akt/cm}$$

$$+T_{\text{av}} + \Delta\frac{AkV}{cm}\left(\frac{Ak}{cm}\sin\frac{\pi t}{12} - \frac{\pi}{12}\cos\frac{\pi t}{12}\right) .$$

A shortcut is to observe that the forced initial-value problem is the same as the original initial-value problem with $T_i = 0$. We need only set $T_i = 0$ in the solution (4.2) of the original initial-value problem to obtain this expression for T_{for}.

FREE INITIAL-VALUE PROBLEM

The free response is defined by the initial-value problem

$$T'_{\text{free}} + \frac{Ak}{cm}T_{\text{free}} = 0, \qquad T_{\text{free}}(0) = T_i ;$$

i.e., by the unforced version of the differential equation subject to the original initial condition. The general solution of this homogeneous equation is just $T_h = Ce^{-Akt/cm}$ found in the previous example. There remains only choosing C in this expression to satisfy the initial condition $T(0) = T_i$. Obviously, $C = T_i$, and the free response is

$$T_{\text{free}}(t) = T_i e^{-Akt/cm} .$$

The forced response is the temperature of the unheated house starting at zero temperature. It contains the oscillatory response induced by the periodic variations in T_{out}.

The free response is the temperature history of the house beginning at its actual initial temperature and driven by zero outside temperature. It is exactly the same as the free response of the original heat-loss model with T_{out} constant. The difference between that original model and the one considered here is the behavior of the forcing function, but the free response neglects all external forcing. ∎

4.3.1 Exercises

EXERCISE GUIDE

To gain experience . . .	Try exercises
Changing variables	1–2
Solving equations	4–5
Finding equilibrium points	7, 9(b), 12(b), 13(a)
Determining stability of steady states	6, 12(d), 13(b)
Analyzing the form of model equations	8, 9(a), 12(a,e), 13(c)
Analyzing and interpreting solution behavior	3, 9(c), 10–11, 12(c)

1. Change the temperature variable T in the heat-loss model

$$\frac{dT}{dt} = -\frac{Ak}{cm}(T - T_{\text{out}}), \qquad T(0) = T_i,$$

from one measured in degrees Kelvin to one measured in degrees Fahrenheit. (If F is temperature in degrees Fahrenheit, then $T = 5(F - 32)/9 + 273$.)

2. Change the temperature variable T in the furnace model

$$\frac{dT}{dt} = -\frac{Ak}{cm}(T - T_{\text{out}}) + \frac{S(t)}{cm}, \qquad T(0) = T_i,$$

from
(a) One measured in degrees Kelvin to one measured in degrees Celsius,
(b) One measured in degrees Kelvin to one measured in degrees Fahrenheit.

3. In example 14, the text finds the furnace heat output rate

$$H = Ak(T_d - T_m)$$

needed to maintain the house at temperature T_d when the outside temperature is T_m. Then it says:

> We should have been able to guess this result! It simply requires that the heat output rate H of the furnace balance the heat-loss rate $Ak(T_d - T_m)$ through the walls and roof when the house is at the desired temperature and the outside temperature is at its minimum.

Explain why $Ak(T_d - T_m)$ represents the rate at which the house is losing heat energy.

4. Use an integrating factor to solve the initial-value problem

$$\frac{dT}{dt} = -\frac{Ak}{cm}(T - T_{\text{out}}) + \frac{H}{cm}, \qquad T(0) = T_i,$$

from the furnace-sizing question.
(a) Verify that you obtain a general solution of the differential equation before you apply the initial condition to determine the arbitrary constant.
(b) Verify that this solution yields the same steady-state solution as in the text.
(c) Why does the integrating factor you use here look familiar?

5. Use an integrating factor to solve the initial value problem

$$T' = -\frac{Ak}{cm}\left(T - T_{\text{av}} - V \sin\frac{\pi t}{12}\right), \qquad T(0) = T_i,$$

modeling a house subject to periodic variations in the outside temperature. Does the integrating factor look familiar? Verify that you obtain the same solution as the text did using undetermined coefficients and a characteristic equation. (*Hint:* To avoid being overcome by the alphabet soup of parameters in this problem, introduce notation for the commonly occurring parameter groups; e.g., let $\alpha = Ak/cm$.)

6. An example in the text uses a solution formula to show that the furnace equation

$$T' = -\frac{Ak}{cm}(T - T_m) + \frac{H}{cm}$$

has a stable equilibrium by showing that all solutions approach the equilibrium. Verify this claim independently by drawing a direction field diagram for this equation.

7. Recall the initial-value problem modeling the house with a furnace,

$$T' = -\frac{Ak}{cm}(T - T_m) + \frac{H}{cm}, \qquad T(0) = T_i.$$

(a) Show that the equilibrium temperature is independent of T_i and that the solution of this initial-value problem approaches that equilibrium regardless of the value of T_i.
(b) Why is it obvious that the value of the equilibrium should be independent of T_i?

8. The initial-value problem

$$\frac{dT}{dt} = -\frac{Ak}{cm}(T - T_{\text{out}}) + \frac{S(t)}{cm}, \qquad T(0) = T_i,$$

is a model for a house containing a furnace. In that model, $S(t)$ represents the heat output rate of the furnace at time t. The note accompanying the model suggests that S is more often a function of temperature T than time t because a furnace is usually controlled by a thermostat that turns the furnace on and off in response to the temperature of the house. Consider this more realistic model

$$\frac{dT}{dt} = -\frac{Ak}{cm}(T - T_{\text{out}}) + \frac{S(T)}{cm}, \qquad T(0) = T_i,$$

(a) Argue that this differential equation is linear only if S has the functional form $S(T) = aT + b$, where a and b are constants; i.e., the differential equation is linear only if the furnace heat source term S is linear in T.

(b) A typical furnace is either on or off, depending on whether the temperature of the house is greater than or less than the thermostat setting. Can such behavior be modeled by a linear function such as $S(T) = aT + b$? What would be an appropriate functional form for S to model this on-off behavior? Could you solve the preceding differential equation above if S had the form you propose?

9. Follow the given steps to choose a central air-conditioning unit for a building that must be maintained at a constant temperature of $25°$ C even when the outside temperature is $46°$.

(a) Derive a model for this situation. It should look very much like the furnace model

$$\frac{dT}{dt} = -\frac{Ak}{cm}(T - T_{\text{out}}) + \frac{S(t)}{cm}, \qquad T(0) = T_i,$$

with the exception of a crucial sign difference. (An air conditioner is a device that promotes heat energy *loss*. It absorbs rather than releases heat at a given rate.)

(b) Find the steady-state solution of your model. Choose the parameter representing the cooling (heat-energy loss) rate of the air conditioner to satisfy the specified conditions.

(c) Does the formula that determines the cooling rate of the air conditioner have a reasonable physical interpretation?

10. Refer to the question of furnace size posed in the second example of this section. If the outside temperature is at its minimum value T_m and the initial temperature of the house is less than T_d, the desired temperature, will the temperature of the house ever reach T_d if the furnace output rate H is chosen by our formula $H = Ak(T_d - T_m)$?

11. The heat-loss model with varying outside temperature

$$T' = -\frac{Ak}{cm}\left(T - T_{\text{av}} - V \sin \frac{\pi t}{12}\right), \qquad T(0) = T_i,$$

has a forcing term whose periodic part is $\sin \pi t/12$, but the solution

$$T(t) = \left(T_i - T_{\text{av}} + \Delta \frac{Ak\pi}{12cm}\right)e^{-Akt/cm}$$

$$+ T_{\text{av}} + \Delta \frac{AkV}{cm}\left(\frac{Ak}{cm}\sin\frac{\pi t}{12} - \frac{\pi}{12}\cos\frac{\pi t}{12}\right)$$

of this initial-value problem involves both the sine and the cosine. Show that the effect of a sum of a sine and a cosine in the solution is still periodic but that the peak of the house temperature does not occur at the same time as the peak in the outside temperature. (This change is called a *phase shift* and the difference is called a *phase angle*.)

(a) Use the formula for the sine of a sum of angles to show that the sine and cosine terms in this solution can be combined to yield a term of the form $\sin(\pi t/12 + \varphi)$, where φ is the phase angle mentioned earlier.

(b) Sketch a graph of both $\sin \pi t/12$ and $\sin(\pi t/12 + \varphi)$ and mark the phase difference φ.

(c) Which hits its peak temperature first each day, the outside or the house? Is this relation physically reasonable?

(*Hint:* In the solution, you have a term of the form $L\cos\alpha + M\sin\alpha$, where L and M are constants and $\alpha = \pi t/12$. Rewrite this expression as $K(P\cos\alpha + Q\sin\alpha)$, where $K = \sqrt{L^2 + M^2}$, $P = L/K$, $Q = M/K$. Find a formula for an angle φ such that $P = \sin\varphi$ and $Q = \cos\varphi$. Note that K is the amplitude of the oscillation in the house temperature, just as V is the amplitude of the oscillation in the outside temperature.)

12. There are various approximate rules for determining the force of air resistance on a falling mass m. One assumes that the force of resistance is proportional to the square of the velocity v of the mass and always acts opposite to the direction of its motion. A model for this situation is

$$\frac{dv}{dt} = -g - \frac{k}{m}v|v|, \qquad v(0) = v_i,$$

where k is the proportionality constant for air resistance. (See the chapter 1 project.)

(a) Does this differential equation succumb to any of the analytic solution techniques you have studied? Explain.

(b) Does this equation have an equilibrium solution?

Explain its physical significance and interpret the formula you obtain for it.

(c) Objects that fall a long distance eventually reach a *terminal velocity*. Could you determine the air resistance coefficient k of an object from a measurement of its terminal velocity?

(d) Draw a direction field diagram for this differential equation for values of v near the equilibrium velocity. Is this equilibrium stable? Is your conclusion physically reasonable? Explain.

(e) Are the concepts of free and forced response applicable here? Explain.

13. The logistic equation $P' = aP - sP^2$ has two equilibrium solutions, $P = 0$ and $P = a/s$. Use a direction field diagram such as the one in figure 4.9, to answer the following.

(a) Which initial values of P yield solutions whose equilibrium value is 0? a/s?

(b) Are there initial values that yield solutions that never reach any equilibrium?

(c) Compare this situation with that of the single equilibrium solution of the *linear* emigration equation $P' = kP - E$, E constant.

4.4 CHAPTER EXERCISES

EXERCISE GUIDE

To gain experience . . .	Try exercises
Finding transients and steady states	1, 3–5, 7–8(b), 10, 12(c), 13, 16
Finding free and forced response	3–6, 7(a), 8(c), 11(a), 12(d)
Finding general solutions	12(a)
Solving initial value problems	8(a), 10, 12(b), 18(b, c)
Changing variables	14
Sketching solution graphs	15(a, c), 16, 17(a, c), 18(a, c)
Analyzing the form of model equations	8(d), 13
Analyzing and interpreting solution behavior	2, 8(c), 9, 11(b), 12(c–d), 13, 15(b), 17(b), 19

1. Find the transient and steady state of the solution of $y' + 2xy = x$, $y(0) = 2$.

2. The differential equation $dy/dt = ky$, k a constant, is sometimes called an *equation of growth and decay*. Explain why this is a good name. What must you know to determine whether this differential equation is describing growth or decay? What solution methods could be applied to finding a general solution of this equation?

3. Find the free and forced response and the transient and steady state of the solution of the initial-value problem

$$y' + xy = e^{-x^2/2}, \qquad y(0) = 0.$$

4. Find the free and forced response and the transient and steady state of

$$y' + \cos x\, y = \frac{1}{x} e^{-\sin x}, \qquad y(\pi) = 0.$$

5. Solve each of the following initial-value problems.

   i. Find transients and steady states if they exist.
   ii. Find the free and forced response of each initial-value problem.

   (a) $y' + y = e^{-x}$, $y(0) = 1$
   (b) $y' - 3y = -3x$, $y(0) = 0$
   (c) $y' + xy = x$, $y(1) = 3$
   (d) $y' + ky = A$, $y(0) = 1$, k, A positive constants

6. Find the free and forced response of $y' + 6ty = t$, $y(0) = 2$.

7. (a) Find the free and forced response of $y' + t^2 y = 2t^2$, $y(0) = 3$.
   (b) Find the transient and steady state of the solution of this initial-value problem.

8. A tank is full of saltwater. Brine in the tank is drained from the bottom so that 10% of the mixture leaves every minute. Water enters the top at an equal rate so that the tank is always full. In addition, salt is stirred into the tank at the rate of 5 lb/min.

Let $s(t)$ denote the total weight of salt in the tank at time t and s_i the total weight of salt in the tank at $t = 0$. Then a model for this situation is the initial-value problem

$$s' + 0.1s = 5, \qquad s(0) = s_i.$$

(a) Solve this initial-value problem.
(b) Find the transient and steady state of the solution of this initial-value problem. Explain the physical significance of each.
(c) Find the free and forced response of this initial-value problem. Explain the physical significance of each.
(d) Argue that the terms in the differential equation do indeed provide a reasonable model for the situation described.

9. The initial-value problem

$$P'(t) - 0.02P(t) = -1,000, \qquad P(0) = P_i,$$

is a model of a population of microbes in which 2% of the current population reproduces per unit time. The term $-1,000$ represents a constant death rate due to the presence of a toxic substance in the microbes' environment.

Argue that a very small initial population will die out while a large initial population will grow. Find the threshold initial population that separates *small* from *large*. That is, find a population P_t with the property that P is increasing if $P_i > P_t$, but P is decreasing if $P_i < P_t$. What happens if $P_i = P_t$?

10. Section 4.3 considered a heat-loss model that includes a term $S(t)$ representing the heat-energy output rate of a furnace,

$$T'(t) = -\frac{Ak}{cm}(T - T_{\text{out}}) + \frac{S(t)}{cm}, \qquad T(0) = T_i.$$

Eventually, that problem was reduced to one in which $S(t) \equiv H$, the (constant) maximum heat output rate of the furnace. Here we wish to consider solutions of this initial-value problem when $S(t)$ is allowed to vary with time.

If $S(t)$ varies with time, we might expect it to range from a minimum of zero, when the furnace is turned off, to a maximum of H. Two possible choices are

i. $S(t) = H(1 + \sin \omega t)/2$,
ii. $S(t) = H \sin^2 \omega t$,

where ω is constant in each case. Find the solution of this initial-value problem for each choice of $S(t)$. Does either exhibit a steady state? What is the transient part of each solution?

11. Consider the differential equation $y' + ky = f(t)$, k a constant.
(a) What range of values of k ensures that the free response of this equation decays?
(b) This equation arises in a model in which y is the concentration of a quantity. For $t > 0$, $f(t) < 0$. Is the function f modeling a source or a sink (loss) of y?

12. A model for the temperature $T(t)$ of a house heated by a wood stove is

$$T'(t) = -\frac{Ak}{cm}(T - T_{\text{out}}) + \frac{S}{cm}, \qquad T(0) = T_i.$$

Here S is the (positive) constant heat release rate of the stove and the other parameters are as in the text. (See exercise 16 of the chapter 2 exercises.)
(a) Find the general solution of the differential equation. Can you use an analysis carried out elsewhere in the text or in other exercises?
(b) Find the solution of the initial-value problem.
(c) Find the transient and steady-state response of this system. Explain the physical significance of the steady state.
(d) Find the free and forced response of this system. Explain the physical significance of the free response.

13. Describe a physical situation of which the initial-value problem

$$T'(t) = -\frac{Ak}{cm}(T - T_{\text{out}}) - \frac{S}{cm}, \qquad T(0) = T_i$$

might be a model. (Note the sign of S, which is opposite that of the usual furnace equation considered in this set of exercises.) How does changing the sign of S affect the transient? The steady state? Are these changes consistent with the physical interpretation you have given this initial-value problem?

14. Introducing *dimensionless* variables can reduce the clutter of parameters with which some models are afflicted.
(a) Argue that the constant k in the growth model $y' = ky$ has units of inverse time (e.g., s^{-1}).
(b) Define the dimensionless time $s = kt$ and the new variable Y, $Y(s) \equiv y(s/k)$. Carefully ex-

press dY/ds in terms of dy/dt and show that $dy/dt = ky$ becomes simply $dY/ds = Y$.

(c) Apply these techniques to the emigration equation $P' = kP - E$ and to the heat-loss equation $T' = -(Ak/cm)(T - T_{out})$.

15. Exercise 14 in the chapter 2 exercises considers the following situation.

An entrepreneur has established a line of credit at a bank. She is charged a daily interest rate of 0.0493% on the total amount she owes the bank at the end of each business day. She borrows at a rate of $b(t)$ dollars per day, and her initial loan was in the amount a_i. The amount she *owes* t days after the initial loan is $a(t)$.

That problem offers four initial-value problems as possible models of this situation.

    i. $a'(t) + 0.000493a = b(t),\ a(0) = a_i$
    ii. $a'(t) - 0.000493a = b(t),\ a(0) = a_i$
    iii. $a'(t) + 0.000493a = -b(t),\ a(0) = a_i$
    iv. $a'(t) - 0.000493a = -b(t),\ a(0) = a_i$

(a) Assume that the function $b(t)$ is a positive constant. Without actually solving these equations, sketch graphs of the solutions of each of these four initial-value problems. Whenever appropriate, comment upon differences in solution behavior caused by different choices of the initial value a_i.

(b) How does your intuition about the situation just described suggest the solution of a correct model should behave? Which initial-value problem gives solution graphs closest to the behavior you expect in a good model?

(c) Which of the initial-value problems possess a steady state? When appropriate, sketch the steady state and transient. Which of these equilibria are stable?

16. Each of the sketches of figure 4.16 represents the solution of some initial-value problem. Sketch the transient and steady state parts of each solution.

17. Exercise 15 of the chapter 2 exercises considers the following situation.

An insect population (denoted by y) is being destroyed by an insecticide at a constant rate d. It is increasing at a rate proportional to the current population; the constant of proportionality is g.

That problem offers four initial-value problems as possible models of this situation.

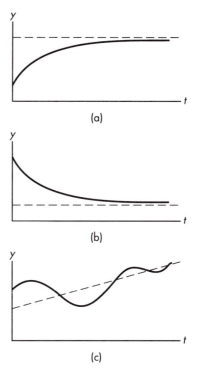

Figure 4.16: Exercise 16.

    i. $y' + gy = d,\ y(0) = y_i$
    ii. $y' - gy = d,\ y(0) = y_i$
    iii. $y' + gy = -d,\ y(0) = y_i$
    iv. $y' - gy = -d,\ y(0) = y_i$

(a) Assume that the parameters g and d are positive constants. Without actually solving these equations, sketch graphs of the solutions of each of these four initial-value problems. Whenever appropriate, comment upon differences in solution behavior caused by different choices of the initial value y_i.

(b) How does your intuition about the situation described here suggest the solution of a correct model should behave? Which initial-value problem gives solution graphs closest to the behavior you expect in a good model?

(c) Which of the initial-value problems possess a steady state? When appropriate, sketch the steady state and transient. Which of these equilibria are stable?

18. (a) Without actually solving the equation, sketch graphs of the solution of $w' = (1 - w)(1 + w)$ for a variety of initial values: positive, negative, and zero.

(b) Solve this equation and confirm that your sketches are correct.

(c) Which is easier, sketching these solutions directly from the equation or sketching them from a solution formula?

19. Argue that the solutions of $y' = -ky$ decay to zero when k is a positive constant. (This is the isotope decay equation derived in exercise 1 of the chapter 2 exercises.) Use the uniqueness theorem 3.3, page 86, to argue that this decay is *asymptotic*; $y(t) \neq 0$ for finite t unless $y(t) \equiv 0$ for *all* t.

4.5 CHAPTER PROJECTS

1. **Heat loss in humans** Think of a human being as a house with a furnace. Heat is lost through the skin to the surroundings, clothing acts as insulation, and metabolism generates heat as if it were a furnace.

   (a) Use these ideas to derive a model for the temperature of a human. Take care to define the meaning of each parameter as it is introduced.

   (b) Suppose an individual weighing 70 kg can choose to wear a 0.5 cm layer of either flannel or copper while venturing into 5° C weather. Use the data of table 2.4, page 39, to determine the heat output rate required to maintain the usual body temperature of 37° C in each of these outfits.

   (c) Show that the normal body temperature is a stable equilibrium of this model. What would be the consequences if this temperature were an *unstable* equilibrium?

2. **General stability analysis** Provide the framework for a general stability analysis of the first-order equation $y' = f(y)$.

   Suppose y_0 is an equilibrium. What can be said about $f(y_0)$? Find the linearized equation that governs a small perturbation of the steady state $y \equiv y_0$. State conditions on f that guarantee stability and instability of this steady state. Are there any undecided situations? Can those be resolved? State your results carefully in the form of a theorem, and organize your arguments into a proof.

3. **Spruce budworm population model** The spruce budworm rapidly defoliates the Canadian balsam fir, destroying entire forests in a few years. Birds are its principal natural enemy.

   (a) Suppose the budworm population $P(t)$ has a population-dependent intrinsic growth rate $k = a - sP$, as in the logistic population model. Further, suppose that predation by birds removes budworms from the population at a rate $R(P)$. Derive the budworm population model

   $$\frac{dP(t)}{dt} = aP - sP^2 - R(P), \qquad P(0) = P_i.$$

   (b) The predation function $R(P)$ exhibits threshold behavior, as illustrated in figure 4.17: once the budworm population reaches a critical level, birds begin consuming them at a nearly constant rate. (Might spraying insecticides by man be modeled similarly?) One functional form that models this behavior is

   $$R(P) = \frac{\beta P^2}{\alpha^2 + P^2},$$

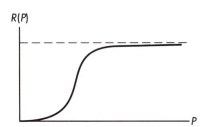

R(P)

Figure 4.17: The predation rate $R(P)$ rises suddenly when the budworm population reaches a critical level.

where α, β are parameters. For example, when α is small, the threshold population is low, but the increase in predation rate is dramatic. The parameter β is the maximum predation rate.

(c) The budworm equation now involves four parameters, a, s, α, and β. To reduce the number of parameters, define the dimensionless variables and parameters

$$u(\tau) = \frac{P}{\alpha}, \qquad A = \frac{\alpha a}{\beta}, \qquad S = \frac{\alpha^2 s}{\beta}, \qquad \tau = \frac{\beta t}{\alpha}.$$

Use the chain rule,

$$\frac{dP(t)}{dt} = \frac{du(\tau)}{d\tau}\frac{d\tau}{dt} = \cdots,$$

to write the budworm model as

$$\frac{du(\tau)}{d\tau} = Au - Su^2 - \frac{u^2}{1+u^2}, \qquad u(0) = u_i,$$

where $u_i = P_i/\alpha$. The differential equation now contains only two parameters, A and S.

(d) Argue that steady states of this dimensionless model occur when

$$A - Su = \frac{u}{1+u^2}.$$

Plot the curve on the right and the line on the left for various (positive) values of A and S. Show that this equation can have one, two, or three steady states, depending on how A and S affect the intersection between the line and curve.

(e) For the cases of one or three steady states, use a phase plane picture like that of figure 4.14, page 150, to ascertain the stability of the steady state(s). Interpret these results physically.

5

OSCILLATORY
PHENOMENA

Whole lotta shakin' goin' on.
Jerry Lee Lewis

Cyclic motions are ubiquitous. We see them in the natural repetitions of life cycles, in the jiggle of an automobile suspension, and in the rhythmic slamming of an old screen door whose spring is too weak.

We will not be so ambitious as to study biological clocks and their cycles, but we will model an intellectual precursor to an automobile suspension, the spring-mass system, as well as an electric circuit, a pendulum, and a population of rabbits and foxes. Many of these inherently oscillatory phenomena will give rise to the same differential equation, for which we shall develop solution techniques in the next chapter.

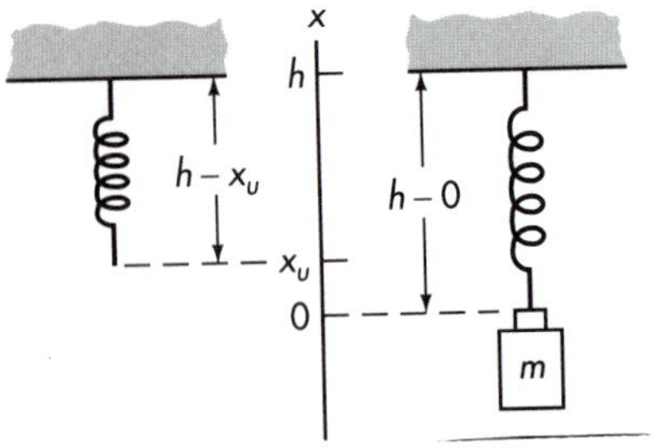

5.1 SPRINGS AND MASSES

Figure 5.1: A vertical spring-mass system showing the location of the free end of the spring at $x = x_u$ before the mass is added and after, when the mass hangs in equilibrium at $x = 0$.

Figure 5.1 shows a typical **spring-mass system**. In the absence of an external disturbance, the mass should hang quietly from the spring, the downward pull of its weight just balanced by the upward pull of the slightly stretched spring. If the mass is pulled further downward, for instance, and released, we expect to it oscillate up and down with slowly decreasing amplitude until it has returned to its equilibrium position.

Which factors determine the frequency of those oscillations? Their amplitude? The speed of the decay back to the rest position? We shall derive several models that begin to answer these questions.

5.1.1   An Undamped System

As in our earlier modeling experiences, we need both a physical law and an experimental fact. The former is the familiar *Newton's Law*,

$$F_t = ma \,,$$

where F_t is the total force acting on the mass m in figure 5.1 and a is its acceleration.

Introducing a signed quantity such as acceleration assumes that we have defined a coordinate axis to specify the positive and negative directions of position, velocity, and acceleration. The axis shown in figure 5.1 takes upward as positive, lets x denote the position of the mass, and chooses its *origin* as the *rest position* of the mass.

> The choice of an origin for a coordinate system is arbitrary. Two different points on the coordinate axes will still be the same absolute distance apart regardless of the location of the origin. However, some choices of origin are more convenient than others. Hindsight and experience are the best guides to good choices.

The experimental fact we shall use relates force in the spring to extension.

Hooke's Law   The force exerted by a spring is proportional to the extension from its unweighted length, and it opposes the extension.

The constant of proportionality k is called the **spring constant**. The extension of the spring is just its current, weighted length minus its original, unweighted length.

> The net extension of the spring is current length minus original length and not vice versa because positive extension corresponds to a lengthened spring.

Figure 5.2 illustrates Hooke's law. A 2 lb weight extends the spring 1 ft from its unstretched position, while a 4 lb weight stretches it 2 ft, twice as far. From the conceptual equation

$$\text{force exerted by spring} = k \times \text{extension of spring},$$

we compute

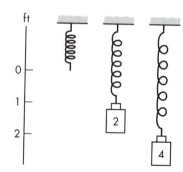

Figure 5.2:   A spring that offers a resisting force of 2 lb when it is extended 1 ft

$$k = \frac{2 \text{ lb}}{1 \text{ ft}} = 2 \text{ lb/ft}.$$

We could equally well compute $k = (4 \text{ lb})/(2 \text{ ft}) = 2 \text{ lb/ft}$.

We require a mathematical formulation of Hooke's Law. If z denotes the net extension of the spring from its *unweighted* length, then downward displacement of the end of the spring from its unweighted position corresponds to positive z. (Negative z would correspond to shortening the spring by pushing the lower end up.) Hence,

$$F_s = kz \,,$$

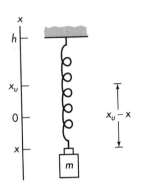

Figure 5.3: The extension of the vertical spring is $z = x_u - x$.

Extension = current length − original length.

HOOKE'S LAW

NEWTON'S LAW

Total force on mass = spring force + gravity.

Law + fact ⇒ equation.

where F_s is the force exerted by the spring on the mass and z is the net extension of the spring from its unweighted length.

This equation certainly restates the experimental fact: The force of the spring is proportional to its extension, and it acts in the direction opposite to the direction of the extension. For example, the spring force acts upward (positive) if we pull the spring downward from its unweighted position.

To use $F_s = kz$, we need an expression for the extension z. When the mass is at an arbitrary coordinate $x(t)$, the lower end of the spring has moved from its unweighted position x_u to x. Hence,

$$z = (h - x(t)) - (h - x_u) = x_u - x(t),$$

where $h - x(t)$ is the current (at time t) length and $h - x_u$ is the original, unweighted length; see figures 5.1 and 5.3.

Now we need to find x_u. The extension z caused by placing the mass at the origin is

$$z = (h - 0) - (h - x_u) = x_u.$$

The length of the spring after adding the weight is $h - 0$; the length before is $h - x_u$. (See figure 5.1.)

When the mass is at rest, its weight and the force of the spring are in equilibrium. Hence, the spring must be exerting a force mg, exactly balancing the force of gravity $-mg$ acting on the mass. Since $F_s = kz = kx_u = mg$, we have

$$x_u = \frac{mg}{k}.$$

As expected, x_u is positive; the end of the spring was above the origin before the weight was added.

Therefore, the force exerted by the spring on the mass when it is at x is

$$F_s = k(x_u - x),$$

where $x_u = mg/k$. This is the form of Hooke's law we shall use; it relates the force exerted by the spring to the position of the mass.

Let us use these laws of Newton and Hooke to model the situation shown in figure 5.1. We shall neglect air resistance, internal friction in the spring, etc. and consider only the force of the spring and the force of gravity acting on the mass.

Since acceleration is the second derivative of displacement, Newton's law is just

$$F_t = mx''(t),$$

where $x(t)$ is the position of the mass at time t. The forces acting on the mass are those of the spring and of gravity: $F_t = F_s + F_g$. From $F_s = k(x_u - x)$ and $F_g = -mg$, we obtain

$$mx''(t) = k(x_u - x(t)) - mg.$$

Substituting $x_u = mg/k$ into this last equation gives the differential equation we sought:

$$mx''(t) + kx(t) = 0.$$

To eliminate the single arbitrary constant in general solutions of first-order equations, we needed one initial condition. That experience suggests that we ought to append *two* initial conditions to a *second*-order equation to specify the motion of the mass uniquely.

What might those two initial conditions be? Physical intuition suggests that the initial position and the initial velocity of the mass are important. We expect the motion to be much different if we lift the mass up slightly and release it from rest than if we give it a strong upward push from the equilibrium position.

If x_i denotes initial position and v_i denotes initial velocity, then such initial conditions would take the form

$$x(0) = x_i, \; x'(0) = v_i.$$

While we have only physical intuition to guide us, the correct initial-value problem to model the situation shown in figure 5.1 is, in fact, the equation we just derived subject to these two initial conditions:

$$mx''(t) + kx(t) = 0. \tag{5.1}$$

$$x(0) = x_i, \; x'(0) = v_i. \tag{5.2}$$

To begin an analysis of this model, note that the differential equation provides a relation between acceleration and position,

$$x''(t) = -\frac{k}{m}x(t).$$

When the mass is above the equilibrium position — i.e., when $x(t) > 0$ — then acceleration is negative and vice versa. The magnitude of acceleration is symmetric across the point $x = 0$. If the mass is not somehow prevented from successively passing back and forth across the equilibrium position, its motion could be symmetric and repetitious.

The relation $x'' = -(k/m)x$ also tells us that, give or take a constant, x must be a function whose second derivative is its own negative, such as a sine or cosine. A little more guessing leads us to conclude that $\sin\sqrt{k/m}\,t$ and $\cos\sqrt{k/m}\,t$ are solutions of $mx'' + kx = 0$. So the spring-mass equation has periodic solutions that oscillate with frequency $\sqrt{k/m}$.

Figure 5.4 shows a graph of $\sin\sqrt{k/m}\,t$, and its shape is consistent with $x'' = -(k/m)x$. When $x(t) > 0$, then $x''(t) < 0$; i.e., the graph is concave down. The graph is concave up when $x(t) < 0$. When the graph crosses the horizontal axis, there is an inflection point.

This system is called **undamped** because, as we shall see, there is no external influence to bring the oscillating mass slowly to a halt.

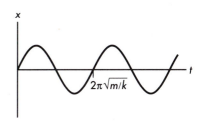

Figure 5.4: A graph of $\sin\sqrt{k/m}\,t$, one solution of $mx'' + kx = 0$.

5.1.2 A Damped System

Let us now turn our system on its side, as shown in figure 5.5, to consider a different problem. The coordinate axis is drawn so that $x = 0$ corresponds to the location of the mass when the spring is unstretched. The new ingredient to consider here is the role of friction between the mass and the surface.

The simplest experimental observation is the following:

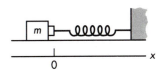

Figure 5.5: A horizontal spring-mass system; the origin is located at the rest point of the free end of the unstretched spring.

Experimental Fact The force of friction is proportional to velocity and acts to oppose the motion.

If F_f is the force of friction exerted by the surface on the mass in figure 5.5 and $v(t) = x'(t)$ is the velocity of the mass, then a mathematical formulation of the experimental fact is

$$F_f = -pv(t).$$

The constant of proportionality p is called the **friction,** or **damping coefficient**. The negative sign ensures that the force of friction opposes the motion; when $v > 0$, then $F_f < 0$ and vice versa.

> That friction should oppose the motion is obvious; that its magnitude should be proportional to velocity is perhaps less clear.
>
> Experience does suggest that the force of friction increases with velocity. (It also depends on the finish of the surfaces in contact, on the force pressing them together, and on the area in contact.) Otherwise, sliding a heavy box across the floor would not become more difficult as we tried to push it faster. (Getting it started is another matter because the force of static friction exceeds that of dynamic friction.)
>
> Choosing the force of friction to be proportional to velocity leads ultimately to a linear differential equation. Although a more complex relation between force and velocity might better fit experimental observations, it would also make our differential equation nonlinear and more difficult to analyze. The relation $F_f = -pv$, like many others, is a compromise between an accurate fit to data and analytic tractability.
>
> The constant of proportionality p has units of force/velocity; e.g., lb/(ft/s) = lb-s/ft.

Since $x = 0$ is the position of the mass when the spring is unstretched, the extension of the spring in this coordinate system is the *negative* of the position x of the mass. Hence, the spring force is

$$F_s = -kx.$$

Now we can proceed just as in the derivation of the undamped model (5.1–5.2). Ignoring all horizontal forces but that of the spring and of friction means that $F_t = F_f + F_s$. (Gravity does not appear because it acts vertically.) Using Newton's Law, $F_t = ma$, the friction force relation $F_f = -pv$, and the spring-force relation $F_s = -kx$, we find

$$mx''(t) = -pv(t) - kx(t).$$

Since $v(t) = x'(t)$, we have the governing differential equation

$$mx''(t) + px'(t) + kx(t) = 0.$$

Since $x'' = v'$, we also could have written this equation as two first-order equations,

$$x' = v,$$

$$mv' = -pv - kx,$$

instead of as a single second-order equation. We will see more of such *first-order systems* later.

Appropriate initial conditions are initial position x_i and initial velocity v_i, as in the suspended spring-mass system. Coupling them with the differential equation gives the complete model for the horizontal spring-mass system with friction shown in figure 5.5:

$$mx''(t) + px'(t) + kx(t) = 0.\qquad(5.3)$$

$$x(0) = x_i, \qquad x'(0) = v_i.\qquad(5.4)$$

This system is called **damped** because friction acts to slow the motion.

5.1.3   A Forced System

Suppose the vertical spring-mass system of figure 5.1 is hung from a ceiling. An energetic grandmother begins rocking vigorously in her chair on the floor directly above. Her motion will certainly cause the mass to move. How is its motion related to the frequency and the strength (amplitude) of the disturbances shaking the ceiling? The model derived here will allow us to explore the effect of such external forcing functions.

Consider the situation shown in figure 5.6, where the motion of grandmother's rocking chair imparts a displacement $h(t)$ to the top of the spring at time t. We proceed as in our first derivation, neglecting air resistance and considering only the force of gravity and the force of the spring.

When the ceiling was stationary, we computed the force of the spring on the mass by finding the extension of the spring. Once again, we compare the current length of the spring with its original unstretched length to find its net extension and, hence, the force it exerts on the mass.

The position of the top of the spring at $t = 0$ is just $h(0)$, and the coordinate of the lower end of the unstretched spring is $x_u = mg/k$, as we found before. Hence, the unstretched length of the spring is

$$h(0) - x_u = h(0) - \frac{mg}{k}.$$

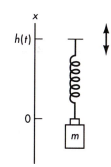

Figure 5.6: The position of the upper end of the vertical spring-mass system varies with time; the coordinate of the upper end of the spring at time t is $h(t)$.

At any instant t, the top of the spring will be at position $h(t)$ while the mass will be at $x(t)$. Thus, the length of the spring at time t is $h(t) - x(t)$, and the net extension of the spring is

$$z = [h(t) - x(t)] - \left[h(0) - \frac{mg}{k}\right].$$

Extension = current length − original length.

From the Hooke's law relation $F_s = kz$, the net force exerted by the spring on the mass is thus

$$F_s = k\left[h(t) - h(0) - x(t) + \frac{mg}{k}\right].$$

The rest of the derivation is straightforward. Since we are considering only the forces of the spring and of gravity, the total force on the mass is

Total force = gravity + spring.

$$F_t = F_g + F_s$$

Using $F_t = mx''(t)$ yields

$$mx''(t) = -mg + k\left[h(t) - h(0) - x(t) + \frac{mg}{k}\right]$$

$$= k[h(t) - h(0)] - kx(t)$$

or

$$mx''(t) + kx(t) = k[h(t) - h(0)].$$

The model for the spring-mass system of figure 5.6 is completed by adding the same set of initial conditions we used before the grandmother's rocker appeared on the scene:

$$mx''(t) + kx(t) = k(h(t) - h(0)). \tag{5.5}$$

$$x(0) = x_i, \qquad x'(0) = v_i. \tag{5.6}$$

FORCED SPRING-MASS MODEL

This system is called **forced** because the movement of the rocker can force the system into motion through the displacement of the upper end of the spring, even if the mass itself is initially in its equilibrium position.

5.1.4 Examples

Suppose we are given the spring shown in Figure 5.2. We observe from that figure that adding a 2-lb weight stretches the spring one foot; its spring constant is $k = (2\text{ lb})/(1\text{ ft}) = 2$ lb/ft.

EXAMPLE 1 *A 6-lb weight is suspended from the spring of figure 5.2. Write an initial-value problem that governs the motion of the mass if it is started from its equilibrium position with a downward velocity of 6 in/s. Neglect air resistance.*

The undamped model (5.1–5.2),

$$x'' + kx = 0, \qquad x(0) = x_i, \qquad x'(0) = v_i,$$

of the first subsection applies; the coordinate system is shown in figure 5.1. We have seen that the spring constant is $k = 2$ lb/ft. The mass is $m = \frac{6}{32} = \frac{3}{16}$ slug. The initial position is $x(0) = 0$. The initial velocity is $v(0) = -6$ in/s $= -0.5$ ft/s. The governing initial-value problem is

$$\frac{3}{16}x''(t) + 2x(t) = 0, \qquad x(0) = 0, \qquad x'(0) = -0.5. \qquad \blacksquare$$

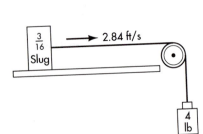

Figure 5.7: The falling 4-lb weight pulls the mass across the table at a constant speed of 2.84 ft/s.

EXAMPLE 2 *The mass of the preceding example is placed on a table as in figure 5.7. After some experimentation, we discover that the falling weight of 4 lb shown there pulls the mass of $\frac{3}{16}$ slug at a steady speed of 2.84 ft/s. The spring of the preceding example is then connected horizontally between this mass and a fixed anchor as in figure 5.5. The spring is stretched 9 in, and the mass is released. Write an initial value problem that governs the resulting motion. Include the effects of friction.*

The damped model (5.3–5.4),

$$x'' + px' + kx = 0, \qquad x(0) = x_i, \qquad x'(0) = v_i,$$

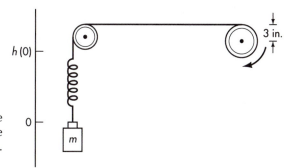

Figure 5.8: The turntable placed on its side oscillates the upper end of the vertical spring. It is shown here at $t = 0$.

of the second subsection applies. We shall use the coordinate system of figure 5.5. We know that $m = \frac{3}{16}$ slug and k = 2 lb/ft. The data of the experiment shown in Figure 5.7 reveals that the friction coefficient is

$$p = (4 \text{ lb})/(2.84 \text{ ft/s}) = 1.41 \text{ lb-s/ft}.$$

The initial position corresponds to stretching the spring 9 in.; i.e., $x(0) = 0.75$ ft. If the mass is simply released, its initial velocity is zero: $x'(0) = 0$.
The governing initial-value problem is

$$\frac{3}{16}x''(t) + 1.41x'(t) + 2x(t) = 0, \qquad x(0) = 0.75, \qquad x'(0) = 0. \quad \blacksquare$$

EXAMPLE 3 *The spring of the first example and a 4-lb weight are suspended as shown in figure 5.8. The end of the cable supporting the spring is connected 3 in. from the center of an old 45-rpm phonograph turntable placed vertically. Figure 5.8 shows the position of the assembly at $t = 0$; the mass is at rest. Neglect air resistance and write an initial-value problem that governs the motion of the mass.*

The forced model (5.5)–(5.6),

$$mx''(t) + kx(t) = k(h(t) - h(0)), \quad x(0) = x_i, \quad x'(0) = v_i,$$

of the third subsection obviously applies. The mass is $m = \frac{1}{8}$ slug, and the spring constant is $k = 2$ lb/ft. The challenge here is finding the position $h(t)$ of the top of the spring at time t.

Figure 5.9 shows the turntable after it has turned through an angle θ. If we neglect the vertical displacement of the cable, then the end attached to the turntable has moved to the right $3 \sin \theta$ in. or $\frac{1}{4} \sin \theta$ ft; i.e., the upper end of the spring has been raised so that its position is $h(t) = h(0) + \frac{1}{4} \sin \theta$. We see that $h(t) = h(0) + \frac{1}{4} \sin \theta(t)$, where $\theta(t)$ is the angular rotation of the turntable at time t.

The turntable is spinning at 45 rpm (revolutions per minute) or 90π rad/min = $3\pi/2$ rad/s, since one revolution covers 2π rad. If t is measured in seconds, then $\theta(t) = 3\pi t/2$, and $h(t) = h(0) + \frac{1}{4} \sin(3\pi t/2)$.

We are told that figure 5.8 shows the position of the mass at $t = 0$, when the mass is at rest. Both initial position and initial velocity must be zero.

Because $k = 2$, the governing initial value problem is

$$\frac{1}{8}x''(t) + 2x(t) = \frac{1}{2} \sin \frac{3\pi t}{2}, \qquad x(0) = 0, \qquad x'(0) = 0.$$

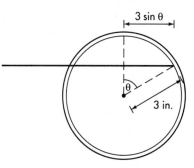

Figure 5.9: After a brief interval of time, the turntable of the previous figure has rotated through an angle θ to the position shown here. The end of the cable connected to the turntable has moved $3 \sin \theta$ in. to the right.

We were careful to convert units of distance to feet and units of time to seconds in this model because we used $g = 32$ ft/s^2 to convert the given weight of 4 lb to a mass of $\frac{1}{8}$ slug. The units in that constant dictated our choice of units for length and time.

Note that the initial location $h(0)$ of the upper end of the spring does not appear in this model. The important quantity is evidently the net displacement of the end of the spring. ∎

5.1.5 Exercises

EXERCISE GUIDE

To gain experience . . .	Try exercises
Deriving spring-mass models	7–9, 10(b–c), 11, 12(c), 13
Writing and using spring-mass models	1, 6, 10(d)
With force relations (Hooke, damping, etc.)	2, 10(a), 12(a–b)
With solutions of spring-mass models	3–4
Analyzing and interpreting solution behavior	5

1. When a mass of 3 kg is suspended from a spring, the end of the spring moves 0.24 m. A force of 0.36 Newtons is required to drag a mass of 7 kg across a table top at a speed of 0.4 m/s.
   (a) Suppose a 2-kg mass is suspended vertically from the spring (as in figure 5.1) and that the mass is set in motion by pulling it down 4 cm and releasing it from rest. Write an initial-value problem whose solution will describe the position of the mass at any time t.
   (b) Suppose the spring is connected to the 7-kg mass on the tabletop, as shown in figure 5.5, and the mass is set in motion by giving it a speed of 0.2 m/s to the right. Write an initial-value problem whose solution will describe the position of the mass at any time t.
   (c) The upper end of the suspended system of part (a) is now connected to the turntable of figure 5.8. Write an initial-value problem whose solution will describe the position of the mass at any time t if the mass is set in motion as in part (a).

2. Does a rubber band obey Hooke's law? Conduct an experiment using coins or similar objects as weights.

3. The undamped spring-mass equation can be written

$$x'' = -\frac{k}{m}x.$$

Since x has to be the negative of its second derivative, the text concludes that x could be a sine or a cosine function. Determine the precise form of those functions by assuming that $x = \sin rt$ and substituting into the differential equation. Obtain a (characteristic) equation for r and solve it to determine a functional form for x. Repeat for the cosine function.

4. The text argues that the undamped spring-mass equation $mx'' + kx = 0$ has the solutions $\sin\sqrt{k/m}\,t$ and $\cos\sqrt{k/m}\,t$.
   (a) Verify this claim by substituting these proposed solutions into the differential equation.
   (b) Show that $C_1 \sin\sqrt{k/m}t + C_2 \cos\sqrt{k/m}t$ is also a solution for any arbitrary constants C_1, C_2. (You are verifying the *principle of superposition* for this second-order equation.)
   (c) Solve the undamped spring-mass model

$$mx'' + kx = 0, \qquad x(0) = x_i, \qquad x'(0) = x_i,$$

   by choosing C_1, C_2 in the solution of part (b) to satisfy the given initial conditions.

5. The text claims that the undamped spring-mass equation $mx'' + kx = 0$ has $\sin\sqrt{k/m}\,t$ as a solution.
   (a) Sketch a graph of this function for $0 \le t \le 2\pi/\sqrt{k/m}$.

(b) How many cycles of the sine function does your graph contain?

(c) What is the period of this function? What does this model predict to be the period of oscillation of the mass?

(d) Does making the spring stiffer cause the mass to oscillate faster or slower? What happens if the mass is changed? If your car bounces too rapidly on a rough road, should you replace its springs with stiffer ones or weaker ones? If you can not change the springs, should you ask your passengers to get out and walk or should you fill the trunk with rocks?

6. The forced spring-mass equation (5.5) has the form

$$mx'' + \cdots = k(h - \cdots).$$

(a) What are the units of kh?

(b) Someone hands you the equation $6x'' + 2x = A \sin \omega t$. What might it model? Describe the physical situation as specifically as you can. What does A represent?

7. The text derives the model

$$mx'' + kx = k(h(t) - h(0)),$$
$$x(0) = x_i, \qquad x'(0) = v_i,$$

for the forced, undamped vertical spring-mass system. It also considered an unforced spring-mass system that was damped by friction. It obtained the model

$$mx'' + px' + kx = 0, \qquad x(0) = x_i, \qquad x'(0) = v_i.$$

In fact, even a freely suspended mass is subject to damping due to air resistance and internal resistance in the spring. These damping forces can be modeled as being proportional to velocity and acting in the opposite direction, just like the force of friction acting on the mass sliding horizontally.

(a) Compare the undamped, forced model with the damped, unforced model. Predict what a model of a forced, damped system would look like.

(b) Verify your prediction by deriving such a model. Specifically, consider the system shown in figure 5.6 and suppose that the mass is subject to air resistance which is proportional to its velocity and acts in the opposite direction.

8. Figure 5.1, which shows the vertical spring-mass system, places the mass below the fixed end of the spring. Place the fixed end of the spring below the mass, like the spring in a stationary pogo stick, and derive a model for this situation.

9. Figure 5.5, which shows the horizontal spring-mass system, has the fixed end of the spring to the right of the mass. Place the fixed end of the spring to the left of the mass and derive a model for this situation.

10. A shock absorber in an automobile can be thought of as a damped version of the vertical spring-mass system of figure 5.1. The mass of the car is m, and it is attached to a piston moving in a chamber filled with hydraulic fluid. See figures 5.10, 5.11. The hydraulic fluid exerts a force on the mass that is proportional to its velocity and *opposes* its motion.

(a) Write a mathematical expression for the force the hydraulic fluid exerts on the moving mass.

(b) Use the expression in part (a) to derive a model of the automobile shock absorber system shown in figure 5.11. Be sure to define a coordinate system.

(c) Derive a model for this same system if the fixed (lower) end of the spring in figure 5.11 is moved to the position $h(t)$ at time t by driving the car over a rough road.

(d) What other model(s) considered in this section have the same mathematical structure as the one you just derived? Would a model of this form be suitable if we wanted to include the effects of air resistance or internal friction in the spring? Would you expect any differences in the relative values of the damping coefficients when you compare a shock absorber model with one involving air resistance?

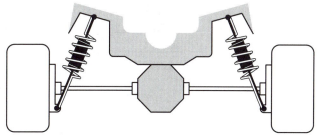

Figure 5.10: A view of a sport car's rear suspension, showing the damping shock absorbers inside coil springs.

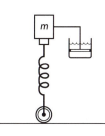

Figure 5.11: A schematic model of an automobile shock absorber system. The lower end of the spring is attached to a wheel of the car. The piston moves in oil to damp the motion of the mass m, which represents the automobile.

11. Suppose the anchor for the horizontal spring-mass system of figure 5.5 is replaced by the turntable of figure 5.8, as shown in figure 5.12. Derive a model for this situation. (Neglect the vertical displacement of the end of the spring, as in the last example in this section.)

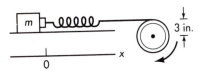

Figure 5.12: The vertical turntable of figure 5.8 is now oscillating the end of the horizontal spring-mass system that was formerly fixed. The spring is attached to the turntable 3 in. from its center.

12. I clamped a 12-in. hacksaw blade in a horizontal position and measured the height of its free end above my work table as I placed quarters on that end. The situation is shown in figure 5.13, and my data appears in the following table.

Number of Quarters	Height Above Table
0	$6\frac{5}{8}$ in.
1	$6\frac{1}{4}$ in.
2	$5\frac{7}{8}$ in.
3	$5\frac{1}{2}$ in.
4	$5\frac{3}{16}$ in.

(a) Is it reasonable to suppose that the displacement of the end of the hacksaw blade is proportional to the force applied? Is such a relationship exact here?

(b) If the displacement is proportional to the force applied, find the proportionality constant.

(c) Derive a model for the position of the end of the hacksaw blade when three quarters are taped to it and it is released from rest from a position 6 in. above the work table. Be sure to define a coordinate system.

13. The left end of the hacksaw blade of figure 5.13 is now being oscillated gently about its old fixed position of $6\frac{5}{8}$ in. above the work table. Specifically, its height as a function of time is

$$h(t) = 6.625 + 0.5 \sin \omega t,$$

where the frequency ω is constant.

(a) How far above its old fixed position does the left end of the hacksaw blade move?

(b) Two quarters are taped to the right end of the blade. Derive a model for the position of this end of the blade if the quarters are at rest before the left end of the blade begins to move. (See exercise 12.)

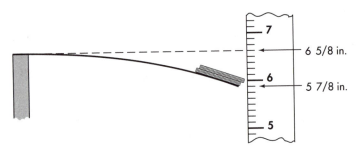

Figure 5.13: A hacksaw blade deflects from its vertical position (shown dashed) when two quarters are laid on its free end.

5.2 RLC CIRCUITS

The mechanical spring-mass systems of the previous section have a precise electrical analogue, an RLC circuit, a circuit containing a resistance R, an inductance L, and a capacitance C. Figure 5.14 illustrates a series RLC circuit, which also includes an *imposed emf* (electromotive force) or an *independent voltage source* E. The charge on the capacitor in this circuit is the analogue of the displacement of a mass attached to a spring.

We begin with a brief summary of the important physical ideas.

Current is the rate of flow of charged particles, and current flows between two points in a circuit because of a potential difference between those two points.

The unit of charge is the *coulomb* (C). The unit of current is the *ampere* (A), 1 C/s.

Electric potential is the ability to move charged particles, and potential differences are measured in *volts* (V). For example, the imposed voltage E in figure 5.14 could be a battery. It lifts or pumps positively charged particles from its negative (lower potential) end to its positive (higher potential) end. The charged particles at its positive end are prepared to flow "downhill" toward its negatively charged end.

The situation is like placing a bucket of water on top of a table, for electrical potential is analogous to mechanical potential energy. The water in the bucket in figure 5.15 flows downhill toward the lower end of the hose, the lower potential part of this fluid circuit. Lifting the bucket onto the table, raising its potential, is like charging a battery.

The narrow constriction in the hose is a resistance. It slows the flow of water and dissipates its potential energy just as the resistor R in the circuit of figure 5.14 slows the flow of charged particles, causing a potential drop. The faster the rate of flow, the greater the potential drop. A precise statement follows.

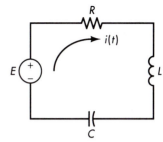

Figure 5.14: A series RLC circuit with an independent voltage source E.

Experimental Fact 1 The potential drop across a resistor is proportional to the rate of flow.

The proportionality constant is called the **resistance** R. If q denotes charge, then dq/dt is the rate of flow of charged particles, or **current** i. We can write the potential change or voltage drop v_R across a resistor as

$$v_R = R\frac{dq}{dt} = iR\,.$$

> The potential that resistors dissipate appears as heat and light energy in toasters, light bulbs, etc.

The capacitor C in figure 5.14 is also a cause of a potential drop. It consists of two conducting plates separated by an insulator, or *dielectric*. The dielectric is a kind of dam for the flow of charged particles. When too much charge builds up on one side of the dam, charged particles back up through the circuit, trying to reach the other side.

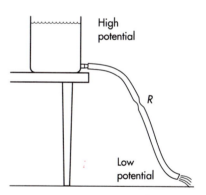

Figure 5.15: The bucket of water on the table is like a charged battery.

Experimental Fact 2 The potential drop across a capacitor is proportional to the charge that has accumulated on the capacitor.

The reciprocal of the constant of proportionality is the **capacitance** C. We can write the potential change or voltage drop v_C across a capacitor as

$$v_C = \frac{1}{C}q\,.$$

> Like a dam holding water in a river, capacitors can store charge. When Ben Franklin flew his kite in a lightning storm, he stored the charge that traveled down the string of his kite in a Leyden jar, a capacitor.

Inductors are coils of wire. They resist the flow of charged particles through a kind of feedback mechanism. Changing currents in the coil induce magnetic fields that in turn induce an opposing current in the inductor. It's a bit like turning some of the flow in a hose back on itself, but it is the *rate of change* of the flow that provides the resistance.

Experimental Fact 3 The potential drop across an inductor is proportional to the rate of change of current in the inductor.

The constant of proportionality is the **inductance** L. We can write the potential change or voltage drop v_L across an inductor as

$$v_L = L \frac{di}{dt} .$$

Since changing electric currents produce magnetic fields, magnetic compasses are useless near electric motors. Indeed, it is precisely those magnetic fields that turn the flow of charged particles into mechanical energy. Conversely, the charged particles that flow in a video display tube sometimes can be affected by the magnetic field from a nearby, powerful loud speaker.

These three experimental facts are summarized in table 5.1.

The basic physical law, the analogue of Newton's law, is actually a statement of conservation of energy in disguise:

Kirchhoff's Voltage Law The sum of the voltage changes around a closed loop is zero.

To use Kirchhoff's voltage law, we simply traverse the loop once in a chosen direction, accounting for each of the voltage changes we encounter.

Traveling the circuit from the negative side of the battery to its positive side, the voltage change is positive $(+E)$ because the battery can pump positively charged particles from its negative side to its positive side. Recalling the water analogy of figure 5.15, we can think of the battery as picking up charged particles from the floor (negative potential) and placing them on the table (positive potential) so that they can flow through the hose. The three experimental facts in Table 5.1 tell us how to account for the voltage drops at the resistor, inductor, and capacitor.

Traveling the circuit of figure 5.14 in a clockwise direction and summing voltage drops ($-E$ at the battery,$\cdots$), Kirchhoff's voltage law yields

$$-E + v_R + v_L + v_C = 0 .$$

Substituting from the experimental facts, we have

$$-E + Ri + L \frac{di}{dt} + \frac{1}{C} q = 0 .$$

TABLE 5.1

Voltage drops across a resistor, a capacitor, and an inductor, as well as the units used to measure the size of each

Circuit Element	Units	Voltage Drop
Resistance R	Ohm (Ω)	$v_R = R dq/dt$
Capacitance C	Farad (F)	$v_C = q/C$
Inductance L	Henry (H)	$v_L = L di/dt$

Since current is the rate of change of charge ($i = dq/dt$), we can write a pair of first-order differential equations:

$$\frac{dq}{dt} = i$$

$$L\frac{di}{dt} = E - Ri - \frac{1}{C}q\,.$$

Alternatively, we can substitute $i = q'$ and replace the first-order system with a single second-order equation

$$Lq'' + Rq' + \frac{1}{C}q = E\,.$$

The mechanical analogy is now evident. Comparing this RLC circuit equation with the equation for a damped, forced spring-mass system,

$$mx'' + px' + kx = k(h(t) - h(0))\,,$$

we see that resistance R is the analogue of the friction coefficient p, charge q is the analogue of displacement x, and so on. The exercises ask you to pursue this analogy in detail.

> We used Kirchhoff's voltage law just like Newton's law. When we modeled spring-mass systems in the previous section, we accumulated all the forces on the mass and set them equal to mass times acceleration. Experimental facts told us how forces due to friction and the spring depended on position and velocity. Here Kirchhoff's voltage law told us how to accumulate the various voltage drops around the circuit (they sum to zero), and the experimental facts told us how to express those voltage drops in terms of charge and current.

Adding a specification of initial charge q_i on the capacitor and initial current i_i completes the model:

$$Lq'' + Rq' + \frac{1}{C}q = E, \qquad q(0) = q_i, \qquad q'(0) = i_i\,.$$

> Since the charge on the capacitor is proportional to the voltage drop across it, an alternate form of the initial condition on charge is $q(0) = Cv_c(0)$.

Since $q' = i$, a derivative of the charge equation yields an equation for the current in this loop,

$$Li'' + Ri' + \frac{1}{C}i = E'\,.$$

Reasonable initial conditions for this equation are the current at $t = 0$, $i(0) = i_i$, and its rate of change at $t = 0$. From the third experimental fact, we can express $i'(0)$ in terms of the initial voltage drop across the inductor,

$$Li'(0) = v_L(0)\,.$$

Hence, a model for the current in the series RLC circuit is

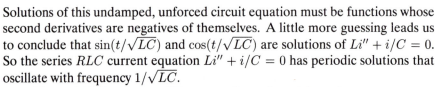

$$Li'' + Ri' + \frac{1}{C}i = E', \qquad i(0) = i_i, \qquad i'(0) = v_L(0)/L.$$

The two models—charge and current—have the same mathematical form.

To gain a feel for the behavior of these two analogous equations, eliminate sources and resistance by setting $E = 0$ and $R = 0$. Then the current equation, for example, can be written

$$i'' = -\frac{1}{LC}i.$$

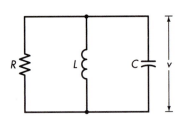

Figure 5.16: A source-free, parallel RLC circuit.

Solutions of this undamped, unforced circuit equation must be functions whose second derivatives are negatives of themselves. A little more guessing leads us to conclude that $\sin(t/\sqrt{LC})$ and $\cos(t/\sqrt{LC})$ are solutions of $Li'' + i/C = 0$. So the series RLC current equation $Li'' + i/C = 0$ has periodic solutions that oscillate with frequency $1/\sqrt{LC}$.

Figure 5.16 illustrates a different configuration of these three components, a source-free, parallel RLC circuit, the dual of the series RLC circuit with $E = 0$. The exercises ask you to derive a governing equation for this circuit using the following statement of conservation of charge.

Kirchhoff's Current Law The sum of the currents into any connection point or *node* of a circuit is zero.

The equation you obtain there is

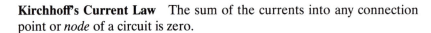

$$Cv'' + \frac{1}{R}v' + \frac{1}{L}v = 0.$$

It is obviously analogous to the series current and charge equations derived before.

> The circuit in figure 5.14 is a *series* circuit because current flows serially from one component to the next; the current flow in each component is the same. The circuit in figure 5.16 is a *parallel* circuit because the current flows in parallel, or simultaneously, through all the components; the voltage drop across all components is the same.
>
> The voltage source in figure 5.14 is called *independent* because the voltage drop across its terminals is independent of the current flowing through it.

5.2.1 Exercises

EXERCISE GUIDE

To gain experience . . .	Try exercises
With electrical and mechanical analogies	1
Deriving circuit models	5–6, 8
With solutions of circuit models	2–4
Analyzing and interpreting solution behavior	7–8

1. The text claims there is an obvious analogy between RLC circuits modeled by the equation

$$Lq'' + Rq' + \frac{1}{C}q = E,$$

and forced, damped spring-mass systems modeled by

$$mx'' + px' + kx = k(h(t) - h(0)).$$

Verify this claim by finding the analogue of L, R, C, E, q, q'. Explain the physical significance of each analogy. Summarize your results in a table.

2. The series LC current equation can be written

$$i'' = -\frac{1}{LC}i.$$

Since i has to be the negative of its second derivative, the text concludes that i could be a sine or a cosine function. Determine the precise form of those functions by assuming that $i = \sin rt$ and substituting into the differential equation. Obtain a (characteristic) equation for r and solve it to determine a functional form for i. Repeat for the cosine function.

3. The text argues that the unforced series LC current equation $Li'' + i/C = 0$ has the solutions $\sin(t/\sqrt{LC})$ and $\cos(t/\sqrt{LC})$.
   (a) Verify this claim by substituting these proposed solutions into the differential equation.
   (b) Show that $C_1 \sin(t/\sqrt{LC}) + C_2 \cos(t/\sqrt{LC})$ is also a solution for any arbitrary constants C_1, C_2. (You are verifying the *principle of superposition* for this second-order equation.)
   (c) Solve the unforced LC circuit model

$$Li'' + \frac{1}{C}i = 0, \qquad i(0) = i_i, \qquad i'(0) = \frac{v_L(0)}{L},$$

   by choosing C_1, C_2 in the solution of part (b) to satisfy the given initial conditions.

4. The text claims that the unforced LC circuit equation $Li'' + i/C = 0$ has $\sin(t/\sqrt{LC})$ as a solution.
   (a) Sketch a graph of this function for $0 \le t \le 2\pi\sqrt{LC}$.
   (b) How many cycles of the sine function does your graph contain?
   (c) What is the period of this function? What does this model predict to be the period of oscillation of the charge on the capacitor in this circuit?

(d) Would increasing the size of the capacitor cause a graph of the oscillations in current versus time to be pushed tighter together or to be more spread out? What if the inductor is enlarged, increasing L by adding more turns to the coil?
(e) Suppose the inductor is actually the magnet coil in a speaker. How would the changes described in the previous part affect the sound you hear?

The following three problems first appeared in the chapter 2 exercises. Now that you know more about RC circuits, you can have a second chance at them.

5. Use the experimental facts in the text and Kirchhoff's voltage law to derive a model for the charge q on the capacitor in the RC circuit shown in figure 5.17. These steps will lead you to the desired model:
   (a) Write a mathematical statement of each experimental fact.
   (b) Write Kirchhoff's voltage law in terms of v_R, v_C.
   (c) Combine the results of the preceding two steps to derive the governing differential equation

$$R\frac{dq}{dt} + \frac{1}{C}q = 0.$$

   Complete the model for a series RC circuit by imposing an appropriate initial condition.

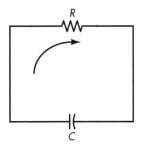

Figure 5.17: A source-free, series RC circuit

6. Suppose the RC circuit of the preceding problem now includes a voltage source (imposed emf) E. Derive the following model for this circuit:

$$R\frac{dq}{dt} + \frac{1}{C}q = E, \ q(0) = q_i.$$

See figure 5.18.

7. Analyze the models of the RC circuits in the two preceding problems by answering the following questions for each:
   (a) By studying the sign of dq/dt in the differential equation, determine whether a given initial charge q_i on the capacitor will grow or decay. (Both R and C are positive constants.)

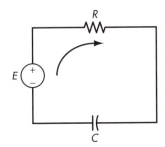

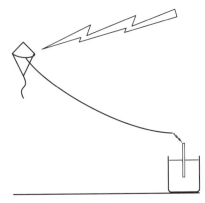

Figure 5.18: A series RC circuit with a voltage source E

(b) By studying the solution of each initial-value problem, determine whether a given initial charge q_i on the capacitor will grow or decay.

(c) If charge stops flowing in either circuit, what is the steady-state charge on the capacitor?

8. The preceding problems deal with RC circuits. Figure 5.19 shows an unusual RC circuit, Ben Franklin's kite flying in a lightning storm. The lightning arises from a voltage difference between the atmosphere and the kite. Current can flow down the wet kite string to the Leyden jar, a kind of capacitor, on the

Figure 5.19: A schematic diagram of Ben Franklin's kite flying in a lightning storm; charge accumulates in the Leyden jar at the lower right.

ground. In principle, you could calculate the capacitance of the Leyden jar and the resistance of the kite string. Derive a model that could help determine the potential difference that caused the lightning bolt. Describe how you would use the model to estimate this potential difference.

5.3   THE PENDULUM

The pendulum in a grandfather clock swings to and fro with comforting regularity, releasing the escapement mechanism at precisely set intervals so the clock can keep time accurately. What determines the period (the elapsed time) of an oscillation? Is it the size of the weight? The length of the pendulum? How hard the weight is pushed when the clock is started? A model of the motion of a pendulum will answer some of those questions.

Figure 5.20 shows a diagram of a pendulum. A mass m is suspended from an arm of length L, whose mass we will neglect. The arm is attached to a pivot which we will assume is frictionless. The angle of the arm from the vertical is denoted θ, and it is measured counter-clockwise in *radians*. (A positive angle is motion to the right; negative, to the left.)

As figure 5.21 suggests, the instantaneous motion of the mass is tangent to its circular path. The displacement of the mass is the distance $s = L\theta$ it has traveled along that circular path.

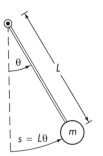

Figure 5.20: A diagram of a pendulum.

If θ is measured in radians, then the distance described by an angle θ along the circumference of a circle of radius L is just $L\theta$; e.g., if $\theta = 2\pi$, then $s = 2\pi L$, the circumference of the circle.

Since L is constant, the velocity of the mass is $v = L d\theta/dt$ and its acceleration is $a = L d^2\theta/dt^2$. We emphasize that velocity and acceleration are *tangent* to the path of the mass.

THE LAW

To derive a model, we will need Newton's Law,

$$F_t = ma\,,$$

THE FACT

and a familiar experimental fact.

Experimental Fact   The force of gravity is proportional to mass.

But since all motion is tangent to the path of the mass, we must consider the components of acceleration and gravity force in the tangential direction.

In the up-is-positive orientation of Figure 5.21, the force of gravity acting *tangent* to the path of the mass is $F_g = -mg\sin\theta$. We have argued before that the acceleration of the mass in the tangential direction is $L\theta''$. Equating total force in Newton's law to the force of gravity, the only force acting on the mass, yields

FACT AND LAW COMBINED

$$mL\theta'' = -mg\sin\theta$$

or

$$L\theta'' + g\sin\theta = 0\,.$$

We can impose two initial conditions, the angle θ_i and the angular velocity ω_i of the pendulum at the instant it was released.

> A common notation for angular velocity is ω, similar to v for linear velocity. Angular velocity is usually measured in radians per unit time; an angular velocity of 2π rad/min is 1 rev/min.

The complete model for the pendulum is

PENDULUM MODEL

$$L\theta'' + g\sin\theta = 0,\qquad \theta(0) = \theta_i,\ \theta'(0) = \omega_i\,.$$

Since the mass m of the pendulum bob appears nowhere in the model, we conclude that the frequency of the ticks in a grandfather clock are independent of the size of the bob. But the length of the arm and the acceleration of gravity do affect the motion.

> If you study a clock's pendulum, you will often find an adjustment screw below the mass to permit small changes in the length of the pendulum arm.
> Since the acceleration of gravity g also affects the motion of a pendulum, grandfather clocks would not be good time-keepers on the space shuttle, even if size were not an issue.

Although we have not formally defined *second-order linear equation*, your experience with first-order equations should persuade you that the pendulum equation $L\theta'' + g\sin\theta = 0$ is probably not linear. But we can linearize this equation using the approximation $\sin\theta \approx \theta$; e.g., $\sin(0.1\,\mathrm{rad}) = 0.0998334\cdots$. Then we obtain the approximate model

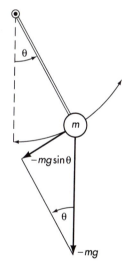

Figure 5.21:   The gravity force $-mg$ $\sin\theta$ acts tangent to the path of the mass when the pendulum is at angle θ.

APPROXIMATE PENDULUM MODEL

$$L\theta'' + g\theta = 0,\qquad \theta(0) = \theta_i,\qquad \theta'(0) = \omega_i\,.$$

> The basis for the approximation $\sin\theta \approx \theta$ is the same as that of linear stability analysis: We hope we can neglect powers of small terms. In this case, we are neglecting the terms $\theta^3, \theta^5, \dots$ in the Taylor polynomial expansion of $\sin\theta$,

$$\sin\theta = \theta - \frac{1}{3!}\theta^3 + \frac{1}{5!}\theta^5 + \cdots.$$

This approximate pendulum model has exactly the same mathematical form as the undamped spring-mass model (5.1–5.2),

$$mx'' + kx = 0, \ x(0) = x_i, \ x'(0) = v_i.$$

The exercises ask you to make precise the analogy between the two.

Like the undamped spring-mass equation, the approximate pendulum equation can be written to suggest what its solutions might be:

$$\theta'' = -\frac{g}{L}\theta.$$

Give or take a constant, θ must be a function whose second derivative is its own negative, such as a sine or cosine. A little more guessing leads us to conclude that $\sin \sqrt{g/L}\, t$ and $\cos \sqrt{g/L}\, t$ are solutions of $L\theta'' + g\theta = 0$. So the approximate pendulum equation has periodic solutions that oscillate with frequency $\sqrt{g/L}$.

By ignoring friction in the pivot and by simplifying the original model equation, we have determined (at least approximately) the frequency of oscillation of a simple pendulum. Are you ready to design a clock?

5.3.1 Exercises

EXERCISE GUIDE

To gain experience . . .	Try exercises
With mechanical analogies	1
Deriving pendulum models	7–9
With solutions of pendulum models	2–4
With the accuracy of the approximate model	5(a), 6
Analyzing and interpreting solution behavior	5(b)

1. The text claims that the approximate pendulum model

$$L\theta'' + g\theta = 0, \qquad \theta(0) = \theta_i, \qquad \theta'(0) = \omega_i$$

and the undamped spring-mass model (5.1–5.2),

$$mx'' + kx = 0, \qquad x(0) = x_i, \qquad x'(0) = v_i$$

are analogous. Verify this claim by finding the analogue of $L, g, \theta, \theta', \theta'', \theta_i$, and ω_i. Explain the physical significance of each analogy. Summarize your results in a table.

2. The approximate pendulum equation can be written

$$\theta'' = -\frac{g}{L}\theta.$$

Since θ has to be the negative of its second derivative, the text concludes that θ could be a sine or a cosine function. Determine the precise form of the sine function by assuming that $\theta = \sin rt$ and substituting into the differential equation. Obtain a (characteristic) equation for r and solve it to determine a functional form for θ. Repeat for the cosine function.

3. The text argues that the approximate pendulum equation $L\theta'' + g\theta = 0$ has the solutions $\sin \sqrt{g/L}\, t$ and $\cos \sqrt{g/L}\, t$.

(a) Verify this claim by substituting these proposed solutions into the differential equation.

(b) Show that $C_1 \sin \sqrt{g/L}\, t + C_2 \cos \sqrt{g/L}\, t$ is also a solution for any arbitrary constants C_1 and C_2. (You are verifying the *principle of superposition* for this second-order equation.)

(c) Solve the approximate pendulum model

$$L\theta'' + g\theta = 0, \qquad \theta(0) = \theta_i, \qquad \theta'(0) = \omega_i.$$

by choosing C_1 and C_2 in the solution of part (b) to satisfy the given initial conditions.

(d) Do the initial values θ_i and ω_i affect the frequency of the pendulum's oscillations? Is the way we start a clock's pendulum important?

4. The text claims that the approximate pendulum equation $L\theta'' + g\theta = 0$ has $\sin\sqrt{g/L}\,t$ as a solution.

(a) Sketch a graph of this function for $0 \le t \le 2\pi/\sqrt{g/L}$.

(b) How many cycles of the sine function does your graph contain?

(c) What is the period of this function? What does this model predict to be the period of oscillation of the pendulum?

(d) Does lengthening its pendulum cause a grandfather clock to run faster or slower? If the clock were moved to the moon where the gravity constant g is smaller, would the clock run faster or slower?

5. A typical grandfather clock might have a pendulum about 30 in. long. There is usually room for it to swing about 6 in. on either side of vertical.

(a) How accurate is the approximation $\sin\theta \approx \theta$ for such a pendulum? (A Taylor polynomial can provide a general expression for the error.)

(b) Use the results of the previous problem to deter-

mine the period of the oscillations of such a pendulum. Is this period consistent with your observations?

6. For what range of values of θ is the approximation $\sin\theta \approx \theta$ accurate to within 1%? 5%? What are these limiting angles in degrees?

7. The text derived the pendulum model using radians to measure the angular deviation of the pendulum from vertical.

(a) Carefully derive the same model with angular deviation measured in *degrees*. (In the interest of good notation, you should give the angle measured in degrees a name other than θ, such as φ.)

(b) Show that your model and the text's are equivalent by changing the dependent variable in your model to an angle measured in radians; e.g., substitute $\varphi = 180\,\theta/\pi$.

8. The text neglected friction in the pivot bearing when it derived the pendulum model. Assume that the force of friction in the pivot is proportional to angular velocity θ' and acts to oppose the motion. Derive the **damped pendulum model**

$$L\theta'' + p\theta' + g\sin\theta = 0, \qquad \theta(0) = \theta_i, \qquad \theta'(0) = \omega_i.$$

What precisely does the term $p\theta'$ represent?

9. The previous problem asks you to derive a model for a pendulum that includes friction in the pivot. Find the corresponding approximate model.

5.4 INTERCONNECTED SYSTEMS

Many of the subjects we have modeled—springs and masses, electric circuits, and even populations—occur not in isolation but as one component of larger systems.

When you ride in a car, you are part of a system that includes your mass and that of the car as well as the springs in the suspension and the springs in the seat. These springs and masses are all interconnected, and the motion of one affects the others.

Populations of foxes and rabbits or owls and field mice are interconnected, too. When prey is plentiful, the predators flourish and the prey is consumed. Then the predator population dwindles because food becomes scarce, and the population of the prey is able to recover.

This section models some of these situations. A key mathematical feature of these models is the appearance of *systems of differential equations*, not the single equation we have seen in most of the previous models.

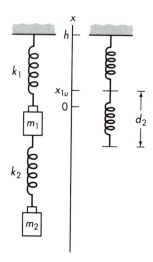

Figure 5.22: A system of two springs and two masses. The origin of the coordinate system is at the equilibrium position of the upper spring when both masses are in place, as shown on the left.

5.4.1 Multiple Springs and Masses

The system of two springs and masses shown in figure 5.22 is a simple multiple spring-mass system. The motion of either mass affects the other through the spring connecting them. We seek a model that describes the coordinates x_1, x_2 of the masses as functions of time.

We use Hooke's law as the basic experimental fact: Spring force F_s is proportional to extension z, or $F_s = kz$. The unifying physical law is Newton's familiar $F_t = ma$.

Figure 5.22 shows the notation and coordinate system we will use. We have chosen the origin to be the rest position of the lower end of the *first* spring when both masses are in place. The position of the lower end of the first spring *before* the weights are added is x_{1u} (again, u for *unweighted*). The unstretched length of the second spring is denoted by d_2.

Just as in the first section of this chapter, we find that x_{1u} is the extension of the first spring needed to balance the added weight. Since the upper spring carries the weight of both masses,

$$x_{1u} = \frac{(m_1 + m_2)g}{k_1} \,.$$

The extension z_1, z_2 of each spring is current length minus original length. When the masses are at x_1, x_2, we have

$$z_1 = (h - x_1) - (h - x_{1u})$$

$$= \frac{(m_1 + m_2)g}{k_1} - x_1 \,,$$

$$z_2 = (x_1 - x_2) - d_2 \,.$$

The forces on the first mass are those of the two springs and that of gravity. The force of the first spring is $k_1 z_1$, but that of the second spring is $-k_2 z_2$ because *positive* extension of the second spring pulls the first mass *down*. The force of gravity on the first mass is $-m_1 g$. Hence, the total force on the first mass is

$$F_{t1} = k_1 z_1 - k_2 z_2 - m_1 g$$

$$= -(k_1 + k_2)x_1 + k_2 x_2 + k_2 d_2 + m_2 g \,.$$

The forces on the second mass are that of the second spring and that of gravity:

$$F_{t2} = k_2 z_2 - m_2 g$$

$$= k_2 (x_1 - x_2 - d_2) - m_2 g \,.$$

Coupling the force expression with Newton's law for each of the masses produces two second-order equations. We can reasonably expect to impose a total of four initial conditions, the initial position and initial velocity of each of the masses. The final model for the multiple spring-mass system of figure 5.22 is the coupled system of differential equations

COUPLED SPRING-MASS MODEL

$$m_1 x_1'' + (k_1 + k_2)x_1 - k_2 x_2 = k_2 d_2 + m_2 g,$$
$$m_2 x_2'' - k_2 x_1 + k_2 x_2 = -k_2 d_2 - m_2 g,$$

subject to the initial conditions

$$x_1(0) = x_{1i}, \qquad x_1'(0) = v_{1i}, \qquad x_2(0) = x_{2i}, \qquad x_2'(0) = v_{2i}.$$

This system is called *coupled* because both unknowns, x_1 and x_2, appear in both equations.

> When we dealt with a single spring, we did not need a parameter such as d_2 for its unweighted length. The length of one of the springs is an unavoidable parameter in this system because it affects the relative location of the two masses. The decision to use the length of the second spring follows naturally from the choice of the origin as the rest position of the first mass.

5.4.2 Predator and Prey

A classic model of an interconnected system is that of predator and prey, such as foxes and rabbits. Foxes consume rabbits as their food, and the fox population grows when rabbits are plentiful. But a larger fox population decimates the rabbits, and the fox population declines from lack of food. Then the rabbit population recovers in the face of smaller numbers of foxes.

To model this situation easily, we will simplify the situation drastically. Rather than confidently advance experimental facts, we will instead propose some tentative ideas.

In the absence of foxes, we can suppose that the growth rate of the rabbits is constant:

Tentative Idea 1 The fraction of the rabbit population that reproduces in a unit time period is constant.

If b_R denotes the constant of proportionality and $R(t)$ is the rabbit population at time t, then approximately

$$b_R R(t) \Delta t$$

Rate $b_R R$ times time Δt.

rabbits are added to the population by reproduction during the time interval from t to $t + \Delta t$.

If foxes are the only predators of rabbits and if we ignore natural causes of death, then the death rate of rabbits should increase in proportion to the number of foxes. Hence, we propose the following.

Tentative Idea 2 The fraction of the *rabbit* population that die in a unit time period is proportional to the *fox* population.

If the constant of proportionality is β and $F(t)$ is the fox population, then the *fraction* of the rabbits that die is $\beta F(t)$. The number of rabbits that die between t and $t + \Delta t$ is approximately

Rate βFR times time Δt.

$$\beta F(t)R(t)\Delta t\,.$$

This last expression can also be interpreted as saying that a certain fraction of the encounters between rabbits and foxes will cause the demise of a rabbit.

To further simplify the situation, suppose that the fraction of foxes that die per unit time is constant:

Tentative Idea 3   The fraction of foxes that die in a unit time period is constant.

If the constant of proportionality is d_F, then the number of foxes lost from t to $t + \Delta t$ is approximately

Rate $d_F F$ times time Δt.

$$d_F F(t)\Delta t\,.$$

If foxes require rabbits for survival, their birth rate should depend on the number of rabbits available.

Tentative Idea 4   The fraction of the *fox* population that reproduce in a unit time period is proportional to the *rabbit* population.

If the constant of proportionality is α, then the *fraction* of foxes that reproduce is $\alpha R(t)$. The number of foxes added to the population from t to $t + \Delta t$ is approximately

Rate αRF times time Δt.

$$\alpha R(t)F(t)\,.$$

This last expression can be interpreted as saying that a certain proportion of the encounters between a rabbit and a fox will provide enough nutrition to permit the birth of a new fox.

We derive our model of foxes and rabbits by combining these tentative ideas with a familiar law.

Law of Conservation of Population   The net change in a population over a given period of time is the number of individuals added less the number of individuals removed.

We use this law twice, once for each of the populations.

The net change in the rabbit population from t to $t+\Delta t$ is $R(t+\Delta t)-R(t)$. Reproduction adds to the number of rabbits, and encounters with foxes reduce the number of rabbits. Hence, the law of conservation of rabbit population yields

Change = number added − number lost.

$$R(t + \Delta t) - R(t) = b_R R(t)\Delta t - \beta F(t)R(t)\Delta t\,.$$

Similarly, conservation of fox population yields

Change = number added − number lost.

$$F(t + \Delta t) - F(t) = \alpha R(t)F(t) - d_F F(t)\Delta t\,.$$

Dividing both equations by Δt and taking the limit as Δt goes to zero produces a pair of first-order equations:

$$F' = -(d_F - \alpha R)F$$
$$R' = (b_R - \beta F)R.$$

The model is completed by imposing two initial conditions, the starting size of each population,

$$F(0) = F_i, \qquad R(0) = R_i.$$

If you speculate that the term $R(t)F(t)$ might make these two equations nonlinear, you're right.

5.4.3 Exercises

EXERCISE GUIDE

To gain experience . . .	Try exercises
With force relations in spring-mass models	1, 3–5
Deriving spring-mass models	2, 6–8
Deriving and analyzing predator-prey models	9–10

1. The text claims the force on the first mass in figure 5.22 is

$$F_{t1} = -(k_1 + k_2)x_1 + k_2x_2 + k_2d_2 + m_2g\,.$$

Check this expression by examining several special cases:

(a) The first mass is moved to x_1, but the second spring retains its original length of d_2. (What is x_2 in this case?) Does the appearance of the term $+m_2g$ have anything to do with our choice of origin for the coordinate system?

(b) Neither spring is extended from its unweighted length. (What are the corresponding values of x_1 and x_2?)

2. Hang a *third* spring with spring constant k_3 from the second mass in figure 5.22. Suspend a third mass m_3 from the end of that spring. Derive a model for the positions of the three masses.

3. The equations for the multiple spring-mass system of figure 5.22 are

$$m_1x_1'' + (k_1 + k_2)x_1 - k_2x_2 = k_2d_2 + m_2g,$$
$$m_2x_2'' - k_2x_1 + k_2x_2 = -k_2d_2 - m_2g.$$

Physically, this system should be identical to that shown in figure 5.1, page 170, if the second spring were perfectly stiff.

What value of k_2 corresponds to making the second spring "perfectly stiff"? What then is the effective mass suspended from the first spring? Do the two equations above reduce to reasonable expressions in this case? (Try dividing by k_2 before it takes the value corresponding to perfect stiffness.)

From this model, you should be able to recover the appropriate version of the undamped spring-mass equation (5.1), page 173,

$$mx'' + kx = 0$$

since one spring is effectively supporting two masses. Add the two differential equations; you will obtain an equation involving both x_1 and x_2. Use the result of the previous paragraph to write an equation in x_1 alone. (Why not use x_2?) What are the effective values of m and k?

4. The equations for the multiple spring-mass system of figure 5.22 are

$$m_1x_1'' + (k_1 + k_2)x_1 - k_2x_2 = k_2d_2 + m_2g\,,$$
$$m_2x_2'' - k_2x_1 + k_2x_2 = -k_2d_2 - m_2g\,.$$

What becomes of these equations if $m_2 = 0$? What is the corresponding physical system? Is the form of these equations when $m_2 = 0$ consistent with an appropriate version of the undamped spring-mass model $mx'' + kx = 0$?

5. The equations for the multiple spring-mass system of figure 5.22 are

$$m_1 x_1'' + (k_1 + k_2)x_1 - k_2 x_2 = k_2 d_2 + m_2 g,$$
$$m_2 x_2'' - k_2 x_1 + k_2 x_2 = -k_2 d_2 - m_2 g.$$

If $m_1 = 0$, then this system corresponds to a single mass suspended from a spring that is a composite of two springs linked together, one with constant k_1 and one with constant k_2. If you can reduce this model to the form

$$m_2 x_2'' + kx_2 = \cdots,$$

then the value of k you obtain will be the effective spring constant of the two linked springs.

Write the coupled spring-mass model when $m_1 = 0$. One differential equation reduces to an algebraic equation involving x_1 and x_2. Use that expression to eliminate x_1 from the remaining differential equation. What is the effective spring constant?

6. Suppose internal resistance in the springs damps the motion of each of the masses shown in figure 5.22; assume the damping force opposes the motion and is proportional to the velocity of the mass. Derive a model for the positions of the masses.

7. Suppose the coordinate of the top of the upper spring in figure 5.22 varies with time; call its location $h(t)$. (Compare with figure 5.6, page 175.) Derive a model for the positions of the two masses.

8. Derive a model for the positions of the masses shown in figure 5.23; take the origin of the coordinate system shown there to be the rest position of m_1. Assume that the force of friction between the horizon-

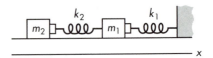

Figure 5.23: The origin of the coordinate system is the rest position of m_1.

tal surface and either of the masses is proportional to velocity and acts to oppose the motion.

9. "Improve" the predator-prey equations

$$F' = -(d_F - \alpha R)F$$
$$R' = (b_R - \beta F)R$$

derived in the text by adding one more effect,

The fraction of rabbits who die due to natural causes in a unit time period is constant.

Call the constant of proportionality d_R and derive the governing equations. Show that the resulting equations have the same mathematical form with b_R replaced by another expression. What is the physical interpretation of the coefficient that replaces b_R?

10. The rabbit population equation from the predator-prey model can be written

$$R' = (b_R - \alpha F)R.$$

The logistic equation can be written

$$P' = (a - sP)P.$$

Argue that both of these equations model populations with grow rates that decline as a certain parameter increases. Explain the significance of that parameter in each case.

5.5 CHAPTER EXERCISES

EXERCISE GUIDE

To gain experience . . .	Try exercises
Deriving and analyzing mechanical models	1–2, 6–7
Deriving and analyzing electrical models	3
Analogies among models	4–5

1. The model (5.1–5.2),

$$x'' + kx = 0, \qquad x(0) = x_i, \qquad x'(0) = v_i,$$

for the vertical spring-mass system was derived using a coordinate system whose origin is at the rest position of the mass; see figure 5.1, page 170.
   (a) Introduce an alternate coordinate system y whose origin is at the lower end of the spring *before* the weight is suspended from it. Derive the model

$$my''(t) + ky(t) = -mg, \; y(0) = y_i, \; y'(0) = v_i,$$

where $y(t)$ is the position of the mass in the new coordinate system at time t, y_i is its initial position, and v_i is its initial velocity.
   (b) Since the model in part (a) and the model (5.1–5.2) describe the same motion, you should be able to convert one initial-value problem into the other. Show that you can by finding a relation between x, the coordinate of figure 5.1, and y, the coordinate used in part (a). (*Hint:* The origin of the y axis is located at x_u in the x-coordinate system.)
       Substitute the expression for y into the model of part(a) and show that it reduces to (5.1–5.2). Reverse the process and obtain the model of part (a) by substituting into (5.1–5.2).

2. Figure 5.24 illustrates a conceptual model of an automobile suspension. The mass is the car, and the "fixed" end of the spring is connected to a wheel

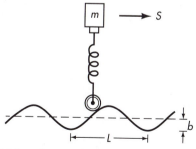

Figure 5.24: This spring-mass system is a conceptual model of an automobile suspension. The mass m represents the automobile. The bumps in the sinusoidal road are L units apart, and each is b units above their average level. The car is moving with speed S.

traversing a sinusoidal road. Derive a model for the vertical motion of the car if it is traveling this road at speed S.

3. As directed in the following, use

   Kirchhoff's Current Law The sum of the currents into any connection point, or *node*, of a circuit is zero.

   and the experimental facts given in the text to derive the equation governing the parallel RLC circuit of figure 5.16, page 184.
   (a) Use Kirchhoff's current law to sum the currents i_R, i_C, i_L through the resistor, capacitor, and inductor where they enter the upper node in figure 5.16.
   (b) Show that the experimental facts governing resistors, capacitors, and inductors can also be written

$$i_R = \frac{1}{R}v, \qquad i_C = C\frac{dv}{dt},$$

$$i_L = \frac{1}{L}\left(\int_0^t v(\tau)\, d\tau - v(0) \right),$$

   where v is the voltage drop across each of the three components.
   (c) Combine the law with the experimental facts to obtain the governing differential equation

$$Cv'' + \frac{1}{R}v' + \frac{1}{L}v = 0.$$

4. Someone hands you the equation $L\theta'' + g\theta = 0$ and says, "This really could be a model of a spring-mass system. The mass is L, the spring constant is g, and θ is the distance the mass has moved from its equilibrium, not an angle." Is such an assertion reasonable? If so, some term in the equation must represent the force of the spring on the mass. Which term is it? Does that term seem to provide a physically reasonable representation of the force exerted by a spring stretched to various lengths?

5. Repeat the previous exercise using the exact pendulum equation $L\theta'' + g\sin\theta = 0$.

6. Figure 5.25 shows a schematic diagram of the passenger-seat suspension system in an automobile; m is the mass of the passenger, k is the spring constant of the spring in the seat, M is the mass of the car, and K is the spring constant of the car's suspension. Derive a model for the positions of the passenger and the automobile.

7. The lower end of the spring in figure 5.25 represents the point where the car's suspension contacts the road. Make the model of the preceding problem more realistic by allowing the coordinate of that point to vary with time.

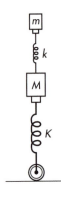

Figure 5.25: A schematic diagram of the passenger-seat suspension system in an automobile

5.6 CHAPTER PROJECTS

1. **Nonlinear spring** The idealization of Hooke's law that force is proportional to extension does not remain valid for large displacements. For sufficiently large extensions, the resisting force increases more rapidly than linearly. For example, as the coils of a spring are stretched nearly straight, the spring will resist even the smallest additional displacement with a substantially increased force. The last entry in the table of exercise 12, page 180, is an example of this phenomena. (What would that entry be if force remained proportional to displacement?) A common modification to accommodate this increase in force is the nonlinear force-extension law

$$F_s = kz + az^3,$$

where a is typically much smaller than k.
   (a) What sign should a have? Sketch the force-extension relationship for both positive and negative values of z and argue that spring force is approximately proportional to extension for small values of z.
   (b) Derive a model similar to (5.1–5.2), page 173, for the vertical, undamped spring-mass system using this force-extension relationship.
   (c) Although *linear* is not yet defined for second-order differential equations, guess an appropriate definition by analogy with the first-order case. Does the differential equation obtained in part (b) appear to be linear or nonlinear?

2. **Bridge cables** Common wisdom holds that the Tacoma Narrows bridge collapsed because of linear resonant vibrations induced by high winds, much as in the classic advertisement, "Is it real or is it Memorex?" in which a glass shatters as it resonates to a singer's voice. (See sections 6.5 and 7.2 for more about resonance.) Recent work by Lazer and McKenna [10] suggests instead that bridges can collapse because of the nonlinear response of the cables that suspend and supposedly stabilize the roadbed.

   The response of the springs we have considered is linear; $F_s = kz$ means that spring force is proportional to extension. Compression or extension are resisted equally, with only the sign of the spring force changing. A cable, however, is a *nonlinear spring*. It resists only extension. That is, its spring force law is

$$F_s = kz^+ = \begin{cases} kz, & z \geq 0 \\ 0, & z < 0. \end{cases}$$

(The superscript $+$ is defined by $z^+ = z$ if $z \geq 0$, $z^+ = 0$ if $z < 0$.)
Figure 5.26 shows a mass m suspended from two cables, a simple prototype of a bridge suspension system. When the cables are modeled so that they resist only extension, the governing equations are

$$m\frac{d^2x}{dt^2} = -k\left(\sqrt{x^2 - y^2} - \ell_0\right)^+ \frac{x}{\sqrt{x^2 + y^2}}$$

$$+k\left(\sqrt{(1/2 - x)^2 + y^2} - \ell_0\right)^+ \frac{1/2 - x}{\sqrt{(1/2 - x)^2 + y^2}}$$

$$-p\frac{dx}{dt} \tag{5.7}$$

$$m\frac{d^2y}{dt^2} = -k\left(\sqrt{x^2 - y^2} - \ell_0\right)^+ \frac{y}{\sqrt{x^2 + y^2}}$$

$$-k\left(\sqrt{(1/2 - x)^2 + y^2} - \ell_0\right)^+ \frac{y}{\sqrt{(1/2 - x)^2 + y^2}}$$

$$-p\frac{dy}{dt} + F\sin\omega t. \tag{5.8}$$

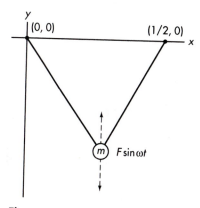

Figure 5.26: A mass m suspended from cables of equal unstretched length ℓ_0

Here k is the coefficient in the spring force law of the cable, p is a damping coefficient, ℓ_0 is the unstretched length of the cables, and F and ω are the magnitude and frequency, respectively, of a vertical force applied directly to the mass.

This system exhibits an extraordinarily rich range of responses. In anticipation of numerical studies of its behavior in project 3 of chapter 7, justify each of the terms in the governing equations (5.7–5.8). In addition, supply appropriate initial conditions.

6

HIGHER-ORDER SOLUTION
METHODS I

To Generalize is to be an Idiot. To Particularize is the Alone Distinction of Merit.
General Knowledges are those Knowledges that Idiots Possess.
William Blake

We saw in chapter 5 that second-order equations arise as models of a variety of oscillatory phenomena. In this chapter, we develop the mathematical concepts and solution procedures these problems require. William Blake notwithstanding, much of the work here is a natural generalization of first-order solution ideas. The next chapter applies these solution techniques to the analysis of several of the oscillation models of the previous chapter.

6.1   BASIC DEFINITIONS

6.1.1   Linear, Homogeneous, etc.

We begin with a series of definitions that apply to second-order equations. Many of these ideas are straightforward generalizations of their first-order analogues.

DEFINITION 1   *The function u is a **solution** on the interval $x_0 \leq x \leq x_1$ of the second-order differential equation $d^2y/dx^2 = f(x, y, dy/dx)$ if u is twice continuously differentiable and if*

$$\frac{d^2u}{dx^2} = f(x, u, \frac{du}{dx}) \text{ for } x_0 \leq x \leq x_1.$$

*Further, u is a **solution of the initial-value problem** $d^2y/dx^2 = f(x, y, dy/dx)$, $y(x_0) = y_i$, $y'(x_0) = z_i$, on the interval $x_0 \leq x \leq x_1$ if u is a solution of the* *differential equation and if*

$$u(x_0) = y_i, \quad \frac{du}{dx}\bigg|_{x=x_0} = z_i.$$

A **linear, second-order equation** is a second-order differential equation that can be written in the form

$$a_2(x)y'' + a_1(x)y' + a_0(x)y = r(x). \qquad (6.1)$$

Of the unknown and its derivatives up to order two, exactly one can appear in any term of the equation, and that unknown must appear raised to the first power only, not as the argument of a nonlinear function. Note the parallel with the general form of the linear first-order equation

$$a_1(x)y' + a_0(x)y = r(x).$$

A second-order equation that is not linear is **nonlinear**.

Recall some definitions from section 3.1: A differential equation is **homogeneous** if the identically zero function is a solution. Otherwise, the equation is **nonhomogeneous**.

A second-order linear equation has **constant coefficients** if the coefficients a_2, a_1, and a_0 in the general form (6.1) are all constants. Otherwise, it has **variable coefficients**.

EXAMPLE 1   *Classify the following as linear or nonlinear, homogeneous or nonhomogeneous, and constant or variable coefficient.*

- The undamped, unforced spring-mass equation of section 5.1,

$$mx''(t) + kx(t) = 0,$$

is linear; choose $a_2 = m$, $a_1 = 0$, $a_0 = k$, $r = 0$, and identify the unknown x in the preceding equation with the dependent variable y in the general form (6.1) and the independent variable t with x in (6.1). This equation is homogeneous because $x(t) = 0$ is a solution, and it has constant coefficients $a_2 = m$, $a_1 = 0$, $a_0 = k$.

- The damped, unforced spring-mass equation of section 5.1,

$$mx''(t) + px'(t) + kx(t) = 0,$$

is linear, homogeneous, and has constant coefficients $a_2 = m$, $a_1 = p$, $a_0 = k$.

- The damped, forced spring-mass equation of section 5.1,

$$mx''(t) + px'(t) + kx(t) = k(h(t) - h(0)),$$

is linear (choose $a_2 = m$, $a_1 = p$, $a_0 = k$, $r = k(h(t) - h(0))$), and it has constant coefficients $a_2 = m$, $a_1 = p$, $a_0 = k$. However, it is nonhomogeneous because substituting the trial solution $x(t) = 0$ in this equation yields

$$0 = k(h(t) - h(0)),$$

an equality that is clearly not valid unless $k = 0$ (no spring) or $h(t) \equiv h(0)$ (no movement of the upper end of the spring).

- The more general spring-mass equation you are asked to derive in exercise 15 of section 5.1 is

$$mx'' + kx + ax^3 = 0.$$

It is certainly homogeneous, for the trial solution $x(t) = 0$ reduces this equation to an equality.

It is nonlinear because of the ax^3 term. Comparing the given equation with the general linear equation (6.1) suggests that we might choose $a_2(t) = m$, $a_1(t) = 0$, $a_0(t) = ax^2$. However, the function a_0 in (6.1) can depend *only* on the independent variable, which is t in this equation. The choice $a_0 = ax^2$ causes a_0 to depend upon the dependent variable x. Similarly, we can not choose $r(t) = ax^3$. The term ax^3 can not be associated with any term in (6.1) because the unknown x is raised to a power other than the first.

- The pendulum equation

$$L\theta'' + g\sin\theta = 0$$

derived in section 5.3, is homogeneous because substituting $\theta = 0$ reduces it to an identity. It is nonlinear because the general linear form (6.1) certainly cannot accommodate the sine of the dependent variable. ∎

6.1.2 Linear Independence and General Solutions

We might expect that second-order equations have two arbitrary constants in their solutions. Roughly speaking, one appears for each antiderivative we need to solve the equation. A **second-order initial-value problem** couples a second-order equation with *two* pieces of initial data. One is the value of the solution, and the other is the value of its derivative at a specified time. (Other types of initial data can occur, but we will seldom encounter such situations.)

For example, the forced, damped spring-mass system model of section 5.1,

$$mx''(t) + px'(t) + kx(t) = k(h(t) - h(0)),$$
$$x(0) = x_i, \qquad x'(0) = v_i,$$

is a linear second-order initial value problem. We will see that the general solution (which we have yet to define) of this differential equation involves two arbitrary constants that are determined by the two initial conditions.

Recall that a general solution of a linear *first*-order homogeneous equation is just a constant multiple of any nontrivial solution of the homogeneous equation. The arbitrary constant is determined by an initial condition, if one is given.

For example, $y = e^{-x}$ is a nontrivial solution of the first-order equation

$$y'(x) + y(x) = 0.$$

A general solution of this homogeneous equation is

$$y_g(x) = Ce^{-x},$$

where C is an arbitrary constant.

The situation is more complicated for second-order equations. The linear homogeneous second-order equation

$$y'' + y' - 12y = 0$$

has two nontrivial solutions

$$y_1(x) = e^{-4x}, \qquad y_2(x) = e^{3x}$$

(which can be found by the characteristic equation method). It's tempting to say simply, "A general solution of the *second*-order homogeneous equation $y'' + y' - 12y = 0$ is

$$y_h(x) = C_1 e^{-4x} + C_2 e^{3x},$$

a sum of constant multiples of *two* nontrivial solutions of the given equation."

But another pair of nontrivial solutions of this differential equation is e^{-4x} and $8e^{-4x}$, and an equally good candidate for a general solution might be

$$C_1 e^{-4x} + C_2 8e^{-4x}.$$

You might object that the nontrivial solutions used here are not "different" in the same way that e^{-4x} and e^{3x} are different. But then you need to be precise about the meaning of *different*.

One sign that the two functions in this last solution are not sufficiently different is the observation that the sum $C_1 e^{-4x} + C_2 8e^{-4x}$ can be written

$$(C_1 + 8C_2)e^{-4x},$$

an expression that really contains only a single arbitrary constant, $C_1 + 8C_2$. The two functions e^{-4x} and $8e^{-4x}$ are not different. They are constant multiples of one another, and any sum containing them will reduce to one involving only a single arbitrary constant. But with only one arbitrary constant we can, in general, satisfy only one initial condition.

To construct a general solution of a linear homogeneous second-order equation, we must use two nontrivial solutions that are *different in the sense that the two arbitrary constants will not collapse into one*. This kind of different is called *linearly independent*: Two functions $y_1(x)$ and $y_2(x)$ are **linearly independent** on the interval $a \leq x \leq b$ if

$$C_1 y_1(x) + C_2 y_2(x) = 0, \qquad a \leq x \leq b,$$

implies $C_1 = C_2 = 0$. Otherwise, they are **linearly dependent** on $a \leq x \leq b$. The expression $C_1 y_1 + C_2 y_2$ is called a **linear combination** of y_1 and y_2. If the interval is obvious from the context, then explicit reference to it can be omitted.

The condition

$$C_1 y_1 + C_2 y_2 = 0 \Rightarrow C_1 = C_2 = 0$$

is just a statement that the two functions can not be multiples of one another. (Read $\Rightarrow$ as "*implies*".) Unfortunately, the idea of avoiding functions that are constant multiples of one another does not generalize easily to more than two functions, although this definition of linear independence does.

EXAMPLE 2   *Show that the functions* $y_1(x) = e^{-4x}$, $y_2(x) = e^{3x}$, *are linearly independent on* $-\infty < x < \infty$.

To establish this claim, assume the contrary; assume that the two functions are linearly *dependent*. Then there are constants C_1 and C_2, not both zero, satisfying

$$C_1 e^{-4x} + C_2 e^{3x} = 0, \qquad -\infty < x < \infty.$$

Suppose $C_2 \neq 0$. Then simple division immediately yields

$$\frac{C_1}{C_2} = e^{7x}.$$

Although the quotient of constants on the left is constant, the exponential on the right is certainly not constant as x ranges over all real numbers. A similar contradiction occurs if we assume $C_1 \neq 0$. These show that e^{-4x} and e^{3x} are linearly independent over the entire real line. ∎

EXAMPLE 3   *Show that the functions* e^{-4x} *and* $8e^{-4x}$ *are linearly dependent on* $-\infty < x < \infty$.

We must exhibit nonzero constants C_1 and C_2 satisfying

$$C_1 e^{-4x} + C_2 8 e^{-4x} = 0$$

for $-\infty < x < \infty$. Dividing this last expression by e^{-4x} yields

$$C_1 = -8C_2.$$

Choosing $C_1 = -8$, $C_2 = 1$, for example, certainly satisfies $C_1 e^{-4x} + C_2 8 e^{-4x} = 0$ for all real x. Since these values of C_1 and C_2 are nonzero, the functions e^{-4x} and $8e^{-4x}$ are linearly dependent on $-\infty < x < \infty$. ∎

The homogeneous solution is a linear combination of independent solutions of the homogeneous equation.

If y_1 and y_2 are linearly independent solutions of a homogeneous linear second-order equation on a given interval, then the **general solution of the homogeneous equation** or the **homogeneous solution** y_h on that interval is just an arbitrary linear combination of those two solutions:

$$y_h(x) = C_1 y_1(x) + C_2 y_2(x).$$

Since y_1 and y_2 are linearly independent, neither of them can be the trivial solution. If one were identically zero, then the constant that multiplies it in the linear combination $C_1 y_1 + C_2 y_2 = 0$ could be any nonzero quantity, thereby rendering the pair linearly dependent.

If y_p is a solution of a nonhomogeneous linear second-order equation (y_p is called a **particular solution**) and if y_h is a general solution of the corresponding homogeneous equation, then

$$y_g(x) = y_p(x) + y_h(x)$$

General solution of *non*homogeneous equation = particular solution + homogeneous solution.

is a **general solution of the nonhomogeneous equation.** The interval of validity of this general solution is the smaller of the intervals upon which y_p is a particular solution of the nonhomogeneous equation and y_h is a general solution of the homogeneous equation.

EXAMPLE 4 *Construct a general solution of the nonhomogeneous equation governing the forced spring-mass system of example 3 of section 5.1,*

$$\frac{1}{8} x''(t) + 2x(t) = \frac{1}{2} \sin \frac{3\pi t}{2},$$

and construct a general solution of the corresponding homogeneous equation,

$$\frac{1}{8} x''(t) + 2x(t) = 0.$$

Although we have not learned to solve second-order equations, we can still steal a few first-order ideas.

We begin with the homogeneous equation $x''/8 + 2x = 0$. Either by using the characteristic equation method to be developed in section 6.3 or by guessing solutions of the form $x = \sin rt$ and $x = \cos rt$ for some constant r, we find two solutions,

$$x_1(t) = \sin 4t, \qquad x_2(t) = \cos 4t.$$

(Rather than just take these for granted, substitute them into the equation and verify that they are indeed solutions.)

To construct a general solution of the homogeneous equation, we must first show that $\sin 4t$ and $\cos 4t$ are linearly independent on the interval of interest for these initial-value problems, $t \geq 0$. Assume there are constants C_1 and C_2 such that

$$C_1 \sin 4t + C_2 \cos 4t = 0$$

for $t \geq 0$. Since the left-hand side of this last expression is identically zero, its derivative is identically zero as well:

$$4C_1 \cos 4t - 4C_2 \sin 4t = 0.$$

These two equations for C_1 and C_2,

$$C_1 \sin 4t + C_2 \cos 4t = 0$$

$$4C_1 \cos 4t - 4C_2 \sin 4t = 0,$$

certainly have the solution $C_1 = C_2 = 0$, as you can verify by direct substitution. The theory of linear equations guarantees that *there is no solution other than this trivial solution if the determinant of coefficients*

$$W = \begin{vmatrix} \sin 4t & \cos 4t \\ 4\cos 4t & -4\sin 4t \end{vmatrix}$$

Value of 2×2 determinant = product of main diagonal − product of off-diagonal entries.

is not zero. (See section A.6.) But this determinant is

$$W = (\sin 4t)(-4\sin 4t) - (4\cos 4t)(\cos 4t)$$
$$= -4\sin^2 4t - 4\cos^2 4t = -4 \neq 0, \qquad t \geq 0.$$

Consequently, the only constants C_1, C_2 satisfying $C_1 \sin 4t + C_2 \cos 4t = 0$ for all $t \geq 0$ are $C_1 = C_2 = 0$. Hence, $\sin 4t$, $\cos 4t$ are linearly independent on $t \geq 0$.

Since these two functions are linearly independent, a general solution of the homogeneous equation $x''/8 + 2x = 0$ is

$$x_h(t) = C_1 \sin 4t + C_2 \cos 4t$$

for $t \geq 0$.

Using undetermined coefficients, we can derive the particular solution

Guess $x_p(t) = A\sin(3\pi t/2)$, substitute, and solve for A.

$$x_p(t) = \frac{16}{64 - 9\pi^2} \sin \frac{3\pi t}{2} \approx -0.644 \sin \frac{3\pi t}{2}$$

of the nonhomogeneous equation $x''/8 + 2x(t) = \frac{1}{2}\sin(3\pi t/2)$. Hence, for $t \geq 0$, a general solution of this nonhomogeneous equation is

$$x_g(t) = x_p(t) + x_h(t)$$
$$\approx -0.644 \sin \frac{3\pi t}{2} + C_1 \sin 4t + C_2 \cos 4t. \qquad \blacksquare$$

6.1.3 Superposition

The last example and the concept of a general solution are both illustrations of an important property of linear equations that was introduced in section 3.1,

Principle of superposition If $y(x)$ is a solution of the linear equation

$$a_2(x)y'' + a_1(x)y' + a_0(x)y = r(x)$$

with forcing term $r(x)$ and if $z(x)$ is a solution of the same differential expression with forcing term $s(x)$,

$$a_2(x)z'' + a_1(x)z' + a_0(x)z = s(x),$$

then the linear combination $u(x) = C_1 y(x) + C_2 z(x)$ is a solution of the differential equation with forcing term $C_1 r(x) + C_2 s(x)$,

$$a_2(x)u'' + a_1(x)u' + a_0(x)u = C_1 r(x) + C_2 s(x).$$

When we constructed a general solution of the homogeneous equation $x''/8 + 2x = 0$, we formed a linear combination of the two functions $\sin 4t$ and $\cos 4t$, each a solution of the equation with a forcing term of zero. Consequently, the principle of superposition asserts that the general solution $u = C_1 \sin 4t + C_2 \cos 4t$ satisfies the equation

$$\frac{1}{8}u'' + 2u = C_1 \cdot 0 + C_2 \cdot 0 = 0 \, ;$$

i.e., the general solution $C_1 \sin 4t + C_2 \cos 4t$ indeed satisfies the homogeneous equation.

Constructing a general solution of the nonhomogeneous equation $x''/8 + 2x = \frac{1}{2}\sin(3\pi t/2)$ follows a similar pattern. The general solution $x_h(t)$ of the homogeneous equation satisfies an equation with zero forcing term, and the particular solution $x_p(t)$ satisfies an equation with forcing term $\frac{1}{2}\sin(3\pi t/2)$. The principle of superposition guarantees that the linear combination

$$u = x_p(t) + x_h(t) \, ,$$

which is the general solution of the nonhomogeneous equation, satisfies the nonhomogeneous equation

$$\frac{1}{8}u''(t) + 2u(t) = \frac{1}{2}\sin\frac{3\pi t}{2} + 0 = \frac{1}{2}\sin\frac{3\pi t}{2} \, ;$$

i.e., the general solution $x_p + x_h$ of the nonhomogeneous equation actually satisfies the nonhomogeneous equation.

EXAMPLE 5 *The preceding discussion used the principle of superposition to show that the general solution $x_h(t) = C_1 \sin 4t + C_2 \cos 4t$ of the unforced spring-mass equation*

$$\frac{1}{8}x''(t) + 2x(t) = 0 \, ,$$

is actually a solution of this equation. Verify by direct substitution that the principle of superposition is correct.

Substituting the general solution $x_h(t) = C_1 \sin 4t + C_2 \cos 4t$ into the differential equation yields

$$\frac{1}{8}x_h''(t) + 2x_h(t) = C_1 \left(-\frac{1}{8} \cdot 16 \sin 4t + 2 \sin 4t \right)$$

$$+ C_2 \left(-\frac{1}{8} \cdot 16 \cos 4t + 2 \cos 4t \right)$$

$$= C_1 \cdot 0 + C_2 \cdot 0 = 0 \, .$$

The principle of superposition is correct: adding two solutions of a homogeneous equation produces a solution of the homogeneous equation. ■

EXAMPLE 6 *Again verify by direct substitution that the principle of superposition is correct, this time using the general solution*

$$x_g(t) = \frac{16}{64 - 9\pi^2} \sin \frac{3\pi t}{2} + x_h(t)$$

of the forced spring-mass equation

$$\frac{1}{8} x''(t) + 2x(t) = \frac{1}{2} \sin \frac{3\pi t}{2},$$

where $x_h(t) = C_1 \sin 4t + C_2 \cos 4t$ is the general solution of the homogeneous equation $x''/8 + 2x = 0$. Recall that $16/(64 - 9\pi^2) \approx -0.644$.

Substituting the general solution x_g into the left side of the differential equation yields

$$\frac{1}{8} x_g''(t) + 2x_g(t) = \frac{0.644 \cdot 9\pi^2}{8 \cdot 4} \sin \frac{3\pi t}{2} - 2 \cdot 0.644 \sin \frac{3\pi t}{2}$$

$$+ \frac{1}{8} x_h''(t) + 2x_h(t)$$

$$= \frac{1}{2} \sin \frac{3\pi t}{2} + 0 = \frac{1}{2} \sin \frac{3\pi t}{2} .$$

The principle of superposition is correct again: Adding a solution of a *non*homogeneous equation to the solution of the corresponding homogeneous equation produces a solution of the *non*homogeneous equation. ∎

Just as with first-order equations, a general solution of a linear differential equation is the starting point for finding a solution of an initial-value problem.

EXAMPLE 7 *Solve the initial-value problem modeling the forced spring-mass system in example 3 of section 5.1,*

$$\frac{1}{8} x''(t) + 2x(t) = \frac{1}{2} \sin \frac{3\pi t}{2},$$

$$x(0) = 0, \qquad x'(0) = 0.$$

We already know that a general solution of the differential equation is

$$x_g(t) = \frac{16}{64 - 9\pi^2} \sin \frac{3\pi t}{2} + C_1 \sin 4t + C_2 \cos 4t$$

$$\approx -0.644 \sin \frac{3\pi t}{2} + C_1 \sin 4t + C_2 \cos 4t .$$

The first initial condition requires

$$x_g(0) = C_2 = 0 .$$

To apply the second initial condition, we compute $x'_g(t)$ and use $C_2 = 0$:

$$x'_g(t) = \frac{16}{64 - 9\pi^2} \frac{3\pi}{2} \cos \frac{3\pi t}{2} + 4C_1 \cos 4t \,.$$

We must choose C_1 to satisfy

$$x'_g(0) = \frac{16}{64 - 9\pi^2} + 4C_1 = 0 \,,$$

or $C_1 = -6\pi/(64 - 9\pi^2) \approx 0.759$.

Hence, the solution of the given initial-value problem is

$$x(t) = \frac{16}{64 - 9\pi^2} \sin \frac{3\pi t}{2} - \frac{6\pi}{64 - p\pi^2} \sin 4t$$

$$\approx -0.644 \sin \frac{3\pi t}{2} + 0.759 \sin 4t \,. \qquad \blacksquare$$

6.1.4   Free and Forced Response

Paralleling first-order linear equations, we can decompose solutions of second-order linear equations into free and forced components and into transient and steady-state parts. The **forced response** is the solution of the original (nonhomogeneous) differential equation subject to zero initial conditions, while the **free response** is the solution of the differential equation with zero forcing subject to the original initial conditions.

Explicitly, if the given initial-value problem is

$$a_2(x)y'' + a_1(x)y' + a_0(x)y = r(x) \,,$$
$$y(0) = y_i, \qquad y'(0) = z_i \,,$$

then the *forced response* y_{for} is the solution of

$$a_2(x)y''_{\text{for}} + a_1(x)y'_{\text{for}} + a_0(x)y_{\text{for}} = r(x) \,,$$
$$y_{\text{for}}(0) = \boxed{0}, \qquad y'_{\text{for}}(0) = \boxed{0} \,,$$

and the *free response* y_{free} is the solution of

$$a_2(x)y''_{\text{free}} + a_1(x)y'_{\text{free}} + a_0(x)y_{\text{free}} = \boxed{0} \,,$$
$$y_{\text{free}}(0) = y_i, \qquad y'_{\text{free}}(0) = z_i \,.$$

(The boxed zero terms emphasize the differences between the original initial-value problem and the free and forced problems.)

The principle of superposition reveals that

$$y(x) = y_{\text{for}}(x) + y_{\text{free}}(x) \,.$$

6.1.5 Equilibria and Transients

As with first-order equations, we define a **steady state**, or **equilibrium**, of a differential equation to be any constant solution. An equilibrium y_{ss} is **stable** if small perturbations of it die out; that is, if solutions whose initial values are near y_{ss} approach y_{ss} in the limit of large time. The **transient** component of a solution is that part (if any) whose limit is zero for large values of the independent variable.

> An equation can also have an oscillating "steady state" or periodic solution. A periodic solution is a solution that is a periodic function of the independent variable. Periodic solutions, like equilibria, can be *stable* or *unstable*, depending upon whether small perturbations of the periodic solution decay or grow with time. A transient can decay to leave a periodic solution.

EXAMPLE 8   *The solution of the initial-value problem*

$$y''(t) + 2y'(t) + 5y(t) = -9.8,$$
$$y(0) = 0, \qquad y'(0) = 0.04$$

is

$$y(t) = -1.96 + e^{-t}(1.96\cos 2t + \sin 2t).$$

Find its transient, if there is one. Does this differential equation have an equilibrium?

> In the solution, we observe the decaying terms that comprise the transient
>
> $$y_{\text{tran}}(t) = e^{-t}(1.96\cos 2t + \sin 2t).$$
>
> If this equation is to have a *constant* steady state y_{ss}, then y_{ss} must satisfy
>
> $$y''_{ss} + 2y'_{ss} + 5y_{ss} = 5y_{ss} = -9.8.$$

Hence, $y_{ss} = 9.8/5 = -1.96$.   ∎

> It is tempting to speculate about the stability of the equilibrium $y_{SS} = -1.96$.
> Since -1.96 is a particular solution of this differential equation, the remaining terms $e^{-t}(1.96\cos 2t + \sin 2t)$ in the solution of the initial-value problem should be a solution of the corresponding homogeneous equation. Perhaps the coefficient 1.96 of $\cos 2t$ and the coefficient 1 of $\sin 2t$ began life as arbitrary constants, before they were assigned these values to satisfy the initial conditions. If so, a general solution of the equation could be $y_g = -1.96 + e^{-t}(C_1\cos 2t + C_2\sin 2t)$.
> The constants C_1, C_2 are determined by the initial conditions, but no matter what values they assume, $\lim_{t\to\infty} y_g(t) = -1.96$. The equilibrium appears to be stable to perturbations of any size; we do not have to start near -1.96 to approach this steady state.

EXAMPLE 9   *The solution of the initial-value problem*

$$y''(t) + 2y'(t) + 5y(t) = 185\sin 4t,$$
$$y(0) = -7, \qquad y'(0) = -47,$$

is

$$y(t) = -8\cos 4t - 11\sin 4t + e^{-t}(\cos 2t - \sin 2t).$$

Find its transient, if there is one. Does this differential equation have an equilibrium?

As before, we identify the transient by finding the decaying terms in the solution,

$$y_{\text{trans}} = e^{-t}(\cos 2t - \sin 2t).$$

To test for an equilibrium, we substitute the constant trial solution y_{ss} into the differential equation. We find

$$y''_{ss} + 2y'_{ss} + 5y_{ss} = 5y_{ss} = 185\sin 4t,$$

or $y_{ss} = 37\sin 4t$. Our assumption that y_{ss} is constant has been contradicted; this equation has no steady state. ∎

Inspecting the solution of the differential equation more carefully shows that it approaches $-8\cos 4t - 11\sin 4t$ as t grows. Hence, this equation has the periodic solution

$$y(t) = -8\cos 4t - 11\sin 4t.$$

Speculating as we did at the end of the previous example suggests that this periodic solution is stable to perturbations of any size.

6.1.6 Exercises

EXERCISE GUIDE

To gain experience . . .	Try exercises
With linear versus nonlinear	1–11
With homogeneous versus nonhomogeneous	1–11, 15(a–b), 18(a)
With constant versus variable coefficient	1–11
Checking solutions of equations	13
With particular solutions	12
With homogeneous solutions	15(b), 18(a, d)
Using general solutions	15(c), 18(b–c)
Solving initial-value problems	15(c), 18(b)
Using the principle of superposition	16, 19
With steady-state and transient solutions	15(a–b, e), 18(a)
With free and forced response	15(b, d), 16, 18(a, c)
With periodic solutions	18(d)
With equations as models	14, 17

Classify the following ten second-order differential equations as linear or nonlinear and as homogeneous or nonhomogeneous. Identify each linear equation as having constant or variable coefficients. Justify each classification.

1. $3x''(t)/16 + 2x(t) = 0$ (See Example 1 of section 5.1.4.)

2. $3x''(t)/16 + 1.41x'(t) + 2x(t) = 0$ (See example 2 of section 5.1.4.)

3. $my''(t) + ky(t) = -mg$, m, k, g constants (See exercise 1 of section 5.5.)

4. $my''(t) + py'(t) = -mg$, m, p, g constants

5. $my''(t) + py'(t) = -ky(t)$, m, p, k constants

6. $x''(t)/8 + 2x(t) = \frac{1}{2}x(t)\sin(3\pi t/2)$

7. $x''(t)/8 + 2x(t) = (\pi t/2)\sin x(t)$

8. $t^3 x''(t) + 2x(t) = \frac{1}{2}\sin(3\pi t/2)$

9. $x''(t) + x'(t)\cos 4t = 2[x(t) - 1]$

10. $x''(t) + x'(t)\cos 4t = 2[x(t) - 1]^2$

11. Verify that the two spring-mass model equations considered in example 4,

$$\frac{1}{8}x''(t) + 2x(t) = \frac{1}{2}\sin\frac{3\pi t}{2},$$

$$\frac{1}{8}x''(t) + 2x(t) = 0,$$

are indeed linear. Verify that the first equation is nonhomogeneous and that the second is homogeneous. Do these equations have constant or variable coefficients?

12. Verify that

$$x_p(t) = \frac{16}{64 - 9\pi^2}\sin\frac{3\pi t}{2} \approx -0.644\sin\frac{3\pi t}{2}$$

is indeed a particular solution of the forced spring-mass equation

$$\frac{1}{8}x''(t) + 2x(t) = \frac{1}{2}\sin\frac{3\pi t}{2}.$$

13. Verify by direct substitution that

$$x(t) = -0.644\sin\frac{3\pi t}{2} + 0.759\sin 4t$$

is a solution of the initial-value problem

$$\frac{1}{8}x''(t) + 2x(t) = \frac{1}{2}\sin\frac{3\pi t}{2},$$

$$x(0) = 0, \qquad x'(0) = 0.$$

14. Identify a particular physical situation of which the initial-value problem

$$y''(t) + 2y'(t) + 5y(t) = -9.8,$$

$$y(0) = 0, \qquad y'(0) = 0.04,$$

of example 8 might be a model. Be sure to specify units.

15. Example 8 considered the initial-value problem

$$y''(t) + 2y'(t) + 5y(t) = -9.8,$$

$$y(0) = 0, \qquad y'(0) = 0.04,$$

whose solution is

$$y(t) = -1.96 + e^{-t}(1.96\cos 2t + \sin 2t).$$

That example found that the transient component of its solution is

$$y_{\text{tran}}(t) = e^{-t}(1.96\cos 2t + \sin 2t)$$

while the equation has the steady state $y_{ss} = -1.96$.

(a) Does the steady state satisfy the homogeneous or the nonhomogeneous differential equation? Could the steady-state solution serve as a particular solution of the nonhomogeneous equation? Is it either the free or the forced response of this initial-value problem? Give a set of initial values that will yield precisely this steady-state solution.

(b) Does the transient component satisfy the homogeneous or the nonhomogeneous differential equation? Is it either the free or the forced response of this initial-value problem?

(c) Show that a general solution of this initial-value problem is

$$y(t) = -1.96 + e^{-t}(C_1\cos 2t + C_2\sin 2t).$$

Derive the solution of this initial-value problem from this general solution.

(d) Use this general solution to find the free and forced response of this initial-value problem.

(e) Use the general solution of part (c) to verify that the steady state $y_{ss} = -1.96$ is stable.

16. Use the principle of superposition to argue that the solution of the forced initial-value problem

$$a_2(x)y'' + a_1(x)y' + a_0(x)y = r(x).$$
$$y(0) = y_i, \qquad y'(0) = z_i,$$

can be written as the sum of its forced and free components,

$$y(x) = y_{\text{for}}(x) + y_{\text{free}}(x).$$

Recall that the forced response y_{for} is the solution of

$$a_2(x)y''_{\text{for}} + a_1(x)y'_{\text{for}} + a_0(x)y_{\text{for}} = r(x),$$
$$y_{\text{for}}(0) = 0, \qquad y'_{\text{for}}(0) = 0,$$

and that the free response y_{free} is the solution of

$$a_2(x)y''_{\text{free}} + a_1(x)y'_{\text{free}} + a_0(x)y_{\text{free}} = 0,$$
$$y_{\text{free}}(0) = y_i, \qquad y'_{\text{free}}(0) = z_i.$$

17. Identify a particular physical situation of which the initial-value problem

$$y''(t) + 2y'(t) + 5y(t) = 185 \sin 4t,$$
$$y(0) = -7, \qquad y'(0) = -47,$$

of example 9 might be a model. Be sure to specify units.

18. Example 9 considered the initial-value problem

$$y''(t) + 2y'(t) + 5y(t) = 185 \sin 4t,$$
$$y(0) = -7, \qquad y'(0) = -47,$$

whose solution is

$$y(t) = -8 \cos 4t - 11 \sin 4t + e^{-t}(\cos 2t - \sin 2t).$$

It found that the transient component of its solution is

$$y_{\text{tran}}(t) = e^{-t}(\cos 2t - \sin 2t)$$

and that it had the periodic solution

$$y_{ps}(t) = 8 \cos 4t - 11 \sin 4t.$$

(a) Does the transient component satisfy the homogeneous or the nonhomogeneous differential equation? Is it either the free or the forced response of this initial-value problem?

(b) Show that a general solution of this initial-value problem is

$$y(t) = -8 \cos 4t - 11 \sin 4t$$
$$+ e^{-t}(C_1 \cos 2t - C_2 \sin 2t).$$

Derive the solution of this initial-value problem from this general solution.

(c) Use the general solution of the preceding part to find the free and forced response of this initial-value problem.

(d) Does the periodic solution satisfy the homogeneous or the nonhomogeneous differential equation? Could it serve as a particular solution of the nonhomogeneous equation? Is it either the free or the forced response of this initial-value problem? Give a set of initial values that will yield precisely this periodic solution.

19. Recall the often-repeated calculus slogan, "The derivative of a sum is the sum of the derivatives." Does this saying apply to second derivatives? Explain why the principle of superposition is a consequence of this statement. (The principle of superposition follows from the linearity of the derivative operation.)

6.2 THE WRONSKIAN AND LINEAR INDEPENDENCE

This section develops a simple linear independence test for solutions of linear equations. To begin, recall the process we used in example 4 of the previous section to show that two functions are linearly independent.

To show that $\sin 4t$, $\cos 4t$ are linearly independent for $t \geq 0$, we seek two constants C_1, C_2 satisfying

$$C_1 \sin 4t + C_2 \cos 4t = 0, \qquad t \geq 0.$$

If the only choice is $C_1 = C_2 = 0$, then $\sin 4t$, $\cos 4t$ are linearly independent on $t \geq 0$.

Since C_1, C_2 are chosen so that $C_1 \sin 4t + C_2 \cos 4t$ is the identically zero function, the derivative of this expression is also identically zero,

$$4C_1 \cos 4t - 4C_2 \sin 4t = 0, \quad t \geq 0.$$

The pair of equations formed from the linear combination and its derivative,

$$C_1 \sin 4t + C_2 \cos 4t = 0 \qquad (6.2)$$

$$4C_1 \cos 4t - 4C_2 \sin 4t = 0, \qquad (6.3)$$

certainly has the solution $C_1 = C_2 = 0$. If there are no other values of C_1 and C_2 satisfying this pair of equations, then $\sin 4t$, $\cos 4t$ are linearly independent.

The theory of linear equations asserts that (6.2–6.3) has *exactly one* solution if the determinant of coefficients

$$W = \begin{vmatrix} \sin 4t & \cos 4t \\ 4\cos 4t & -4\sin 4t \end{vmatrix}$$

is not zero. By subtracting the product of the off-diagonal entries from the product of those on the diagonal, we find this determinant to be

$$W = (\sin 4t)(-4\sin 4t) - (4\cos 4t)(\cos 4t)$$

$$= -4\sin^2 4t - 4\cos^2 4t = -4 \neq 0, \quad t \geq 0.$$

Hence, the obvious solution $C_1 = C_2 = 0$ of (6.2–6.3) is the only solution of this pair of equations, and we conclude that $\sin 4t$, $\cos 4t$ are linearly independent on $t \geq 0$.

You may be more familiar with these determinant ideas in the context of Cramer's rule for solving systems of linear equations. This result states that if $W \neq 0$, then the *unique* solution of the pair of equations (6.2–6.3) is

$$C_1 = \frac{R_1}{W}, \qquad C_2 = \frac{R_2}{W}.$$

Here R_1 and R_2 are the determinants obtained by replacing the first and second columns of W, respectively, by the zero terms appearing on the right side of (6.2–6.3),

$$R_1 = \begin{vmatrix} 0 & \cos 4t \\ 0 & -4\sin 4t \end{vmatrix},$$

$$R_2 = \begin{vmatrix} \sin 4t & 0 \\ 4\cos 4t & 0 \end{vmatrix}.$$

But the columns of zeros in these determinants force $R_1 = R_2 = 0$, and the only solution of (6.2–6.3) is $C_1 = C_2 = 0$. See section A.6.

This approach extends quite nicely to pairs of solutions of second-order equations. We describe that extension here.

Let $y_1(x)$, $y_2(x)$ be two functions that are continuously differentiable on the interval $a \leq x \leq b$. Define the **Wronskian** $W(y_1, y_2)$ of y_1, y_2 by

DEFINITION OF THE WRONSKIAN

$$W(y_1, y_2) = \begin{vmatrix} y_1(x) & y_2(x) \\ y_1'(x) & y_2'(x) \end{vmatrix}$$

(This determinant is named after the Polish mathematician H. Wronski.) We may expand the determinant to write

EXPANDED WRONSKIAN

$$W(y_1, y_2) = y_1(x)y_2'(x) - y_2(x)y_1'(x).$$

The Wronskian of two functions is evidently a function of x.

To establish that the two functions y_1, y_2 are linearly independent on the interval $a \leq x \leq b$, we must show that the *only* constants C_1, C_2 satisfying

$$C_1 y_1(x) + C_2 y_2(x) = 0, \qquad a \leq x \leq b,$$

are $C_1 = C_2 = 0$.

We can proceed exactly as before, with y_1, y_2 replacing $\sin 4t$, $\cos 4t$. Since the linear combination $C_1 y_1 + C_2 y_2$ is identically zero on the interval $[a, b]$, its derivative must be zero there as well,

$$\frac{d}{dx}[C_1 y_1(x) + C_2 y_2(x)] = C_1 y_1'(x) + C_2 y_2'(x) = 0, \qquad a \leq x \leq b.$$

We now have a pair of simultaneous linear equations for the unknown constants C_1, C_2,

$$C_1 y_1(x) + C_2 y_2(x) = 0$$
$$C_1 y_1'(x) + C_2 y_2'(x) = 0,$$

the analogue of (6.2–6.3). The first equation is the original linear combination, and the second is its derivative. This pair of equations has no other solution than $C_1 = C_2 = 0$ if its determinant of coefficients is not zero on $[a, b]$; that is, $C_1 = C_2 = 0$ if

$$W(y_1, y_2) \neq 0, \qquad a \leq x \leq b,$$

because the Wronskian is exactly the determinant of coefficients of this pair of linear equations. This argument proves the first property of the Wronskian.

Theorem 6.1 The pair of continuously differentiable functions y_1, y_2 is linearly independent on the interval $a \leq x \leq b$ if $W(y_1, y_2) \neq 0$ there.

$W \neq 0$ guarantees linear independence.

EXAMPLE 10   *Show that the pair of functions* $\sin 4t$, $\cos 4t$ *is linearly independent on* $t \geq 0$.

Since these two functions are certainly continuously differentiable, we can immediately compute the Wronskian,

$$W(\sin 4t, \cos 4t) = \sin 4t (\cos 4t)' - \cos 4t (\sin 4t)'$$
$$= -4 \sin^2 4t - 4 \cos^2 4t = -4 \neq 0, \qquad t \geq 0.$$

Hence, theorem 6.1 guarantees that $\sin 4t$, $\cos 4t$ are linearly independent on $t \geq 0$.   ∎

Note that this theorem tells us nothing if the Wronskian happens to be zero at one or more points scattered through the interval. The following two results together rectify this deficiency when the functions in question are both solutions of a second-order homogeneous linear differential equation. Further, they show that we need check the Wronskian only at a single point of the interval.

The Wronskian of two solutions is either always zero or never zero.

Theorem 6.2   Let y_1, y_2 be solutions of the second-order homogeneous linear differential equation

$$y''(x) + P(x)y' + Q(x)y = 0, \qquad a \leq x \leq b.$$

Assume that y_1, y_2 are defined and twice continuously differentiable on the interval $[a, b]$ and that the coefficient $P(x)$ is continuous on $[a, b]$. Then either $W(y_1, y_2) \neq 0$ for every x in $[a, b]$ or $W(y_1, y_2) = 0$ for every x in $[a, b]$.

Technically, the requirement that y_1, y_2 be twice continuously differentiable is redundant; see the definition of solution of a second-order equation, definition 1, page 198.

The essence of the proof of this theorem is the observation that W (as a function of x) is a solution of the first-order linear equation

$$\frac{dW}{dx} + P(x)W = 0.$$

Hence, W may be written

$$W = C \exp\left(-\int_a^x P(s)\, ds \right),$$

where C is an arbitrary constant. Since the exponential is never zero, either $W \neq 0$ on $[a, b]$ because $C \neq 0$ or $W \equiv 0$ on $[a, b]$ because $C = 0$. The exercises lead you through this proof step by step.

Theorem 6.2 asserts that *if the Wronskian of two solutions of a linear equation is zero at one point of the interval of interest, then it is zero throughout that interval.* However, we have yet to show that a zero Wronskian implies linear dependence. The next theorem disposes of that case.

Theorem 6.3 Let y_1, y_2 be twice continuously differentiable solutions of the linear equation

If the Wronskian of two solutions is zero at one point, then the solutions are linearly dependent.

$$y''(x) + P(x)y' + Q(x)y = 0, \qquad a \le x \le b.$$

Let the coefficient P be continuous on $[a, b]$. If $W(y_1, y_2) = 0$ at any point of $[a, b]$, then y_1, y_2 are linearly dependent on $[a, b]$.

The proof of this theorem is outlined in the exercises as well.

Together, these theorems provide the following linear independence test. Recall that the functions in question must be twice continuously differentiable solutions of $y''(x) + P(x)y' + Q(x)y = 0$ and that the coefficient P must be continuous on the interval of interest.

Linear Independence Test Two solutions of the linear equation $y''(x) + P(x)y' + Q(x)y = 0$ are linearly *independent* on their domain of definition if their Wronskian is *not zero* for at least one point of the interval. The two solutions are linearly *dependent* if their Wronskian is *zero* for at least one point of the interval.

EXAMPLE 11 *Use the linear independence test to show that e^{-4x}, e^{3x} are linearly independent solutions of*

$$y'' + y' - 12y = 0, \qquad -\infty < x < \infty.$$

Substitution immediately verifies that these two functions are solutions of this equation on the given interval.

This differential equation is certainly of the form required by the linear independence test; take $P(x) = 1$, $Q(x) = -12$.

The Wronskian of these two solutions is

$$W(e^{-4x}, e^{3x}) = e^{-4x}(e^{3x})' - e^{3x}(e^{-4x})'$$

$$= 7e^{-x}.$$

Since $e^{-x} \ne 0$ for $-\infty < x < \infty$, the solutions e^{-4x}, e^{3x} are linearly independent there. ∎

EXAMPLE 12 *Use the linear independence test to show that e^{-4x}, $8e^{-4x}$ are linearly dependent solutions of*

$$y'' + y' - 12y = 0, \quad -\infty < x < \infty.$$

Obviously, both functions are solutions of the given equation. Their Wronskian is

$$W(e^{-4x}, 8e^{-4x}) = e^{-4x}(8e^{-4x})' - 8e^{-4x}(e^{-4x})' = 0.$$

Hence, these two solutions are linearly dependent because $W = 0$ for all x. ∎

EXAMPLE 13 *Two solutions of*

$$x^2 y'' + xy' - 4y = 0$$

are x^2 and x^{-2}. Are they linearly independent on any nontrivial intervals?

To apply the linear independence test, the given equation must be put into the form $y''(x) + P(x)y' + Q(x)y = 0$. To perform the necessary division by x^2, we can consider only $x \neq 0$. We obtain

$$y'' + x^{-1}y' - 4x^{-2}y = 0.$$

We now have the form $y''(x) + P(x)y' + Q(x)y = 0$ with $P(x) = x^{-1}$ and $Q(x) = -4x^{-2}$. This form of P is certainly continuous for $x \neq 0$.

Evaluating the Wronskian leads to

$$W(x^2, x^{-2}) = x^2(x^{-2})' - x^{-2}(x^2)'$$
$$= -4x^{-1} \neq 0, \ x \neq 0.$$

Hence, x^2 and x^{-2} are linearly independent solutions of $x^2y'' + xy' - 4y = 0$ on any interval that excludes $x = 0$.

6.2.1 Exercises

EXERCISE GUIDE

To gain experience . . .	Try exercises
Verifying linear independence	1, 2–3, 12–13
Constructing general solutions	2–3
Verifying linearity of equations	4
Constructing general solutions	5
With proofs of Wronskian relations and linear independence	6–11

1. Determine whether each of the given pairs of functions is linearly independent for all x. When the two functions are not linearly independent for all x, give an interval where they are linearly independent, if possible. If they are not linearly independent, explain why.
   (a) $\cos \omega x$, $\sin \omega x$, ω a positive constant
   (b) $\cos 2x$, $\cos x$
   (c) $\cos 2x$, $\sin 2(x - \pi/4)$
   (d) e^{rx}, e^{sx}, r, s constants, $r \neq s$
   (e) e^{rx}, xe^{rx}, r constant
   (f) e^{rx}, $e^{r(x-1)}$, r constant

   In exercises 2–3:

   i. Determine whether the two functions are linearly independent or linearly dependent solutions of the given differential equation on the interval shown.

   ii. Explain the independence/dependence behavior you observe.

   iii. Write a general solution of the differential equation each time you find a linearly independent pair of solutions.

2. (a) $y'' + y = 0$; $\sin x$, $\cos x$ on $-\infty < x < \infty$
   (b) $y'' + y = 0$; $\sin x$, $\cos(x - \pi/2)$ on $-\infty < x < \infty$
   (c) $y'' + y = 0$; $\sin x$, $\cos(x - \pi/4)$ on $-\infty < x < \infty$

3. (a) $y'' - y = 0$; e^x, e^{-x} on $-\infty < x < \infty$
   (b) $y'' - y = 0$; e^x, e^{1-x} on $-\infty < x < \infty$
   (c) $y'' - y = 0$; e^{-x}, e^{1-x} on $-\infty < x < \infty$

4. Verify that the differential equation

$$y''(x) + P(x)y' + Q(x)y = 0$$

appearing in theorem 6.2 is indeed linear. Can the general linear second-order differential equation

$$a_2(x)y'' + a_1(x)y' + a_0(x)y = r(x)$$

always be written in this form?

5. Example 13 considers the solutions x^2 and x^{-2} of

$$x^2 y'' + xy' - 4y = 0.$$

Write a general solution of this equation for $x \neq 0$. Can you guess a general solution of the nonhomogeneous equation

$$x^2 y'' + xy' - 4y = -4?$$

(*Hint:* Does this equation have a steady state?)

6. To contribute to the proof of theorem 6.2, show by completing the following steps that the Wronskian $W(y_1, y_2)$ satisfies the differential equation

$$\frac{dW}{dx} + P(x)W = 0$$

as a function of x. Recall that y_1, y_2 are twice continuously differentiable solutions of

$$y''(x) + P(x)y' + Q(x)y = 0.$$

(a) Show directly from the definition of W that $dW/dx = y_1 y_2'' - y_2 y_1''$.
(b) Each of the functions y_1 and y_2 satisfies $y'' + Py' + Qy = 0$. Write out the mathematical statements of these facts, multiply the equation for y_1 by y_2 and that for y_2 by y_1. Subtract the two expressions and show that the sum reduces to $dW/dx + PW = 0$.
(c) The preceding steps derived the differential equation for W from the definition of W. Describe a series of steps like these that would allow you to work backwards, establishing that W is a solution of $dW/dx + PW = 0$ by directly substituting in this equation.

7. The proof of theorem 6.2 depends upon the claim that a solution of

$$\frac{dW}{dx} + P(x)W = 0$$

is

$$W = C \exp\left(-\int_a^x P(s)\, ds\right).$$

(a) Verify by direct substitution that the expression $W = C \exp\left(-\int_a^x P(s)\, ds\right)$ is indeed a solution of $dW/dx + PW = 0$.

(b) Use a solution method you learned earlier to derive the solution $W = C \exp\left(\int_a^x P(s)\, ds\right)$ from $dW/dx + PW = 0$.

8. The hypotheses of theorem 6.2 require that the two solutions y_1 and y_2 be twice continuously differentiable and that the coefficient $P(x)$ be continuous. Explain why each of these conditions is imposed.

9. Combine the ideas of the preceding three exercises to write out a complete proof of theorem 6.2.

10. This exercise outlines a proof of theorem 6.3. Suppose that y_1, y_2 are two solutions of

$$y''(x) + P(x)y' + Q(x)y = 0$$

and that their Wronskian $W(y_1, y_2)$ is zero throughout the interval $[a, b]$ upon which the solutions y_1, y_2 are defined.

(a) Argue that if both y_1 and y_2 were identically zero, then they would be linearly dependent.
(b) Now consider the case when at least one of the two is not identically zero; suppose it is y_1. Argue that

$$\frac{W}{y_1^2} = \frac{d}{dx}\left(\frac{y_2}{y_1}\right) = 0$$

and, hence, that y_2/y_1 is constant on any subinterval of $[a, b]$ where $y_1 \neq 0$. Does this complete the proof that y_1 and y_2 are linearly dependent? If not, add the necessary details.

11. State precisely how the linear independence test follows from the three theorems stated in this section.

12. Suppose y_1, y_2 are twice continuously differentiable solutions of

$$y''(x) + P(x)y' + Q(x)y = 0$$

on some interval $0 \leq x \leq x_1$, that P is continuous on that interval, and that $y_1(0) = 1$, $y_1'(0) = 0$, $y_2(0) = 0$, $y_2'(0) = 1$. Show that y_1, y_2 are linearly independent on $0 \leq x \leq x_1$.

13. Suppose y_1, y_2 are twice continuously differentiable solutions of

$$y''(x) + P(x)y' + Q(x)y = 0$$

on some interval $0 \leq x \leq x_1$, that P is continuous on that interval, and that $y_1(0) = a$, $y_1'(0) = b$, $y_2(0) = c$, $y_2'(0) = d$. Give a condition involving the four constants a, b, c, d which guarantees that y_1, y_2 are linearly independent on $0 \leq x \leq x_1$.

6.3 CHARACTERISTIC EQUATIONS: REAL ROOTS

6.3.1 The Possibilities

We saw in section 3.7 that a constant-coefficient, homogeneous, first-order linear equation such as the population equation

$$P' - kP = 0$$

could be solved by finding its *characteristic equation*: Assume a solution of the form $P(t) = e^{rt}$, r a constant, substitute into the differential equation, and divide by the nonzero factor e^{rt} to obtain the characteristic equation

$$r - k = 0.$$

Then we use $r = k$ to write a nontrivial solution $P(t) = e^{kt}$ of $P' = kP$. A general solution is $P_g(t) = Ce^{kt}$, where C is an arbitrary constant.

The same procedure applies to constant-coefficient, homogeneous, second-order, linear equations, but the increase in the order of the equation leads to a few new wrinkles.

The general constant-coefficient, homogeneous, second-order, linear differential equation has the form

$$b_2 y'' + b_1 y' + b_0 y = 0, \tag{6.4}$$

where the coefficients b_2, b_1, b_0 are all constants. Assuming a solution of the form $y = e^{rx}$ and substituting into this equation yields

$$b_2 r^2 e^{rx} + b_1 r e^{rx} + b_0 e^{rx} = 0.$$

Since e^{rx} is never zero, we can divide it out of each term to obtain the **characteristic equation**

$$b_2 r^2 + b_1 r + b_0 = 0.$$

The roots of the characteristic equation are called **characteristic roots** of the differential equation.

EXAMPLE 14 *Find the characteristic equation for the constant-coefficient, homogeneous, second-order linear differential equation*

$$y'' + y' - 12y = 0.$$

Use that characteristic equation to find two solutions of this differential equation.

> A careful reader would argue that this request is carelessly worded. We should ask for two *nontrivial* solutions to prevent the answer, "Well, one solution is $y = 0$. You find the other."

The equation $y'' + y' - 12y = 0$ fits the general form (6.4) with $b_2 = 1$, b_1 and $b_0 = -12$. Assuming $y = e^{rx}$, substituting in the differential equation, and dividing out the nonzero factor e^{rx} leads to

$$r^2 e^{rx} + r e^{rx} - 12 e^{rx} = 0,$$
$$r^2 + r - 12 = 0.$$

The characteristic equation for the given differential equation is $r^2 + r - 12 = 0$. This quadratic may be solved by factoring,

$$r^2 + r - 12 = (r+4)(r-3) = 0.$$

Hence, its roots are $r = -4, 3$, and two solutions of the differential equation $y'' + y' - 12y = 0$ are

$$y_1 = e^{-4x}, \qquad y_2 = e^{3x}. \qquad\qquad\blacksquare$$

The characteristic equation $b_2 r^2 + b_1 r + b_0 = 0$ of the second-order, constant-coefficient differential equation (6.4) is a quadratic equation with real coefficients. Consequently, its roots fall into one of three categories:

1. Two distinct real roots,
2. A single repeated real root, or
3. A pair of complex conjugate roots.

> Recall that a *complex number* is one of the form $\alpha + i\beta$, where α, β are real numbers and $i^2 = -1$; e.g., $3 - 2i$, $-1 + i$, and πi are all complex numbers. The *real part* of the complex number $z = \alpha + i\beta$ is α; we write Re $z = \alpha$. The *imaginary part* of z is β; we write Im $z = \beta$. If $z = 3 - 2i$, then Re $z = 3$ and Im $z = -2i$.
> The complex conjugate of $\alpha + i\beta$ is $\alpha - i\beta$. The complex conjugate of $3 - 2i$ is $3 + 2i$, and the complex conjugate of πi is $-\pi i$.

Examples of these three case follow:

1. $r^2 + r - 12 = (r+4)(r-3) = 0$ has the *distinct real roots* $r = -4, 3$ (as we just saw in the previous example).
2. $r^2 + 8r + 16 = (r+4)(r+4) = 0$ has the single *repeated real root* $r = -4$.
3. $r^2 + 8r + 20 = 0$ has the pair of *complex conjugate roots* $r = 4 + 2i, 4 - 2i$.

In the last case, solve $r^2 + 8r + 20 = 0$ using the quadratic formula:

$$r = \frac{-8 \pm \sqrt{8^2 - 4 \cdot 20}}{2} = 4 \pm i.$$

Note that the negative *discriminant* (the term inside the square root),

$$8^2 - 4 \cdot 20 = -16,$$

guarantees complex roots of this equation because $\sqrt{-16} = 4i$.

In case 1, when the characteristic equation $b_2 r^2 + b_1 r + b_0 = 0$ has two *distinct real roots* $r_1, r_2, r_1 \neq r_2$, then two solutions of $b_2 y'' + b_1 y' + b_0 y = 0$ are

$$e^{r_1 x}, \qquad e^{r_2 x}.$$

6.3 Characteristic Equations: Real Roots

We shall see that these two solutions are linearly independent; they are all we need to construct a general solution of the differential equation (6.4).

In case 2, when there is only *one real root*, we cannot hope to construct a general solution until we have somehow found a second, linearly independent solution. In case 3, when there is a pair of *complex conjugate roots*, we again have two solutions that can be shown to be linearly independent, but we have the problem of interpreting an exponential with a complex exponent.

Resolving the two issues we identified in the preceding paragraph will occupy most of this section and the next. Notice that we avoided these problems with first-order equations because a linear characteristic equation such as $r - k = 0$ from the simple population equation can have only one real solution. We need only one nontrivial solution of a homogeneous first-order differential equation such as $P' - kP = 0$ to construct a general solution.

EXAMPLE 15   *Find the characteristic equation and one solution of*

$$y'' + 8y' + 16y = 0.$$

Assume $y = e^{rx}$ and substitute in $y'' + 8y' + 16y = 0$ to obtain

$$r^2 e^{rx} + 8re^{rx} + 16e^{rx} = 0.$$

Dividing out the exponential factor yields the characteristic equation,

$$r^2 + 8r + 16 = 0.$$

As illustrated in case 2, this characteristic equation has the single real root $r = -4$. Hence, one solution of $y'' + 8y' + 16y = 0$ is

$$y_1 = e^{-4x}.$$

We need other techniques to obtain a second solution.   ∎

EXAMPLE 16   *Find the characteristic equation of*

$$y'' + 8y' + 20y = 0.$$

Assume $y = e^{rx}$ and substitute in $y'' + 8y' + 16y = 0$ to obtain

$$r^2 e^{rx} + 8re^{rx} + 20e^{rx} = 0.$$

Dividing out the exponential factor yields the characteristic equation,

$$r^2 + 8r + 20 = 0.$$

As illustrated in case 3, this characteristic equation has the pair of complex conjugate roots $r = 4 \pm 2i$.   ∎

Based on the work in the last example, we can properly claim that the differential equation $y'' + 8y' + 20y = 0$ has the two solutions $e^{(4+2i)x}$, $e^{(4-2i)x}$, but we will postpone the question of interpreting these complex exponentials.

6.3.2 Distinct Real Roots

When the characteristic equation $b_2 r^2 + b_1 r + b_0 = 0$ has distinct real roots r_1, $r_2, r_1 \neq r_2$, then the two solutions

$$y_1 = e^{r_1 x}, \qquad y_2 = e^{r_2 x}$$

are linearly independent. The general solution of the constant-coefficient, homogeneous equation $b_2 y'' + b_1 y' + b_0 y = 0$ is

$$y_g(x) = C_1 e^{r_1 x} + C_2 e^{r_2 x}.$$

The exercises ask you to verify the claim of linear independence.

For example, we saw in example 1 that two solutions of

$$y'' + y' - 12y = 0$$

are

$$y_1 = e^{-4x}, \qquad y_2 = e^{3x}$$

because its characteristic equation

$$r^2 + r - 12 = 0$$

has the distinct real roots $r = -4, 3$. In example 11, we verified that these two solutions are indeed linearly independent. Hence, a general solution of $y'' + y' - 12y = 0$ is

$$y_g(x) = C_1 e^{-4x} + C_2 e^{3x}.$$

EXAMPLE 17 *Find a general solution of*

$$6y'' - 12y' - 90y = 0$$

that is valid for all x.

From the assumption $y = e^{rx}$ and the usual manipulations, we obtain the characteristic equation

$$6r^2 - 12r - 90 = (3r + 9)(2r - 10) = 0.$$

Its roots are $r = -3, 5$, and two solutions of $6y'' - 12y' - 90y = 0$ are

$$y_1 = e^{-3x}, \qquad y_2 = e^{5x}.$$

To test for linear independence, we evaluate the Wronskian

$$W(e^{-3x}, e^{5x}) = e^{-3x}(e^{5x})' - e^{5x}(e^{-3x})'$$
$$= 5e^{2x} + 3e^{2x} = 8e^{2x}.$$

Since $8e^x \neq 0$ for all x, the solutions e^{-3x}, e^{5x} are linearly independent for all x. A general solution of $6y'' - 12y' - 90y = 0$ valid for all x is

$$y_g(x) = C_1 e^{-3x} + C_2 e^{5x}.$$

∎

6.3.3   Repeated Real Roots

We saw in example 15 that the characteristic equation

$$r^2 + 8r + 16 = 0$$

of the homogeneous differential equation

$$y'' + 8y' + 16y = 0$$

has the single repeated real root $r = -4$. Consequently, this differential equation has one solution,

$$y_1 = e^{-4x}.$$

How can we get a second linearly independent solution from which to construct a general solution?

To understand why the characteristic equation has given us only one solution, let us factor the derivatives appearing in the differential equation:

$$y'' + 8y' + 16y = \frac{d^2y}{dx^2} + 8\frac{dy}{dx} + 16y$$

$$= \left(\frac{d}{dx} + 4\right)\left(\frac{d}{dx} + 4\right)y = 0.$$

What happens when we substitute the solution $y_1 = e^{-4x}$ into this expression ? The two factors $(\frac{d}{dx} + 4)$ appearing in the equation will play different roles. We have boxed the second one to emphasize the difference:

$$\left(\frac{d}{dx} + 4\right)\boxed{\left(\frac{d}{dx} + 4\right)}e^{-4x} = \left(\frac{d}{dx} + 4\right)\boxed{\left(\frac{d}{dx} + 4\right)e^{-4x}}$$

$$= \left(\frac{d}{dx} + 4\right)\boxed{\left(-4e^{-4x} + 4e^{-4x}\right)}$$

$$= \left(\frac{d}{dx} + 4\right)\boxed{0} = 0.$$

The zero appears in the preceding box because e^{-4x} satisfies

$$\left(\frac{d}{dx} + 4\right)e^{-4x} = 0.$$

That is, e^{-4x} is a solution of the original second-order differential equation because it satisfies the *first*-order differential equation $(\frac{d}{dx} + 4)e^{-4x} = 0$. The left-hand factor $(\frac{d}{dx} + 4)$ in the differential equation never had any work to do because the right-hand one reduced e^{-4x} to zero entirely on its own.

This observation suggests that we could obtain a second solution y_2 if we could construct y_2 so that

$$\boxed{\left(\frac{d}{dx} + 4\right) y_2}$$

is a nontrivial function (rather than the zero function obtained previously) which the left-hand factor $(\frac{d}{dx} + 4)$ could reduce to zero. That is, if we let

$$s(x) = \boxed{\left(\frac{d}{dx} + 4\right) y_2(x)}, \qquad (6.5)$$

then we would like to choose y_2 so that

$$y_2'' + 8y_2' + 16y_2 = \left(\frac{d}{dx} + 4\right)\boxed{\left(\frac{d}{dx} + 4\right) y_2(x)}$$

$$= \left(\frac{d}{dx} + 4\right) s(x) = 0.$$

The last line tells us what $s(x)$ can be: Solve $s' + 4s = 0$ and choose $s(x) = e^{-4x}$.

Using $s = e^{-4x}$ in (6.5) provides a first-order differential equation for the second solution y_2,

$$e^{-4x} = \boxed{\left(\frac{d}{dx} + 4\right) y_2(x)},$$

or

$$y_2' + 4y_2 = e^{-4x}.$$

We can solve this first-order, nonhomogeneous, linear, constant-coefficient equation by the method of undetermined coefficients. The solution y_2 has the form

$$y_2 = Axe^{-4x}.$$

The first guess for the particular solution is $y_2 = Ae^{-4x}$, but it satisfies the homogeneous equation. Hence, we multiply it by x.

Since the actual value of A is unimportant here, we can simply take our second solution to be

$$y_2 = xe^{-4x}.$$

The exercises ask you to verify by direct substitution that y_2 is indeed a solution of $y'' + 8y' + 16y = 0$. They also ask you to show that xe^{-4x} is linearly independent of e^{-4x} for all x.

The pattern that this analysis suggests is one of multiplying the first solution by the independent variable to obtain the second. The exercises ask you to verify in general that:

> If the characteristic equation for a constant-coefficient, homogeneous, second-order linear equation has the single *repeated real root* r, then a pair of linearly independent solutions is
>
> $$e^{rx}, \qquad xe^{rx}.$$
>
> A general solution is $C_1 e^{rx} + C_2 xe^{rx}$.

EXAMPLE 18 *Find a general solution of*

$$4y'' - 12y' + 9y = 0.$$

To find the characteristic equation, substitute the assumed solution $y = e^{rx}$ into the differential equation. We obtain

$$4r^2 - 12r + 9 = 4(r - \frac{3}{2})^2 = 0.$$

The characteristic equation evidently has the single repeated real root $r = \frac{3}{2}$. The preceding discussion indicates that two solutions of $4y'' - 12y' + 9y = 0$ are

$$y_1 = e^{3x/2}, \qquad y_2 = xe^{3x/2}.$$

To assess their linear independence, we evaluate the Wronskian of these two solutions:

$$W(e^{3x/2}, xe^{3x/2}) = e^{3x/2}(xe^{3x/2})' - xe^{3x/2}(e^{3x/2})'$$
$$= e^{3x} \neq 0$$

for all x. Hence, these two solutions are linearly independent for all x.

A general solution of $4y'' - 12y' + 9y = 0$ is

$$y_g = C_1 e^{3x/2} + C_2 xe^{3x/2}. \qquad \blacksquare$$

EXAMPLE 19 *Solve the initial-value problem*

$$4y'' - 12y' + 9y = 0, \qquad y(0) = 0, \qquad y'(0) = -3.$$

From the previous example, we have the general solution $y_g = C_1 e^{3x/2} + C_2 xe^{3x/2}$. The initial condition $y(0) = 0$ forces $C_1 = 0$. Then we calculate

$$y_g' = C_2 \left(\frac{3xe^{3x/2}}{2} + e^{3x/2} \right).$$

The second initial condition $y'(0) = -3$ forces $C_2 = -3$. The solution of the initial-value problem is

$$y(x) = -3xe^{3x/2}. \qquad \blacksquare$$

EXERCISE GUIDE

To gain experience . . .	Try exercises
Using the characteristic equation method	1–10, 11–17, 23
Verifying linearly independence of solutions	1–10, 18–19, 20–21(b)
Verifying given functions are solutions	20–22(a)
Finding general solutions	1–10, 18–19, 20–21(c), 23
Solving initial value problems	1–10
With the foundations of the characteristic equation method	11–17, 22(b), 23–24
Understanding repeated roots	22, 24–26

In exercises 1–10:

i. Find the characteristic equation of the given differential equation.

ii. Find two linearly independent solutions of the differential equation.

iii. Verify that these solutions are linearly independent for all values of the independent variable.

iv. Write a general solution of the differential equation.

v. Solve the differential equation subject to the given initial conditions.

1. $2y'' - 2y' - 4y = 0$, $y(0) = 1$, $y'(0) = -4$

2. $y'' = y$, $y(0) = 0$, $y'(0) = 2$

3. $y'' = -2y' + y$, $y(0) = 1$, $y'(0) = -1$

4. $x''(t) + x'(t) - 6x(t) = 0$, $x(1) = 6$, $x'(1) = 0$

5. $2w''(x) - 3w'(x) + w(x) = 0$, $w(0) = 2$, $w'(0) = 1$

6. $8z(t) - 10z'(t) - 3z''(t) = 0$, $z(0) = 2$, $z'(0) = 1$

7. $4y'' - 8y' + 4y = 0$, $y(0) = -3$, $y'(0) = 4$

8. $x''/3 - 2x' - 9x = 0$, $x(0) = 0$, $x'(0) = 1$

9. $12u'' + 5u' - 2u = 0$, $u(1) = 1$, $u'(1) = -1$

10. $y'' + 2ay' + a^2y = 0$, $y(0) = m$, $y'(0) = n$, a, m, n constants

In exercises 11–17, write a constant-coefficient, second-order differential equation that has the given pair of functions as solutions. Write the most general equation you can.

11. e^x, e^{-x}

12. e^{2x}, e^{-3x}

13. e^x, xe^x

14. $-e^{-x}, xe^{-x}$

15. e^{x+1}, e^{2x-1}

16. e^x, π

17. e^x

18. Verify by direct substitution that

$$y_1 = e^{-4x}, \quad y_{3y} = e^{3x}$$

are solutions of the differential equation

$$y'' + y' - 12y = 0$$

of the first example in this section. Show that y_1, y_2 are linearly independent. Construct a general solution of this differential equation.

19. If the characteristic equation

$$b_2 r^2 + b_1 r + b_0 = 0$$

of the general constant-coefficient, homogeneous equation

$$b_2 y'' + b_1 y' + b_0 y = 0$$

has the distinct real roots r_1, r_2, $r_1 \neq r_2$, then the differential equation has two solutions:

$$y_1 = e^{r_1 x}, \qquad y_2 = e^{r_2 x}.$$

Use the linear independence test to verify that these two solutions are linearly independent and, hence, that a general solution of this differential equation is

$$y_g(x) = C_1 e^{r_1 x} + C_2 e^{r_2 x}.$$

Is the condition $r_1 \neq r_2$ essential for linear independence? Is the condition that the roots be real essential? (*Hint:* example 17 demonstrates linear independence of solutions when the roots of the characteristic equation are $r = -3, 5$. Can you use it as a guide to the general situation considered here?)

20. The discussion at the end of this section considers the constant-coefficient, homogeneous equation

$$y'' + 8y' + 16y = 0,$$

whose characteristic equation has the single repeated real root -4. Hence, one solution of this equation is $y_1 = e^{-4x}$. The text claims that a second linearly independent solution is $y_2 = xe^{-4x}$.
   (a) Verify by substitution in $y'' + 8y' + 16y = 0$ that $y_2 = xe^{-4x}$ is indeed a solution.
   (b) Verify that these two solutions are linearly independent for all x.
   (c) Write a general solution of $y'' + 8y' + 16y = 0$.

21. The discussion at the end of this section claims

   If the characteristic equation for a constant-coefficient, homogeneous, second-order linear equation has the single repeated real root r, then a pair of linearly independent solutions is e^{rx}, xe^{rx}. A general solution is $C_1 e^{rx} + C_2 xe^{rx}$.

Verify this claim by completing the following steps for the general equation

$$b_2 y'' + b_1 y' + b_0 y = 0.$$

You may wish to generalize your solution of the preceding exercise, which asks for much the same analysis in the context of a specific equation.
   (a) Verify by direct substitution that y_2 is a solution of $b_2 y'' + b_1 y' + b_0 y = 0$. (*Hint:* Use the quadratic equation to find r in terms of the coefficients b_i. Remember that r is a repeated root. What does that fact tell you about the discriminant $b_1^2 - 4b_2 b_0$?)
   (b) Verify that y_1 and y_2 are linearly independent for all x.
   (c) Write a general solution of $b_2 y'' + b_1 y' + b_0 y = 0$.

22. (a) Verify by direct substitution that

$$y_1 = e^{3x/2}, \qquad y_2 = xe^{3x/2}$$

   are solutions of

$$4y'' - 12y' + 9y = 0,$$

the differential equation considered in example 18.
   (b) Write the differential equation $4y'' - 12y' + 9y = 0$ in the factored form used in the text,

$$4 \left(\frac{d}{dx} - \frac{3}{2} \right) \boxed{\left(\frac{d}{dx} - \frac{3}{2} \right) y} = 0.$$

Mimic the analysis that precedes example 18 to answer the following questions:

   i. When $y = e^{3x/2}$, what expression does the factor $\left(\frac{d}{dx} - \frac{3}{2} \right)$ outside the box operate on? Is that expression a solution of $dy/dx - 3y/2 = 0$?

   ii. When $y = xe^{3x/2}$, what expression does the factor $\left(\frac{d}{dx} - \frac{3}{2} \right)$ outside the box operate upon? Is that expression a solution of $dy/dx - 3y/2 = 0$?

23. By studying the discriminant $b_1^2 - 4b_2 b_0$ of the characteristic equation $b_2 r^2 + b_1 r + b_0 = 0$, give conditions on the coefficients b_2, b_1, b_0 that guarantee that the constant-coefficient differential equation

$$b_2 y'' + b_1 y' + b_0 = 0$$

has two linearly independent solutions of the given form.
   (a) $e^{r_1 x}$, $e^{r_2 x}$,
   (b) e^{rx}, xe^{rx}.
For each case, write out the general solution in terms of the coefficients b_2, b_1, b_0.

24. (a) Find the undetermined coefficient A in the assumed form

$$y_2 = Axe^{-4x}$$

   of a particular solution of

$$y_2' + 4y_2 = e^{-4x},$$

the first-order equation we encountered when we were trying to find a second solution of the equation

$$y'' + 8y' + 16y = 0,$$

whose characteristic equation has a single repeated real root.
   (b) Verify that y_2 is a solution of $y'' + 8y' + 16y = 0$ for *any* choice of the constant A, including the

value you find. This last argument justifies taking $A = 1$ in the discussion in the text.

25. The differential equation

$$y'' + 8y' + 16y = 0$$

has the single repeated characteristic root $r = -4$, and one solution is obviously $y_1 = e^{-4x}$. The last part of this section argues that a second solution is xe^{-4x}. This exercise outlines an alternate approach, known as *reduction of order*, to deriving that second solution.

(a) Assume that a second solution has the form $y_2 = u(x)e^{-4x}$, where $u(x)$ is an unknown function. Substitute y_2 into the differential equation and show that it reduces to $u'' = 0$.

(b) Argue that one solution of $u'' = 0$ is $u(x) = x$ and, hence, that a second solution of the differential equation is $y_2(x) = xe^{-4x}$.

26. Repeat the preceding exercise for the general equation $b_2 y'' + b_1 y' + b_0 y = 0$ with a single repeated characteristic root. (*Hint:* What is the value of the discriminant $b_1^2 - 4b_2 b_0$ when the characteristic equation has a single repeated root?)

6.4 CHARACTERISTIC EQUATIONS: COMPLEX ROOTS

The preceding section explored the solutions of constant-coefficient, linear, homogeneous differential equations whose characteristic equations had distinct or repeated real roots. This section determines the form of the solution when the characteristic equation has complex conjugate roots.

6.4.1 Complex Conjugate Roots

We saw earlier that the characteristic equation

$$r^2 + 8r + 20 = 0$$

for the differential equation

$$y'' + 8y' + 20y = 0$$

has the pair of complex conjugate roots $r = 4 + 2i, 4 - 2i$. While we can certainly write $e^{(4+2i)x}$, $e^{(4-2i)x}$ as solutions of $y'' + 8y' + 20y = 0$, we do not know the significance of such complex exponentials. We seek a way of expressing these complex exponentials in terms of real-valued functions.

To make such a conversion, recall the Taylor series about the origin for the exponential, the cosine, and the sine

$$e^x = 1 + x + \frac{x^2}{2} + \frac{x^3}{6} + \frac{x^4}{24} + \frac{x^5}{120} + \cdots,$$

$$\cos x = 1 - \frac{x^2}{2} + \frac{x^4}{24} + \cdots,$$

$$\sin x = x - \frac{x^3}{6} + \frac{x^5}{120} + \cdots,$$

Now let β be a real number, and write the series for the exponential with $x = i\beta$,

$$e^{i\beta} = 1 + i\beta + \frac{(i\beta)^2}{2} + \frac{(i\beta)^3}{6} + \frac{(i\beta)^4}{24} + \frac{(i\beta)^5}{120} + \cdots$$

$$= 1 - \frac{\beta^2}{2} + \frac{\beta^4}{24} + \cdots$$

$$+ i\left(\beta + \frac{\beta^3}{6} + \frac{\beta^5}{120} + \cdots\right)$$

$$= \cos\beta + i\sin\beta.$$

The last equality follows from comparing the preceding series for $e^{i\beta}$ with those for $\cos x$ and $\sin x$ using $x = \beta$.

This relationship is called **Euler's formula**:

$$e^{i\beta} = \cos\beta + i\sin\beta.$$

EXAMPLE 20   *Use Euler's formula to write two linearly independent,* real-valued *solutions of*

$$y'' + 8y' + 20y = 0.$$

We already know that the characteristic equation $r^2 + 8r + 20 = 0$ has the complex conjugate roots $4 \pm 2i$. Hence, the differential equation has the complex exponential solutions

$$z_1 = e^{(4+2i)x}, \qquad z_2 = e^{(4-2i)x}.$$

Using Euler's formula with $\beta = 2x$, we may rewrite z_1 as

$$z_1 = e^{(4+2i)x} = e^{4x}e^{i(2x)}$$

$$= e^{4x}(\cos 2x + i\sin 2x).$$

In the same way, we may write

$$z_2 = e^{4x}(\cos 2x - i\sin 2x).$$

Since both z_1 and z_2 are solutions of the homogeneous linear differential equation $y'' + 8y' + 20y = 0$, the principle of superposition guarantees that any linear combination of them will also be a solution. Is there a linear combination that would eliminate the imaginary terms from z_1 and z_2? Indeed, the sum of these two expressions is real-valued:

$$z_1 + z_2 = 2e^{4x}\cos 2x.$$

We remind you that the function $2e^{4x}\cos 2x$ is still a solution of $y'' + 8y' + 20y = 0$ because it is a linear combination of solutions of this homogeneous linear equation.

Since any constant multiple of a solution of a homogeneous linear equation is again a solution of that equation, we can multiply this new solution by $\frac{1}{2}$ to obtain a real-valued solution of $y'' + 8y' + 20y = 0$:

$$y_1 = e^{4x}\cos 2x.$$

Another linear combination gives a purely imaginary solution,

$$z_1 - z_2 = 2ie^{4x}\sin 2x.$$

Using the same reasoning as in the preceding paragraph, we can multiply this homogeneous solution by $1/2i$ to obtain a second real-valued solution of $y'' + 8y' + 20y = 0$:

$$y_2 = e^{4x}\sin 2x.$$

The exercises ask you to verify by direct substitution that y_1 and y_2 are indeed solutions of the differential equation.

To show that y_1, y_2 are linearly independent solutions, we evaluate their Wronskian,

$$
\begin{aligned}
W(e^{4x}\cos 2x, e^{4x}\sin 2x) &= e^{4x}\cos 2x(e^{4x}\sin 2x)' - e^{4x}\sin 2x(e^{4x}\cos 2x)' \\
&= 2e^{4x}(\cos^2 2x + \sin^2 2x) = 2e^{4x} \neq 0
\end{aligned}
$$

for all x. Hence, for all x, a real-valued general solution of $y'' + 8y' + 20y = 0$ is

$$y_g = C_1 e^{4x}\cos 2x + C_2 e^{4x}\sin 2x. \qquad \blacksquare$$

Obviously (meaning you can do it in the exercises), this analysis can be carried over without change to the general case:

A differential equation whose characteristic equation has the complex conjugate roots $r = \alpha + i\beta$, $\alpha - i\beta$ has the general solution

$$y_g = C_1 e^{\alpha x}\cos \beta x + C_2 e^{\alpha x}\sin \beta x.$$

The real part of the characteristic root appears in the exponential, and the complex part appears as the angular frequency in the cosine and sine.

> The constant ω in either $\cos \omega x$ or $\sin \omega x$ is called the **angular frequency**, or **circular frequency**, because it measures the number of radians by which the angle ωx in either the cosine or the sine increases when x increases by one unit. The quotient $\omega/2\pi$ is the **cyclic frequency**, or simply **frequency**; it measures the number of complete cycles executed by the cosine or sine when x increases by one unit. The inverse $2\pi/\omega$ of cyclic frequency is the **period**, the time required to complete one full cycle.
>
> Angular frequency is measured in radians per unit of x — e.g., in radians per second if x is measured in seconds. Frequency is measured in cycles per unit of x — e.g., in cycles per second, or hertz, if x is measured in seconds. Period has units of time, such as seconds.

6.4.2 More Examples

Examples of equations with both real and complex characteristic roots follow.

EXAMPLE 21 *Solve the initial-value problem*

$$\frac{3}{16}x''(t) + 1.41x'(t) + 2x(t) = 0, \qquad x(0) = 0.75, \qquad x'(0) = 0,$$

which models the damped spring-mass system of example 2, section 5.1.4.

The usual assumption $x(t) = e^{rt}$ leads to the characteristic equation

$$\frac{3}{16}r^2 + 1.41r + 2 = 0.$$

The quadratic formula reveals that its roots are approximately $r = -1.90, -5.62$. Hence, two linearly independent solutions of the differential equation are $e^{-1.87t}$, $e^{-5.65t}$, and a general solution is

$$x_g(t) = C_1 e^{-1.90t} + C_2 e^{-5.62t}.$$

To solve the initial-value problem, we use the initial conditions to determine C_1, C_2. These conditions require

$$x_g(0) = C_1 + C_2 = 0.75,$$
$$x_g'(0) = -1.90C_1 - 5.62C_2 = 0.$$

Solving this pair of simultaneous equations yields

$$C_1 = 1.13, \qquad C_2 = -0.38.$$

The solution of the given initial-value problem is

$$x(t) = 1.13e^{-1.90t} - 0.38e^{-5.62t}. \qquad \blacksquare$$

It appears that this spring-mass system does not oscillate and that it approaches a steady state of $x = 0$. The solution itself is the transient component of the response.

Such a system is called *overdamped*. We will explore these physical interpretations more deeply in the next chapter.

EXAMPLE 22 *Consider the spring-mass system of the preceding example with a slightly stronger spring, one with a spring constant $k = 2.67$ lb/ft. The governing initial-value problem is*

$$\frac{3}{16}x''(t) + 1.41x'(t) + 2.67x(t) = 0, \qquad x(0) = 0.75, \qquad x'(0) = 0.$$

Find its solution.

As before, we find the characteristic equation

$$\frac{3}{16}r^2 + 1.41r + 2.67 = 0.$$

Since its discriminant

$$(1.41)^2 - 4\frac{3}{16}2.67$$

is zero, this characteristic equation has the single repeated real root $r = -3.76$. Two linearly independent solutions of the differential equation are thus

$e^{-3.76t}, te^{-3.76t}$. A general solution is

$$x_g(t) = C_1 e^{-3.76t} + C_2 t e^{-3.76t}.$$

The initial conditions lead to two equations for the constants C_1, C_2,

$$x_g(0) = C_1 = 0.75,$$
$$x'_g(0) = -3.76C_1 + C_2 = 0.$$

Their solution is $C_1 = 0.75$, $C_2 = 2.82$.

The solution of the given initial-value problem is

$$x(t) = 0.75e^{-3.76t} + 2.82te^{-3.76t}. \qquad \blacksquare$$

This solution also decays to a steady state of $x = 0$ without oscillating. A system such as this with a single characteristic root is called *critically damped* because, as we shall see, it separates overdamped systems such as that of the preceding example from damped, oscillating systems such as the one in the next example.

EXAMPLE 23   *Now consider the spring-mass system of the last example with a still stronger spring, one with spring constant $k = 4$ lb/ft. The governing initial-value problem is*

$$\frac{3}{16}x''(t) + 1.41x'(t) + 4x(t) = 0, \qquad x(0) = 0.75, \qquad x'(0) = 0.$$

Find its solution.

The usual argument yields the characteristic equation

$$\frac{3}{16}r^2 + 1.41r + 4 = 0.$$

It roots are $r = -3.76 + 2.68i, -3.76 - 2.68i$.

Rather than write the complex exponential form of the solutions arising from these roots and form linear combinations to obtain real-valued solutions, we proceed directly to the real-valued solutions. The real part -3.76 of the root governs the exponential, and the imaginary part 2.68 is the frequency in the cosine and sine terms. The general solution is

$$x_g(t) = e^{-3.76t}(C_1\cos 2.68t + C_2\sin 2.68t).$$

The initial conditions require that C_1, C_2 be the solution of

$$x_g(0) = C_1 = 0.75,$$

$$x'_g(0) = -3.76C_1 + 2.68C_2 = 0.$$

The solution of this pair of equations is $C_1 = 0.75$, $C_2 = 1.05$.

The solution of the given initial-value problem is

$$x(t) = e^{-3.76t}(0.75\cos 2.68t + 1.05\sin 2.68t). \qquad \blacksquare$$

The solution oscillates because of the cosine and sine terms, but the decaying exponential still leads to a steady state of $x = 0$. Such a system is called *underdamped*.

EXAMPLE 24 *Solve the initial-value problem*

$$\frac{3}{16}x''(t) + 2x(t) = 0, \qquad x(0) = 0, \qquad x'(0) = -0.5,$$

which models the undamped spring-mass system of example 1, section 5.1.

The characteristic equation is

$$\frac{3}{16}r^2 + 2 = 0.$$

It has the pure imaginary roots $r = \pm\sqrt{32/3}\,i \approx 3.27i$. Since the real part of these complex roots is zero, the general solution contains no exponential term. However, the imaginary part 3.27 enters as the frequency in the cosine and sine terms. The general solution of this differential equation is

$$x_g(t) = C_1 \cos 3.27t + C_2 \sin 3.27t.$$

The initial conditions require

$$x_g(0) = C_1 = 0,$$
$$x_g'(0) = 3.27C_2 = -0.5.$$

The solution of this pair of equations is obviously $C_1 = 0$, $C_2 = -0.15$.
The solution of the given initial-value problem is

$$x(t) = -0.15 \sin 3.27t. \qquad\qquad \blacksquare$$

This solution exhibits no decay whatsoever; this system is *undamped*. It has a periodic solution with period $2\pi/3.27 = 1.92$ s.

> Equivalently, the circular frequency of this solution is 3.27 rad/s. Its cyclic frequency is $3.27/2\pi = 0.52$ Hz. In other words, it takes 1.92 s for the mass to complete a round trip, bouncing down and then back up to its starting position, or it completes 0.52 round trips every second.

6.4.3 Summary of Characteristic Equations

The constant-coefficient, homogeneous, second-order linear equation

$$b_2 y'' + b_1 y' + b_0 y = 0$$

can be solved by assuming $y = e^{rx}$, r a constant. Substituting this proposed solution in the differential equation and dividing by the exponential leaves the *characteristic equation*

$$b_2 r^2 + b_1 r + b_0 = 0.$$

Since the coefficients b_i are real, this quadratic equation may have

1. Distinct real roots r_1, r_2 if the discriminant $b_1^2 - 4b_2b_0 > 0$.
2. A single real root r if the discriminant $b_1^2 - 4b_2b_0 = 0$.
3. Complex conjugate roots $\alpha \pm i\beta$ if the discriminant $b_1^2 - 4b_2b_0 < 0$.

The corresponding real-valued linearly independent solutions of $b_2 y'' + b_1 y' + b_0 y = 0$ are

Distinct real roots $r_1, r_2:$ $y_1 = e^{r_1 x}, \; y_2 = e^{r_2 x}$
Repeated real root $r:$ $y_1 = e^{rx}, \; y_2 = xe^{rx}$
Complex conjugate roots $\alpha \pm i\beta:$ $y_1 = e^{\alpha t} \sin \beta t, \; y_2 = e^{\alpha t} \cos \beta t$

6.4.4 Exercises

EXERCISE GUIDE

To gain experience . . .	Try exercises
Using the characteristic equation method	1–16
Verifying linearly independence of solutions	1–16, 28(d)
Verifying given functions are solutions	26(b)
Deriving real-valued solutions from complex-valued solutions	26–29
Finding general solutions	1–23, 24–25(a), 27(a), 28(e)
Solving initial-value problems	17–23, 24(e), 25(d), 27(b)
Finding steady states	30, 31
Analyzing solution behavior	24(b–d, f), 25(b, c), 33–34
With the foundations of the characteristic equation method	32–33

In exercises 1–16:

i. Find the characteristic equation of the given differential equation. If the concept of characteristic equation is not appropriate, explain why and go on to the next exercise.

ii. Find two linearly independent solutions of the differential equation.

iii. Verify that these solutions are linearly independent for all values of the independent variable.

iv. Write a general solution of the differential equation.

1. $y'' - 9y = 0$

2. $4x''(t) + 8x(t) + 5x(t) = 0$

3. $w''(x) + 9w(x) = 0$

4. $2z''(t) - 7z'(t) + 3z(t) = 0$

5. $y'' - 4y' + 40y = 0$

6. $y'' - 4y' + 40y = \cos \pi x$

7. $x'' + x' - 6x = 0$

8. $9y'' + 36y' + 4y = 0$

9. $9y'' + 36y' + 4y^2 = 0$

10. $mx''(t) + kx(t) = 0$, m, k constants

11. $mx''(t) + kx(t) = k[h(t) - h(0)]$, m, k constants

12. $L\theta'' + g\theta = 0$, L, g constants

13. $L\theta'' + g \sin \theta = 0$, L, g constants

14. $mx'' + px' + kx = 0$, m, p, k constants, $p^2 - 4mk > 0$

15. $mx'' + px' + kx = 0$, m, p, k constants, $p^2 - 4mk = 0$

16. $mx'' + px' + kx = 0$, m, p, k constants, $p^2 - 4mk < 0$

In exercises 17–23:

> **i.** Write a general solution of the differential equation.
>
> **ii.** Solve the initial-value problem.

17. $u'' - 4u' + 40u = 0$, $u(\pi/6) = 1$, $u'(\pi/6) = -1$

18. $9x'' + 36x' + 4x = 0$, $x(0) = 2$, $x'(0) = 3$

19. $4y'' + 8y' + 5y = 0$, $y(\pi) = 0$, $y'(\pi) = -4$

20. $z'' + z' - 6z = 0$, $z(0) = 4$, $z'(0) = -6$

21. $x'' + 9x = 0$, $x(\pi) = 3$, $x'(\pi) = -9$

22. $2w'' - 7w' + 3w = 0$, $w(0) = 8$, $w'(0) = 12$

23. $x'' - 9x = 0$, $x(0) = 18$, $x'(0) = 18$

24. In the undamped spring-mass system model derived in section 5.1,

$$mx''(t) + kx(t) = 0, \qquad x(0) = x_i, \qquad x'(0) = v_i,$$

the parameters m, k in the differential equation are positive constants.

(a) Find a general solution of the differential equation.

(b) Argue that your solution is purely oscillatory, neither growing nor decaying with time.

(c) Find the angular frequency and period of these oscillations.

(d) Argue that the period of the oscillations in this spring-mass system depends only on the mass m and the spring constant k, not upon the initial conditions; i.e., the period of the oscillations is independent of how the motion is initiated.

(e) Find the solution of this initial-value problem.

(f) Find the maximum displacement of the mass from its equilibrium position. Does it reduce to the value you expect when the initial velocity v_i is zero?

25. In the damped spring-mass system model derived in section 5.1,

$$mx''(t) + px'(t) + kx(t) = 0,$$

$$x(0) = x_i, \qquad x'(0) = v_i,$$

the parameters m, p, k in the differential equation are positive constants.

(a) Find a general solution of this differential equation. If you need to consider different ranges of the parameters m, p, k, state them clearly.

(b) Argue that your solution is decaying toward a steady state of zero, regardless of the values of $m, p,$ and k. Which parameter(s) in the equation determine(s) the rate of decay? Are they the one(s) you expect?

(c) For what ranges of the parameters does the solution of this equation exhibit oscillations? Are the underlying oscillations of the same frequency as those of the undamped spring-mass model? (See exercise 24.)

(d) Find the solution of the initial-value problem.

26. The first example in this section begins with two complex exponential solutions

$$z_1 = e^{(4+2i)x}, \qquad z_2 = e^{(4-2i)x}$$

of the differential equation

$$y'' + 8y' + 20y = 0.$$

It then obtains two new solutions

$$y_1 = e^{4x}\cos 2x, \qquad y_2 = e^{4x}\sin 2x$$

as linear combinations of z_1, z_2.

(a) Write y_1 and y_2 as linear combinations of z_1, z_2.

(b) Verify by substitution that y_1, y_2 are indeed solutions of $y'' + 8y' + 20y = 0$. Why is this direct verification actually unnecessary?

27. The last example in this section shows that the differential equation

$$\frac{3}{16}x''(t) + 2x(t) = 0,$$

from the undamped spring-mass model of example 1, section 5.1, has the characteristic equation

$$\frac{3}{16}r^2 + 2 = 0.$$

The roots of this equation are $r = \pm\sqrt{32/3}\,i \approx \pm3.27i$. Consequently, $3x''/16 + 2x = 0$ has the linearly independent solutions

$$z_1 = e^{3.27it}, \qquad z_2 = e^{-3.27it}.$$

(a) Write a general solution of $3x''/16 + 2x = 0$ in terms of the complex exponential solutions z_1, z_2.

(b) Choose the constants in your answer to part (a) so that your solution satisfies the initial conditions

$$x(0) = 0, \qquad x'(0) = -0.5.$$

(Your constants will be complex numbers.)

(c) Use Euler's formula to reduce your (apparently) complex-valued solution of this initial-value problem to that obtained in the example in the text,

$$x(t) = -0.15 \sin 3.27t.$$

28. The differential equation

$$\frac{3}{16}x''(t) + 1.41x'(t) + 4x(t) = 0,$$

considered in example 23 has the characteristic equation

$$\frac{3}{16}r^2 + 1.41r + 4 = 0.$$

Its roots are $r = -3.76+2.68i, -3.76-2.68i$. The example proceeded from these roots directly to a real-valued general solution of the differential equation. This exercise asks you to derive that real-valued solution in more detail.

(a) Write a general solution of $3x''/16 + 1.41x' + 4x = 0$ in terms of complex exponentials.

(b) Choose a set of values for the arbitrary constants in your general solution that reduces it, via Euler's formula, to the real-valued function $e^{-3.76t} \cos 2.68t$.

(c) Choose another set of values for the arbitrary constants in your general solution that reduces it, via Euler's formula, to the real-valued function $e^{-3.76t} \sin 2.68t$.

(d) Show that the solutions obtained in parts (b) and (c) are linearly independent.

(e) Write a general solution of $3x''/16 + 1.41x' + 4x = 0$ using these real-valued solutions.

29. Suppose that the characteristic equation for a homogeneous, constant-coefficient, second-order, linear differential equation has the complex conjugate roots $r = \alpha \pm i\beta$. Using the preceding exercise as a guide, argue that a general solution of such a differential equation is

$$e^{\alpha x}(C_1 \cos \beta x + C_2 \sin \beta x).$$

30. Find the steady state, if it exists, of each of the initial-value problems in exercises 17–23.

31. The preceding exercise asks you to find steady states, if they exist, for a series of initial-value problems. Can you determine if a steady state exists simply by looking at the roots of the characteristic polynomial? What root must the polynomial possess to give a (nonzero) steady-state solution? Could you predict the steady state of an unforced initial value problem from knowledge of the roots of the characteristic

polynomial alone? Justify your reasoning. Try your ideas out on some of the initial-value problems of exercises 17–23.

32. The characteristic equation for the general homogeneous, constant coefficient, linear, second-order equation

$$b_2 y'' + b_1 y' + b_0 y = 0$$

is

$$b_2 r^2 + b_1 r + b_0 = 0.$$

The discriminant of this quadratic equation is

$$d = b_1^2 - 4b_2 b_0.$$

To justify the statements in the summary at the end of this section, argue that the sign of d determines which of three cases (real distinct, single repeated, or complex conjugate roots of $b_2 r^2 + b_1 r + b_0 = 0$) occurs. Specifically, show each of the following.

(a) If $d > 0$, then the characteristic equation has real, distinct roots.

(b) If $d = 0$, then the characteristic equation has a single repeated real root.

(c) If $d < 0$, then the characteristic equation has a pair of complex conjugate roots.

33. Suppose the discriminant of a characteristic equation is negative. (See the previous exercise.) What condition(s) on the coefficients of the homogeneous differential equation guarantee(s) the solution will decay to a steady state of zero?

34. Example 22 considers the initial-value problem

$$\frac{3}{16}x''(t) + 1.41x'(t) + 2.67x(t) = 0,$$

$$x(0) = 0.75, \qquad x'(0) = 0,$$

and claims that it models the same spring-mass system as does the initial-value problem

$$\frac{3}{16}x''(t) + 1.41x'(t) + 2x(t) = 0,$$

$$x(0) = 0.75, \qquad x'(0) = 0,$$

but for a slightly stiffer spring. Justify this claim. How is the term *stiffer spring* reflected in relative values of spring constants?

6.5 UNDETERMINED COEFFICIENTS

Section 3.8 showed that the method of undetermined coefficients can find particular solutions of *constant-coefficient, linear*, first-order equations whose forcing terms are polynomials, sines, cosines, or exponentials. (See table 3.3 for a quick review.) Those ideas extend immediately to constant-coefficient, second-order, linear equations

$$b_2 y'' + b_1 y' + b_0 y = f(x)$$

with forcing terms $f(x)$ from the same family.

The essential idea of undetermined coefficients is to assume a trial particular solution that involves the same sorts of functions as appear in the forcing term. For example, if the forcing term is

$$\sin 2\pi t,$$

then the trial particular solution is

$$A \cos 2\pi t + B \sin 2\pi t.$$

The coefficients A and B are determined by substituting the trial solution into the differential equation.

If part of a trial particular solution is a solution of the *homogeneous* equation, then the coefficient of that part will disappear when it is substituted into the differential equation. Hence, its coefficient will remain undetermined. We discovered that this difficulty could be circumvented by multiplying the first trial solution by the independent variable. See the examples in section 3.8.2 for illustrations of this process.

In this situation lies the only real difference between undetermined coefficients for first-order equations and second-order equations. Multiplying a trial solution of a second-order equation by the independent variable may yet leave a solution of the homogeneous equation in the trial solution. In that case, we simply multiply by the independent variable one more time.

The rules for efficiently guessing a form of the particular solution (or, to be more elegant, constructing a trial solution) are summarized in table 6.1. When

TABLE 6.1

Particular solutions via undetermined coefficients for the constant-coefficent, second-order linear equation $b_2 y'' + b_1 y' + b_0 y = f(x)$

Forcing Term $f(x)$	Trial Particular Solution
$a_n x^n + \ldots + a_1 x + a_0$	$A_n x^n + \ldots + A_1 x + A_0$
$(a_n x^n + \ldots + a_1 x + a_0) e^{qx}$	$(A_n x^n + \ldots + A_1 x + A_0) e^{qx}$
$(a_n x^n + \ldots + a_1 x + a_0) \cos px$	$(A_n x^n + \ldots + A_1 x + A_0) \cos px$
$+ (b_n x^n + \ldots + b_1 x + b_0) \sin px$	$+ (B_n x^n + \ldots + B_1 x + B_0) \sin px$
$a e^{qx} \cos px + b e^{qx} \sin px$	$A e^{qx} \cos px + B e^{qx} \sin px$

If the assumed form of the particular solution solves the corresponding homogeneous equation, multiply the assumed form by x. Repeat if necessary.

Lowercase letters a, a_i, b, b_i are constants given in the forcing function. The coefficients to be determined are denoted by uppercase letters A, A_i, B, B_i.

you compare them with the rules for first-order equations in table 3.1, you will see that we have added several natural extensions, such as forcing terms which are the product of a polynomial and an exponential.

In simplest terms, the important points for efficiently forming trial particular solutions are these:

- Guess a trial particular solution of the same form as the forcing function.
- If the forcing term involves a sine *or* a cosine, your trial solution should contain *both* a sine *and* a cosine.
- If any part of your trial solution is a solution of the homogeneous differential equation, multiply the trial solution by the independent variable. Repeat if necessary.

We illustrate these rules by examples, beginning with a situation in which our first trial solution includes a term that is a solution of the homogeneous equation.

EXAMPLE 25   *Find a particular solution of*

$$x''(t) = -g,$$

the equation for the position of a rock moving vertically; g is the constant of gravitation. (See exercise 5 of the chapter 1 exercises.)

This equation is certainly a constant-coefficient, linear, second-order equation. Since its forcing term is a constant (i.e., a polynomial of degree $n = 0$), table 6.1 directs us to use the trial solution

The first trial particular solution ...

$$x_p(t) = A_0,$$

where A_0 is a coefficient to be determined. But substituting x_p into $x'' = -g$ immediately yields

... fails.

$$x_p'' = (A_0)'' = 0.$$

Our trial solution is a solution of the homogeneous equation. We can not determine A_0.

Abandoning table 6.1 for the moment, we can solve $x'' = -g$ directly by two successive antiderivatives,

$$x'(t) = -\int g\, dt = -gt + C_1,$$

$$x(t) = \int (-gt + C_1)\, dt = -gt^2/2 + C_1 t + C_2,$$

where C_1, C_2 are arbitrary constants. Substituting this expression for $x(t)$ into the left side of $x'' = -g$ reveals

$$x''(t) = \left(\frac{-gt^2}{2}\right)'' + (C_1 t)'' + (C_2)'' = -g + 0 + 0.$$

6.5   Undetermined Coefficients

That is, a particular solution of $x'' = -g$ is

$$x_p(t) = \frac{-gt^2}{2}.$$

The other terms in the solution, $C_1 t + C_2$, satisfy the homogeneous equation $x''(t) = 0$.

We could have arrived at this same particular solution by sticking with the guidelines of Table 6.1. We would begin by finding the solutions of the homogeneous version of this equation,

$$x''(t) = 0.$$

Since its characteristic equation $r^2 = 0$ has the repeated real root $r = 0$, a pair of linearly independent solutions of $x''(t) = 0$ is

LINEARLY INDEPENDENT HOMOGENEOUS SOLUTIONS

$$x_1 = e^{0t} = 1, \qquad x_2 = te^{0t} = t.$$

We see immediately that our first trial solution $x_p(t) = A_0$ is just a multiple of the homogeneous solution $x_1 = 1$. Following table 6.1, we multiply the trial solution by the independent variable to obtain a new trial solution,

SECOND TRIAL PARTICULAR SOLUTION

$$x_p(t) = A_0 t.$$

But this trial solution is just a multiple of the second homogeneous solution $x_2 = t$. Hence, we again multiply by the independent variable to obtain yet another trial solution

THIRD TRIAL SOLUTION

$$x_p(t) = A_0 t^2,$$

that, at last, does not contain a solution of the homogeneous differential equation.

Substituting this trial solution into the original differential equation

Substitute the trial solution. $x'' = -g$ yields

$$x_p''(t) = (A_0 t^2)'' = 2A_0 = -g$$

or

Find the undetermined coefficient.

$$A_0 = \frac{-g}{2}.$$

Hence, the method of undetermined coefficients yields a particular solution of $x'' = -g$,

PARTICULAR SOLUTION

$$x_p(t) = \frac{-gt^2}{2}.$$

With no additional effort, we can couple this particular solution with the homogeneous solutions found above to obtain the general solution

GENERAL SOLUTION

$$x_g(t) = \frac{-gt^2}{2} + C_1 t + C_2,$$

the solution we obtained by antidifferentiation.   ∎

This example motivates the strategy of multiplying a trial particular solution that satisfies the homogeneous equation by the independent variable. Do you see any similarities between this example and example 36 of section 3.8, which motivated the same strategy for first-order equations?

The next example treats a similar situation, but for a differential equation that can not be solved by antidifferentiation.

EXAMPLE 26 *Find a particular solution of*

$$y'' + 4y' + 4y = 8e^{-2x}.$$

To find linearly independent solutions of the corresponding homogeneous equation,

$$y'' + 4y' + 4y = 0,$$

we find the roots of its characteristic equation,

$$r^2 + 4r + 4 = (r + 2)^2 = 0.$$

Since the characteristic equation has the repeated real root $r = -2$, linearly independent solutions of the homogeneous equation are

$$y_1 = e^{-2x}, \qquad y_2 = xe^{-2x}.$$

LINEARLY INDEPENDENT
HOMOGENEOUS SOLUTIONS

Because the forcing term in $y'' + 4y' + 4y = 8e^{-2x}$ is an exponential, table 6.1 suggests a trial particular solution of the same form,

$$y_p(x) = A_0 e^{-2x}.$$

FIRST TRIAL PARTICULAR SOLUTION

But this trial solution satisfies the homogeneous equation $y'' + 4y' + 4y = 0$ because it is a constant multiple of the homogeneous solution $y_1 = e^{-2x}$. We multiply by the independent variable to obtain a new trial solution,

$$y_p(x) = A_0 x e^{-2x}.$$

SECOND TRIAL SOLUTION

Now we have a multiple of the homogeneous solution $y_2 = xe^{-2x}$. One more multiplication of the trial solution by x yields

$$y_p(x) = A_0 x^2 e^{-2x}.$$

THIRD TRIAL SOLUTION

We accept this expression as our final trial solution because it contains no solutions of the homogeneous differential equation.

Substituting the trial particular solution $y_p(x) = A_0 x^2 e^{-2x}$ into the nonhomogeneous equation $y'' + 4y' + 4y = 8e^{-2x}$ yields

Substitute the trial solution.

$$y_p'' + 4y_p' + 4y_p = A_0 e^{-2x}(2 - 8x + 4x^2) + 4A_0 e^{-2x}(2x - 2x^2) + 4A_0 x^2 e^{-2x}$$

$$= 2A_0 e^{-2x} = 8e^{-2x}.$$

From the last equality, we have $2A_0 = 8$, or $A_0 = 4$. A particular solution of $y'' + 4y' + 4y = 8e^{-2x}$ is

PARTICULAR SOLUTION

$$y_p(x) = 4x^2 e^{-2x}.$$

Since we know the homogeneous solutions y_1, y_2, we can also write a general solution of $y'' + 4y' + 4y = 8e^{-2x}$,

GENERAL SOLUTION

$$y_g(x) = 4x^2 e^{-2x} + C_1 e^{-2x} + C_2 x e^{-2x}. \qquad \blacksquare$$

EXAMPLE 27 *Find a general solution of*

$$y'' + 4y' + 4y = 6x e^{-2x}.$$

The homogeneous version of this equation is $y'' + 4y' + 4y = 0$. We have already found linearly independent solutions of that homogeneous equation,

LINEARLY INDEPENDENT HOMOGENEOUS SOLUTIONS

$y_1 = e^{-2x}$, $y_2 = x e^{-2x}$.

Since the forcing term in $y'' + 4y' + 4y = 6x e^{-2x}$ is a polynomial of first-order multiplying an exponential, table 6.1 directs us to consider a trial particular solution that is also a product of a first-order polynomial with the same exponential,

FIRST TRIAL PARTICULAR SOLUTION

$$y_p(x) = (A_1 x + A_0) e^{-2x}.$$

But both the terms in this trial particular solution are solutions of the homogeneous equation $y'' + 4y' + 4y = 0$: $A_0 e^{-2x} = A_0 y_1$ and $A_1 x e^{-2x} = A_1 y_2$.

To obtain a new trial solution, we multiply our first guess by x and obtain

SECOND TRIAL SOLUTION

$$y_p(x) = x(A_1 x + A_0) e^{-2x} = (A_1 x^2 + A_0 x) e^{-2x}.$$

The last term in this trial solution, $A_0 x e^{-2x}$, is still a solution of the homogeneous differential equation, so we must again multiply by x to produce yet another trial solution

THIRD TRIAL SOLUTION

$$y_p(x) = x(A_1 x^2 + A_0 x) e^{-2x} = (A_1 x^3 + A_0 x^2) e^{-2x}.$$

This trial solution contains no solutions of the homogeneous equation $y'' + 4y' + 4y = 0$.

Substitute the trial solution.

Substituting the trial particular solution $y_p(x) = (A_1 x^3 + A_0 x^2) e^{-2x}$ into the differential equation $y'' + 4y' + 4y = 6x e^{-2x}$ and simplifying yields

$$(-6A_0 + 6A_1) x e^{-2x} + 2A_0 e^{-2x} = 6x e^{-2x}.$$

Equate coefficients.

Equating coefficients of $x e^{-2x}$ and e^{-2x} on both sides of this last expression gives two equations for the two undetermined coefficients A_0, A_1,

$$
\begin{aligned}
e^{-2x}: \qquad & 2A_0 = 0 \\
x e^{-2x}: \qquad & -6A_0 + 6A_1 = 6
\end{aligned}
$$

The solution of this pair of simultaneous equations is obviously $A_0 = 0$, $A_1 = 1$. A particular solution of $y'' + 4y' + 4y = 6x e^{-2x}$ is thus

PARTICULAR SOLUTION

$$y_p(x) = x^3 e^{-2x}.$$

A general solution of $y'' + 4y' + 4y = 6xe^{-2x}$ is

$$y_g(x) = x^3 e^{-2x} + C_1 e^{-2x} + C_2 x e^{-2x}.$$

EXAMPLE 28   *Find a particular solution of*

$$y'' + 4y' + 4y = 18xe^{-2x} - \sin \pi x - 2e^{-2x}.$$

The homogeneous version of this equation is $y'' + 4y' + 4y = 0$. Linearly independent solutions of this homogeneous equation are e^{-2x}, xe^{-2x}.

The forcing term $18xe^{-2x} - \sin \pi x - 2e^{-2x}$ can be separated into two groups, those involving the exponential e^{-2x} and those involving the $\sin \pi x$ term. To simplify the algebra, consider two separate problems, one for each group of forcing terms:

Problem 1: $y'' + 4y' + 4y = 18xe^{-2x} - 2e^{-2x}$,

Problem 2: $y'' + 4y' + 4y = -\sin \pi x$.

The principle of superposition guarantees that a sum of particular solutions of problems 1 and 2 will be a solution of the original equation $y'' + 4y' + 4y = 18xe^{-2x} - \sin \pi x - 2e^{-2x}$.

The forcing term in problem 1, $(18x - 2)e^{-2x}$, is a polynomial of first-order multiplying the exponential e^{-2x}. Since this is precisely the same form as the forcing term in the previous example, we can mimic that analysis exactly. In particular, we find that our initial trial solution

$$y_p(x) = (A_1 x + A_0)e^{-2x}$$

must be multiplied twice by the independent variable before we obtain the final trial solution, which is free of homogeneous solutions:

$$y_p(x) = (A_1 x^3 + A_0 x^2)e^{-2x}.$$

Substituting $y_p(x) = (A_1 x^3 + A_0 x^2)e^{-2x}$ into problem 1 yields a slight variant of an expression in the previous example,

$$(-6A_0 + 6A_1)xe^{-2x} + 2A_0 e^{-2x} = 18xe^{-2x} - 2e^{-2x}.$$

Equating coefficients of xe^{-2x} and e^{-2x} yields two simultaneous equations for the two undetermined coefficients A_0, A_1,

$$e^{-2x}: \qquad 2A_0 = -2$$
$$xe^{-2x}: \quad -6A_0 + 6A_1 = 18$$

The solution of this pair of equations is $A_0 = -1$, $A_1 = 2$. A particular solution of problem 1 is

$$y_{p1}(x) = (2x^3 - x^2)e^{-2x}.$$

To find a particular solution of problem 2, table 6.1 suggests the trial solution

$$y_p(x) = A \cos \pi x + B \sin \pi x.$$

Since none of the terms in this trial solution is a solution of the homogeneous equation $y'' + 4y' + 4y = 0$, we can substitute the trial solution directly into problem 2. A little algebra yields

$$[(4 - \pi^2)A + 4\pi B] \cos \pi x + [-4\pi A + (4 - \pi^2)B] \sin \pi x = -\sin \pi x.$$

Equating coefficients of $\cos \pi x$, $\sin \pi x$ gives two equations for the two undetermined coefficients A, B:

$$\cos \pi x : \quad (4 - \pi^2)A + 4\pi B = 0$$
$$\sin \pi x : \quad -4\pi A + (4 - \pi^2)B = -1$$

The solution of this system is

$$A = -\frac{4\pi}{(4 + \pi^2)^2}, \qquad B = \frac{4 - \pi^2}{(4 + \pi^2)^2}.$$

A particular solution of problem 2 is thus

PARTICULAR SOLUTION OF PROBLEM 2

$$y_{p2}(x) = -\frac{4\pi}{(4 + \pi^2)^2} \cos \pi x + \frac{4 - \pi^2}{(4 + \pi^2)^2} \sin \pi x.$$

The particular solution of the original problem is the sum of particular solutions 1 and 2.

The principle of superposition allows us to combine the particular solutions y_{p1} and y_{p2} of problems 1 and 2 to obtain a particular solution of the original equation, $y'' + 4y' + 4y = 18xe^{-2x} - \sin \pi x - 2e^{-2x}$:

$$y_p(x) = y_{p1}(x) + y_{p2}(x)$$

$$= (2x^3 - x^2)e^{-2x} - \frac{4\pi}{(4 + \pi^2)^2} \cos \pi x + \frac{4 - \pi^2}{(4 + \pi^2)^2} \sin \pi x. \qquad \blacksquare$$

EXAMPLE 29 *Solve the initial-value problem*

$$\frac{1}{8}x''(t) + 2x(t) = \frac{1}{2} \sin \frac{3\pi t}{2}, \qquad x(0) = 0, \qquad x'(0) = 0,$$

that governs the forced spring-mass system of example 3, section 5.1.

The characteristic equation of this differential equation is

$$\frac{1}{8}r^2 + 2 = 0.$$

LINEARLY INDEPENDENT HOMOGENEOUS SOLUTIONS

Its roots are $r = \pm\sqrt{-16} = \pm 4i$. Linearly independent solutions of the homogeneous equation $x''/8 + 2x = 0$ are $\cos 4t$, $\sin 4t$. Table 6.1 suggests the trial particular solution

TRIAL PARTICULAR SOLUTION

$$x_p(t) = A \cos \frac{3\pi t}{2} + B \sin \frac{3\pi t}{2}.$$

Since these particular cosine and sine functions are not solutions of the homogeneous equation $x''/8 + 2x = 0$, we can substitute this trial solution directly into the given nonhomogeneous equation:

SUBSTITUTE TRIAL SOLUTION

$$\left(-\left(\frac{3\pi}{2}\right)^2 8 + 2\right) A \cos \frac{3\pi t}{2} + \left(-\left(\frac{3\pi}{2}\right)^2 8 + 2\right) B \sin \frac{3\pi t}{2} = \frac{1}{2} \sin \frac{3\pi t}{2}.$$

Equating coefficients of $\cos(3\pi t/2)$ and $\sin(3\pi t/2)$ yields

Equate coefficients.

$$A = 0, \qquad B = \frac{16}{64 - 9\pi^2}.$$

A particular solution of the given nonhomogeneous equation is

$$x_p(t) = \frac{16}{64 - 9\pi^2} \sin \frac{3\pi t}{2}.$$

PARTICULAR SOLUTION

Combining this particular solution with the homogeneous solutions found before allows us to construct the general solution

$$x_g(t) = \frac{16}{64 - 9\pi^2} \sin \frac{3\pi t}{2} + C_1 \cos 4t + C_2 \sin 4t.$$

GENERAL SOLUTION

The first initial condition requires

$$x(0) = C_1 = 0.$$

Using $C_1 = 0$ and imposing the second initial condition lead to

Find C_1, C_2.

$$x'(0) = \frac{24\pi}{64 - 9\pi^2} + 4C_2 = 0,$$

or

$$C_2 = \frac{-6\pi}{64 - 9\pi^2}.$$

The solution of the given initial value problem is

$$x(t) = \frac{16}{64 - 9\pi^2} \sin \frac{3\pi t}{2} - \frac{6\pi}{64 - 9\pi^2} \sin 4t$$

SOLUTION OF INITIAL–VALUE PROBLEM

$$\approx -0.64 \sin \frac{3\pi t}{2} + 0.76 \sin 4t. \qquad \blacksquare$$

Evidently, the motion of this undamped system neither grows nor decays with time.

EXAMPLE 30   *Suppose the angular frequency of the forcing function in the differential equation of the preceding example is changed from $3\pi/2$ to 4. Find the position of the mass as a function of time if it is started with the initial conditions $x(0) = x'(0) = 0.$*

6.5   Undetermined Coefficients

We are asked to solve the initial-value problem

$$\frac{1}{8}x''(t) + 2x(t) = \frac{1}{2}\sin 4t, \qquad x(0) = 0, \qquad x'(0) = 0.$$

Since only the forcing term has changed from the previous example, we know that linearly independent solutions of the homogeneous equation $x''/8 + 2x = 0$ are $\cos 4t$, $\sin 4t$.

We can begin finding a particular solution with the usual trial solution for a sine or cosine forcing term,

$$x_p(t) = A\cos 4t + B\sin 4t.$$

But for the change in angular frequency, this trial solution is the same as that of the previous example.

Obviously, both terms in this trial solution are multiples of the homogeneous solutions $\cos 4t$, $\sin 4t$. We can create a new trial solution by multiplying by the independent variable:

$$x_p(t) = At\cos 4t + Bt\sin 4t.$$

Since none of the terms in this trial solution is a solution of $x''/8 + 2x = 0$, we can substitute it into the nonhomogeneous differential equation $x''/8 + 2x = (\sin 4t)/2$:

$$\frac{1}{8}x_p''(t) + 2x_p(t) = -A\sin 4t + B\cos 4t = \frac{1}{2}\sin 4t,$$

Equating coefficients of the $\sin 4t$ and $\cos 4t$ terms yields equations for A and B:

$$-A = \frac{1}{2},$$

$$B = 0.$$

Obviously, $A = -\frac{1}{2}$, $B = 0$, and a particular solution of $x''/8 + 2x = (\sin 4t)/2$ is

$$x_p(t) = -\frac{1}{2}t\sin 4t.$$

A general solution of $x''/8 + 2x = (\sin 4t)/2$ is

$$x_g(t) = -\frac{1}{2}t\sin 4t + C_1\cos 4t + C_2\sin 4t.$$

The initial conditions $x(0) = x'(0) = 0$ require

$$x_g(0) = C_1 = 0,$$

$$x_g'(0) = 4C_2 = 0.$$

Hence, the solution of the given initial-value problem is

$$x(t) = -\frac{1}{2}t \sin 4t.$$

■

The mass in this system *never* oscillates periodically. Indeed, the factor t in this solution causes the displacement to swing to ever greater extremes until finally the spring breaks from overextension. This system is in *resonance*, a phenomenon we study in the next chapter.

EXAMPLE 31   *Determine an appropriate trial particular solution for*

$$y'' - y' - 6y = 3x \sin 2x - 5 \cos 3x + x^3(e^{-2x} + e^{2x}) + \pi - x^4.$$

As in the fourth example, we shall decompose the given problem into subproblems, each of which can be solved individually. The principle of superposition guarantees that the sum of particular solutions of the subproblems is a particular solution of the original equation.

We form each subproblem by collecting in each all forcing terms of similar type — i.e., all exponentials with the same exponent, all sines and cosines with the same frequency, all polynomials, and so forth. One such set of subproblems is

Separate into subproblems.

- All sine, cosine terms with angular frequency 2:

   Problem 1: $y'' - y' - 6y = 3x \sin 2x$

- All sine, cosine terms with angular frequency 3:

   Problem 2: $y'' - y' - 6y = 5 \cos 3x$

- All exponentials with exponent $-2x$:

   Problem 3: $y'' - y' - 6y = x^3 e^{-2x}$

- All exponentials with exponent $2x$:

   Problem 4: $y'' - y' - 6y = x^3 e^{2x}$

- All polynomials:

   Problem 5: $y'' - y' - 6y = \pi - x^4$

To construct trial solutions for each of these subproblems, we need to know the solutions of the homogeneous equation

$$y'' - y' - 6y = 0.$$

From the characteristic equation $r^2 - r + 6 = (r - 3)(r + 2) = 0$, we find the homogeneous solutions e^{3x}, e^{-2x}.

For problem 1, a trial solution is

TRIAL SOLUTION FOR PROBLEM 1

$$y_{p1}(x) = (A_1 x + A_0) \cos 2x + (B_1 x + B_0) \sin 2x.$$

None of the terms in this expression is a solution of the homogeneous equation $y'' - y' - 6y = 0$. Both the cosine and the sine in the trial solution are multiplied by a first-order polynomial because the first-order polynomial $3x$ multiplies the sine term in problem 1.

For problem 2, a trial solution is

TRIAL SOLUTION FOR PROBLEM 2

$$y_{p2}(x) = C \cos 3x + D \sin 3x.$$

The logic behind the form of this trial solution is identical with that for problem 1.

For problem 3, an initial trial solution is

FIRST TRIAL SOLUTION FOR PROBLEM 3

$$y_{p3}(x) = (E_3 x^3 + E_2 x^2 + E_1 x + E_0)e^{-2x}.$$

The exponential is multiplied by a third-order polynomial because the forcing term in problem 3 is the third-order polynomial x^3 multiplying an exponential.

But the term $E_0 e^{-2x}$ in this trial solution is a solution of the homogeneous equation $y'' - y' - 6y = 0$. Hence, we must multiply this entire expression by the independent variable x to obtain an improved trial solution

SECOND TRIAL SOLUTION FOR PROBLEM 3

$$y_{p3}(x) = x(E_3 x^3 + E_2 x^2 + E_1 x + E_0)e^{-2x}.$$

This expression is an acceptable trial solution for problem 3 because no terms in it satisfy the homogeneous equation $y'' - y' - 6y = 0$.

For problem 4, a trial solution is

TRIAL SOLUTION FOR PROBLEM 4

$$y_{p4}(x) = (F_3 x^3 + F_2 x^2 + F_1 x + F_0)e^{2x}.$$

The argument justifying its form is the same as that for the first trial solution for problem 3. We accept this trial solution as it stands because it contains no solutions of the homogeneous equation $y'' - y' - 6y = 0$.

For problem 5, a trial solution is

TRIAL SOLUTION FOR PROBLEM 5

$$y_{p5}(x) = G_4 x^4 + G_3 x^3 + G_2 x^2 + G_1 x + G_0,$$

a polynomial of fourth order because the forcing term in problem 5 is a polynomial of fourth order, $\pi - x^4$. No terms in this trial solution satisfy the homogeneous equation $y'' - y' - 6y = 0$.

A sum of these particular solutions gives the form of a particular solution of the given nonhomogeneous equation,

Superimpose to obtain a complete particular solution.

$$y_p(x) = y_{p1} + y_{p2} + y_{p3} + y_{p4} + y_{p5}.$$

The undetermined coefficients in each of the trial solutions $y_{p1}, \ldots, y_{p5}$ are best found separately before the subproblem solutions are summed. We leave the determination of those coefficients to the exercises. ∎

EXERCISE GUIDE

To gain experience . . .	Try exercises
Forming trial particular solutions	1–27, 32(a–b), 33(b), 35
Finding values of undetermined coefficients	1–27, 35
Determining when to use undetermined coefficients	28
Finding general solutions	1–27, 29–31, 39–40
Solving initial-value problems	32(a–b), 33(a)
With the foundations of undetermined coefficients	36–38, 41
Using superposition	34–35
Analyzing solution behavior	32(c), 33(b)

For each of the differential equations in exercises 1–27,

    i. Give an efficient form of a trial particular solution.

    ii. Find a particular solution.

    iii. Find a general solution.

1. $y'' - 9y = 14 - 2x$

2. $y'' - 9y = x^2 - 1$

3. $y'' - 9y = e^{-2x}$

4. $y'' - 9y = xe^{-2x}$

5. $y'' - 9y = 4 - e^{-3x}$

6. $y'' - 9y = xe^{-3x}$

7. $y'' - 9y = 2e^{3x} - e^{-3x} - 5x$

8. $4x''(t) + 8x'(t) + 5x(t) = e^t(\sin t/2 - 6)$

9. $4x''(t) + 8x'(t) + 5x(t) = 6\sin t/2$

10. $4x''(t) + 8x'(t) + 5x(t) = t - 2$

11. $w''(x) + 9w(x) = x\sin 3x + 2\cos 4x$

12. $w''(x) + 9w(x) = e^{3x}$

13. $w''(x) + 9w(x) = xe^{-x}$

14. $w''(x) + 9w(x) = x^2 - 3$

15. $w''(x) + 9w(x) = 2 - \cos 3x$

16. $2z''(t) - 7z'(t) + 3z(t) = e^{t/2}\sin \pi t$

17. $2z''(t) - 7z'(t) + 3z(t) = 6e^{3t}$

18. $2z''(t) - 7z'(t) + 3z(t) = -1$

19. $2z''(t) - 7z'(t) + 3z(t) = e^{t/2}\sin \pi t + 6e^{3t} - 1$

20. $y'' + 4y' + 40y = xe^{-2x}\cos 6x$

21. $y'' + 4y' + 40y = xe^{-2x}$

22. $y'' + 4y' + 40y = -\sin 6x$

23. $x'' + x' - 6x = -3t$

24. $x'' + x' - 6x = te^{-3t} - \cos 2t$

25. $9y'' + 36y' + 4y = xe^{-2x/3}$

26. $4y''(x) + 8y'(x) + 5y(x) = \cos x/2 + x^3$

27. $w'' + w' - 6w = e^{2x} - e^{-3x}$

28. Verify that

$$x''(t) = -g$$

is indeed a constant-coefficient, linear, second-order equation, as the text claims in example 25.

29. Verify that the solution

$$x_g(t) = -gt^2/2 + C_1 t + C_2$$

of

$$x''(t) = -g$$

obtained in example 25 is indeed a *general* solution.

30. Verify that the solution

$$y_g(x) = 4x^2 e^{-2x} + C_1 e^{-2x} + C_2 x e^{-2x}$$

of

$$y'' + 4y' + 4y = 8e^{-2x}$$

obtained in example 26 is indeed a *general* solution.

31. Verify that the solution

$$y_g(x) = x^3 e^{-2x} + C_1 e^{-2x} + C_2 x e^{-2x}$$

of

$$y'' + 4y' + 4y = 6x e^{-2x}$$

obtained in example 27 is indeed a *general* solution. If you completed the previous exercise, can you use some of its analysis here? Explain.

32. Consider the undamped, forced spring-mass system model

$$mx''(t) + kx(t) = A \sin \omega t, \quad x(0) = x_i, \quad x'(0) = v_i,$$

where A, ω are fixed parameters. (Recall that m is mass, k is the spring constant, and x_i, v_i are the initial position and velocity.)

(a) Assume $\omega \neq \sqrt{k/m}$ and solve the initial-value problem.

(b) Assume $\omega = \sqrt{k/m}$ and solve the initial-value problem.

(c) How does the behavior of the solutions obtained in parts (a) and (b) differ? How does the value of ω affect the solution procedure you use?

33. (a) Solve the damped, forced spring-mass system model

$$mx''(t) + px'(t) + kx(t) = A \sin \omega t,$$

$$x(0) = x_i, \qquad x'(0) = v_i,$$

where A, ω are fixed parameters. (Recall that m is mass, p is the damping or friction coefficient, k is the spring constant, and x_i, v_i are the initial position and velocity.)

(b) The previous exercise considered the undamped ($p = 0$) version of this model, and two cases arose, $\omega = \sqrt{k/m}$ and $\omega \neq \sqrt{k/m}$. Do such considerations arise here? Does the solution of this problem remain bounded for all time?

34. Example 28 obtained the particular solution

$$y_p(x) = (2x^3 - 2x^2)e^{-2x} - \frac{4\pi}{(4 + \pi^2)^2} \cos \pi x$$

$$+ \frac{4 - \pi^2}{(4 + \pi^2)^2} \sin \pi x$$

of

$$y'' + 4y' + 4y = 18x e^{-2x} - \sin \pi x - 2e^{-2x}$$

by decomposing the original problem into two subproblems:

Problem 1: $y'' + 4y' + 4y = 18x e^{-2x} - 2e^{-2x}$,

Problem 2: $y'' + 4y' + 4y = -\sin \pi x$.

Verify by direct substitution that y_p, which is a sum of a particular solution of problem 1 and a particular solution of problem 2, is indeed a solution of the full problem. Which part of y_p is a solution of problem 1? Of problem 2? Do these parts each yield the expected forcing terms when they are substituted into the left side of the equation? Show how your calculations support your answer to this last question.

35. Example 31 broke the problem of finding the form of a particular solution of

$$y'' - y' - 6y = 3x \sin 2x - 5 \cos 3x + x^3 (e^{-2x} + e^{2x})$$

$$+ \pi - x^4.$$

into a series of subproblems:

Problem 1: $y'' - y' - 6y = 3x \sin 2x$

Problem 2: $y'' - y' - 6y = 5 \cos 3x$

Problem 3: $y'' - y' - 6y = x^3 e^{-2x}$

Problem 4: $y'' - y' - 6y = x^3 e^{2x}$

Problem 5: $y'' - y' - 6y = \pi - x^4$

Find a particular solution of the original differential equation by finding a particular solution of each of these subproblems and summing. You may use as much of the information in example 31 as you find useful.

36. In its search for a particular solution of

$$x''(t) = -g,$$

example 25 proposed three forms for a trial solution of $x'' = -g$. They are

$$x_{p1}(t) = A_0,$$

$$x_{p2}(t) = A_0 t,$$

$$x_{p3}(t) = A_0 t^2.$$

Confirm the wisdom of finally using x_{p3} by direct substitution of the first two trial solutions in $x''(t) = -g$. Explain why these two trial solutions are unsatisfactory. Show that the third trial solution is not a solution of the homogeneous equation $x'' = 0$.

37. In its search for a particular solution of

$$y'' + 4y' + 4y = 8e^{-2x},$$

example 26 proposed three forms for a trial solution of $y'' + 4y' + 4y = 8e^{-2x}$:

$$y_{p1}(x) = A_0 e^{-2x},$$
$$y_{p2}(x) = A_0 x e^{-2x},$$
$$y_{p3}(x) = A_0 x^2 e^{-2x}.$$

Confirm the wisdom of finally using y_{p3} by direct substitution of y_{p1} and y_{p2} in $y'' + 4y' + 4y = 8e^{-2x}$. Explain why these two trial solutions are unsatisfactory. Show that x_{p3} is not a solution of the homogeneous equation $y'' + 4y' + 4y = 0$.

38. In its search for a particular solution of

$$y'' + 4y' + 4y = 6xe^{-2x},$$

example 27 proposed three forms for a trial solution of $y'' + 4y' + 4y = 6xe^{-2x}$:

$$y_{p1}(x) = (A_1 x + A_0)e^{-2x},$$
$$y_{p2}(x) = (A_1 x^2 + A_0 x)e^{-2x},$$
$$y_{p3}(x) = (A_1 x^3 + A_0 x^2)e^{-2x}.$$

Confirm the wisdom of finally using y_{p3} by direct substitution of the first two trial solutions in $y'' + 4y' + 4y = 6xe^{-2x}$. Explain why these two trial solutions are unsatisfactory. Show that the third trial solution is not a solution of the homogeneous equation $y'' + 4y' + 4y = 0$.

39. Example 28 found that a particular solution of

$$y'' + 4y' + 4y = 18xe^{-2x} - \sin \pi x - 2e^{-2x}$$

is

$$y_p(x) = (2x^3 - 2x^2)e^{-2x} - \frac{4\pi}{(4 + \pi^2)^2} \cos \pi x$$
$$+ \frac{4 - \pi^2}{(4 + \pi^2)^2} \sin \pi x.$$

Find a general solution of this differential equation.

40. Example 29 found that a solution of the initial value problem

$$\frac{1}{8}x''(t) + 2x(t) = \frac{1}{2}\sin \frac{3\pi t}{2},$$

$$x(0) = 0, \qquad x'(0) = 0,$$

is

$$x(t) = \frac{16}{64 - 9\pi^2}\sin \frac{3\pi t}{2} - \frac{6\pi}{64 - 9\pi^2}\sin 4t$$

$$\approx -0.64 \sin \frac{3\pi t}{2} + 0.76 \sin 4t.$$

Use this solution as a starting point for the construction of a general solution of the differential equation. How does the general solution you obtain compare with that obtained in example 29? Explain the differences and similarities. Is one general solution better than the other in any sense?

41. Table 6.1, which summarizes the form of the trial particular solution for various forcing functions, does not include the hyperbolic sine or cosine among its forcing terms. Correct this omission by adding entries for forcing terms of the form
   (a) $a \cosh px + b \sinh px$
   (b) $(a_n x^n + \cdots + a_1 x + a_0) \cosh px + (b_n x^n + \cdots + b_1 x + b_0) \sinh px$
   (c) $ae^{qx} \cosh px + be^{qx} \sinh px$
   (Recall that $\cosh px = (e^{px} + e^{-px})/2$, $\sinh px = (e^{px} - e^{-px})/2$.) Do these entries add any really new information?

6.6 NUMERICAL METHODS

Recall the idea behind the Euler method for finding a numerical approximation to the solution of a single first-order equation such as the heat-loss equation

$$\frac{dT}{dt} = -\frac{Ak}{cm}(T - T_{\text{out}}).$$

Suppose we have found an approximate temperature T_n at some time t_n. We want to step forward to the time $t_{n+1} = t_n + \Delta t$ to find an approximation T_{n+1} to the temperature at that time.

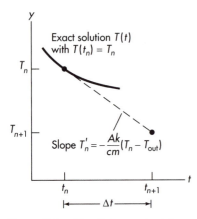

Figure 6.1: The Euler method for a single first-order equation: The slope at the current point directs us to the next point.

Figure 6.1 shows the Euler strategy: *Use the slope given by the differential equation to locate the next approximate solution point.* Analytically, we write

$$T_{n+1} = T_n + T'_n \Delta t = T_n + \left(-\frac{Ak}{cm}(T_n - T_{\text{out}}) \right) \Delta t.$$

The slope T'_n is just the right-hand side of the differential equation evaluated at t_n. In words, the Euler method is

$$\boxed{\text{new value}} = \boxed{\text{old value}} + \boxed{\text{slope}} \times \boxed{\text{step size}}$$

How do these ideas extend to a *second*-order equation? A good prototype is the forced spring-mass equation of example 3, section 5.1:

$$\frac{1}{8}x'' + 2x = \frac{1}{2}\sin\frac{3\pi t}{2}. \tag{6.6}$$

The slope x' does not even appear in this equation, although we do know that $x' = v$, the velocity of the mass.

How can we rewrite this equation so that velocity v appears explicitly? Since $x' = v$, we can replace x'' with v' to obtain

$$\frac{1}{8}v' + 2x = \frac{1}{2}\sin\frac{3\pi t}{2}.$$

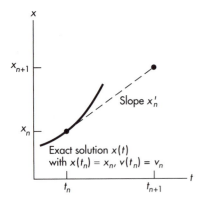

But now there are two unknowns, x and v. We seem to need a second equation. We get it from the definition of v,

$$x' = v.$$

With just a bit of algebra we can rewrite these two equations as a *system of first-order equations*.

$$x' = v$$

$$v' = -16x + 4\sin 3\pi t/2.$$

This system is equivalent to the original second-order equation (6.6) in the sense that a solution of one leads to a solution of the other.

If the second-order equation has the initial conditions $x(0) = x_i$, $x'(0) = v_i$, then equivalent initial conditions for the first-order system are $x(0) = x_i$, $v(0) = v_i$, since x' and v are analogous dependent variables.

Now we can simply apply the Euler method simultaneously to both equations of this first-order system. Instead of having a single equation such as $T' = \cdots$ whose slope information steers us across the (T, t)-plane, we have two equations, $x' = \cdots$ to steer us across the (x, t)-plane and $v' = \cdots$ to steer us across the (v, t)-plane. Figure 6.2 suggests how this process works for these two equations.

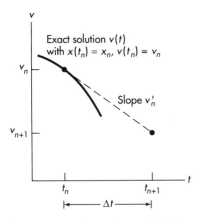

Figure 6.2: The Euler method for a system: Working simultaneously in two planes, the slope at the current point directs us to the next value of each dependent variable.

Analytically, we write the **Euler method** for this system as

$$x_{n+1} = x_n + x'_n \Delta t = x_n + v_n \Delta t$$

$$v_{n+1} = v_n + v'_n \Delta t = v_n + (-16x_n + 4\sin 3\pi t_n/2)\Delta t.$$

The value x_{n+1} approximates the exact position $x(t_{n+1})$ at time $t_{n+1} = t_n + \Delta t$; i.e., $x_{n+1} \approx x(t_{n+1})$. Similarly, $v_{n+1} \approx v(t_{n+1})$. Given the initial values $x(0) = x_i$, $x'(0) = v_i$, we take $x_0 = x_i$, $v_0 = v_i$, as starting values for Euler's method.

Before we illustrate this method, we summarize the general situation. A general second-order equation of the form

$$y'' = g(t, y, y')$$

can always be converted to a first-order system by introducing a variable z to account for the derivative y':

$$y' = z$$
$$z' = g(t, y, z)$$

The last equation just uses $z' = (y')' = y''$. (In the spring-mass equation considered previously, we introduced v to account for x'.) For example, the constant-coefficient second-order equation

$$b_2 y'' + b_1 y' + b_0 y = h(t)$$

is equivalent to the first-order system

$$y' = z$$
$$b_2 z' = -b_0 y - b_1 z + h(t).$$

We can easily state the Euler method for a general system of two first-order equations.

Euler method   Given the initial-value problem

$$y' = f(t, y, z), \qquad y(t_0) = y_i$$
$$z' = g(t, y, z), \qquad z(t_0) = z_i,$$

then the **Euler method** approximation to the solution of this initial-value problem is the sequence $(y_1, z_1), (y_2, z_2), (y_3, z_3), \ldots$ computed using

$$\boxed{\begin{aligned} y_{n+1} &= y_n + f(t_n, y_n, z_n)\Delta t \\ z_{n+1} &= z_n + g(t_n, y_n, z_n)\Delta t, \end{aligned}}$$

EULER METHOD

beginning with $(y_0, z_0) = (y_i, z_i)$.

The pair (y_n, z_n) approximates the exact solution $(y(t), z(t))$ at $t = t_n$.

We have just stated the Euler method for a system that is even more general than one obtained from a single second-order equation. For example, with such a general statement of the Euler method, we could solve the predator-prey (fox-rabbit) system

$$F' = -(d_F - \beta R)F$$
$$R' = (b_R - \alpha F)R$$

of section 5.4.2 by identifying y with F, z with R, and setting $f(t, F, R) = -(d_F - \beta R)F$ and $g(t, F, R) = (b_R - \alpha F)R$.

EXAMPLE 32 *Using $\Delta t = 0.5$, find the first three steps of the Euler method approximation to the solution of*

$$\frac{1}{8}x'' + 2x = \frac{1}{2}\sin\frac{3\pi t}{2}, \qquad x(0) = 0, \qquad x'(0) = 0.$$

From the preceding discussion, we know that an equivalent initial-value problem for a first-order system is

$$x' = v$$

$$v' = -16x + 4\sin\frac{3\pi t}{2},$$

$x(0) = 0$, $v(0) = 0$. Using the general notation introduced earlier, this system has the form

$$x' = f(t, x, v)$$

$$v' = g(t, x, v),$$

with

$$f(t, x, v) = v, \qquad g(t, x, v) = -16x + 4\sin\frac{3\pi t}{2}.$$

Using $\Delta t = 0.5$ and the given initial values $x_0 = x(0) = 0$ and $v_0 = v(0) = 0$, we find successively that

$$x_1 = x_0 + v_0\Delta t$$

$$= 0 + 0 \cdot 0.5 = 0$$

$$v_1 = v_0 + [-16x_0 + 4\sin 3\pi t_0/2]\Delta t$$

$$= 0 + [-16 \cdot 0 + 4\sin 3\pi \cdot 0/2]0.5 = 0,$$

$$x_2 = x_1 + v_1\Delta t$$

$$= 0 + 0 \cdot 0.5 = 0$$

$$v_2 = v_1 + [-16x_1 + 4\sin 3\pi t_1/2]\Delta t$$

$$= 0 + [-16 \cdot 0 + 4\sin 3\pi \cdot 0.5/2]0.5 = 1.4142,$$

$$x_3 = 0 + 1.4142 \cdot 0.5 = 0.7071$$

$$v_3 = 1.4142 + [-16 \cdot 0 + 4\sin 3\pi \cdot 1.0/2]0.5 = -6.2426.$$

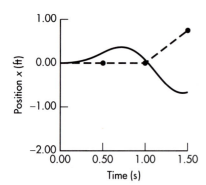

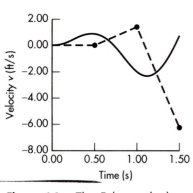

These approximations are $(x(0.5), v(0.5)) \approx (0, 0)$, $(x(1.0), v(1.0)) \approx (0, 1.4142)$, $(x(1.5), v(1.5)) \approx (0.7071, -6.2426)$. They are illustrated in figure 6.3. ∎

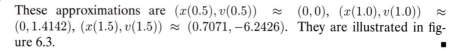

Figure 6.3: The Euler method approximation (closed circles) using $\Delta t = 0.5$ and the exact solution (solid curve) of $x' = v$, $v' = -16x + 4\sin 3\pi t/2$, $x(0) = 0$, $v(0) = 0$.

The exact solution shown in figure 6.3 and subsequent graphs was derived in example 29 of section 6.5:

$$x(t) = \frac{16}{64 - 9\pi^2} \sin \frac{3\pi t}{2} - \frac{6\pi}{64 - 9\pi^2} \sin 4t.$$

The corresponding velocity component $v(t)$ is obtained by differentiating this expression.

Note that we have approximate values of both position and velocity; we couldn't find one without the other. But we know we need both position and velocity to specify the motion of the mass: The mass could be in a given position at an instant of time moving upward ($v' > 0$) or downward ($v' < 0$). A still photograph, which can record only the position of the mass, wouldn't distinguish between the two cases. From this perspective, an equivalent first-order system seems a natural way to approach a second-order equation.

Of course, the *Heun*, or *modified Euler, method* can be carried over to systems as well. This method modifies Euler by first taking an Euler step, using the slope at the current point to determine the direction of the step. Then it steps back to the current point. The final approximate value is obtained from a step whose direction is the average of the slope at the current point and the slope at the end of that first, tentative Euler step.

As we saw in chapter 3, that geometric explanation of the Heun method is not the most efficient way of implementing it. An efficient version follows.

Heun Method   For the initial-value problem

$$y' = f(t, y, z), \qquad y(t_0) = y_i$$
$$z' = g(t, y, z), \qquad z(t_0) = z_i,$$

the **Heun Method** computes

$$k_1 = f(t_n, y_n, z_n)\Delta t,$$
$$l_1 = g(t_n, y_n, z_n)\Delta t,$$
$$k_2 = f(t_{n+1}, y_n + k_1, z_n + l_1)\Delta t,$$
$$l_2 = g(t_{n+1}, y_n + k_1, z_n + l_1)\Delta t,$$

$$\boxed{\begin{aligned} y_{n+1} &= y_n + \frac{k_1 + k_2}{2}, \\[1em] z_{n+1} &= z_n + \frac{l_1 + l_2}{2}, \end{aligned}}$$

HEUN METHOD

beginning with $(y_0, z_0) = (y_i, z_i)$.

EXAMPLE 33   *Using $\Delta t = 0.5$, find the first three steps of the Heun method approximation to the solution of*

$$\frac{1}{8}x'' + 2x = \frac{1}{2}\sin\frac{3\pi t}{2}, \qquad x(0) = 0, \qquad x'(0) = 0.$$

We again use the equivalent first-order system

$$x' = v$$

$$v' = -16x + 4\sin\frac{3\pi t}{2},$$

$x(0) = 0$, $v(0) = 0$.

Using the initial values $x_0 = v_0 = 0$, we find that the approximation at $t_1 = 0.5$ is (x_1, v_1), where

$$k_1 = 0 \cdot 0.5 = 0$$

$$l_1 = \left(-16 \cdot 0 + 4\sin 3\pi \cdot \frac{0}{2}\right)0.5 = 0,$$

$$k_2 = 0 \cdot 0.5 = 0$$

$$l_2 = \left(-16 \cdot 0 + 4\sin 3\pi \cdot \frac{0.5}{2}\right)0.5 = 1.4142.,$$

$$x_1 = 0 + \frac{0+0}{2} = 0$$

$$v_1 = 0 + \frac{0 + 1.4142}{2} = 0.7071.$$

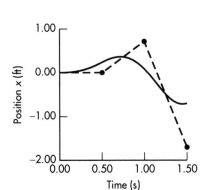

From $(x_1, v_1) = (0, 0.7071)$, we compute (x_2, v_2),

$$k_1 = 0.7071 \cdot 0.5 = 0.3536$$

$$l_1 = \left(-16 \cdot 0 + 4\sin 3\pi \cdot \frac{0.5}{2}\right)0.5 = 1.4142,$$

$$k_2 = (0.7071 + 1.4142) \cdot 0.5 = 1.0607$$

$$l_2 = \left[-16(0 + 0.3536) + 4\sin 3\pi \cdot \frac{0.5}{2}\right]0.5 = -4.8284$$

$$x_2 = 0 + \frac{0.3536 + 1.0607}{2} = 0.7071$$

$$v_2 = 0.7071 + \frac{1.4142 - 4.8284}{2} = -1.0000$$

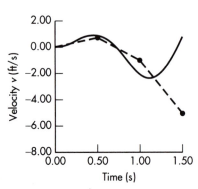

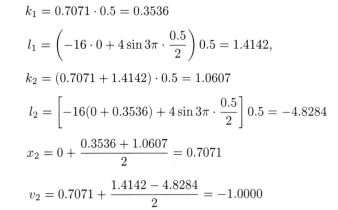

Figure 6.4: The Heun method approximation (closed circles) using $\Delta t = 0.5$ and the exact solution (solid curve) of $x' = v$, $v' = -16x + 4\sin 3\pi t/2$, $x(0) = 0$, $v(0) = 0$.

The last step follows the same pattern. We find $k_1 = -0.50000$, $l_1 = -7.6569$, $k_2 = -4.3284$, $l_2 = -0.2426$ and $(x_3, v_3) = (-1.7071, -4.9497)$. Figure 6.4 illustrates these results. ∎

The fourth-order Runge-Kutta method extends easily to systems, too.

Fourth-order Runge-Kutta Method   For the initial-value problem

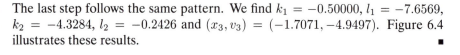

$$y' = f(t, y, z), \qquad y(t_0) = y_i$$

$$z' = g(t, y, z), \qquad z(t_0) = z_i,$$

the **fourth-order Runge-Kutta method** successively computes the intermediate quantities

$$k_1 = f(t_n, y_n, z_n)\Delta t$$

$$l_1 = g(t_n, y_n, z_n)\Delta t,$$

$$k_2 = f(t_n + \Delta t/2, y_n + k_1/2, z_n + l_1/2)\,\Delta t$$

$$l_2 = g(t_n + \Delta t/2, y_n + k_1/2, z_n + l_1/2)\Delta t,$$

$$k_3 = f(t_n + \Delta t/2, y_n + k_2/2, z_n + l_2/2)\Delta t$$

$$l_3 = g(t_n + \Delta t/2, y_n + k_2/2, z_n + l_2/2)\Delta t,$$

$$k_4 = f(t_n + \Delta t, y_n + k_3, z_n + l_3)\Delta t$$

$$l_4 = g(t_n + \Delta t, y_n + k_3, z_n + l_3)\Delta t,$$

to find

$$y_{n+1} = \frac{y_n + k_1 + 2k_2 + 2k_3 + k_4}{6}$$

$$z_{n+1} = z_n + \frac{l_1 + 2l_2 + 2l_3 + l_4}{6}.$$

RUNGE-KUTTA METHOD

It begins with $(y_0, z_0) = (y_i, z_i)$.

EXAMPLE 34 *Using $\Delta t = 0.5$, find the first three steps of the fourth-order Runge-Kutta approximation to the solution of*

$$\frac{1}{8}x'' + 2x = \frac{1}{2}\sin\frac{3\pi t}{2}, \qquad x(0) = 0, \qquad x'(0) = 0.$$

Again, we replace the second-order initial-value problem with an equivalent initial-value problem for a first-order system:

$$x' = v,$$

$$v' = -16x + 4\sin\frac{3\pi t}{2},$$

$x(0) = 0, v(0) = 0.$

Beginning with $t_0 = 0$, $x_0 = x(0) = 0$ and $v_0 = v(0) = 0$, we compute

$$k_1 = v_0\Delta t = 0 \cdot 0.5 = 0$$

$$l_1 = \left(-16x_0 + 4\sin\frac{3\pi t_0}{2}\right)\Delta t = (0 + 0)0.5 = 0$$

$$k_2 = \left(v_0 + \frac{l_0}{2}\right)\Delta t = (0 + 0)0.5 = 0$$

$$l_2 = \left[-16\left(x_0 + \frac{k_1}{2}\right) + 4\sin\frac{3\pi(t_0 + \Delta t/2)}{2}\right]\Delta t = 1.8478$$

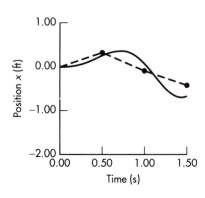

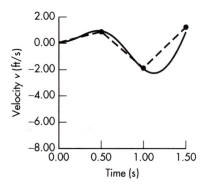

Figure 6.5: The fourth-order Runge-Kutta approximation (closed circles) using $\Delta t = 0.5$ and the exact solution (solid curve) of $x' = v$, $v' = -16x + 4\sin 3\pi t/2$, $x(0) = 0$, $v(0) = 0$

$$k_3 = \left(v_0 + \frac{l_2}{2}\right)\Delta t = 0.4619$$

$$l_3 = \left[-16\left(x_0 + \frac{k_2}{2}\right) + 4\sin\frac{3\pi(t_0 + \Delta t/2)}{2}\right]\Delta t = 1.8478$$

$$k_4 = (v_0 + l_3)\Delta t = 0.9239$$

$$l_4 = \left[-16(x_0 + k_3) + 4\sin\frac{3\pi(t_0 + \Delta t)}{2}\right]\Delta t = -2.2813$$

to find

$$x_1 = x_0 + \frac{k_1 + 2k_2 + 2k_3 + k_4}{6} = 0.3080$$

$$v_1 = v_0 + \frac{l_1 + 2l_2 + 2l_3 + l_4}{6} = 0.8516.$$

That is, $x(0.5) \approx 0.3080$, and $v(0.5) \approx 0.8516$. These computations and the balance of those required to approximate the solution at $t = 1.0$ and $t = 1.5$ are summarized in table 6.2.

Figure 6.5 illustrates graphically the remarkable accuracy of this method. Compare it with the corresponding figures for the Euler method (figure 6.3) and the Heun method (figure 6.4). ∎

EXAMPLE 35 *Using available software, compare the Euler method approximation, the Heun method approximation, and the exact solution of*

$$\frac{1}{8}x'' + 2x = \frac{1}{2}\sin\frac{3\pi t}{2}, \qquad x(0) = 0, \qquad x'(0) = 0.$$

Take 20 time steps to cover three full periods of the forcing function.

We apply each of these methods to the first-order system considered in the previous examples.

TABLE 6.2

The fourth-order Runge-Kutta calculations required to approximate the solution of $x' = v$, $v' = -16x + 4\sin 3\pi t/2$, $x(0) = 0$, $v(0) = 0$, using $\Delta t = 0.5$

n	t_n	x_n	v_n				
0	0	0	0				
			$k_i =$	0	0	0.4619	0.9239
			$l_i =$	0	1.8478	1.8478	−2.2813
1	0.5	0.3080	0.8516				
			$k_i =$	0.4258	0.1634	−0.8073	−1.5156
			$l_i =$	−1.0495	−4.9323	−3.8828	1.9944
2	1.0	−0.0883	−1.9293				
			$k_i =$	−0.9646	−1.2881	−0.0148	1.5819
			$l_i =$	−1.2938	3.7994	5.0931	2.2387
3	1.5	−0.4197	1.1924				

TABLE 6.3

A comparison of the Euler, Heun, and Runge-Kutta
approximate solutions of $x' = v$, $v' = -16x + 4 \sin 3\pi t/2$, $x(0) = 0$, $v(0) = 0$ using $\Delta t = 0.2$

t	n		Euler	Heun	Runge-Kutta	Exact
1.0	5	x_n	0.3664	0.0809	0.0684	0.0696
		v_n	−1.4518	−2.4361	−1.9697	−1.9871
2.0	10	x_n	0.1615	1.2565	0.7378	0.7509
		v_n	3.6439	2.7315	2.5739	2.5927
3.0	15	x_n	−1.3373	−1.4858	−1.0310	−1.0513
		v_n	1.3004	7.0941	2.4751	2.5653
4.0	20	x_n	0.7691	−1.9695	−0.2002	−0.2186
		v_n	−7.1175	−11.5977	−5.7682	−5.9454

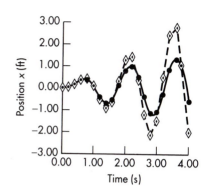

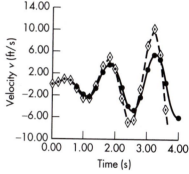

Figure 6.6: The fourth-order Runge-Kutta approximations (closed circles) using $\Delta t = 0.2$ lie almost precisely on the exact solution (solid curve) of $x' = v$, $v' = -16x + 4 \sin 3\pi t/2$, $x(0) = 0$, $v(0) = 0$. The Heun approximations (diamonds connected by dashed lines) drift further away as time increases.

Since the forcing function $4 \sin 3\pi t/2$ first repeats itself when $3\pi t/2 = 2\pi$, its period is

$$t = \frac{2\pi}{3\pi/2} = \frac{4}{3} \text{ s}.$$

Hence, covering three periods occupies the time interval $0 \le t \le 4$, and taking 20 twenty time steps requires $\Delta t = \frac{4}{20} = 0.2$ s.

Rather than list all the approximate solution values obtained at $t = 0.2, 0.4, 0.6, \ldots, 4$ s, we list in table 6.3 just the values at $t = 1, 2, 3, 4$ s. Obviously, Euler is least accurate, Heun is intermediate in accuracy, and Runge-Kutta is most accurate.

Figure 6.6 illustrates these accuracy claims. The closed circles representing the Runge-Kutta approximation fall almost exactly on the solid curve of the exact solution for both position x and velocity v. The Heun approximations, represented by diamonds connected by a dashed line, drift farther from the exact solutions as time grows. The Euler approximation is too far gone to even include in the graphs. ∎

EXAMPLE 36   *Solve the predator-prey (fox-rabbit) model*

$$F' = -(0.04 - 0.0011R)F$$
$$R' = (0.06 - 0.0009F)R,$$

$F(0) = 40$, $R(0) = 60$, *on the interval $0 \le t \le 200$ using available software and the fourth-order Runge-Kutta method. Choose a step size that seems to give about 5% accuracy.*

The mechanics of this process should be straightforward by now. The accuracy requirement is another matter. We do not know an exact solution with which to compare the numerical solution. Indeed, this is the hard nut of any numerical computation: How do you know your answer is any good if you do not know the exact answer? (And you would not bother with the computation if you already knew the answer!)

TABLE 6.4

Values of the approximate solution of the predator-prey model $F' = -(0.04 - .0011R)F$, $R' = (0.06 - 0.0009F)R$, $F(0) = 40$, $R(0) = 60$, obtained using fourth-order Runge-Kutta with $\Delta t = 20$

t	n	F_n	R_n
0.0000	0	40.0000	60.0000
20.0000	1	82.7367	70.4604
40.0000	2	118.9134	33.9187
60.0000	3	89.2980	16.9523
80.0000	4	56.2978	15.4215
100.0000	5	37.7815	22.3828
120.0000	6	33.0832	40.0997
140.0000	7	48.5883	66.8903
160.0000	8	98.1605	60.8590
180.0000	9	113.5144	27.0033
200.0000	10	78.8115	15.8129

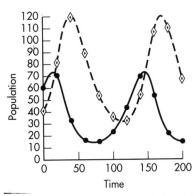

Figure 6.7: Approximate solution of the predator-prey model $F' = -(0.04 -0.0011R)F$, $R' = (0.06 - 0.0009F) R$, $F(0) = 40$, $R(0) = 60$, using fourth-order Runge-Kutta with $\Delta t = 5$ (solid and dashed line segments) and with $\Delta t = 20$ (closed circles and diamonds). The solid segments connect approximate values of r, the dashed segments, f.

One approach is solving the system twice, once with a value of Δt that seems reasonable and then again with a value of Δt that is significantly smaller. If the two sets of approximate solutions are similar, then we can have some confidence that they are reasonably accurate. To estimate error, we can treat the approximate solution with the smaller Δt as if it were "exact" (even though it is not).

Figure 6.7 illustrates approximate solutions obtained using $\Delta t = 20$ (closed circles and diamonds) and $\Delta t = 5$ (solid and dashed line segments). The close proximity of the two solutions suggests that $\Delta t = 20$ is small enough to give a decent approximate solution on this interval. The $\Delta t = 20$ approximate solution values appear in table 6.4.

Since the greatest deviation between the two approximate solutions in figure 6.7 seems to occur at $t = 200$, we compare the two sets of approximate solutions at that time:

$$\Delta t = 20 : \quad F(200) \approx F_{10} = 78.8115, \quad R(200) \approx R_{10} = 15.8129,$$

$$\Delta t = 5 : \quad F(200) \approx F_{40} = 77.4718, \quad R(200) \approx R_{40} = 15.0100.$$

Treating the $\Delta t = 5$ approximate solution as if it were exact suggests the error estimates

$$\left| \frac{F_{10} - F_{40}}{F_{40}} \right| = \frac{78.8115 - 77.4718}{77.4718} \approx 0.0173,$$

$$\left| \frac{R_{10} - R_{40}}{R_{40}} \right| = \frac{15.8129 - 15.0100}{15.0100} \approx 0.0535.$$

The larger estimated error occurs in the rabbit population, and the error is just over 5%, the level we sought. ∎

EXERCISE GUIDE

To gain experience . . .	Try exercises
Converting second order equations to systems	1–11
Using the Euler method	3, 6, 9, 12, 20
Using the Heun method	4, 7, 10, 12, 20
Using the Runge-Kutta method	5, 8, 11, 12, 20
Analyzing error in the methods	13–15
With geometrical interpretations of the methods	16–18
Analyzing and interpreting behavior of solutions	19–20

1. The text claims that

$$x(t) = \frac{16}{64 - 9\pi^2} \sin \frac{3\pi t}{2} - \frac{6\pi}{64 - 9\pi^2} \sin 4t$$

is the solution of the initial-value problem

$$\frac{1}{8}x'' + 2x = \frac{1}{2} \sin \frac{3\pi t}{2}, \qquad x(0) = 0, \qquad x'(0) = 0$$

considered in several of the examples in the text.
(a) Verify this claim.
(b) Use the solution expression given to write a solution $(x(t), v(t))$ of the first-order initial-value problem

$$x' = v$$

$$v' = -16x + 4 \sin \frac{3\pi t}{2},$$

$$x(0) = 0, v(0) = 0.$$

2. The text claims that the constant-coefficient, second-order equation

$$b_2 y'' + b_1 y' + b_0 y = h(t)$$

is equivalent to the first-order system

$$y' = z$$

$$b_2 z' = -b_0 y - b_1 z + h(t).$$

Verify this claim as follows.

(a) Suppose $y(t)$ is a solution of the second-order equation. Show that $(y(t), y'(t))$ solves the first-order system.

(b) Suppose $(y(t), z(t))$ solves the first-order system. Show that $y(t)$ solves the second-order equation.

3. Write a first-order system that is equivalent to the second-order equation $y'' = f(t, y, y')$. Write out the recipe for the Euler method for such a system.

4. Write a first-order system that is equivalent to the second-order equation $y'' = f(t, y, y')$. Write out the recipe for the Heun method for such a system.

5. Write a first-order system that is equivalent to the second-order equation $y'' = f(t, y, y')$. Write out the recipe for the fourth-order Runge-Kutta method for such a system.

6. After writing it as a system of two first-order equations, use four steps of the Euler method to approximate the solution at $t = 2$ of $x'' + 4x' + 40x = -\sin 6t$, $x(0) = 0, x'(0) = 3/74$. Find the exact solution of the initial-value problem, and determine the error in the Euler method solution.

7. Repeat exercise 6 using the Heun method.

8. Repeat exercise 6 using the fourth-order Runge-Kutta method.

9. When a mass of 0.2 kg is suspended from a spring, it extends 0.08 m. The mass is set in motion by raising it 0.1 m above its equilibrium position and releasing it. Use four steps of the Euler method to approximate the position and the velocity of the mass at 1 s. Find the exact solution of the governing initial-value problem, and determine the error in the approximate position and velocity values.

10. Repeat exercise 9 using the Heun method.

11. Repeat exercise 9 using the fourth-order Runge-Kutta method.

12. Using the numerical method of your choice and available software, estimate the period of the solution of the nonlinear initial-value problem $\theta'' + 4\sin\theta = 0$, $\theta(0) = 1.5$, $\theta'(0) = 0$. How does that estimated period compare with the period of the solution of the corresponding approximate linear problem? Of what mechanical system might this problem be a model?

13. Using 10, 20, and 40 steps of the Euler method and available software, approximate the solution at $t = 2$ of the initial-value problem $x'' + \pi^2 x = 0$, $x(0) = 1$, $x'(0) = 1$. Determine the error in each case, and show that it is proportional to Δt.

14. Repeat exercise 13 using the Heun method. Show that the error is proportional to Δt^2.

15. Repeat exercise 13 using the fourth-order Runge-Kutta method. Show that the error is proportional to Δt^4.

16. Speaking of the Heun method, the text says

> This method modifies Euler by first taking an Euler step, using the slope at the current point to determine the direction of the step. Then it steps back to the current point. The final approximate value is obtained from a step whose direction is the average of the slope at the current point and the slope at the end of that first, tentative Euler step.

Expand the expressions for $k_1, \ldots, l_2$ in the Heun method,

$$y_{n+1} = y_n + (k_1 + k_2)/2,$$

$$z_{n+1} = z_n + (l_1 + l_2)/2,$$

and show which contains the Euler step using the slope at t_n and which contains the step taken with the slope at the end of the Euler step. Point out the term that contains the average of the two slopes.

17. We were able to interpret the Euler method as taking a step forward in time using the slope of the solution curve through the current point. The Heun method has a similar interpretation involving an average of the slope at the beginning of the step and at the end of an Euler step. By interpreting each of $k_1, \ldots, k_4$, $l_1, \ldots, l_4$ in the Runge-Kutta method as a slope multiplying a time step, provide a similar interpretation of that method.

18. Example 36 of this section compares two approximate solutions of a version of the predator-prey model at $t = 200$. One approximation has $(F(200), R(200)) \approx (F_{10}, R_{10})$, and the other has $(F(200), R(200)) \approx (F_{40}, R_{40})$. Why are the subscripts on the approximate solutions different?

19. Figure 6.7 shows approximate solutions of the predator-prey model

$$F' = -(0.04 - 0.0011R)F,$$

$$R' = (0.06 - 0.0009F)R,$$

$F(0) = 40$, $R(0) = 60$. The response seems to be periodic. The peaks of the fox population and the rabbit population are out of phase. Is this behavior consistent with the physical situation being modeled?

20. Suppose the initial conditions in the predator-prey model of example 36 can vary by $\pm 20\%$. Does the period of the oscillations in the populations change? Use a numerical method of your choice and available software to answer this question.

6.7 CHAPTER EXERCISES

EXERCISE GUIDE

To gain experience . . .	Try exercises
With steady-state and periodic solutions	1–2
Determining linear independence	3–5
Constructing general solutions	4–14
Solving initial-value problems	10–12
Analyzing and interpreting solution behavior	15–18

1. Argue that a steady-state solution of a differential equation can *always* be used as a particular solution.

2. Argue that a periodic solution of a differential equation can *always* be used as a particular solution.

3. Determine whether each of the given pairs of functions is linearly independent for all x. When the two functions are not linearly independent for all x, give an interval where they are linearly independent, if possible. If they are not linearly independent, explain why.
   (a) x, x^2
   (b) $1 + x, 1 - x$
   (c) $\ln|x|, \ln x^2$

In exercises 4 and 5:

   i. Determine whether the two functions are linearly independent or linearly dependent solutions of the given differential equation on the interval shown.

   ii. Explain the independence/dependence behavior you observe.

   iii. Write a general solution of the differential equation each time you find a linearly independent pair of solutions.

4. (a) $y'' + xy' = 0$; $e^{-x^2/2}, 1$ on $1 \leq x$
   (b) $y'' + xy' = 0$; $e^{(1-x^2)/2}, \pi$ on $1 \leq x$
   (c) $y'' + xy' = 0$; $e^{-x^2/2}, 2 - e^{-x^2/2}$ on $1 \leq x$

5. (a) $x^2 y'' - 3xy' + 3y = 0$; x, x^3 on $x \leq -2$
   (b) $x^2 y'' - 3xy' + 3y = 0$; $x(1 - x)(1 + x), x^3$ on $x \leq -2$
   (c) $x^2 y'' - 3xy' + 3y = 0$; $x, x(4 + x^2)$ on $x \leq -2$

In exercises 6–14, find a general solution of the given equation or solve the given initial-value problem.

6. $4x''(t) + 8x'(t) + 5x(t) = t(t - 2)$

7. $y'' - 9y = e^{-3x}$

8. $y'' + 4y' + 40y = -2x^2 + e^{-2x}$

9. $y'' - 4y' + 40y = e^{2x} - e^{-2x}$

10. $z'' + 9z = 2\cos 4t$, $z(\pi) = 3$, $z'(\pi) = -9$

11. $z'' + 9z = t\sin 3t$, $z(\pi) = 3$, $z'(\pi) = -9$

12. $2w'' - 7w' + 3w = 6e^{3t}$, $w(0) = 8$, $w'(0) = -12$

13. $2w'' - 7w' + 3w = -1$

14. $y'' + 4y' + 4y = 16e^{2x} + 12\sin 2x$

15. Find conditions on the parameters α, β that guarantee that all solutions of

$$x'' + \alpha x' + \beta x = 0$$

are
(a) Purely oscillatory,
(b) Decaying to zero for large t,
(c) Contain *no* oscillatory terms,
(d) Growing in amplitude for all t.
What might α, β represent if this differential equation arose in a spring-mass model? Are all of the values you found physically reasonable in such a setting?

16. Find initial values x_i, v_i that guarantee the solution of

$$x'' - 12x = 0, \qquad x(0) = x_i, \qquad x'(0) = v_i,$$

is
(a) Decaying to zero,
(b) Growing without bound,
(c) Constant for all t.
Could this initial-value problem be a model of a spring-mass system?

17. Find conditions on the parameters a, b that guarantee

$$x'' + ax' + bx = 4\sin 2t$$

has solutions that
(a) Have constant amplitude for all time,
(b) Have period π in the limit as $t \to \infty$,
(c) Are oscillatory and include a component with period 4,
(d) Have amplitude growing linearly with time.
What might a, b represent if this differential equation arose in a spring mass model? Are all of the values you found physically reasonable in such a setting?

18. Find a linear, constant-coefficient, homogeneous differential equation having the given pair of functions as solutions. For each function, find an initial-value problem of which it is a solution.
(a) $\cos x, 2\sin x$
(b) $3x, 2e^{-x}$

Use characteristic equations and undetermined coefficients to find general solutions of the third- and fourth-order equations in exercises 19–24.

19. $y''' - 4y' = 0$

20. $y''' + y'' - y - 1 = 0$

21. $y''' - 3y'' + 4y' - 12y = 0$

22. $y^{iv} - 16y = 0$

23. $y''' - 4y' = 48e^{4t}$

24. $y^{iv} - 16y = -30\cos t$

6.8 CHAPTER PROJECTS

1. **Basic definitions for nth order equations** A linear differential equation of order n is one of the form

$$a_n(x)\frac{d^n y}{dx^n} + \cdots + a_1(x)\frac{dy}{dx} + a_0(x)y = r(x).$$

(a) For such an equation, define *solution, homogeneous, nonhomogeneous, initial-value problem, solution of initial-value problem.*
(b) Complete the definition: *The functions* $y_1, \ldots, y_n$ *are linearly independent on the interval* .... Define *general solution* of a linear, homogeneous equation of order n. Define *general solution* of a linear, nonhomogeneous equation.
(c) Formulate an analogue of the Wronskian condition for the linear independence of n solutions of a linear, homogeneous differential equation of order n. Conjecture what the analogues of the theorems in section 6.2 might be. (These relate the Wronskian and linear independence.)
(d) State and prove a principle of superposition for linear equations of order n.

2. **Solution methods for constant-coefficient equations of order n** Consider the constant-coefficient linear equation of order n,

$$b_n y^{(n)} + \cdots + b_1 y' + b_0 y = r(x)$$

where $b_n, \ldots, b_0$ are (real) constants.
(a) Develop the characteristic equation solution method for the homogeneous form of this equation. Summarize your method in a form similar to the second-order summary on page 232. What linear independence results are needed to put this method on a rigorous foundation?
(b) Develop the method of undetermined coefficients for this equation when the forcing term $r(x)$ is a polynomial, a linear combination of sines and cosines, or an exponential. Provide a table similar to table 6.1 for these cases.

3. **The van der Pol equation** The van der Pol equation

$$\frac{d^2 x}{dt^2} - \epsilon\left(1 - x^2\right)\frac{dx}{dt} + x = 0$$

is a classic source of nonlinear oscillatory phenomena.
(a) B. van der Pol first described and analyzed this equation in the mid-1920s in terms of "negative resistance" in vacuum tube circuits. (See [19], for example.) Explain this terminology in terms of a hypothetical mechanical system or electrical circuit. Specifically, show the variation in the damping-force coefficient with amplitude.
(b) Using available software and a numerical method of your choice, show that this equation achieves the same periodic limiting state for a wide range of initial conditions. Consider $\epsilon = 0.1, 1.0, 10.0$.
(c) Compare the shape of the periodic oscillations for $\epsilon = 0.1$ with those for $\epsilon = 10$. Explain why the former are reasonably sinusoidal. The latter are called *relaxation oscillations*. Suggest why that name is appropriate.

7

QUALITATIVE BEHAVIOR OF HIGHER-ORDER SYSTEMS

And put himself upon his good behavior.
George Noel Gordon, Lord Byron

The preceding two chapters introduced models of various oscillatory phenomena and developed some solution techniques for the second-order equations that appear so frequently in those models. But the real question is the behavior of the solutions of those models. Now we will use those solution techniques to analyze the behavior of such systems. We will explore the relation between initial conditions and system response, examine the effect of external forcing, and study the relation between equilibrium points and periodic solutions.

7.1 UNFORCED SPRING-MASS SYSTEMS

Recall that in section 5.1 we derived initial-value problems such as

$$mx''(t) + px'(t) + kx(t) = 0, \qquad x(0) = x_i, \qquad x'(0) = v_i,$$

as models of spring-mass systems. Here m is the mass, x_i is the initial position in a coordinate system whose origin is at the rest position of the mass, and v_i is the initial velocity given to the mass. The spring constant k is the proportionality constant in Hooke's Law, the experimental observation that force exerted by the spring is proportional to its extension. The damping coefficient p serves a parallel purpose for friction or other damping forces. It is the proportionality constant relating the force of (sliding) friction to velocity.

The differential equation $mx''(t) + px'(t) + kx(t) = 0$ is simply a statement of Newton's Law. The term mx'' is mass times acceleration. It is balanced by the sum of the force of friction and the force of the spring, $-px' - kx$.

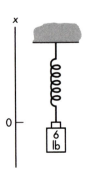

Figure 7.1: The vertical spring-mass system of example 1.

7.1.1 Undamped Systems

In an undamped system, the damping coefficient p is zero. A typical model is then

$$mx''(t) + kx(t) = 0, \qquad x(0) = x_i, \qquad x'(0) = v_i.$$

EXAMPLE 1 *A spring suspended vertically extends one foot when a 2-lb weight is added to it. (Hence, its spring constant is $k = 2$ lb/1 ft $= 2$ lb/ft.) A 6-lb weight is suspended from this spring, and the mass is started in motion from its equilibrium position with a downward velocity of 0.5 ft/s. (See figure 7.1.) Will the mass continue oscillating or will the oscillations decay to zero? If the mass continues oscillating, determine the period of the oscillations of the mass.*

Example 1 of section 5.1, showed that the governing initial-value problem is

$$\frac{3}{16}x''(t) + 2x(t) = 0, \qquad x(0) = 0, \qquad x'(0) = -0.5.$$

Note that the mass m is $\frac{6}{32} = \frac{3}{16}$ slug.

Example 24 of section 6.4 solved this initial-value problem. Briefly, the constant coefficient differential equation $3x''/16 + 2x = 0$ has the characteristic equation

$$\frac{3}{16}r^2 + 2 = 0,$$

which has the pure imaginary roots $r = \pm\sqrt{32/3}i \approx \pm 3.27i$. The general solution of $3x''/16 + 2x = 0$ is

$$x_g(t) = C_1 \cos 3.27t + C_2 \sin 3.27t.$$

The initial conditions $x(0) = 0$, $x'(0) = -0.5$, require $C_1 = 0$, $C_2 = -0.15$. The solution of the initial value problem is

$$x(t) = -0.15 \sin 3.27t.$$

Because the roots of the characteristic equation are pure imaginary, the solution exhibits no exponential decay (or growth). It oscillates steadily with period $2\pi/3.27 = 1.92$ s, as illustrated in figure 7.2. ∎

Do *all* undamped spring-mass systems oscillate periodically without decay? How do the initial conditions affect the response? Can they cause oscillations to grow or decay?

EXAMPLE 2 *Answer the preceding questions by showing that the general solution of the undamped spring-mass equation*

$$mx''(t) + kx(t) = 0$$

never contains exponential factors to cause growth or decay.

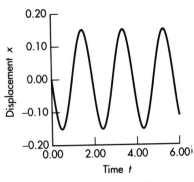

Figure 7.2: The steady oscillations of the solution of the undamped spring-mass model $3x''/16 + 2x = 0$, $x(0) = 0$, $x'(0) = -0.5$.

The characteristic equation of $mx''(t) + kx(t) = 0$ is

$$mr^2 + k = 0.$$

Its roots are $r = \pm\sqrt{-k/m}$. Since $k > 0$ and $m > 0$, these roots are pure imaginary; $r = \pm i\sqrt{k/m}$. The general solution of $mx''(t) + kx(t) = 0$ is

$$x_g(t) = C_1 \cos \omega_n t + C_2 \sin \omega_n t,$$

where $\omega_n = \sqrt{k/m}$.

Regardless of the values of the constants C_1 and C_2 (which are determined by the initial conditions), the solution will be a sum of the simple periodic functions $\cos \omega_n t$ and $\sin \omega_n t$. No exponential decay occurs because the roots of the characteristic equation are pure imaginary. The roots of $mr^2 + k = 0$ will always be pure imaginary as long as m and k are positive. ∎

The frequency $\omega_n = \sqrt{k/m}$ appearing in the general solution $x_g(t) = C_1 \cos \omega_n t + C_2 \sin \omega_n t$ is the **natural (angular) frequency** of the undamped spring-mass system with mass m and spring constant k. We use the term *angular* (or *circular*) frequency because the angle $\omega_n t$, the argument of the cosine and sine terms in this solution, passes through an angle of ω_n radians in one unit of time. For example, the natural frequency of the system appearing in example 1 is $\sqrt{2/(3/16)} \approx 3.27$ rad/s, or a little more than π rad/s.

The **natural (cyclic) frequency** is $\omega_n/2\pi$. We use the term *cyclic* frequency because $\omega_n/2\pi$ is the number of round trips the mass makes in one unit of time. (Each round trip, or complete cycle, corresponds to an angle of 2π rad.) The natural cyclic frequency of the system of example 1 is $3.27/2\pi \approx 0.52$ Hz (cycles/s); about one-half of a complete cycle is executed each second. Compare this cyclic frequency with the angular frequency of approximately π rad/s. How many radians correspond to a complete cycle?

The **natural period**, or the time for one complete cycle, is $2\pi/\omega_n$. When the natural frequency $\omega_n/2\pi$ has units of cycles per second, or hertz, the period $2\pi/\omega_n$ has units of seconds (per cycle). We saw in example 1 that the natural period of that system is $2\pi/3.27 = 1.92$ s, or about 2 s. If the mass is executing about 1/2 cycle each second, it should require about 2 s for one complete cycle.

EXAMPLE 3   *Show that an undamped spring-mass system always oscillates at its natural frequency, regardless of how the motion is started. Show that the graph of those oscillations is a simple sinusoid.*

The motion of a mass m suspended from a spring with spring constant k in the absence of damping is governed by the initial-value problem

$$mx''(t) + kx(t) = 0, \qquad x(0) = x_i, \qquad x'(0) = v_i.$$

We saw in Example 2 that the general solution of $mx''(t) + kx(t) = 0$ is

$$x_g(t) = C_1 \cos \omega_n t + C_2 \sin \omega_n t,$$

where ω_n is the natural frequency $\sqrt{k/m}$.

The initial conditions $x(0) = x_i$, $x'(0) = v_i$, require that $C_1 = x_i$, $C_2 = v_i/\omega_n$. Hence, the solution of this initial value problem is

$$x(t) = x_i \cos \omega_n t + \frac{v_i}{\omega_n} \sin \omega_n t. \tag{7.1}$$

The undamped spring-mass system always oscillates at its natural frequency ω_n, but it is not yet clear that the graph of the sum of the sine and cosine term is simply a sinusoid.

To argue that (7.1) is, in fact, a sinusoid, we must combine the two terms in (7.1) into a single expression involving a sine function. If we could find an angle φ and a positive constant A such that

$$x_i = A \sin \varphi, \quad \frac{v_i}{\omega_n} = A \cos \varphi, \tag{7.2}$$

then a simple trigonometric identity would suffice, for (7.1) would become

$$x(t) = A \sin \varphi \cos \omega_n t + A \cos \varphi \sin \omega_n t$$
$$= A \sin(\omega_n t + \varphi).$$

This form of the solution of the initial-value problem is certainly a simple sinusoid of frequency ω_n. It has simply been shifted by the **phase angle** φ. The factor A is evidently the maximum displacement or **amplitude** of the oscillations. See figure 7.3.

> The graph of $\sin(\omega t + \varphi)$ is just the graph of $\sin \omega t$ shifted to the *left* by φ radians. When $t = 0$, $\sin \omega t = 0$, but $\sin(\omega t + \varphi) = \sin \varphi$. The phase shift φ has given the graph of $\sin(\omega t + \varphi)$ a "head start" of φ radians.

However, we must still show that we can actually find a phase angle φ and a positive amplitude A satisfying $x_i = A \sin \varphi$, $v_i/\omega_n = A \cos \varphi$. To find A, square and sum these two expressions:

$$x_i^2 + \left(\frac{v_i}{\omega_n}\right)^2 = A^2 \cos^2 \varphi + A^2 \sin^2 \varphi = A^2.$$

Hence,

$$A = \sqrt{x_i^2 + \left(\frac{v_i}{\omega_n}\right)^2}. \tag{7.3}$$

Since the motion was begun with nontrivial initial conditions, either $x_i \neq 0$ or $v_i \neq 0$. Assuming that the latter holds, we can divide the first expression in (7.2) by the second to obtain

$$\tan \varphi = \frac{\omega_n x_i}{v_i}.$$

Since the inverse tangent relation is not single valued, we have two choices for φ in the interval $0 \leq \varphi < 2\pi$ which satisfy this last relation. We choose the

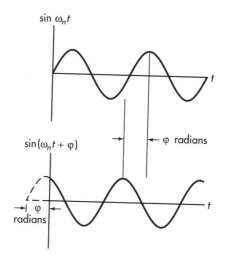

Figure 7.3: The upper graph is $\sin \omega t$. The lower graph is $\sin(\omega t + \varphi)$ — that is, $\sin \omega t$ shifted by φ radians. The corresponding time shift is φ/ω.

one consistent with the two equations (7.2) and with $A > 0$. (The next example illustrates this process.)

Regardless of the (nontrivial) values of the initial displacement and velocity, the motion described by $A\sin(\omega_n t + \varphi)$ has a simple sinusoidal graph with frequency ω_n.

∎

The work of the previous example shows that we have an *alternate form of the general solution* of the undamped spring-mass equation $mx'' + kx = 0$. That alternate general solution,

$$x_g(t) = A\sin(\omega_n t + \varphi),$$

is called the **phase angle-amplitude** form. The amplitude A is determined by $A = \sqrt{x_i^2 + (v_i/\omega_n)^2}$. The phase angle φ is chosen to satisfy simultaneously $\tan\varphi = \omega_n x_i / v_i$, equations (7.2), and the condition $A > 0$.

The expression $A\sin(\omega_n t + \varphi)$ has arbitrary constants A, φ in place of C_1, C_2. While expressing A, φ in terms of the initial conditions is harder than finding C_1, C_2, the phase angle-amplitude form of the general solution immediately gives us the amplitude of the motion and its phase relative to a sine function.

EXAMPLE 4   *Find the solution of the initial-value problem*

$$\frac{3}{16}x''(t) + 2x(t) = 0, \qquad x(0) = 0, \qquad x'(0) = -0.5$$

in phase angle–amplitude form.

From our work in example 1, we know $\omega_n = \sqrt{2/\frac{3}{16}} \approx 3.27$. Hence, we seek values of A and φ that make the alternate general solution

$$x_g(t) = A\sin(3.27t + \varphi)$$

of this differential equation satisfy the initial conditions $x(0) = 0$, $x'(0) = -0.5$.

7.1 Unforced Spring-Mass Systems

These require

$$A \sin \varphi = 0, \qquad 3.27A \cos \varphi = -0.5.$$

These equations are the analogue of (7.2).

Solving each of the preceding equations for the trigonometric function, squaring, and adding gives

$$A^2 = \left(\frac{-0.5}{3.27} \right)^2.$$

Hence, $A = 0.5/3.27 \approx 0.15$. (Note that A is positive.)

To find φ, divide $A \sin \varphi = 0$ by $3.27A \cos \varphi = -0.5$ to obtain

$$\tan \varphi = 0.$$

Either $\varphi = 0$ or $\varphi = \pi$. Both choices satisfy $A \sin \varphi = 0$, but only $\varphi = \pi$ satisfies $3.27A \cos \varphi = -0.5$. Hence, $\varphi = \pi$, and the solution of the initial-value problem is

$$x(t) = 0.15 \sin(3.27t + \pi).$$

This solution is the same as that obtained in example 1, for $\sin(3.27t + \pi) = -\sin 3.27t$. ∎

7.1.2 Damped Systems

We have seen that damped spring-mass systems are modeled by initial-value problems of the form

$$mx''(t) + px'(t) + kx(t) = 0, \qquad x(0) = x_i, \qquad x'(0) = v_i.$$

The positive constant p is the damping coefficient.

To see what sorts of behavior we might find in these systems, we examine the characteristic roots of $mx''(t) + px'(t) + kx(t) = 0$. The characteristic equation is

$$mr^2 + pr + k = 0.$$

Its roots are

$$r_1, r_2 = \frac{-p \pm \sqrt{p^2 - 4mk}}{2m}.$$

Evidently, three cases arise, depending upon the sign of the *discriminant* $p^2 - 4mk$. If p is small enough, then the discriminant is negative. The characteristic roots have an imaginary part, and the solution of $mx''(t) + px'(t) + kx(t) = 0$ contains oscillatory terms. That is, the mass oscillates if the damping is not too large. Such a system is said to be *underdamped*.

If the damping coefficient is large enough to make $p^2 - 4mk$ nonnegative, then the characteristic roots are real and the solution of the differential equation contains no oscillatory terms. The mass can not oscillate if the damping is sufficiently large. Such a system is said to be *overdamped*.

More precisely, if $p^2 - 4mk < 0$, then the system is **underdamped**. If $p^2 - 4mk > 0$, then the system is **overdamped**. At the transition $p^2 - 4mk = 0$, the system is said to be **critically damped**.

EXAMPLE 5 *Example 2, section 5.1.4, showed that the horizontal spring-mass system illustrated in figure 7.4 is modeled by the initial-value problem*

$$\frac{3}{16}x''(t) + 1.41x'(t) + 2x(t) = 0, \qquad x(0) = 0.75, \qquad x'(0) = 0.$$

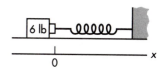

Figure 7.4: The horizontal spring-mass system of example 5.

(The weight of 6 lb has mass of $\frac{3}{16}$ slug. The spring constant is 2 lb/ft, and the friction coefficient is 1.41 lb-s/ft.) Is the force of friction between the mass and the table sufficient to overdamp this system?

This initial-value problem was solved in example 21 of Section 6.4. But we don't need the complete solution. We can answer the damping question simply by examining the sign of the discriminant of the characteristic equation of $3x''/16 + 1.41x' + 2x = 0$,

$$\frac{3}{16}r^2 + 1.41r + 2 = 0.$$

The discriminant is

$$(1.41)^2 - 4\left(\frac{3}{16}\right)2 = 0.49 > 0.$$

Hence, the characteristic roots are real and distinct. The system is overdamped. The initial conditions have nothing whatsoever to do with whether or not the system is damped.

Indeed, in the course of solving this initial-value problem in section 6.4.2, we found that the general solution of the differential equation is

$$x_g(t) = C_1 e^{-1.90t} + C_2 e^{-5.62t}.$$

The absence of oscillatory terms is simply a reflection of the real characteristic roots dictated by the positive discriminant. Because $\lim_{t\to\infty} x_g(t) = 0$, this system always attains a steady state of zero.

The graph of the solution of this initial value problem is the solid curve in figure 7.5; it decays to zero without oscillating. (With a negative initial velocity of sufficient magnitude, the solution will cross the x-axis once before it decays to zero.) ∎

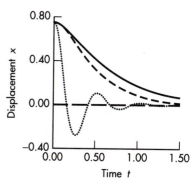

Figure 7.5: Solutions of the damped spring-mass model $3x''/16 + 1.41x' + kx = 0$, $x(0) = 0.75$, $x'(0) = 0$ for $k = 2$ (overdamped: heavy curve), $k = 8/3$ (critically damped: dashed curve), and $k = 200$ (underdamped: light curve).

EXAMPLE 6 *Can the spring in the system of the preceding example be replaced by one that will allow the system to oscillate? If so, specify a range of acceptable spring constant values.*

If the mass and friction coefficient are fixed and the spring has constant k, then the system is governed by the differential equation

$$\frac{3}{16}x'' + 1.41x' + kx = 0.$$

The system will oscillate if this differential equation has a characteristic equation with complex roots.

The discriminant of the characteristic equation is

$$(1.41)^2 - 4\left(\frac{3}{16}\right)k \approx 2 - \frac{3k}{4}.$$

A negative discriminant leads to complex roots. Hence, the system will oscillate if $2 - 3k/4 < 0$, or if

$$k > \frac{8}{3}.$$

The finer dashed curve in figure 7.5 illustrates the decaying oscillations typical of an underdamped system ($k = 200$ lb/ft).

The system is critically damped if $k = 8/3$. Displacement decays to zero without oscillating, as the broader dashed curve of figure 7.5 shows. Compare these conclusions with the solutions of the initial-value problems in some of the other examples of section 6.4.2, in which this system is considered with springs having $k = \frac{8}{3}$ and $k = 4$ lb/ft. ∎

EXAMPLE 7   *Show that every damped spring-mass system achieves a steady state of zero as $t \to \infty$. In other words, show that the rest state is a stable equilibrium of a damped system.*

The characteristic roots of the damped spring-mass equation $mx'' + px' + kx = 0$ are

$$r_1, r_2 = \frac{-p \pm \sqrt{p^2 - 4mk}}{2m}.$$

We have to consider three cases: underdamped, overdamped, and critically damped.

If the system is underdamped, then these roots are a complex conjugate pair. They may be written

$$r_1, r_2 = \alpha \pm i\beta,$$

where

$$\alpha = -\frac{p}{2m}, \qquad \beta = \frac{\sqrt{4mk - p^2}}{2m}.$$

The corresponding general solution of $mx'' + px' + kx = 0$ is

$$x_g(t) = e^{\alpha t}(C_1 \cos \beta t + C_2 \sin \beta t).$$

Since $\alpha < 0$, $\lim_{t \to \infty} x_g(t) = 0$ for any choice of C_1, C_2; i.e., for any choice of initial values x_i, v_i, the solution of the initial-value problem modeling the underdamped spring-mass system tends toward a steady state of zero.

If the system is overdamped, then the differential equation has the general solution

$$x_g(t) = C_1 e^{r_1 t} + C_2 e^{r_2 t},$$

where r_1, r_2 are the distinct real roots

$$r_1, r_2 = \frac{-p \pm \sqrt{p^2 - 4mk}}{2m}.$$

Evidently, both roots are negative, and for any set of initial values, the solution of the differential equation must again decay to a steady state of zero, regardless of the initial conditions.

Obviously, the root $-p - \sqrt{\cdots}$ is negative. The root with the positive sign is negative because $\sqrt{p^2 - 4mk} < p$ as long as $p^2 - 4mk > 0$ and $m, k > 0$.

If the system is critically damped, then the single repeated characteristic root is $r = -p/2m$, and the general solution of the differential equation is

$$x_g(t) = C_1 e^{rt} + C_2 t e^{rt}.$$

Since $r < 0$, $\lim_{t \to \infty} x_g(t) = 0$.

You can use l' Hôpital's rule to show $\lim_{t \to \infty} e^{rt} = 0$.

Hence, damped spring-mass systems decay to a zero steady state. ∎

The analysis of the last example also shows that the transient response of a damped, unforced spring-mass system model is simply the solution itself.

7.1.3 Exercises

EXERCISE GUIDE

To gain experience . . .	Try exercises
With phase angle–amplitude solutions	1, 13–14, 16, 19–21
With relations among amplitude, phase, period, initial conditions, etc.	2–4, 15, 16–18
Analyzing and interpreting damped response	4, 5(a), 6–11
Analyzing and interpreting periodic response	2–3, 5(b–c), 8, 12, 15–16, 18

1. Derive the phase angle-amplitude form of the solution of the initial-value problem modeling each of the undamped systems described here. Assume a general solution of the form $x_g(t) = A \sin(\omega t + \varphi)$, and determine the constants A and φ directly from the initial conditions.

   (a) A mass of 600 g is suspended from a spring that extended 4 cm when a mass of 100 g was hung from it. The 600 g mass is set in motion by releasing it from rest 6 cm above its equilibrium position.

   (b) The spring of part (a) is now set horizontally, attached to a fixed anchor as in figure 7.4. A 2-kg air puck is attached to the free end of the spring. (An air puck rides on a cushion of air so that it can slide with essentially no friction.) The puck is pushed 7 cm toward the fixed anchor and released.

2. Consider the system described in part (a) of exercise 1.

   (a) Suppose the mass is released from rest from an arbitrary position x_i. How does the amplitude of the resulting motion depend upon x_i? How does the phase angle depend upon x_i? Do these relationships seem physically reasonable?

(b) Suppose the mass is started in motion from its equilibrium position with a velocity v_i. How does the amplitude of the resulting motion depend upon v_i? How does the phase angle depend upon v_i? Do these relationships seem physically reasonable?

3. The air-puck system described in part (b) of exercise 1 is set in motion from its equilibrium position with an unknown initial velocity.

(a) The period of its oscillations is observed to be 6.39 s. Can you determine the initial velocity of the puck?

(b) The maximum displacement of the puck is observed to be 14.1 cm. Can you determine its initial velocity?

4. The system described in part (b) of exercise 1 is not truly undamped, although that exercise suggested that you could model it that way. In fact, during a period of 4 min, the maximum displacement of the oscillations of the puck was observed to decrease from 10 cm to 5.6 cm. What is the damping coefficient?

5. Example 2 shows that the general solution of the undamped model equation

$$mx''(t) + kx(t) = 0$$

contains no exponential factors to cause either growth or decay.

(a) Explain why no solution of the initial-value problem

$$mx''(t) + kx(t) = 0, \quad x(0) = x_i, \quad x'(0) = v_i,$$

will exhibit growth or decay, regardless of the choice of initial values x_i, v_i.

(b) Show that all solutions of this initial-value problem oscillate with period $2\pi\sqrt{m/k}$.

(c) What happens when $x_i = v_i = 0$?

6. A 12-kg mass is suspended from a spring. Air resistance and internal friction in the spring resist the motion of the mass with a force whose magnitude is 0.02 times the velocity of the mass. Find the range of spring-constant values for which the mass will *not* oscillate.

7. A mass of 4 kg is suspended from a spring in a liquid that offers a resistance force whose magnitude is eight times the velocity of the mass.

(a) Does decreasing the magnitude of the spring constant cause the mass to oscillate slower or faster?

(b) What range of values of the spring constant will prevent the mass oscillating?

8. You are given the characteristic roots r_1, r_2 of a model of a spring-mass system. Give a step by step recipe (an algorithm) for determining from r_1, r_2 whether the system is damped or undamped. If it is damped, determine whether it is underdamped, critically damped, or overdamped.

9. Example 2 of section 5.1.4, derives a model for an unforced, horizontal spring-mass system,

$$\frac{3}{16}x''(t) + 1.41x'(t) + 2x(t) = 0,$$

$$x(0) = 0.75, \qquad x'(0) = 0.$$

Example 21 of section 6.4.2 finds that the solution of this initial value problem is

$$x(t) = 1.13e^{-1.90t} - 0.38e^{-2.83t}.$$

(a) In what way(s) does the form of this solution tell you that this system is overdamped? Verify that it is indeed overdamped.

(b) Does the position $x(t)$ ever become negative? (*Hint:* Examine the local extrema of the function x.)

(c) Review the physical situation this initial-value problem models. Might you be able to start the mass moving in such a way that its position could be negative at some time? Describe such a set of initial conditions qualitatively.

(d) Provide a mathematical statement for the set of initial conditions you described in part (c). Solve the given differential equation subject to those initial conditions and show that you do indeed obtain negative values of $x(t)$. (If you don't obtain the desired results, revise your guess about an appropriate set of initial conditions.)

(e) For a given initial position x_i, give the range of initial velocities that will always yield negative values of $x(t)$ at some time.

10. Example 7 argues that any solution of the overdamped spring-mass system model

$$mx''(t) + px'(t) + kx(t) = 0, \quad x(0) = x_i, \quad x'(0) = v_i$$

must decay to zero because the real, distinct characteristic roots of $mx'' + px' + kx = 0$ are both negative. Justify the claim that those roots, which are defined by

$$r_1, r_2 = \frac{-p \pm \sqrt{p^2 - 4mk}}{2m},$$

are both negative.

11. Example 7 claims that the critically damped spring-mass system model

$$mx''(t) + px'(t) + kx(t) = 0, \; x(0) = x_i, \; x'(0) = v_i,$$

possesses a stable zero steady state because its general solution

$$x_g(t) = C_1 e^{rt} + C_2 t e^{rt}$$

decays to zero. (The repeated characteristic root r is $-p/2m$.) Verify that $\lim_{t \to \infty} x_g(t) = 0$.

12. Show that the general solution

$$x_g(t) = C_1 \cos \omega_n t + C_2 \sin \omega_n t$$

of

$$mx''(t) + kx(t) = 0$$

does indeed have period $2\pi/\omega_n = 2\pi/\sqrt{k/m}$; i.e., show that $x_g(t + 2\pi/\omega_n) = x_g(t)$ for all t.

13. To make more plausible the claim of example 3 that the general solution

$$x_g(t) = C_1 \cos \omega_n t + C_2 \sin \omega_n t$$

can be replaced by the phase angle-amplitude form

$$x(t) = A \sin(\omega_n t + \varphi),$$

expand $A \sin(\omega_n t + \varphi)$ using the sum-of-angles identity for the sine function. Find C_1, C_2 in terms of A, φ by equating the two forms of the general solution. Express C_1, C_2 in terms of A in the four special cases $\varphi = 0, \pi/2, \pi, 3\pi/2$. Can the phase angle-amplitude form duplicate behavior that involves just a sine or just a cosine in the familiar general solution form $x_g = C_1 \cos \omega_n t + C_2 \sin \omega_n t$?

(The next exercise asks you to go the other way, finding A, φ in terms of C_1, C_2.)

14. Express the arbitrary constants A, φ in the phase an-

gle form of the general solution

$$x_g(t) = A \sin(\omega t + \varphi),$$

of the undamped spring-mass equation $mx'' + kx = 0$ in terms of the arbitrary constants C_1, C_2 that appear in the usual form of the general solution

$$x_g(t) = C_1 \cos \omega t + C_2 \sin \omega t.$$

(The previous exercise asks you to go the other way, finding C_1, C_2 in terms of A, φ.)

15. Show that the maximum value of the expression

$$C_1 \cos \omega t + C_2 \sin \omega t$$

is $\sqrt{C_1^2 + C_2^2}$. That is, find the amplitude of this typical solution expression in terms of the constants C_1, C_2.

16. Example 3 solves the initial-value problem

$$mx''(t) + kx(t) = 0, \; x(0) = x_i, \; x'(0) = v_i,$$

by finding the constants A, φ in the phase angle-amplitude form of the general solution,

$$x(t) = A \sin(\omega_n t + \varphi).$$

Those constants are determined by

$$A = \sqrt{x_i^2 + \left(\frac{v_i}{\omega_n}\right)^2}, \qquad \tan \varphi = \frac{\omega_n x_i}{v_i},$$

where φ, $0 \leq \varphi < 2\pi$, is chosen to satisfy $\tan \varphi = \omega_n/v_i$, equation (7.2), and the condition $A > 0$.

(a) Does the formula $A = \sqrt{x_i^2 + (v_i/\omega_n)^2}$ give a reasonable value for amplitude when $v_i = 0$? (Recall that $\sqrt{x^2} = |x|$.)

(b) What are possible values of φ when $v_i = 0$, $x_i \neq 0$? Does the sign of x_i determine φ uniquely? Is the resulting solution physically reasonable?

(c) What are possible values of φ when $x_i = 0$, $v_i \neq 0$? Does the sign of v_i determine φ uniquely? Is the resulting solution physically reasonable?

(Hint: See example 4.)

17. Figure 7.3 shows that the graph of $\sin(\omega t + \varphi)$ is just the graph of $\sin \omega t$ shifted to the left by φ radians; i.e., the graph of $\sin(\omega t + \varphi)$ starts φ radians before the graph of $\sin \omega t$. But the horizontal axis in figure 7.3 is measured in units of time. By how many *time* units is the graph of $\sin(\omega t + \varphi)$ ahead of the graph of $\sin \omega t$?

That is, find a formula that translates a phase shift of φ radians into units of time.

(*Hint:* A phase shift of π radians corresponds to a phase shift of one-half period. Can you establish a similar proportionality relationship for arbitrary phase angles?)

18. An undamped, vertical spring-mass system is set in motion by a variety of combinations of initial displacements and velocities. Each choice is required to produce a maximum displacement of M units from the equilibrium position of the mass.
    (a) Given the initial position, prescribe an initial velocity that will produce the desired maximum displacement. Is the choice of initial velocity unique? Is there a maximum initial displacement?
    (b) As the initial displacement is increased to larger and larger positive values and the initial velocity is given the corresponding positive value you found in part (a), will the mass reach its peak displacement sooner or later? (*Hint:* examine the phase angle-amplitude form.) Is your analytic conclusion consistent with your physical intuition?

19. Show that the underdamped problem considered in example 23 of section 6.4.2,

$$\frac{3}{16}x''(t) + 1.41x'(t) + 4x(t) = 0,$$

$$x(0) = 0.75, \qquad x'(0) = 0,$$

can be solved by beginning with a general solution in

phase angle-amplitude form

$$x_g(t) = Ae^{-3.76t}\sin(2.68t + \varphi)$$

and using the initial conditions $x(0) = 0.75$, $x'(0) = 0$, to determine the arbitrary constants A and φ. Explain the presence of the numbers 3.76 and 2.68 in this general solution. Show that your phase angle-amplitude solution is equivalent to that obtained from the familiar form of the general solution,

$$x(t) = e^{-3.76t}(0.75\cos 2.68t + 1.05\sin 2.68t)$$

20. Show that the general underdamped initial-value problem

$$mx''(t) + px'(t) + kx(t) = 0, \ x(0) = x_i, \ x'(0) = 0,$$

can be solved by beginning with a general solution in the phase angle-amplitude form

$$x_g(t) = Ae^{\alpha t}\sin(\beta t + \varphi),$$

where A, φ are arbitrary constants to be determined by the initial conditions and α, β are determined by the characteristic equation.

21. The preceding exercise describes a phase angle-amplitude form for a general solution of an underdamped initial-value problem. Find the values of the arbitrary constants A, φ in that solution in terms of the arbitrary constants C_1, C_2 that appear in the usual general solution form

$$x_g(t) = e^{\alpha t}(C_1\cos\beta t + C_2 sin\beta t).$$

7.2 FORCED SPRING-MASS SYSTEMS

Example 3 of section 5.1, derived the initial-value problem

$$\frac{1}{8}x''(t) + 2x(t) = \frac{1}{2}\sin\frac{3\pi t}{2}, \qquad x(0) = 0, \qquad x'(0) = 0, \qquad (7.4)$$

as a model of a vertical spring-mass system whose upper end is being oscillated with angular frequency $3\pi/2$ rad/s. (See figure 7.6.) Example 29 of section 6.5 found that the solution of this initial-value problem is

$$x(t) = \frac{16}{64 - 9\pi^2}\sin\frac{3\pi t}{2} - \frac{6\pi}{64 - 9\pi^2}\sin 4t \qquad (7.5)$$

$$\approx -0.64\sin\frac{3\pi t}{2} + 0.76\sin 4t.$$

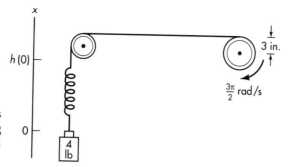

Figure 7.6: This spring-mass system is forced by oscillating the upper end of the spring at a frequency of $3\pi/2$ rad/s.

Graphs of this solution and of the forcing term are shown in figure 7.7. Note the presence in this solution of oscillations at both the frequencies $3\pi/2$ and 4 rad/s. The forcing frequency $3\pi/2$ rad/s corresponds to a period of $2\pi/(3\pi/2) = 4/3 = 1.33$ s, and the natural frequency of 4 rad/s corresponds to a period of $2\pi/4 = 1.57$ s.

The term involving $\sin(3\pi t/2)$ evidently arises from the external forcing. To explain the other term, which has angular frequency 4, note that the natural frequency of this system is

$$\omega_n = \sqrt{\frac{2}{1/8}} = 4 \text{ rad/s.}$$

The $\sin 4t$ term has been stimulated by the external forcing, but it represents the natural oscillations of the system.

How does the forcing frequency affect the amplitude of the response? What determines the balance between natural frequency and forced frequency in the system's response?

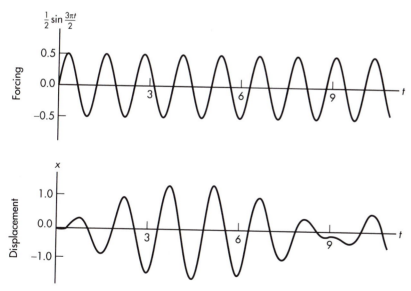

Figure 7.7: A plot of the forcing term (upper curve) and the displacement (lower curve) of the forced spring-mass system shown in figure 7.6. Note the presence in the displacement curve of two separate oscillations, one at the natural frequency and one at the forcing frequency.

To begin answering these questions, consider this same system with the forcing frequency changed from $3\pi/2$ rad/s to 4 rad/s, the natural frequency of this system. In example 29 of section 6.5, we used undetermined coefficients to solve the governing initial-value problem,

$$\frac{1}{8}x''(t) + 2x(t) = \frac{1}{2}\sin 4t, \qquad x(0) = 0, \qquad x'(0) = 0. \tag{7.6}$$

Recall that our first guess for a particular solution was the linear combination

$$x_p(t) = A\cos 4t + B\sin 4t.$$

Since both of the terms in this proposed particular solution are solutions of the homogeneous equation $x''/8 + 2x = 0$, we multiplied by t. The particular solution we finally obtained is

$$x_p(t) = -\frac{1}{16}t\sin 4t.$$

We were forced to add the factor t in this solution because the forcing frequency of 4 rad/s is identical to this undamped system's natural frequency. Because of this added factor, the magnitude of the solution grows without bound, in contrast to the solution (7.5) of the system (7.4), whose forcing frequency is different from its natural frequency.

The undamped system (7.6) is said to be in **resonance** because it is being forced at its natural frequency. The natural frequency of an *undamped* system is also called its **resonant frequency**. Evidently, the *response of an undamped system in resonance grows without bound.*

> Although the solution of the mathematical model grows without bound for all time, in reality displacement is limited by two factors. At very large displacements, the simple linear relationship of Hooke's law fails, and the governing differential equation must be replaced by one containing a nonlinear spring force-displacement relationship; see project 1 of the chapter 5 projects. Or the spring breaks because it is stretched too far.

We have just seen that the response of a system can change dramatically as the forcing frequency varies. The next example explores this phenomena in more detail.

EXAMPLE 8 *Examine the behavior of the response of the undamped, forced system*

$$\frac{1}{8}x''(t) + 2x(t) = F\sin\omega t, \qquad x(0) = 0, \qquad x'(0) = 0,$$

as the forcing frequency ω varies. Do not allow resonance to occur.

Forbidding resonance is equivalent to requiring $\omega \neq \omega_n$, where

$$\omega_n = \sqrt{\frac{2}{1/8}} = 4 \text{ rad/s}.$$

To find a solution of this initial-value problem, we can mimic the analysis of example 29 of section 6.5.

The method of undetermined coefficients suggests a particular solution of the form

$$x_p(t) = A \cos \omega t + B \sin \omega t.$$

Since $\omega \neq \omega_n$, none of the terms in this trial particular solution is a solution of the homogeneous equation $x''/8 + 2x = 0$; no added factor of t is required. Substituting into $x''/8 + 2x = \sin \omega t$ leads to

$$A = 0, \qquad B = \frac{8F}{16 - \omega^2}.$$

Hence, a general solution of $x''/8 + 2x = F \sin \omega t$ is

$$x_g(t) = \frac{8F}{16 - \omega^2} \sin \omega t + C_1 \cos 4t + C_2 \sin 4t.$$

The initial conditions $x(0) = x'(0) = 0$ require $C_1 = 0, C_2 = -2\omega F/(16 - \omega^2)$. The solution of the initial-value problem is

$$x(t) = \frac{8F}{16 - \omega^2} \sin \omega t - \frac{2\omega F}{16 - \omega^2} \sin 4t$$

$$= \frac{F}{16 - \omega^2} (8 \sin \omega t - 2\omega \sin 4t).$$

The extremes of displacement occur at values of t for which $x'(t) = 0$. Rather than compute those values explicitly, we note that the magnitude of such extremes will be controlled by the leading factor

$$\frac{F}{16 - \omega^2}.$$

As ω approaches the resonant frequency of this system, 4 rad/s, this factor grows without bound. At resonance ($\omega = 4$) the solution is no longer valid, and we

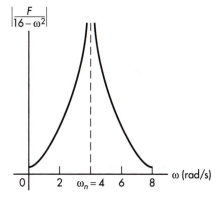

Figure 7.8: A plot of the amplitude of the displacement of the undamped system $x''/8 + 2x = F \sin \omega t$ considered in example 8 as the forcing frequency ω approaches 4 rad/s, the natural frequency of this system.

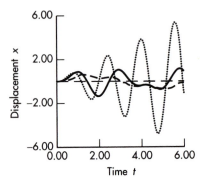

Figure 7.9: The response of $x''/8 + 2x = \sin\omega t$, $x(0) = 0.75$, $x'(0) = 0$, for $\omega = 1$ rad/s (dashed curve), $\omega = 2.5$ rad/s (heavy curve), and $\omega = 3.9$ rad/s (light curve). Note that the amplitude of the response increases as ω approaches the resonant frequency $\omega_n = 4$ rad/s, although the forcing amplitude remains constant at 1. When the forcing frequency is near the resonant frequency ($\omega = 3.9$), the peak amplitude increases nearly linearly with t.

must turn to the solution

$$x(t) = -\frac{1}{16}t\sin 4t$$

discussed earlier, which grows with time because of the factor t multiplying the sine function. Figure 7.8 shows the qualitative behavior of the maximum displacement as ω approaches the resonant frequency $\omega_n = 4$ with F fixed. Figure 7.9 shows plots of the displacement for three different forcing frequencies, $\omega = 1, 2.5, 3.9$, all with forcing amplitude $F = 1$. Note how the amplitude of the response grows as the frequency approaches the resonant frequency, 4 rad/s.

This solution also shows that the extreme displacements are proportional to the amplitude F of the forcing function. ∎

EXAMPLE 9 *Can the solution of the forced, underdamped system*

$$\frac{3}{16}x''(t) + 1.41x'(t) + 8x(t) = F\sin\omega t, \qquad x(0) = 0, \qquad x'(0) = 0$$

grow without bound? What value of the forcing frequency ω will produce the largest displacements over time?

The condition $k > \frac{8}{3}$ derived in example 6 can be used to verify that this system is indeed underdamped. Alternatively, the presence of complex characteristic roots confirms that it is underdamped.

The characteristic roots of $3x''/16 + 1.41x' + 8x = 0$ are $-3.76 \pm 5.34i$. Hence, the homogeneous part of the general solution of this differential equation is

$$e^{-3.76t}(C_1 \cos 5.34t + C_2 \sin 5.34t).$$

Obviously, this part of the solution decays with time for any values of C_1, C_2. It does not produce unbounded growth, regardless of the forcing frequency (or initial conditions), and it contributes next to nothing to the maximum displacement after a sufficient period of time has passed. (For example, at $t = 10$ s, $e^{-3.76t} \approx 4.38 \times 10^{-14}$.)

To examine the contributions of the particular solution to the displacement, we use undetermined coefficients and assume a particular solution of the form

$$x_p(t) = B\cos\omega t + C\sin\omega t.$$

After the usual substitution in the differential equation, we find that

$$B = -\frac{1.41\omega^2 F}{2\omega^2 + D^2}, \qquad C = \frac{DF}{2\omega^2 + D^2},$$

where $D = k - 3\omega^2/16$.

Our real interest here is in the amplitude, or maximum displacement, of the particular solution x_p. Just as in example 3, we could represent this particular

278

solution in the phase angle–amplitude form

$$x_p(t) = A \sin(\omega t + \varphi).$$

The calculations that derived the amplitude expression (7.3) from the solution (7.1) considered in example 3 could also lead us to the amplitude A of this particular solution,

$$A = \sqrt{B^2 + C^2} = \frac{F}{\sqrt{2\omega^2 + D^2}}.$$

We can now answer the question, What value of forcing frequency ω makes the amplitude A of the response a maximum? by seeking the maximum of A as a function of ω. Using the chain rule, we calculate

$$\frac{dA}{d\omega} = -\frac{F}{2(2\omega^2 + D^2)^{3/2}} \left(4\omega + 2D\frac{dD}{d\omega} \right).$$

Since only the term in the large parentheses can make $dA/d\omega$ zero, we evaluate it to find

$$4\omega + 2D\frac{dD}{d\omega} = \omega \left(\frac{9\omega^2}{64} - 2 \right).$$

Hence, $dA/d\omega = 0$ when $\omega = 0$, a value we discard (why?), and when $9\omega^2/64 - 2 = 0$, or

$$\omega = \sqrt{\frac{128}{9}} = \frac{8\sqrt{2}}{3} \approx 3.77 \text{ rad/s.} \qquad \blacksquare$$

The particular solution whose amplitude we have just examined is, in fact, the *periodic solution* of this underdamped system, and the homogeneous solution is its transient. We have shown that this limiting periodic state has maximum amplitude at a forcing frequency of 3.77 rad/s. The forcing frequency at which a periodic solution of an *under*damped system has its maximum is called the **resonant frequency** of that system.

Note that the resonant frequency of this damped system is *not* the natural frequency of the corresponding undamped system. That natural frequency is

$$\omega_n = \sqrt{\frac{8}{\frac{3}{16}}} \approx 6.53 \text{ rad/s}$$

while we have seen that the resonant frequency is 3.77 rad/s. Without damping, the resonant frequency would be ω_n.

Figure 7.10 shows the variation of the amplitude of the forced response with frequency for the damped system of example 9 and for its undamped analogue. The resonant frequency in the damped case is $\omega_r = 3.77$ rad/s, as calculated earlier, while the resonant frequency without damping is the natural frequency $\omega_n = 6.53$ rad/s. With damping, the amplitude of the response reaches a finite maximum when the system is forced at its resonant frequency. Without damping, the amplitude of the response grows without bound as the forcing frequency approaches the resonant (natural) frequency.

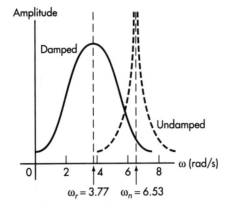

Figure 7.10: Amplitude of response versus forcing frequency for the system of example 2 (solid curve) and its undamped analogue (dashed curve)

The exercises ask you to show that the resonant frequency of the underdamped spring-mass system

$$mx'' + px' + kx = F \sin \omega t$$

is

$$\omega_r = \sqrt{\frac{k}{m} - \frac{p^2}{2m^2}}.$$

That is, the amplitude of the periodic solution of this equation is a maximum when $\omega = \omega_r$.

> Adjusting the forcing frequency for maximum amplitude is the complement of tuning a radio, in which the natural frequency of the tuner circuit is adjusted for maximum amplitude at the fixed forcing frequency of the desired station. Resonance in the moderately damped tuner circuit provides the strongest signal.
>
> Resonance was long believed to account for the collapse of the Tacoma Narrows Bridge in 1940. The bridge began twisting and undulating in moderate winds that might have excited certain resonant modes. Ultimately, the amplitude of the response grew so large that the bridge collapsed. However, more recent research suggests that nonlinearity rather than resonance may have been the culprit; see project 3 at the end of this chapter.

Why do we not speak of the resonant frequency of an *over*damped system? (Exercise 17 requests an answer.)

7.2.1 Exercises

EXERCISE GUIDE

To gain experience . . .	Try exercises
Relating physical systems to equations	1
Finding resonant frequency	2–4, 13(c), 14
With relations among system parameters and resonance	3–5, 14, 17
Analyzing solution behavior as forcing frequency varies	7, 13(c), 15–16
Solving forced equations	8–9, 11
Finding maxima of solutions	10, 13(b–c)
With phase angle–amplitude solutions	11–12

1. *Completely* describe a physical situation of which each of the following initial-value problems could be a model.
   (a) $2x'' + 3x' + x = 0$, $x(0) = 1$, $x'(0) = -2$,
   (b) $2x'' + 3x' + x = 5 \sin \omega t$, $x(0) = 1$, $x'(0) = -2$.

2. Find the resonant frequency of each of the following systems.
   (a) A mass of 600 g is suspended from a spring that extended 4 cm when a mass of 100 g was hung from it. The 600-g mass is set in motion by releasing it from rest 6 cm above its equilibrium position.
   (b) The spring of part (a) is now set horizontally, attached to a fixed anchor as in figure 7.4. A 2-kg air puck is attached to the free end of the spring. (An air puck rides on a cushion of air so that it can slide with essentially no friction.) The puck is pushed 7 cm toward the fixed anchor and released.

   Do you have enough information in each case to determine the resonant frequency? Are you given any extra information? List both missing and extraneous information, if any.

3. Consider the system described in part (a) of exercise 2.
   (a) Suppose the mass is released from rest from an arbitrary position x_i. How does the resonant frequency of the system depend upon x_i?
   (b) Suppose the mass is started in motion from its equilibrium position with a velocity v_i. How does the resonant frequency of the system depend upon v_i?

4. An air-puck system similar to the one described in part (b) of exercise 2 is set in motion from its equilibrium position with an unknown initial velocity.
   (a) The maximum displacement of the puck is observed to be 14.1 cm. Can you determine the resonant frequency of the system?
   (b) The period of its oscillations is observed to be 6.39 s. Can you determine the resonant frequency of the system?

5. The system described in part (b) of exercise 2 is not truly undamped, although that exercise suggested that you could model it that way. In fact, during a period of 4 min, the maximum displacement of the oscillations of the puck was observed to decrease from 10 cm to 5.6 cm. What is the system's resonant frequency?

6. Following the style of example 8, examine the behavior of the response of the general undamped, forced system

$$mx''(t) + kx(t) = F \sin \omega t,$$

$$x(0) = 0, \qquad x'(0) = 0,$$

as the forcing frequency ω varies. Do not allow resonance to occur.

7. Following the style of example 8, examine the behavior of the response of the general damped, forced system

$$mx''(t) + px' + kx(t) = F \sin \omega t,$$

$$x(0) = x_i, \qquad x'(0) = v_i,$$

as the forcing frequency ω varies. Does the form of the solution change at resonance?

8. Example 8 solves the initial-value problem

$$\frac{1}{8}x''(t) + 2x(t) = F \sin \omega t,$$

$$x(0) = 0, \qquad x'(0) = 0,$$

of which the problem solved in example 29 of section 6.5,

$$\frac{1}{8}x''(t) + 2x(t) = \frac{1}{2} \sin \frac{3\pi t}{2},$$

$$x(0) = 0, \qquad x'(0) = 0,$$

is a special case. Show that the solution obtained in example 8 reduces to the solution found in example 29, section 6.5,

$$x(t) = \frac{16}{64 - 9\pi^2} \sin \frac{3\pi t}{2} - \frac{6\pi}{64 - 9\pi^2} \sin 4t,$$

when the parameters F, ω appearing in the first equation are given the appropriate values. Similarly, show that the undetermined coefficients used to find the particular solution and the arbitrary constants in the general solution reduce to the appropriate values.

9. Supply the details of the derivation of the solution

$$x(t) = \frac{F}{16 - \omega^2}(8 \sin \omega t - 2\omega \sin 4t)$$

of the initial-value problem

$$\frac{1}{8}x''(t) + 2x(t) = F \sin \omega t, \quad x(0) = 0, \quad x'(0) = 0,$$

considered in example 8.

10. Example 8 suggests that extreme displacements of the solution of the forced initial-value problem

$$\frac{1}{8}x''(t) + 2x(t) = F \sin \omega t,$$

$$x(0) = 0, \qquad x'(0) = 0,$$

occur at those times t for which $x'(t) = 0$, where

$$x(t) = \frac{F}{16 - \omega^2}(8 \sin \omega t - 2\omega \sin 4t)$$

is the solution of this initial-value problem. Find an equation whose solutions are these values of t. In the case $F = \frac{1}{2}$, $\omega = 3\pi/2$, find the first two such values of t graphically, numerically, or by other means. What are the corresponding extreme values of displacement?

11. Find the solution of the system of example 8

$$\frac{1}{8}x''(t) + 2x(t) = F \sin \omega t,$$

$$x(0) = 0, \qquad x'(0) = 0,$$

with both the particular and homogeneous parts of the solution expressed in phase angle-amplitude form. That is, express the homogeneous part in a form such as

$$A \sin(\omega_n t + \varphi)$$

and the particular part in a form such as

$$B \sin(\omega t + \theta),$$

where B, θ are the amplitude and phase angle, respectively, of the particular solution.

12. Example 9 claims that the particular solution

$$x_p(t) = B \cos \omega t + C \sin \omega t$$

of the underdamped differential equation

$$\frac{3}{16}x''(t) + 1.41x'(t) + 8x(t) = F \sin \omega t$$

can be represented in the phase angle–amplitude form

$$x_p(t) = A \sin(\omega t + \varphi)$$

with amplitude

$$A = \frac{F}{\sqrt{2\omega^2 + D^2}},$$

where $D = k - 3\omega^2/16$. It suggests that the calculations are identical with those of example 3.
(a) Verify this claim by deriving the given expression for A.
(b) Obtain an expression for the phase angle φ. Does the response represented by the particular solution x_p given above lead (reach a peak before) or lag (reach a peak after) the forcing function? Does it seem physically reasonable that the parameters you find appearing in φ should control the phase relationship between the forcing and the response?

13. Consider a forced, damped system described by the differential equation

$$\frac{3}{16}x''(t) + 1.41x'(t) + kx(t) = F \sin \omega t.$$

(This equation with $k = 8$ was considered in example 9.)
(a) For what range of values of the spring constant k is this system *underdamped*? Assume in the following that k is restricted to that range of values.
(b) Find the amplitude of the particular solution of this system.
(c) Show that for $k > 16/3$ this system can exhibit resonance; i.e., show that there is a value of the forcing frequency ω for which the amplitude of the periodic solution of this equation exhibits a local maximum. Find the corresponding resonant frequency. Show that for $k \leq 16/3$, the amplitude does not exhibit a local maximum. Do the values of k for which resonance does not occur correspond to the system being nearly overdamped or very far from being overdamped? Is this relation physically reasonable?

14. The resonant frequency of a damped, forced spring-mass system is the forcing frequency which gives the periodic solution of the equation its largest amplitude. Is the resonant frequency of a damped, forced system the same as the frequency of the decaying os-

cillations in the unforced system? Answer using information obtained in example 9.

15. Using available software, conduct some numerical experiments to estimate the amplitude of the periodic solution of

$$\frac{1}{8}x'' + 2x = \sin \omega t$$

for ω equal to 0.1, 0.5, 0.9, 1.1, 2, and 4 times the resonant frequency of this system. Plot the amplitude values versus frequency and comment on the connection with the graph of figure 7.8. What might happen were you to force this system at precisely its resonant frequency?

16. Repeat exercise 15 for a system with the same mass and spring constant but with a damping coefficient one-quarter that required to critically damp the system. Can you force this system exactly at its resonant frequency for an extended period of time?

17. Answer the question that closes section 7.2:

Why do we not speak of the resonant frequency of an *over*damped system?

7.3   RLC CIRCUITS

Section 5.2 showed that the series RLC circuit shown in figure 7.11 is modeled by the initial-value problem

$$Lq'' + Rq' + \frac{1}{C}q = E, \qquad q(0) = q_i, \qquad q'(0) = i_i.$$

Here q is the charge on the capacitor C and E is a voltage source which can vary with time. The initial charge q_i on the capacitor can also be expressed in terms of the initial voltage drop $v_C(0)$ across the capacitor: $q_i = Cv_C(0)$. The initial current in the circuit i_i is dq/dt evaluated at $t = 0$.

This model has precisely the same mathematical form as the spring-mass models we have just analyzed,

$$mx'' + px' + kx = f(t), \qquad x(0) = x_i, \qquad x'(0) = v_i.$$

From the mathematical analogy between spring-mass systems and RLC circuits, we can derive physical analogies, and we can adapt the analysis of one physical problem to the other. This transfer of understanding between two apparently different physical situations illustrates the power of mathematical abstraction.

The mathematical analogy follows from a comparison of the two equations. For example, mass m corresponds to inductance L because these parameters are the coefficients of the second derivative terms in their respective equations. These analogies are summarized in table 7.1.

Likewise, we can construct physical analogies from the mathematical parallels. For example, the inertia term mx'' in the spring-mass equation, the "mass times acceleration" expression from Newton's law, is analogous to the inductive potential term Lq'' in the RLC circuit equation. Additional physical analogies are listed in table 7.2.

The ideas we developed in the previous section—underdamped, overdamped, resonance, etc.—apply equally well to RLC circuits. The analogies guide their development, but we leave most of the details to the exercises.

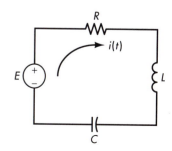

Figure 7.11: A series RLC circuit with an independent voltage source E.

EXAMPLE 10   *The series RLC circuit of figure 7.11 has its inductance and capacitance fixed. The fixed resistance R is replaced by a potentiometer, a variable resistor. For what range of values of R is the response of the circuit underdamped?*

TABLE 7.1

Mathematical analogies between the spring-mass model
$mx'' + px' + kx = f(t)$, $x(0) = x_i$, $x'(0) = v_i$, and the series
RLC circuit model $Lq'' + Rq' + q/C = E$, $q(0) = q_i$, $q'(0) = i_i$

Spring-Mass System	Series RLC Circuit
Mass m	Inductance L
Friction coefficient p	Resistance R
Spring constant k	Reciprocal capacitance $1/C$
Force $f(t)$ applied to spring	Voltage source E
Displacement x of mass	Charge q on capacitor
Velocity x'	Current $q' = i$
Acceleration x''	Rate of change of current di/dt
Initial position x_i of mass	Initial charge q_i on capacitor
Initial velocity v_i	Initial current i_i

From the work of section 7.1, we know that a spring-mass system is under-damped when $p^2 - 4mk < 0$. Using the analogies of table 7.1, we immediately conclude that the circuit is underdamped if $R^2 - 4L(1/C) < 0$ or if

$$R < 2\sqrt{\frac{L}{C}}$$

Alternatively, we could argue that a spring-mass equation is underdamped when its characteristic roots are complex, for then it exhibits decaying oscillations. We can apply the same criterion to the characteristic roots of $Lq'' + Rq' + q/C = 0$, the equation governing the series RLC circuit. The discriminant of the characteristic equation $Lr^2 + Rr + 1/C = 0$ is $R^2 - 4L/C$. Complex characteristic roots occur when $R^2 - 4L/C < 0$, and we reach the same conclusion as before. ∎

EXAMPLE 11   *What is the natural frequency of an undamped RLC circuit?*

First we should decide what makes a series RLC circuit undamped. Consulting the tables reveals that resistance R is analogous to the friction or damping coefficient p; each is the coefficient of the first derivative term in its respective differential equation. Such a circuit is undamped when $R = 0$.

The natural frequency of the undamped spring-mass equation $mx'' + kx = 0$ is the real part of the imaginary characteristic root: $\omega_n = \sqrt{k/m}$. Since the spring constant k is analogous to the reciprocal capacitance $1/C$ and m is analogous to L, the natural (circular) frequency of the undamped series RLC circuit is $\omega_n = 1/\sqrt{LC}$.

TABLE 7.2

Physical analogies between the spring-mass model
$mx'' + px' + kx = f(t)$, $x(0) = x_i$, $x'(0) = v_i$, and the series RLC
circuit model $Lq'' + Rq' + q/C = E$, $q(0) = q_i$, $q'(0) = i_i$

Spring-mass system	Series RLC Circuit
Inertia mx''	Inductive potential $Lq'' = Li'$
Damping force px'	Voltage drop $Rq' = Ri$ across resistor
Restoring force kx	Capacitive potential q/C

Again, a direct attack is just as good. The characteristic roots of $Lq'' + q/C = 0$ are $\pm 1/\sqrt{-LC}$. Hence, the natural frequency, the frequency of the sine and cosine terms in the general solution of this equation, is $1/\sqrt{LC}$. ∎

Oscillations in electrical circuits often occur on a much faster time scale than mechanical oscillations, although the 60 hertz oscillations of the power lines in your house are a ubiquitous exception. We can take advantage of that short time scale, commonly microseconds, to explore the interaction between forcing frequency, natural frequency, and damping in a circuit.

> The inductors and capacitors that appear in household electronic devices such as radios have values that are on the *milli* (10^{-3}, prefix *m*), *micro* (10^{-6}, prefix μ) scale, or *pico* (10^{-12}, prefix *p*); e.g., one might find an inductor of 1 mH = 10^{-3} H and a capacitor of 1 μF = 10^{-6} F.

EXAMPLE 12 *Suppose the elements in the circuit of figure 7.11 have the values $R = 200\Omega$, $L = 100$ μH, $C = 100$ pF, $E \equiv 0$. (The circuit has no voltage source.) What is the natural frequency of the corresponding undamped circuit? What is the frequency of the oscillations in the damped circuit? How much time is required for the amplitude of the oscillations in the charge on the capacitor to die to one-half its initial value?*

The natural frequency of the undamped ($R = 0$) circuit is

$$\omega_n = \frac{1}{\sqrt{LC}} = \frac{1}{\sqrt{100 \times 10^{-6} \cdot 100 \times 10^{-12}}}$$

$$= 1 \times 10^7 \text{ rad/s} = 10 \text{ rad/}\mu\text{s}.$$

The corresponding cyclic frequency is 1.5915 cycles/μs.

The frequency of the oscillations in the damped circuit is the imaginary part of the complex characteristic root of $Lr^2 + Rr + 1/C = 0$. Using the given circuit parameters, those roots are

$$r = -\frac{R}{2L} \pm i\sqrt{\frac{1}{LC} - \left(\frac{R}{2L}\right)^2} = -2 \times 10^6 \pm i9.798 \times 10^6.$$

Hence, the relatively small damping of 200 Ω has reduced the frequency only slightly, to 9.798×10^6 rad/s, or 9.798 rad/μs. The corresponding cyclic frequency is 1.5594 cycles/μs.

The real part of this characteristic root, $-R/2L = -2 \times 10^6$, controls the exponential decay of these oscillations. The oscillations will have one-half their initial amplitude when

$$e^{-2 \times 10^6 t} = \frac{1}{2}.$$

Using logarithms, we solve for t, the time for the amplitude to decay by one-half:

$$-2 \times 10^6 t = -\ln 2$$

$$t = (2 \times 10^{-6}) \ln 2 \approx 1.3863 \times 10^{-6} \text{ s} = 1.3863 \ \mu\text{s}.$$

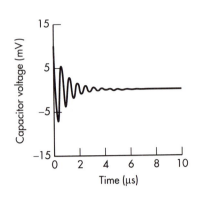

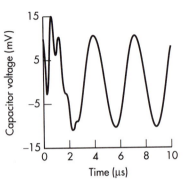

Figure 7.12: Capacitor voltage versus time for the unforced (upper graph) and forced (lower graph) RLC circuit of Figure 7.11 with $R = 200\ \Omega$, $L = 100\ \mu H$, $C = 100\ pF$. The forcing function for the lower graph is $E = V \sin \omega t$ with $V = 10\ mV$, $\omega = 2\ rad/\mu s$. Initial conditions are $v_C(0) = 10\ mV$ (corresponding to $q(0) = 10^{-12}$ C) and $i(0) = q'(0) = -1\ \mu A$.

This decay is illustrated in the upper graph of figure 7.12. Is that graph consistent with these calculations? ∎

The vertical axes of the graphs in figures 7.12 and 7.13 plot the voltage drop across the capacitor rather than charge on the capacitor because voltage drop is easily measured by attaching oscilloscope probes to either side of the capacitor. From experimental fact 2 of section 5.2, the relation between capacitor voltage and charge is $v_C = q/C$.

EXAMPLE 13 *Suppose the RLC circuit of the previous example is now subject to a voltage source $E = V \sin \omega t$ with $V = 10\ mV$, $\omega = 2\ rad/\mu s$. What is the amplitude of the periodic response to the forcing that appears after the transient dies away? About how much time must pass before that periodic solution establishes itself?*

To answer the second question first, we recall from the previous example that the transient decays to half its amplitude in about 1.4 μs. So 2 or 3 μs should be enough time for the transient to be effectively dissipated.

To determine the amplitude of the periodic solution, we must first find a particular solution of

$$Lq'' + Rq' + q/C = V \sin \omega t$$

that does not involve any homogeneous solution terms. That particular solution will be the desired periodic solution.

Since we are concerned about amplitude, we seek a particular solution in the phase angle-amplitude form

$$q_p = A \sin(\omega t + \varphi),$$

where A, φ are undetermined coefficients. (The exercises use the more conventional form $q_p = B_1 \cos \omega t + B_2 \sin \omega t$.)

Substituting in the differential equation, expanding $\sin(\omega t + \varphi)$ using the sum of angles identity, and equating coefficients of $\sin \omega t$ and $\cos \omega t$ lead, respectively, to

$$A \left[\left(\frac{1}{C} - L\omega^2 \right) \cos \varphi - R\omega \sin \varphi \right] = V \tag{7.7}$$

$$\left(\frac{1}{C} - L\omega^2 \right) \sin \varphi + R\omega \cos \varphi = 0.$$

From the second equation, we have

$$\tan \varphi = -\frac{R\omega}{1/C - L\omega^2}.$$

We conclude that

$$\cos \varphi = \pm \frac{1/C - L\omega^2}{D}, \qquad \sin \varphi = \mp \frac{R\omega}{D}, \tag{7.8}$$

where

7   Qualitative Behavior of Higher-Order Systems

$$D = \sqrt{\left(\frac{1}{C} - L\omega^2\right)^2 + R^2\omega^2}.$$

Using these expressions for $\cos\varphi$ and $\sin\varphi$ in (7.7), we find

$$A = \pm\frac{V}{D}.$$

To ensure $A > 0$, we choose the positive sign above (and hence the positive sign with the cosine and the negative sign with the sine in (7.8)) if $V > 0$. All the signs are reversed if $V < 0$.

For the given circuit parameters $R = 200\ \Omega$, $L = 100\ \mu$H, $C = 100$ pF and $V = 10$ mV, we calculate $A = 1.0408 \times 10^{-12}$ C. This charge on a 100 pF capacitor corresponds to a voltage drop across the capacitor of 10.408 mV, because $v_C = q/C$.

The lower graph of figure 7.12 (which was computed using a fourth-order Runge-Kutta method with $\Delta t = 0.001\mu$s) illustrates this periodic response with its amplitude of approximately 10 mV. That same graph shows that the transient has decayed to insignificance in about 3 μs.

Compare the decay of the transient shown in the upper graph with the emergence of the periodic solution in the lower graph. Note the appearance of the higher frequency unforced oscillations in the early part of the forced response, when the forced and the free response are struggling for dominance. ∎

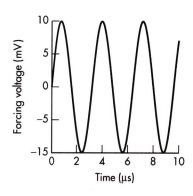

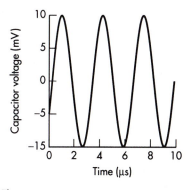

Figure 7.13: Voltage source (upper graph) and periodic response (capacitor voltage — lower graph) versus time for the RLC circuit of figure 7.11 with $R = 4,500\ \Omega$, $L = 100\ \mu$H, $C = 100$ pF. The forcing function is $E = V\sin\omega t$ with $V = 10$ mV, $\omega = 2$ rad/μs.

EXAMPLE 14 *Increase the resistance in the circuit of the previous example to $R = 4,500\ \Omega$. What is the phase relationship between the forcing function and the periodic response?*

Because the forcing amplitude $V = 10$ mV is positive, the analysis of the previous example and equation (7.8) force the phase angle φ to satisfy

$$\cos\varphi = \frac{1/C - L\omega^2}{D}, \qquad \sin\varphi = -\frac{R\omega}{D},$$

where D is as before. For the given circuit parameters, we find

$$\cos\varphi = 0.7295, \qquad \sin\varphi = -0.6839.$$

Since φ must lie in the fourth quadrant, we have $\varphi = 5.5300$ rad, or about 317°.

When $t = 0$, the argument of the periodic response $A\sin(\omega t + \varphi)$ is about 317° ahead of the argument of the forcing function $V\sin\omega t$. At an angle of 317°, the sine function in the periodic solution will have already passed through its first maximum at 90° ($\pi/2$ rad), its second zero at 180° (π rad), and its first minimum at 270° ($3\pi/2$ rad). The periodic solution will be climbing back towards zero on its way to another maximum when the forcing function is only beginning its climb from zero toward its first maximum.

Figure 7.13 compares the periodic response (lower graph) with the forcing function (upper graph) to illustrate this phase difference. The 317° phase difference is nearly 90% of a full period of 360°. Do the graphs show a time difference between peaks that is about 90% of the period of the forcing function? ∎

1. Find conditions on the parameters R, L, C that guarantee that a series RLC circuit such as that of figure 7.11 is

   i. Underdamped

   ii. Critically damped

   iii. Overdamped

   The voltage drop v across a *parallel RLC* circuit such as that shown in figure 7.14 is governed by

   $$Cv'' + \frac{1}{R}v' + \frac{1}{L}v = 0.$$

   Exercises 2–9 refer to such circuits.

2. Construct a table similar to table 7.1 illustrating the mathematical analogies between a parallel RLC circuit and a spring-mass system.

3. Construct a table similar to table 7.2 illustrating the physical analogies between a parallel RLC circuit and a spring-mass system.

4. Construct a table similar to table 7.1 illustrating the mathematical analogies between a parallel RLC circuit and a *series RLC* circuit.

5. Construct a table similar to table 7.2 illustrating the physical analogies between a parallel RLC circuit and a *series RLC* circuit.

6. Find the natural frequency of an undamped parallel RLC circuit.

7. Find conditions on the parameters R, L, C that make a parallel RLC circuit

   (a) Underdamped

   (b) Critically damped

   (c) Overdamped

8. Find the value of forcing frequency ω that maximizes the amplitude of the periodic response of the forced parallel RLC circuit modeled by

   $$Cv'' + \frac{1}{R}v' + \frac{1}{L}v = F\sin\omega t.$$

9. Find the phase angle between the forcing function and the periodic response of the parallel RLC circuit of exercise 8.

10. The oscillations of the charge on the capacitor in a series RLC circuit without a voltage source are governed by $Lq'' + Rq' + q/C = 0$.

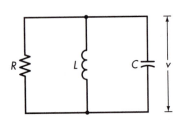

Figure 7.14: A parallel RLC circuit.

(a) Write the frequency of the oscillations in this circuit in terms of the natural frequency of the undamped circuit. Which of the three circuit parameters R, L, C cause the two frequencies to differ?

(b) Find an expression for the time required for the amplitude of the oscillations in this circuit to decrease to one-half their initial-value. Which of the three circuit parameters R, L, C control this characteristic decay time?

11. Referring to the series RLC circuit of figure 7.11 with $R = 200\ \Omega$, $L = 100\ \mu H$, $C = 100$ pF, the text says glibly, "So two or three microseconds should be enough time for the transient to be effectively dissipated." But "effectively" is not very precise.

(a) How much time must elapse before the amplitude of the transient has decayed to 1% of its original size?

(b) How much time must elapse before the amplitude of the transient has decayed to δ times its original size, where $0 < \delta < 1$?

(c) Answer the preceding question for the general RLC circuit governed by $Lq'' + Rq' + q/C = 0$.

12. The text finds the amplitude of the periodic response of the series RLC circuit equation

$$Lq'' + Rq' + q/C = V \sin \omega t$$

by seeking a particular solution in the phase angle-amplitude form $q_p = A \sin(\omega t + \varphi)$. Instead, derive the amplitude as follows.

(a) Assume an undetermined coefficients solution of the form $q_p = B_1 \cos \omega t + B_2 \sin \omega t$. Under what conditions on R, L, and/or C can you be sure that this proposed particular solution contains no solutions of the corresponding homogeneous equation?

(b) Find B_1, B_2.

(c) Argue that the amplitude of this solution, its maximum value, is $\sqrt{B_1^2 + B_2^2}$, thereby verifying the amplitude expression obtained in the text.

13. Example 13 derives the amplitude of the periodic response of

$$Lq'' + Rq' + q/C = V \sin \omega t$$

using the phase angle-amplitude form $A \sin(\omega t + \varphi)$ of the particular solution. Supply the details of that derivation.

14. Example 14 determines that the periodic response of a certain series RLC circuit leads the forcing function by 317°. That circuit has $R = 4,500\ \Omega$, $L = 100\ \mu H$, and $C = 100$ pF, and it is subject to a voltage source $E = V \sin \omega t$ with $V = 10$ mV and $\omega = 2$ rad/μs. The response and the forcing are shown in figure 7.13. Answer the question the text asks:

The 317° phase difference is about 90% of a full period of 360°. Do the graphs show a time difference between peaks that is about 90% of the period of the forcing function?

Compute the period of the forcing function and compare it with the time differences between peaks you read from the graphs of figure 7.13.

15. Could example 13 have concluded that the phase angle of the periodic response is $\varphi = -43°$ rather than $\varphi = 317°$? Verify that the equations defining φ are satisfied by this choice. Would one then say that the periodic response *lags* the forcing by 43°? Sketch a version of figure 7.13 and show this phase difference.

16. Example 13 concludes that the periodic response of the circuit considered there leads the forcing by 317°. Express this lead in seconds. Is the time lead you obtain consistent with that shown in figure 7.13?

17. Find the amplitude of the periodic response of the series RLC circuit considered in example 13. That circuit has $R = 4,500\ \Omega$, $L = 100\ \mu H$, and $C = 100$ pF, and it is subject to a voltage source $E = V \sin \omega t$ with $V = 10$ mV and $\omega = 2$ rad/μs. Compare that computed amplitude with the one shown in the appropriate graph of figure 7.13. Which of the parameters R, L, C, ω, V affect the amplitude? Does doubling the strength V of the voltage source change the amplitude? How about doubling the frequency of the voltage source?

18. Find the phase angle between the periodic response and the forcing function in a series RLC circuit with $R = 200\ \Omega$, $L = 100\ \mu H$, and $C = 100$ pF. The voltage source is $E = V \sin \omega t$ with $V = 10$ mV and $\omega = 2$ rad/μs. Which of the parameters R, L, C, ω, V affect the phase angle? Does doubling the strength V of the voltage source change the phase angle? How about doubling the frequency of the voltage source? 7.13.

Audio purists worry when phase changes with frequency because sound quality suffers. The apparent location of an instrument could change as the notes it played rose or fell in frequency.

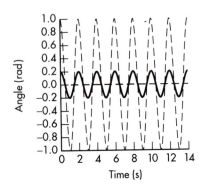

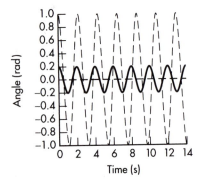

Figure 7.15: Graphs of solutions of the linear (upper) and the nonlinear (lower) pendulum equations for $\theta_i = 0.15$ rad (solid curve) and $\theta_i = 1$ rad (dashed curve). In addition, $L = 0.9930$ m, $\omega_i = 0$.

7.4 LINEAR VERSUS NONLINEAR

In what ways are nonlinear models of oscillatory phenomena different from linear models? How does nonlinearity change the relations among frequency, amplitude, initial conditions, and the other parameters of the problem? What relationships do we find between nonlinear equations and linear approximations when we study the stability of equilibrium points?

A natural vehicle for such comparisons is the nonlinear pendulum model of section 5.3,

$$L\theta'' + g\sin\theta = 0, \qquad \theta(0) = \theta_i, \qquad \theta'(0) = \omega_i,$$

and its linear approximation

$$L\theta'' + g\theta = 0, \qquad \theta(0) = \theta_i, \qquad \theta'(0) = \omega_i.$$

Recall that L is the length of the pendulum arm, g is the acceleration of gravity, θ is the angle in radians of the pendulum from the vertical, θ_i is the initial angular displacement of the pendulum, and ω_i is its initial angular velocity.

7.4.1 Amplitude and Period

To simplify our study of the relation between the amplitude and the period of the pendulum's oscillations, let us always start the pendulum by displacing it to one side and releasing it from rest: $\omega_i = 0$.

In this case, the solution of the linear problem $L\theta'' + g\theta = 0$, $\theta(0) = \theta_i$, $\theta'(0) = 0$ is

$$\theta(t) = \theta_i \cos\sqrt{\frac{g}{L}}t.$$

We make two observations:

1. The natural frequency $\omega_n = \sqrt{g/L}$ is independent of initial position θ_i.
2. The amplitude θ_i of the response is independent of length L and the acceleration of gravity g.

The upper graph of figure 7.15 illustrates the first observation for a pendulum of length $L = 0.9930$ m. The solid curve is the graph of angular displacement for $\theta_i = 0.15$ rad, and the dotted curve is that for $\theta_i = 1$ rad. The periods, and hence the frequencies, of these two curves are identical.

What is the relation between period and amplitude for the nonlinear equation $L\theta'' + g\sin\theta = 0$? Unfortunately, analytic solution tools fail, but fourth-order Runge-Kutta enables us to draw the graphs shown in the lower half of figure 7.15.

The solid curves in figure 7.15 seem to have the same frequency, but the dotted curve in the lower graph clearly has a longer period. With each swing, it crosses through the vertical, or $\theta = 0$, position slightly later than the motion described by the solid curve. *The period of the nonlinear pendulum has increased with amplitude, while that of the linear pendulum remained fixed.*

Why is the difference between the two equations pronounced at the larger amplitude but not at the smaller amplitude? At the smaller amplitude of 0.15

rad ($\approx 8.6°$), the approximation $\sin \theta \approx \theta$ that connects the nonlinear pendulum equation

$$L\theta'' + g \sin \theta = 0$$

and the linear pendulum equation

$$L\theta'' + g\theta = 0$$

is more nearly valid than it is at 1 rad ($\approx 57°$); $\sin 0.15 = 0.14944$, while $\sin 1 = 0.84147$. As the nonlinearity asserts itself, predictions based on the solution of a linear approximation lose validity.

We leave it to you to conduct a parallel investigation of the relation between amplitude and the parameters L and g. (Does conservation of energy offer any clues?)

7.4.2 Equilibria

What are the steady states of the nonlinear pendulum equation? They are the constant solutions θ_{ss} of

$$L\theta''_{ss} + g \sin \theta_{ss} = g \sin \theta_{ss} = 0.$$

From $\sin \theta_{ss} = 0$, we obtain

$$\theta_{ss} = 0, \pm\pi, \pm 2\pi, \ldots .$$

This endless list of possible steady states is at first surprising, for the only apparent stationary position of the pendulum is hanging straight down. But figure 7.16 explains the list.

Steady state values such as $\theta_{ss} = 0, \pm 2\pi, \ldots$, which involve zero or *even* multiples of π, correspond to the familiar straight-down position, perhaps after the pendulum has passed through one or more complete revolutions around its pivot. (A full revolution increments θ by 2π. The increment is positive if the revolution is counterclockwise and negative if the rotation is clockwise.)

The equilibria at *odd* multiples of π occur when the pendulum is positioned straight *up*. Knowing that this inverted position is difficult to attain in practice suggests that these equilibria are unstable, although their straight-down counterparts ought to be stable. The linear stability analysis that follows confirms our intuition in several cases. (Compare these examples with the analysis of the

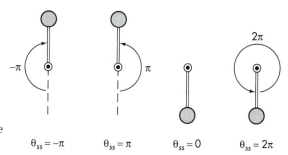

Figure 7.16: Some possible steady states of a pendulum.

$\theta_{ss} = -\pi$ $\theta_{ss} = \pi$ $\theta_{ss} = 0$ $\theta_{ss} = 2\pi$

stability of the steady states of the logistic equation $P' = aP - sP^2$ in section 4.2.)

EXAMPLE 15 *Determine the stability of the equilibrium* $\theta_{ss} = 2\pi$.

Identify the steady state.

What becomes of a small perturbation of the steady state $\theta_{ss} = 2\pi$? We suspect it does not grow, for this steady state corresponds to the pendulum hanging straight-down. If we can confirm our intuition mathematically, we will have established the stability of this steady state.

Perturb the steady state.

Call this small perturbation $p(t)$; it is defined by

$$\theta(t) = 2\pi + p(t),$$

where $\theta(t)$ is a solution of the nonlinear pendulum equation $L\theta'' + g\sin\theta = 0$.

The steady state plus its perturbation must satisfy the governing equation because that sum describes the motion of the pendulum. We believe that every possible motion of the pendulum is a solution of that equation.

Substitute into differential equation.

Substituting θ into the differential equation leads to an equation for the unknown perturbation p,

$$L(2\pi + p(t))'' + g\sin(2\pi + p(t)) = 0$$

$$Lp''(t) + g\sin(2\pi + p(t)) = 0$$

Expand the nonlinearity.

To find a *linear* (but approximate) equation for p, we expand $\sin x$ in a Taylor polynomial about $x = 2\pi$:

$$\sin(2\pi + p) = \sin x\big|_{x=2\pi} + \left(\frac{d}{dx} \sin x \bigg|_{x=2\pi} \right) p + \cdots$$

$$= \sin 2\pi + (\cos 2\pi)p + \cdots$$

$$= p + \cdots.$$

Linearize.

Hence, the approximation

$$\sin(2\pi + p) \approx p$$

gives us a linear equation for p. This approximation fails if the perturbation p becomes large, but by then stability is already out the window.

The resulting *linear* equation for the perturbation p is

LINEAR EQUATION FOR PERTURBATION

$$Lp'' + gp = 0,$$

just the familiar approximate pendulum equation. We know that the general solution of this equation is

$$p(t) = C_1 \cos \sqrt{\frac{g}{L}}t + C_2 \sin \sqrt{\frac{g}{L}}t.$$

It neither grows nor decays.

If the perturbation can be kept arbitrarily close to the equilibrium by making the initial perturbation sufficiently small, then the equilibrium is called **neutrally stable**. In the absence of damping, oscillations of the pendulum continue indefinitely. A small perturbation of the straight-down steady state will lead to small oscillations around that steady state. Neutral stability is the next best thing to stability.

The steady state $\theta_{ss} = 2\pi$ is *neutrally stable*. ■

The approximate, *linear* pendulum equation seems to govern the perturbations of the neutrally stable steady states of the original nonlinear pendulum equation.

What happens to a steady state we believe to be unstable?

EXAMPLE 16 *Determine the stability of the equilibrium $\theta_{ss} = \pi$.*

Proceeding as before, we wish to determine the fate of the small perturbation p of the steady state $\theta_{ss} = \pi$. Hence, we define p by

Perturb the steady state.

$$\theta(t) = \pi + p(t).$$

where $\theta(t)$ is a solution of $L\theta'' + g\sin\theta = 0$. Substituting into the differential equation leads to

Substitute into differential equation.

$$L(\pi + p(t))'' + g\sin(\pi + p(t)) = 0$$
$$Lp''(t) + g\sin(\pi + p(t)) = 0$$

Now we expand $\sin x$ in a Taylor polynomial about $x = \pi$:

Expand nonlinearity.

$$\sin(\pi + p) = \sin x|_{x=\pi} + \left(\frac{d}{dx} \sin x \Big|_{x=\pi} \right) p + \cdots$$

$$= \sin\pi + (\cos\pi)p + \cdots$$

$$= -p + \cdots.$$

The linearizing approximation is

Linearize.

$$\sin(\pi + p) \approx -p.$$

The resulting *linear* equation for the perturbation p is

$$Lp'' - gp = 0,$$

LINEAR EQUATION FOR PERTURBATION

only slightly different from the approximate pendulum equation. But that negative sign before the second term has profound consequences. The general solution of this equation is

$$p = C_1 e^{\sqrt{g/L}\, t} + C_2 e^{-\sqrt{g/L}\, t}.$$

Except in the very special circumstances that render $C_1 = 0$, this perturbation grows exponentially.

The straight-up equilibrium $\theta_{ss} = \pi$ is unstable. ■

7.4 Linear versus Nonlinear

The linearizations

$$\sin(2\pi + p) \approx p, \ \sin(\pi + p) \approx -p,$$

used in these two examples are also consequences of familiar trig identities and our old friend, $\sin p \approx p$. We used Taylor polynomial expansions in these examples because they can be applied to a wide variety of nonlinearities. But, of course, $\sin p \approx p$ was first derived using a Taylor polynomial.

7.4.3 Exercises

EXERCISE GUIDE

To gain experience . . .	Try exercises
Identifying linear and nonlinear equations	1, 5, 6–7(c–d), 11(b)
Solving linear equations	2–3, 5
With validity of linear models	6(b), 11(b), 12–13
Relating period, amplitude, etc. to system parameters	3–4, 6–10, 11(a), 12–13(a–b)
With damped pendulum model	17–19
Determining stability of equilibria	14–18, 20–24
Analyzing and interpreting solution behavior	6–9, 14–15, 18, 21

1. Verify that the pendulum equation

$$L\theta'' + g\sin\theta = 0$$

is nonlinear and that its approximate counterpart

$$L\theta'' + g\theta = 0$$

is linear. Are these equations homogeneous or non-homogeneous?

2. Derive the solution

$$\theta(t) = \theta_i \cos\sqrt{\frac{k}{m}}\,t$$

of the linear initial-value problem $L\theta'' + g\theta = 0$, $\theta(0) = \theta_i$, $\theta'(0) = 0$, thereby verifying the solution formula used in the text.

3. Solve the general initial-value problem of the approximate pendulum model,

$$L\theta'' + g\theta = 0, \quad \theta(0) = \theta_i, \quad \theta'(0) = \omega_i.$$

Argue that the frequency of this solution is *always* $\sqrt{g/L}$, regardless of the initial values θ_i, ω_i.

4. The text makes the following observations about the solution

$$\theta(t) = \theta_i \cos\sqrt{\frac{g}{L}}\,t$$

of the initial-value problem $L\theta'' + g\theta = 0$, $\theta(0) = \theta_i$, $\theta'(0) = 0$:

   i. The natural frequency $\omega_n = \sqrt{g/L}$ is independent of initial position θ_i.

   ii. The amplitude θ_i of the response is independent of length L and the acceleration of gravity g.

Verify each claim.

5. The text claims, "Unfortunately, our analytic solution tools fail when we apply them to the nonlinear equation $L\theta'' + g\sin\theta = 0$." Demonstrate this assertion by attempting to solve this equation by the characteristic equation method. To what types of equations is this method applicable? What fails here?

6. The text makes two observations about the behavior of the solution of the linear pendulum problem $L\theta'' + g\theta = 0$, $\theta(0) = \theta_i$, $\theta'(0) = 0$:

  i. The natural frequency $\omega_n = \sqrt{g/L}$ is independent of initial position θ_i.

  ii. The amplitude of the response is independent of length L and the acceleration of gravity g.

  (a) Construct a parallel set of observations for the initial-value problem $L\theta'' + g\theta = 0$, $\theta(0) = 0$, $\theta'(0) = \omega_i$.

  (b) Describe the physical difference between this problem and the one considered in the text.

  (c) Is your first observation valid for the corresponding nonlinear problem? Investigate numerically.

  (d) Is your second observation valid for the corresponding nonlinear problem? Investigate numerically.

7. Repeat parts (a) and (b) of exercise 6 for the initial-value problem $L\theta'' + g\theta = 0$, $\theta(0) = \theta_i$, $\theta'(0) = \omega_i$.

8. The text observes that the frequency of the response of the linear pendulum equation $L\theta'' + g\theta = 0$ is independent of the initial position θ_i. It then goes on to show that the frequency of the response of the nonlinear pendulum equation $L\theta'' + g\sin\theta = 0$ is dependent upon θ_i.

    Conduct a parallel numerical investigation to determine if the nonlinear pendulum equation obeys the second property the text observed for the linear equation,

    The amplitude θ_i of the response is independent of length L and the acceleration of gravity g.

9. A pendulum executing small oscillations is transported to the moon, where the force of gravity is less than that on the earth. What becomes of the period of the pendulum's oscillations? What becomes of the amplitude of those oscillations? What can you say if the pendulum is executing large oscillations?

10. The numerical examples graphed in figure 7.15 were computed for a pendulum with length 0.9930 m. Why do you think that unusual length was chosen?

11. The upper graph in figure 7.15 shows solutions of

$$0.9930\theta'' + 9.8\theta = 0, \qquad \theta(0) = \theta_i, \qquad \theta'(0) = 0$$

for $\theta_i = 0.15$ and $\theta = 1$.

  (a) Read the period of the solutions from the graph. Are the values you read consistent with the solution of this initial-value problem?

  (b) Describe the pendulum modeled by this initial-

value problem. Does the choice of the value of θ_i affect the validity of this model?

12. The *dashed* curves in figure 7.15 show the response of the approximate pendulum equation $L\theta'' + g\theta = 0$ (upper graph) and the nonlinear pendulum equation $L\theta'' + g\sin\theta = 0$ (lower graph) when the mass is released from rest from an angle of 1 rad. (Recall $L = 0.9930$ m.)

  (a) For the motion shown, how good or bad is the approximation $\sin\theta \approx \theta$ that relates these two equations?

  (b) What difference in period, if any, can you detect in the graph between the responses of the two equations?

  (c) Are your answers to the preceding two questions consistent?

13. The *solid* curves in figure 7.15 show the response of the approximate pendulum equation $L\theta'' + g\theta = 0$ (upper graph) and the nonlinear pendulum equation $L\theta'' + g\sin\theta = 0$ (lower graph) when the mass is released from rest from an angle of 0.15 rad. (Recall $L = 0.9930$ m.)

  (a) For the motion shown, how good or bad is the approximation $\sin\theta \approx \theta$ that relates these two equations?

  (b) What difference in period, if any, can you detect in the graph between the responses of the two equations?

  (c) Are your answers to the preceding two questions consistent?

14. Determine the stability of the equilibrium $\theta_{ss} = 0$ of the pendulum equation $L\theta'' + g\sin\theta = 0$. Are your results consistent with your intuition?

15. Determine the stability of the equilibrium $\theta_{ss} = -\pi$ of the pendulum equation $L\theta'' + g\sin\theta = 0$. Are your results consistent with your intuition?

16. Find the equilibrium points of the approximate pendulum equation $L\theta'' + g\theta = 0$. Determine their stability.

17. Find the equilibrium points of the *damped* approximate pendulum equation $L\theta'' + p\theta' + g\theta = 0$. Determine their stability.

18. **(a)** Find the equilibrium points of the *damped* pendulum equation

$$L\theta'' + p\theta' + g\sin\theta = 0.$$

  (b) Which of these correspond to a straight-up position of the pendulum? To a straight-down position? Are there any other equilibrium positions?

(c) Determine the stability of one of the straight-up equilibria. Are your results consistent with your intuition?

(d) Determine the stability of one of the straight-down equilibria. Are your results consistent with your intuition?

19. The preceding exercise glibly refers to the *damped pendulum equation* $L\theta'' + p\theta' + g\sin\theta = 0$. Examine the derivation of the pendulum equation in section 5.3. What physical effects might the damping term $p\theta'$ represent? When the equation is written in this form, which of the parameters L, g, or the mass m of the pendulum bob might the coefficient p involve?

20. In conducting its linear stability analysis of the equilibrium point $\theta_{ss} = 2\pi$ of the nonlinear pendulum equation, the text expands $\sin x$ in a Taylor polynomial about $x = 2\pi$:

$$\sin(2\pi + p) = \sin x\big|_{x=2\pi} + \left(\frac{d}{dx}\sin x\bigg|_{x=2\pi}\right)p + \cdots$$

$$= \sin 2\pi + (\cos 2\pi)p + \cdots.$$

(a) What is the lowest power of p that appears in the terms represented by the ellipsis $\cdots$?

(b) Are these terms really smaller than the term involving p? Specifically compare the term involving p with the first term represented by the ellipsis $\cdots$ when $p = 0.1$.

(c) If you kept any of the terms represented by the ellipsis $\cdots$, would the resulting equation in p be linear? Explain.

21. The linearized stability analyses conducted in the text lead to the equations

$$Lp'' + gp = 0$$

and

$$Lp'' - gp = 0$$

for the perturbation p of the steady state in question.

(a) Find the general solution of each of these equations.

(b) Argue that the solutions of the first are always bounded.

(c) Argue that the solutions of the second grow exponentially unless p and p' obey a special initial condition. State that condition.

In exercises 22–24, find the equilibrium points of the given equation and determine their stability.

22. $x'' + x^2 - 1 = 0$

23. $x'' + x^2 - x = 0$

24. $x'' + x^3 - x = 0$

7.5 INTRODUCTION TO THE PHASE PLANE

To understand the behavior of devices such as a pendulum or a spring-mass system, we need information about both position and velocity. That information can be presented in two separate graphs, position versus time and velocity versus time, as in figure 7.17. But in the absence of forcing, it is really the relation between velocity and position that matters. For example, a pendulum with *positive* displacement and *positive* velocity will continue to swing even further from its equilibrium, while one with *negative* position and *positive* velocity must be moving toward its equilibrium.

For a second-order equation such as the pendulum equation, the natural way to capture the relation between velocity and position is in a graph of velocity versus position. In general, a plot of the derivative of the dependent variable (velocity) against the dependent variable (position) in a second-order equation is known as a **phase plane**. For a system of two first-order equations, the phase plane is a graph of one dependent variable against the other. A curve in the phase plane that represents the evolution of the two variables θ and ω with time is called a **trajectory**.

For example, a phase plane plot of the pendulum equation

$$L\theta'' + g\sin\theta = 0$$

is a graph of $\omega = \theta'$ versus θ — that is, a graph of velocity versus position. To make angular velocity ω more visible, we could write this second-order equation as a first-order system

$$\theta' = \omega$$

$$L\omega' = -g\sin\theta.$$

Recall that in section 6.6, we converted second-order equations to first-order systems by introducing a new dependent variable— ω in this case—to replace the first derivative of the unknown. The first equation in the system is just the definition of the new dependent variable.

Figure 7.18 shows the phase plane plot of ω versus θ that corresponds to the graphs of ω versus t and θ versus t shown in figure 7.17. Notice that *a simple closed curve in the phase plane corresponds to a periodic solution*. Each time the trajectory is traced, position and velocity return to the same values. Because time never appears explicitly in the pendulum equations $\theta' = \omega$, $L\omega' = -g\sin\theta$, the time elapsed in traversing a section of the trajectory depends only on θ and ω, not on any external influence that could change from cycle to cycle. We could choose any point on the trajectory as initial-values of θ and θ', and we would always retrace this trajectory.

Systems such as the pendulum equation, in which the only functions of time are the unknowns themselves, are called **autonomous**. The phase plane perspective is particularly appropriate for autonomous systems because the response of the system depends upon its state, the current values of the dependent variables, and not upon time. The phase plane picture is invariant in time.

Another autonomous system is the predator-prey (fox-rabbit) model of section 5.4,

$$F' = -(d_F - \alpha R)F$$

$$R' = (b_R - \beta F)R.$$

The *unforced* spring-mass equation is also autonomous.

Which direction on the trajectory of Figure 7.18 corresponds to increasing time, clockwise or counterclockwise? Starting at $\theta = 0.15\,\text{rad}$, $\omega = 0\,\text{rad/s}$ places us on the positive θ-axis at the three o'clock position. Releasing the pendulum causes it to drop; θ will decrease. At the same time, its velocity will *decrease* in the sense of moving from zero to negative values; ω is negative because θ is decreasing with time. We trace the trajectory in the clockwise direction.

These physical arguments are a reflection of the governing differential equations,

$$\theta' = \omega$$

$$L\omega' = -g\sin\theta.$$

If $\theta > 0$, then $L\omega' = -g\sin\theta$ forces $\omega' < 0$. If initially $\omega = 0$ and $\omega' < 0$, then ω must become negative, forcing the trajectory from its starting point on the positive θ-axis into the fourth quadrant.

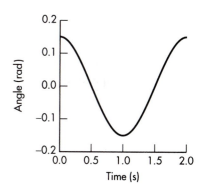

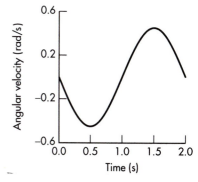

Figure 7.17: Graphs of angular position θ (upper graph) and angular velocity $\omega = \theta'$ (lower graph) versus time for the nonlinear pendulum model $L\theta'' + g\sin\theta = 0$, $\theta(0) = 0.15$, $\theta'(0) = 0$ with $L = 0.9930$ m.

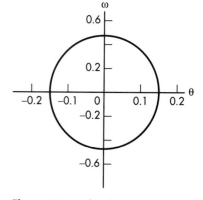

Figure 7.18: The phase plane plot of angular velocity ω versus angular position θ corresponding to the velocity- and position-versus-time graphs of figure 7.17.

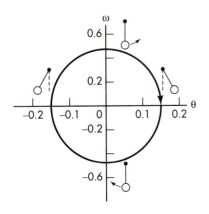

Figure 7.19: The phase plane of figure 7.18 marked with the position of the pendulum. The trajectory is traversed in the clockwise direction as time increases.

Figure 7.19 redraws this same trajectory with diagrams of the pendulum added to emphasize the connection between location in the phase plane and motion of the pendulum.

At a steady state, both θ and ω are constant. Hence, the trajectory of a steady state is simply a point in the phase plane. The equilibria of the pendulum equation must all lie on the horizontal axis in the phase plane, because the velocity ω is zero at such a point.

Since steady states are constant solutions of the governing equations, we can find them by solving the differential equations with the derivatives set to zero:

$$0 = \theta'_{ss} = \omega_{ss}$$

$$0 = L\omega'_{ss} = -g \sin \theta_{ss}.$$

The first equation confirms the observation of the preceding paragraph: $\omega_{ss} = 0$. From the second equation, we find $\sin \theta_{ss} = 0$ and

$$\theta_{ss} = 0, \pm\pi, \pm 2\pi, \ldots.$$

These equilibrium or **stationary** points are shown in figure 7.20.

What becomes of trajectories that start near these stationary points? From the stability analysis of section 7.4, we know the essence of the answer. The stationary points $0, \pm 2\pi, \ldots$, which correspond to the pendulum's usual straight-down rest position, are neutrally stable. Perturbations of these equilibria result in small periodic oscillations around the equilibrium point. That is, near the stationary points $\theta = 0, \pm 2\pi, \ldots$, we should find trajectories representing those small periodic oscillations that are closed curves.

The closed, elliptically shaped curves surrounding $\theta = 0, \pm 2\pi$ in figure 7.21 are these neutrally stable oscillations. Indeed, these oscillations occur even for the relatively large perturbations shown in this figure.

> Since any point on a trajectory could serve as an initial point, the distance between points on the trajectory and the equilibrium gives an idea of the size of the perturbation. For example, the elliptically shaped curve closest to the origin in figure 7.21 could have been started from the point on the positive θ-axis with $\theta = 1, \omega = 0$. Of course, the change from $\theta = 0$ to $\theta = 1$ is hardly a small perturbation.
>
> If the closed curve of figure 7.18 were scaled to fit these axes, it would represent a small perturbation of the equilibrium at $\theta = 0$. That curve would be a tiny ellipse around the origin passing through the point $\theta = 0.15, \omega = 0$. (The curve in figure 7.18 appears more circular than the closed curves in figure 7.21 because of the relative scaling of the axes.)

The stationary points $\theta = \pm\pi, \ldots$ are unstable because they correspond to attempting to stop the pendulum in an inverted position. We saw in section 7.4 that small perturbations of these equilibria grow with time. The linear analysis used there can not tell us anything once these perturbations become large, for it is accurate only for small perturbations.

The trajectory labeled U (for *unstable*) in figure 7.21 begins close to $\theta = -\pi, \omega = 0$; it is a small perturbation of that unstable equilibrium point. Although it eventually returns to its starting point, it also cycles far from that equilibrium point, demonstrating the instability of the equilibrium $\theta = -\pi$.

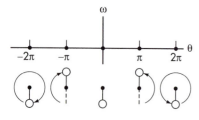

Figure 7.20: The equilibrium points of the nonlinear pendulum equation $L\theta'' + g \sin \theta = 0$ lie on the θ-axis in the phase plane.

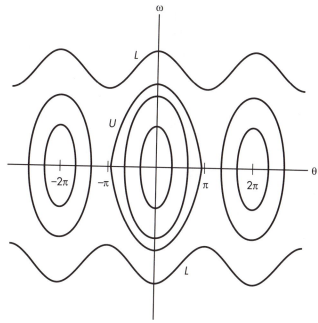

Figure 7.21: The phase plane for the undamped pendulum equation $L\theta'' + g\sin\theta = 0$.

Physically, this motion corresponds to releasing the pendulum from rest from the position shown in figure 7.22, a negative angular displacement just short of vertical. The pendulum swings down and back up the other side, not quite reaching the vertical position $\theta = \pi$ before it returns to its starting point. The trajectory in figure 7.21 likewise falls just short of $\theta = \pi$.

The curves labeled L (for *looping*) in Figure 7.21 represent continual looping motion of the pendulum. It has been displaced and pushed so hard that it spins continuously around its pivot like a propeller. Exercise 18 asks you to decide which of the curves labeled L represents clockwise motion and which represents counterclockwise motion.

If damping is added to the pendulum equation, then the phase plane picture of figure 7.21 should change. Eventually, any motion of the pendulum must come to rest as friction in the pivot or air resistance exhausts its energy. That is, the stationary points $\theta = 0, \pm2\pi, \ldots$ change from neutrally stable to stable. They become **attracting points** or **attractors**, drawing to them trajectories that start near enough. Physical intuition suggests that the stationary points $\theta = \pm\pi, \pm3\pi, \ldots$ corresponding to the inverted position will remain unstable.

Figure 7.23 illustrates the phase plane for the damped pendulum equation

$$\theta' = \omega$$

$$L\omega' = -g\sin\theta - p\omega.$$

The closed curves around the stable equilibria at $\theta = 0, \pm2\pi$ have become spirals; the exercises ask you to verify that they are spiraling toward the stable equilibria (not away), justifying our calling a stable equilibrium an *attracting point*.

Figure 7.22: The position of the pendulum at the start of the trajectory labeled U in figure 7.21. The pendulum is displaced just to the left of vertical, to a negative angle slightly smaller in magnitude than $-\pi$.

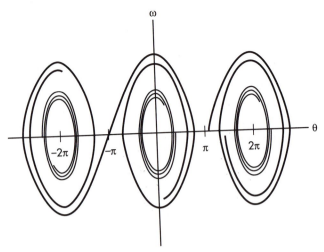

Figure 7.23: The phase plane for the damped pendulum equation $L\theta'' + p\theta' + g\sin\theta = 0$.

Perturbations of the unstable equilibria at $\theta = \pm\pi$ also spiral toward the nearest *stable* equilibrium points. Physically, the inverted pendulum winds or unwinds once around its pivot when it is perturbed. Then it oscillates back and forth until it eventually settles to rest in a straight-down position.

The trajectories shown in these phase plane diagrams were computed using a fourth-order Runge-Kutta method to solve the nonlinear pendulum equations. But the real point is that we can deduce from the equations themselves much of the general shape of the trajectories in the phase plane. The exercises take you through such analyses.

Any autonomous system such as the ones mentioned in this section can be written in the general form

$$y' = f(y, z)$$
$$z' = g(y, z),$$

where prime (') denotes differentiation with respect to an independent variable such as t. For example, in the pendulum equation, $y = \theta$, $z = \omega$, and so forth.

The stationary points or equilibria of such systems are simply solutions (y_{ss}, z_{ss}) of

$$f(y_{ss}, z_{ss}) = 0$$
$$g(y_{ss}, z_{ss}) = 0.$$

Finally, a phase plane trajectory can be thought of as a solution of the single first-order equation

$$\frac{dy}{dz} = \frac{f(y, z)}{g(y, z)},$$

provided $g \neq 0$.

7 Qualitative Behavior of Higher-Order Systems

EXERCISE GUIDE

To gain experience . . .	Try exercises
Relating second-order equations to autonomous systems	1–2, 10–12(a)
Finding stationary points	3
Drawing phase plane plots	4–7(a–c), 9, 10–12(b)
Analyzing phase plane plots	4–7(d–e), 8–9, 10–12(b), 13–22

1. Obtain the damped pendulum system

$$\theta' = \omega$$
$$L\omega' = -g\sin\theta - p\omega$$

from the second-order damped pendulum equation $L\theta'' + p\theta' + g\sin\theta = 0$.

2. Identify the dependent variables y, z and the functions f, g so that each of the following systems or equations can be written in the autonomous form

$$y' = f(y, z)$$
$$z' = g(y, z).$$

If necessary, convert second-order equations to first-order systems.
(a) undamped pendulum system

$$\theta' = \omega$$
$$L\omega' = -g\sin\theta.$$

(b) damped pendulum equation $L\theta'' + p\theta' + g\sin\theta = 0$
(c) approximate pendulum equation $L\theta'' + g\theta = 0$
(d) unforced spring-mass equation $mx'' + px' + kx = 0$
(e) predator-prey system

$$F' = -(d_F - \alpha R)F$$
$$R' = (b_R - \beta F)R.$$

3. Find the stationary points of each of the equations listed in the previous exercise.

4. Solve the approximate pendulum model

$$L\theta'' + g\theta = 0, \qquad \theta(0) = \theta_i, \qquad \theta'(0) = 0,$$

with $L = 0.9930$ m, $g = 9.8$ m/s^2, $\theta_i = 0.15$ rad.
(a) Write formulas for $\theta(t)$ and $\omega(t) = \theta'(t)$. Use these to construct a phase plane plot of ω versus θ. (You will be plotting a curve defined by parametric equations with t as the parameter.)
(b) Mark the times $t = 0.5, 1.0, 1.5, 2.0$ s on the trajectory. Which direction, clockwise or counterclockwise, corresponds to t increasing?
(c) Solve the θ and ω equations for the trigonometric function each contains. Square and add to show that ω and θ satisfy an equation whose graph is a simple closed curve. Identify that curve.
(d) Mark the amplitude or maximum value of $\theta(t)$ on your phase plane graph. How is the amplitude related to the initial conditions?
(e) Compare your results with figure 7.18, a trajectory of the solution of the nonlinear pendulum equation $L\theta'' + g\sin\theta = 0$ subject to the same initial conditions.

5. Repeat exercise 4 using arbitrary values of the parameters L, g, and θ_i.

6. Repeat exercise 4 using the initial conditions $\theta(0) = 0$, $\theta'(0) = \omega_i$, and arbitrary values of the parameters L, g, and ω_i. What value(s) must ω_i have if the motion is to have a specified amplitude?

7. Repeat exercise 4 using the general initial conditions $\theta(0) = \theta_i, \theta'(0) = \omega_i$, and arbitrary values of the parameters L, g, θ_i, and ω_i. What value(s) must θ_i and ω_i have if the motion is to have a specified amplitude?

8. Referring to the pendulum trajectory shown in figure 7.18, the text claims

We could choose any point on the trajectory as initial values of θ and θ', and we would always retrace this trajectory.

Read from the graph the initial conditions appropriate for starting on the trajectory at

(a) the 3 o'clock position (positive θ-axis),
(b) the 6 o'clock position (negative ω-axis),
(c) the 9 o'clock position (negative θ-axis),
(d) the 12 o'clock position (positive ω-axis).

9. The text argues that the trajectory of figure 7.18 is traced in the clockwise direction starting from $\theta = 0.15$, $\omega = 0$, the 3 o'clock position in the phase plane.

(a) What happens when the pendulum is started with initial conditions that correspond to the 9 o'clock position? Argue both from your physical intuition about the behavior of the pendulum and from the governing equations,

$$\theta' = \omega$$
$$L\omega' = -g \sin \theta.$$

(b) Starting on the phase plane trajectory of figure 7.18 at the 12 o'clock position corresponds to beginning the motion with the pendulum hanging straight down. What is the sign of the initial velocity of the pendulum? In which direction must the pendulum move as it begins its motion, to the right or to the left? In which direction is the trajectory traced, clockwise or counterclockwise? Argue both from your physical intuition about the behavior of the pendulum and from the governing equations.

(c) Answer the question posed in part (b) when the starting point is the 6 o'clock position on the trajectory.

10. (a) Write a first-order system corresponding to the approximate pendulum equation $L\theta'' + g\theta = 0$.

(b) Draw a complete phase plane diagram for this system. Identify both stationary points and periodic solutions. Is it reasonable physically to consider large values of θ and ω?

11. (a) Letting $v = x'$, write a first-order system corresponding to the undamped spring-mass equation $mx'' + kx = 0$.

(b) Draw a complete phase plane diagram for this system. Identify both stationary points and periodic solutions. Are there physical limitations on the size of the values of x and v you can consider?

12. (a) Letting $v = x'$, write a first-order system cor-

responding to the damped spring-mass equation $mx'' + px' + kx = 0$.

(b) Draw a complete phase plane diagram for this system. Identify stationary points, attractors, and periodic solutions, if any. Are there physical limitations on the size of the values of x and v you can consider?

Exercises 13–18 ask about figure 7.21, the phase plane diagram of the undamped pendulum. In answering each question provide both a *physical* and a *mathematical* justification for your answer. Base your mathematical arguments on the governing equations,

$$\theta' = \omega$$
$$L\omega' = -g \sin \theta.$$

13. Are the elliptically shaped curves that represent oscillations around the stable equilibrium points $\theta = 0, \pm 2\pi$ traversed in a clockwise or a counterclockwise direction?

14. The trajectory labeled U is a perturbation of the unstable equilibrium at $\theta = -\pi$. Is it traversed in a clockwise or a counterclockwise direction?

15. Suppose the unstable equilibrium at $\theta = -\pi$ were perturbed to a point slightly to the left of $-\pi$—i.e., $\theta(0) < -\pi$. Sketch the resulting trajectory and give the direction in which it is traversed.

16. Suppose the unstable equilibrium at $\theta = \pi$ were perturbed to a point slightly to the left of π—i.e., $\theta(0) < \pi$. How would the resulting trajectory compare with the curve labeled U?

17. Suppose the unstable equilibrium at $\theta = \pi$ were perturbed to a point slightly to the right of π—i.e., $\theta(0) > \pi$. Sketch the resulting trajectory and give the direction in which it is traversed.

18. The text claims that the trajectories labeled L represent looping motion of the pendulum; it is spinning continuously in one direction about its pivot.

(a) Justify this claim.

(b) In which direction is the pendulum spinning on the upper trajectory labeled L? On the lower trajectory?

(c) Might you find a trajectory like the upper one labeled L with valleys so deep they nearly touched the θ-axis at odd multiples of π? If so, how would the initial conditions for such a trajectory compare with those for the one shown in figure 7.21? Could such a trajectory actually cross the θ-axis?

Exercises 19–22 ask about figure 7.23, the phase plane diagram of the damped pendulum. In answering each question provide both a *physical* and a *mathematical* justification for your answer. Base your mathematical arguments on the governing equations,

$$\theta' = \omega$$
$$L\omega' = -g \sin \theta - p\omega.$$

19. Are the spiral trajectories around the stable equilibrium points $\theta = 0, \pm 2\pi$ being traversed in a clockwise or a counterclockwise direction?

20. Sketch a trajectory that begins on the θ-axis slightly to the left of $\theta = \pi$. Such a trajectory represents a small perturbation of the unstable equilibrium at $\theta = \pi$.

21. Figure 7.21, the phase plane for the undamped pendulum, includes two trajectories marked L. Sketch their analogues for the damped pendulum.

22. Argue on physical grounds that no trajectory in the phase plane of figure 7.21 for the damped pendulum can be a closed curve unless it is a single point.

7.6 CHAPTER EXERCISES

1. A weight of 32 lb is suspended from a spring so that it hangs in a container of light oil. The oil resists the motion of the weight with a force equal in magnitude to three times the velocity of the weight. The spring exerts a resisting force of 4 lb for every foot it is extended. The weight is raised 1 ft and released. How much time is required for the amplitude of the motion to decrease by 50%?

2. Suppose the weight in the system described in exercise 1 can be varied arbitrarily without affecting the damping force exerted by the oil. (For example, a hollow container could be filled with varying amounts of lead shot.) For what range of *weights* will this system be underdamped? Overdamped? Critically damped? Is this variation with weight physically reasonable?

3. A 2-kg mass is suspended from a spring that extended 7 cm when the mass was hung from it. That mass is suspended in a variety of different fluids, ranging from air to heavy motor oil. Argue that after a disturbance of its position, the mass will always tend to return to its original equilibrium position, *no matter*

what the numerical value of the damping constant for the various fluids.

4. The differential equation

$$2x'' - x' + 4x = 0$$

is proposed as a model of a damped spring-mass system. What is wrong with it? What is the effect of the proposed damping term in this equation?

5. A certain spring extends 4 cm when a mass of 100 g is suspended from it. A 600-g mass is hanging at rest from this spring when it is suddenly struck from below. The mass moves upward 13 cm. What initial velocity was given to it by this blow? Would the mass have moved just as far if the blow had been directed downward?

6. A horizontal spring-mass system such as that of figure 7.4, page 269, can be taken as a simplified model of the energy absorbing barriers that are now placed around bridge abutments on some interstate highways. The fixed end of the spring is attached to the bridge abutment on the right, and the mass is that of the car and the restraint system combined.

(a) If the speed limit were raised from 55 mi/h to 65 mi/h, how would the spring have to be changed to limit maximum displacement from a collision at the higher speed to that at the lower speed? (Ignore damping.)

(b) How would the presence of damping alter your conclusion in part (a)? Should such systems be underdamped or overdamped?

7. Under what initial conditions is the maximum displacement of an undamped spring-mass system independent of both the mass and the spring constant?

8. A screen door on a porch is often pulled shut by an attached spring. It is usually set into motion from its equilibrium (closed) position by pushing it open. Friction in the hinges and between the door and the door frame damps the motion. Should such a system be overdamped, underdamped, or critically damped? In theory, will an overdamped screen door ever shut tightly?

9. Describe key difference(s) between the response of an undamped spring-mass system and a damped system when each is forced at its resonant frequency.

10. You are given the characteristic roots r_1, r_2 of a model of a spring mass system. Give a step-by-step recipe (an algorithm) for determining from r_1, r_2 whether or not resonance is possible. If resonance can occur, determine the resonant frequency of the system.

11. The resonant frequency of an underdamped spring-mass system is the value of the forcing frequency ω that makes the amplitude of the periodic solution of

$$mx'' + px' + kx = F \sin \omega t$$

a maximum.

(a) Find a particular solution (of frequency ω) of this equation and show that its amplitude A is F/D, where

$$D = \sqrt{(k - m\omega^2)^2 + p^2\omega^2}.$$

(b) Argue that the resonant frequency must be the value of ω that makes D a *minimum*. Find that minimum and show that the resonant frequency is

$$\omega_r = \sqrt{\frac{k}{m} - \frac{p^2}{2m^2}}.$$

(*Hint:* It is easier to minimize D^2 than D.)

(c) How does this resonant frequency ω_r compare with the natural frequency of the *undamped* system as damping decreases?

(d) Use the formula for ω_r to check the resonant frequency value found in example 9 of section 7.2.

12. Using the result of part (a) of exercise 11 and the ideas outlined there, show that the forcing frequency that gives the *velocity* of the mass its maximum amplitude is the undamped natural frequency $\omega_n = \sqrt{k/m}$. How does this frequency compare with the resonant frequency—the forcing frequency that produces maximum amplitude of displacement—found in exercise 11 when damping is small?

13. Using the amplitude expression derived in example 13, find the forcing frequency ω that maximizes the amplitude of the periodic response (capacitor voltage or capacitor charge) of a series RLC circuit with voltage source $E = V \sin \omega t$.

14. Show that the periodic response of the *current* in a series RLC circuit has maximum amplitude when the voltage source $E = V \sin \omega t$ has frequency $\omega = 1/\sqrt{LC}$.

15. Generalize the text's analysis of the stability of the equilibria $\theta_{ss} = \pi, 2\pi$ of $L\theta'' + g \sin \theta = 0$ to show that

(a) The equilibrium points $0, \pm 2\pi, \pm 4\pi, \ldots$ of $L\theta'' + g \sin \theta = 0$ are neutrally stable,

(b) The equilibrium points $\pm \pi, \pm 3\pi, \ldots$ of $L\theta'' + g \sin \theta = 0$ are unstable.

16. Determine the stability of the nontrivial equilibrium solution of the predator-prey (fox-rabbit) equations

$$F' = -(d_F - \alpha R)F$$
$$R' = (b_R - \beta F)R,$$

that were derived in section 5.4. Can you recognize the linear equation that governs perturbations of this equilibrium? Have you seen a similar equation in other contexts? Can you suggest a linear population model that is valid near the equilibrium?

17. Under suitable conditions, the uniqueness theorem for a single first-order equation prohibits a first-order initial-value problem from having more than one distinct solution. Similar theorems are available for second-order equations: Under suitable conditions, a second-order initial-value problem can have no more than one distinct solution. What does such a

theorem say about the possibility of two distinct trajectories of an autonomous system meeting in a common point?

18. The quantity $2\sqrt{mk}$ is sometimes called the *critical damping* value for the system $mx'' + px' + kx = 0$, and the ratio $d = p/2\sqrt{mk}$ is called the *damping ratio*. Explain these names by showing that the system is underdamped, critically damped, or overdamped according to whether $d < 1$, $d = 1$, $d > 1$. Show that the resonant frequency of an underdamped system can be written $\omega_n\sqrt{1 - d^2}$, where $\omega_n = \sqrt{k/m}$ is the natural frequency.

7.7 CHAPTER PROJECTS

1. **Period of a nonlinear spring** Project 1 of the chapter 5 projects asks for a derivation of a model of a spring-mass system with the nonlinear force-extension relation

$$F_s = kz + az^3,$$

where $a > 0$ is much smaller than k. Using available software, determine the variation with a of the period of such a system (undamped). (Keep k, initial displacement, and mass fixed.) For fixed spring parameters, determine the variation of the period with amplitude. What are the corresponding relations for a linear spring-mass system?

   Use some of your computed solution values to construct phase plane diagrams for the solutions of this nonlinear system. How do they compare with the corresponding trajectories for the undamped linear system?

2. **Predator-prey phase plane** Sketch the main qualitative features of a phase plane diagram for the predator-prey equations

$$F' = -(d_F - \alpha R)F$$
$$R' = (b_R - \beta F)R$$

derived in section 5.4. Mark equilibrium points and describe their stability. Use the slope expression

$$\frac{dF}{dR} = \frac{-(d_F - \alpha R)F}{(b_R - \beta F)R}$$

to determine the direction of trajectories as they cross the axes and other lines where dF/dR is easily evaluated. Comment on the physical significance of this diagram. Using available software, confirm these qualitative sketches numerically and add any detail that was omitted.

3. **More bridge cables** Project 2 of chapter 5 introduced in (5.7–5.8) a prototype model for the nonlinear response of the cables in a suspension bridge such as the Golden Gate. This project explores the great variety of periodic responses of this system. In the following, use $m = 1$, $g = 9.8$, $k = 10$, $p = 0.05$, $\omega = 1$, $F = 8.5$, and $\ell_0 = 0.3135$.
   (a) Write the system (5.7–5.8) of two second-order equations as a system of four first-order equations.

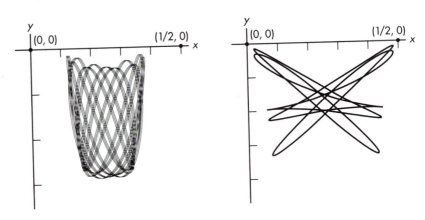

Figure 7.24: Two trajectories from the prototype bridge suspension model (5.7–5.8).

(b) Using available software with graphical output, show that the mass simply oscillates vertically if the initial conditions do not displace it to the left or right. Plot y versus x so that you can see the motion of the mass. (Try a second- or fourth-order Runge-Kutta method. Allow the system to run long enough, say 30 or more periods, for the transient to decay.)

(c) Now impose initial conditions that displace the mass to the side. Compare the long-time motion you find with that of others who complete this same project. Figure 7.24 shows two examples of the motion that can result. Do you find what might be chaotic motion, trajectories without any apparent regular pattern?

8

SYSTEMS OF EQUATIONS

. . . nature's system must be a real beauty. . .
Buckminster Fuller (From ''Profiles: In the Outlaw Area'' by Calvin Tomkins,
The New Yorker, January 8, 1966.)

Systems of differential equations provide a general setting that can encompass
apparently diverse problems. We begin with the usual definitions, develop solution techniques from previous ideas, and finally display these concepts in the
unifying light of matrices.

8.1 BASIC DEFINITIONS: SYSTEMS

Equations of order two (or higher) can be converted to systems of first-order
equations by introducing a new variable to account for the first derivative. For
example, the damped pendulum equation

$$L\theta'' + g\sin\theta = 0$$

can be converted to a *first-order system* by introducing a variable $\omega = \theta'$,

$$\theta' = \omega$$

$$L\omega' = -g\sin\theta.$$

The first equation in the system is just the definition of the new variable, and the
second equation is the original second-order equation rewritten with ω' replacing θ''. The first-order system and the second-order equation are equivalent in
the sense that a solution of one leads to a solution of the other.

Systems of first-order equations can arise naturally in models, too, as in our study of the interaction of predator and prey (foxes and rabbits) in section 5.4.2:

$$F' = -(d_F - \alpha R)F$$
$$R' = (b_R - \beta F)R$$

Indeed, systems are the most general framework for the study of differential equations, for they incorporate higher-order equations as well as systems themselves.

For simplicity, we will confine ourselves to systems of two first-order equations, but all the ideas we present generalize naturally to three or more equations. The exercises start you on some of those generalizations.

The generic first-order system of two differential equations has the form

$$y' = f(t, y, z) \tag{8.1}$$
$$z' = g(t, y, z) \tag{8.2}$$

where y and z are the **dependent variables** and t is the **independent variable**. A *solution* of a system is a pair of continuously differentiable functions $(y(t), z(t))$ that reduce both equations to an identity when substituted for the corresponding dependent variables. An **initial-value problem** for a first-order system specifies a value for each of the dependent variables at a fixed value of the independent variable, and a *solution of an initial-value problem* is a pair of functions that solves the system and satisfies the initial conditions.

More precisely, we state:

DEFINITION 1   *The pair of functions $(u(t), v(t))$ is a* **solution** *on the interval $t_0 \leq t \leq t_1$ of the first-order system (8.1-8.2) if u, v are continuously differentiable and if*

$$u' = f(t, u, v)$$
$$v' = g(t, u, v),$$

for $t_0 \leq t \leq t_1$. Further, (u, v) is a **solution of the initial-value problem**

$$y' = f(t, y, z), \qquad y(t_0) = y_i,$$
$$z' = g(t, y, z), \qquad z(t_0) = z_i,$$

if (u, v) is a solution of the system (8.1-8.2) and, in addition,

$$u(t_0) = y_i, \qquad v(t_0) = z_i.$$

The equations $y(t_0) = y_i$, $z(t_0) = z_i$, are the **initial conditions**.

We write the solution $(y(t), z(t))$ as an ordered pair to emphasize that we need two functions to solve a system of two first-order equations and that the first function goes with the first equation and the second function with the second equation.

For example, if $\theta(t)$ is a solution of the pendulum equation $L\theta'' + g\sin\theta = 0$, then a solution of the equivalent first order system $\theta' = \omega$, $L\omega' = -g\sin\theta$ is the pair $(\theta(t), \theta'(t))$.

EXAMPLE 1 *Write the spring-mass equation $mx'' + kx = 0$ as a first-order system. Use a solution of this second-order equation to write a solution of the equivalent system.*

Introduce the additional dependent variable $v = x'$; v is the velocity of the mass. Then the equivalent system is

$$x' = v$$
$$mv' = -kx.$$

One solution of $mx'' + kx = 0$ is $x(t) = \sin \sqrt{k/m}\, t$. Then a solution of the system is $x(t) = \sin \sqrt{k/m}\, t$ and

$$v(t) = \left(\sin \sqrt{\frac{k}{m}}\, t \right)' = \sqrt{\frac{k}{m}} \cos \sqrt{\frac{k}{m}}\, t.$$

We can also work backward. Given the solution pair

$$(x, v) = \left(\sin \sqrt{\frac{k}{m}}\, t, \sqrt{\frac{k}{m}} \cos \sqrt{\frac{k}{m}}\, t \right)$$

for the system, we can identify the first component $x = \sin \sqrt{k/m}\, t$ as a solution of the original second order equation. ∎

The system is **linear** if f and g are linear in the dependent variables; i.e., if f, g can be written in the form

$$f(t, y, z) = k_1(t)y + k_2(t)z + F(t)$$
$$g(t, y, z) = \ell_1(t)y + \ell_2(t)z + G(t).$$

The dependent variables can appear raised only to the first power and multiplied only by a function of the independent variable. Otherwise, the system is **nonlinear**.

The system is **homogeneous** if it has the **trivial solution** $y \equiv 0$, $z \equiv 0$. Otherwise, it is **nonhomogeneous**.

A linear system has **constant coefficients** if the coefficients k_1, k_2, and ℓ_1, ℓ_2 in the expansions $f(t, y, z) = k_1(t)y + k_2(t)z + F(t)$, $g(t, y, z) = \ell_1(t)y + \ell_2(t)z + G(t)$ are constant. Otherwise, it has **variable coefficients**.

EXAMPLE 2 *Are the pendulum and spring-mass systems given above linear or nonlinear, homogeneous or nonhomogeneous, constant or variable coefficient?*

The pendulum system

$$\theta' = \omega$$
$$L\omega' = -g \sin \theta.$$

is nonlinear because of the term $\sin \theta$. It is homogeneous because substituting the solution pair $(\theta, \omega) = (0, 0)$ reduces both equations to identities. Since *constant-coefficient* is defined only for linear systems, it has no meaning for this system.

The spring-mass system

$$x' = v$$

$$mv' = -kx.$$

is linear. We can make the identifications $y = x$, $z = v$, $k_1 = 0$, $k_2 = 1$, $\ell_1 = -k/m$, $\ell_2 = 0$. It is a constant-coefficient system because each of the coefficients k_1, k_2, ℓ_1, ℓ_2 is constant. It is homogeneous because it possesses the trivial solution $(x, v) = (0, 0)$.

Had we been given a forced spring-mass system, such as

$$x' = v$$

$$mv' = -kx - F \sin \pi t,$$

then we would have concluded that it is nonhomogeneous because substituting $(x, v) = (0, 0)$ fails to yield equality:

$$0 = 0$$

$$0 \neq 0 - F \sin \pi t.$$

However, this forced system is linear and has constant coefficients, just like its homogeneous counterpart in the preceding paragraph. ∎

Two solutions $(y_1(t), z_1(t))$ and $(y_2(t), z_2(t))$ of a first order system are **linearly independent** on the interval $t_0 \leq t \leq t_1$ if

$$C_1 y_1(t) + C_2 y_2(t) = 0$$

$$C_1 z_1(t) + C_2 z_2(t) = 0, \ t_0 \leq t \leq t_1,$$

for any constants C_1, C_2 forces $C_1 = C_2 = 0$. Otherwise, they are **linearly dependent**.

A sum of multiples of two solutions,

$$(C_1 y_1(t) + C_2 y_2(t), \ C_1 z_1(t) + C_2 z_2(t))$$

is called a **linear combination** of the solutions $(y_1, z_1), (y_2, z_2)$. Any linear combination of two solutions of a linear, homogeneous system is again a solution of that system. Two solutions of a linear system (homogeneous or nonhomogeneous) are **linearly independent** if the only linear combination that yields the trivial solution is the one with $C_1 = C_2 = 0$.

Paralleling our study in chapter 6 of the Wronskian for second-order equations, Cramer's rule reveals that the solutions $(y_1(t), z_1(t))$ and $(y_2(t), z_2(t))$ are linearly independent on $t_0 \leq t \leq t_1$ if the determinant

$$W(t) = \begin{vmatrix} y_1 & y_2 \\ z_1 & z_2 \end{vmatrix}$$

is not zero on that interval. The determinant $W(t)$ is known as the **Wronskian**. As with the Wronskian of two solutions of a second-order equation, under suitable conditions, either $W(t)$ is identically zero or it is never zero on the interval of definition of the solutions $(y_1, z_1), (y_2, z_2)$. Hence, we have the following.

Linear Independence Test The solutions (y_1, z_1) and (y_2, z_2) are *linearly independent* if their Wronskian $W(t)$ is not zero for one point in their interval of definition.

Suppose $(y_1(t), z_1(t))$ and $(y_2(t), z_2(t))$ are linearly independent solutions of the homogeneous system

$$y' = k_1(t)y + k_2(t)z$$
$$z' = \ell_1(t)y + \ell_2(t)z$$

and that C_1, C_2 are arbitrary constants. Then the **general solution of the homogeneous linear system** or the **homogeneous solution** is a solution (y_h, z_h) of the form

$$y_h = C_1 y_1 + C_2 y_2$$
$$z_h = C_1 z_1 + C_2 z_2.$$

Equivalently, we can write

$$\boxed{\begin{aligned}(y_h, z_h) &= C_1(y_1, z_1) + C_2(y_2, z_2) \\ &= (C_1 y_1 + C_2 y_2, C_1 z_1 + C_2 z_2).\end{aligned}}$$

HOMOGENEOUS SOLUTION

A given solution (y_p, z_p) of a nonhomogeneous linear system

$$y' = k_1(t)y + k_2(t)z + F(t)$$
$$z' = \ell_1(t)y + \ell_2(t)z + G(t)$$

is often called a **particular solution**. A **general solution of a nonhomogeneous system** is a solution (y_g, z_g) of the form

$$y_g = y_p + y_h$$
$$z_g = z_p + z_h$$

where $(y_p(t), z_p(t))$ is a particular solution of the nonhomogeneous system and $(y_h(t), z_h(t))$ is a general solution of the homogeneous equation. To emphasize the analogy with a single second-order equation, we can also write this general solution in the form

$$\boxed{\begin{aligned}(y_g, z_g) &= (y_p, z_p) + (y_h, z_h) \\ &= (y_p, z_p) + C_1(y_1, z_1) + C_2(y_2, z_2).\end{aligned}}$$

GENERAL SOLUTION

The general solution construction is applied in parallel to both components of the solution.

EXAMPLE 3 *The linear system*

$$y' = z$$
$$z' = 16y + 8t$$

is equivalent to the second-order equation $y'' - 16y = 8t$. Use that equivalence to find a general solution of this equation.

Linearly independent solutions of the homogeneous equation $y'' - 16y = 0$ are e^{4t} and e^{-4t}. Using $z = y'$, we can construct from them two pairs of solutions

$$(y_1, z_1) = (e^{4t}, 4e^{4t})$$
$$(y_2, z_2) = (e^{-4t}, -4e^{-4t})$$

of the homogeneous system

$$y' = z$$
$$z' = 16y.$$

To verify that these solutions are linearly independent, we evaluate the determinant:

$$\begin{vmatrix} y_1 & y_2 \\ z_1 & z_2 \end{vmatrix} = \begin{vmatrix} e^{4t} & e^{-4t} \\ 4e^{4t} & -4e^{-4t} \end{vmatrix} = -8 \neq 0.$$

(This expression is just the Wronskian $W(e^{4t}, e^{-4t})$ of the two solutions of the equivalent second-order equation $y'' - 16y = 0$.) A general solution of the homogeneous system is the solution pair (y_h, z_h), where

$$y_h = C_1 y_1 + C_2 y_2 = C_1 e^{4t} + C_2 e^{-4t}$$
$$z_h = C_1 z_1 + C_2 z_2 = 4C_1 e^{4t} - 4C_2 e^{-4t}.$$

To obtain a particular solution of the nonhomogeneous system, we note that a particular solution of the equivalent second-order equation $y'' - 16y = 8t$ is $t/2$. Again using $z = y'$, we find the particular solution pair

$$(y_p, z_p) = \left(\frac{t}{2}, \frac{1}{2} \right).$$

Combining this solution pair with the homogeneous solution above, we have a general solution of the original system

$$(y_g, z_g) = (y_p, z_p) + C_1(y_1, z_1) + C_2(y_2, z_2)$$

$$= (y_p, z_p) + (y_h, z_h)$$

$$= \left(\frac{t}{2} + y_h, \frac{1}{2} + z_h \right)$$

$$= \left(\frac{t}{2} + C_1 e^{4t} + C_2 e^{-4t}, \frac{1}{2} + 4C_1 e^{4t} - 4C_2 e^{-4t} \right) \qquad \blacksquare$$

A **principle of superposition** can also be stated for linear systems; we leave it for the exercises.

8.1.1 Exercises

1. Write each of the following second-order equations as an equivalent first-order system. If initial conditions are given, write an equivalent initial-value problem for the system. When appropriate, give a physical interpretation of the new variable you introduce.
   (a) $4y'' - 16y' + 2y = 0$
   (b) $4y'' - 16y' + 2y = e^{2t}$, $y(0) = 1$, $y'(0) = -2$
   (c) $4y'' + 16y'(1 - y^2) = y'$
   (d) the damped spring-mass equation $mx'' + px' + kx = 0$, $x(0) = x_i$, $x'(0) = v_i$
   (e) the forced, undamped spring-mass equation $mx'' + kx = A \sin \omega t$
   (f) the series RLC equation $Rq'' + Rq' + q/C = E(t)$
   (g) the approximate pendulum equation $L\theta'' + g\theta = 0$, $\theta(0) = \theta_i$, $\theta'(0) = \omega_i$
   (h) the damped pendulum equation $L\theta'' + p\theta' + g \sin \theta = 0$,

2. Write a first-order system that is equivalent to the generic linear second-order equation $a_2 y'' + a_1 y' + a_0 y = h(t)$.

3. Write a first-order system that is equivalent to the generic second-order equation $y'' = f(t, y, y')$.

4. Verify that the second-order equation

$$a_2(t)y'' + a_1(t)y' + a_0(t)y = h(t)$$

is equivalent to the first-order system

$$y' = z$$
$$a_2(t)z' = -a_0(t)y - a_1(t)z + h(t).$$

as follows:
   (a) Suppose $y(t)$ is a solution of the second-order equation. Show that $(y(t), y'(t))$ solves the first-order system.
   (b) Suppose $(y(t), z(t))$ solves the first-order system. Show that $y(t)$ solves the second-order equation.

5. Classify each of the following equations as
   i. Linear or nonlinear,
   ii. Homogeneous or nonhomogeneous,
   iii. Constant- or variable-coefficient.

   If necessary, convert second-order equations to first-order systems.
   (a) $4y'' - 16y' + 2y = 0$
   (b) $4y'' - 16y' + 2y = e^{2t}$, $y(0) = 1$, $y'(0) = -2$
   (c) $4y'' + 16y'(1 - y^2) = y'$
   (d) the damped spring-mass equation $mx'' + px' + kx = 0$,
   (e) the forced, damped spring-mass equation $mx'' + px' + kx = A \sin \omega t$

(f) the forced, *un*damped spring-mass equation $mx'' + kx = A \sin \omega t$

(g) the predator-prey system

$$F' = -(d_F - \alpha R)F$$
$$R' = (b_R - \beta F)R$$

(h) the damped pendulum system

$$\theta' = \omega$$
$$L\omega' = -g \sin \theta - p\omega,$$

(i) the forced pendulum equation $L\theta'' + p\theta' + g \sin \theta = A \sin \pi t$

(j) the series RLC equation $Rq'' + Rq' + q/C = E(t)$

(k) the approximate pendulum equation $L\theta'' + g\theta = 0$

6. Example 3 considers the systems

$$y' = z$$
$$z' = 16y + 8t$$

and

$$y' = z$$
$$z' = 16y.$$

Verify that both systems are linear, that the first is nonhomogeneous, that the second is homogeneous, and that both have constant coefficients.

7. Verify by substitution that

$$(y_1, z_1) = (e^{4t}, 4e^{4t})$$
$$(y_2, z_2) = (e^{-4t}, -4e^{-4t})$$

are both pairs of solutions of the system

$$y' = z$$
$$z' = 16y,$$

as claimed in example 3. Explain how you could find a solution of the equivalent second-order equation $y'' - 16y = 0$ from each of these solution pairs.

8. Example 3 obtains the general solution

$$(y_g, z_g) = \left(y_h + \frac{t}{2}, z_h + \frac{1}{2} \right)$$
$$= \left(C_1 e^{4t} + C_2 e^{-4t} + \frac{t}{2}, \right.$$
$$\left. 4C_1 e^{4t} - 4C_2 e^{-4t} + \frac{1}{2} \right)$$

of the linear system

$$y' = z$$
$$z' = 16y + 8t.$$

(a) Verify by direct substitution that (y_g, z_g) is indeed a solution of this system.

(b) Use this general solution to solve this system subject to the initial conditions $(y(0), z(0)) = (2, \frac{3}{2})$.

9. Show that the linear system

$$y' = k_1(t)y + k_2(t)z + F(t)$$
$$z' = \ell_1(t)y + \ell_2(t)z + G(t)$$

is homogeneous if and only if $F = G = 0$.

10. Show that a homogeneous system has the steady state $(0, 0)$.

11. For each of the given systems,

i. Verify that each of the following pairs of functions is a solution of the system:

$$y' = 2z$$
$$z' = -2y.$$

ii. Determine which pairs are linearly independent.

iii. Use each linearly independent pair to write a general solution of this system.

(a) $(y_1, z_1) = (\sin 2t, 2 \cos 2t)$
(b) $(y_2, z_2) = (\sin 2(t - \pi/2), 2 \cos 2(t - \pi/2))$
(c) $(y_3, z_3) = (\sin 2(t - \pi/4), 2 \cos 2(t - \pi/4))$

12. Suppose (y_1, z_1) and (y_2, z_2) are solutions of the homogeneous system

$$y' = k_1(t)y + k_2(t)z$$
$$z' = \ell_1(t)y + \ell_2(t)z.$$

Verify by substitution that the combination (y_h, z_h),

$$y_h = C_1 y_1 + C_2 y_2$$
$$z_h = C_1 z_1 + C_2 z_2,$$

is also a solution of the homogeneous system. Does it matter whether or not (y_1, z_1) and (y_2, z_2) are linearly independent?

13. The text claims

Any linear combination of two solutions of a linear, homogeneous system is again a solution of that system.

(a) Prove that this claim is correct.
(b) Is the result of this exercise the same as the result of exercise 12? Explain.

14. (a) Show that a constant multiple of a solution of a linear homogeneous system is again a solution of that system.
(b) Is this result different from the result of the preceding two exercises or is it a special case? Explain.

15. Suppose (y_h, z_h) is a solution of the homogeneous system

$$y' = k_1(t)y + k_2(t)z$$
$$z' = \ell_1(t)y + \ell_2(t)z$$

and that (y_p, z_p) is a particular solution of the nonhomogeneous system

$$y' = k_1(t)y + k_2(t)z + F(t)$$
$$z' = \ell_1(t)y + \ell_2(t)z + G(t).$$

Verify by substitution that the combination (y_g, z_g),

$$y_g = y_p + y_h$$
$$z_g = z_p + z_h,$$

is also a solution of the nonhomogeneous system. Does it matter whether or not (y_h, z_h) is nontrivial?

16. Example 1 uses the solution $x = \sin\sqrt{k/m}\,t$ of the spring-mass equation $mx'' + kx = 0$ to construct a solution of the equivalent system

$$x' = v$$
$$mv' = -kx.$$

(a) Find a solution of $mx'' + kx = 0$ that is linearly independent of $\sin\sqrt{k/m}\,t$. Use it to construct another solution of this system.
(b) Show that the solution you construct in part (a) and the solution

$$(x, v) = \left(\sin\sqrt{\frac{k}{m}}\,t, \sqrt{\frac{k}{m}}\cos\sqrt{\frac{k}{m}}\,t\right)$$

obtained in the text are linearly independent.

17. (a) Find a nontrivial solution of the second-order homogeneous equation $x'' + 4x = 0$.

(b) Write this second-order equation as a first-order system. Use the solution you just found to write a solution of this first-order system.
(c) Find a *general* solution of $x'' + 4x = 0$.
(d) Use the general solution you just found to write a general solution of the equivalent first-order system of part (b).
(e) Verify that you have a general solution of the system. Do you evaluate a determinant that looks like the Wronskian of two solutions of the second-order equation $x'' + 4x = 8t$?
(f) Find a particular solution of $x'' + 4x = 8t$.
(g) Write $x'' + 4x = 8t$ as a first-order system. Use the particular solution you just found to write a particular solution of this first-order system.
(h) Combine the analysis of the preceding steps to write a general solution of the nonhomogeneous system.

18. Repeat exercise 17 for the homogeneous equation $x'' - 9x = 0$ and the nonhomogeneous equation $x'' - 9x = -10\sin t$.

19. The text claims, "Cramer's rule reveals that the solutions $(y_1(t), z_1(t))$ and $(y_2(t), z_2(t))$ are linearly independent on $t_0 \leq t \leq t_1$ if the determinant

$$\begin{vmatrix} y_1 & y_2 \\ z_1 & z_2 \end{vmatrix}$$

is not zero on that interval." Prove this assertion.

20. The text states that the linear independence of the solutions $(y_1(t), z_1(t))$ and $(y_2(t), z_2(t))$ of a linear system can be tested by examining the determinant

$$\begin{vmatrix} y_1 & y_2 \\ z_1 & z_2 \end{vmatrix}.$$

Show that if the linear system was obtained from a linear, second-order equation, then this determinant is just the Wronskian of two solutions of the second-order equation.

21. (a) One solution of a linear, homogeneous system satisfies the initial conditions $(y_1(0), z_1(0)) = (1, 0)$. Another satisfies the initial conditions $(y_2(0), z_2(0)) = (0, 1)$. Show that the two solutions must be linearly independent.
(b) Give a general relation between two sets of initial conditions that will guarantee that the corresponding solutions of a homogeneous system are linearly independent.

22. State a *principle of superposition* for linear systems. It might begin

> Let (y_j, z_j) be solutions of the nonhomogeneous systems
>
> $$y' = k_1(t)y + k_2(t)z + F_j(t)$$
> $$z' = \ell_1(t)y + \ell_2(t)z + G_j(t)$$
>
> for $j = 1, 2$. Then for any constants C_1, C_2, the combination (y_c, z_c), where
>
> $$y_c = C_1 y_1 + C_2 y_2$$
>
> $\vdots$

23. Give definitions of *free response* and *forced response* for the linear first-order system

$$y' = k_1(t)y + k_2(t)z + F(t), \qquad y(0) = y_i$$
$$z' = \ell_1(t)y + \ell_2(t)z + G(t), \qquad z(0) = z_i$$

that are consistent with those given for a single second-order equation.

24. The text suggests in passing that an equation of *any* order can be reduced to a first-order system.
   (a) By introducing the variables $y_1 = u$, $y_2 = u'$, and

$y_3 = u''$, find a system of three first-order equations in y_1, y_2, y_3 that is equivalent to the linear third-order equation

$$b_3 u^{(3)} + b_2 u'' + b_1 u' + b_0 u = h(t).$$

   (b) Repeat these same steps for the generic third-order equation

$$u^{(3)} = f(t, u, u', u'').$$

   (c) Does one equation include the other as a special case?

25. Repeat exercise 24 for the fourth-order equations

$$b_4 u^{(4)} + b_3 u^{(3)} + b_2 u'' + b_1 u' + b_0 u = h(t)$$

and

$$u^{(4)} = f(t, u, u', u'', u^{(3)}).$$

26. Repeat exercise 24 for the n-th order equations

$$b_n u^{(n)} + \cdots + b_2 u'' + b_1 u' + b_0 u = h(t)$$

and

$$u^{(n)} = f(t, u, u', u'', \ldots, u^{(n-1)}).$$

8.2 CONSTANT-COEFFICIENT HOMOGENEOUS SYSTEMS

We want to find a general solution of the generic constant-coefficient, homogeneous linear system of two first-order equations,

$$y' = ay + bz$$
$$z' = cy + dz.$$

Our earlier experience with constant-coefficient equations immediately suggests guessing exponential solutions of the form e^{rt}, where the parameter r will be determined by substituting the guess into the equations. A concrete example from section 8.1 can give a little more guidance.

In example 3, we found the linearly independent solutions

$$(y_1, z_1) = (e^{4t}, 4e^{4t})$$
$$(y_2, z_2) = (e^{-4t}, -4e^{-4t})$$

of the homogeneous system

$$y' = z$$
$$z' = 16y.$$

There we were able to solve the equivalent second-order equation $y'' - 16y = 0$ for $y_1 = e^{4t}$ and $y_2 = e^{-4t}$. Then we used the first equation in the system, $z = y'$, to find z_1 and z_2. We need an approach to guide us when we can not find an equivalent second-order equation.

These solutions certainly involve exponentials of the form e^{rt}. For example, the pair (y_1, z_1) has $r = 4$. But $y_1 = e^{4t}$, while $z_1 = 4e^{4t}$; they have different coefficients. This last observation suggests we should guess solutions of the form

$$(y_1, z_1) = (pe^{rt}, qe^{rt}),$$

where the coefficients p, q and the growth rate r are constants to be determined by substituting into the system. The following example explores this process.

EXAMPLE 4 *Use the assumed solution form $(y_1, z_1) = (pe^{rt}, qe^{rt})$ to find linearly independent solutions of*

$$y' = z$$
$$z' = 16y.$$

Substituting the guess for (y_1, z_1) into the system yields

$$rpe^{rt} = qe^{rt}$$
$$rqe^{rt} = 16pe^{rt}.$$

Dividing by the exponential and collecting terms leaves

$$rp - q = 0 \tag{8.3}$$

$$-16p + rq = 0. \tag{8.4}$$

For any value of r, we always have the solution $p = q = 0$. But choosing p and q to be zero gives us only a trivial solution of the differential equation, hardly a building block for a general solution.

Are there values of r for which the pair of equations (8.3-8.4) has *nontrivial* solutions for p and q? If we tried to solve for p using Cramer's rule, we would have

$$p = \Delta^{-1} \begin{vmatrix} 0 & -1 \\ 0 & r \end{vmatrix} = \frac{0}{\Delta}$$

where Δ is the determinant of coefficients

$$\Delta = \begin{vmatrix} r & -1 \\ -16 & r \end{vmatrix} = r^2 - 16.$$

We can obtain a nonzero value of p only if $\Delta = 0$. (See section A.6.) But requiring that the determinant of coefficients be zero gives an equation for r,

$$\Delta = r^2 - 16 = 0,$$

the *characteristic equation* of this system of differential equations.

> Of course, this is the characteristic equation of the equivalent second-order equation $y'' - 16y = 0$ as well.

The solutions of the characteristic equation $r^2 - 16 = 0$ are $r = \pm 4$. The corresponding values of p and q must satisfy (8.3 – 8.4) with either $r = 4$ or $r = -4$.

When $r = 4$, we have

$$4p - q = 0$$

$$-16p + 4q = 0.$$

But the second equation is redundant; it is the first multiplied by -4. Hence, we can determine p and q only to within a constant multiple:

$$q = 4p, \qquad p \text{ arbitrary.}$$

Choosing $p = 1$ for simplicity, we find a solution corresponding to $r = +4$,

$$(y_1, z_1) = (pe^{4t}, qe^{4t}) = (e^{4t}, 4e^{4t}).$$

Similarly, the values of p and q corresponding to $r = -4$ must satisfy

$$-4p - q = 0$$

$$-16p - 4q = 0.$$

The second equation is again a multiple of the first, but now $q = -4p$. Choosing $p = 1$ leads to a solution pair corresponding to $r = -4$,

$$(y_2, z_2) = (pe^{-4t}, qe^{-4t}) = (e^{-4t}, -4e^{-4t}).$$

Finally, to verify linear independence, we evaluate the Wronskian of these two solution pairs:

$$W(t) = \begin{vmatrix} y_1 & y_2 \\ z_1 & z_2 \end{vmatrix} = \begin{vmatrix} e^{4t} & e^{-4t} \\ 4e^{4t} & -4e^{-4t} \end{vmatrix} = -8.$$

The solutions are linearly independent because $W(t) \neq 0$.

A general solution of this system is

$$(y_g, z_g) = C_1(y_1, z_1) + C_2(y_2, z_2)$$

$$= C_1(e^{4t}, 4e^{4t}) + C_2(e^{-4t}, -4e^{-4t}).$$

As with second-order equations, the general solution is a linear combination of two linearly independent solutions. ∎

In retrospect, we should not be surprised to find the coefficients p and q just to within a constant multiple. We can always multiply a solution of a homogeneous linear system by a constant and obtain a new solution.

The next example considers a slightly different system, one with imaginary characteristic roots.

EXAMPLE 5 *Use the assumed solution form* $(y_1, z_1) = (pe^{rt}, qe^{rt})$ *to find linearly independent solutions of*

$$y' = -z$$
$$z' = 16y.$$

We proceed exactly as in the preceding example. Substitute the guess for (y_1, z_1) into the system,

$$rpe^{rt} = -qe^{rt}$$
$$rqe^{rt} = 16pe^{rt},$$

and divide by the exponential,

$$rp - q = 0 \tag{8.5}$$

$$-16p + rq = 0. \tag{8.6}$$

To avoid trivial solutions of (8.5–8.6), require that the determinant of coefficients be zero to obtain the characteristic equation:

$$\begin{vmatrix} r & -1 \\ -16 & r \end{vmatrix} = r^2 + 16 = 0$$

The roots of the characteristic equation $r^2 + 16 = 0$ are $r = \pm 4i$. The corresponding values of p and q must satisfy (8.5–8.6) with either $r = 4i$ or $r = -4i$.

When $r = 4i$, we have

$$4i\,p - q = 0$$
$$-16p + 4i\,q = 0.$$

Ignoring the second redundant equation and choosing $p = 1$ in the first, we find a solution corresponding to $r = +4i$,

$$(y_+, z_+) = (pe^{4it}, qe^{4it}) = (e^{4it}, 4ie^{4it}).$$

Similarly, the values of p and q corresponding to $r = -4i$ must satisfy

$$-4i\,p - q = 0$$
$$-16p - 4i\,q = 0.$$

Choosing $p = 1$ in the first equation leads to a solution pair corresponding to $r = -4i$,

$$(y_-, z_-) = (pe^{-4it}, qe^{-4it}) = (e^{-4it}, -4ie^{-4it}).$$

Finally, to verify linear independence, evaluate the Wronskian

$$W(t) = \begin{vmatrix} y_+ & y_- \\ z_+ & z_- \end{vmatrix} = \begin{vmatrix} e^{4it} & e^{-4it} \\ 4i\,e^{4it} & -4i\,e^{-4it} \end{vmatrix} = -8i \neq 0.$$

A general solution is

$$(y_g, z_g) = C_1(y_+, z_+) + C_2(y_-, z_-)$$
$$= C_1(e^{4it}, 4ie^{4it}) + C_2(e^{-4it}, -4ie^{-4it}). \qquad \blacksquare$$

Appropriate choices of C_1 and C_2 lead to real-valued solutions, as the next example illustrates.

EXAMPLE 6 *From the pair of solutions*

$$(y_+, z_+) = (e^{4it}, 4ie^{4it})$$
$$(y_-, z_-) = (e^{-4it}, -4ie^{-4it})$$

derived in example 5, obtain a pair of real-valued, linearly independent solutions.

We need Euler's formula for the complex exponential,

$$e^{\pm 4it} = \cos 4t \pm i \sin 4t.$$

Since we are solving a linear, homogeneous system, any linear combination of solutions is again a solution. Writing the two solutions one above the other will suggest linear combinations that can eliminate the imaginary terms:

$$(y_+, z_+) = (\cos 4t + i \sin 4t, 4i \cos 4t - 4 \sin 4t)$$
$$(y_-, z_-) = (\cos 4t - i \sin 4t, -4i \cos 4t - 4 \sin 4t)$$

Adding the two solution pairs and dividing by 2 is one possibility; that is, use a linear combination with $C_1 = C_2 = 1/2$. The first real-valued solution pair is

$$(y_1, z_1) = \frac{1}{2}(y_+, z_+) + \frac{1}{2}(y_-, z_-) = (\cos 4t, -4 \sin 4t).$$

Subtracting the first solution pair from the second and dividing by $2i$ (using a linear combination with $C_1 = 1/2i$, $C_2 = -1/2i$) gives the second real-valued solution pair,

$$(y_2, z_2) = \frac{1}{2i}(y_+, z_+) - \frac{1}{2i}(y_-, z_-) = (\sin 4t, 4 \cos 4t).$$

Note that the conversion of complex-valued exponentials into real-valued solutions involving the sine and cosine followed the same pattern as with constant-coefficient, second-order equations: Add and divide by 2 or subtract and divide by $2i$.

We verify linear independence by testing the Wronskian:

$$W(t) = \begin{vmatrix} y_1 & y_2 \\ z_1 & z_2 \end{vmatrix} = \begin{vmatrix} \cos 4t & \sin 4t \\ -4\sin 4t & 4\cos 4t \end{vmatrix} = 4(\cos^2 4t + \sin^2 4t) = 4 \neq 0.$$

Since $W(t) \neq 0$, the solution pairs (y_1, z_1) and (y_2, z_2) are linearly independent. ∎

Observe that these two real-valued solutions are simply the real part and the imaginary part of the complex solution (y_+, z_+). The next example shows that this occurrence is no coincidence.

EXAMPLE 7 *Show that the real part and the imaginary part of the complex-valued solution*

$$(y_+, z_+) = (e^{4it}, 4ie^{4it})$$

obtained in example 5 also form a pair of linearly independent real-valued solutions of the system

$$y' = -z$$
$$z' = 16y.$$

Using Euler's formula, we expand this solution into its real and imaginary parts,

$$\begin{aligned}(y_+, z_+) &= (e^{4it}, 4ie^{4it}) \\ &= (\cos 4t + i\sin 4t, 4i\cos 4t - 4\sin 4t) \\ &= (\cos 4t, -4\sin 4t) + i(\sin 4t, 4\cos 4t) \\ &= (y_r, z_r) + i(y_i, z_i),\end{aligned}$$

where the *real part* of this solution is

$$(y_r, z_r) = (\cos 4t, -4\sin 4t)$$

and the *imaginary part* is

$$(y_i, z_i) = (\sin 4t, 4\cos 4t).$$

Rather than substitute these candidate solutions directly into the system, we observe that the complex-valued solution (y_+, z_+) is a linear combination of two real-valued pairs,

$$(y_+, z_+) = C_1(y_r, z_r) + C_2(y_i, z_i)$$

with $C_1 = 1$, a real number, and $C_2 = i$, an imaginary number. Since the system of differential equations has real coefficients, substituting this combination will preserve real and imaginary terms; no real term will be multiplied by i to convert it to imaginary or vice versa. Likewise, no derivative will convert a real expression to imaginary, since the two components (y_r, z_r) and (y_i, z_i) are each real-valued. Hence, the real part and the imaginary part must individually be solutions of the system.

Symbolically, this argument is

$$y'_+ = z_+ \Rightarrow y'_r + iy'_i = z_r + iz_i$$

$$z'_+ = 16y_+ \Rightarrow z'_r + iz'_i = 16y_r + i16y_i$$

Equating the real parts and the imaginary parts shows that (y_r, z_r) and (y_i, z_i) are, individually, solutions of the original system.

Linear independence follows from evaluating the Wronskian, exactly as in the preceding example. ∎

These examples show many of the basic ideas of the **characteristic equation method**. The following steps summarize their application to the generic constant-coefficient, linear system

$$y' = ay + bz$$

$$z' = cy + dz.$$

CHARACTERISTIC EQUATION METHOD

1. Assume a solution pair of the form $(y, z) = (pe^{rt}, qe^{rt})$, where p, q, r are constants.
   Substitute into the system of differential equations and divide by the exponential to obtain

   $$(a - r)p + bq = 0 \tag{8.7}$$

   $$cp + (d - r)q = 0. \tag{8.8}$$

2. The system of linear algebraic equations (8.7–8.8) has a nontrivial solution for p and q only if the determinant of coefficients is zero:

   $$\begin{vmatrix} a - r & b \\ c & d - r \end{vmatrix} = (a - r)(d - r) - bc = 0.$$

   Solve the resulting quadratic **characteristic equation**

   $$r^2 - (a + d)r + ad - bc = 0$$

   for the two **characteristic roots** r_1, r_2.

3. Set $r = r_1$ in (8.7–8.8). One of the equations will be redundant. Solve the remaining equation for q_1 in terms of p_1 (or vice versa) to obtain one solution $(p_1 e^{r_1 t}, q_1 e^{r_1 t})$.
   (a) If $r_2 \neq r_1$, repeat the preceding with $r = r_2$.
   (b) If $r_2 = r_1$, then seek a solution of the form $((P_1 + P_2 t)e^{r_1 t}, (Q_1 + Q_2 t)e^{r_1 t})$.

4. Linearly independent solutions are
   (a) $(p_1 e^{r_1 t}, q_1 e^{r_1 t})$, $(p_2 e^{r_2 t}, q_2 e^{r_2 t})$ if the characteristic roots r_1, r_2 are *real and distinct*,

(b) The *real part* and the *imaginary part* of the complex-valued solution $(p_1 e^{r_1 t}, q_1 e^{r_1 t})$ if r_1, r_2 are a *complex conjugate pair*,

(c) Of the form $(p e^{rt}, q e^{rt})$ and $((P_1 + P_2 t)e^{rt}, (Q_1 + Q_2 t)e^{rt})$ if there is a *single characteristic root* r.

General verification of most of the statements just given is left for the exercises. Step 4(a) is illustrated in example 4 and in example 8, which follows. Using the real part and the imaginary part of a complex solution to construct linearly independent solutions, as in step 4(b), was illustrated earlier in example 7. The repeated root case 4(c) is shown below in example 11.

EXAMPLE 8   *Find a general solution of*

$$y' = -\frac{y}{2} + \frac{3z}{2}$$

$$z' = \frac{3y}{2} - \frac{z}{2}.$$

Substituting the assumed solution form $(y, z) = (p e^{rt}, q e^{rt})$ into this system and dividing by the exponential leads to

$$\left(-\frac{1}{2} - r\right) p + \frac{3q}{2} = 0 \tag{8.9}$$

SYSTEM FOR p, q

$$\frac{3p}{2} + \left(-\frac{1}{2} - r\right) q = 0. \tag{8.10}$$

This system of algebraic equations has nontrivial solutions in p and q only if its determinant of coefficients is zero. Hence, we have the characteristic equation

$$\begin{vmatrix} -\frac{1}{2} - r & \frac{3}{2} \\ \frac{3}{2} & -\frac{1}{2} - r \end{vmatrix} = \left(-\frac{1}{2} - r\right)^2 - \frac{9}{4} = 0,$$

$$(r - 1)(r + 2) = 0.$$

CHARACTERISTIC EQUATION

The characteristic roots are $r_1, r_2 = 1, -2$.

To find the coefficients p, q when $r = r_1 = 1$, we substitute into (8.9–8.10) to obtain

Find p, q for the first root.

$$\frac{-3p}{2} + \frac{3q}{2} = 0$$

$$\frac{3p}{2} - \frac{3q}{2} = 0.$$

The second equation is redundant, and from the first we find $q = p$. Choosing $p = 1$ for simplicity, one solution of this system is

$$(y_1, z_1) = (e^t, e^t).$$

FIRST SOLUTION

To find the coefficients p, q when $r = -2$, we again substitute into (8.9)–(8.10) to obtain

Find p, q for the second root.

$$\frac{3p}{2} + \frac{3q}{2} = 0$$

$$\frac{3p}{2} + \frac{3q}{2} = 0.$$

We find $q = -p$. Another solution of this system is

SECOND SOLUTION

$$(y_2, z_2) = (e^{-2t}, -e^{-2t}).$$

Verification of linear independence is left to the exercises. A general solution is

$$(y_h, z_h) = C_1(y_1, z_1) + C_2(y_2, z_2)$$

GENERAL SOLUTION
$$= (C_1 e^t + C_2 e^{-2t}, C_1 e^t - C_2 e^{-2t}). \qquad \blacksquare$$

EXAMPLE 9 *Find the solution of the initial-value problem*

$$y' = -\frac{y}{2} + \frac{3z}{2}, \qquad y(0) = 3$$

$$z' = \frac{3y}{2} - \frac{z}{2}, \qquad z(0) = -1.$$

We need to choose the constants C_1, C_2 in the general solution found in example 8,

$$(y_h, z_h) = (C_1 e^t + C_2 e^{-2t}, C_1 e^t - C_2 e^{-2t}),$$

to satisfy the initial conditions $(y(0), z(0)) = (2, -1)$.

Apply initial conditions.

Setting $t = 0$ in the general solution yields

$$(y_h(0), z_h(0)) = (C_1 + C_2, C_1 - C_2) = (3, -1).$$

Equating the first and second components of the solution pair gives us two equations for the two unknown constants C_1, C_2,

$$C_1 + C_2 = 3$$
$$C_1 - C_2 = -1.$$

Solve for C_1, C_2.
Write the solution of the initial-value problem.

Solving yields $C_1 = 1$, $C_2 = 2$. The solution of the initial-value problem is $(e^t + 2e^{-2t}, e^t - 2e^{-2t})$ or

$$y(t) = e^t + 2e^{-2t}$$
$$z(t) = e^t - 2e^{-2t}. \qquad \blacksquare$$

EXAMPLE 10 *Find linearly independent solutions of the system*

$$y' = y + 3z$$
$$z' = -2z.$$

Substituting the assumed solution form $(y, z) = (pe^{rt}, qe^{rt})$ into the system yields

$$(1 - r)p + 3q = 0$$

SYSTEM FOR p, q

$$(-2 - r)q = 0.$$

Setting the determinant of coefficients to zero (to permit nontrivial solutions for p, q) yields the characteristic equation

$$(1 - r)(2 + r) = 0.$$

CHARACTERISTIC EQUATION

Its roots are $r = 1, -2$.

Setting $r = 1$, we find p, q must satisfy

Find p, q for the first root.

$$3q = 0$$

$$-3q = 0.$$

Evidently, $q = 0$ but p is arbitrary. Choosing $p = 1$ gives the first solution,

$$(y_1, z_1) = (e^t, 0).$$

FIRST SOLUTION

To find the coefficients p, q of the second solution, set $r = -2$ in the system of algebraic equations:

Find p, q for the second root.

$$3p + 3q = 0$$

$$0 = 0.$$

The single useful equation yields $q = -p$. Setting $p = 1$ yields

$$(y_2, z_2) = (e^{-2t}, -e^{-2t})$$

SECOND SOLUTION

Verification of linear independence is left to the exercises. ∎

The system in example 10 can be uncoupled by first solving the second equation for $z(t)$, and then using that solution as a forcing term in the first equation to find $y(t)$. The two solutions we just found correspond to the trivial solution $z = 0$ and to the nontrivial solution $z = -e^{-2t}$.

EXAMPLE 11 *Find linearly independent solutions of the system*

$$y' = z$$

$$z' = -9y + 6z.$$

Substituting $(y, z) = (pe^{rt}, qe^{rt})$ and dividing by the exponential leads to

$$-rp + q = 0$$

SYSTEM FOR p, q

$$-9p + (6 - r)q = 0.$$

Setting the determinant of coefficients to zero gives the characteristic equation

$$r^2 - 6r + 9 = (r - 3)^2 = 0,$$

CHARACTERISTIC EQUATION

which has the single repeated root $r = 3$.

To find p, q, we set $r = 3$ in the algebraic system:

$$-3p + q = 0 \tag{8.11}$$

$$-9p + 3q = 0 \tag{8.12}$$

FIRST SOLUTION

We obtain $q = 3p$. Choosing $p = 1$ gives one solution of the system,

$$(y_1, z_1) = (e^{3t}, 3e^{3t}).$$

SPECIAL FORM FOR SECOND SOLUTION

As directed in step 3(b) on page 322, we seek a second solution in the form

$$(y, z) = \left((P_1 + P_2 t)e^{3t}, (Q_1 + Q_2 t)e^{3t}\right).$$

Substituting this assumed form into the system of differential equations yields

$$3P_1 e^{3t} + \boxed{t(P_2 e^{3t})'} + P_2 e^{3t} = Q_1 e^{3t} + \boxed{t(Q_2 e^{3t})}$$

$$3Q_1 e^{3t} + \boxed{t(Q_2 e^{3t})'} + Q_2 e^{3t} = -9P_1 e^{3t} + \boxed{t(-9P_2 e^{3t})} + 6Q_1 e^{3t} + \boxed{t(6Q_2 e^{3t})}.$$

We can equate all the terms multiplying te^{3t}, which are boxed, and all the terms multiplying just e^{3t} to obtain two separate sets of equations. The equations coming from the boxed terms reduce to (8.11–8.12) with $p = P_2$ and $q = Q_2$; we conclude that $Q_2 = 3P_2$.

Using $P_2 = 1$ and $Q_2 = 3P_2 = 3$, the equations from the terms that are not boxed become

$$3P_1 - Q_1 = -1$$
$$9P_1 - 3Q_1 = -3,$$

from which we conclude $Q_1 = 1 + 3P_1$. With $P_1 = 1$, a second solution of this system is

SECOND SOLUTION

$$(y_2, z_2) = \left((1 + t)e^{3t}, (4 + 3t)e^{3t}\right).$$

Verification of linear independence is left for the exercises. ∎

8.2.1 Exercises

EXERCISE GUIDE

To gain experience . . .	*Try exercises*
Finding general solutions	1–12, 16(a), 17, 22–23
Solving initial-value problems	1–12
Verifying linear independence	13–15, 20–21
Analyzing characteristic roots	16(b)
Obtaining real-valued solutions from complex characteristic roots	1, 5–6, 9, 11, 20–21
With the foundations of the characteristic equation method	16, 18–21

In exercises 1–12,

    i. Find a general solution of the given system,

    ii. Find a solution of the system that satisfies the given initial conditions.

1. $y' = 2y + 4z,\ y(0) = 2$
$z' = -4y + 2z,\ z(0) = -2$

2. $y' = -y + 3z,\ y(0) = 0$
$z' = 3y - z,\ z(0) = 4$

3. $y' = 4y + 4z,\ y(0) = 4$
$z' = 3y - 4z,\ z(0) = 1$

4. $y' = 5y/2 + 3\sqrt{3}\,z/2,\ y(0) = 6$
$z' = 3\sqrt{3}\,y/2 - z/2,\ z(0) = 0$

5. $y' = -4y - 5z/2,\ y(0) = 0$
$z' = y - 2z,\ z(0) = 0$

6. $y' = -4y - 5z/2,\ y(0) = 1$
$z' = y - 2z,\ z(0) = -6$

7. $y' = y/2 + \sqrt{3}\,z/2,\ y(0) = 6$
$z' = \sqrt{3}\,y/2 - z/2,\ z(0) = 4$

8. $y' = z,\ y(0) = 1$
$z' = y,\ z(0) = 0$

9. $x' = v,\ x(0) = x_i$
$mv' = -kx,\ v(0) = 0$

10. $x' = v,\ x(0) = 0$
$mv' = -kx,\ v(0) = v_i$

11. $x' = v,\ x(0) = x_i$
$mv' = -kx - pv,\ v(0) = 0$

12. $\theta' = \omega,\ \theta(0) = \theta_i$
$L\omega' = -g\theta,\ \omega(0) = 0$

13. Verify that the solutions

$$(y_1, z_1) = (e^t, e^t),\ (y_2, z_2) = (e^{-2t}, -e^{-2t})$$

of the linear system considered in example 8 are linearly independent.

14. Verify that the solutions

$$(y_1, z_1) = (e^t, 0),\ (y_2, z_2) = (e^{-2t}, -e^{-2t}),$$

of the linear system considered in example 10 are linearly independent.

15. Verify that the solutions

$$(y_1, z_1) = (e^{3t}, 3e^{3t}),\ (y_2, z_2)$$
$$= \big((1+t)e^{3t},\ (4+3t)e^{3t}\big),$$

of the linear system considered in example 11 are linearly independent.

16. The text claims that the generic constant-coefficient system

$$y' = ay + bz$$
$$z' = cy + dz$$

has the characteristic equation

$$r^2 - (a + d)r + ad - bc = 0.$$

  (a) Verify this claim, beginning with the assumed solution form $(y, z) = (pe^{rt}, qe^{rt})$.

  (b) Find conditions on the coefficients a, b, c, d that guarantee that the characteristic roots are

    i. Real and distinct,

    ii. Real and equal,

    iii. Complex conjugates.

17. Supply the details of the derivation of the second solution

$$(y_-, z_-) = (e^{-4it}, -4ie^{-4it})$$

of the system

$$y' = z$$
$$z' = 16y.$$

that were omitted from example 5 .

18. The text's general description of the characteristic equation method begins

  Assume a solution pair of the form $(y, z) = (pe^{rt}, qe^{rt})$, where p, q, r are constants. Substitute into the system of differential equations

$$y' = ay + bz$$
$$z' = cy + dz$$

and divide out the exponential to obtain

$$(a - r)p + bq = 0$$
$$cp + (d - r)q = 0.$$

Follow these steps to derive this system of algebraic equations.

19. While describing the characteristic equation method, the text claims

  The system of linear algebraic equations

$$(a - r)p + bq = 0$$
$$cp + (d - r)q = 0.$$

has a nontrivial solution for p and q only if the determinant of coefficients is zero:

$$\begin{vmatrix} a-r & b \\ c & d-r \end{vmatrix} = (a-r)(d-r) - bc = 0.$$

Use Cramer's rule to justify this statement.

20. Example 7 shows that the real part and the imaginary part of a complex-valued solution of a certain linear homogeneous system with real coefficients are each solutions of that same system. Generalize that argument to show the following:

Suppose

$$(y_c, z_c) = (y_1, z_1) + i(y_2, z_2)$$

is a complex-valued solution of the linear homogeneous system

$$y' = k_1(t)y + k_2(t)z$$
$$z' = \ell_1(t)y + \ell_2(t)z,$$

where $y_1, z_1, y_2, z_2, k_1, k_2, \ell_1$, and ℓ_2 are all real-valued functions. Then (y_1, z_1) and (y_2, z_2) are each real-valued solutions of the same system.

Are the two real-valued solutions necessarily linearly independent?

21. The text claims linearly independent solutions of

$$y' = ay + bz$$
$$z' = cy + dz$$

are

i. $(p_1 e^{r_1 t}, q_1 e^{r_1 t}), (p_2 e^{r_2 t}, q_2 e^{r_2 t})$ if the characteristic roots r_1, r_2 are real and distinct,

ii. The real part and the imaginary part of the complex-valued solution $(p_1 e^{r_1 t}, q_1 e^{r_1 t})$ if r_1, r_2 are a complex conjugate pair.

Verify the claims of linear independence as follows.

(a) Evaluate the Wronskian $W(t)$ of the solutions $(p_1 e^{r_1 t}, q_1 e^{r_1 t}), (p_2 e^{r_2 t}, q_2 e^{r_2 t})$. What conditions must you impose on p_1, q_1, p_2, q_2? How can you be sure these conditions are satisfied?

(b) Write $r_1, r_2 = \alpha \pm i\beta$, $p_1 = P_1 + iP_2$, $q_1 = Q_1 + iQ_2$, where all parameters are real. Use this notation to write the real part and the imaginary part of the solution $(p_1 e^{r_1 t}, q_1 e^{r_1 t})$. Evaluate the Wronskian and proceed as in the previous part.

22. Fill in the details of the computation of the second solution in example 11. Start with the paragraph that opens, "As directed in step 3(b) ... "

23. In a remark following example 10 the text says

The system in example 10 can be uncoupled by first solving the second equation for $z(t)$ and then using that solution as a forcing term in the first equation to find $y(t)$.

Carry out this process.
The system considered in example 10 is

$$y' = y + 3z$$
$$z' = -2z.$$

24. Write the first-order system that is equivalent to $b_2 y'' + b_1 y' + b_0 y = 0$, b_i constant. Show that the characteristic equation of this system is $b_2 r^2 + b_1 r + b_0 = 0$.

8.3 NONHOMOGENEOUS SYSTEMS: VARIATION OF PARAMETERS

We seek particular solutions of nonhomogeneous linear systems using the method of *variation of parameters*, which we briefly examined for first-order equations in chapter 3, section 3. We discovered there that we could find particular solutions of first-order equations in the form $u(t)y_h(t)$, where y_h was a nontrivial solution of the *homogeneous* equation and u was a function to be determined by substituting into the *non*homogeneous equation.

One motivation for this approach is an observation about the form of particular solutions of first-order equations that are found using an integrating factor. For example, the integrating factor $\mu(t) = e^t$ helps us solve

$$y' + y = f(t)$$

by converting the differential expression on the left into the derivative of a product:

$$\mu(t)(y' + y) = \mu(t)f(t),$$

$$(\mu y)' = \mu f,$$

$$y = \frac{1}{\mu} \int \mu f.$$

The important observation here is that $1/\mu = e^{-t}$ is a solution of the homogeneous equation $y' + y = 0$. The particular solution has the form $y_h u$ with $u = \int \mu f$ in this case.

To extend variation of parameters to the generic linear nonhomogeneous system

$$y' = k_1(t)y + k_2(t)z + F(t)$$
$$z' = \ell_1(t)y + \ell_2(t)z + G(t),$$

we first suppose we have linearly independent solution pairs (y_1, z_1) and (y_2, z_2) of the corresponding homogeneous system,

$$y' = k_1(t)y + k_2(t)z$$
$$z' = \ell_1(t)y + \ell_2(t)z.$$

We shall seek a particular solution (y_p, z_p) of the nonhomogeneous system in the form

$$(y_p, z_p) = u_1(t)(y_1, z_1) + u_2(t)(y_2, z_2),$$

or, equivalently,

$$y_p = u_1(t)y_1 + u_2(t)y_2$$
$$z_p = u_1(t)z_1 + u_2(t)z_2$$

where u_1, u_2 are functions to be determined by substituting the proposed particular solution into the nonhomogeneous system.

> It may not be obvious that this form of the particular solution is the correct generalization of the first-order form $y_p = u y_h$. The way to discover the correct form is to try the possibilities. We omit the ideas that fail.

Substituting the proposed form of y_p into the first equation goes as follows:

$$y_p' = k_1 y_p + k_2 z_p + F$$
$$u_1' y_1 + u_2' y_2 + u_1 y_1' + u_2 y_2' = k_1 u_1 y_1 + k_1 u_2 y_2 + k_2 u_1 z_1 + k_2 u_2 z_2 + F$$
$$u_1' y_1 + u_2' y_2 = F + u_1(k_1 y_1 + k_2 z_1 - y_1') + u_2(k_1 y_2 + k_2 z_2 - y_2')$$

The terms in parentheses in the last line are zero because y_1, y_2 satisfy the first equation in the *homogeneous* system. Hence,

$$y_1 u_1' + y_2 u_2' = F.$$

We obtain a similar result from the second equation. (The details are left for the exercises.) Consequently, u_1', u_2' are determined by the pair of linear equations

$$y_1 u_1' + y_2 u_2' = F$$
$$z_1 u_1' + z_2 u_2' = G.$$

We can solve this system using Cramer's rule. The determinant of coefficients is just the Wronskian of the homogeneous solutions,

$$W(t) = y_1 z_2 - z_1 y_2.$$

Since these homogeneous solutions are linearly independent, we know $W \neq 0$ and the system defining u_1', u_2' has exactly one solution. It is

$$u_1' = \frac{F z_2 - G y_2}{W(t)}, \qquad u_2' = \frac{G y_1 - F z_1}{W(t)}.$$

The steps in the **variation of parameters** process for solving a nonhomogeneous linear system are as follows:

VARIATION OF PARAMETERS

1. Assume a particular solution of the form

$$\boxed{\begin{aligned} y_p &= u_1(t)y_1 + u_2(t)y_2 \\ z_p &= u_1(t)z_1 + u_2(t)z_2 \end{aligned}}$$

where $(y_1, z_1), (y_2, z_2)$ are linearly independent solutions of the homogeneous system.

Substitute (y_p, z_p) into the nonhomogeneous system and solve the resulting system of linear equations for u_1', u_2':

$$\boxed{u_1' = \frac{F z_2 - G y_2}{W(t)}, \qquad u_2' = \frac{G y_1 - F z_1}{W(t)}} \tag{8.13}$$

2. Integrate the expressions for u_1', u_2' and write the resulting particular solution.

The last step is often easier said then done. Frequently, these antiderivatives must be written as definite integrals using the fundamental theorem of calculus. Indeed, much of the value of variation of parameters is its ability to provide a representation of a particular solution, not a practical formula.

EXAMPLE 12 *Find a particular solution of*

$$y' = -4y + 2z + 10$$
$$z' = -3y + 3z + 5t.$$

Using the characteristic equation method of the previous section, we find that

$$(y_1, z_1) = (2e^{-3t}, e^{-3t}), \quad (y_2, z_2) = (e^{2t}, 3e^{2t})$$

are linearly independent solutions of the homogeneous system

$$y' = -4y + 2z$$
$$z' = -3y + 3z.$$

We note that the Wronskian of these solutions is

$$W(t) = 5e^{-t}.$$

To use variation of parameters, we assume a particular solution of the form

$$y_p = u_1(t)y_1 + u_2(t)y_2 = 2e^{-3t}u_1(t) + e^{2t}u_2(t) \tag{8.14}$$
$$z_p = u_1(t)z_1 + u_2(t)z_2 = e^{-3t}u_1(t) + 3e^{2t}u_2(t). \tag{8.15}$$

Substituting this form into the system of differential equations leads to a system of linear equations for u_1', u_2'. The solution of that system is given by equations (8.13) which, with $W(t) = 5e^{-t}$, yield

$$u_1' = 6e^{3t} - te^{3t}$$
$$u_2' = 2te^{-2t} - 2e^{-2t}.$$

Integrating and setting the constant of integration to zero gives

$$u_1 = \left(\frac{19}{9} - \frac{t}{3}\right)e^{3t}$$

$$u_2 = \left(t - \frac{1}{2}\right)e^{-2t}.$$

Combining these expressions with those for the homogeneous solutions in (8.14–8.15) leads to a particular solution of this system,

$$y_p = \frac{85}{18} - \frac{5t}{3}$$

$$z_p = \frac{65}{18} - \frac{10t}{3}. \qquad \blacksquare$$

Had we retained the constants of integration, they would have multiplied solutions of the homogeneous system. Since those solutions are linearly independent, we would have obtained a *general solution* of the nonhomogeneous system,

$$y_g = \frac{85}{18} - \frac{5t}{3} + C_1y_1 + C_2y_2$$

$$= \frac{85}{18} - \frac{5t}{3} + 2C_1e^{-3t} + C_2e^{2t}$$

$$z_g = \frac{65}{18} - \frac{10t}{3} + C_1z_1 + C_2z_2$$

$$= \frac{65}{18} - \frac{10t}{3} + C_1e^{-3t} + 3C_2e^{2t}.$$

We could satisfy initial conditions, if any had been given, with appropriate choices of C_1, C_2. The exercises ask you to explore these ideas.

The proverbial alert reader will have noticed that variation of parameters can be applied to any linear system, constant- or variable-coefficient. But homogeneous solutions are the building blocks of variation of parameters, and we know how to find them only for constant-coefficient systems. Consequently, our discussion has been limited to constant-coefficient systems.

Even when explicit solutions of the homogeneous system can not be found, variation of parameters provides a way of representing a particular solution; see the discussion of fundamental matrices in the next section. When actual solution values are needed, numerical techniques such as Runge-Kutta are often best for variable-coefficient systems.

The method of undetermined coefficients is alive and well for systems, and it works much as you might guess: Given a forcing term involving an exponential, a sine or a cosine, or a polynomial in t, seek a particular solution of the same form. For example, if e^t appears in the forcing term, guess a particular solution of the form (pe^t, qe^t) and substitute to find (p, q). See exercise 18.

8.3.1 Exercises

EXERCISE GUIDE

To gain experience . . .	Try exercises
Finding homogeneous solutions	1–11
With variation of parameters and particular solutions	1–11, 14
Finding and using general solutions	1–11, 14, 15(a), 16
Solving initial-value problems	1–11, 15(b)
With the foundations of variation of parameters	12–13, 17
Extending undetermined coefficients to systems	18

In exercises 1–11, . . .

i. Find two linearly independent solutions (y_1, z_1), (y_2, z_2) of the homogeneous system.

ii. Begin a variation of parameters solution of the nonhomogeneous system by finding u_1, u_2 in the assumed form $(y_p, z_p) = u_1(t)(y_1, z_1) + u_2(t)(y_2, z_2)$. If necessary, express antiderivatives as definite integrals using the fundamental theorem of calculus.

iii. Complete the variation of parameters solution by writing (y_p, z_p).

iv. Write a general solution of the nonhomogeneous system.

v. Find the solution of the nonhomogeneous system that satisfies the given initial conditions.

If a homogeneous solution is available from some other source (e.g., a previous exercise), use it rather than re-deriving a solution.

1. $y' = -4y + 2z + 15t$, $y(0) = 3$
   $z' = -3y + 3z - 20$, $z(0) = 4$

2. $y' = -y + 3z - 8e^{-2t}$, $y(0) = 0$
   $z' = 3y - z + 12$, $z(0) = 4$

3. $y' = 2y + 4z - 2$, $y(0) = 2$
   $z' = -4y + 2z + 4$, $z(0) = -2$

4. $y' = 4y + 4z - 24t$, $y(0) = 4$
   $z' = 3y - 4z + 8\cos 2t$, $z(0) = 1$

5. $y' = 5y/2 + 3\sqrt{3}z/2 + 4$, $y(0) = 6$
   $z' = 3\sqrt{3}y/2 - z/2 - 6t$, $z(0) = 0$

6. $y' = -4y - 5z/2 + 1$, $y(0) = 1$
   $z' = y - 2z - 1$, $z(0) = -6$

7. $y' = -4y - 5z/2 + 1$, $y(0) = 0$
   $z' = y - 2z - 1$, $z(0) = 0$

8. $x' = v$, $x(0) = x_i$
   $mv' = -kx + F \sin \omega t$, $v(0) = 0$

9. $x' = v$, $x(0) = 0$
   $mv' = -kx + F \sin \omega t$, $v(0) = v_i$

10. $x' = v$, $x(0) = x_i$
    $mv' = -kx - pv + Fe^{-t}$, $v(0) = 0$

11. $\theta' = \omega$, $\theta(0) = \theta_i$
    $L\omega' = -g\theta + 2 \cos \pi t$, $\omega(0) = 0$

12. Complete the variation of parameters derivation of the equations for u_1', u_2' by substituting the second component $z_p = u_1 z_1 + u_2 z_2$ of the proposed particular solution into the second equation

$$z' = \ell_1(t)y + \ell_2(t)z + G(t)$$

in the nonhomogeneous system to show that u_1', u_2' must satisfy

$$z_1 u_1' + z_2 u_2' = G.$$

13. Example 12 solves the nonhomogeneous system

$$y' = -4y + 2z + 10$$
$$z' = -3y + 3z + 5t$$

using variation of parameters and the particular solution form

$$y_p = u_1(t)y_1 + u_2(t)y_2 = 2e^{-3t}u_1(t) + e^{2t}u_2(t)$$
$$z_p = u_1(t)z_1 + u_2(t)z_2 = e^{-3t}u_1(t) + 3e^{2t}u_2(t).$$

It finds u_1', u_2' from equation (8.13). Instead, substitute this particular solution form directly into the system of differential equations, find the system of linear equation defining u_1', u_2', and solve for u_1', u_2'. Show that you obtain the same result as the example.

14. Example 12 obtains the particular solution

$$y_p = \frac{85}{18} - \frac{5t}{3}$$
$$z_p = \frac{65}{18} - \frac{10t}{3}$$

of the nonhomogeneous system

$$y' = -4y + 2z + 10$$
$$z' = -3y + 3z + 5t.$$

Verify by substitution that (y_p, z_p) is indeed a particular solution of this system. Explain why it is *not* a general solution.

15. The text claims after example 12 that

$$y_g = \frac{85}{18} - \frac{5t}{3} + 2C_1 e^{-3t} + C_2 e^{2t}$$
$$z_g = \frac{65}{18} - \frac{10t}{3} + C_1 e^{-3t} + 3C_2 e^{2t}$$

is a general solution of the nonhomogeneous system

$$y' = -4y + 2z + 10$$
$$z' = -3y + 3z + 5t.$$

(a) Verify that (y_g, z_g) is indeed a general solution.

(b) Find the solution of this system that satisfies the initial conditions $(y(0), z(0)) = (\frac{139}{18}, \frac{137}{18})$.

16. Example 12 obtains the particular solution

$$y_p = \frac{85}{18} - \frac{5t}{3}$$
$$z_p = \frac{65}{18} - \frac{10t}{3}$$

of the nonhomogeneous system

$$y' = -4y + 2z + 10$$
$$z' = -3y + 3z + 5t$$

by setting to zero the constants of integration it finds in computing u_1, u_2 from

$$u_1' = 6e^{3t} - te^{3t}$$
$$u_2' = 2te^{-2t} - 2e^{-2t}.$$

Keep those constants (you might call them C_1, C_2) and show that you actually obtain a *general solution* of the nonhomogeneous system.

17. What fails in variation of parameters if the homogeneous solutions are *not* linearly independent?

18. Formulate an extension of the method of undetermined coefficients to systems and apply it to exercises 1–11, as assigned.

8.4 THE MATRIX PERSPECTIVE

Using matrices to reformulate the results of the previous sections provides simpler notation, clearer insight into fundamental mathematical issues, and a direct route to extensions to systems of arbitrary order. We assume familiarity with the matrix ideas reviewed in section A.5.

8.4.1 General Linear Systems

The general linear system of two equations,

$$y' = k_1(t)y + k_2(t)z + F(t)$$
$$z' = \ell_1(t)y + \ell_2(t)z + G(t),$$

can be written as the matrix system

MATRIX NOTATION FOR LINEAR SYSTEM

$$y' = \mathbf{A}(t)\mathbf{y} + \mathbf{f}(t)$$

by introducing the vector functions

$$\mathbf{y}(t) = \begin{pmatrix} y(t) \\ z(t) \end{pmatrix}, \qquad \mathbf{f}(t) = \begin{pmatrix} F(t) \\ G(t) \end{pmatrix},$$

and the matrix function

$$\mathbf{A}(t) = \begin{pmatrix} k_1(t) & k_2(t) \\ \ell_1(t) & \ell_2(t) \end{pmatrix}.$$

Where we once spoke of a solution as the ordered pair $(y(t), z(t))$, we can now speak of the *solution vector* $\mathbf{y}(t)$, which was just defined.

> The formalism of the notation is irrelevant, as long as it conveys the order of the solution functions; the first solution function $y(t)$ is either the one on the left in the ordered pair or the one on the top in the vector.

Of course, two solution vectors $\mathbf{y}_1(t), \mathbf{y}_2(t)$ are *linearly independent* if their Wronskian is not zero at one point of their interval of definition. But the Wronskian is just the determinant of the matrix whose columns are $\mathbf{y}_1(t), \mathbf{y}_2(t)$:

WRONSKIAN

$$W(\mathbf{y}_1(t), \mathbf{y}_2(t)) = \det (\ \mathbf{y}_1(t) \quad \mathbf{y}_2(t)\ )$$

That is, these two solutions are linearly independent if and only if the matrix

FUNDAMENTAL MATRIX

$$\mathbf{Y}(t) = (\ \mathbf{y}_1(t) \quad \mathbf{y}_2(t)\ )$$

is *nonsingular*. A square matrix whose columns are linearly independent solutions of the system of linear differential equations is called a **fundamental matrix** of that system. The corresponding solution vectors are said to form a set of **fundamental solutions** of the system.

A general solution of a homogeneous system is a linear combination of fundamental solutions,

$$\mathbf{y}_g(t) = c_1\mathbf{y}_1(t) + c_2\mathbf{y}_2(t) = \mathbf{Y}(t)\mathbf{c},$$

where $\mathbf{Y}(t) = (\ \mathbf{y}_1(t)\ \ \mathbf{y}_2(t)\ )$ and $\mathbf{c}$ is a vector of arbitrary constants,

$$\mathbf{c} = \begin{pmatrix} c_1 \\ c_2 \end{pmatrix}.$$

EXAMPLE 13 *Using solutions obtained earlier, construct a fundamental matrix and a general solution of the homogeneous system*

$$y' = -\frac{y}{2} + \frac{3z}{2}$$

$$z' = \frac{3y}{2} - \frac{z}{2}.$$

In matrix form, this system is $\mathbf{y}'(t) = \mathbf{A}\mathbf{y}(t)$, where

$$\mathbf{y}(t) = \begin{pmatrix} y(t) \\ z(t) \end{pmatrix}, \qquad \mathbf{A} = \begin{pmatrix} -\frac{1}{2} & \frac{3}{2} \\ \frac{3}{2} & -\frac{1}{2} \end{pmatrix}.$$

In example 8 of section 8.2, we found the solutions $(y_1, z_1) = (e^t, e^t)$, $(y_2, z_2) = (e^{-2t}, -e^{-2t})$, or in vector notation

$$\mathbf{y}_1(t) = \begin{pmatrix} e^t \\ e^t \end{pmatrix}, \qquad \mathbf{y}_2(t) = \begin{pmatrix} e^{-2t} \\ -e^{-2t} \end{pmatrix}.$$

These solutions are linearly independent for all t because

$$W(\mathbf{y}_1(t), \mathbf{y}_2(t)) = \det (\ \mathbf{y}_1(t)\ \ \mathbf{y}_2(t)\ ) = \det \begin{pmatrix} e^t & e^{-2t} \\ e^t & -e^{-2t} \end{pmatrix} = -2e^{-t} \neq 0.$$

That is,

$$\mathbf{Y}(t) = (\ \mathbf{y}_1(t)\ \ \mathbf{y}_2(t)\ ) = \begin{pmatrix} e^t & e^{-2t} \\ e^t & -e^{-2t} \end{pmatrix}$$

is a fundamental matrix for this system.

A general solution is

$$\mathbf{y}_g(t) = \mathbf{Y}(t)\mathbf{c} = c_1 \begin{pmatrix} e^t \\ e^t \end{pmatrix} + c_2 \begin{pmatrix} e^{-2t} \\ -e^{-2t} \end{pmatrix}. \qquad \blacksquare$$

If $\mathbf{Y}(t)$ is a fundamental matrix of $\mathbf{y}'(t) = \mathbf{A}(t)\mathbf{y}(t)$ and if $\mathbf{y}_p(t)$ is a particular solution of $\mathbf{y}'(t) = \mathbf{A}(t)\mathbf{y}(t) + \mathbf{f}(t)$, then a *general solution* of the nonhomogeneous equation $\mathbf{y}'(t) = \mathbf{A}(t)\mathbf{y}(t) + \mathbf{f}(t)$ is

$$\mathbf{y}_g(t) = \mathbf{y}_p(t) + \mathbf{Y}(t)\mathbf{c}.$$

This construction is the same as for any other linear problem: A particular solution plus a linear combination of linearly independent homogeneous solutions.

The solution of the initial-value problem

$$\mathbf{y}'(t) = \mathbf{A}(t)\mathbf{y}(t) + \mathbf{f}(t), \qquad \mathbf{y}(t_0) = \mathbf{y}_i, \qquad \blacksquare$$

is easily represented using a fundamental matrix $\mathbf{Y}(t)$ for this system. The constant vector $\mathbf{c}$ in the general solution is determined by the initial conditions,

Impose initial conditions.

$$\mathbf{y}_g(t_0) = \mathbf{y}_p(t_0) + \mathbf{Y}(t_0)\mathbf{c} = \mathbf{y}_i,$$

or

$$\mathbf{Y}(t_0)\mathbf{c} = \mathbf{y}_i - \mathbf{y}_p(t_0).$$

Since the columns of $\mathbf{Y}(t)$ are linearly independent (or since $\det \mathbf{Y}(t) \neq 0$), the matrix $\mathbf{Y}(t_0)$ is certainly invertible, and we can write

Solve for **c**.

$$\mathbf{c} = \mathbf{Y}(t_0)^{-1}(\mathbf{y}_i - \mathbf{y}_p(t_0)).$$

Use **c** in the general solution.

The solution of the initial-value problem is thus

$$\mathbf{y}(t) = \mathbf{Y}(t)\mathbf{Y}(t_0)^{-1}(\mathbf{y}_i - \mathbf{y}_p(t_0)) + \mathbf{y}_p(t).$$

Practically speaking, the inverse matrix $\mathbf{Y}(t_0)^{-1}$ is not computed. Instead, the linear system $\mathbf{Y}(t_0)\mathbf{c} = \mathbf{y}_i - \mathbf{y}_p(t_0)$ is solved for the unknown $\mathbf{c}$.

As a general rule, *never* compute the inverse of a matrix and multiply to solve a system. Elimination leads to the solution in fewer operations. Cramer's rule is quick for 2×2 systems, but it is too inefficient for much larger systems.

EXAMPLE 14 *Using solutions obtained earlier, write a general solution of the nonhomogeneous system*

$$\mathbf{y}'(t) = \begin{pmatrix} -4 & 2 \\ -3 & 3 \end{pmatrix} \mathbf{y}(t) + \begin{pmatrix} 10 \\ 5t \end{pmatrix},$$

and find the solution of this initial-value problem subject to the initial condition

$$\mathbf{y}(0) = \mathbf{y}_i = \begin{pmatrix} \frac{98}{18} \\ \frac{29}{18} \end{pmatrix}.$$

FUNDAMENTAL MATRIX

From example 12 of section 8.3, we have the linearly independent homogeneous solutions $(2e^{-3t}, e^{-3t})$, $(e^{2t}, 3e^{2t})$. Hence, a fundamental matrix is

$$\mathbf{Y}(t) = \begin{pmatrix} 2e^{-3t} & e^{2t} \\ e^{-3t} & 3e^{2t} \end{pmatrix}.$$

From that same example, a particular solution is

$$\mathbf{y}_p(t) = \begin{pmatrix} \frac{85}{18} - \frac{5t}{3} \\ \frac{65}{18} - \frac{10t}{3} \end{pmatrix}.$$

GENERAL SOLUTION

A general solution is $\mathbf{y}_g(t) = \mathbf{Y}(t)\mathbf{c} + \mathbf{y}_p(t)$, or

$$\mathbf{y}_g(t) = \begin{pmatrix} 2e^{-3t} & e^{2t} \\ e^{-3t} & 3e^{2t} \end{pmatrix} \begin{pmatrix} c_1 \\ c_2 \end{pmatrix} + \begin{pmatrix} \frac{85}{18} - \frac{5t}{3} \\ \frac{65}{18} - \frac{10t}{3} \end{pmatrix}.$$

To determine $\mathbf{c}$, set $t = 0$ and solve $\mathbf{y}_g(0) = \mathbf{Y}(0)\mathbf{c} + \mathbf{y}_p(0) = \mathbf{y}_i$; i.e., solve

Impose initial conditions to determine **c**.

$$\begin{pmatrix} 2 & 1 \\ 1 & 3 \end{pmatrix} \begin{pmatrix} c_1 \\ c_2 \end{pmatrix} = \begin{pmatrix} \frac{93}{18} \\ \frac{29}{18} \end{pmatrix} - \begin{pmatrix} \frac{85}{18} \\ \frac{65}{18} \end{pmatrix} = \begin{pmatrix} 1 \\ -2 \end{pmatrix}.$$

Using Cramer's rule or elimination, we find $c_1 = 1$, $c_2 = -1$. The solution of the initial-value problem is

Use **c** in the general solution to write the solution of the initial-value problem.

$$\mathbf{y}(t) = \mathbf{Y}(t) \begin{pmatrix} 1 \\ -1 \end{pmatrix} + \mathbf{y}_p(t),$$

where $\mathbf{Y}(t)$ and $\mathbf{y}_p(t)$ are as given before. ∎

Although the preceding discussion considered only 2×2 matrices, the extension to $n \times n$ matrices should be obvious. The general matrix notation is unchanged as the size of the system increases.

8.4.2 Constant-Coefficient Homogeneous Systems

Now restrict consideration to constant-coefficient ($\mathbf{A}(t) = \mathbf{B} \equiv$ constant), homogeneous ($\mathbf{f}(t) \equiv 0$) systems. How is the characteristic equation process implemented using matrix notation?

To solve the constant-coefficient, homogeneous system $\mathbf{y}' = \mathbf{By}$, we began in section 8.2 with the assumed solution form $(y, z) = (pe^{rt}, qe^{rt})$ where p, q, and r were constants to be determined. That form corresponds to the proposed solution vector

$$\mathbf{y} = \mathbf{p}e^{rt} = \begin{pmatrix} p \\ q \end{pmatrix} e^{rt}.$$

FORM OF CHARACTERISTIC EQUATION SOLUTION

Since $\mathbf{y}' = r\mathbf{p}e^{rt}$, substituting in $\mathbf{y}' = \mathbf{By}$ leads to

$$r\mathbf{p}e^{rt} = \mathbf{Bp}e^{rt}.$$

Canceling the nonzero exponential leaves the relation $\mathbf{Bp} = r\mathbf{p}$ or

$$(\mathbf{B} - r\mathbf{I})\mathbf{p} = 0$$

between the characteristic root r and the unknown vector $\mathbf{p}$. ($\mathbf{I}$ is the identity matrix.)

This form of the characteristic equation puts this process in a new light. Solving the equation $\mathbf{Bp} = r\mathbf{p}$ is asking for a (nonzero) vector $\mathbf{p}$ whose *direction* is unaffected when it is multiplied by the matrix $\mathbf{B}$, while its *magnitude* is altered by the factor r. The nonzero vector $\mathbf{p}$ is called a **characteristic vector** or **eigenvector** of the matrix $\mathbf{B}$, and r is its corresponding **characteristic value** or **eigenvalue**.

To find the eigenvalues of the matrix $\mathbf{B}$, note that the defining relation $(\mathbf{B} - r\mathbf{I})\mathbf{p} = 0$ is a *homogeneous* system of linear equations; it always has the solution $\mathbf{p} \equiv 0$. From Cramer's rule, it can have nontrivial solutions $\mathbf{p} \neq 0$ only when the determinant of coefficients is zero – i.e., only when

$$\boxed{\det(\mathbf{B} - r\mathbf{I}) = 0.}$$

This expression is the **characteristic equation** of the matrix $\mathbf{B}$, the equation whose solutions are the eigenvalues of $\mathbf{B}$.

> These circumstances are the complement of those we encountered in studying the Wronskian: The columns of the matrix $\mathbf{Y}(t)$ are linearly independent if and only if the system of equations $\mathbf{Y}(t)\mathbf{c} = 0$ has only the trivial solution $\mathbf{c} = 0$. That system has only the trivial solution if $W = \det \mathbf{Y} \neq 0$. In contrast, we require here that $\det(\mathbf{B} - r\mathbf{I}) = 0$ to obtain *non*trivial solutions in addition to the trivial solution.

EXAMPLE 15   *Find the eigenvalues and eigenvectors of the matrix*

$$\mathbf{B} = \begin{pmatrix} -\frac{1}{2} & \frac{3}{2} \\ \frac{3}{2} & -\frac{1}{2} \end{pmatrix}.$$

The characteristic equation defining the eigenvalues r is

$$\det(\mathbf{B} - r\mathbf{I}) = \begin{pmatrix} -\frac{1}{2} - r & \frac{3}{2} \\ \frac{3}{2} & -\frac{1}{2} - r \end{pmatrix} = 0.$$

Expanding the determinant gives the *characteristic polynomial* of the matrix $\mathbf{B}$,

$$(-1/2 - r)^2 - 9/4 = r^2 + r - 2 = (r + 2)(r - 1) = 0.$$

The roots of this polynomial are $r = -2, 1$; that is, the eigenvalues of the matrix $\mathbf{B}$ are $r = -2, 1$.

Substitute the first eigenvalue to find the eigen vector.

To find the eigenvector corresponding to the eigenvalue $r_1 = 1$, we return to the defining relation $(\mathbf{B} - r\mathbf{I})\mathbf{p} = 0$, substitute $r_1 = 1$, and solve for the components p, q of the eigenvector $\mathbf{p}_1$:

$$(\mathbf{B} - \mathbf{I})\mathbf{p}_1 = \begin{pmatrix} -\frac{3}{2} & \frac{3}{2} \\ \frac{3}{2} & -\frac{3}{2} \end{pmatrix} \begin{pmatrix} p \\ q \end{pmatrix} = 0.$$

Multiplying by $\frac{2}{3}$, we have two equations,

$$-p + q = 0$$

$$p - q = 0,$$

which are not independent; one of the unknowns is left free to vary. Subtracting the two equations gives

$$q = p.$$

Arbitrarily choosing $p = 1$, we find $q = 1$. An eigenvector of $\mathbf{B}$ associated with the eigenvalue $r_1 = 1$ is

$$\mathbf{p}_1 = \begin{pmatrix} 1 \\ 1 \end{pmatrix}.$$

Substitute the second eigenvalue to find the eigenvector.

To find the eigenvector of $\mathbf{B}$ associated with the eigenvalue $r_2 = -2$, substitute $r_2 = -2$ into the defining relation $(\mathbf{B} - r_2\mathbf{I})\mathbf{p} = 0$ and solve for the components of $\mathbf{p}$:

$$(\mathbf{B} + 2\mathbf{I})\mathbf{p}_2 = \begin{pmatrix} \frac{3}{2} & \frac{3}{2} \\ \frac{3}{2} & \frac{3}{2} \end{pmatrix} \begin{pmatrix} p \\ q \end{pmatrix} = 0.$$

A nontrivial solution of the resulting equations,

$$p + q = 0$$

$$p + q = 0,$$

requires $q = -p \neq 0$; choose $p = 1$, $q = -1$. An eigenvector of $\mathbf{B}$ associated with the eigenvalue $r_2 = -2$ is

$$\mathbf{p}_2 = \begin{pmatrix} 1 \\ -1 \end{pmatrix}.$$

■

Note that eigenvectors can be specified only to within a constant multiple; indeed, if a is a constant and $\mathbf{B}\mathbf{p} = r\mathbf{p}$, then $\mathbf{B}(a\mathbf{p}) = r(a\mathbf{p})$. If $\mathbf{p}$ is a eigenvector, then so is $a\mathbf{p}$.

To summarize the preceding example, the steps in finding the eigenvalues r_i and eigenvectors $\mathbf{p}_i$ of a matrix $\mathbf{B}$ are as follows:

1. Solve the characteristic polynomial

FINDING EIGENVALUES AND EIGENVECTORS

$$\det(\mathbf{B} - r\mathbf{I}) = 0$$

for the eigenvalues r_i.

2. Determine the components of the eigenvector $\mathbf{p}_i$ by finding a nontrivial solution of the homogeneous system $(\mathbf{B} - r_i\mathbf{I})\mathbf{p}_i = 0$. (One or more components of $\mathbf{p}_i$ will be arbitrary.)

The next example illustrates how the eigenvalues and eigenvectors of the coefficient matrix $\mathbf{B}$ are combined to construct a solution of $\mathbf{y}' = \mathbf{B}\mathbf{y}$.

EXAMPLE 16   *Find a general solution of*

$$\mathbf{y}' = \begin{pmatrix} -\frac{1}{2} & \frac{3}{2} \\ \frac{3}{2} & -\frac{1}{2} \end{pmatrix} \mathbf{y}.$$

Assume solutions of the form $\mathbf{y} = \mathbf{p}e^{rt}$. Substitute in the homogeneous differential equation and divide out e^{rt} to obtain the eigenvalue relation $r\mathbf{p} = \mathbf{B}\mathbf{p}$, or

$$(\mathbf{B} - r\mathbf{I})\mathbf{p} = 0,$$

where $\mathbf{B}$ is the coefficient matrix

$$\mathbf{B} = \begin{pmatrix} -\frac{1}{2} & \frac{3}{2} \\ \frac{3}{2} & -\frac{1}{2} \end{pmatrix}.$$

Evidently, r must be a root of the characteristic polynomial

CHARACTERISTIC POLYNOMIAL

$$\det(\mathbf{B} - r\mathbf{I}) = 0;$$

i.e., r is an eigenvalue of $\mathbf{B}$ and $\mathbf{p}$ is a corresponding eigenvector.

From the preceding example, the eigenvalues and eigenvectors of the coefficient matrix $\mathbf{B}$ are

$$r_1 = 1, \quad \mathbf{p}_1 = \begin{pmatrix} 1 \\ 1 \end{pmatrix} \quad \text{and} \quad r_2 = -2, \quad \mathbf{p}_2 = \begin{pmatrix} 1 \\ -1 \end{pmatrix}.$$

Since each eigenvalue-eigenvector pair generates a solution vector $\mathbf{p}_i e^{r_i t}$, a candidate fundamental matrix is $\mathbf{Y}(t) = (\ \mathbf{p}_1 e^t \quad \mathbf{p}_2 e^{-2t}\ )$. To test linear independence, evaluate the Wronskian at $t = 0$:

$$W(\mathbf{p}_1 e^t, \mathbf{p}_2 e^{-2t})\big|_{t=0} = \det \mathbf{Y}(0) = \det (\ \mathbf{p}_1\ \mathbf{p}_2\ ) = -2 \neq 0.$$

We see that $\mathbf{Y}(t)$ is indeed a fundamental matrix and, furthermore, that *the linear independence of the solutions is a direct consequence of the linear independence of the eigenvectors* of the coefficient matrix.

Hence, a general solution of this system is

$$\mathbf{y}_g(t) = \mathbf{Y}(t)\mathbf{c} = c_1 \begin{pmatrix} 1 \\ 1 \end{pmatrix} e^t + c_2 \begin{pmatrix} 1 \\ -1 \end{pmatrix} e^{-2t}. \qquad \blacksquare$$

As this example illustrates, the key to the eigenvector construction of a fundamental matrix for a system of n constant-coefficient differential equations is the existence of n linearly independent eigenvectors of the coefficient matrix. Unfortunately, not every matrix is blessed with a complete set of linearly independent eigenvectors; see exercise 8. However, *symmetric* matrices ($\mathbf{B} = \mathbf{B}^T$) and *Hermitian* matrices ($\mathbf{B} = \overline{\mathbf{B}}^T$) always have a complete set of linearly independent eigenvectors.

> Recall that overbar denotes complex conjugate: $\overline{\alpha + i\beta} = \alpha - i\beta$. The complex conjugate of a matrix is the matrix of complex conjugates; i.e., if the entries of $\mathbf{B}$ are b_{ij}, then the entries of $\overline{\mathbf{B}}$ are $\overline{b}_{ij}$.

To emphasize the importance of linearly independent eigenvectors, we solve an initial-value problem involving the system of the previous example.

EXAMPLE 17 *Solve the system of the previous example subject to the initial condition*

$$\mathbf{y}(0) = \mathbf{y}_i = \begin{pmatrix} 1 \\ 5 \end{pmatrix}.$$

From the previous example, a general solution is $\mathbf{y}_g(t) = \mathbf{Y}(t)\mathbf{c}$, where $\mathbf{Y}(t)$ is the fundamental matrix given there. The initial condition demands that the vector $\mathbf{c}$ of arbitrary constants be chosen to satisfy

$$\mathbf{y}_g(0) = \mathbf{Y}(0)\mathbf{c} = \mathbf{y}_i.$$

Symbolically, $\mathbf{c} = \mathbf{Y}(0)^{-1}\mathbf{y}_i$; the inverse matrix $\mathbf{Y}(0)^{-1} = (\ \mathbf{p}_1\ \mathbf{p}_2\ )^{-1}$ exists precisely because the eigenvectors of the coefficient matrix are linearly independent.

Practically, we solve the system of equations defining $\mathbf{c}$ rather than compute the inverse matrix and multiply:

$$\mathbf{Y}(0)\mathbf{c} = \begin{pmatrix} 1 & 1 \\ 1 & -1 \end{pmatrix} \mathbf{c} = \begin{pmatrix} 1 \\ 5 \end{pmatrix}.$$

Elimination yields $c_1 = 3$, $c_2 = -2$. The solution of the initial value problem is

Use **c** to write solution of initial-value problem.

$$\mathbf{y}(t) = \mathbf{Y}(t) \begin{pmatrix} 3 \\ -2 \end{pmatrix} = 3 \begin{pmatrix} 1 \\ 1 \end{pmatrix} e^t + -2 \begin{pmatrix} 1 \\ -1 \end{pmatrix} e^{-2t}. \qquad \blacksquare$$

The preceding ideas work equally well when the eigenvalues are complex. Just a little additional effort is needed to obtain real-valued solutions.

EXAMPLE 18 *Find a real-valued fundamental matrix for the system* $\mathbf{y}' = \mathbf{B}\mathbf{y}$ *where*

$$\mathbf{B} = \begin{pmatrix} 0 & -1 \\ 16 & 0 \end{pmatrix}.$$

The characteristic polynomial of this matrix is

CHARACTERISTIC POLYNOMIAL

$$\det(\mathbf{B} - r\mathbf{I}) = \det \begin{pmatrix} -r & -1 \\ 16 & -r \end{pmatrix} = r^2 + 16 = 0.$$

Its roots, the eigenvalues of $\mathbf{B}$, are evidently $r = \pm 4i$.
To find an eigenvector associated with $r_1 = 4i$, solve

COMPLEX EIGENVALUES

Substitute first eigenvalue to find eigenvector.

$$(\mathbf{B} - 4i\mathbf{I})\mathbf{p} = \begin{pmatrix} -4i & -1 \\ 16 & -4i \end{pmatrix} \begin{pmatrix} p \\ q \end{pmatrix} = 0.$$

Both equations provide the same relation between p and q,

$$q = 4ip.$$

Choosing $p = 1$, an eigenvector associated with the eigenvalues $r_1 = 4i$ is

FIRST EIGENVECTOR

$$\mathbf{p}_1 = \begin{pmatrix} 1 \\ 4i \end{pmatrix}.$$

Parallel calculations find the eigenvector associated with $r_2 = -4i$,

Use the second eigenvalue to find a second eigenvector.

$$\mathbf{p}_2 = \begin{pmatrix} 1 \\ -4i \end{pmatrix}.$$

After verifying the linear independence of the two eigenvectors $\mathbf{p}_1$ and $\mathbf{p}_2$, we can certainly form the general solution

COMPLEX-VALUED GENERAL SOLUTION

$$c_1 \begin{pmatrix} 1 \\ 4i \end{pmatrix} e^{4it} + c_2 \begin{pmatrix} 1 \\ -4i \end{pmatrix} e^{-4it}. \qquad (8.16)$$

How do we obtain a *real*-valued solution?
The key lies in two observations:

1. Eigenvectors and eigenvalues of real matrices occur in complex conjugate pairs: If $r, \mathbf{p}$ is an eigenvalue, eigenvector pair for the real-valued matrix $\mathbf{B}$, then so is $\bar{r}, \bar{\mathbf{p}}$.

The proof is simple. Suppose $\mathbf{Bp} = r\mathbf{p}$. Form the complex conjugate of both sides of the equation. Since $\mathbf{B}$ is real-valued, $\overline{\mathbf{B}} = \mathbf{B}$. Hence, $\mathbf{B}\bar{\mathbf{p}} = \bar{r}\bar{\mathbf{p}}$, and the conjugates form an eigenvalue, eigenvector pair as well.

2. If $z = \alpha + i\beta$ is a complex quantity, then

$$\frac{z + \bar{z}}{2} = \alpha = \operatorname{Re} z, \qquad \frac{z - \bar{z}}{2i} = \beta = \operatorname{Im} z.$$

Exercise 10 requests a verification.

The first statement certainly holds in this case: $r_2 = -4i = \overline{4i} = \bar{r}_1$. And the eigenvectors are clearly complex conjugate pairs as well, $\mathbf{p}_2 = \bar{\mathbf{p}}_1$. Consequently, the choice of arbitrary constants $c_1 = c_2 = 1/2$ in the general solution (8.16) leads to the real-valued solution

$$\mathbf{y}_1 = \frac{1}{2}\left(\mathbf{p}_1 e^{4it} + \mathbf{p}_2 e^{-4it}\right) = \operatorname{Re}\left(\mathbf{p}_1 e^{4it}\right).$$

Using Euler's formula, $e^{4it} = \cos 4t + i \sin 4t$, we find

$$\mathbf{p}_1 e^{4it} = \begin{pmatrix} \cos 4t + i \sin 4t \\ 4i \cos 4t - 4 \sin 4t \end{pmatrix}.$$

Hence, the real part of this vector function is

$$\mathbf{y}_1 = \operatorname{Re}\left(\mathbf{p}_1 e^{4it}\right) = \begin{pmatrix} \cos 4t \\ -4 \sin 4t \end{pmatrix}.$$

Likewise, the choice $c_1 = -c_2 = 1/2i$ in the general solution (8.16) leads to another real-valued solution,

$$\mathbf{y}_2 = \frac{1}{2i}\left(\mathbf{p}_1 e^{4it} - \mathbf{p}_2 e^{-4it}\right) = \operatorname{Im}\left(\mathbf{p}_1 e^{4it}\right) = \begin{pmatrix} \sin 4t \\ 4 \cos 4t \end{pmatrix}.$$

A real-valued fundamental matrix is

$$\mathbf{Y}(t) = \begin{pmatrix} \cos 4t & \sin 4t \\ -4 \sin 4t & 4 \cos 4t \end{pmatrix} \mathbf{c}.$$

(Verification of linear independence is left to exercise 11.) ∎

To summarize from the matrix perspective, the **characteristic equation method** depends entirely on the eigenvectors of the (constant) coefficient matrix $\mathbf{B}$. If the $n \times n$ matrix $\mathbf{B}$ has n linearly independent eigenvectors $\mathbf{p}_1, \ldots, \mathbf{p}_n$ with associated eigenvalues $r_1, \ldots, r_n$, then a *fundamental matrix* for

$$\mathbf{y}' = \mathbf{By}$$

is

$$\mathbf{Y}(t) = \left(\ \mathbf{p}_1 e^{r_1 t} \cdots \mathbf{p}_n e^{r_n t} \ \right),$$

FUNDAMENTAL MATRIX

and a *general solution* of the homogeneous system is

$$\mathbf{y}_g(t) = \mathbf{Y}(t)\mathbf{c},$$

GENERAL SOLUTION

where $\mathbf{c}$ is the constant vector $(c_1 \cdots c_n)^T$.

If $\mathbf{B}$ is a real matrix, then complex eigenvectors and eigenvalues occur in conjugate pairs $\mathbf{p}, \overline{\mathbf{p}}$ and $r, \overline{r}$. The complex-valued columns

$$\mathbf{Y}(t) = \left(\cdots \mathbf{p}e^{rt} \quad \overline{\mathbf{p}}e^{\overline{r}t} \cdots \right)$$

in the fundamental matrix can be replaced by the real columns

REAL-VALUED FUNDAMENTAL SOLUTIONS WHEN EIGENVECTORS AND EIGENVALUES ARE COMPLEX

$$\mathbf{Y}(t) = \left(\cdots \text{Re}(\mathbf{p}e^{rt}) \ \ \text{Im}(\mathbf{p}e^{rt}) \cdots \right).$$

If $\mathbf{B}$ does not have a complete set of linearly independent eigenvectors, then these ideas will not produce a fundamental matrix. However, we will not explore that possibility here.

8.4.3 Variation of Parameters

Given a fundamental matrix $\mathbf{Y}(t)$ for the homogeneous system $\mathbf{y}' = \mathbf{A}(t)\mathbf{y}$, we can find a particular solution of the nonhomogeneous equation

$$\mathbf{y}' = \mathbf{A}(t)\mathbf{y} + \mathbf{f}(t) \tag{8.17}$$

via *variation of parameters*.

Variation of parameters assumes a particular solution can be written by replacing the constant vector $\mathbf{c}$ in the homogeneous general solution $\mathbf{Y}(t)\mathbf{c}$ with a variable vector $\mathbf{u}(t)$,

$$\mathbf{y}_p(t) = \mathbf{Y}(t)\mathbf{u}(t),$$

VARIATION OF PARAMETERS FORM OF PARTICULAR SOLUTION

where $\mathbf{u}(t)$ is to be determined by substituting into the nonhomogeneous system (8.17).

Substituting this assumed form of $\mathbf{y}_p(t)$ yields

$$\mathbf{y}_p' = \mathbf{Y}'(t)\mathbf{u}(t) + \mathbf{Y}(t)\mathbf{u}'(t) = \mathbf{A}(t)\mathbf{Y}(t)\mathbf{u}(t) + \mathbf{f}(t).$$

Now each of the columns of the fundamental matrix $\mathbf{Y}(t)$ is a solution of $\mathbf{y}' = \mathbf{A}\mathbf{y}$; that is, $\mathbf{Y}' = \mathbf{A}\mathbf{Y}$. So the previous expression reduces to

$$\mathbf{Y}(t)\mathbf{u}'(t) = \mathbf{f}(t),$$

EQUATION DEFINING u'

the basic equation of variation of parameters, the equation defining the derivative of the unknown vector function $\mathbf{u}(t)$. We need only solve this system of equations for the components of $\mathbf{u}'$, integrate to obtain $\mathbf{u}$ (often a difficult chore), and write $\mathbf{y}_p(t) = \mathbf{Y}(t)\mathbf{u}(t)$ for the particular solution.

Since we have learned to solve only constant-coefficient, homogeneous systems, we limit ourselves to constant-coefficient, nonhomogeneous systems as well.

EXAMPLE 19   *Find a particular solution of* $\mathbf{y}'(t) = \mathbf{B}\mathbf{y}(t) + \mathbf{f}(t)$, *where*

$$\mathbf{B} = \begin{pmatrix} -4 & 2 \\ -3 & 3 \end{pmatrix}, \qquad \mathbf{f}(t) = \begin{pmatrix} 10 \\ 5t \end{pmatrix}.$$

From example 14, a fundamental matrix for this system is

FUNDAMENTAL MATRIX

$$\mathbf{Y}(t) = \begin{pmatrix} 2e^{-3t} & e^{2t} \\ e^{-3t} & 3e^{2t} \end{pmatrix}.$$

Then using the basic equation of variation of parameters, $\mathbf{Y}(t)\mathbf{u}'(t) = \mathbf{f}(t)$, we can write

EQUATION FOR u′

$$\begin{pmatrix} 2e^{-3t} & e^{2t} \\ e^{-3t} & 3e^{2t} \end{pmatrix} \begin{pmatrix} u_1'(t) \\ u_2'(t) \end{pmatrix} = \mathbf{f}(t) = \begin{pmatrix} 10 \\ 5t \end{pmatrix}.$$

Solving this system by Cramer's rule requires calculation of the determinant of coefficients, which is just the Wronskian

$$W = \det \begin{pmatrix} 2e^{-3t} & e^{2t} \\ e^{-3t} & 3e^{2t} \end{pmatrix} = 5e^{-t}.$$

Solve for **u′**.

Then

$$u_1' = \frac{1}{W} \det \begin{pmatrix} 10 & e^{2t} \\ 5t & 3e^{2t} \end{pmatrix} = 6e^{3t} - te^{3t},$$

$$u_2' = \frac{1}{W} \det \begin{pmatrix} 2e^{-3t} & 10 \\ e^{-3t} & 5t \end{pmatrix} = 2te^{-2t} - 2e^{-2t}.$$

Integrate.

Integrating yields

$$u_1 = \left(\frac{19}{9} - \frac{t}{3} \right) e^{3t},$$

$$u_2 = \left(t - \frac{1}{2} \right) e^{-2t}.$$

Write **y**_p.

The particular solution is

$$\mathbf{y}_p = \mathbf{Y}(t)\mathbf{u} = \begin{pmatrix} 2e^{-3t} & e^{2t} \\ e^{-3t} & 3e^{2t} \end{pmatrix} \begin{pmatrix} u_1(t) \\ u_2(t) \end{pmatrix} = \begin{pmatrix} \frac{85}{18} - \frac{5t}{3} \\ \frac{65}{18} - \frac{10t}{3} \end{pmatrix}. \qquad \blacksquare$$

Clearly, the integrations required by variation of parameters can become formidable. Much of the value of this method lies in the representation formula it provides for a particular solution.

To obtain such a representation, apply the fundamental theorem of calculus to the basic equation $\mathbf{u}'(t) = \mathbf{Y}(t)^{-1}\mathbf{f}(t)$ to write

EXPRESSION FOR u

$$\mathbf{u}(t) = \int_a^t \mathbf{Y}(s)^{-1}\mathbf{f}(s)\, ds,$$

where a can be chosen arbitrarily. Hence, given a fundamental matrix $\mathbf{Y}(t)$, a general solution of $\mathbf{y}' = \mathbf{A}(t)\mathbf{y} + \mathbf{f}(t)$ is

$$\mathbf{y}_g(t) = \mathbf{Y}(t) \int_a^t \mathbf{Y}(s)^{-1}\mathbf{f}(s)\, ds + \mathbf{Y}(t)\mathbf{c}.$$

GENERAL SOLUTION OF
$\mathbf{y}' = \mathbf{A}(t)\mathbf{y} + \mathbf{f}(t)$

The first term is the particular solution of the nonhomogeneous equation, and the second term is the general solution of the corresponding homogeneous system $\mathbf{y}' = \mathbf{A}(t)\mathbf{y}$.

Of course, the method of undetermined coefficients extends to constant-coefficient matrix systems. For example, if the forcing term includes e^t, then a particular solution would have the form $\mathbf{p}e^t$ for some undetermined coefficient vector $\mathbf{p}$. If $\mathbf{p}e^t$ could solve the homogeneous system, then the appropriate form for a particular solution is $\mathbf{p}_1 t e^t + \mathbf{p}_2$, where the vectors $\mathbf{p}_i$ are again determined by substituting in the nonhomogeneous system. See exercise 26 in this section as well as exercise 29 of the chapter exercises for chapter 8.

8.4.4 Exercises

EXERCISE GUIDE

To gain experience . . .	Try exercises
Finding eigenvalues and eigenvectors	1–6, 7–9
Finding fundamental matrices	1–6, 11
Using variation of parameters to find particular solutions	15–20
Finding general solutions	1–6, 17–21
Solving initial-value problems	1–6, 22
Obtaining real-valued solutions from complex eigenvalues	5, 10–11
With the foundations of variation of parameters	13–14
With the relation between characteristic equations and eigenvalues	23
Analyzing behavior of solutions	24
With eigenvalue ideas	24–25
Extending undetermined coefficients to systems	26

In exercises 1–6, for the given matrix **B** . . .

i. Find the eigenvalues and associated eigenvectors of **B**,

ii. Find linearly independent solutions of $\mathbf{y}' = \mathbf{B}\mathbf{y}$,

iii. Find a fundamental matrix of $\mathbf{y}' = \mathbf{B}\mathbf{y}$,

iv. Write a general solution of $\mathbf{y}' = \mathbf{B}\mathbf{y}$,

v. Find a solution of $\mathbf{y}' = \mathbf{B}\mathbf{y}$, $\mathbf{y}(0) = (\,4 \quad -4\,)^T$.

1. $\mathbf{B} = \begin{pmatrix} 1 & 0 \\ 4 & 3 \end{pmatrix}$

2. $\mathbf{B} = \begin{pmatrix} -2 & 0 \\ 2 & -1 \end{pmatrix}$

3. $\mathbf{B} = \begin{pmatrix} 11 & 2 \\ 1 & 10 \end{pmatrix}$

4. $\mathbf{B} = \begin{pmatrix} 1 & 2 \\ -1 & 1 \end{pmatrix}$

5. $\mathbf{B} = \begin{pmatrix} 4 & 3 \\ -3 & 4 \end{pmatrix}$

6. $\mathbf{B} = \begin{pmatrix} \frac{3}{2} & \frac{\sqrt{3}}{2} \\ \frac{-\sqrt{3}}{2} & \frac{3}{2} \end{pmatrix}$

7. Verify the details omitted from example 18 by showing that $\mathbf{p}_2 = (\,1 \quad -4i\,)^T$ is indeed an eigenvector of the matrix

$$\mathbf{B} = \begin{pmatrix} 0 & -1 \\ 16 & 0 \end{pmatrix}$$

associated with the eigenvalue $r_2 = -4i$; that is, verify that $\mathbf{B}\mathbf{p}_2 = -4i\mathbf{p}_2$.

8. Show that the matrix

$$\mathbf{B} = \begin{pmatrix} 1 & 1 \\ 0 & 1 \end{pmatrix}$$

does not have two linearly independent eigenvectors. Find the (single) eigenvalue of $\mathbf{B}$ and show that any eigenvector $\mathbf{p} = (\,p \quad q\,)^T$ associated with it must have $q = 0$. Hence, every eigenvector of this matrix is a scalar multiple of the vector $(\,1 \quad 0\,)^T$.

9. Show that the diagonal matrix

$$\mathbf{B} = \begin{pmatrix} a & 0 \\ 0 & a \end{pmatrix},$$

$a \neq 0$, has two linearly independent eigenvectors even though it has the single repeated eigenvalue $r = a$. Show that the two independent eigenvectors can be chosen to be multiples of $(\,1 \quad 0\,)^T$ and $(\,0 \quad 1\,)^T$.

10. Using $\bar{z} = \alpha - i\beta$, verify the claim made in the text:

If $z = \alpha + i\beta$ is a complex quantity, then

$$\frac{z + \bar{z}}{2} = \alpha = \mathrm{Re}\ z, \quad \frac{z - \bar{z}}{2i} = \beta = \mathrm{Im}\ z.$$

11. Verify the result of example 18 that

$$\mathbf{Y}(t) = \begin{pmatrix} \cos 4t & \sin 4t \\ -4\sin 4t & 4\cos 4t \end{pmatrix} \mathbf{c}.$$

is a fundamental matrix for the system $\mathbf{y}' = \mathbf{B}\mathbf{y}$ where

$$\mathbf{B} = \begin{pmatrix} 0 & -1 \\ 16 & 0 \end{pmatrix}.$$

Also show that $\mathbf{Y}' = \mathbf{B}\mathbf{Y}$.

12. The text shows that the unknown vector function $\mathbf{u}(t)$ in the variation of parameters solution $\mathbf{y}_p(t) = \mathbf{Y}(t)\mathbf{u}(t)$ must satisfy

$$\mathbf{Y}(t)\mathbf{u}'(t) = \mathbf{f}(t).$$

Show that in the case of a 2×2 system, solving this set of equations by Cramer's rule leads to exactly the same equations (8.13) for u_1', u_2' as in the text's earlier treatment of variation of parameters.

13. In example 14, the text constructs a fundamental matrix for the system

$$\mathbf{y}'(t) = \begin{pmatrix} -4 & 2 \\ -3 & 3 \end{pmatrix} \mathbf{y}(t)$$

using solutions derived in a previous section. Instead, find the eigenvalues and eigenvectors of the coefficient matrix and use them to construct a fundamental matrix. How does your fundamental matrix compare with the one obtained in example 14? Must the two fundamental matrices be identical?

Use your fundamental matrix to find the solution of this system subject to the initial condition $\mathbf{y}(0) = (\,1 \quad -2\,)^T$.

14. The text shows in general that substituting the particular solution form $\mathbf{y}_p(t) = \mathbf{Y}(t)\mathbf{u}(t)$ into $\mathbf{y}' = \mathbf{A}\mathbf{y} + \mathbf{f}$ leads to the system of equations

$$\mathbf{Y}(t)\mathbf{u}'(t) = \mathbf{f}(t)$$

for the derivative of the unknown vector function $\mathbf{u}(t)$. Verify this claim in the following special case by substituting directly into the system $\mathbf{y}' = \mathbf{A}\mathbf{y} + \mathbf{f}$ considered in example 19, where

$$\mathbf{A} = \begin{pmatrix} -4 & 2 \\ -3 & 3 \end{pmatrix}, \quad \mathbf{f}(t) = \begin{pmatrix} 10 \\ 5t \end{pmatrix}.$$

A fundamental matrix for this system is

$$\mathbf{Y}(t) = \begin{pmatrix} 2e^{-3t} & e^{2t} \\ e^{-3t} & 3e^{2t} \end{pmatrix}.$$

In exercises 15–16, use the given solutions $\mathbf{y}_1, \mathbf{y}_2$ of the homogeneous equation to construct a particular solution of the nonhomogeneous equation.

15. $\mathbf{y}_1 = \begin{pmatrix} -e^{4t} \\ e^{4t} \end{pmatrix}$, $\mathbf{y}_2 = \begin{pmatrix} e^{-4t} \\ e^{-4t} \end{pmatrix}$,

$$\mathbf{y}' = \begin{pmatrix} 0 & -4 \\ -4 & 0 \end{pmatrix} \mathbf{y} + \begin{pmatrix} 4t \\ -2e^{2t} \end{pmatrix}$$

16. $\mathbf{y}_1 = \begin{pmatrix} e^t \cos t \\ -e^t \sin t \end{pmatrix}$, $\mathbf{y}_2 = \begin{pmatrix} e^t \sin t \\ e^t \cos t \end{pmatrix}$,

$$\mathbf{y}' = \begin{pmatrix} 1 & 1 \\ -1 & 1 \end{pmatrix} \mathbf{y} + \begin{pmatrix} 2e^t \\ -e^t \end{pmatrix}$$

In exercises 17–20, find a general solution of the given system. If you can use work from earlier exercises, do so

17. $\mathbf{y}' = \begin{pmatrix} 1 & 0 \\ 4 & 3 \end{pmatrix} \mathbf{y} + \begin{pmatrix} 4e^{-2t} \\ 6e^{2t} \end{pmatrix}$

18. $\mathbf{y}' = \begin{pmatrix} -2 & 0 \\ 2 & 1 \end{pmatrix} \mathbf{y} + \begin{pmatrix} -6t \\ 2e^{3t} \end{pmatrix}$

19. $\mathbf{y}' = \begin{pmatrix} 11 & 2 \\ 1 & 10 \end{pmatrix} \mathbf{y} + \begin{pmatrix} 162 \\ -108t \end{pmatrix}$

20. $\mathbf{y}' = \begin{pmatrix} 1 & 2 \\ -1 & 1 \end{pmatrix} \mathbf{y} + \begin{pmatrix} 4\sqrt{2}e^{2t} \\ -4\sqrt{2}e^{-2t} \end{pmatrix}$

21. Verify that $\mathbf{Y}(t)\mathbf{c}$ is, in fact, a linear combination of the columns of $\mathbf{Y}(t)$ in the 2×2 case $\mathbf{Y} = (\begin{array}{cc} \mathbf{y}_1 & \mathbf{y}_2 \end{array})$, where

$$\mathbf{y}_i = \begin{pmatrix} y_i \\ z_i \end{pmatrix}, \qquad \mathbf{c} = \begin{pmatrix} c_1 \\ c_2 \end{pmatrix}.$$

Carry out the matrix multiplication and show that $\mathbf{Y}(t)\mathbf{c} = c_1\mathbf{y}_1 + c_2\mathbf{y}_2$.

22. Use the general solution expression

$$\mathbf{y}_g(t) = \mathbf{Y}(t) \int_a^t \mathbf{Y}(s)^{-1}\mathbf{f}(s)\, ds + \mathbf{Y}(t)\mathbf{c}$$

for the system $\mathbf{y}' = \mathbf{A}(t)\mathbf{y} + \mathbf{f}(t)$ to write a solution of this equation subject to the initial condition $\mathbf{y}(0) = \mathbf{y}_i$. (Choose a to simplify the integral at $t = 0$.) How would the expression for $\mathbf{c}$ simplify if $\mathbf{Y}(0) = \mathbf{I}$?

23. Write the characteristic equation of the scalar second-order equation $b_2 y'' + b_1 y' + b_0 y = 0$, b_i constants. Now write this equation as a first-order system and find the characteristic equation of the corresponding (constant) coefficient matrix. Show that both characteristic equations are identical.

24. Suppose you have conducted a linear stability analysis of an equilibrium point of a system of nonlinear differential equations. You have found that the perturbations $\mathbf{y}(t)$ of that equilibrium satisfy the constant coefficient system $\mathbf{y}' = \mathbf{B}\mathbf{y}$, where $\mathbf{B}$ has a complete set of linearly independent eigenvectors. By examining the general solution of this system, justify the following statements:
  (a) The equilibrium is stable if the real part of *every* eigenvalue of $\mathbf{B}$ is negative.
  (b) The equilibrium is unstable if the real part of *any* eigenvalue of $\mathbf{B}$ is positive.
  (c) The equilibrium is neutrally stable if the real part of every eigenvalue of $\mathbf{B}$ is zero.

25. Given an eigenvalue r of the matrix $\mathbf{B}$, why can Cramer's rule not be used to solve $(\mathbf{B} - r\mathbf{I})\mathbf{p} = 0$ for the components of the eigenvector $\mathbf{p}$?

26. Formulate an extension of the method of undetermined coefficients to systems and apply it to exercises 17–20, as assigned.

8.5   CHAPTER EXERCISES

EXERCISE GUIDE

To gain experience . . .	Try exercises
Generalizing concepts and methods to higher-order equations	1, 6, 11, 22–27
Solving homogeneous systems	2–5, 17–22, 24–27
Finding particular solutions by variation of parameters	7–10, 12–16, 23
Finding general solutions	2–5, 7–10
Solving initial-value problems	2–5, 7–10
Comparing with numerical solutions	28
Extending undetermined coefficients to systems	29

1. Provide a definition for each of the following terms for

   i. A system of three first-order equations,

   ii. A system of n first-order equations,

   that is consistent with the definition given in the text for a system of two equations.
   - **(a)** Solution
   - **(b)** Homogeneous system, nonhomogeneous system,
   - **(c)** Linear system, nonlinear system,
   - **(d)** Constant-coefficient linear system, variable-coefficient linear system,
   - **(e)** Linearly independent solutions, linearly dependent solutions,
   - **(f)** General solution of homogeneous linear system, particular solution of nonhomogeneous linear system, general solution of nonhomogeneous linear system.

In exercises 2–5,

   i. Find a general solution of the given system.

   ii. Find a solution of the system that satisfies the given initial conditions.

2. $y' = z, \ y(0) = 0$
$z' = y, \ z(0) = 1$

3. $y' = -4y + 2z, \ y(0) = 3$
$z' = -3y + 3z, \ z(0) = 4$

4. $y' = -4z, \ y(0) = 0$
$z' = -4y, \ z(0) = 2$

5. $y' = y + z, \ y(0) = 1$
$z' = -y + z, \ z(0) = 2$

6. Begin a generalization of the characteristic equation method to a homogeneous system of three constant-coefficient linear equations by completing the following steps.
   - **(a)** Write a generic homogeneous system of three constant-coefficient linear equations.
   - **(b)** Postulate a form for the solution of this system involving the exponential factor e^{rt}.
   - **(c)** Substitute your assumed solution form into the system of differential equations. Find a set of three linear algebraic equations involving r and the coefficients in your proposed solution.
   - **(d)** Use Cramer's rule to find a condition that guarantees that the three equations you found in the previous part have a nontrivial solution. State the *characteristic equation* for this system.
   - **(e)** If the coefficients in your system of three differential equations are real, then the characteris-

tic equation will be a cubic equation in r with real coefficients. What sorts of roots can such an equation have? Ignoring repeated roots, speculate on the types of solutions such roots would generate.

In exercises 7–10,

   i. Find two linearly independent solutions (y_1, z_1), (y_2, z_2) of the homogeneous system.

   ii. Begin a variation of parameters solution of the nonhomogeneous system by finding u_1, u_2 in the assumed form $(y_p, z_p) = u_1(t)(y_1, z_1) + u_2(t)(y_2, z_2)$. If necessary, express antiderivatives as definite integrals using the fundamental theorem of calculus.

   iii. Complete the variation of parameters solution by writing (y_p, z_p).

   iv. Write a general solution of the nonhomogeneous system.

   v. Find the solution of the nonhomogeneous system that satisfies the given initial conditions.

If a homogeneous solution is available from some other source (e.g., a previous exercise), use it rather than re-derive it.

7. $y' = y/2 + \sqrt{3}\,z/2 + 2t, \ y(0) = 6$
$z' = \sqrt{3}\,y/2 - z/2 - 4t^2, \ z(0) = 4$

8. $y' = z + 4\cos t, \ y(0) = 1$
$z' = y, \ z(0) = 0$

9. $y' = y + 3z + e^{2t}, \ y(0) = 1$
$z' = 3y + z - 1, \ z(0) = 0$

10. $y' = y - z - 2, \ y(0) = 1$
$z' = 2y + 4z - 2e^{-2t}, \ z(0) = 0$

11. Describe an extension of variation of parameters to a system of *three* linear, first-order equations. Find the system of equations that define u'_1, u'_2, u'_3. (You need not solve for u'_1, u'_2, u'_3. Could you if you had to?)

Find a general solution of the systems given in exercises 12–27.

12. $\mathbf{y}' = \begin{pmatrix} 5 & -1 \\ -1 & 5 \end{pmatrix} \mathbf{y} + \begin{pmatrix} -12 \\ 2e-6t \end{pmatrix}$

13. $\mathbf{y}' = \begin{pmatrix} -7 & -3 \\ 6 & 2 \end{pmatrix} \mathbf{y} + \begin{pmatrix} 6t \\ -9 \end{pmatrix}$

14. $\mathbf{y}' = \begin{pmatrix} 11 & 6 \\ -30 & -16 \end{pmatrix} \mathbf{y} + \begin{pmatrix} 2e^t \\ -2 \end{pmatrix}$

15. $\mathbf{y}' = \begin{pmatrix} -9 & -4 \\ 20 & 9 \end{pmatrix} \mathbf{y} + \begin{pmatrix} e^{-t} \\ -e^t \end{pmatrix}$

16. $\mathbf{y'} = \begin{pmatrix} 0 & -4 \\ 4 & 0 \end{pmatrix} \mathbf{y} + \begin{pmatrix} \cos 4t \\ -\sin 4t \end{pmatrix}$

17. $\mathbf{y'} = \begin{pmatrix} -4 & -4 \\ 8 & 4 \end{pmatrix} \mathbf{y}$

18. $\mathbf{y'} = \begin{pmatrix} 1 & 1 \\ -2 & -1 \end{pmatrix} \mathbf{y}$

19. $\mathbf{y'} = \begin{pmatrix} -17 & -13 \\ 18 & 13 \end{pmatrix} \mathbf{y}$

20. $\mathbf{y'} = \begin{pmatrix} -8 & -5 \\ 9 & 4 \end{pmatrix} \mathbf{y}$

21. $\mathbf{y'} = \begin{pmatrix} -3 & -2 \\ 4 & 1 \end{pmatrix} \mathbf{y}$

22. $\mathbf{y'} = \begin{pmatrix} 2 & 2 & 0 \\ 2 & 1 & 1 \\ -7 & 2 & 3 \end{pmatrix} \mathbf{y}$

23. $\mathbf{y'} = \begin{pmatrix} 2 & 2 & 0 \\ 2 & 1 & 1 \\ -7 & 2 & 3 \end{pmatrix} \mathbf{y} + \begin{pmatrix} 0 \\ 0 \\ -1 \end{pmatrix}$

24. $\mathbf{y'} = \begin{pmatrix} 1 & 0 & -2 \\ 2 & 2 & 4 \\ 0 & 0 & 2 \end{pmatrix} \mathbf{y}$

25. $\mathbf{y'} = \begin{pmatrix} 0 & 1 & 0 \\ 0 & 0 & 1 \\ -12 & 13 & 0 \end{pmatrix} \mathbf{y}$

26. $\mathbf{y'} = \begin{pmatrix} 0 & 1 & 0 \\ 0 & 0 & 1 \\ 12 & 16 & 3 \end{pmatrix} \mathbf{y}$

27. $\mathbf{y'} = \begin{pmatrix} 8 & 2 & 12 \\ 5 & 2 & 9 \\ -5 & -1 & -8 \end{pmatrix} \mathbf{y}$

28. Choose a homogeneous system and a nonhomogeneous system from the exercises at the end of section 8.2 and 8.3. Solve each system, and confirm the correctness of your analytic solution by comparing it with a numerical solution obtained using available software and a method of your choice.

29. Formulate an extension of the method of undetermined coefficients to systems and apply it to exercises 12–16 and 23, as assigned.

8.6 CHAPTER PROJECTS

1. **Linear systems of order** n Extend the basic definitions and solution methods of this chapter to systems of n linear, constant-coefficient differential equations. Illustrate your ideas on some systems of three equations.

2. **Linear equations of order** n Propose a reasonable definition for a linear differential equation of order n. Show that any such equation can be written as an equivalent system of n linear, first-order differential equations. Describe an extension of the solution methods of this chapter to such equations. Illustrate them on a fourth-order example.

3. **Stability analysis of predator-prey system** Show that a linear stability analysis of the equilibrium points of the predator-prey equations (see section 5.4)

$$F' = -(d_F - \alpha R)F$$
$$R' = (b_R - \beta F)R$$

leads to a system of constant-coefficient, linear, homogeneous equations; i.e., if (F_{ss}, R_{ss}) is an equilibrium point of this system and $(F_{ss} + p_F(t), R_{ss} + p_R(t))$ is a nearby solution, then (p_F, p_R) is (approximately) the solution of a linear system.

Analyze the behavior of perturbations of each equilibrium point, and explain the physical significance of your results. Sketch the behavior of the perturbations on a phase plane plot of F versus R.

You will find that the equilibrium at the origin is a *saddle point*, a point at which some solutions increase while others decrease. Can some perturbations return to the origin? On the average, do perturbations of this equilibrium return to the origin? Would you characterize a saddle point as stable or unstable?

Do you find any shortcomings in this model?

4. **Equilibria of autonomous system** Suppose (y_0, z_0) is a steady state of the autonomous system

$$y' = f(y, z)$$
$$z' = g(y, z).$$

Show that a linear stability analysis of a perturbation $(y(t) + p_y(t), z(t) + p_z(t))$ of this equilibrium point leads to the constant-coefficient linear system

$$p'_y = f_y(y_0, z_0)p_y + f_z(y_0, z_0)p_z$$
$$p'_z = g_y(y_0, z_0)p_y + g_z(y_0, z_0)p_z.$$

Here $f_y = \partial f / \partial y$, etc. (*Hint:* Since (y_0, z_0) is an equilibrium point, $f(y_0, z_0) = \cdots$. Use a Taylor expansion.)

Find conditions on the partial derivatives f_y, f_z, g_y, g_z which guarantee that this equilibrium is stable, neutrally stable, or unstable. Does your analysis cover all possibilities? Are there other cases that you might reasonably classify as unstable because either p_y or p_z could grow?

5. **Existence theorem for systems** For a system of two equations, state and prove an existence theorem that is an extension of the existence theorem of section 3.9.2. Show carefully how the bounding arguments in the proof of the theorem for a single equation must be modified to account for two equations.

9

DIFFUSION MODELS AND BOUNDARY-VALUE PROBLEMS

Here the boundaries meet and all contradictions exist side by side.
Fedor Mikhailovich Dostoevski

Boundary-value problems give the data needed to specify a solution uniquely at two different points, not just at a single initial point. They arise naturally in models of diffusion, among many other important applications.

9.1 DIFFUSION MODELS

Diffusion is the process that causes a dark drop of ink to spread from one end of a narrow pan of water to the other until all of the water is uniformly tinted by the diluted droplet. Likewise, diffusion is the process that carries heat through the walls of a house and pollution from one side of a quiet lake to another. It is central to many equilibrium phenomena in the natural world.

The rate at which a blob of ink spreads through the water is determined by the change in concentration of ink between adjacent points in the fluid. When the ink is uniformly spread through the water, then there is no change in concentration of ink from one point to another. Diffusion is effectively halted, as shown in the right side of figure 9.1. When the droplet of ink first enters the water, there is a sharp concentration change at the boundary between the ink and the water. The ink diffuses rapidly toward the clear water.

The natural measure of the change in concentration from point to point is the derivative of concentration with respect to position, the **concentration gradient**. If $c(x)$ is the concentration at point x, then the concentration gradient at x is $c'(x)$. Since diffusion carries material from high concentration toward low concentration, the diffusive flow will be in the *positive* x direction if c' is *negative* and vice versa.

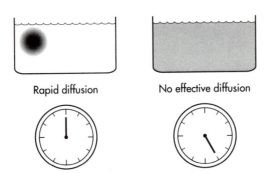

Figure 9.1: When a drop of ink first enters the water, it diffuses rapidly at the boundary of the droplet, where the concentration gradient is greatest. Later, as shown on the right, diffusion is effectively halted because the concentration is uniform and the concentration gradient is zero.

One way to characterize diffusion is to find the rate of flow of ink (or salt or any other dissolved substance) through an imaginary unit surface placed in the fluid. The experimental fact that governs diffusion uses that idea.

> **Experimental Fact** The rate of diffusion of a substance dissolved in a liquid through a unit area is proportional to the negative of its *concentration gradient*.

The constant of proportionality is the **diffusion coefficient**. If $c(x)$ denotes concentration, then this experimental fact can be written:

Flow rate due to diffusion through unit area at x is $-Dc'(x)$.

This idea is illustrated in figure 9.2.

> This experimental fact is sometimes known as Fick's law, but it is a "law" in the same sense as Hooke's law, which relates force in a spring to its extension. If the quantity that is diffusing is heat energy, then it is called Fourier's law.

Let us use these ideas to model the transfer of salt from the ocean to an estuary through a tidal inlet, as illustrated in figure 9.3. The ocean on the left contains salt in a concentration of c_0 mass per unit volume (say, g/m^3). Tidal flow can carry the salty water through the connecting channel at a velocity of $V(x)$ m/s into the fresher water on the right, where the salt concentration is c_L. Even without the flow of saltwater, diffusion would carry salt from the ocean through the channel toward the lower concentration in the estuary, just as the ink droplet in water spreads, seeking lower concentrations of itself.

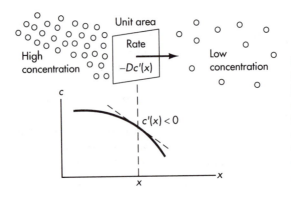

Figure 9.2: An illustration of the experimental fact governing diffusion. Ink (or any other dissolved substance) will diffuse from left to right because the concentration gradient is negative at the point shown; concentration is higher on the left than on the right.

9 Diffusion Models and Boundary Value Problems

We want to formulate a model that will describe the equilibrium distribution of salt in the connecting channel. The model should incorporate transport of salt due both to flow of water in the channel and to diffusion through the water in the channel. The governing law is that of conservation of mass (of salt, in this case).

Law of Conservation of Mass   The net change in mass over a given period of time is the mass added less the mass removed.

Suppose we focus our attention on a section of the channel between the coordinates x and $x + \Delta x$, as shown in figure 9.4. Since our concern is the equilibrium concentration of salt in this imaginary box, or *control volume*, the net change in the mass of salt in this box over time is zero. Consequently, the conservation law tells us:

At equilibrium, the rate of flow *into* the control volume equals the rate of flow *out of* the control volume.

The net volume of water per unit time that flows through either wall is the product of the area of the wall and the flow rate of the water in the channel, AV; see figure 9.5. The total rate at which *salt* enters due to this flow is just the product of the volume flow rate of water with the concentration of salt in the water, AVc. Hence, due to the flow in the channel, salt *enters* from the left at the rate

$$A(x)V(x)c(x)$$

and it *leaves* from the right at the rate

$$A(x + \Delta x)V(x + \Delta x)c(x + \Delta x).$$

(Our *enter* and *leave* labels suppose the water is flowing from left to right, as drawn in figure 9.4.)

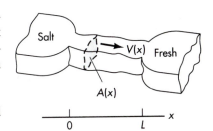

Figure 9.3: A schematic diagram of a channel connecting two bodies of water, salt on the left and fresh on the right. The length of the connecting channel is L. The area of the cross section at coordinate x is $A(x)$.

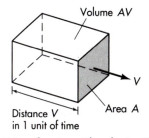

Figure 9.5: If water with velocity V flows through an area A, then in 1 unit of time a volume AV of water will pass.

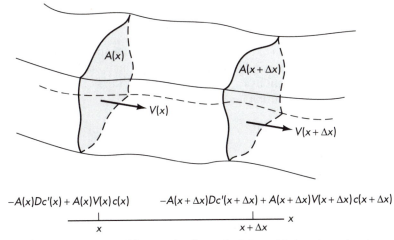

$$-A(x)Dc'(x) + A(x)V(x)c(x) \qquad -A(x + \Delta x)Dc'(x + \Delta x) + A(x + \Delta x)V(x + \Delta x)c(x + \Delta x)$$

Figure 9.4: At equilibrium, the flow of salt into this imaginary box in the channel equals the net flow out. This picture is drawn as if salt flows in through the left wall and out through the right wall.

From the experimental fact, the rate of transport of salt *through a unit area* due to diffusion is $-Dc'$. Hence, the net rate of diffusion *in* through the left wall of the control volume is

$$-A(x)Dc'(x),$$

and the net rate *out* through the right wall is

$$-A(x + \Delta x)Dc'(x + \Delta x).$$

As figure 9.4 illustrates, the *total* flow rate *in* through the left is

$$-A(x)Dc'(x) + A(x)V(x)c(x),$$

and the *total out* through the right is

$$-A(x + \Delta x)Dc'(x + \Delta x) + A(x + \Delta x)V(x + \Delta x)c(x + \Delta x).$$

Since the mass of salt in the control volume cannot change over time at equilibrium, conservation of mass demands that the rate of salt flow in equal the rate of salt flow out:

Rate in = rate out.

$$-A(x)Dc'(x) + A(x)V(x)c(x) = -A(x + \Delta x)Dc'(x + \Delta x)$$
$$+A(x + \Delta x)V(x + \Delta x)c(x + \Delta x).$$

Collecting terms and dividing by Δx lead to

$$-D\frac{A(x + \Delta x)c'(x + \Delta x) - A(x)c'(x)}{\Delta x}$$

$$+\frac{A(x + \Delta x)V(x + \Delta x)c(x + \Delta x) - A(x)V(x)c(x)}{\Delta x} = 0.$$

In the limit as Δx goes to zero, we obtain the second-order differential equation

$$-D(A(x)c'(x))' + (A(x)V(x)c(x))' = 0.$$

We expect a solution of this second-order equation to contain two arbitrary constants. We need two conditions to determine them, and the natural ones to use are the concentrations c_0, c_L at the inlet and the outlet of the channel. Hence, our complete *diffusion model* is

DIFFUSION MODEL

$$-D(A(x)c'(x))' + (A(x)V(x)c(x))' = 0, \qquad c(0) = c_0, \qquad c(L) = c_L.$$

The diffusion model is an example of a *boundary-value problem*, and the auxiliary conditions

$$c(0) = c_0, \qquad c(L) = c_L$$

are called *boundary conditions*. We solve boundary-value problems as you might expect: Find a general solution of the differential equation and use the boundary conditions to determine the arbitrary constants. We carry out this process in the next section.

9 Diffusion Models and Boundary Value Problems

EXAMPLE 1 *Suppose the body of water on the left in figure 9.3 is the ocean, which has a salt concentration of about 28 g/m³, and the body of water on the right is a large freshwater lake. The channel is approximately semicircular in cross section with a diameter of 10 m, and the water in it is flowing toward the lake at 1 km/h. The channel is 4 km long. Write a model for the salt concentration in the channel.*

A little library research reveals that the diffusion coefficient of salt in water is about $D = 1.09 \times 10^{-9}$ m²/s. Adopting meters and seconds as our units of length and time, we find that

$$V = 1 \text{ km/h} = 0.2778 \text{ m/s}.$$

The length of the channel is $L = 4,000$ m.

Since the area of the channel is constant, it can be canceled from each of the terms in the diffusion equation $-D(A(x)c'(x))' + (A(x)V(x)c(x))' = 0$, leaving

$$-Dc'' + Vc' = 0,$$

or, when we incorporate the values for D, V,

$$-1.09 \times 10^{-9} c'' + 0.2778 c' = 0.$$

The boundary value at $x = 0$ is determined by the concentration in the ocean. Hence, $c_0 = 28$. The concentration of salt in the lake is claimed to be zero, suggesting $c_L = 0$. You can reasonably argue that there must be some salt at the outlet of the channel, even if the lake is so large that it eventually dilutes what enters to vanishingly small proportions. Nonetheless, we will take the simplifying approach that the salt entering the lake from the channel disperses immediately. Then we can take $c(L) = c(4,000) = 0$.

The complete model with boundary conditions is

$$-1.09 \times 10^{-9} c'' + 0.2778 c' = 0, \qquad c(0) = 28, \qquad c(4,000) = 0. \qquad \blacksquare$$

EXAMPLE 2 *A quiet circular pond (figure 9.6) has radius L and depth d. Measurements at several points around its perimeter show that the concentration of salt is c_L. Find a model for the concentration of salt in the pond.*

The exercises let you tackle this modeling job from scratch. Here we take a shortcut.

Since the concentration of salt measured at the edge of the pond does not vary as we move around the pond, it seems safe to assume that concentration varies only with distance from the center of the pond. Because salt can flow only in the radial direction, away from or toward the center, the appropriate control volume is bounded by two concentric cylinders, as shown in figure 9.7. Any salt reaching the boundary must pass through those cylinders. The inner cylinder has radius x and height d. Hence, the surface area of its side is $A(x) = 2\pi x d$, and that is precisely the area term in the diffusion equation. Since there is no current, $V = 0$.

Using $A = 2\pi x d$ in the diffusion equation, we have

$$-D(2\pi x d c'(x))' = 0.$$

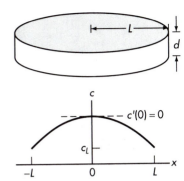

Figure 9.6: A circular pond of radius L and depth d. To be symmetric across the center of the pond, any hypothetical concentration profile like the one shown must have $c'(0) = 0$.

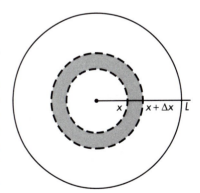

Figure 9.7: The pond of figure 9.6 viewed from above showing the tops of the concentric cylinders of radius x and $x+\Delta x$ that bound the control volume.

After simplifying, the governing equation is

$$xc''(x) + c'(x) = 0, \qquad c(0) = c_0, \qquad c(L) = c_L.$$

The boundary condition at the perimeter of the pond is easy: $c(L) = c_L$. The center is more difficult because the salt concentration there is unknown.

However, we do expect the concentration profile to be symmetric across the center of the pond, as suggested by the hypothetical concentration graph shown in figure 9.6. If the concentration function $c(x)$ is to be symmetric about $x = 0$ and have a continuous first derivative, then $c'(0) = 0$. This is the second boundary condition we sought.

The complete model for axially symmetric diffusion in a circle is

$$xc''(x) + c'(x) = 0, \qquad c'(0) = 0, \qquad c(L) = c_L.$$

The outer boundary condition comes from measurement. The boundary condition at the center comes from symmetry. ∎

> When we solve this boundary-value problem in the next section, we will discover that the hypothetical concentration profile of figure 9.6 is incorrect. In fact, the concentration is constant: $c(x) \equiv c_L$. Of course, $c'(0) = 0$ is still satisfied.

Although we derived the diffusion model thinking of salt in water, we could have equally well dealt with any substance that moves by convection, the action of the flowing water, as well as by diffusion, the tendency to move down concentration gradients. Almost any small particle suspended in water behaves this way, including many forms of industrial and domestic pollutants. The same ideas apply to heat flow, too, a topic you can explore in the exercises.

EXAMPLE 3 *Suppose the quantity of interest in the channel of figure 9.3 is now the concentration of particles of an industrial waste that are small enough to diffuse according to the experimental fact. The body of water on the left in that figure is badly polluted; it has a concentration c_0 of the undesirable particles. To protect the other body of water, the right end of the channel has been dammed. No water is flowing in the channel. Derive a model for the concentration of the pollutant in the channel under these circumstances.*

The diffusion equation $-D(A(x)c'(x))' + (A(x)V(x)c(x))' = 0$ applies because flow of water in the channel and diffusion through the water are still the only means of transporting the pollutant, the concentration of which we are denoting by $c(t)$. However, $V = 0$ because no water is flowing in the channel.

Evidently $c(0) = c_0$ because water with that concentration of pollutant is at the left end of the channel.

At the right end of the channel, the boundary condition must reflect the fact that all flow of the pollutant is prohibited. Setting $V = 0$ has already eliminated flow due to the movement of the water. Since the flow of salt due to diffusion is given by $-Dc'$, we will have no diffusion if $c' = 0$. Hence, the boundary condition at $x = L$ is

$$c'(L) = 0,$$

and the complete model is the boundary value problem

$$-D(A(x)c'(x))' = 0, \qquad c(0) = c_0, \qquad c'(L) = 0. \qquad ∎$$

A no-flow boundary condition like $c'(L) = 0$ is known as a *Neumann* boundary condition or a boundary condition of the *second kind*. A boundary condition like $c(0) = c_0$ that specifies the value itself is a *Dirichlet* boundary condition or a boundary condition of the *first kind*.

EXAMPLE 4 *The bodies of water in figure 9.3 have been cleared of pollutant, but years of neglect have contaminated the bed and walls of the channel. They release pollutant at a concentration rate of $P(x)$ g/m³ · s into the water in the channel. Derive a model for the concentration of pollutant in the channel.*

If we return to the control volume of figure 9.4, we now have *three* ways to add pollutant to that imaginary box, transport of pollutant through the wall of the box by the flowing water, diffusion through the wall due to a nonzero concentration gradient, and the added effect of release into the volume of the box at the concentration rate $P(x)$. Since we are at equilibrium, conservation of mass still demands that the rate at which pollution enters the control volume equal the rate at which it exits.

Release of pollutant from the bed and walls occurs at a concentration rate of $P(x)$. Hence, the total mass of pollutant released into the control volume per unit time is the product of the box's volume with the concentration rate P. Since the area of a side wall of the control volume is $A(x)$ and its width is Δx, its volume is about $A(x)\Delta x$. Hence, a mass of $A(x)P(x)\Delta x$ of salt is added to the control volume per unit time by the polluted bed and walls of the channel.

Combining this source with the flow and diffusion terms shown in figure 9.4, our "rate in equals rate out" conservation equation is

$$-A(x)Dc'(x) + A(x)V(x)c(x) + A(x)P(x)\Delta x = -A(x + \Delta x)Dc'(x + \Delta x)$$
$$+A(x + \Delta x)V(x + \Delta x)c(x + \Delta x).$$

Collecting terms and dividing by Δx lead to

$$-D\frac{A(x + \Delta x)c'(x + \Delta x) - A(x)c'(x)}{\Delta x}$$

$$+\frac{A(x + \Delta x)V(x + \Delta x)c(x + \Delta x) - A(x)V(x)c(x)}{\Delta x} = A(x)P(x).$$

In the limit as Δx approaches zero, we have the *diffusion equation with source term*

$$-D(A(x)c'(x))' + (A(x)V(x)c(x))' = A(x)P(x).$$

Since the large bodies of water at either end of the channel are free of pollutant, we can suppose that any pollutant which enters them from the channel is dispersed completely so that the ends of the channel are effectively pollution free. Hence, reasonable boundary conditions are $c(0) = c(L) = 0$. The complete model is

$$-D(A(x)c'(x))' + (A(x)V(x)c(x))' = A(x)P(x), \qquad c(0) = c(L) = 0. \quad \blacksquare$$

EXERCISE GUIDE

To gain experience . . .	Try exercises
With concepts underlying the derivation of diffusion models	1–2
Classifying diffusion equations	4–5
Deriving material diffusion models	3, 6–7
Deriving thermal diffusion models	8–14

1. The text derives the model for the distribution of salt in the channel as if the higher concentration of salt were on the left. Reverse the place of the ocean and the estuary and show that the model remains unchanged. The boundary values of c_0, c_L do not affect the derivation of the governing differential equation.

2. The text derives the model for the distribution of salt in the channel as if the higher concentration of salt were on the left —, that is, as if diffusion were carrying the salt from left to right. It also supposes that the flow of water is in the same direction. Argue that neither the direction of diffusion nor the direction of water flow matter. The basic conservation balance,

$$-A(x)Dc'(x)+A(x)V(x)c(x) =$$
$$- A(x + \Delta x)Dc'(x + \Delta x)$$
$$+ A(x + \Delta x)V(x + \Delta x)c(x + \Delta x).$$

still holds. If either the concentration gradient c' or the velocity V is such that one of the "in" terms on the left is actually removing salt from the control volume, then the sign of that term will be negative. Likewise, "out" terms on the right will change sign if they are actually contributing to the salt in the box rather than removing it.

3. The text derives the diffusion model

$$-D(A(x)c'(x))' + (A(x)V(x)c(x))' = 0$$

under the assumption that the diffusion coefficient D is constant. Show that if the diffusion coefficient varies with position, then the governing equation is

$$-(D(x)A(x)c'(x))' + (A(x)V(x)c(x))' = 0.$$

4. Is the diffusion equation

$$-D(A(x)c'(x))' + (A(x)V(x)c(x))' = 0$$

linear or nonlinear? Homogeneous or nonhomogeneous? Under what conditions would it have constant coefficients?

5. Is the diffusion equation with source term

$$-D(A(x)c'(x))' + (A(x)V(x)c(x))' = A(x)P(x)$$

linear or nonlinear? Homogeneous or nonhomogeneous? Under what conditions would it have constant coefficients?

6. Suppose the cross section of the channel in example 1 is approximately semicircular throughout its length, but that it is 90 m in diameter at the ocean and tapers roughly linearly to a diameter of 10 m at the lake. Furthermore, suppose water is flowing *from* the lake *to* the ocean at $62.5L^2/(45L - 40x)^2$ km/h, where x is distance from the end of the channel connected to the ocean. Write a model for the concentration of salt in the channel.

7. Example 2 took some short cuts to derive a model for the situation shown in figure 9.6. Proceed from first principles, using the conservation law as in the text and the control volume shown in figure 9.7, to derive the model

$$xc''(x) + c'(x) = 0, \qquad c(0) = c_0, \qquad c(L) = c_L.$$

Exercises 8–14 ask you to derive models for the *equilibrium* temperature distribution in various bodies. The central experimental fact is as follows:

Experimental fact The rate of diffusion of heat energy through a unit area is proportional to the negative of the *temperature gradient* at that point.

The constant of proportionality is the *thermal conductivity* k. If $T(x)$ denotes temperature, then:

Heat flow rate through a unit area at x is $-kT'(x)$.

The basic conservation law is as follows:

> **Law of Conservation of Heat Energy**   The net change in the heat energy of a body over a given period of time is the amount of heat energy added less the amount of heat energy removed.

At equilibrium, this conservation law demands that the rate at which heat is added to a control volume equal the rate at which heat is lost.

8. The ends of a thin insulated rod of length L are maintained at temperatures T_0, T_L. Suppose the rod has constant cross-sectional area A. Since the rod is thin and its sides are insulated, you can assume that heat flows only along its length. Derive a model for the temperature distribution in this rod as follows.

(a) Argue that the rate at which heat energy flows through the left wall of the control volume shown in figure 9.8 is $-kAT'(x)$. Similarly, argue that the heat flow rate through the wall on the right is $-kAT'(x + \Delta x)$.

(b) Why does the conservation law let you conclude that at equilibrium $-kAT'(x + \Delta x)+ kAT'(x) = 0$?

(c) Use a limiting argument to derive the differential equation $-kT''(x) = 0$.

(d) Argue that the complete *temperature model* is

$$-T''(x) = 0, \qquad T(0) = T_0, \qquad T(L) = T_L.$$

(e) Can you find the temperature distribution in the rod? Does it seem reasonable?

9. Repeat exercise 8 when the right end of the bar is perfectly insulated so that no heat energy can diffuse through it. Argue that you obtain almost the same model as before; only the boundary condition at $x = L$ is changed to

$$T'(L) = 0.$$

10. Repeat exercise 9 under each of the following conditions:

(a) Cross-sectional area A varies with position. Obtain the model

$$-(A(x)T'(x))' = 0,$$
$$T(0) = T_0, \qquad T(L) = T_L.$$

(b) Both cross-sectional area A and thermal conductivity k vary with position. Obtain the model

$$-(k(x)A(x)T'(x))' = 0,$$
$$T(0) = T_0, \qquad T(L) = T_L.$$

(c) Use the results of part (a) to write a model of the temperature distribution in a rod whose area varies uniformly from left to right so that the area at $x = L$ is twice that at $x = 0$.

11. The top and the bottom of a thin circular plate of thickness d and radius L are insulated; see figure 9.9. Its boundary is at temperature T_L. Because of symmetry and insulation, heat can flow only in the radial direction. Using a control volume such as that in figure 9.6 derive the model

$$-(xT'(x))' = 0, \qquad T'(0) = 0, \qquad T(L) = T_L.$$

(*Hint:* The area of the surface of the control volume at radius x is $A(x) = 2\pi x d$.)

12. Repeat exercise 11 when the outer edge of the disk is perfectly insulated so that no heat energy can diffuse through it. Argue that you obtain almost the same model as before; only the boundary condition at $x = L$ is changed to

$$T'(L) = 0.$$

Try to guess a solution of the boundary-value problem in this case. Does your guess seem physically reasonable? Do you seem to need more information?

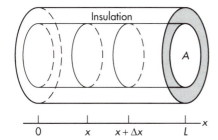

Figure 9.8: An insulated rod with cross-sectional area A. The control volume has its left end at x and its right end at $x + \Delta x$.

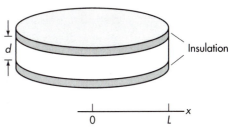

Figure 9.9: The top and bottom of a thin circular plate of thickness d and radius L are insulated.

13. By passing a current through the thin rod shown in figure 9.8, it is possible to add heat to it at a given rate. Suppose that heat energy per unit volume is being added at the rate $H(x)$. Derive the model

$$-kAT''(x) = H(x), \qquad T(0) = T_0, \qquad T(L) = T_L.$$

14. How is the preceding model changed if the left end of the bar is perfectly insulated so that no heat can diffuse through it?

9.2 BOUNDARY-VALUE PROBLEMS

The diffusion model

$$-D(A(x)c'(x))' + (A(x)c(x))' = 0, \qquad c(0) = c_0, \qquad c(L) = c_L.$$

derived in section 9.1 is an example of a **boundary-value problem**, a differential equation coupled with specifications of the solution at two different values of the independent variable.

Auxiliary conditions such as

$$c(0) = c_0, \qquad c(L) = c_L,$$

or

$$c(0) = c_0, \qquad c'(L) = 0,$$

are called **boundary conditions**. In contrast to initial conditions, which are all imposed at a single point, boundary conditions are imposed at two different points. Those points define the interval upon which we must solve the boundary-value problem. In the diffusion models of section 9.1, that interval was always $0 \le x \le L$.

A **solution of a boundary-value problem** is a solution of the differential equation which satisfies the boundary conditions and which is defined throughout the interval at whose ends the boundary conditions are given. To solve a boundary-value problem (at least for a linear equation):

1. Find a general solution of the differential equation.
2. Choose the arbitrary constants in the general solution so that the boundary conditions are satisfied.

The question of uniqueness, the number of solutions a boundary-value problem can possess, is addressed in project 2 of this chapter's projects.

EXAMPLE 5   *Solve the diffusion model*

$$-Dc'' + Vc' = 0, \qquad c(0) = 28, \qquad c(L) = 0.$$

derived in example 1. Here $D = 1.09 \times 10^{-9}$ m²/s, $V = 0.2778$ m/s, $L = 4,000$ m, and concentrations are measured in g/m³.

Find a general solution.

We can solve the constant-coefficient, homogeneous differential equation

$$-Dc'' + Vc' = 0$$

using the characteristic equation method. The characteristic equation $-Dr^2 + Vr = 0$ has roots

$$r = 0, \qquad \frac{V}{D},$$

and the general solution of the differential equation is

$$c_g = C_1 + C_2 e^{Vx/D}.$$

At $x = 0$, the boundary condition requires

$$c_g(0) = C_1 + C_2 = 28.$$

Determine C_1, C_2 from the boundary conditions.

The boundary condition at $x = L$ requires

$$c_g(L) = C_1 + C_2 e^{VL/D} = 0.$$

To find C_1, C_2, we solve these two simultaneous equations,

$$C_1 + C_2 = 28$$

$$C_1 + C_2 e^{VL/D} = 0,$$

and find

$$C_1 = \frac{28 e^{VL/D}}{e^{VL/D} - 1}, \qquad C_2 = \frac{-28}{e^{VL/D} - 1}.$$

Hence, the solution of the boundary-value problem is the general solution with C_1, C_2 replaced by these values:

$$c(x) = \frac{28 e^{VL/D}}{e^{VL/D} - 1} - \frac{28 e^{Vx/D}}{e^{VL/D} - 1}.$$

∎

EXAMPLE 6 *Describe the concentration profile given by the solution obtained in example 5.*

With the given values of V, L, and D, we find $VL/D \approx 1.02 \times 10^{12}$. Hence,

$$e^{VL/D} - 1 \approx e^{VL/D},$$

and we can write the solution found in example 5 as

$$c(x) = \frac{28 e^{VL/D}}{e^{VL/D} - 1} - \frac{28 e^{Vx/D}}{e^{VL/D} - 1}$$

$$\approx 28 - 28 e^{-V(L-x)/D}.$$

What can we learn from this approximate expression? When x is near zero, the second term is negligible; $e^{-VL/D} \approx 10^{-12}$. So $c \approx 28$ unless x is close enough to L to make $e^{-V(L-x)/D}$ significant, say, 0.1. If $e^{-V(L-x)/D} = 0.1$, then we find

$$x = L - \frac{D \ln 10}{V} \approx L - 9.05 \times 10^{-9}.$$

In other words, the concentration remains at its inlet level of 28 g/m^3 throughout the length of the channel, dropping to its ending value of zero only in the last fraction of a millimeter. This precipitous drop is sketched (though not to scale) in figure 9.10. ∎

EXAMPLE 7   *Example 3 proposes the model*

$$-D(A(x)c'(x)))' = 0, \qquad c(0) = c_0, \qquad c'(L) = 0,$$

for diffusion in a channel of length L when the end at $x = L$ is dammed to stop diffusion. Solve this boundary value problem when D and A are constant.

When D and A are constant, the differential equation simplifies to the constant-coefficient equation

$$c''(x) = 0.$$

Either from the characteristic equation $r^2 = 0$ with repeated root $r = 0$ or from direct integration, we find the general solution

Find a general solution.

$$c_g = C_1 + C_2 x.$$

Determine C_1, C_2 from the boundary conditions.

The boundary conditions require

$$c_g(0) = C_1 = c_0$$
$$c'_g(L) = C_2 = 0.$$

Hence, the solution of the boundary-value problem is

$$c(x) = c_0.$$

The concentration is constant throughout the length of the channel. ∎

EXAMPLE 8   *Example 2 proposes the model*

$$xc''(x) + c'(x) = 0, \qquad c'(0) = 0, \qquad c(L) = c_L,$$

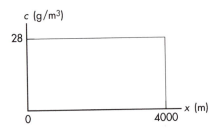

c (g/m³)

28

0   4000   x (m)

Figure 9.10: A graph of the solution of the boundary-value problem $-Dc'' + Vc' = 0$, $c(0) = 28$, $c(L) = 0$, analyzed in example 6. The rapid decline from $c = 28$ to $c = 0$ near $x = 4,000$ occurs much more precipitously than can be drawn here.

for the diffusion of salt in a circular pond of radius L. Solve this boundary-value problem.

Unfortunately, this second-order equation has variable coefficients, and we are equipped to solve only constant-coefficient, second-order equations. Were we to skip ahead to section 10.2, however, we would discover that a general solution of this equation is

Find a general solution.

$$c_g = C_1 + C_2 \ln x.$$

Accepting this general solution as given, we proceed to find the constants C_1 and C_2. To evaluate the boundary condition at $x = 0$, we compute

Determine C_1, C_2 from the boundary conditions.

$$c_g'(x) = \frac{C_2}{x},$$

a function that is not even defined at $x = 0$ *unless* $C_2 = 0$. If $C_2 = 0$, then

$$c_g'(0) = 0,$$

and the boundary condition is satisfied.

We choose the constant C_2 not to satisfy the boundary condition explicitly but simply to permit the solution to be defined throughout the interval $0 \le x \le L$. What remains of the general solution does in fact satisfy the symmetry boundary condition $c'(0) = 0$.

The second boundary condition requires

$$c_g(L) = C_1 = c_L.$$

Hence, the solution of the boundary-value problem is

$$c(x) = c_L.$$

The salt concentration is constant throughout the pond. ∎

EXAMPLE 9   *Example 4 proposes a model for the level of pollutant in a channel when the undesirable material is being released from the walls and bed of the channel at a specified rate; the channel connects two clean bodies of water. If the release rate of the pollutant is a constant P, if the cross-sectional area of the channel is a constant, and if there is no flow in the channel, then the model reduces to*

$$-Dc''(x) = P, \qquad c(0) = c(L) = 0.$$

Solve this boundary-value problem.

A general solution of this differential equation is easily obtained by direct integration, but we will exercise the constant-coefficient machinery we learned earlier.

As in example 7, we find from the characteristic equation $r^2 = 0$ that the homogeneous equation $c''(x) = 0$ has the general solution

Find a general solution.

$$c_h = C_1 + C_2 x.$$

To solve the forced equation $c''(x) = -P/D$ with P and D constant, we use undetermined coefficients with the initial guess

$$c_p = B,$$

where B is the undetermined coefficient. But this guess satisfies the homogeneous equation. Multiplying by the independent variable leads to

$$c_p = Bx,$$

which again solves the homogeneous equation. Another multiplication leads to the final form of the particular solution,

$$c_p = Bx^2.$$

Substituting in $c''(x) = -P/D$ gives

$$2B = -\frac{P}{D},$$

and

$$c_p = -\frac{Px^2}{2D}.$$

The general solution of the differential equation is

$$c_g = C_1 + C_2 x - \frac{Px^2}{2D}.$$

Determine C_1, C_2 from the boundary conditions.

The boundary conditions require

$$c_g(0) = C_1 = 0$$

$$c_g(L) = C_2 L - \frac{PL^2}{2D} = 0.$$

We conclude that $C_1 = 0$, $C_2 = \frac{PL}{2D}$. The solution of the boundary-value problem is

$$c(x) = \frac{Px}{2D}(L - x).$$

The concentration of pollutant reaches a peak value of $PL^2/8D$ at $x = L/2$, the midpoint of the channel. ∎

9   Diffusion Models and Boundary Value Problems

9.2.1 Exercises

EXERCISE GUIDE

To gain experience . . .	Try exercises
Finding general solutions	1–7, 18
Solving boundary value problems	1–15
Analyzing and interpreting solution behavior	8–17

In exercises 1–7,

    i. Find a general solution of the differential equation.

    ii. Find a solution of the given boundary-value problem.

1. $y'' - y = 0$, $y(0) = 1$, $y(2) = 0$

2. $y'' - y = 0$, $y(0) = 1$, $y'(2) = 4$

3. $y'' - y = \sin \pi x$, $y(0) = y(2) = 0$

4. $y'' - y = 0$, $y(0) = y(2) = 0$

5. $y'' - y = \sin \pi x$, $y(0) = 1$, $y'(2) = 4$

6. $y'' - 2y' + y = t^2$, $y(0) = -2$, $y(1) = 1$

7. $y'' - 2y' + y = e^t$, $y(0) = 0$, $y(1) = 1$

Each of exercises 8–15 displays a boundary-value problem based on a model for equilibrium temperature derived in the indicated exercise from the previous section.

    i. Solve the boundary-value problem.

    ii. Comment on the physical significance of the solution in light of the situation being modeled.

8. From exercise 8,

$$-T''(x) = 0, \qquad T(0) = T_0, \qquad T(L) = T_L.$$

9. From exercise 9,

$$-T''(x) = 0, \qquad T(0) = T_0, \qquad T'(L) = 0.$$

10. From exercise 11,

$$-(xT'(x))' = 0, \qquad T'(0) = 0, \qquad T(L) = T_L.$$

(*Hint:* See example 8 for a general solution.)

11. From exercise 12,

$$-(xT'(x))' = 0, \qquad T'(0) = 0, \qquad T'(L) = 0.$$

(*Hint:* See example 8 for a general solution.) Have you found one or many solutions of this boundary-value problem? Do you seem to need more information?

12. From exercise 13 with $H(x) = x(L - x)$, k, A constant,

$$-kAT''(x) = x(L - x),$$
$$T(0) = T_0, \qquad T(L) = T_L.$$

13. From exercise 14 with $H(x) = x(L - x)$, k, A constant,

$$-kAT''(x) = x(L - x),$$
$$T'(0) = 0, \qquad T(L) = T_L.$$

14. From part (a) of chapter exercise 1,

$$T''(x) = 0, \qquad T(0) = T_{in}, \qquad T(L) = T_{out}.$$

15. Example 6 claims

$$e^{VL/D} - 1 \approx e^{VL/D}$$

when VL/D is large. How large must this quantity be to keep the error in this approximation under 1%? Do the values used in the text, $D = 1.09 \times 10^{-9}$ m²/s, $V = 0.2778$ m/s, and $L = 4{,}000$ m, meet this criterion?

16. Verify the claim made in example 6 while analyzing the approximate solution $c \approx 28 - 28e^{-V(L-x)/D}$: If $e^{-V(L-x)/D} = 0.1$, then we find

$$x = L - \frac{D \ln 10}{V}.$$

17. Example 8 solved the boundary-value problem

$$xc''(x) + c'(x) = 0, \qquad c'(0) = 0, \qquad c(L) = c_L,$$

modeling the concentration of salt in the circular pond of figure 9.6. That figure also shows a hypothetical graph of the concentration profile in the pond. Redraw figure 9.6 with the correct concentration profile, as found in example 8.

18. Find a general solution of the differential equation considered in example 9,

$$-Dc''(x) = P,$$

by integration. Verify that you obtain the same general solution as in the example. (Recall that D, P are constants.)

9.3 TIME-DEPENDENT DIFFUSION

The diffusion models considered so far in this chapter suppose that the distribution of salt or pollutant or heat energy is in *equilibrium*; that is, the value of concentration of pollutant at a given point is constant in time. This section derives models that relax that assumption, permitting a study of the approach to equilibrium as well as the equilibrium itself.

The models derived here use several independent variables, those for spatial position as well as for time. Consequently, partial derivatives appear in the resulting models, which involve *partial differential equations*.

Section 9.1 studied the diffusion of salt and pollutants in water. The models here consider an analogous phenomena, the diffusion of heat energy in a solid such as the thin, insulated rod shown in figure 9.11. The qualifiers *thin* and *insulated* mean that heat energy flow is effectively limited to the axial direction, the left-right direction in the figure.

The governing experimental fact is gradient-dependent diffusion, just as with substances dissolved in water:

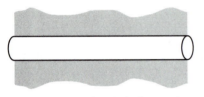

Figure 9.11: A thin rod whose sides are insulated.

> **Experimental Fact** The rate of diffusion of heat energy through a unit area is proportional to the negative of the *temperature gradient*.

The constant of proportionality is the *thermal conductivity* k. If $T(x)$ denotes temperature, then

$$\text{heat flow rate through a unit area at } x = -k\frac{\partial T(x,t)}{\partial x}.$$

Figure 9.12 illustrates this idea.

> The two-variable notation $T(x,t)$ indicates that temperature depends both on position x along the rod's axis and on time t. Consequently, the temperature gradient is denoted by the partial derivative $\partial T/\partial x$.

The basic conservation law is as follows.

> **Law of Conservation of Heat Energy** The net change in the heat energy of a body over a given period of time is the amount of heat energy added less the amount of heat energy removed.

Away from equilibrium, this conservation law demands that the net change of heat energy within the control volume equal the rate at which heat is added to a control volume less the rate at which heat is lost.

9 Diffusion Models and Boundary Value Problems

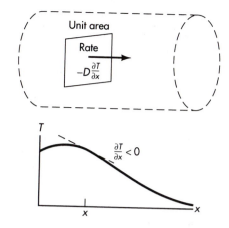

Figure 9.12: Heat is flowing through the unit area at the rate $-k\partial T(x,t)/\partial x$. The flow is from left to right because the temperature gradient is negative at that point.

This conservation law is a precise analogue of the law of conservation of population, among others: The amount added less the amount lost equals the net change over a fixed time interval.

To derive a model, we account for heat energy gained and lost in a control volume such as that in figure 9.13. Since heat is flowing only in the axial direction (horizontally in figure 9.13), we need account only for heat flow through the left and right sides of the control volume. The quantities required by conservation of energy for a change over Δt are:

$$\text{change in heat energy over } \Delta t = Q(t + \Delta t) - Q(t)$$

$$\text{heat energy } in \text{ at } x = -kA(x)\frac{\partial T(x,t)}{\partial x}\Delta t$$

$$\text{heat energy } out \text{ at } x + \Delta x = -kA(x + \Delta x)\frac{\partial T(x + \Delta x, t)}{\partial x}\Delta t$$

Here $A(x)$ is the area of the cross section of the rod at coordinate x, and $Q(t)$ is the heat energy in the control volume at time t. Having assumed that heat energy is flowing only in the axial direction, we need account only for heat energy flow through the circular ends of the control volume at x and $x + \Delta x$.

The gradient expression $-k\partial T/\partial x$ gives heat flow *in* on the left side of the control volume and *out* on the right side because x is increasing from left to right and that is the direction in which the gradient is computed. On the left, the direction of increasing x points into the control volume, and on the right it points out.

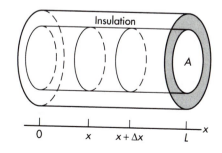

Figure 9.13: An insulated rod with cross-sectional area A. The control volume has its left end at x and its right end at $x + \Delta x$.

Conservation of energy requires that the net change over Δt in heat energy within the control volume equal the amount of heat energy that flowed in less the amount that flowed out. In words, conservation of energy demands

change over Δt of heat energy in control volume

= amount gained − amount lost.

The corresponding mathematical statement is

$$Q(t + \Delta t) - Q(t) = -kA(x)\frac{\partial T(x,t)}{\partial x}\Delta t + kA(x + \Delta x)\frac{\partial T(x + \Delta x, t)}{\partial x}\Delta t.$$

To relate heat energy Q to temperature T, use $Q = cmT$, where c is the specific heat of the material in the rod and m is the mass of the control volume. (See page 39.) Now the mass of the control volume is just its volume multiplied by the density ρ of the rod material. The volume of the slice of the rod between x and $x + \Delta x$ is approximately

$$A(x)\Delta x,$$

so the mass of that piece is $\rho A(x)\Delta x$. Consequently, the heat energy in the control volume at time t is $c\rho A(x)\Delta x T(x,t)$.

With this substitution, the conservation of energy relation becomes

$$Q(t + \Delta t) - Q(t) = c\rho A(x)T(x, t + \Delta t)\Delta x - c\rho A(x)T(x,t)\Delta x$$

$$= -kA(x)\frac{\partial T(x,t)}{\partial x}\Delta t + kA(x + \Delta x)\frac{\partial T(x + \Delta x)}{\Delta x}\Delta t.$$

Division by both Δt and Δx yields

$$c\rho A(x)\frac{T(x, t + \Delta t) - T(x,t)}{\Delta t}$$

$$= k\frac{A(x + \Delta x)T_x(x + \Delta x, t) - A(x)T_x(x,t)}{\Delta x},$$

where T_x denotes $\partial T/\partial x$. With the limits

$$\lim_{\Delta t \to 0} \frac{T(x, t + \Delta t) - T(x,t)}{\Delta t} = \frac{\partial T(x,t)}{\partial t}$$

and

$$\lim_{\Delta x \to 0} \frac{A(x + \Delta x)T_x(x + \Delta x, t) - A(x)T_x(x,t)}{\Delta x} = \frac{\partial(A(x)T(x,t))}{\partial x}$$

the conservation relation yields the governing equation, the **heat equation**

$$c\rho A(x)\frac{\partial T(x,t)}{\partial t} = k\frac{\partial(A(x)T(x,t))}{\partial x}.$$

9   Diffusion Models and Boundary Value Problems

The heat equation is a **partial differential equation** because it involves partial derivatives, in contrast with the ordinary derivatives in the *ordinary differential equations* studied earlier.

If the rod is in equilibrium, then $\partial T/\partial t = 0$ and the heat equation reduces to the diffusion equation considered in section 9.1.

To complete the heat-flow model, we need appropriate auxiliary data. The spatial derivative is of second-order ($\partial(AT_x)/\partial x = A\partial^2 T/\partial x^2 + \cdots$), and it represents the effects of diffusion. Boundary conditions at the left and right ends of the bar might be appropriate; for example, the ends might be kept at zero temperature:

$$T(0,t) = 0, \qquad T(L,t) = 0, \qquad t > 0.$$

The time derivative $\partial T/\partial t$ is of first order and should demand an initial condition. But that initial condition must be specified at each value of x. For example, if the initial temperature distribution in the rod followed the sine curve shown in figure 9.14, the corresponding initial condition would be

$$T(x,0) = M\sin\frac{\pi x}{L}, \qquad 0 < x < L.$$

Using these auxiliary conditions, the complete heat-flow model for the rod is the **initial-boundary value problem**

$$c\rho A(x)\frac{\partial T(x,t)}{\partial t} = k\frac{\partial(A(x)T(x,t))}{\partial x}$$

$$T(x,0) = M\sin\frac{\pi x}{L}, \qquad 0 < x < L$$

$$T(0,t) = 0, \qquad T(L,t) = 0, \qquad t > 0.$$

MODEL FOR TEMPERATURE IN A THIN ROD

The first line is the heat equation, the second is the initial condition, and the third contains the boundary conditions.

EXAMPLE 10   *Write a model for the temperature distribution in a thin rod of constant cross-sectional area, such as that shown in figure 9.14. Use the initial and boundary conditions shown there.*

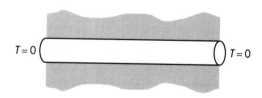

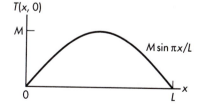

Figure 9.14:   The ends of a thin rod of length L are maintained at zero temperature. At $t = 0$, the temperature in the rod is given by $M\sin\pi x/L$.

If area A is constant, then the heat equation simplifies to

$$\frac{\partial T(x,t)}{\partial t} = \kappa \frac{\partial^2 T(x,t)}{\partial x^2}, \qquad (9.1)$$

where $\kappa = k/c\rho$ is the *thermal diffusivity*. The initial and boundary conditions are

$$T(x,0) = M \sin \frac{\pi x}{L}, \qquad 0 < x < L \qquad (9.2)$$

$$T(0,t) = 0, \qquad T(L,t) = 0, \qquad t > 0. \qquad \blacksquare$$

The fundamental idea in the derivation of this model is the application of conservation of energy to a control volume. The terms involving the change over time of the net heat energy in the box lead in the limit to the time derivative T_t. The terms involving the difference in the temperature gradient on either side of the control volume lead to the second-order spatial derivative $(AT_x)_x$.

To emphasize the role of the control volume, the next example examines the case of a wire heated by a current passing through it. The heat generated by the current adds another term to the conservation of energy balance.

EXAMPLE 11 *Suppose a piece of insulated wire of constant cross-sectional area A and length L is heated by a current passing through it. Assume the ends of the wire are maintained at zero temperature and that the wire is initially at some temperature $f(x)$, $0 \le x \le L$. Derive a model for the variation with position and time of the temperature in the wire.*

The situation is essentially that shown in figure 9.14, but the initial temperature is given by an arbitrary function $f(x)$.

To derive the governing partial differential equation, employ a control volume such as that illustrated in figure 9.13. Using the ideas given earlier, we can write expressions for the heat energy flow through the sides of the control volume and for the net change in heat energy within the control volume. The challenge is incorporating the heat generated by the current in the wire.

Let H denote the *rate* at which the current generates heat energy *per unit volume* of wire. Then over a time period Δt, the current will add $H\Delta V\Delta t$ units of heat energy (e.g., calories) to a volume ΔV of wire. If the case of the control volume show in figure 9.13, the volume is $\Delta V = A\Delta x$.

Conservation of heat energy demands

CONSERVATION OF ENERGY

change over Δt of heat energy in control volume = heat flow in through left

$-$ heat flow out through right $+$ heat added by current.

The terms involved in this accounting are

change in heat energy over $\Delta t = Q(t + \Delta t) - Q(t)$

heat energy *in* at $x = -kA\dfrac{\partial T(x,t)}{\partial x}\Delta t$

heat energy *out* at $x + \Delta x = -kA\dfrac{\partial T(x + \Delta x, t)}{\partial x}\Delta t$

$$\text{heat generated by current} = HA\Delta x \Delta t$$

Equating the terms as required by conservation of energy leads to

$$Q(t + \Delta t) - Q(t)$$

$$= -kA\frac{\partial T(x,t)}{\partial x}\Delta t + kA\frac{\partial T(x + \Delta x, t)}{\partial x}\Delta t + HA\Delta x \Delta t.$$

CONSERVATION OF ENERGY

Once again, we can relate heat energy Q in the control volume to its temperature using $Q = cmT$, where c is the specific heat of the wire and $m = \rho\Delta V = \rho A\Delta x$ expresses the mass of the control volume in terms of the density ρ of the wire. Making this substitution and dividing out the constant factor A lead to

$$c\rho\frac{T(x, t + \Delta t) - T(x,t)}{\Delta t} = k\frac{T_x(x + \Delta x, t) - T_x(x,t)}{\Delta x} + H.$$

Letting Δt and Δx go to zero yields the governing partial differential equation:

$$\frac{\partial T(x,t)}{\partial t} = \kappa\frac{\partial^2 T(x,t)}{\partial x^2} + \frac{H}{c\rho},$$

NONHOMOGENEOUS HEAT EQUATION

where $\kappa = k/c\rho$. This is a nonhomogeneous version of the heat equation (9.1). To complete the model, impose both the boundary conditions corresponding to zero end temperature:

$$T(0,t) = T(L,t) = 0, \qquad t > 0,$$

BOUNDARY CONDITIONS

and the initial condition,

$$T(x,0) = f(x), \qquad 0 < x < L. \qquad \blacksquare$$

INITIAL CONDITIONS

The dependence of the rate of heat flow on the gradient of the temperature can also be used to develop alternate boundary conditions.

EXAMPLE 12 *Suppose a thin insulated rod of constant cross-sectional area has its left end insulated against heat flow while its right end is maintained at temperature T_L, as shown in figure 9.15. Its initial temperature varies linearly from zero at the left end to T_L at the right end. Write a model for the temperature in the rod.*

Since cross-sectional area is constant and no heat is generated within the rod, the governing equation is the heat equation (9.1),

$$\frac{\partial T(x,t)}{\partial t} = \kappa\frac{\partial^2 T(x,t)}{\partial x^2}.$$

HEAT EQUATION

The initial condition is evidently

$$T(x,0) = \frac{T_L x}{L}, \qquad 0 < x < L,$$

INITIAL CONDITION

and the boundary condition at $x = L$ must be

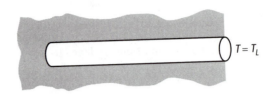

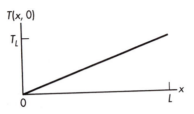

Figure 9.15: The left end of a thin rod is insulated. Its right end is kept at temperature T_L. At $t = 0$, the temperature increases linearly from zero at the left to T_L at the right.

BOUNDARY CONDITION AT $x = L$

$$T(L, t) = T_L, \qquad t > 0.$$

Insulating the left end of the bar means that the rate of heat energy flow through that surface is zero. Since rate of heat flow is proportional to T_x, the insulated boundary condition is

INSULATED BOUNDARY CONDITION AT $x = 0$

$$T_x(0, t) = 0, \qquad t > 0. \qquad\blacksquare$$

The insulated boundary condition is another example of a *Neumann* boundary condition.

EXAMPLE 13 *Many of the processor chips in personal computers are narrow, thin rectangular blocks, as illustrated in figure 9.16. They are mounted on circuit boards, which in a poor design could trap heat generated by the power supply below. Viewing the chip from its narrow end, derive a model that will capture the variation in temperature of the chip from its hot underside to its cool top. Assume the chip starts at temperature zero and that its top remains at that temperature while its bottom is at temperature T_0.*

Since the height of the chip is relatively small compared with its width and depth, simplify the problem by neglecting the variation in temperature from left to right and from front to back in the end view of figure 9.17. That is, assume that heat flows only in the vertical direction from the lower surface $y = 0$ at temperature T_0 to the upper surface $y = h$ at temperature zero.

An appropriate control volume is the box with width Δx, height Δy, and depth Δz shown in that figure. Neglecting temperature variations in all but the vertical direction means no heat flows through the sides of the box. We need account only for heat flow through the top and bottom of the box.

If the control volume is Δz units deep, then the area of its top and bottom is $\Delta z \Delta x$. The accounting required by conservation of energy is

$$\text{change in heat energy over } \Delta t = Q(t + \Delta t) - Q(t)$$

$$\text{heat energy } in \text{ at } y = -k(\Delta z \Delta x)\frac{\partial T(y, t)}{\partial y}\Delta t$$

Figure 9.16: A computer chip mounted on a circuit board.

$$\text{heat energy } out \text{ at } y + \Delta y = -k(\Delta z \Delta x)\frac{\partial T(y + \Delta y, t)}{\partial y}\Delta t$$

9 Diffusion Models and Boundary Value Problems

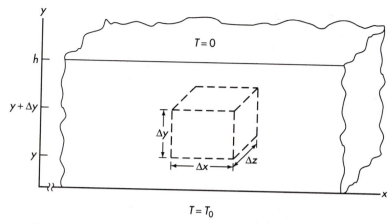

Figure 9.17: An end view of the computer chip showing a control volume with height Δy, width Δx, and depth Δz. The lower surface is at temperature T_0, the upper surface at zero temperature.

Equating the terms as required by conservation of energy leads to

$$Q(t + \Delta t) - Q(t) = -k(\Delta z \Delta x)\frac{\partial T(y,t)}{\partial y}\Delta t + k(\Delta z \Delta x)\frac{\partial T(y + \Delta y, t)}{\partial y}\Delta t.$$

CONSERVATION OF ENERGY

Once again, we can relate heat energy Q in the control volume to its temperature using $Q = cmT$, where c is the specific heat of the chip and $m = \rho A \Delta x \Delta y \Delta z$ expresses the mass of the control volume in terms of the density ρ of the chip. Making this substitution and dividing out the common factor $\Delta x \Delta z$ lead to

$$c\rho\frac{T(y, t + \Delta t) - T(y,t)}{\Delta t} = k\frac{T_y(y + \Delta x, t) - T_y(y,t)}{\Delta y}.$$

Letting Δt and Δy go to zero yields the governing partial differential equation,

$$\frac{\partial T(y,t)}{\partial t} = \kappa\frac{\partial^2 T(y,t)}{\partial y^2},$$

HEAT EQUATION

where $\kappa = k/c\rho$. This is the heat equation (9.1) with y replacing x. To complete the model, impose both the boundary conditions giving the temperature on the bottom and top of the chip,

$$T(0,t) = T_0, \qquad T(L,t) = 0, \qquad t > 0,$$

BOUNDARY CONDITIONS

and the zero temperature initial condition,

$$T(y,0) = 0, \qquad 0 < y < h. \qquad \blacksquare$$

INITIAL CONDITION

9.3.1 Summary

Rate of heat flow through a unit area is proportional to the *negative* temperature gradient perpendicular to the area—e.g., to $-T_x(x,t)$ for the unit area shown in figure 9.12. The proportionality constant is k, the thermal conductivity.

Conservation of heat energy demands that net heat energy flowing into a control volume such as that in figure 9.13 or 9.17 be balanced by the net increase in energy within the control volume. An example of the appropriate accounting appears on page 373. Relating heat energy Q to temperature T via $Q = cmT$ leads to governing equations such as the heat equation (9.1). A complete model includes boundary and initial conditions, as in (9.2).

9.3.2 Exercises

EXERCISE GUIDE

To gain experience . . .	Try exercises
Finding thermal diffusivity κ	1–2
Defining *homogeneous*, etc for the heat equation	3
Deriving models of thermal diffusion	4–11
Interpreting terms in the heat equation	12–13

1. According to table 2.4, page 39, the specific heat of copper is 0.09 cal/g·K and its thermal conductivity is 0.908 cal/s·K·cm²/cm. (The thermal conductivity units include the last "/cm" because the data in table 2.4 are for samples 1 cm thick.) Its density is 8.85 g/cm³. Compute the value (including units) of its thermal diffusivity.

2. Repeat exercise 1 for eastern white pine. Thermal conductivity is 0.0003 cal/s·K·cm, specific heat is 0.42 cal/g·K, and density is 0.373 g/cm³.

3. The text casually refers to $T_t = \kappa T_{xx}$ as the *homogeneous* heat equation and $T_t = \kappa T_{xx} + H/c\rho$ as the *nonhomogeneous* heat equation. Define *homogeneous* and *nonhomogeneous* in this setting and verify by substitution that the text's claims are correct.

4. Example 11 derives a model for the temperature in a wire of constant cross-sectional area when it is heated by passing a current through it. Derive a model for the same situation, but allow the cross-sectional area of the wire to vary with position.

5. The sides and both ends of a thin rod of length L are insulated. Initially, its temperature is given by $T_L x/L$, $0 < x < L$. Write an initial-boundary-value problem that models the temperature in the rod.

6. Two copper rods of length L are heated to temperatures T_1 and T_2, respectively. Their ends are quickly joined, the entire assembly is wrapped with insulation, and the outer ends of the double length rod are maintained at the initial temperature of the corresponding segment. Derive a model for the variation with time and position of the temperature of the complete assembly. Assume heat flows unhindered across the junction of the two rods.

7. Suppose the thermal conductivity k of the rod considered in example 10 varies with position because of manufacturing irregularities. Derive a model for the temperature variations with time in the rod.

8. The top and bottom of a thin circular plate of thickness d and radius L are insulated; see figure 9.9. The boundary of the disk is maintained at the uniform temperature T_L, and its initial temperature is a constant T_i. Since there is no angular variation, we can suppose that heat flows only in the radial (center to circumference) direction.

   (a) Using the doughnut-shaped control volume shown in figure 9.7 derive the following equation for the temperature of the plate as a function of time t and radius x:

   $$xT_t(x,t) = \kappa(xT_x(x,t))_x.$$

   (b) Argue that circular symmetry demands $T_x(0,t) = 0$. Add a boundary condition at $x = L$ and an initial condition to complete the model.

9. Suppose the computer chip considered in example 13 is generating heat internally at a rate of H cal/cm³·s.

9   Diffusion Models and Boundary Value Problems

Derive a model for the vertical temperature profile of the chip as a function of time. Neglect temperature variations in all but the vertical direction.

10. The model of the temperature profile in the computer chip of example 13 neglected temperature variation in the horizontal direction in the end view shown in figure 9.17. Improve the model by considering temperature variations in both the horizontal and vertical directions as follows. (Let w denote the width of the chip.)

   (a) Consider the control volume in figure 9.17. Argue that the heat flow in through the left side over time Δt is $-k \Delta y T_x(y, t) \Delta t$. Construct a similar expression for the heat flow out through the right side.

   (b) Conservation of energy demands equality between change in heat energy within the control volume and heat flow through the walls of the control volume. Write out the terms in that energy balance.

   (c) Collect terms and take appropriate limits to obtain

   $$T_t(x, y, t) = \kappa \left[T_{xx}(x, y, t) + T_{yy}(x, y, t) \right].$$

   (d) The bottom of the chip is at temperature T_0 and its sides and top are at temperature zero. Initially, the chip's temperature is zero. Complete the model by using this information to specify boundary conditions on $T(x, y, t)$ at $x = 0, w$ for $0 < y < h$ and at $y = 0, h$ for $0 < x < w$, as well as initial conditions at $t = 0$ for $0 < x < w$, $0 < y < h$.

11. How would the boundary conditions in exercise 10 change if the left and right sides of the computer chip were insulated? If the top were insulated instead? (See example 12.)

12. Example 11 showed that

   $$T_t - \kappa T_{xx} = H/c\rho$$

   governs heat flow in the presence of a heat source $H > 0$. (The heat source in that example is a current passing through the wire.) In the absence of thermal diffusion ($\kappa = 0$), the heat source obviously acts to increase temperature. Using similar logic, argue that the effect of thermal diffusion is to lower temperature; that is, diffusion is dissipative. Is this observation physically reasonable? For example, would a hot spot in the center of an otherwise cool rod diminish by spreading or intensify and become hotter?

13. Examine example 11. What is happening in the rod if $T_t - \kappa T_{xx} = H/c\rho$ is the governing equation and $H < 0$?

9.4 FOURIER METHODS

The partial differential equation models of the previous section are solved by combining *separation of variables* for partial differential equations with *eigenfunction expansions*, a technique for representing certain functions as infinite sums of sines and cosines, for example.

The separation method provides a family of solutions of the partial differential equation which also satisfy the (homogeneous) boundary conditions. Representing the initial condition in an eigenfunction expansion leads to a superposition of solutions that satisfies the partial differential equation, the boundary conditions, and the initial condition.

9.4.1 Separation of Variables

To illustrate the separation idea, consider the model for heat flow in a thin rod derived in example 10. (See figure 9.14.) The heat equation

$$\frac{\partial T(x, t)}{\partial t} = \kappa \frac{\partial^2 T(x, t)}{\partial x^2}, \tag{9.3}$$

is subject to the initial and boundary conditions

$$T(x, 0) = M \sin \pi x / L, \qquad 0 < x < L$$

$$T(0, t) = 0, \qquad T(L, t) = 0, \qquad t > 0. \tag{9.4}$$

Separation of variables begins with the assumption that the unknown temperature can be written as a product of two functions, one dependent only on x, the other only on t:

$$T(x, t) = X(x)\Theta(t).$$

Substituting this assumed solution form into the heat equation (9.3) yields

$$X(x)\Theta'(t) = \kappa X''(x)\Theta(t).$$

Prime denotes the derivative with respect to the argument shown: $\Theta'(t) = d\Theta(t)/dt$, whereas $X'(x) = dX(x)/dx$.

Collecting all functions of t on one side of the equation and all functions of x on the other leads to

$$\frac{\Theta'(t)}{\kappa\Theta(t)} = \frac{X''(x)}{X(x)}.$$

Now fix t and vary x. Since the quotient on the left is constant as long as t is fixed, the quotient on the right must be constant. Reversing the roles of x and t forces us to conclude that *the two quotients are constant*. That observation is the key to separation of variables.

Denote that common constant by $-\lambda$:

$$\frac{\Theta'(t)}{\kappa\Theta(t)} = \frac{X''(x)}{X(x)} = -\lambda.$$

(The negative sign is for convenience later.) Two *ordinary* differential equations for $\Theta(t)$ and $X(x)$ result:

$$\Theta'(t) + \lambda\kappa\Theta(t) = 0$$

$$X''(x) + \lambda X(x) = 0,$$

where λ remains to be determined.

The homogeneous boundary conditions $T(0, t) = X(0)\Theta(t) = 0$ and $T(L, t) = X(L)\Theta(t) = 0$, $t > 0$, lead immediately to boundary conditions on

$X(x)$:

$$X(0) = X(L) = 0.$$

To determine the spatial dependence of the temperature (and to evaluate the separation constant λ), we must solve the boundary-value problem

$$X''(x) + \lambda X(x) = 0, \qquad X(0) = X(L) = 0. \tag{9.5}$$

For any value of λ, one obvious solution is $X(x) \equiv 0$. Unfortunately, choosing the trivial solution for X leads to $T(x, t) \equiv 0$, a solution of the heat equation that can never satisfy the given initial condition.

Hence, determining X requires finding *nontrivial solutions* of the homogeneous boundary-value problem (9.5). More precisely, we seek those values of λ for which

$$X''(x) + \lambda X(x) = 0, \qquad X(0) = X(L) = 0$$

EIGENVALUE PROBLEM

has *nontrivial solutions*. The values of λ for which there are nontrivial solutions are **eigenvalues**, and the corresponding nontrivial solutions are **eigenfunctions**. The homogeneous boundary-value problem containing the parameter λ is an **eigenvalue problem**.

> An eigenvalue problem couples a *homogeneous* differential equation with *homogeneous* boundary conditions. That combination *always* has the trivial solution. The trick is finding the values of λ, the eigenvalues, for which the problem has *non*trivial solutions.
>
> The separation process must always lead to a (homogeneous) eigenvalue problem. If necessary, the given initial-boundary value problem is transformed to obtain homogeneous boundary conditions and a homogeneous differential equation; see example 17 below.

To find the eigenvalues and eigenfunctions, we begin with a general solution of $X'' + \lambda X = 0$ and impose the boundary conditions. The characteristic equation of $X'' + \lambda X = 0$ is

Find the eigenfunctions and eigenvalues.

$$r^2 + \lambda = 0.$$

Its roots are $r = \pm\sqrt{-\lambda}$. Three cases arise, corresponding to λ negative, zero, and positive. The corresponding solutions are:

$$\lambda < 0: \quad X(x) = C_1 e^{\sqrt{-\lambda}x} + C_2 e^{-\sqrt{-\lambda}x}$$

$$\lambda = 0: \quad X(x) = C_1 + C_2 x$$

$$\lambda > 0: \quad X(x) = C_1 \cos\sqrt{\lambda}x + C_2 \sin\sqrt{\lambda}x.$$

When $\lambda < 0$, the boundary conditions $X(0) = X(L) = 0$ require

$$x = 0: \qquad\qquad C_1 + C_2 = 0$$

$$x = L: \quad C_1 e^{\sqrt{-\lambda}L} + C_2 e^{-\sqrt{-\lambda}L} = 0.$$

Using Cramer's rule to solve this pair of equations for C_1, C_2, we find that the determinant of coefficients is

$$\det \begin{bmatrix} 1 & 1 \\ e^{\sqrt{-\lambda}L} & e^{-\sqrt{-\lambda}L} \end{bmatrix} = e^{\sqrt{-\lambda}L} - e^{-\sqrt{-\lambda}L} \neq 0.$$

Hence, $C_1 = C_2 = 0$. The eigenvalue problem (9.5) has no negative eigenvalues. A similar analysis, which is left to exercise 5, shows that $\lambda = 0$ is not an eigenvalue.

If $\lambda > 0$, then the boundary condition $X(0) = 0$ requires

$$X(0) = C_1 = 0.$$

The other boundary condition, $X(L) = 0$, then demands

$$X(L) = C_2 \sin \sqrt{\lambda} L = 0.$$

Choosing $C_2 = 0$ is possible but undesirable, for we seek nontrivial solutions. Alternatively, we can choose λ to satisfy

$$\sin \sqrt{\lambda} L = 0,$$

which requires that

$$\sqrt{\lambda} L = 0, \ \pm\pi, \pm 2\pi, \ldots$$

or

$$\lambda = \frac{\pi^2}{L^2}, \frac{4\pi^2}{L^2}, \ldots.$$

(The value $\lambda = 0$ is dropped because it leads again to the trivial solution.) The constant C_2 remains arbitrary.

Hence, the boundary-value problem

$$X''(x) + \lambda X(x) = 0, \qquad X(0) = X(L) = 0$$

has the nontrivial solutions, or *eigenfunctions*,

EIGENFUNCTIONS $X_n(x)$

$$X_n(x) = C \sin \frac{n\pi x}{L},$$

where C is arbitrary, corresponding to the *eigenvalues*

EIGENVALUES λ_n

$$\lambda_n = \frac{n^2 \pi^2}{L^2}, \qquad n = 1, 2, \ldots.$$

For convenience, let $C = 1$.

For each value of λ, there is a corresponding solution of the Θ equation. A general solution of

$$\Theta'(t) + \lambda_n \kappa \Theta(t) = 0$$

is obviously

CORRESPONDING $\Theta_n(t)$

$$\Theta_n(t) = c_n e^{-\lambda_n \kappa t}, \qquad n = 1, 2, \ldots.$$

For each integer n, the product $T_n(x, t) = X_n(x)\Theta_n(t)$ is a solution of the governing equation (9.3). That is, for $n = 1, 2, \ldots$

$T_n = X_n(x)\Theta_n(t)$ solves the partial differential equation . . .

$$T_n(x, t) = c_n e^{-n^2 \pi^2 \kappa t / L^2} \sin \frac{n\pi x}{L}$$

is a solution of

$$\frac{\partial T(x,t)}{\partial t} = \kappa \frac{\partial^2 T(x,t)}{\partial x^2}.$$

Furthermore, for each n, $T_n(x,t)$ satisfies the boundary conditions,

... and T_n satisfies the boundary conditions.

$$T_n(0,t) = T_n(L,t) = 0, \qquad n = 1, 2. \ldots .$$

Satisfying the initial condition demands

Impose initial conditions.

$$T_n(x,0) = c_n \sin \frac{n\pi x}{L} = M \sin \frac{\pi x}{L}.$$

Evidently, $n = 1$ and $c_1 = M$. The solution of the initial-boundary value problem (9.3–9.4) is

SOLUTION OF INITIAL-BOUNDARY VALUE PROBLEM

$$T(x,t) = M e^{-\pi^2 \kappa t / L^2} \sin \frac{\pi x}{L}.$$

Note that $T(x,t) \to 0$ as $t \to \infty$ for each x, $0 \le x \le L$. This solution is decaying to the equilibrium zero temperature we would expect from the zero boundary conditions. That equilibrium temperature is the solution of the governing problem (9.3–9.4) with $\partial T/\partial t = 0$; i.e., $T_{xx} = 0$ subject to $T = 0$ at $x = 0, L$.

The hottest point in the rod is always in its interior, at $x = L/2$ in this case. The rate of decay toward the equilibrium is governed by $\pi^2 \kappa / L^2$. The temperature decays more quickly in a shorter rod (L smaller) or in one that conducts heat better (κ larger).

9.4.2 Eigenfunction Expansions

What if the initial conditions were other than a sine curve that conveniently matches the spatial dependence of one of the solutions $T_n(x,t)$? The following example explores this possibility.

EXAMPLE 14   *Suppose the initial temperature distribution in the rod shown in figure 9.14 is*

$$T(x,0) = Mx(L - x).$$

Find the resulting temperature in the rod.

In spite of the change in the initial condition, the separation of variables solution process follows the steps illustrated previously:

1. Assume the solution has the product form $T(x,t) = X(x)\Theta(t)$.

2. Substitute in the governing heat equation (9.3), separate variables, and introduce a separation constant $-\lambda$,

$$\frac{\Theta'(t)}{\kappa\Theta(t)} = \frac{X''(x)}{X(x)} = -\lambda.$$

3. Use the boundary conditions $T(0, t) = T(L, t) = 0$ to obtain an eigenvalue problem for $X(x)$:

$$X''(x) + \lambda X(x) = 0, \qquad X(0) = X(L) = 0.$$

4. Solve the eigenvalue problem to obtain the eigenfunction, eigenvalue pairs

$$X_n(x) = \sin\frac{n\pi x}{L}, \qquad \lambda_n = \frac{n^2\pi^2}{L^2}, \qquad n = 1, 2, \ldots,$$

and using the eigenvalues λ_n, solve the equation $\Theta' + \kappa\lambda\Theta = 0$ to find

$$\Theta_n(t) = c_n e^{-\lambda_n \kappa t}, \qquad n = 1, 2, \ldots.$$

5. Obtain a family

$$T_n(x, t) = c_n e^{-n^2\pi^2\kappa t/L^2} \sin\frac{n\pi x}{L}$$

of solutions of the heat equation (9.3) that also satisfy the boundary conditions $T(0, t) = T(L, t) = 0$.

Unfortunately, none of the individual solutions $T_n(x, t)$ can satisfy the initial condition

$$T(x, 0) = Mx(L - x),$$

because

$$T_n(x, 0) = c_n \sin\frac{n\pi x}{L}.$$

The individual sine curves of $T_n(x, 0)$ and the parabola can never match exactly.

Since each of the $T_n(x, t)$ satisfies both the *homogeneous*, linear heat equation and the *homogeneous* boundary conditions, any finite sum of these functions is also a solution. Furthermore, the infinite sum

$$T(x, t) = \sum_{n=1}^{\infty} T_n(x, t) = \sum_{n=1}^{\infty} c_n e^{-n^2\pi^2\kappa t/L^2} \sin\frac{n\pi x}{L}$$

is also a solution, if it converges.

> Claiming that a finite sum of solutions of the homogeneous equation is again a solution of the equation is an application of the principle of superposition, a concept that is easily extended from ordinary to partial differential equations. Would a reasonable definition of *linear partial differential equation* include the heat equation?
>
> Although it is a serious mathematical question, we will take for granted the convergence of the infinite series in order to concentrate on the construction of a solution in this form.

If this infinite sum is to satisfy the initial condition, then we must find constants c_n so that

$$T(x,0) = \sum_{n=1}^{\infty} c_n \sin \frac{n\pi x}{L} = Mx(L-x), \qquad 0 < x < L. \qquad (9.6)$$

This request may not be completely preposterous, for one-half cycle of a sine function does resemble the parabolic arc $Mx(L-x), 0 < x < L$.

The key to finding the constants c_n is the **orthogonality** on $0 \le x \le L$ of the family of functions $\sin n\pi x/L$,

$$\int_0^L \sin \frac{m\pi x}{L} \sin \frac{n\pi x}{L} \, dx = \begin{cases} 0, & m \ne n \\ L/2, & m = n. \end{cases} \qquad (9.7)$$

The adjective *orthogonal* is used because the integral of the product of the two functions is a natural generalization of the dot product of two vectors when the two sine functions are regarded as members of the vector space of continuous functions. We will see later that this family of functions is orthogonal because it is a family of eigenfunctions.

To use orthogonality to find c_n, choose an integer m and multiply both sides of (9.6) by $\sin m\pi x/L$. If integration and the infinite sum can be interchanged, then

$$\sum_{n=1}^{\infty} c_n \int_0^L \sin \frac{m\pi x}{L} \sin \frac{n\pi x}{L} \, dx = \int_0^L \sin \frac{m\pi x}{L} Mx(L-x) \, dx.$$

Because of the orthogonality of the family of sine functions, every integral on the left is zero except for one case, $n = m$, when the integral is $L/2$. The infinite sum collapses to a single term, and we have

$$c_m = \frac{2}{L} \int_0^L \sin \frac{m\pi x}{L} Mx(L-x) \, dx.$$

Evaluating the integral yields

$$c_m = \frac{4ML^2}{m^3 \pi^3} \left(1 - (-1)^m \right),$$

or

$$c_m = \begin{cases} 8ML^2/m^3\pi^3, & m \text{ odd} \\ 0 & m \text{ even}. \end{cases} \qquad (9.8)$$

With the choice of c_n given by (9.8), we have

$$\sum_{n=1}^{\infty} c_n \sin \frac{n\pi x}{L} = Mx(L-x), \qquad 0 \le x \le L. \qquad (9.9)$$

The meaning of equality will be made precise in the eigenfunction expansion theorem stated shortly. This series is an example of an *eigenfunction expansion*

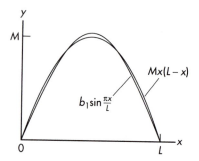

Figure 9.18: A comparison of the function $Mx(L - x)$, $0 \leq x \leq L$, and the first term of its eigenfunction expansion, $(8ML^2/\pi^3)\sin(\pi x/L)$.

of $Mx(L - x)$ in terms of the eigenfunctions $X_n = \sin n\pi x/L$ of the eigenvalue problem

$$X''(x) + \lambda X(x) = 0, \qquad X(0) = X(L) = 0.$$

To suggest the power of the eigenfunction expansion of $Mx(L - x)$, figure 9.18 compares the graph of this function with the first term in its expansion. The graph of the expansion which includes higher-order terms is indistinguishable from that of $Mx(L - x)$.

Hence, we have the formal solution

$$T(x, t) = \sum_{n=1}^{\infty} c_n X_n(x)\Theta_n(t)$$

$$= \sum_{n=1,3,\ldots}^{\infty} \frac{8ML^2}{n^3\pi^3} e^{-n^2\pi^2\kappa t/L^2} \sin\frac{n\pi x}{L}$$

of the heat equation (9.3) subject to the boundary conditions

$$T(0, t) = T(L, t) = 0, \qquad t > 0,$$

and to the initial condition

$$T(x, 0) = Mx(L - x), \qquad 0 < x < L.$$

Here c_n is given by (9.8). ∎

Figure 9.18 suggests that the eigenfunction expansion (9.9) is converging rapidly to $Mx(L - x)$ on $0 \leq x \leq L$. What function does this expansion represent for other values of x?

The answer is surprisingly easy. For arbitrary x, let

$$f(x) = \sum_{n=1}^{\infty} c_n \sin\frac{n\pi x}{L},$$

where c_n is given by (9.8). Since $f(x)$ is a sum of sine functions, it must have the same symmetry about the y-axis as the sine function; that is, it must be an **odd function**, one which satisfies

$$f(x) = -f(-x).$$

An arbitrary odd function is anti-symmetric about the origin, as illustrated in figure 9.19.

Since $f(x) = Mx(L - x)$, $0 \leq x \leq L$, its definition for $-L \leq x < 0$ is immediate from $f(x) = -f(-x) = -M(-x)(L - (-x)) = Mx(L + x)$,

$$f(x) = \begin{cases} Mx(L - x), & 0 \leq x \leq L, \\ Mx(L + x), & -L \leq x < 0. \end{cases}$$

The *odd extension* of $Mx(L - x)$ is illustrated in figure 9.20.

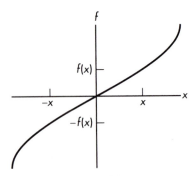

Figure 9.19: An odd function, $f(x) = -f(-x)$. Its graph for $x < 0$ is the reflection of the graph for $x > 0$ across the y-axis and then the x-axis.

To reach beyond the interval $-L \leq x \leq L$, note that the term of lowest frequency (largest period) in the expansion for f is $\sin \pi x / L$, a function with period $2L$; i.e., $\sin \pi x / L = \sin \pi(x+2L)/L$ for all x. The sum itself—and, consequently, $f(x)$—must have period $2L$. The eigenfunction expansion converges to the *2L-periodic extension* of $f(x)$.

Finally, a bit of house cleaning. Since the odd integers $n = 1, 3, 5, \ldots$ can be written as $n = 2p - 1$, $p = 1, 2, 3, \ldots$, the change of index $n = 2p - 1$ transforms the sum in the solution expression to

$$T(x,t) = \sum_{p=1}^{\infty} \frac{8ML^2}{(2p-1)^3 \pi^3} e^{-(2p-1)^2 \pi^2 \kappa t / L^2} \sin \frac{(2p-1)\pi x}{L}.$$ ∎

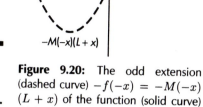

Figure 9.20: The odd extension (dashed curve) $-f(-x) = -M(-x)$ $(L + x)$ of the function (solid curve) $f(x) = Mx(L - x)$, $x > 0$.

The construction illustrated in this example is an example of an **eigenfunction expansion**, a series of the form

$$\sum^{\infty} c_n X_n(x),$$

where $\{X_n(x)\}$ are the eigenfunctions of an eigenvalue problem of the form

$$X''(x) + \lambda X(x) = 0$$
$$aX(0) + bX'(0) = 0 \qquad (9.10)$$
$$cX(L) + dX'(L) = 0.$$

In the previous example, $a = c = 1$, $b = d = 0$, and $X_n(x) = \sin n\pi x / L$.

The eigenvalue problem (9.10) is an example of a *self-adjoint* boundary-value problem. See exercise 14 for a more general example.

The important property of eigenfunctions of (9.10) that correspond to distinct eigenvalues is their *orthogonality*,

$$\int_0^L X_n(x) X_m(x) \, dx = 0, \qquad m \neq n,$$

as in (9.7), for example, where $X_n(x) = \sin n\pi x / L$.

Theorem 9.1 (Eigenfunction Expansion) Let $\{X_n(x)\}$ be a family of eigenfunctions of the eigenvalue problem (9.10), where neither a and b nor c and d are simultaneously zero.

i. Eigenfunctions corresponding to distinct eigenvalues are orthogonal on $0 \leq x \leq L$; i.e.,

$$\int_0^L X_n(x) X_m(x) \, dx = 0, \qquad m \neq n.$$

ORTHOGONAL EIGENFUNCTIONS

ii. If $f(x)$ is square integrable on $0 \leq x \leq L$, that is, if

$$\int_0^L [f(x)]^2 \, dx < \infty,$$

then its eigenfunction expansion converges in mean square,

CONVERGENCE OF EIGENFUNCTION EXPANSION

$$\lim_{p \to \infty} \int_0^L \left(f(x) - \sum^p c_n X_n(x) \right)^2 dx = 0,$$

where

GENERALIZED FOURIER COEFFICIENT

$$c_n = \left(\int_0^L [X_n(x)]^2 \, dx \right)^{-1} \int_0^L f(x) X_n(x) \, dx. \qquad (9.11)$$

In the previous example with $X_n(x) = \sin n\pi x/L$, the orthogonality could be verified by direct integration:

$$\int_0^L \sin \frac{n\pi x}{L} \sin \frac{m\pi x}{L} \, dx = 0, \qquad m \neq n.$$

Since

$$\int_0^L \left(\sin \frac{n\pi x}{L} \right)^2 dx = \frac{L}{2},$$

the (generalized) **Fourier coefficient** c_n given in part (ii) of the theorem is just the coefficient c_n we computed in (9.8). The eigenfunction expansion $\sum c_n X_n(x)$ is sometimes called a generalized *Fourier expansion*.

The *mean square convergence* provided by this theorem permits the infinite sum and $f(x)$ to differ at a large number of points, provided their total "length" on the interval $0 \leq x \leq L$ is zero. For our purposes, these differences may occur where f suffers a jump discontinuity, and we will consider only functions with a finite number of such jumps.

The proof of the mean square convergence of the eigenfunction expansion is rather delicate and will be omitted. We will illustrate the idea of the proof of orthogonality and the use of the orthogonality relation to compute the generalized Fourier coefficient c_n.

Proof (Sketch)

Consider part (i) in the special case of the boundary conditions $X(0) = X(L) = 0$. Suppose $X(x)$ is an eigenfunction of (9.10) corresponding to the eigenvalue λ, and $Y(x)$ is an eigenfunction corresponding to the eigenvalue μ; that is,

$$X''(x) + \lambda X(x) = 0, \qquad X(0) = X(L) = 0,$$

and

$$Y''(x) + \mu Y(x) = 0, \qquad Y(0) = Y(L) = 0.$$

Further, suppose $\lambda \neq \mu$. To establish orthogonality, we must show

$$\int_0^L X(x) Y(x) \, dx = 0.$$

Substitute $Y = -Y''/\mu$ in this integral, integrate by parts using $u = X$, $v = Y'$, and apply the boundary conditions $X(0) = X(L) = 0$:

$$\int_0^L X(x)Y(x)\,dx = -\frac{1}{\mu}\int_0^L X(x)Y''(x)\,dx$$

$$= -\frac{1}{\mu}\left(X(x)Y'(x)\big|_{x=0}^{x=L} - \int_0^L X'(x)Y'(x)\,dx \right)$$

$$= \frac{1}{\mu}\int_0^L X'(x)Y'(x)\,dx.$$

One more integration by parts using $u = X'$, $v = Y'$ and the boundary conditions $Y(0) = Y(L) = 0$ yields

$$\int_0^L X(x)Y(x)\,dx = -\frac{1}{\mu}\int_0^L X''(x)Y(x)\,dx.$$

Now substitute $X'' = -\lambda X$ and move all terms to the left to obtain

$$\left(1 - \frac{\lambda}{\mu}\right)\int_0^L X(x)Y(x)\,dx = 0.$$

Since $\lambda/\mu \neq 1$, the integral must be zero, establishing the orthogonality condition.

The argument for more general boundary conditions is similar; see exercise 22 for the boundary conditions $X'(0) = X(L) = 0$.

To show the plausibility of the formula for the generalized Fourier coefficient c_n given in part (ii), suppose that the eigenfunction expansion converges to $f(x)$ at each point of $0 \leq x \leq L$:

$$f(x) = \sum_{n=1}^{\infty} c_n X_n(x), \qquad 0 \leq x \leq L.$$

Mimicking the calculations in the previous example, choose a fixed integer m, multiply through by $X_m(x)$, and assume integration and summation can be interchanged:

$$\int_0^L f(x)X_m(x)\,dx = \sum_{n=1}^{\infty} c_n \int_0^L X_n(x)X_m(x)\,dx.$$

Because of orthogonality, the integral inside the summation is zero except when $n = m$, leaving

$$\int_0^L f(x)X_m(x)\,dx = c_m \int_0^L [X_m(x)]^2\,dx,$$

from which the formula (9.11) for c_m follows immediately. ∎

9.4   Fourier Methods

EXAMPLE 15   *Expand the function*

$$\frac{T_L x}{L}, \qquad 0 \le x \le L,$$

in the eigenfunctions of the eigenvalue problem

$$X''(x) + \lambda X(x) = 0, \qquad X'(0) = X'(L) = 0.$$

We leave to exercise 6 the task of showing that this eigenvalue problem has no nontrivial solutions for $\lambda \le 0$.

For $\lambda > 0$, the general solution is $X_g = C_1 \cos \sqrt{\lambda} x + C_2 \sin \sqrt{\lambda} x$. The boundary condition $X'(0) = 0$ immediately forces $C_2 = 0$. The second boundary condition $X'(L) = 0$ then requires

$$-C_1 \sqrt{\lambda} \sin \sqrt{\lambda} L = 0.$$

We reject the choice $C_1 = 0$ because we want nontrivial solutions. The alternative is

$$\sin \sqrt{\lambda} L = 0 \Rightarrow \sqrt{\lambda} L = n\pi, \qquad n = 0, 1, 2, \dots .$$

EIGENVALUES AND EIGENFUNCTIONS   That is, the eigenvalues and corresponding eigenfunctions are

$$\lambda_n = \frac{n^2 \pi^2}{L^2}, \qquad X_n(x) = \cos \frac{n\pi x}{L}, \qquad n = 0, 1, 2, \dots .$$

To complete the expansion, we need constants c_n such that

$$\sum_{n=0}^{\infty} c_n \cos \frac{n\pi x}{L} = \frac{T_L x}{L}, \qquad 0 \le x \le L,$$

where the equality is understood in the sense of mean square convergence, as in part (ii) of the eigenfunction expansion theorem. To use formula (9.11) for c_n, compute

$$\int_0^L [X_n(x)]^2 \, dx = \int_0^L \cos^2 \frac{n\pi x}{L} \, dx = \begin{cases} L, & n = 0 \\ L/2, & n = 1, 2, \dots \end{cases}$$

For $n = 0$, (9.11) yields

$$c_0 = \frac{1}{L} \int_0^L \frac{T_L x}{L} \, dx = \frac{T_L}{2},$$

and for $n = 1, 2, \dots$,

$$c_n = \frac{2}{L} \int_0^L \frac{T_L x}{L} \cos \frac{n\pi x}{L} \, dx$$

$$= \frac{2T_L}{n^2\pi^2}\left((-1)^n - 1\right), \qquad n = 1, 2, \dots,$$

or

$$c_n = \begin{cases} -4T_L/n^2\pi^2, & n = 1, 3, \dots \\ 0 & n = 2, 4, \dots. \end{cases}$$

Hence, the eigenfunction expansion is

$$\frac{T_L}{2} - \sum_{n=1,3,\dots} \frac{4T_L}{n^2\pi^2} \cos\frac{n\pi x}{L} = \frac{T_L x}{L}, \qquad 0 \le x \le L, \qquad (9.12)$$

where equality is again understood in the sense of mean square convergence. ∎

To what function $f(x)$ does this eigenfunction expansion converge when x is not in $0 \le x \le L$? Since f is a sum of a constant and cosines, it must have the same symmetry across the y-axis as those two functions; that is, f must be an **even** function,

$$f(x) = f(-x).$$

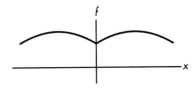

Figure 9.21: An even function.

Figure 9.21 illustrates a generic even function, which is its own reflection across the y-axis.

For $-L \le x < 0$, the eigenfunction expansion above evidently must converge in mean square to the even extension of $T_L x/L$, the dashed line shown in figure 9.22. Since the terms in the eigenfunction expansion each have period $2L$, the expansion converges in mean square to the $2L$-periodic extension of this function for x outside of $-L \le x \le L$.

The rate of convergence of the expansion is illustrated in figure 9.23. ∎

Solving the initial-boundary value problem of the next example will put that eigenfunction expansion to use.

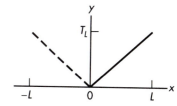

Figure 9.22: The function $T_L x/L$, $0 \le x \le L$ (solid line), and its even extension (dashed line) to $-L \le x < 0$.

EXAMPLE 16 *Suppose both ends of a thin, insulated rod of length L are insulated and its initial temperature varies linearly from zero at $x = 0$ to T_L at $x = L$. Using the model derived in exercise 5 of section 9.3,*

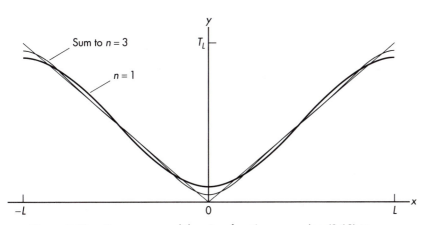

Figure 9.23: Convergence of the eigenfunction expansion (9.12) to the even extension of $T_L x/L$ from $0 \le x \le L$.

$$\frac{\partial T(x,t)}{\partial t} = \kappa \frac{\partial^2 T(x,t)}{\partial x^2},$$

$$T_x(0,t) = T_x(L,t) = 0, \qquad t > 0,$$

$$T(x,0) = T_L x/L, \qquad 0 < x < L,$$

find the variation with position and time of the temperature of the rod.

Assuming $T(x,t) = X(x)\Theta(t)$, substituting in the partial differential equation, and separating variables leads to

Separate variables.

$$\frac{\Theta'(t)}{\kappa\Theta(t)} = \frac{X''(x)}{X(x)} = -\lambda,$$

where the separation constant λ arises because the variables x and t are independent. The corresponding ordinary differential equations are

$$\Theta'(t) + \kappa\lambda\Theta(t) = 0,$$

$$X''(x) + \lambda X(x) = 0.$$

Applying the boundary condition $T_x(0,t) = X'(0)\Theta(t) = 0$ leads to $X'(0) = 0$. Similarly, $X'(L) = 0$. Coupling these boundary conditions with the differential equation for X, we have the eigenvalue problem

EIGENVALUE PROBLEM

$$X''(x) + \lambda X(x) = 0, \qquad X'(0) = X'(L) = 0.$$

From the preceding example, its eigenvalues and eigenfunctions are

EIGENVALUES AND EIGENFUNCTIONS

$$\lambda_n = \frac{n^2\pi^2}{L^2}, \qquad X_n(x) = \cos\frac{n\pi x}{L}, \qquad n = 0, 1, 2, \ldots.$$

The corresponding solutions of the Θ equation are

$$\Theta_n(t) = c_n e^{-\kappa n^2\pi^2 t/L^2}.$$

Since every member of the family of functions

$$T_n(x,t) = X_n(x)\Theta_n(t) = c_n e^{-\kappa n^2\pi^2 t/L^2}\cos\frac{n\pi x}{L}, \qquad n = 0, 1, 2, \ldots$$

solves the homogeneous partial differential equation and satisfies the homogeneous boundary conditions, so does the sum

$$T(x,t) = \sum_{n=0}^{\infty} T_n(x,t) = \sum_{n=0}^{\infty} c_n e^{-\kappa n^2\pi^2 t/L^2}\cos\frac{n\pi x}{L},$$

provided the infinite sum converges uniformly.

Satisfying the initial condition requires choosing the constants c_n so that

$$\sum_{n=0}^{\infty} c_n \cos\frac{n\pi x}{L} = \frac{T_L x}{L}, \qquad 0 < x < L;$$

that is, the constants c_n are the coefficients of the eigenfunction expansion (9.12) of $T_L x/L$ on $0 \le x \le L$. These were found in the preceding example.

Using the expression for c_n from that example, we have the (formal) solution of the initial-boundary value problem,

$$T(x,t) = \frac{T_L}{2} - \sum_{n=1,3,\ldots} \frac{4T_L}{n^2 \pi^2} e^{-\kappa n^2 \pi^2 t/L^2} \cos \frac{n\pi x}{L}.$$

Note that this solution decays to a steady state of $T_L/2$, the average of the initial temperature distribution, as the heat energy trapped in the rod by the insulation redistributes itself uniformly. Chapter exercise 13 asks you to illustrate this property in general. ∎

EXAMPLE 17 *Example 11 of section 9.3 derived a model for an insulated wire being heated by a current passing through it. With the particular choice of initial condition given here, that model is*

$$\frac{\partial T(x,t)}{\partial t} = \kappa \frac{\partial^2 T(x,t)}{\partial x^2} + \frac{H}{c\rho},$$

$$T(0,t) = T(L,t) = 0, \qquad t > 0,$$

$$T(x,0) = M + \frac{Hx(L-x)}{2k}, \qquad 0 < x < L.$$

The term H arises from the heat generated by the current. Assume H is constant and find a solution of this initial-boundary value problem.

Separation of variables assumes a solution of the form

$$T(x,t) = X(x)\Theta(t)$$

and substitutes in the differential equation,

$$X(x)\Theta'(t) = \kappa X''(x)\Theta(t) + H/c\rho.$$

But the nonhomogeneous term $H/c\rho$ prevents separation into expressions involving functions of x alone and functions of t alone. We must find a new problem in which the differential equation is *homogeneous*.

The key is the solution $E(x)$ of the *equilibrium problem*,

$$0 = \kappa E''(x) + H/c\rho, \qquad E(0) = E(L) = 0,$$

the model that governs when temperature is constant in time. We readily find that the equilibrium temperature is

$$E(x) = \frac{Hx(L-x)}{2\kappa}.$$

Since the equilibrium temperature accounts for the effects of the nonhomogeneous term $H/c\rho$, we expect that the difference

$$u(x,t) = T(x,t) - E(x)$$

should satisfy a *homogeneous* differential equation. Indeed, substituting $T(x,t) = u(x,t) + E(x)$ into $T_t = \kappa T_{xx} + H$ yields

$$\frac{\partial u(x,t)}{\partial t} = \kappa \frac{\partial^2 u(x,t)}{\partial x^2}, \tag{9.13}$$

a homogeneous heat equation. Furthermore, we find the boundary conditions

$$u(0,t) = u(L,t) = 0, \qquad t > 0, \tag{9.14}$$

and the initial condition

$$u(x,0) = T(x,0) - E(x) = M, \qquad 0 \le x \le L.$$

The new dependent variable u satisfies a homogeneous partial differential equations subject to homogeneous boundary conditions. To apply separation of variables, assume

$$u(x,t) = X(x)\Theta(t).$$

Then *precisely* the same calculations as in example 14 lead to the eigenvalue problem (9.5). Its eigenfunctions are

$$X_n(x) = \sin \frac{n\pi x}{L}, \qquad n = 1, 2, \ldots .$$

From the eigenvalues $\lambda_n = n^2 \pi^2 / L^2$, the corresponding solutions of the $\Theta(t)$ equation are

$$\Theta_n(t) = c_n e^{-n^2 \pi^2 \kappa t / L^2}, \qquad n = 1, 2, \ldots .$$

If it converges appropriately, the sum

$$u(x,t) = \sum_{n=1}^{\infty} c_n e^{-n^2 \pi^2 \kappa t / L^2} \sin \frac{n\pi x}{L}$$

satisfies the homogeneous differential equation (9.13) and boundary conditions (9.14). The initial condition demands

$$\sum_{n=1}^{\infty} c_n \sin \frac{n\pi x}{L} = M, \qquad 0 < x < L. \tag{9.15}$$

Using the orthogonality relation (9.7) and the formula (9.11) for c_n, we find

$$c_n = \frac{2}{L} \int_0^L M \sin \frac{n\pi x}{L} \, dx$$

$$= \frac{2M}{n\pi} (1 - (-1)^n) = \begin{cases} 4M/n\pi, & n = 1, 3, \ldots \\ 0, & n = 2, 4, \ldots . \end{cases}$$

With this value of c_n, the solution of the original initial-boundary value problem is

$$T(x,t) = E(x) + u(x,t)$$

$$= \frac{Hx(L-x)}{2k} + \sum_{n=1}^{\infty} c_n e^{-n^2\pi^2\kappa t/L^2} \sin\frac{n\pi x}{L}. \qquad \blacksquare$$

For $-L \le x \le 0$, the series given by (9.15) should converge to the odd extension of $f(x) = M, \, 0 \le x \le L$. The resulting *square wave* is shown in figure 9.24 along with the first few terms of the corresponding expansion. Note that

- The convergence is slower for this discontinuous function than for either the continuously differentiable function of figures 9.18 and 9.20 or the merely continuous triangular wave of figure 9.23,
- At the jump discontinuity at $x = 0$, the eigenfunction expansion is converging to the average of the limiting values from the left and the right.

9.4.3 Fourier Series

The eigenfunction expansion ideas discussed previously also apply to eigenvalue problems with **periodic boundary conditions**

$$X''(x) + \lambda X(x) = 0, \qquad -L < x < L,$$

$$X(-L) - X(L) = 0$$

$$X'(-L) - X'(L) = 0.$$

The eigenfunctions of this problem are 1, $\sin\pi x/L$, $\cos\pi x/L$, $\ldots$, $\sin n\pi x/L$, $\cos n\pi x/L$, $\ldots$. (See exercise 26.) The corresponding eigenfunction expansion is a **Fourier series**

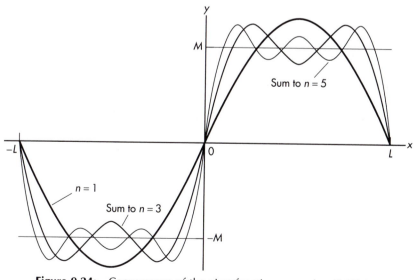

Figure 9.24: Convergence of the eigenfunction expansion (9.15) to the odd extension of $f(x) = M$ from $0 \le x \le L$.

$$a_0 + \sum_{n=1}^{\infty} \left(a_n \cos \frac{n\pi x}{L} + c_n \sin \frac{n\pi x}{L} \right), \qquad (9.16)$$

where the **Fourier coefficients** a_n, b_n are defined by

$$a_n = \begin{cases} \dfrac{1}{2L} \displaystyle\int_{-L}^{L} f(x)\,dx, & n = 0, \\[2ex] \dfrac{1}{L} \displaystyle\int_{-L}^{L} f(x) \cos \dfrac{n\pi x}{L}\,dx, & n = 1, 2, \ldots, \end{cases}$$

$$b_n = \frac{1}{L} \int_{-L}^{L} f(x) \sin \frac{n\pi x}{L}\,dx, \qquad n = 1, 2, \ldots. \qquad (9.17)$$

To state (without proof) a stronger convergence result for this particular set of eigenfunctions, we require some terminology. A function is **piecewise continuous** on the interval $-L \le x \le L$ if it is continuous at all but a finite number of points, called *jump discontinuities*, where the limits from the left and the right exist but differ in value. (See figure 11.2 of section 11.1.) The limit from the left of f at x is

$$\lim_{h \to 0^+} f(x - h),$$

where $h \to 0^+$ indicates that h approaches zero through positive values. The limit from the right is similar with $f(x + h)$ replacing $f(x - h)$.

Theorem 9.2 (Fourier Expansion Theorem)   Let f and f' each be piecewise continuous on $-L \le x \le L$. Then at every point of $-L \le x \le L$, the Fourier expansion (9.16) converges to

$$\frac{1}{2} \left[\lim_{h \to 0^+} f(x - h) + \lim_{h \to 0^+} f(x + h) \right],$$

where f is defined by its periodic extension $f(x) = f(x \pm 2L)$ for values of x outside of $-L \le x \le L$.

In other words, the Fourier expansion converges to the function at points of continuity of f and to the midpoint of the jump at points of discontinuity.

The Fourier coefficient formulas are again consequences of orthogonality. The appropriate relation for the cosine is

$$\int_{-L}^{L} \cos \frac{m\pi x}{L} \cos \frac{n\pi x}{L}\,dx = \begin{cases} 0, & m \ne n \\ L, & m = n. \end{cases}$$

There is a corresponding relation with sine replacing cosine as well as an orthogonality condition for mixed products,

$$\int_{-L}^{L} \sin \frac{m\pi x}{L} \cos \frac{n\pi x}{L}\,dx = 0, \qquad \text{for all } m, n.$$

See exercise 28.

If the Fourier series (9.16) contains only sine functions (i.e., $a_n \equiv 0$), it is called a *sine series*. As we saw in example 14, an odd function is expanded in a sine series because it has the same symmetry about the origin as the sine function, $f(x) = -f(-x)$, and vice versa.

Similarly, an even function, $f(x) = f(-x)$, is expanded in a series containing only cosines (and the constant term a_0) because it has the same symmetry about the origin as the cosine function. That series is called a *cosine series*. See figure 9.21.

Since the eigenfunction expansions we derived earlier are either sine or cosine series, the stronger convergence result of the Fourier expansion theorem applies, as shown in figure 9.24 for the square wave function. (Note the jump discontinuity at $x = 0$.) However, not all eigenfunction expansions are special cases of Fourier series, as the exercises illustrate.

9.4.4   Summary

To solve a linear, *homogeneous* partial differential equation by separation of variables:

1. Assume the solution has the product form $T(x, t) = X(x)\Theta(t)$.

2. Substitute in the partial differential equation, separate variables, and introduce a separation constant $-\lambda$.

3. Use the *homogeneous* boundary conditions to obtain an eigenvalue problem for $X(x)$.

4. Solve the eigenvalue problem to obtain eigenfunction, eigenvalue pairs $X_n(x)$, λ_n. Using the eigenvalues λ_n, solve the Θ equation for Θ_n. The family of functions $X_n(x)\Theta_n(t)$ satisfies both the differential equation and the boundary conditions, as does $\sum c_n X_n(x)\Theta(t)$.

5. To satisfy the initial condition $T(x, 0) = f(x)$, require that c_n be the coefficients in an eigenfunction expansion for $f(x)$, $\sum c_n X_n(x) = f(x)$.

6. Compute c_n from (9.11) and write the complete solution of the initial-boundary value problem as $\sum c_n X_n(x)\Theta_n(t)$.

If the differential equation or the boundary conditions are not homogeneous, subtract the solution of the equilibrium boundary-value problem. All these manipulations assume that differentiation and the infinite sum can be interchanged.

> The eigenfunction expansion theorem stated here is essentially that of Coddington and Levinson [4]. For a more accessible introduction to eigenfunction expansions, see Weinberger's text [20].

EXERCISE GUIDE

To gain experience . . .	Try exercises
Separating variables	1–2
Verifying general solutions of eigenvalue problems	4
Finding eigenvalues and eigenfunctions	5–6, 11, 26
Finding eigenfunction expansions	7–12
With superposition in the heat equation	13
With odd and even extensions	7–10, 14, 29
With orthogonality and generalized Fourier coefficients	7–12, 15, 22, 27–29
Solving initial boundary-value problems	16–21
With physical interpretations of initial boundary-value problems	18–21
With equilibrium boundary value problems	23–25
With nonhomogeneous equations and boundary conditions	25
Analyzing and interpreting solution behavior	3

In exercises 1 and 2, suppose $T(x, t) = X(x)\Theta(t)$, where $T(x, t)$ satisfies the given partial differential equation and boundary conditions.

   i. Find an eigenvalue problem for $X(x)$ and an ordinary differential equation for $\Theta(t)$.

   ii. Solve the eigenvalue problem. Solve the equation defining $\Theta(t)$.

   iii. Write a sum which satisfies both the differential equation and the boundary conditions.

1. $T_t = \kappa T_{xx}$, $T_x(0, t) = T(L, t) = 0$

2. $T_t = 4T_{xx}$, $T(0, t) = T_x(L, t) = 0$

3. The thermal diffusivity values for copper and wood are $\kappa = 0.11$ cm²/s and $\kappa = 0.0019$ cm²/s, respectively. Compare the rate of temperature decay in identical thin rods with insulated sides made from these two materials; e.g., if the ends of the rods are at temperature zero and the initial temperatures are identical, how do the midpoint temperatures compare as time passes?

4. The book claims that the general solution of $X''(x) + \lambda X(x) = 0$ is

$$\lambda < 0: \quad X(x) = C_1 e^{\sqrt{-\lambda}x} + C_2 e^{-\sqrt{-\lambda}x}$$

$$\lambda = 0: \quad X(x) = C_1 + C_2 x$$

$$\lambda > 0: \quad X(x) = C_1 \cos \sqrt{\lambda}x + C_2 \sin \sqrt{\lambda}x.$$

Verify this claim.

5. Show that $\lambda = 0$ is not an eigenvalue of

$$X''(x) + \lambda X(x) = 0, \qquad X(0) = X(L) = 0.$$

That is, show that the boundary conditions force $C_1 = C_2 = 0$ in the general solution of this equation when $\lambda = 0$.

6. Show that the eigenvalue problem

$$X''(x) + \lambda X(x) = 0, \qquad X'(0) = X'(L) = 0,$$

which arose in example 16, has no nontrivial solutions for $\lambda \leq 0$.

7. Find the expansion of $f(x) = T_L x/L$, $0 \leq x \leq L$, using the eigenfunctions of example 14, $X_n(x) =$

$\sin n\pi x/L$, $n = 1, 2, \dots$. To what function does the eigenfunction expansion converge for $-L \le x \le L$?

8. Repeat the previous problem for $f(x) = L - x$.

9. Find the expansion of $f(x) = L - x$, $0 \le x \le L$, using the eigenfunctions of example 15, $X_n(x) = \cos n\pi x/L$, $n = 0, 1, 2, \dots$. To what function does the eigenfunction expansion converge for $-L \le x \le L$?

10. Repeat exercise 9 for $f(x) = 4$.

11. Solve the eigenvalue problem $X''(x) + \lambda X(x) = 0$, $X'(0) = X(\pi) = 0$. Using those eigenfunctions, expand $f(x) = x/\pi$, $0 \le x \le \pi$. To what function does the eigenfunction expansion converge for $-\pi \le x \le \pi$?

12. Using the eigenfunctions found in the previous problem, expand $f(x) = x^2 - \pi x$, $0 \le x \le \pi$. To what function does the eigenfunction expansion converge for $-\pi \le x \le \pi$?

13. The text claims that any finite sum of solutions

$$T_n(x,t) = c_n e^{-n^2 \pi^2 \kappa t/L} \sin \frac{n\pi x}{L}$$

of the heat equation

$$\frac{\partial T(x,t)}{\partial t} = \kappa \frac{\partial^2 T(x,t)}{\partial x^2}$$

subject to the boundary conditions

$$T(0,t) = T(L,t) = 0$$

again satisfies both the partial differential equation and the boundary conditions. Explicitly verify this claim for the sum $T_1(x,t) + T_2(x,t)$ by substituting $T_1 + T_2$ into the heat equation and into the boundary conditions.
(a) Use the explicit formulas for T_1, T_2.
(b) Use only the knowledge that T_1, T_2 individually satisfy the heat equation and the boundary conditions.

14. Sketch and write a formula for the odd extension and the even extension of the following functions:
(a) $\cos \pi x$, $0 < x < 1$
(b) $\sin \pi x$, $0 < x < 1$
(c) $\sqrt{x}$, $0 < x < 4$
(d) $|x|$, $0 < x < L$
(e) $\begin{cases} -1, & 0 < x \le 2 \\ 1, & 2 < x < 4 \end{cases}$

15. Verify the text's calculation in example 14 of the generalized Fourier coefficient c_n for $f(x) = Mx(L-x)$, $0 \le x < L$. (See (9.8).)

16. Solve the initial–boundary value problem

$$T_t(x,t) = \kappa T_{xx}(x,t)$$
$$T(0,t) = T(4,t) = 0, \ t > 0$$
$$T(x,0) = -2, \ 0 < x < 4.$$

17. Solve the initial–boundary value problem

$$T_t(x,t) = \kappa T_{xx}(x,t)$$
$$T_x(0,t) = T_x(4,t) = 0, \ t > 0$$
$$T(x,0) = 8 - 2x, \ 0 < x < 4.$$

18. Solve the initial–boundary value problem

$$T_t(x,t) = \kappa T_{xx}(x,t)$$
$$T_x(0,t) = T(\pi,t) = 0, \ t > 0$$
$$T(x,0) = x^2 - \pi x, \ 0 < x < \pi.$$

Describe a physical situation which this problem might model. As time passes, does the solution appear to approach the proper equilibrium solution?

19. Repeat exercise 18 with the initial condition $T(x,0) = x/\pi$.

20. Solve the initial–boundary value problem

$$T_t(x,t) = \kappa T_{xx}(x,t)$$
$$T(0,t) = 4, \ T(L,t) = 2, \ t > 0$$
$$T(x,0) = -2, \ 0 < x < L.$$

Describe a physical situation which this problem might model.
Hint: To solve by separation of variables, find an equilibrium solution $E(x)$ which satisfies the nonhomogeneous boundary conditions. Show that $u(x,t) = T(x,t) - E(x)$ satisfies a problem with homogeneous boundary conditions.

21. Repeat exercise 20 with the boundary conditions $T_x(0,t) = 0, T(L,t) = 8$.

22. Verify the orthogonality relation $\int_0^L X(x)Y(x)\,dx = 0$ when X, Y are eigenfunctions of

$$X''(x) + \lambda X(x) = 0, \ X'(0) = X(L) = 0,$$

corresponding to distinct eigenvalues.

23. Solve the equilibrium boundary-value problem

$$0 = \kappa E''(x) + H/c\rho, \qquad E(0) = E(L) = 0,$$

encountered in example 17.

24. Example 16 shows that the limit in time of the solution of the initial-boundary value problem

$$\frac{\partial T(x,t)}{\partial t} = \kappa \frac{\partial^2 T(x,t)}{\partial x^2},$$

$$T_x(0,t) = T_x(L,t) = 0, \qquad t > 0,$$

$$T(x,0) = \frac{T_L x}{L}, \qquad 0 < x < L,$$

is $T_L/2$. Obtain this equilibrium solution directly from the model that governs at equilibrium.

25. Verify the computation in example 17: $u(x,t) = T(x,t) - E(x)$ satisfies the homogeneous equation $u_t = \kappa u_{xx}$ if T solves the nonhomogeneous equation $T_t = \kappa T_{xx} + H/c\rho$ and E solves the equilibrium equation $0 = \kappa E''(x) + H/c\rho$. Do you need a formula for $E(x)$? Does the result of this exercise depend on whether H is constant?

26. Verify that 1, $\sin \pi x/L$, $\cos \pi x/L$, ..., $\sin n\pi x/L$, $\cos \pi x/L$, ... are eigenfunctions of the periodic eigenvalue problem

$$X''(x) + \lambda X(x) = 0$$

$$X(-L) - X(L) = 0$$

$$X'(-L) - X'(L) = 0.$$

Find the corresponding eigenvalues.

27. Using formulas involving the sum and difference of angles, evaluate the integrals and verify the following orthogonality relations:

(a) $\displaystyle\int_{-L}^{L} \sin \frac{m\pi x}{L} \sin \frac{n\pi x}{L}\, dx = \begin{cases} 0, & m \neq n \\ L, & m = n. \end{cases}$

(b) $\displaystyle\int_{-L}^{L} \cos \frac{m\pi x}{L} \cos \frac{n\pi x}{L}\, dx = \begin{cases} 0, & m \neq n \\ L, & m = n \neq 0. \end{cases}$

(c) $\displaystyle\int_{-L}^{L} \cos \frac{m\pi x}{L} \sin \frac{n\pi x}{L}\, dx = 0$ for all m, n

28. Using the orthogonality relations of the previous problem, derive the Fourier coefficient formulas (9.17). Assume that the Fourier expansion converges to $f(x)$ for $-L \leq x \leq L$,

$$f(x) = \frac{a_0}{2} + \sum_{n=1}^{\infty} \left(a_n \cos \frac{n\pi x}{L} + b_n \sin \frac{n\pi x}{L} \right),$$

and that summation and integration can be interchanged.

At a jump discontinuity, the Fourier expansion does not converge to $f(x)$, but if there are only a finite number of such points, they can not affect the values of the integrals in the formulas for the coefficients a_n, b_n.

29. Example 14 constructed a Fourier sine expansion of the odd function

$$f(x) = \begin{cases} Mx(L-x), & 0 \leq x < L, \\ Mx(L+x), & -L < x < 0. \end{cases}$$

(See figure 9.20.) Using the Fourier coefficient formula (9.17), verify directly that $a_n = 0$, $n = 0, 1, 2, \ldots$.

9.5 CHAPTER EXERCISES

EXERCISE GUIDE

To gain experience . . .	Try exercises
Deriving models and boundary conditions	1–2, 9–11
Solving boundary-value problems	2–8
Separating variables	12
With orthogonality and eigenfunction expansion	14–16
Solving eigenvalue problems	15
Solving initial boundary-value problems	17–19
Analyzing and interpreting solution behavior	13

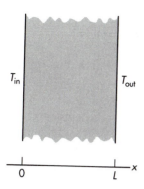

T_{in} T_{out}

0   L x

Figure 9.25: A cross section of the wall of a house.

1. Figure 9.25 shows a cross section of the wall of a house. Suppose that the temperature in the wall varies only from left to right, not vertically, as would happen if the inner and outer temperatures T_{in} and T_{out} were uniform.
   (a) Assuming that the thermal conductivity k of the material in the wall is constant, derive the model

   $$T''(x) = 0, \ T(0) = T_{\text{in}}, \ T(L) = T_{\text{out}}$$

   for the temperature profile $T(x)$ in the wall. Sketch the wall and show the control volume you use.
   (b) In fact, the thermal conductivity of the wall of a house is not constant. It varies with the material (wall board, insulation, sheathing, siding, etc.) one encounters at various positions within the wall. Derive the model

   $$(k(x)T'(x))' = 0, \ T(0) = T_{\text{in}}, \ T(L) = T_{\text{out}}$$

   for the temperature profile $T(x)$ in the wall. Can you use the same control volume as in part (a)?

2. Is the heat loss model developed in section 2.3 an equilibrium-temperature model? What is the equilibrium temperature of the house in that model?

In exercises 3–8, solve the given boundary-value problem.

3. $y'' - y = 4, \ y'(0) = 4, \ y(2) = -1$

4. $2y'' - 4y' + 10y = 0, \ y(-1) = 1, \ y(1) = 0$

5. $y'' + y = 0, \ y(0) + y'(0) = 0, \ y(\pi/2) - y'(\pi/2) = 0$

6. $y'' + y = 2\cos t, \ y(0) = 0, \ y(\pi/2) = 0$

7. $x^2 y'' - 2xy' + 2y = 0, \ y(1) = 2, \ y(2) + y'(2) = 11$
   (Homogeneous solutions are x, x^2.)

8. $x^2 y'' - 2xy' + 2y = 4x^2, \ y(1) = 2, \ y(2) = 4$
   (Homogeneous solutions are x, x^2.)

9. Suppose the wall of a house has cooled uniformly to the outside temperature T_{out}. Derive the model

   $$T_t = \kappa T_{xx},$$
   $$T(0,t) = T_{\text{in}}, \qquad T(L,t) = T_{\text{out}}, \qquad t > 0,$$
   $$T(x,0) = T_{\text{out}},$$

   that describes how the temperature profile through the wall changes when the inner side of the wall is suddenly raised to the temperature T_{in} by turning on the furnace in the house. See figure 9.25. (Since the wall's height is much greater than its width, the only significant heat flow is from left to right. Construct a control volume within the wall, but consider only heat flow through its left and right boundaries.)

10. The house heat-loss model considered in section 2.3 supposed that temperature was uniform throughout the house. This exercise asks you to improve that model by incorporating variations with position.
    (a) Sketch the cross section of a flat roofed (for simplicity) house. Using a rectangular control volume and the law of conservation of energy, derive a governing partial differential equation.
    (b) Again for simplicity, suppose the walls of the house are at the outside temperature T_{out}. Complete the model by coupling the differential equation of part (a) with appropriate boundary and initial conditions.

11. Suppose the right end of the rod in figure 9.11, page 366, is covered with a poor insulator having thermal conductivity δ. Argue that the rate of heat flow through this poor insulator is $A(L)\delta(T(L,t) - T_L)$. But conservation of energy requires that the flow of heat energy through the end cap must equal the rate of heat flow through the end of the rod, $-kA(L)T_x(L,t)$. Equate these two expressions to derive a boundary condition of the form

    $$T(L,t) + aT_x(L,t) = b, \qquad t > 0.$$

    Express the constants a, b in terms of $k, \delta, T_{\text{out}}$. What would be the form of the boundary condition if the insulated cap were applied at $x = 0$?

    This boundary condition is known as a boundary condition of the *third kind*.

12. Suppose $T(x,t) = X(x)\Theta(t)$, where $T(x,t)$ satisfies

    $$xT_t = \kappa (xT_x)_x, \ T_x(0,t) = T(L,t) = 0.$$

(The equation and boundary conditions arise in a model of heat flow in a thin circular disk whose top and bottom are insulated; see exercise 8 of section 9.3.) Find an eigenvalue problem for $X(x)$ and an ordinary differential equation for $\Theta(t)$. Can the equation for $X(x)$ be solved by any methods you have studied?

13. The results of example 16 suggest that the initial heat energy in a rod whose ends are insulated simply redistributes itself uniformly at equilibrium. Verify this observation in general by showing that as $t \to \infty$, the solution of the initial-boundary value problem

$$\frac{\partial T(x,t)}{\partial t} = \kappa \frac{\partial^2 T(x,t)}{\partial x^2},$$

$$T_x(0,t) = T_x(L,t) = 0, \qquad t > 0,$$

$$T(x,0) = g(x), \qquad 0 < x < L,$$

tends to

$$\frac{1}{L} \int_0^L g(x)\, dx.$$

14. The eigenfunction expansion theorem can be extended to the more general *self-adjoint* problem

$$-(p(x)X'(x))' + q(x)X(x) = \lambda X(x)$$

$$aX(0) + bX'(0) = 0$$

$$cX(L) + dX'(L) = 0,$$

where p is continuously differential and nonzero on $0 \le x \le L$ and q is continuous there. The eigenfunctions are then *orthogonal with weight* p,

$$\int_0^L p(x)X_n(x)X_m(x)\, dx = 0, \qquad m \neq n.$$

(a) Verify the orthogonality relation for the boundary conditions $X(0) = X(L) = 0$.

(b) Verify the orthogonality relation for the boundary conditions $X'(0) = X'(L) = 0$.

(c) Verify the orthogonality relation for the general boundary conditions given in the problem statement.

15. Solve the eigenvalue problem $X''(x) + \lambda X(x) = 0$, $X(0) = X'(L) = 0$. Using those eigenfunctions, expand $f(x) = 2, 0 \le x \le L$. To what function does the eigenfunction expansion converge for $-L \le x \le L$?

16. Using the eigenfunctions found in the previous problem, expand $f(x) = x/L$. To what function does the eigenfunction expansion converge for $-L \le x \le L$?

17. Solve the initial–boundary value problem

$$T_t(x,t) = \kappa T_{xx}(x,t)$$

$$T(0,t) = T_x(L,t) = 0, \qquad t > 0$$

$$T(x,0) = 2, \qquad 0 < x < \pi.$$

Describe a physical situation that this problem might model. As time passes, does the solution appear to approach the proper equilibrium solution?

18. Repeat exercise 17 with the initial condition $T(x,0) = x/L$.

19. Find the vertical temperature profile of the computer chip modeled in example 13, page 372. That is, solve

$$\frac{\partial T(y,t)}{\partial t} = \kappa \frac{\partial^2 T(y,t)}{\partial y^2}$$

$$T(0,t) = T_0, \qquad T(L,t) = 0, \qquad t > 0$$

$$T(y,0) = 0, \qquad 0 < y < h.$$

20. A long circular pipe of inner radius r and outer radius R is carrying steam at temperature T_s. Its outer surface is at temperature T_{out}. Show that the equilibrium temperature in the wall of the pipe is modeled by

$$-\kappa(xT'(x))' = 0, \qquad T(r) = T_s, \qquad T(R) = T_{\text{out}},$$

where x is the radial coordinate. What assumptions must you make about the direction of heat flow?

9.6 CHAPTER PROJECTS

1. **Shooting method** Numerical methods for solving initial-value problems, such as the Euler method or Runge-Kutta methods, can be adapted to solve boundary-value problems by using the *shooting method*: the solution

of a boundary-value problem is the solution of an initial-value problem involving the same differential equation with an appropriate but unknown initial slope. For example, the solution of the boundary-value problem

$$y'' + y = 0, \qquad y(0) = 1, \qquad y(1) = 2, \qquad (9.18)$$

is also a solution of the initial-value problem

$$y'' + y = 0, \qquad y(0) = 1, \qquad y'(0) = v_i, \qquad (9.19)$$

for an appropriate choice of initial slope v_i. The shooting method applies a numerical method to the initial-value problem and varies v_i until the second boundary condition $y(1) = 2$ is satisfied.

Verify the claims of the preceding paragraph by solving the boundary-value problem (9.18), finding the slope of that solution at $x = 0$, solving the initial-value problem (9.19) using the value of v_i just found, and showing that the two solutions are identical.

To implement the shooting method analytically, write the solution of the initial-value problem (9.19) using v_i as a parameter. Require that the solution of the initial-value problem satisfy the second boundary condition $y(1) = 2$ to obtain an equation for the desired value of v_i. Show that when v_i has this value, the solution of the initial-value problem is again the solution of the boundary-value problem.

Implement the shooting method numerically. Use available software for an initial-value solver to compute the numerical solution of several boundary-value problems of your choice from the exercises of section 9.2. Vary the initial slope until the second boundary condition is satisfied to sufficient accuracy. How sensitive is the solution of the boundary-value problem to the value of the initial slope? Can you develop an algorithm for varying the initial slope that leads methodically to a solution of the boundary-value problem?

The bellicose name *shooting* comes from thinking of the graph of the solution of the boundary-value problem as the trajectory of a cannon ball fired from a gun located at $x = 0$, $y = 1$. The angle of the cannon must be adjusted until the cannon ball hits the point $x = 1, y = 2$.

2. **The minimum principle** At equilibrium the internal temperature of a thin insulated rod is never less than the smaller of its end temperatures; i.e., its minimum temperature occurs at an end. That physical observation is the basis of a theorem known as the *minimum principle* that can be used to establish uniqueness of solutions of boundary-value problems.

   (a) Suppose a thin insulated rod is heated by a current passing through it while its ends are maintained at zero temperature. Show that its equilibrium temperature is modeled by

   $$-\kappa A(x)T''(x) - \kappa A'(x)T'(x) = f(x), \qquad T(0) = T(L) = 0,$$

   where $A(x)$ is the cross-sectional area of the rod, $\kappa > 0$ is its thermal diffusivity, L is its length, and $f(x) > 0$ is a function proportional to the rate at which heat is generated by the current.

(b) Prove that the minimum of $T(x)$ must occur at $x = 0$ or $x = L$. Argue by contradiction. Suppose a negative minimum of $T(x)$ occurs at x_0, $0 < x_0 < L$. Then $T'(x_0) = 0$ and $T''(x_0) \geq 0$. How is the differential equation contradicted?

(c) Using these same ideas, prove the following theorem.

Theorem 9.3 (Minimum Principle) Let $y(x)$ be a solution of the boundary-value problem

$$-y''(x) + P(x)y' + Q(x)y = f(x), \qquad y(0) = y(L) = 0,$$

where $f(x) \geq 0, Q(x) > 0, 0 \leq x \leq L$. Then $y(x) \geq 0, 0 \leq x \leq L$.

State and prove the corresponding maximum principle for $f(x) \leq 0$, $0 \leq x \leq L$.

(d) Prove the following theorem.

Theorem 9.4 (Uniqueness for Boundary-Value Problems) If $Q(x) > 0, 0 \leq x \leq L$, then the boundary-value problem

$$-y''(x) + P(x)y' + Q(x)y = f(x), \ y(0) = a, y(L) = b,$$

has at most one solution.

To begin the proof, suppose the boundary-value problem has two solutions. Of what boundary-value problem is their difference a solution? What do the maximum and minimum principles assert about the solutions of that boundary-value problem?

10

HIGHER-ORDER SOLUTION METHODS II

We shall reach greater and greater platitudes of achievement.
Mayor Richard J. Daley

The solution techniques we developed for constant-coefficient equations solved a great many problems of interest, but there are also many equations with variable coefficients, such as the diffusion equation in a circle. Their solutions are still out of reach. This chapter will develop analytic solution techniques for these variable-coefficient equations.

10.1 REDUCTION OF ORDER

Reduction of order is a technique for finding a second solution of a homogeneous linear equation when we have somehow found a first solution. The steps in the method are straightforward:

1. Given one solution y_1 of a linear, homogeneous, second-order equation, assume there is a second solution in the form $y_2(x) = u(x)y_1(x)$. Substitute this form into the homogeneous equation.

2. Simplify to find a *first-order* equation for u', the derivative of the unknown function. Solve and integrate to obtain u and the second solution $y_2 = uy_1$.

To motivate this technique, suppose we naively applied the characteristic equation method to

$$4y'' - 12y' + 9y = 0,$$

which was solved completely in example 18 of section 6.3.3. We would assume $y = e^{rx}$, substitute in the differential equation, and find the characteristic equation

$$4r^2 - 12r + 9 = 4\left(r - \frac{3}{2}\right)^2 = 0.$$

Its single root is $r = 3/2$, and one solution of $4y'' - 12y' + 9y = 0$ is $y_1 = e^{3x/2}$.

We know the characteristic equation formalism for repeated real roots: Obtain the second solution as x times the first, $y_2 = xe^{rx}$. Generalizing this idea suggests seeking a second solution in the form

Assume $y_2 = uy_1$.

$$y_2(x) = u(x)y_1(x) = u(x)e^{3x/2},$$

where we must determine the unknown function u. We find that

$$y_2' = u'e^{3x/2} + u\frac{3}{2}e^{3x/2},$$

$$y_2'' = u''e^{3x/2} + 3u'e^{3x/2} + \frac{9u}{4}e^{3x/2}.$$

Substitute y_2 into the equation.

Substituting into the differential equation yields

$$4y_2'' - 12y_2' + 9y_2 = 4u''e^{3x/2} + u'(12e^{3x/2} - 12e^{3x/2})$$

$$+ u\left(4\frac{9}{4}e^{3x/2} - 12\frac{3}{2}e^{3x/2} + 9e^{3x/2}\right) \qquad (10.1)$$

The last term in parentheses is zero because $e^{3x/2}$ is a solution of the homogeneous equation. To summarize, we have

$$4y_2'' - 12y_2' + 9y_2 = (4u'')e^{3x/2} = 0,$$

or

$$4u'' = 0.$$

At first glance, the second-order differential equation $4u'' = 0$ for the unknown factor u does not seem to be much of an advance over the original second-order equation (although we could easily integrate this one). A closer look reveals, however, that we actually have a *first-order* equation in the unknown u',

$$4(u')' = 0.$$

Find a first-order equation for $v = u'$.

To formalize that observation, define $v(x) = u'(x)$ and rewrite $4u'' = 0$ as

$$4v' = 0,$$

a *first-order* equation. We have reduced the problem of finding another solution of the second-order equation to the problem of solving the first-order equation $4v' = 0$. Hence, the name *reduction of order*.

The first-order equation $4v' = 0$ is easily solved by a simple anti-differentiation:

Solve for v.

$$v(x) = C,$$

where C is an arbitrary constant. Since $u' = v = C$, another antidifferentiation yields $u(x) = Cx + D$, where D is another arbitrary constant. Because we need only the simplest form of u that meets our needs, not a general family of such functions, we choose $C = 1$ and $D = 0$. From $u(x) = x$, we have the second solution,

Use u to write y_2.

$$y_2(x) = xy_1(x) = xe^{3x/2}.$$

To better appreciate why this reduction of order occurs, let us repeat the calculation that leads from the original second-order equation to the first-order equation $4v' = 0$, this time using the symbol y_1 in place of $e^{3x/2}$. We begin with $y_2 = uy_1$ and find

$$y_2' = u'y_1 + uy_1',$$
$$y_2'' = u''y_1 + 2u'u_1' + uy_1''.$$

Substituting in $4y'' - 12y' + 9y = 0$ yields

$$4y_2'' - 12y_2' + 9y_2 = 4y_1u'' + (8y_1' - 12y_1)u' + u(4y_1'' - 12y_1' + 9y_1).$$

As in equation (10.1), the last expression in parentheses, the coefficient of u, is zero precisely because y_1 is a solution of the original homogeneous differential equation $4y'' - 12y' + 9y = 0$. Hence, we have a homogeneous equation of first order for the unknown u',

$$4y_2'' - 12y_2' + 9y_2 = (4y_1)u'' + (8y_1' - 12y_1)u' = 0.$$

It is easy to argue in general that the substitution $y_2 = uy_1$ reduces *any* homogeneous second-order linear differential equation of which y_1 is a solution to a first-order equation for u', providing the coefficient of y'' is never zero. (See exercise 12.)

EXAMPLE 1   *One solution of*

$$y'' + y' - 12y = 0$$

is $y_1 = e^{-4x}$. Find a second linearly independent solution.

We assume the form $y_2 = u(x)e^{-4x}$ for the second solution, substitute in $y'' + y' - 12y = 0$, and obtain

Assume $y_2 = uy_1$.

Substitute y_2 into the equation.

$$y_2'' + y_2' - 12y_2 = u''e^{-4x} + u'(-8e^{-4x} + e^{-4x})$$
$$+ u(16e^{-4x} - 4e^{-4x} - 12e^{-4x})$$
$$= u''e^{-4x} - 7u'e^{-4x} = 0.$$

As expected, the coefficient of u vanishes because y_1 is a solution of the homogeneous equation $y'' + y' - 12y = 0$. We have

$$u'' - 7u' = 0,$$

Find a first-order equation for $v = u'$. or, if $v = u'$, the first-order equation

$$v' - 7v = 0.$$

Solve for v. Find $u = \int v$. One solution of the reduced-order equation $v' - 7v = 0$ is $v(x) = e^{7x}$. Since $u = \int v \, dx$,

$$u(x) = \int e^{7x} \, dx = \frac{e^{7x}}{7},$$

where we dropped the constant of integration. A second solution of $y'' + y' - 12y = 0$ is

$$uy_1 = \frac{e^{7x}}{7} e^{-4x} = \frac{e^{3x}}{7}.$$

Use u to write y_2. Since a constant multiple of a solution of a linear homogeneous equation is again a solution of that equation, we can ignore the factor $\frac{1}{7}$ and take

$$y_2(x) = e^{3x}.$$

A straightforward computation shows that $W(e^{-4x}, e^{3x}) \neq 0$ for all x, verifying the linear independence of this second solution. ∎

Is this the same pair of solutions we would have obtained by the characteristic equation method? See example 14 of section 6.3.1.

EXAMPLE 2   *One solution of*

$$x^2 y'' - 3xy' + 3y = 0$$

is $y_1(x) = x$. Find a second solution.

Since the given equation is linear and homogeneous, we can use reduction of order to find a second solution in the form $y_2(x) = u(x)x$. Substituting y_2 in $x^2 y'' - 3xy' + 3y = 0$ and simplifying yields

Assume $y_2 = uy_1$.
Substitute y_2 into the equation.

$$x^2 y_2'' - 3xy_2' + 3y_2 = x^3 u'' - x^2 u' = 0.$$

Find a first-order equation for $v = u'$. The usual reduction-of-order substitution $v = u'$ reduces the second-order equation $x^3 u'' - x^2 u'$ to the first-order equation

$$x^3 v' - x^2 v = 0.$$

We solve $x^3 v' - x^2 v = 0$ by separating variables,

$$\frac{dv}{v} = \frac{dx}{x},$$

and integrating to obtain $v(x) = Cx$. Since we need only one solution of $x^3 v' - x^2 v = 0$, we may choose the arbitrary constant C for convenience; e.g., $C = 1$. Then from $u' = v = x$, we obtain

Solve for v.

Find $u = \int v$.

$$u(x) = \frac{x^2}{2},$$

where we have again chosen an arbitrary constant of integration to give us the simplest possible expression.

Hence, a second solution of $x^2 y'' - 3xy' + 3y = 0$ is

Use u to write y_2.

$$uy_1 = \frac{x^2}{2} x = \frac{x^3}{2}.$$

Since we are solving a homogeneous equation, we could multiply this solution by a constant to obtain the simplified second solution

$$y_2 = x^3.$$

$\blacksquare$

10.1.1 Summary

The method of reduction of order obtains a second solution y_2 of a linear, homogeneous, second-order differential equation given a first solution y_1.

We assume $y_2(x) = u(x)y_1(x)$ and substitute y_2 into the given equation. After the simplification that results because y_1 solves the given homogeneous equation, we obtain a second-order equation for the unknown function u that contains only u'' and u'; i.e., we really have a first-order equation for the unknown u'. Letting $u' = v$, we solve the first-order equation for v, integrate once to obtain u, and write the new solution as $u(x)y_1(x)$. These steps are given concisely at the beginning of this section.

10.1.2 Exercises

EXERCISE GUIDE

To gain experience . . .	Try exercises
Using reduction of order	1–11
Solving initial-value problems	8–10
With the foundations of reduction of order	12–14

In exercises 1–7, use the given solution and reduction of order to find a second solution of each of the following differential equations. If you believe that the method is not applicable to a given equation, explain why.

1. $x^2 y'' - 3xy' + 3y = 0$, $y_1 = x^3$

2. $y'' - y = 0$, $y_1 = e^{1-x}$

3. $y'' + y' - 12y = 0$, $y_1 = e^{3x}$

4. $y'' + y = \cos \pi x$, $y_1 = (1 - \pi^2)^{-1} \cos \pi x$

5. $x^2 y'' + 2xy' - 6y = 0$, $y_1 = x^{-3}$

6. $x^2 y'' - 3xy' + 4y = 0$, $y_1 = x^2$

7. $3x^2 y'' - 3xy' + 6y = 0$, $y_1 = x \cos(\ln x)$

In exercises 8–10, use the information given to solve the initial-value problem.

8. $t^2 x'' - 3tx' + 4x = 0$, $x(1) = 1$, $x'(1) = 0$. One solution of this equation is t^2.

9. $x'' + x' - 12x = 2t$, $x(0) = 0$, $x'(0) = 4$. One solution of the corresponding homogeneous equation is e^{3t}.

10. $3t^2 w''(t) - 3tw'(t) + 6w(t) = 0$, $w(1) = 2$, $w'(1) = 1$. One solution of this equation is $t \cos(\ln t)$.

11. Example 2 sought a second solution of

$$x^2 y'' - 3xy' + 3y = 0$$

in the form $y_2 = u(x)x$, given the first solution $y_1 = x$.

(a) Supply the details of the calculations that led from the substitution of y_2 into $x^2 y'' - 3xy' + 3y = 0$ to a second-order equation for u,

$$x^2 y_2'' - 3xy_2' + 3y_2 = x^3 u'' - x^2 u' = 0.$$

(b) Make the substitution $u' = v$ in $x^3 u'' - x^2 u' = 0$ and solve the resulting first-order equation by an integrating factor. (Example 2 used separation of variables.)

(c) Obtain an expression for $u(x)$. Show that you can ultimately derive the same second solution as in example 2, $y_2 = x^3$.

12. The text claims

It is easy to argue in general that the substitution $y_2 = uy_1$ reduces *any* homogeneous second-order linear differential equation of which y_1 is a solution to a first-order equation for u', providing the coefficient of y'' is never zero.

Verify this claim by showing that if y_1 is a solution of

$$y'' + P(x)y' + Q(x)y = 0,$$

then the substitution $y_2 = uy_1$ reduces the given equation to

$$y_1 u'' + (2y_1' + Py_1)u' = 0.$$

Explain why this demonstration verifies the claim. Make explicit note of the step where you use the fact that y_1 is a solution of the original equation.

13. The preceding problem asks you to show that the substitution $y_2 = uy_1$ reduces the equation

$$y'' + P(x)y' + Q(x)y = 0$$

to

$$y_1 u'' + (2y_1' + Py_1)u' = 0,$$

where y_1 is a solution of the given equation $y'' + P(x)y' + Q(x)y = 0$. Follow the steps given here to show that a second solution of this differential equation is

$$y_2(x) = y_1(x) \int^x y_1^{-2}(r) \exp\left(-\int^r P(s)\, ds\right) dr.$$

(a) Use the substitution $u' = v$ to rewrite $y_1 u'' + (2y_1' + Py_1)u' = 0$ as a first-order equation.

(b) Solve the resulting first-order equation using an integrating factor.

(c) Integrate once to obtain u from the expression for v.

Compare the complexity of this formula with the relative simplicity of the steps in the method of reduction of order. It is certainly simpler and more reliable to understand the method and apply it to each problem than it is to memorize this formula.

14. Exercise 13 derived a general expression for a second solution y_2 of a linear, second-order, homogeneous equation. Use the linear independence test to show that y_2 is linearly independent of y_1 on any interval on which $y_1 \neq 0$. (*Hint:* Compute the Wronskian using $y_2 = uy_1$, not the formula obtained here. You will find $W(y_1, y_2) = u'y_1^2$. Can you guarantee $u' \neq 0$? Why would y_1 and y_2 automatically be linearly dependent if $u'(x) = 0$ throughout the interval of interest?)

Why is the restriction $y_1 \neq 0$ consistent with the formula obtained in the preceding problem?

10.2 CAUCHY-EULER EQUATIONS

In example 2 of the preceding section we used reduction of order to find that a second solution of

$$x^2 y'' - 3xy' + 3y = 0$$

is x^3, given that a first solution is x. How was that first solution derived?

One approach is to observe that the power of x in each coefficient is the same as the order of the derivative in that term; x^2 multiplies y'', and so on. Consequently, assuming a solution of the form

$$y(x) = x^r$$

would lead to a kind of characteristic equation for the unknown constant r. Each derivative of y would reduce the exponent by 1, but that reduction would be compensated by the power of x in the coefficient. Every term would still contain x^r, and we could divide out that common factor if $x > 0$.

Explicitly, substituting $y = x^r$ into $x^2 y'' - 3xy' + 3y = 0$ leads to

$$x^2 y'' - 3xy' + 3y = x^2 r(r-1)x^{r-2} - 3xrx^{r-1} + 3x^r$$

$$= r(r-1)x^r - 3rx^r + 3x^r = 0.$$

As long as $x > 0$, we can divide x^r out of the last equation to obtain a quadratic equation for r,

$$r(r-1) - 3r + 3 = r^2 - 4r + 3 = (r-3)(r-1) = 0.$$

The roots of this *indicial equation* are obviously $r = 1, 3$. Hence, solutions of the differential equation $x^2 y'' - 3xy' + 3y = 0$ are x, x^3, $x > 0$.

We have found a pair of linearly independent (verify!) solutions of the given equation by exploiting the observation that the power of x matches the order of the derivative in each term.

Equations of this form are called **Cauchy-Euler** equations. (They are also known as equidimensional equations.)

Specifically, Cauchy-Euler equations are linear, second-order homogeneous equations that can be written in the form

$$x^2 y'' + b_1 xy' + b_0 y = 0,$$

CAUCHY-EULER EQUATION

where b_1, b_0 are constants. To solve such equations,

1. Assume a solution of the form $y = x^r$ and substitute into the differential equation $x^2 y'' + b_1 xy' + b_0 y = 0$.

2. Divide out x^r, $x > 0$, and solve the resulting quadratic equation for r:

$$r(r-1) + b_1 r + b_0 = 0.$$

INDICIAL EQUATION

If the **indicial** (or *auxiliary*) **equation** $r(r-1) + b_1 r + b_0 = 0$ has two distinct real roots r_1, r_2, then the differential equation has the pair of linearly independent solutions

$$x^{r_1}, \ x^{r_2}, \qquad x > 0$$

Linearly-independent solutions for distinct real roots.

These are easily shown to be linearly independent for $x > 0$. (See exercise 15.) Later in this section we explore the other possibilities: A single repeated root and complex conjugate roots of the indicial equation.

EXAMPLE 3 *Find a general solution valid for $x > 0$ of*

$$x^2y'' + xy' - 4y = 0.$$

This equation is certainly of Cauchy-Euler type, for the power of x in each coefficient matches the order of the derivative of the unknown y. Assuming a solution in the form $y = x^r$ and substituting leads to

Assume $y = x^r$.

$$x^2y'' + xy' - 4y = r(r-1)x^r + rx^r - 4x^r = 0.$$

Since $x > 0$, we may divide by x^r to obtain

INDICIAL EQUATION

$$r(r-1) + r - 4 = r^2 - 4 = (r-2)(r+2) = 0.$$

Obviously, $r = \pm 2$, and two solutions of the differential equation are x^2, x^{-2}. We find $W(x^2, x^{-2}) = -4/x$; these solutions are linearly independent for $x > 0$. A general solution of $x^2y'' + xy' - 4y = 0$ is

GENERAL SOLUTION

$$y_g(x) = C_1x^2 + C_2x^{-2}, \qquad x > 0. \qquad \blacksquare$$

EXAMPLE 4   *Find a general solution valid for $x > 0$ of*

$$x^2y'' - xy' + y = 0.$$

We seek solutions of this Cauchy-Euler equation in the form $y = x^r$. The usual substitution of x^r in $x^2y'' - xy' + y = 0$ and division by x^r leads to

Assume $y = x^r$.

INDICIAL EQUATION

$$r(r-1) - r + 1 = r^2 - 2r + 1 = (r-1)^2 = 0.$$

The quadratic has the single repeated root $r = 1$. One solution of $x^2y'' - xy' + y = 0$ is $y_1 = x$. What is another?

We turn to reduction of order and assume there is a second solution of the form $y_2 = u(x)x$. Substituting this assumed solution in $x^2y'' - xy' + y = 0$ yields

$$x^2y_2'' - xy_2' + y_2 = x^2(u''x + 2u') - x(u'x + u) + ux$$
$$= x^3u'' + x^2u' = 0.$$

The reduction of order substitution $u' = v$ yields the usual first-order equation

$$x^3v' + x^2v = 0,$$

whose solution by separation of variables is

$$\frac{dv}{v} = -\frac{dx}{x},$$

$$v = Ce^{-\ln x} = Ce^{\ln(1/x)} = \frac{C}{x}.$$

We choose $C = 1$ and obtain

$$u(x) = \int v(x)\,dx = \int \frac{1}{x}\,dx = \ln x + D.$$

Choosing the arbitrary constant $D = 0$ finally yields the second solution

$$y_2(x) = x \ln x.$$

In this example with repeated roots of the quadratic equation defining r, we find that the second solution is simply the first multiplied by $\ln x$.

Since $W(x, x \ln x) = x$, the pair of solutions $x, x \ln x$ is linearly independent for $x > 0$. A general solution of $x^2 y'' - xy' + y = 0$ is

$$y_g(x) = C_1 x + C_2 x \ln x, \qquad x > 0. \qquad \blacksquare$$

GENERAL SOLUTION

The situation observed in this example holds in general. If the quadratic equation $r(r-1) + b_1 r + b_0 = 0$ has the single repeated root r, then two linearly independent solutions of the differential equation $x^2 y'' + b_1 xy' + b_0 y = 0$ are

$$x^r, \qquad x^r \ln x, \qquad x > 0.$$

Linearly independent solutions for repeated root

See exercise 16.

EXAMPLE 5 *Find a general solution of the diffusion equation in a circle,*

$$xc'' + c' = 0,$$

which was derived in example 2, page 355, to model the concentration $c(x)$ of salt in a pond.

To see that this is indeed a Cauchy-Euler equation, multiply the given expression by x: $x^2 c'' + xc' = 0$.

Assuming a solution of the form $c(x) = x^r$ and substituting into $x^2 c'' + xc' = 0$, we find that r must satisfy

Assume $c = x^r$.

$$r(r-1) + r = r^2 = 0.$$

INDICIAL EQUATION

The (repeated) roots of this indicial equation are $r = 0$. Linearly independent solutions are $x^0 = 1$ and $x^0 \ln x = \ln x$, $x > 0$. A general solution of $xc'' + c' = 0$ is

$$c_g(x) = C_1 + C_2 \ln x, \qquad x > 0. \qquad \blacksquare$$

GENERAL SOLUTION

EXAMPLE 6 *Find two solutions of*

$$x^2 y'' + 5xy' + 13y = 0$$

that are linearly independent for $x > 0$.

Since $x^2 y'' + 5xy' + 13y = 0$ has the form of a Cauchy-Euler equation, we assume $y = x^r$ and substitute. We obtain

Assume $y = x^r$.

$$x^2 y'' + 5xy' + 13y = r(r-1)x^r + 5rx^r + 13x^r = 0.$$

The unknown constant r evidently must satisfy

$$r(r-1) + 5r + 13 = r^2 + 4r + 13 = 0.$$

INDICIAL EQUATION

The quadratic formula reveals that the roots of this equation are $r = -2 \pm 3i$.

Two solutions of $x^2 y'' + 5xy' + 13y = 0$ are certainly x^{-2+3i}, x^{-2-3i}, but they are not real-valued. Using the identity

$$x = e^{\ln x}$$

and Euler's formula,

$$e^{i\beta} = \cos\beta + i\sin\beta,$$

we can write the first of these two solutions as

$$x^{-2+3i} = x^{-2} x^{3i} = x^{-2}(e^{\ln x})^{3i}$$
$$= x^{-2} e^{i(3\ln x)} = x^{-2}\left[\cos(3\ln x) + i\sin(3\ln x)\right].$$

Similarly, the second may be written

$$x^{-2-3i} = x^{-2}\left[\cos(3\ln x) - i\sin(3\ln x)\right].$$

Now the principle of superposition guarantees that linear combinations of these two solutions will again be a solution of the homogeneous linear equation $x^2 y'' + 5xy' + 13y = 0$. We can form linear combinations of x^{-2+3i} and x^{-2-3i} that will give us real-valued solutions. If we take $C_1 = C_2 = \frac{1}{2}$ in the linear combination

$$C_1 x^{-2+3i} + C_2 x^{-2-3i} = e^{-2}\left[(C_1 + C_2)\cos(3\ln x) + i(C_1 - C_2)\sin(3\ln x)\right],$$

we obtain

FIRST REAL-VALUED SOLUTION

$$y_1(x) = x^{-2}\cos(3\ln x).$$

The choice $C_1 = 1/2i, C_2 = -1/2i$ yields

SECOND REAL-VALUED SOLUTION

$$y_2(x) = x^{-2}\sin(3\ln x).$$

We leave it to you to verify that y_1, y_2 are indeed linearly independent for $x > 0$. ∎

The arguments of example 6 obviously extend directly to the general situation in which the roots of the quadratic equation $r(r-1) + b_1 r + b_0 = 0$ are the complex conjugate pair $r = \alpha \pm i\beta$. In analogy with the result of the last example, the linearly independent solutions of $x^2 y'' + b_1 xy' + b_0 y = 0$ are

Linearly independent solutions for complex roots

$$x^\alpha \cos(\beta\ln x), \qquad x^\alpha \sin(\beta\ln x), \qquad x > 0.$$

See Exercise 17. ∎

EXAMPLE 7 *Find two linearly independent solutions of*

$$(t-4)^2 y'' - (t-4)y' + 5y = 0.$$

State the values of t for which the solutions you find are linearly independent.

Although this equation is not precisely in the Cauchy-Euler form $x^2 y'' +$

$b_1 xy' + b_0 y = 0$, it obviously could be transformed by the substitution $x = t - 4$. Alternatively, we could seek directly solutions in the form $y(t) = (t - 4)^r$.

To pursue the latter approach, substitute $y(t) = (t - 4)^r$ in $(t - 4)^2 y'' - (t - 4)y' + 5y = 0$ and simplify in the usual manner. We obtain

$$(t - 4)^2 y'' - (t - 4)y' + 5y = r(r - 1)(t - 4)^r - r(t - 4)^r + 5(t - 4)^r.$$

If $t > 4$, then r must be a root of

$$r(r - 1) - r + 5 = r^2 - 2r + 5 = 0.$$

The quadratic formula reveals $r = 1 \pm 2i$.

A pair of solutions of $(t - 4)^2 y'' - (t - 4)y' + 5y = 0$ is evidently

$$(t - 4)^{1+2i}, \qquad (t - 4)^{1-2i},$$

but they are not real-valued. However, they may be rewritten using Euler's formula, just as in example 6. We obtain

$$(t - 4)^{1+2i} = (t - 4)\left[\cos(2\ln(t - 4)) + i\sin(2\ln(t - 4))\right]$$

and

$$(t - 4)^{1-2i} = (t - 4)\left[\cos(2\ln(t - 4)) - i\sin(2\ln(t - 4))\right].$$

Mimicking the linear combinations that lead to the real-valued solutions in example 6, we combine the two preceding expressions to obtain the real-valued solutions

$$(t - 4)\cos\left[2\ln(t - 4)\right], \qquad (t - 4)\sin\left[2\ln(t - 4)\right], \qquad t > 4.$$

We leave it to you to prove that this pair of solutions is linearly independent for $t > 4$. ∎

All the preceding work has treated x^r with $x > 0$. How do we solve

$$x^2 y'' + b_1 xy' + b_0 y = 0$$

when $x < 0$? In chapter project 1, we ask you to show that the appropriate solution form is $y = (-x)^r$ for $x < 0$. The same solution forms appear but with x replaced by $-x$. To encompass both $x > 0$ and $x < 0$ in the following summary, we replace x with $|x|$.

10.2.1 Summary

A Cauchy-Euler equation is a second-order homogeneous linear differential equation in which the power of the independent variable in the coefficient matches the order of the derivative—i.e., an equation of the form

$$x^2 y'' + b_1 xy' + b_0 y = 0.$$

To solve, assume a solution of the form $|x|^r$, $x \neq 0$, substitute in the differential

---

Side margin notes:

Assume $y = (t - 4)^r$.

INDICIAL EQUATION

REAL-VALUED SOLUTIONS

TABLE 10.1

General solution of the Cauchy-Euler equation $x_2 y'' + b_1 xy' + b_0 y = 0$

Type of Root	General Solution for $x \neq 0$
Real $r_1 \neq r_2$	$C_1 |x|^{r_1} + C_2 |x|^{r_2}$
Real repeated r	$C_1 |x|^r + C_2 |x|^r \ln |x|$
Complex conjugate pair $\alpha \pm i\beta$	$|x|^\alpha (C_1 \cos (\beta \ln |x|) + C_2 \sin(\beta \ln |x|))$

$y = |x|^r \Rightarrow r(r-1) + b_1 r + b_0 = 0.$

equation $x^2 y'' + b_1 xy' + b_0 y = 0$, and find that r must satisfy the quadratic indicial equation

$$r(r - 1) + b_1 r + b_0 = 0.$$

If this quadratic has a pair of distinct real roots r_1, r_2, then

$$|x|^{r_1}, \qquad |x|^{r_2}$$

are linearly independent solutions of $x^2 y'' + b_1 xy' + b_0 y = 0$ for $x \neq 0$. If $r(r - 1) + b_1 r + b_0 = 0$ has the single repeated root r, then

$$|x|^r, \qquad |x|^r \ln |x|$$

are linearly independent solutions of $x^2 y'' + b_1 xy' + b_0 y = 0$ for $x \neq 0$. If the quadratic indicial equation has the pair of complex conjugate roots $r = \alpha \pm i\beta$, then

$$|x|^\alpha \cos(\beta \ln |x|), \qquad |x|^\alpha \sin(\beta \ln |x|)$$

are linearly independent solutions of $x^2 y'' + b_1 xy' + b_0 y = 0$ for $x \neq 0$.

These solution forms are summarized in table 10.1.

The pair x^r, $x^r \ln x$ is derived from the single solution x^r by reduction of order. The real-valued pair $x^\alpha \cos(\beta \ln x)$, $x^\alpha \sin(\beta \ln x)$ is obtained by rewriting the pair of complex-valued solutions $x^{\alpha \pm i\beta}$ using Euler's formula and forming linear combinations.

10.2.2   Exercises

EXERCISE GUIDE

To gain experience . . .	Try exercises
Solving Cauchy-Euler equations	1–10, 13, 18
Verifying linear independence	1–10, 15(a), 16(c), 18
Checking solutions	9–12
Finding equations, given solutions	9–12
Obtaining real-valued solutions from complex roots	3, 5, 17–18
With the foundations of the Cauchy-Euler method	13–17

In exercises 1–8 find, if possible, a pair of linearly independent solutions of each of the following equations. Verify linear independence, and state the interval upon which it holds. If you cannot find such a pair of solutions, explain why each solution method you tried failed.

1. $x^2 y'' - 2xy' + 2y = 0$

2. $x^2 y'' - 3xy' + 4y = 0$

3. $x^2 y'' - xy' + 5y = 0$

4. $x^2 y'' + 2xy' - 6y = 0$

5. $3x^2 y'' - 3xy' + 6y = 0$

6. $x^2 y'' - 2x^2 y' + 2y = 0$

7. $4(2t - 3)^2 u''(t) - 2(2t - 3)u'(t) + 5u(t) = 0$

8. $(1 - x)^2 y'' + 2(1 - x)y' + 2y = 0$

In exercises 9–12, ...

    i. Write Cauchy-Euler equations that have the given pair of functions as solutions.

    ii. Verify by substitution that the equation you found is indeed solved by the pair of functions.

9. $x, 1/x$

10. $(2 + x)^3, 2 + x$

11. $x^2 \cos(4 \ln x), x^2 \sin(4 \ln x)$

12. $x^2 \ln x, x^2$

13. Show by substituting the assumed solution form $y = x^r$ into the general Cauchy-Euler equation

$$x^2 y'' + b_1 xy' + b_0 y = 0,$$

that the unknown constant r is a solution of the quadratic equation

$$r(r - 1) + b_1 r + b_0 = 0.$$

What restrictions must you place on x?

14. The preceding problem shows that the general Cauchy-Euler equation

$$x^2 y'' + b_1 xy' + b_0 y = 0$$

has a solution of the form x^r if r is a root of

$$r(r - 1) + b_1 r + b_0 = 0.$$

Find conditions on the coefficients b_1, b_0 which guarantee that this quadratic equation has

    (a) Distinct real roots,

    (b) A single repeated real root,

    (c) A pair of complex conjugate roots.

15. **(a)** Verify the text's claim that x^{r_1}, x^{r_2} are linearly independent on any interval not including $x = 0$ if r_1, r_2 are distinct real numbers. Why must you exclude $x = 0$?

    (b) Is the result still true if r_1, r_2 are complex conjugate pairs?

16. The text claims

    If the quadratic equation

$$r(r - 1) + b_1 r + b_0 = 0$$

    defining r has the single repeated root r, then two linearly independent solutions of the differential equation

$$x^2 y'' + b_1 xy' + b_0 y = 0$$

    for $x > 0$ are x^r, $x^r \ln x$.

Verify this claim by completing the following steps.

    (a) Show that if $r(r - 1) + b_1 r + b_0 = 0$ possess a single repeated real root, then b_1 can be expressed in terms of b_0. (*Hint:* What is the value of the discriminant?)

    (b) Using x^r as the first solution, construct a second solution of the differential equation of the form $u(x)x^r$ via reduction of order. Using part (a), show that $u(x) = \ln x$ by solving the resulting differential equation for u.

    (c) Verify that x^r and $x^r \ln x$ are linearly independent for $x > 0$.

17. Following example 6, the text claims

    The arguments of example 6 obviously extend directly to the general situation in which the roots of the quadratic equation $r(r - 1) + b_1 r + b_0 = 0$ are the complex conjugate pair $r = \alpha \pm i\beta$. ... (T)he linearly independent solutions of $x^2 y'' + b_1 xy' + b_0 y = 0$ for $x > 0$ are

$$x^\alpha \cos(\beta \ln x), \qquad x^\alpha \sin(\beta \ln x).$$

Verify this claim by deriving this pair of real-valued solutions from the pair of complex-valued solutions $x^{\alpha + i\beta}, x^{\alpha - i\beta}$. Show that these real-valued solutions form a linearly independent pair for $x > 0$.

18. Fill in the details in example 7 of the derivation of the solutions

$$(t - 4) \cos [2 \ln(t - 4)], \qquad (t - 4) \sin [2 \ln(t - 4)]$$

of

$$(t-4)^2 y'' - (t-4)y' + 5y = 0.$$

Verify that this pair of solutions is linearly independent for $t > 4$.

19. The boundary-value problem

$$-(xT'(x))' = 0, \qquad T'(0) = 0, \qquad T(L) = T_L,$$

models the equilibrium temperature in a thin disk of radius L whose top and bottom are insulated while its circumference is maintained at temperature T_L; see exercise 11, page 359, of section 9.1.
  (a) Solve this boundary-value problem.
  (b) Explain the physical significance of the solution you obtain.

20. The boundary-value problem

$$-(xT'(x))' = 0, \qquad T'(0) = 0, \qquad T'(L) = 0,$$

models the equilibrium temperature in a thin disk of radius L whose top, bottom and circumference are insulated; see exercise 12, page 359, of section 9.1.
  (a) Solve this boundary-value problem.
  (b) Explain the physical significance of the solution you obtain. What additional physical information is required to specify the solution uniquely?

21. Solve the boundary-value problem governing the equilibrium temperature in the wall of a steam pipe:

$$-\kappa(xT'(x))' = 0, \qquad T(r) = T_s, \qquad T(R) = T_{\text{out}}.$$

(See exercise 20, page 398, of the chapter exercises for chapter 9.) Sketch a graph of the resulting temperature profile. Why is the temperature variation through the wall of the pipe different than that through the wall of a house?

10.3   VARIATION OF PARAMETERS

Variation of parameters is a method for finding particular solutions of second-order equations without the restrictions of constant coefficients and special forcing terms that the method of undetermined coefficients imposes. In return for increased generality, we are required to know two solutions of the corresponding *homogeneous* equation in order to construct the desired particular solution.

10.3.1   From the Systems Perspective

If you have not studied variation of parameters for systems of first-order differential equations, skip this subsection. The next subsection provides a detailed introduction to variation of parameters that is meant for you.

Recall that in section 8.3 we used variation of parameters to construct a particular solution of the first-order system

$$y' = k_1(t)y + k_2(t)z + F(t)$$
$$z' = \ell_1(t)y + \ell_2(t)z + G(t).$$

We assumed we knew linearly independent solution pairs (y_1, z_1) and (y_2, z_2) of the corresponding homogeneous system. We sought a particular solution (y_p, z_p) of the nonhomogeneous system in the form

$$y_p = u_1(t)y_1 + u_2(t)y_2$$
$$z_p = u_1(t)z_1 + u_2(t)z_2$$

where u_1, u_2 were functions to be determined.

Substituting this proposed particular solution into the nonhomogeneous system led to two simultaneous equations for u_1', u_2',

$$y_1 u_1' + y_2 u_2' = F$$

$$z_1 u_1' + z_2 u_2' = G.$$

We solved this system using Cramer's rule to obtain

$$u_1' = \frac{Fz_2 - Gy_2}{W}, \qquad u_2' = \frac{Gy_1 - Fz_1}{W},$$

where W is just the Wronskian of the homogeneous solutions,

$$W = y_1 z_2 - z_1 y_2.$$

To specialize this process for the second-order linear equation

$$y'' + P(x)y' + Q(x)y = R(x),$$

we write it as a first-order system by introducing the variable $z = y'$:

$$y' = z$$

$$z' = -Q(x)y - P(x)z + R(x)$$

These equations correspond to the generic linear system with $F = 0$ and $G = R$.

If we suppose we have two linearly independent solutions y_1, y_2 of the homogeneous equation $y'' + P(x)y' + Q(x)y = 0$, then we can seek a particular solution in the form

$$y_p = u_1(x)y_1 + u_2(x)y_2.$$

The results for the first-order system tell us that u_1', u_2' satisfy the system of equations

$$y_1 u_1' + y_2 u_2' = F = 0$$

$$y_1' u_1' + y_2' u_2' = G = R.$$

The solution of this system is

$$u_1' = -\frac{Ry_2}{W}, \qquad u_2' = \frac{Ry_1}{W},$$

where W is the Wronskian $W(y_1, y_2) = y_1 y_2' - y_2 y_1'$. We need only find u_1, u_2 to write a particular solution of $y'' + P(x)y' + Q(x)y = R(x)$.

10.3.2 Without the Systems Perspective

If you read the previous subsection and understood it, skip this section. Otherwise, you should study the detailed example presented here to see the central ideas behind variation of parameters.

Recall that for a linear, first-order equation, variation of parameters seeks a particular solution in the form

$$y_p(x) = u(x)y_h(x),$$

where u is an unknown function and y_h is a known solution of the homogeneous equation. Substituting this form into the nonhomogeneous equation led to a simple equation for u'; see section 3.3.2.

To generalize this structure to second-order equations, assume that we know two linearly independent solutions y_1, y_2 of the homogeneous equation

$$y'' + P(x)y' + Q(x)y = 0.$$

Then we can speculate that a particular solution of the nonhomogeneous equation

$$y'' + P(x)y' + Q(x)y = R(x)$$

is a sum of functions of x multiplying those solutions of the homogeneous equation,

$$y_p(x) = u_1(x)y_1(x) + u_2(x)y_2(x).$$

> Variation of parameters is sometimes called variation of constants because this expression looks like a general solution of the homogeneous equation with C_1, C_2 replaced by the functions u_1, u_2.

The major task of variation of parameters is determining the unknown functions u_1, u_2. An example will guide us.

EXAMPLE 8 *Find a particular solution of*

$$y'' + 4y' + 4y = 8e^{-2x}$$

in the form $y_p = u_1 y_1 + u_2 y_2$.

Using the characteristic equation method, we find that linearly independent solutions of the homogeneous equation $y'' + 4y' + 4y = 0$ are

$$y_1(x) = e^{-2x}, \qquad y_2(x) = xe^{-2x}.$$

Now we seek a particular solution of $y'' + 4y' + 4y = 8e^{-2x}$ in the form

Assume $y_p = u_1 y_1 + u_2 y_2$.

$$y_p = u_1(x)y_1(x) + u_2(x)y_2(x)$$
$$= u_1(x)e^{-2x} + u_2(x)xe^{-2x}.$$

Given an assumed form for a solution, our usual procedure is to substitute that form into the differential equation to find relations that will determine the unknowns, the functions u_1, u_2 in this case. However, we anticipate that this procedure will give us only one relation defining two unknown functions. We expect to need a second relation defining u_1, u_2.

As we consider forms that second relation might take, we should also realize that substituting this proposed solution in the differential equation will lead to an expression involving not only u_1 and u_2 but their first and second derivatives as well. Can we simplify these derivative expressions by our choice of the second relation that u_1, u_2 satisfy?

With these ideas in mind, we begin computing derivatives of the solution form $y_p = u_1 y_1 + u_2 y_2$:

$$y_p' = u_1'e^{-2x} + u_2'xe^{-2x} + u_1(e^{-2x})' + u_2(xe^{-2x})'.$$

The complexity of the expression will grow as we compute y_p'' unless we can impose some simplifying assumptions now.

One possibility is the requirement

$$u_1'e^{-2x} + u_2'xe^{-2x} = 0,$$

Requiring that $u_1'y_1 + u_2'y_2 = 0$.

which eliminates the first two terms in the expression for y_p'. It certainly prevents second derivatives of u_1, u_2 appearing in the expression for y_p'', as we see in the following:

$$y_p'' = u_1(e^{-2x})'' + u_2(xe^{-2x})'' + u_1'(e^{-2x})' + u_2'(xe^{-2x})'.$$

We now substitute into the differential equation $y'' + 4y' + 4y = 8e^{-2x}$ the preceding expressions for y_p, for y_p', and for y_p'', using the simplified expression for the latter. After collecting terms, we obtain

$$u_1 \boxed{[(e^{-2x})'' + 4(e^{-2x})' + 4e^{-2x}]}$$
$$+ u_2 \boxed{[(xe^{-2x})'' + 4(xe^{-2x})' + 4xe^{-2x}]} + u_1'(e^{-2x})' + u_2'(xe^{-2x})'$$
$$= 8e^{-2x}.$$

Each of the boxed terms, the coefficients of u_1, u_2, are zero because they are precisely the *homogeneous* form of the differential equation $y'' + 4y' + 4y = 0$ with the homogeneous solutions e^{-2x} and xe^{-2x} substituted for the unknown. Hence, this last expression simplifies to

$$u_1'(e^{-2x})' + u_2'(xe^{-2x})' = 8e^{-2x}.$$

We have obtained two equations for the two unknowns u_1', u_2',

$$e^{-2x}u_1' + xe^{-2x}u_2' = 0$$
$$(e^{-2x})'u_1' + (xe^{-2x})'u_2' = 8e^{-2x}.$$

Two equations for u_1', u_2'.

The first is the condition we imposed in hopes of simplifying the second derivative, and the second is the result of substituting the assumed solution form in the differential equation.

We can solve this pair of equations for u_1', u_2' using Cramer's rule. The determinant of coefficients is just the Wronskian,

$$W(e^{-2x}, xe^{-2x}) = e^{-2x}(xe^{-2x})' - xe^{-2x}(e^{-2x})',$$

and the solution of the system of equations is

Solve for u_1', u_2'.

$$u_1' = -\frac{8e^{-2x}xe^{-2x}}{W(e^{-2x}, xe^{-2x})}$$

$$u_2' = \frac{8e^{-2x}e^{-2x}}{W(e^{-2x}, xe^{-2x})}.$$

We have carried these expressions without simplifying so that we can identify the pattern of this analysis for later generalization.

Since

$$W(e^{-2x}, xe^{-2x}) = e^{-4x},$$

we can simplify these expressions to

$$u_1' = -8x, \qquad u_2' = 8.$$

Integrate to find u_1, u_2.
A single antidifferentiation yields

$$u_1 = -4x^2, \qquad u_2 = 8x.$$

We set the constant of integration to zero in both cases because we seek *any* pair of nontrivial functions u_1, u_2 satisfying the equations for u_1', u_2'.

Using these expressions for u_1, u_2 in the solution form $y_p = u_1 y_1 + u_2 y_2$ yields

Use u_1, u_2 to write y_p.

$$y_p(x) = -4x^2 e^{-2x} + 8x(xe^{-2x}) = 4x^2 e^{-2x}.$$

Compare this particular solution with that obtained by undetermined coefficients in example 26, section 6.5. ∎

To generalize this process to other linear equations, let us summarize the steps in the last example for a general linear equation

$$y'' + P(x)y' + Q(x)y = R(x).$$

We begin with two solutions of the homogeneous equation; call them y_1, y_2. We assume that a particular solution of the nonhomogeneous equation has the form

Assumed form for y_p

$$y_p(x) = u_1(x)y_1(x) + u_2(x)y_2(x),$$

where u_1, u_2 are unknown functions. To simplify the calculation of the derivatives of y_p and to prevent second derivatives of u_1, u_2 appearing in y_p'', we impose the condition

Imposed condition

$$u_1' y_1 + u_2' y_2 = 0.$$

Substituting the assumed solution form y_p in the differential equation, using the preceding relation, and exploiting the fact that y_1, y_2 are solutions of the *homogeneous* equation lead to a second equation for u_1', u_2',

Consequence of differential equation

$$u_1' y_1' + u_2' y_2' = R(x).$$

Solving these simultaneous equations for u_1' and u_2' leads to

Solve for u_1', u_2'.

$$u_1' = -\frac{Ry_2}{W}, \qquad u_2' = \frac{Ry_1}{W},$$

In principle, an antiderivative of each of these expressions leads to expressions for u_1 and u_2 and then to the particular solution $y_p = u_1 y_1 + u_2 y_2$.

10.3.3 Examples

From either of the perspectives described in the previous sections, we arrive at the same recipe for finding a particular solution of the second-order linear equation

$$y'' + P(x)y' + Q(x)y = R(x)$$

by the method of **variation of parameters**:

1. Given two linearly independent solutions y_1, y_2 of the homogeneous equation $y'' + P(x)y' + Q(x)y = 0$, assume a particular solution of the form

$$\boxed{y_p(x) = u_1(x)y_1(x) + u_2(x)y_2(x),}$$

VARIATION OF PARAMETERS FORM OF PARTICULAR SOLUTION

where u_1, u_2 are unknown functions. Require that u_1', u_2' satisfy

$$u_1'y_1 + u_2'y_2 = 0.$$

Substitute the assumed solution form y_p into $y'' + Py' + Qy = R$, use the preceding relation, and find a second equation for u_1', u_2',

$$u_1'y_1' + u_2'y_2' = R(x).$$

Solve these two equations to obtain

$$\boxed{u_1' = -\frac{Ry_2}{W}, \quad u_2' = \frac{Ry_1}{W},}$$

where W is the Wronskian $W(y_1, y_2) = y_1y_2' - y_2y_1'$.

2. Integrate the expressions for u_1', u_2' and write the resulting particular solution.

EXAMPLE 9 *Find a general solution of*

$$x^2y'' + xy' - 4y = \ln x$$

valid for $x > 0$.

In example 3 of section 10.2, we found that solutions of the homogeneous equation $x^2y'' + xy' - 4y = 0$, a Cauchy-Euler equation, are $y_1 = x^2$, $y_2 = x^{-2}$. These are linearly independent for $x \neq 0$.

To use variation of parameters to find a particular solution of this equation, assume

$$y_p(x) = u_1(x)x^2 + u_2(x)x^{-2}.$$

Assume $y_p = u_1y_1 + u_2y_2$.

For consistency with the present notation, divide the entire equation by x^2 to make the coefficient of y'' be unity; that is, $P(x) = 1/x$, $Q(x) = -4/x^2$, and $R(x) = (\ln x)/x^2$.

The functions u_1', u_2' must satisfy

$$u_1'y_1 + u_2'y_2 = x^2u_1' + x^{-2}u_2' = 0,$$

as well as the equation that comes from substituting y_p into the differential equation,

$$u_1' y_1' + u_2' y_2' = 2x u_1' - 2x^{-3} u_2' = \frac{\ln x}{x^2}.$$

Solve for u_1', u_2'.

Solving this pair of equations yields

$$u_1' = -\frac{x^2 \ln x}{W(x^2, x^{-2})} = \frac{\ln x}{4x^3}$$

$$u_2' = \frac{x^{-2} \ln x}{W(x^2, x^{-2})} = -\frac{x \ln x}{4},$$

Integrate to find u_1, u_2.

where we have used $W(x^2, x^{-2}) = -4/x$. Integrating these expressions by parts and setting all constants of integration to zero yield

$$u_1(x) = -\frac{2 \ln x + 1}{16x^2}$$

$$u_2(x) = -\frac{x^2(2 \ln x - 1)}{16}.$$

Use u_1, u_2 to write y_p.

Substituting these expression for u_1, u_2 into $y_p = u_1 y_1 + u_2 y_2$ yields

$$y_p(x) = -\frac{\ln x}{4}.$$

Hence, a general solution of $x^2 y'' + xy' - 4y = \ln x$ is

GENERAL SOLUTION

$$y_g(x) = -\frac{\ln x}{4} + C_1 x^2 + C_2 x^{-2}, \qquad x > 0. \qquad \blacksquare$$

EXAMPLE 10   *Find a solution of the initial value problem*

$$x^2 y'' - xy' + y = \pi x^2 \cos \pi x, \qquad y(1) = 0, \qquad y'(1) = 3.$$

Example 4 of section 10.2 found that a pair of linearly independent solutions of the homogeneous equation $x^2 y'' - xy' + y = 0$ is $y_1(x) = x$, $y_2(x) = x \ln x$, $x > 0$. Consequently, we assume a particular solution of the form

Assume $y_p = u_1 y_1 + u_2 y_2$.

$$y_p(x) = u_1(x) y_1(x) + u_2(x) y_2(x) = x u_1(x) + x \ln x \, u_2(x).$$

To put the equation into standard form, divide through by x^2. Then $R(x) = \pi \cos \pi x$.

Proceeding directly to

Solve for u_1', u_2'.

$$u_1' = -\frac{R y_2}{W}, \qquad u_2' = \frac{R y_1}{W},$$

the expression obtained by solving the equations for u_1', u_2', we find

$$u_1' = -\frac{(\pi \cos \pi x)(x \ln x)}{W(x, x \ln x)}$$

$$u_2' = \frac{x \pi \cos \pi x}{W(x, x \ln x)}.$$

Since $W(x, x \ln x) = x$, we have

$$u_1' = -\pi \cos \pi x \ln x, \qquad u_2' = \pi \cos \pi x.$$

From the second expression, we easily obtain

Integrate to find u_1, u_2.

$$u_2(x) = \sin \pi x.$$

Since the first expression lacks an explicit antiderivative, we employ the fundamental theorem of calculus (see section A.2) to write

$$u_1(x) = -\pi \int_1^x \cos \pi s \, \ln s \, ds.$$

The lower limit of integration was chosen to make this expression easy to evaluate at the initial point $x = 1$.

Using these expressions for u_1, u_2 in the particular solution form $y_p = u_1 y_1 + u_2 y_2$ yields

Use u_1, u_2 to write y_p.

$$y_p(x) = -\pi x \int_1^x \cos \pi s \, \ln s \, ds + x \sin \pi x \, \ln x.$$

A general solution of $x^2 y'' - xy' + y = \cos \pi x$ is obviously

$$y_g(x) = -\pi x \int_1^x \cos \pi s \, \ln s \, ds + x \sin \pi x \, \ln x + C_1 x + C_2 x \ln x.$$

GENERAL SOLUTION

To find C_1, and C_2, we impose the initial conditions $y(1) = 0$, $y'(1) = 3$:

$$y_g(1) = C_1 = 0$$
$$y_g'(1) = C_1 + C_2 = 3.$$

Impose initial conditions.

We find $C_1 = 0$, $C_2 = 3$. The solution of the initial-value problem is

$$y(x) = -\pi x \int_1^x \cos \pi s \, \ln s \, ds + (x \sin \pi x + 3x) \ln x. \qquad \blacksquare$$

SOLUTION OF INITIAL-VALUE PROBLEM

10.3.4 Summary

Variation of parameters finds a particular solution of a second-order linear equation

$$y'' + P(x)y' + Q(x)y = R(x)$$

by assuming a particular solution of the form

$$y_p(x) = u_1(x)y_1(x) + u_2(x)y_2(x),$$

where u_1, u_2 are unknown functions and y_1, y_2 are two *known* solutions of the corresponding homogeneous equation. We find that u_1', u_2' must satisfy the simultaneous equations

$$u_1'y_1 + u_2'y_2 = 0$$
$$u_1'y_1' + u_2'y_2' = R(x),$$

whose solution is

$$u_1' = -\frac{Ry_2}{W}, \qquad u_2' = \frac{Ry_1}{W}.$$

Here W is the Wronskian $W(y_1, y_2) = y_1y_2' - y_1'y_2$.

These expressions for u_1', u_2' are integrated (not necessarily in closed form) and substituted in $y_p = u_1y_1 + u_2y_2$ to obtain the desired particular solution.

The first of the two equations defining u_1' and u_2' can be thought of as a condition imposed to simplify the calculation of y_p''. The second equation is a consequence of substituting y_p into the differential equation.

10.3.5   Exercises

EXERCISE GUIDE

To gain experience . . .	Try exercises
Using variation of parameters	1–12, 17, 20
Finding general solutions	1–12, 20
Finding free and forced response	16
With the foundations of variation of parameters	13–15
Generalizing variation of parameters	18–19

Use variation of parameters to find a general solution of the differential equations in exercises 1–12.

1. $y'' - y = 1$

2. $y'' - y = -2t$

3. $y'' - 9y = e^{-3x}$

4. $y'' - 9y = e^{3x}$

5. $y'' - 9y = \cosh 3x$

6. $y'' + 16y = 2\tan 4x$

7. $w'' + 9w = 2e^{3x}$

8. $y'' + y' - 6y = e^{2t}$

9. $y'' + y = 4t$

10. $t^2u + tu' - 4u = t\ln t$

11. $t^2u'' + 2tu' - 6u = 1 + 1/t$

12. $x^2c'' + 2xc' = Px$, P a constant

13. Why would variation of parameters fail if the two homogeneous solutions y_1 and y_2 were not linearly independent?

14. Find an example in section 3.3.3 in which the particular solution obtained by an integrating factor is in the form of a function of the independent variable multiplying a solution of the homogeneous equation—that is, in the form $u(x)y_h(x)$.

15. Verify directly that the boxed terms, the coefficients of u_1 and u_2, in

$$u_1 \boxed{[(e^{-2x})'' + 4(e^{-2x})' + 4e^{-2x}]}$$

$$+ u_2 \boxed{[(xe^{-2x})'' + 4(xe^{-2x})' + 4xe^{-2x}]}$$

$$+ u_1'(e^{-2x})' + u_2'(xe^{-2x})'$$

$$= 8e^{-2x}$$

are zero. This expression was obtained in example 8 when the assumed solution $y_p = u_1 e^{-2x} + u_2 xe^{-2x}$ was substituted into the differential equation $y'' + 4y' + 4y = 8e^{-2x}$.

16. Find the free and forced response of the initial-value problem considered in example 10,

$$x^2 y'' - xy' + y = \cos \pi x, \qquad y(1) = 0, \qquad y'(1) = 1.$$

Does this system possess a steady state?

17. Example 8 solves the problem it considers in great detail, while example 9 finds a particular solution of

$$x^2 y'' + xy' - 4y = \ln x$$

18. but omits many steps. Fill in the details of example 9 in the manner of example 8. In particular, derive the equations that u_1', and u_2' satisfy.

18. The text suggests that the arguments of example 8 can be applied to the general linear equation

$$y'' + P(x)y' + Q(x)y = R(x)$$

to find the equations

$$u_1' = -\frac{Ry_2}{W(y_1, y_2)}, \qquad u_2' = \frac{Ry_1}{W(y_1, y_2)}.$$

Supply the details of this argument. In particular, derive the equations that u_1' and u_2' satisfy. (*Hint:* The preceding exercise asks you to perform this same task for a specific equation. Mimic the appropriate steps here.)

19. Suppose y_1, y_2 are linearly independent solutions of

$$a_2(x)y'' + a_1(x)y' + a_0(x) = r(x)$$

and that $a_2(x) \neq 0$. Find equations for the functions u_1' and u_2' in the particular solution form $y_p = u_1 y_1 + u_2 y_2$.

20. The functions x and $x \ln x$ are solutions of

$$x^2 y'' - 4xy' + 6y = \frac{1}{x}$$

for $x > 0$. Find a general solution of this equation.

10.4 POWER SERIES METHODS

In this and earlier chapters, we have followed a pattern of assuming solutions of a particular form and substituting the solution form into the differential equation to determine the unknown constants or functions in that form. We are now about to make what is apparently the most general possible assumption, that the solution can be expanded in a power series whose coefficients we can determine. We proceed formally at first; the last part of this section explores the validity of this bold assumption.

Recall that a **power series about** x_0 is an infinite series of the form

$$\sum_{n=0}^{\infty} c_n (x - x_0)^n,$$

where the c_n are constants. Only integer powers of x appear in a power series. Within the interval of convergence of this series, the function it represents can be integrated or differentiated by integrating or differentiating the series term by term. That is, if

$$f(x) = \sum_{n=0}^{\infty} c_n (x - x_0)^n$$

$$= c_0 + c_1(x - x_0) + c_2(x - x_0)^2 + c_3(x - x_0)^3 + \cdots,$$

then

$$f'(x) = \sum_{n=0}^{\infty} \left(c_n (x - x_0)^n \right)'$$

$$= \sum_{n=1}^{\infty} c_n n (x - x_0)^{n-1}$$

and

$$\int f(x)\,dx = \sum_{n=0}^{\infty} c_n \int (x - x_0)^n\,dx$$

$$= C + \sum_{n=0}^{\infty} c_n \frac{(x - x_0)^{n+1}}{n + 1}$$

whenever the original series for f converges. (In the series for $\int f$, C is just a constant of integration.) See section A.4.

> Why did the starting value of the summation index n in the series for f' change from $n = 0$ to $n = 1$? (See exercise 1.)

For the moment, we will proceed formally with such manipulations.

The constants c_n in the power series for f are just the **Taylor coefficients**

$$c_n = \frac{f^{(n)}(x_0)}{n!}.$$

To see this relation for $n = 0$, evaluate $f(x) = \sum_{n=0}^{\infty} c_n (x - x_0)^n$ at $x = x_0$:

$$f(x_0) = c_0.$$

Similarly, evaluating $f'(x) = \sum_{n=1}^{\infty} c_n n (x - x_0)^{n-1}$ at $x = x_0$ yields

$$f'(x_0) = c_1,$$

and so on for $f''(x_0)$, $f^{(3)}(x_0)$, …. Evidently, a formal power series like $f(x) = \sum_{n=0}^{\infty} c_n (x - x_0)^n$ can be constructed for any function that has an infinite number of derivatives about the point x_0.

> But if the series does not converge to the value $f(x)$, constructing the series is a waste.

To solve differential equations using power series, we will proceed in the other direction. Rather than construct the series from the known function f, we will attempt to construct as much of the solution function as we can by finding the coefficients in its power series directly from the differential equation. That is, given a linear differential equation and an expansion point x_0, we will attempt to find the coefficients c_n in the power series expansion for the solution about x_0. *We substitute the assumed solution form into the equation and equate coefficients of powers of x to find the unknown constants c_n.*

This method is not as general as you might like to believe, for not every solution of every linear differential equation has a power series expansion about an arbitrary point. For example, the Cauchy-Euler equation

$$x^2 y'' + 7xy' + 9y = 0$$

has the solutions x^{-3}, $x^{-3} \ln |x|$. Neither of these functions has a power series expansion about $x_0 = 0$.

EXAMPLE 11 *Find a power series solution about $x_0 = 0$ of*

$$y'' + y = 0.$$

We assume the solution has the form of a power series in x about $x_0 = 0$,

Assume a power series solution.

$$
\begin{aligned}
y(x) &= \sum_{n=0}^{\infty} c_n x^n \\
&= c_0 + c_1 x + c_2 x^2 + c_3 x^3 + \cdots.
\end{aligned}
$$

Then we differentiate term by term to compute

$$
\begin{aligned}
y'(x) &= c_1 + 2c_2 x + 3c_3 x^2 + 4c_4 x^3 + \cdots \\
&= \sum_{n=1}^{\infty} nc_n x^{n-1},
\end{aligned}
$$

$$
\begin{aligned}
y''(x) &= 2c_2 + 6c_3 x + 12c_4 x^2 + 20c_5 x^3 + \cdots \\
&= \sum_{n=2}^{\infty} n(n-1)c_n x^{n-2}.
\end{aligned}
$$

Note the change in the starting value of the summation index n each time the series is differentiated.

Substituting the power series expansions for y, y'' in the differential equation $y'' + y = 0$ yields

$$
\begin{aligned}
y'' + y &= 2c_2 + 6c_3 x + 12c_4 x^2 + 20c_5 x^3 + \cdots + c_0 + c_1 x + c_2 x^2 + c_3 x^3 + \cdots \\
&= (2c_2 + c_0) + (6c_3 + c_1)x + (12c_4 + c_2)x^2 + (20c_5 + c_3)x^3 + \cdots \\
&= 0.
\end{aligned}
$$

Since the right hand side is zero, each of the coefficients of a power of x on the left must be zero. Hence, from the successive powers of x, we obtain

Equate coefficients of like powers of x.

10.4 Power Series Methods

$$x^0: \quad 2c_2 + c_0 = 0$$
$$x^1: \quad 6c_3 + c_1 = 0$$
$$x^2: \quad 12c_4 + c_2 = 0$$
$$x^3: \quad 20c_5 + c_3 = 0$$
$$\vdots$$

Solve for the c's.

We solve each of these equations for the coefficient with the higher index:

$$x^0: \quad c_2 = -\frac{c_0}{2}$$

$$x^1: \quad c_3 = -\frac{c_1}{6}$$

$$x^2: \quad c_4 = -\frac{c_2}{12}$$

$$x^3: \quad c_5 = -\frac{c_3}{20}$$

$$\vdots$$

Evidently, the even-indexed coefficients can be expressed in terms of c_0,

$$c_2 = -\frac{c_0}{2}$$

$$c_4 = -\frac{c_2}{12} = \frac{c_0}{24}$$

$$\vdots$$

while the odd ones can be written in terms of c_1:

$$c_3 = -\frac{c_1}{6}$$

$$c_5 = -\frac{c_3}{20} = \frac{c_1}{120}$$

$$\vdots$$

Using these values of c_n in the assumed power series solution form $y(x) = \sum_{n=0}^{\infty} c_n x^n$, we obtain

GENERAL SOLUTION

$$y(x) = c_0 \left(1 - \frac{x^2}{2} + \frac{x^4}{24} + \cdots \right) + c_1 \left(x - \frac{x^3}{20} + \frac{x^5}{120} + \cdots \right),$$

the power series solution we sought. ∎

The power series solution obtained in example 11 contains two apparently arbitrary constants, c_0 and c_1. Are they the two constants we usually encounter in a general solution of a linear, second-order equation?

The strong suggestion that they are comes from the observation that evaluating the power series solution at $x = 0$ gives

$$y(0) = c_0, \qquad y'(0) = c_1.$$

For further confirmation, note that $\cos x$ and $\sin x$ are linearly independent solutions of $y'' + y = 0$. Since these functions have the power series expansions

$$\cos x = 1 - \frac{x^2}{2} + \frac{x^4}{24} + \cdots$$

$$\sin x = x - \frac{x^3}{20} + \frac{x^5}{120} + \cdots,$$

a general solution of $y'' + y = 0$ is

$$y(x) = C_1 \cos x + C_2 \sin x$$

$$= C_1 \left(1 - \frac{x^2}{2} + \frac{x^4}{24} + \cdots \right) + C_2 \left(x - \frac{x^3}{20} + \frac{x^5}{120} + \cdots \right),$$

just a restatement of the power series solution we obtained in example 11 directly from the differential equation.

EXAMPLE 12   *Find the general expression for the coefficients c_n in the power series about zero of the solution of*

$$y'' + y = 0.$$

That is, find a formula for c_n in terms of n.

We begin with the same assumed solution form as in example 11,

Assume a power series solution.

$$y(x) = \sum_{n=0}^{\infty} c_n x^n,$$

but we shall preserve the general index n throughout in order to find a formula for c_n. Taking care to adjust the starting index of the sum to account for terms lost to differentiation, we find

$$y'(x) = \sum_{n=1}^{\infty} n c_n x^{n-1}$$

$$y''(x) = \sum_{n=2}^{\infty} n(n-1) c_n x^{n-2}.$$

Substituting the expressions for y and y'' in the differential equation $y'' + y = 0$ yields

Substitute.

$$y'' + y = \sum_{2}^{\infty} n(n-1) c_n x^{n-2} + \sum_{n=0}^{\infty} c_n x^n$$

$$= 0.$$

To parallel example 11, we must collect coefficients of like powers of x. However, the two sums in the preceding equation contain different powers of x; the first has x^n and the second, x^{n-2}.

To rewrite the second sum so that the exponent of x is the index itself, as in the first sum, rather than the index less 2, as in the second, we make the change of index $m = n - 2$ in the second sum. Since $n = m + 2$, we obtain

$$\sum_{n=2}^{\infty} n(n-1)c_n x^{n-2} = \sum_{m=0}^{\infty} (m+2)(m+1)c_{m+2} x^m.$$

Note that the starting value of the lower index has changed to $m = 0$ because $m = 0$ when $n = 2$, the old starting value.

Using the altered form of the second sum gives us

$$y'' + y = \sum_{m=0}^{\infty} (m+2)(m+1)c_{m+2} x^m + \sum_{n=0}^{\infty} c_n x^n.$$

But the notation m and n for the summation indices in this last expression is entirely arbitrary. We could relabel them both k, for instance. Then the result of substituting in the differential equation would read

$$y'' + y = \sum_{k=0}^{\infty} (k+2)(k+1)c_{k+2} x^k + \sum_{k=0}^{\infty} c_k x^k.$$

Since both summations start at the same index, we may combine them into one. Furthermore, since the power of x in both is the same, we may factor it out of the combined sum to obtain

$$y'' + y = \sum_{k=0}^{\infty} [(k+2)(k+1)c_{k+2} + c_k] x^k = 0.$$

Because this power series sums to zero, each of the coefficients of x^k must be zero. (The power series for the identically zero function is the one that has every coefficient equal to zero.) Hence, we obtain the **recurrence relation**

Equate coefficients of like powers of x.

$$(k+2)(k+1)c_{k+2} + c_k = 0, \qquad k = 0, 1, 2, \ldots,$$

which successively expresses each coefficient in terms of one or more of its predecessors. Solving for the coefficient with the higher index, we find

Solve for the c's.

$$c_{k+2} = -\frac{c_k}{(k+2)(k+1)}.$$

Working our way back toward the coefficient with lowest index, we calculate c_k in terms of c_{k-2}, then c_{k-2} in terms of c_{k-4}, and so forth. That is, we use the preceding formula with k replaced by $k - 2$, then with k replaced by $k - 4$, and so on.

$$c_k = -\frac{c_{k-2}}{k(k-1)}$$

$$= \frac{c_{k-4}}{k(k-1)(k-2)(k-3)} = \cdots.$$

If k is even, then the sequence terminates after $k/2$ steps with c_0:

$$c_k = (-1)^{k/2} \frac{c_0}{k!}, \qquad k \text{ even}.$$

If k is odd, then the sequence terminates after $(k-1)/2$ steps with c_1:

$$c_k = (-1)^{(k-1)/2} \frac{c_1}{k!}, \qquad k \text{ odd}.$$

To see the pattern of these general calculations, try a few specific cases, such as $k = 6, 7, 8, 9$.

Although we could substitute these expressions for c_k as they stand in the power series solution form, the result has an awkward appearance,

$$y(x) = c_0 \sum_{k=0,2,\ldots} (-1)^{k/2} \frac{x^k}{k!} + c_1 \sum_{k=1,3,\ldots} (-1)^{(k-1)/2} \frac{x^k}{k!}.$$

GENERAL SOLUTION

Alternatively, we can write the even indices as $k = 2n$ and the odd indices as $k = 2n+1$, where $n = 0, 1, 2, \ldots$. Now we can use the common index n to write this solution in a prettier form,

$$y(x) = c_0 \sum_{n=0}^{\infty} (-1)^n \frac{x^{2n}}{(2n)!} + c_1 \sum_{n=0}^{\infty} (-1)^n \frac{x^{2n+1}}{(2n+1)!}.$$

We immediately identify the two series in this expression as the power series expansions about $x = 0$ of $\cos x$ and $\sin x$. Hence, we have obtained the same solution of $y'' + y = 0$ as the characteristic equation method. The leading coefficients c_0, c_1 are indeed the arbitrary constants in the general solution. ∎

EXAMPLE 13 *Find the first five nonzero terms in a power series solution of the initial-value problem*

$$y'' + (x-1)^2 y' - 2(x-1)^3 y = 0, \qquad y(1) = 1, \qquad y'(1) = -8.$$

Since the initial conditions are given at $x = 1$ and a power series is most easily evaluated at its expansion point, we assume a power series solution of $y'' + (x-1)^2 y' - 2(x-1)^3 y = 0$ that is expanded about $x_0 = 1$,

Assume a power series solution.

$$y(x) = \sum_{0}^{\infty} c_n (x-1)^n$$

$$= c_0 + c_1(x-1) + c_2(x-1)^2$$
$$+ c_3(x-1)^3 + c_4(x-1)^4 + c_5(x-1)^5 + c_6(x-1)^6 + \cdots.$$

At the beginning of a power series solution, we usually do not know how many terms to keep in order to finally obtain the desired number of nonzero terms. We added two extra to be safe.

We compute

$$y'(x) = c_1 + 2c_2(x-1) + 3c_3(x-1)^2 + 4c_4(x-1)^3$$
$$+ 5c_5(x-1)^4 + 6c_6(x-1)^5 + \cdots.$$
$$y''(x) = 2c_2 + 6c_3(x-1) + 12c_4(x-1)^2$$
$$+ 20c_5(x-1)^3 + 30c_6(x-1)^4 + \cdots.$$

Substitute.

Substituting these expression for y, y', and y'' into the differential equation $y'' + (x-1)^2 y' - 2(x-1)^3 y = 0$ leads to

$$y'' + (x-1)^2 y' - 2(x-1)^3 y = 2c_2 + 6c_3(x-1) + 12c_4(x-1)^2$$
$$+ 20c_5(x-1)^3 + 30c_6(x-1)^4 + \cdots$$
$$+ c_1(x-1)^2 + 2c_2(x-1)^3 + 3c_3(x-1)^4 + \cdots$$
$$- 2c_0(x-1)^3 - 2c_1(x-1)^4 + \cdots$$
$$= 0.$$

Each expression has been expanded to the same order (four in this case) as it was substituted in the differential equation. There is no point in keeping a particular power of $x-1$ in one expression if we have not kept it in another, because we will be required to collect *all* coefficients of a given power of $x-1$ as we solve for the coefficients c_n.

We collect coefficients of powers of $x-1$ to obtain

$$2c_2 + 6c_3(x-1) + (12c_4 + c_1)(x-1)^2 + (20c_5 + 2c_2 - 2c_0)(x-1)^3$$
$$+ (30c_6 + 3c_3 - 2c_1)(x-1)^4 + \cdots = 0.$$

Equate coefficients of like powers of $x-1$.

Setting each coefficient to zero yields a series of equations for the c_n,

$$
\begin{aligned}
(x-1)^0 &: & 2c_2 &= 0 \\
(x-1)^1 &: & 6c_3 &= 0 \\
(x-1)^2 &: & 12c_4 + c_1 &= 0 \\
(x-1)^3 &: & 20c_5 + 2c_2 - 2c_0 &= 0 \\
(x-1)^4 &: & 30c_6 + 3c_3 - 2c_1 &= 0.
\end{aligned}
$$

Solve for the c's.

The solution of these equations is

$$c_2 = 0, \qquad c_3 = 0, \qquad c_4 = -\frac{c_1}{12},$$
$$c_5 = \frac{c_0 - c_2}{10} = \frac{c_0}{10},$$
$$c_6 = \frac{c_1}{15} - \frac{c_3}{10} = \frac{c_1}{15}.$$

Using these values of $c_2, \ldots, c_6$, we find that our solution is

$$y(x) = c_0 + c_1(x-1) - c_1 \frac{(x-1)^4}{12} + c_0 \frac{(x-1)^5}{10} + c_1 \frac{(x-1)^6}{15} + \cdots$$

$$= c_0 \left(1 + \frac{(x-1)^5}{10} + \cdots\right) + c_1 \left((x-1) - \frac{(x-1)^4}{12} + \frac{(x-1)^6}{15} + \cdots\right).$$

GENERAL SOLUTION

To determine the remaining coefficients c_0 and c_1, we turn to the initial conditions, as we have in solving every other initial value problem. Combining the first initial condition with the solution form $y(x) = \sum_{n=0}^{\infty} c_n(x-1)^n$, we have

Impose initial conditions.

$$y(1) = c_0 = 1.$$

Similarly, from the second initial condition and

$$y'(x) = c_0 \left(\frac{(x-1)^4}{2} + \cdots\right) + c_1 \left(1 - \frac{(x-1)^3}{3} + \frac{2(x-1)^5}{5} + \cdots\right),$$

we obtain

$$y'(1) = c_1 = -8.$$

Although the coefficients c_2, c_3, $\ldots$ are found in terms of c_0 and c_1 by substituting in the differential equation, the values of c_0 and c_1 depend *only* upon the given initial conditions. We can choose to apply the initial conditions at any convenient point in the solution process. See exercise 18.

When these values of c_0 and c_1 are substituted in the solution $y(x) = \sum_{n=0}^{\infty} c_n(x-1)^n$, we obtain the solution of the initial-value problem,

$$y(x) = 1 - 8(x-1) + \frac{(x-1)^5}{10} - \frac{2(x-1)^4}{3} - \frac{8(x-1)^6}{15} + \cdots. \quad \blacksquare$$

SOLUTION OF INITIAL-VALUE PROBLEM

These ideas apply equally well to nonhomogeneous equations. We only need to expand the forcing term in a series about the same point as the solution.

EXAMPLE 14 *Find the first five nonzero terms in a series expansion of the solution of*

$$y'' - 2xy = \ln(1+x), \qquad y(0) = y'(0) = 1.$$

We require a series about the initial point $x = 0$. The Taylor series for the forcing term is

$$\ln(1+x) = x - \frac{x^2}{2} + \frac{x^3}{3} + \cdots, \qquad |x| < 1.$$

The appropriate form for the solution is

Assume a power series solution.

$$y(x) = c_0 + c_1 x + c_2 x^2 + c_3 x^3 + \cdots.$$

The initial conditions require $c_0 = c_1 = 1$.

Impose initial conditions.

10.4 *Power Series Methods*

Substituting the proposed solution in the equation and using the series expansion of $\ln(1 + x)$ lead to

$$y'' - 2xy = 2c_2 + 6c_3x + 12c_4x^2 + 20c_5x^3 + \cdots$$

$$- 2x - 2x^2 - 2c_2x^3 + \cdots$$

$$= x - \frac{x^2}{2} + \frac{x^3}{3} + \cdots.$$

Equating like powers of x yields

$$x^0 : \qquad 2c_2 = 0$$

$$x^1 : \qquad 6c_3 - 1 = 1$$

$$x^2 : \qquad 12c_4 - 2 = -\frac{1}{2}$$

$$x^3 : \qquad 20c_5 - 2c_2 = \frac{1}{3}.$$

We conclude that

$$c_2 = 0$$

$$c_3 = \frac{1}{3}$$

$$c_4 = \frac{1}{8}$$

$$c_5 = \frac{1}{60}.$$

Hence, the solution we seek is

$$y(x) = 1 + x + \frac{x^3}{3} + \frac{x^4}{8} + \frac{x^5}{60} + \cdots. \qquad \blacksquare$$

10.4.1   Existence of a Power Series Solution

We remarked earlier that the Cauchy-Euler equation

$$x^2y'' + 7xy' + 9y = 0$$

is an example of a linear differential equation that does *not* have power series solutions about $x = 0$. Its linearly independent solutions x^{-3} and $x^{-3} \ln x$ both involve *negative* powers of x; a power series about $x = 0$ could only use non-negative powers. Here we describe conditions that guarantee the existence of power series solutions.

A function $f(x)$ is **analytic** about the point x_0 if it has a convergent power series representation about x_0; i.e., if for some $\delta > 0$,

$$f(x) = \sum_{n=0}^{\infty} c_n(x - x_0)^n, \qquad |x - x_0| < \delta.$$

We are asking the question, Which linear differential equations have a pair of linearly independent analytic solutions?

If a function is analytic at x_0, then the function itself and each of its derivatives must certainly be defined at x_0; we need only evaluate the power series or its term-by-term derivative at $x = x_0$. (And convergent power series can be differentiated term-by-term.) That is,

$$f(x_0) = \sum_{n=0}^{\infty} c_n(x_0 - x_0)^n = c_0$$

$$f'(x_0) = \sum_{n=1}^{\infty} n c_n(x_0 - x_0)^{n-1} = c_1$$

$$f''(x_0) = \sum_{n=2}^{\infty} n(n-1) c_n(x_0 - x_0)^{n-2} = 2c_2$$

$$\vdots$$

Consequently, a function that is not defined at x_0 is not analytic there; e.g.,

$$f(x) = \frac{1}{x^2}$$

is not analytic at $x_0 = 0$. Likewise, a function that is not differentiable at x_0 is not analytic there; e.g.,

$$f(x) = |x|$$

is not analytic at $x_0 = 0$ because $|x|$ has no derivative at $x = 0$. (See figure 10.1.) Likewise, $f(x) = (x + 1)^{3/2}$ is not analytic at $x_0 = -1$ because

$$f'' = \frac{3}{4(x+1)^{1/2}}$$

is not defined there.

On the other hand, a function that is a polynomial (a finite sum of nonnegative powers) in $x - x_0$ or has a known power series is analytic at x_0. For example

$$f(x) = 4 - (x + 2)^3$$

is analytic at $x_0 = -2$ (or any other point) because it can be written as a finite power series in $x + 2$. The function

$$\sin x = x - \frac{x^3}{3!} + \frac{x^5}{5!} + \cdots$$

is certainly analytic at $x = 0$ because of its well known convergent Taylor series expansion. Other familiar functions that are analytic for all x are $\cos x$ and e^x.

Definition 1 The point x_0 is an **ordinary point** of

$$y'' + P(x)y' + Q(x)y = 0$$

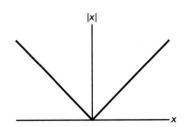

Figure 10.1 Because of the "corner" at $x = 0$, the function $|x|$ is not differentiable at $x = 0$. Hence, it is not analytic at $x = 0$.

if P and Q are analytic at x_0. Otherwise, x_0 is a **singular point** of the equation. ∎

EXAMPLE 15   *The point $x_0 = 0$ is an ordinary point of*

$$y'' + (\cos \pi x)y' + e^x y = 0.$$

The coefficients $P(x) = \cos \pi x$ and $Q(x) = e^x$ have convergent power series expansions about $x_0 = 0$:

$$\cos \pi x = 1 - \frac{x^2}{2!} + \frac{x^4}{4!} + \cdots,$$

$$e^x = 1 + x + \frac{x^2}{2} + \cdots. \qquad ∎$$

EXAMPLE 16   *The point $x_0 = 1$ is a singular point of*

$$y'' + (\cos \pi x)y' + \frac{y}{1-x} = 0,$$

but the point $x_0 = 0$ is an ordinary point of the same equation.

Although $P(x) = \cos \pi x$ is certainly analytic at $x_0 = 1$, the coefficient $Q(x) = 1/(1-x)$ is not because it is not even defined there.

On the other hand, at $x_0 = 0$ both $P(x) = \cos \pi x$ and $Q(x) = 1/(1-x)$ are analytic. The cosine has its familiar Taylor series expansion, and for $1/(1-x)$ we can construct the Taylor series

$$\frac{1}{1-x} = 1 + x + x^2 + x^3 + \cdots + x^n + \cdots, \qquad |x| < 1,$$

using

$$f(x) = (1-x)^{-1} \Rightarrow f(0) = 1$$
$$f'(x) = (1-x)^{-2} \Rightarrow f'(0) = 1$$
$$f''(x) = 2(1-x)^{-3} \Rightarrow f''(0) = 2$$
$$\vdots$$
$$f^{(n)}(x) = n!(1-x)^{-n-1} \Rightarrow f^{(n)}(0) = n!$$

and the ratio test (or geometric series) to determine the radius of convergence. Hence, $x_0 = 0$ is an ordinary point of this equation. ∎

Since both of the coefficients P and Q in the previous example have power series expansions about the ordinary point $x_0 = 0$, we might reasonably expect that solutions of this equation would have convergent power series expansions about $x_0 = 0$ as well. However, it seems equally likely that such series solutions would converge only for $|x| < 1$, the radius of convergence of $Q(x) = 1/(1 -$

x), for at $x = 1$ this function is not analytic. The next theorem confirms this suspicion.

Theorem 10.1 (Existence of Analytic Solutions) Let $P(x), Q(x)$ be functions that are analytic at x_0 with radii of convergence R_P, R_Q. Then the differential equation

$$y'' + P(x)y' + Q(x)y = 0$$

has a pair of linearly independent power series solutions with radius of convergence $R = \min(R_P, R_Q)$.

Rather than prove this theorem, we outline the main steps in the argument:

1. To construct linearly independent solutions, find a pair of solutions that satisfy the two sets of initial conditions $y(x_0) = 1$, $y'(x_0) = 0$ and $y(x_0) = 0, y'(x_0) = 1$. If such solutions exist, then at $x = 0$ the Wronskian $W = 1 \neq 0$.

2. Assume a solution of the form

$$y(x) = \sum_{n=0}^{\infty} c_n (x - x_0)^n.$$

The first set of initial conditions requires $c_0 = 1, c_1 = 0$. Substitute into the differential equation to obtain a recursion formula for $c_2, c_3, \ldots$ in terms of the coefficients of the power series for P, Q.

3. Use the convergence of the power series expansions for P, Q to show that the power series for y converges for $|x - x_0| < R$. The convergence of that series then justifies the formal manipulations (interchange of differentiation and summation, etc.) of the previous step.

4. Repeat the preceding two steps for the second set of initial conditions, for which $c_0 = 0, c_1 = 1$.

For example, this theorem tells us that the equation

$$y'' + (\cos \pi x \, y) + \frac{y}{1 - x} = 0$$

will have power series solutions about the ordinary point $x_0 = 0$ whose radius of convergence is the smaller of infinity, the radius of convergence of the series for $\cos \pi x$, and 1, the radius of convergence of the series for $1/(1 - x)$.

> Full understanding of power series requires the perspective of the complex plane. For example, the power series about $x_0 = 0$ for $1/(1 + x^2)$ convergences only for $|x| < 1$, even though this function is continuous and differentiable for all x. The problem is a "hidden" singularity at $x = i$, a point off the real axis, that limits the radius of convergence to $R = 1$.

10.4.2   Summary

The steps in finding a **power series solution** of a linear differential equation follow.

1. Assume a solution of the form

$$\sum_{n=0}^{\infty} c_n(x - x_0)^n$$

where x_0 is an ordinary point of the equation. If initial conditions are given, choose x_0 to be the initial point.

2. Compute the derivatives of the power series term by term. Substitute in the differential equation and collect coefficients of like powers of x. Take care to account for all terms containing a given power of x. If necessary, adjust summation indices so that all series are written with the same exponent of x.

3. Set the coefficient of each power of x to zero. Solve the resulting equations for the highest numbered coefficient c_n in terms of the lower numbered coefficient(s) in each equation. Working recursively, express each coefficient in terms of c_0 and c_1 and write the series solution using just c_0 and c_1.

4. If initial conditions are given, use them to find c_0 and c_1.

Project 3 asks you to provide a proof of theorem 10.1 following the outline on page 435.

10.4.3 Exercises

EXERCISE GUIDE

To gain experience . . .	Try exercises
Differentiating, integrating, manipulating power series	1–3, 11, 14–15, 16, 19
Locating ordinary points	4–10, 16
Finding general solutions	4–10, 13 (a)–(c)
Solving initial-value problems	6–7, 12, 13(e), 18
Using Taylor series	11, 13(a), 16, 19
With validity of power series solutions	13(d), 16
With recurrence relations	17(a)
With analytic functions	19–20
Finding radius of convergence	2–3(c), 12(b), 17(b)

1. After writing $f(x) = \sum_{n=0}^{\infty} c_n(x - x_0)^n$, the text asks

Why did the starting value of the summation index n in

$$f'(x) = \sum_{n=0}^{\infty} (c_n(x - x_0)^n)'$$

$$= \sum_{n=1}^{\infty} c_n n(x - x_0)^{n-1}$$

change from $n = 0$ to $n = 1$?

Answer this question.

2. Given the convergent power series expansion

$$e^x = \sum_{n=0}^{\infty} \frac{x^n}{n!},$$

verify the following.

(a) $d(e^x)/dx = e^x$ using term by term differentiation,

(b) $\int_0^t e^x \, dx = e^t - 1$ using term by term integration.

(c) Determine the radius of convergence of this power series.

3. Given the convergent power series expansions

$$\sin x = \sum_{n=1}^{\infty} (-1)^{n+1} \frac{x^{2n-1}}{(2n-1)!},$$

$$\cos x = \sum_{n=0}^{\infty} (-1)^n \frac{x^{2n}}{(2n)!},$$

verify each of the following.

(a) $d(\cos x)/dx = -\sin x$ using term-by-term differentiation,

(b) $d(\sin x)/dx = \cos x$ using term-by-term differentiation.

(c) Determine the radius of convergence of each power series.

For Exercises 4–10,

   i. Verify that the given expansion point or initial point is an ordinary point of the equation.

   ii. Find the first four nonzero terms in power series solutions of the differential equation or initial-value problem.

4. $y'' + xy = 0$ (about $x = 0$)

5. $y'' + (x + 2)y = 0$ (about $x = -2$)

6. $y'' - y = 0$, $y(0) = 2$, $y'(0) = 2$

7. $y'' - y = 0$, $y(1) = -1$, $y'(1) = 4$

8. $x'' - (t - 2)x' - 3(t - 2)x = 0$ (about $t = 2$)

9. $y' + y = x^2$ (about $x = 0$)

10. $y'' - 3y' + 2x^2 y = e^x$ (about $x = 0$)

11. Formally compute the derivatives f'', $f^{(3)}$, and $f^{(4)}$ of the power series function

$$f(x) = \sum_{n=0}^{\infty} c_n (x - x_0)^n,$$

evaluate each at $x = x_0$, and verify the Taylor coefficient formula

$$c_n = \frac{f^{(n)}(x_0)}{n!}$$

for $n = 2, 3, 4$.

12. In example 11, the text claims that the constants c_0 and c_1 that remained undetermined at the end of the power series solution of

$$y'' + y = 0$$

satisfy

$$y(0) = c_0, \qquad y'(0) = c_1.$$

(a) Use the power series solution

$$y(x) = c_0 \left(1 - \frac{x^2}{2} + \frac{x^4}{24} + \cdots \right)$$

$$+ c_1 \left(x - \frac{x^3}{20} + \frac{x^5}{120} + \cdots \right)$$

obtained in example 11 to verify this claim.

(b) Verify this same claim for the alternate form of this solution,

$$y(x) = c_0 \sum_{n=0}^{\infty} (-1)^n \frac{x^{2n}}{(2n)!}$$

$$+ c_1 \sum_{n=0}^{\infty} (-1)^n \frac{x^{2n+1}}{(2n+1)!},$$

which was obtained in example 12, and find its radius of convergence.

(c) Verify in general that if $y(x)$ is defined by the convergent power series

$$y(x) = \sum_{n=0}^{\infty} c_n (x - x_0)^n,$$

then $y(x_0) = c_0$, $y'(x_0) = c_1$.

13. (a) Use the Taylor coefficient formula

$$c_n = \frac{f^{(n)}(x_0)}{n!}$$

to verify that power series expansions of $\cos x$ and $\sin x$ about $x_0 = 0$ are

$$\cos x = 1 - \frac{x^2}{2} + \frac{x^4}{24} + \cdots,$$

$$\sin x = x - \frac{x^3}{6} + \frac{x^5}{120} + \cdots.$$

(b) Verify by substitution that $\cos x$, $\sin x$ are indeed solutions of the differential equation considered in example 11,

$$y'' + y = 0.$$

(c) Use the results of parts (a) and (b) to show that a general solution of $y'' + y = 0$ may be written in the power series form

$$y(x) = C_1 \left(1 - \frac{x^2}{2} + \frac{x^4}{24} + \cdots \right)$$

$$+ C_2 \left(x - \frac{x^3}{6} + \frac{x^5}{120} + \cdots \right),$$

where C_1, C_2 are the usual arbitrary constants.

(d) Verify by direct substitution in the differential equation $y'' + y = 0$ that the series expression in part (c) is indeed a solution.

(e) Express the arbitrary constants C_1, C_2 appearing in part (c) in terms of the initial values $y(0), y'(0)$.

14. Verify that the first three terms in each of the sums in

$$y(x) = c_0 \sum_{n=0}^{\infty} (-1)^n \frac{x^{2n}}{(2n)!} + c_1 \sum_{n=0}^{\infty} (-1)^n \frac{x^{2n+1}}{(2n+1)!},$$

the solution of $y'' + y = 0$ obtained in example 12, are the same as those in the solution of the same equation found in example 11,

$$y(x) = c_0 \left(1 - \frac{x^2}{2} + \frac{x^4}{24} + \cdots \right)$$

$$+ c_1 \left(x - \frac{x^3}{20} + \frac{x^5}{120} + \cdots \right).$$

15. Examples 11 and 12 both find power series solutions about $x_0 = 0$ of

$$y'' + y = 0.$$

Example 11 works with the first few terms of the proposed series solution, while example 12 uses the summation notation $\sum$ throughout. Write out the two ex-

amples side by side, matching the equations in each example that represent identical steps. (For example, match the equations that represent the calculation of y' in each example.) Add any computational details omitted from either example.

What are the major differences between the two approaches?

16. Verify that the Cauchy-Euler equation

$$x^2 y'' + 7xy' + 9y = 0$$

has the solutions x^{-3}, $|x|^{-3} \ln x$. Show that neither of these functions has a convergent power series expansion about $x_0 = 0$. (Recall that if a function has a convergent power series expansion, then its coefficients must be defined by the Taylor coefficient formula $c_n = f^{(n)}(x_0)/n!$.) Is $x = 0$ an ordinary point of this equation?

17. In example 12, the text derives from the recurrence relation

$$c_k + (k+2)(k+1)c_{k+2} = 0, \qquad k = 0, 1, 2, \ldots$$

the expressions

$$c_k = (-1)^{k/2} \frac{c_0}{k!}, \qquad k \text{ even}$$

and

$$c_k = (-1)^{(k-1)/2} \frac{c_1}{k!}, \qquad k \text{ odd}$$

for the coefficients in a power series expansion. Then the text says

To get the pattern of these general calculations, try a few specific cases, such as $k = 6, 7, 8, 9$.

(a) Do just that; e.g., write c_6 in terms of c_4, then in terms of c_2 and so on. Do enough concrete cases to convince yourself that the exponent of -1 in the two expressions for c_k is correct.

(b) Find the radius of convergence of this series.

18. Example 13 found the first five nonzero terms in a power series solution of the initial value problem

$$y'' + (x-1)^2 y' - 2(x-1)^3 y = 0,$$

$$y(1) = 1, \qquad y'(1) = -8.$$

It began with the series expansion

$$y(x) = \sum_{0}^{\infty} c_n (x - 1)^n$$

$$f(x_0) = \sum_{n=0}^{\infty} c_n (x_0 - x_0)^n = c_0$$

and found expressions for $c_2, c_3, \ldots$ in terms of c_0, c_1. Then it used the initial conditions to determine c_0, c_1. Reverse this process: Find the first five nonzero terms in a power series expansion of the solution of this initial-value problem by *first* evaluating c_0, c_1 in this series expansion, and then substituting in the differential equation to evaluate the remaining coefficients in the series. Show that you obtain the same solution as example 13.

19. Suppose $f(x)$ is analytic at x_0. Verify that

$$f'(x_0) = \sum_{n=1}^{\infty} n c_n (x_0 - x_0)^{n-1} = c_1$$

$$f''(x_0) = \sum_{n=2}^{\infty} n(n-1) c_n (x_0 - x_0)^{n-2} = 2c_2$$

$$\vdots$$

20. Show that a polynomial in x (a finite sum of non-negative powers of x) is analytic at every finite value of x.

10.5 REGULAR SINGULAR POINTS

The previous section constructed power series solutions of linear equations

$$y'' + P(x)y' + Q(x)y = 0$$

about *ordinary* points, those where the coefficients P and Q are both analytic. But what happens at a *singular* point? For example, $x = 0$ is a singular point of the equation

$$y'' + \frac{3}{2x}y' - \frac{1}{2x^2}y = 0$$

because neither $P = 3/2x$ nor $Q = -1/2x^2$ is defined, much less analytic, at $x = 0$. What is the behavior at $x = 0$ of the solutions of such an equation? Is the singularity in the coefficients reflected in the solutions as well?

We can solve this particular example. Multiplying by $2x^2$ reduces it to the Cauchy-Euler equation

$$2x^2 y'' + 3xy - y = 0,$$

whose indicial equation

$$2r(r - 1) + 3r - 1 = 2r^2 + r - 1 = (2r - 1)(r + 1) = 0$$

has roots $r = -1, \frac{1}{2}$. Linearly independent solutions of this equation are

$$y_1 = x^{-1}, \qquad y_2 = x^{1/2}.$$

Neither solution is analytic at $x = 0$.

But these solutions could be thought of as nearly analytic in the sense that they are either a negative or a noninteger power of x multiplying a (trivial) power series

$$x^{-1} = x^{-1}(1 + 0x + 0x^2 + \cdots)$$

$$x^{1/2} = x^{1/2}(1 + 0x + 0x^2 + \cdots).$$

That observation is the key to the technique of this section, the *method of Frobenius*, which seeks solutions of linear equations in the form of an arbitrary power of x multiplying a power series. Identifying the leading powers of x (x^{-1} and $x^{1/2}$ in this example) is important because those leading terms capture the behavior of the solutions near the singular point.

The method of Frobenius will be restricted to linear equations whose coefficients suffer only a mild singularity at the expansion point. In the preceding example, both $xP(x) = \frac{3}{2}$ and $x^2Q(x) = -\frac{1}{2}$ are analytic at $x = 0$; the singularity in P is no worse than $1/x$ and that in Q is no worse than $1/x^2$. This weak singularlity in P and Q is an example of a *regular singular point*.

Definition 2 The point x_0 is a **regular singular point** of

$$y'' + P(x)y' + Q(x)y = 0$$

if x_0 is a singular point (P and/or Q are not analytic at x_0) but both

$$(x - x_0)P(x) \quad \text{and} \quad (x - x_0)^2 Q(x)$$

are analytic there. A singular point that is not regular is called **irregular**.

To construct a series solution of a linear equation in a neighborhood of a *regular singular point* x_0, the **method of Frobenius** seeks a solution in the form

An arbitrary power of x multiplies the power series.

$$y(x) = (x - x_0)^r \sum_{n=0}^{\infty} c_n(x - x_0)^n,$$

where both the exponent r and the power series coefficients c_n are to be determined by substitution in the given differential equation. To avoid ambiguity in determining r, we assume that $c_0 \neq 0$; that is, we have factored the highest-possible power of $x - x_0$ from the series and placed it in $(x - x_0)^r$.

We illustrate the method of Frobenius with an example. The techniques are much like those of finding power series expansions with the added wrinkle of determining r as well.

EXAMPLE 17 *Find two linearly independent series solutions about $x_0 = 0$ of*

$$2x^2y'' + 3x(1 - x)y' - y = 0.$$

Written in the usual form, this equation is

$$y'' + \frac{3(1 - x)}{2x}y' - \frac{1}{2x^2}y = 0.$$

Evidently, $x_0 = 0$ is a singular point of this equation. But both $xP(x) = 3(1 - x)/2$ and $x^2Q(x) = 1/2$ are analytic at the singular point, so we seek a solution about a *regular* singular point.

To use the method of Frobenius, we seek a solution in the form

Assume Frobenius series solution.

$$y(x) = x^r \sum_{n=0}^{\infty} c_n x^n$$

$$= c_0 x^r + c_1 x^{r+1} + c_2 x^{r+2} + \cdots,$$

where we assume $c_0 \neq 0$. The derivatives we need to substitute into the differential equation are

$$y'(x) = c_0 r x^{r-1} + c_1 (r+1) x^r + c_2 (r+2) x^{r+1} + \cdots$$

$$y''(x) = c_0 r (r-1) x^{r-2} + c_1 (r+1) r x^{r-1} + c_2 (r+2)(r+1) x^r + \cdots.$$

Substituting the expressions for y, y', and y'' into $2x^2 y'' + 3x(1-x)y' - y = 0$ and keeping terms up to x^{r+2} yields

Substitute.

$$2c_0 r(r-1) x^r + 2c_1 (r+1) r x^{r+1} + 2c_2 (r+2)(r+1) x^{r+2} + \cdots$$

$$+3c_0 r x^r + 3c_1 (r+1) x^{r+1} \qquad + 3c_2 (r+2) x^{r+2} + \cdots$$

$$-3c_0 r x^{r+1} \qquad - 3c_1 (r+1) x^{r+2} - \cdots$$

$$-c_0 x^r - c_1 x^{r+1} \qquad - c_2 x^{r+2} - \cdots = 0$$

The top line is $x^2 y''$, the next two are $3x(1-x)y'$, and the last is $-y$. To ease the task of collecting coefficients of like powers of x, corresponding powers appear in the same column, x^r, x^{r+1}, and x^{r+2} from left to right. Try a similar technique to organize your own work.

Since the sum is zero, we require that the coefficient of each power of x be zero:

Equate like powers of x.

$$x^r: \qquad c_0[2r(r-1) + 3r - 1] \qquad = 0$$

$$x^{r+1}: \qquad c_1[2r(r+1) + 3(r+1) - 1] - 3c_0 \qquad = 0$$

$$x^{r+2}: \quad c_2[2(r+2)(r+1) + 3(r+2) - 1] - 3c_1(r+1) = 0 \qquad (10.2)$$

$$\vdots$$

The first equation,

$$c_0[2r(r-1) + 3r - 1] = 0,$$

the one coming from the lowest power of x, requires either

$$c_0 = 0 \quad \text{or} \quad 2r(r-1) + 3r - 1 = 0.$$

Since we assumed $c_0 \neq 0$, we must choose r to satisfy the **indicial equation**

$$2r(r-1) + 3r - 1 = (2r-1)(r+1) = 0.$$

Indicial equation from $c_0 \neq 0$.

Hence,

$$r = -1, \frac{1}{2}.$$

The indicial equation determines r.

To ensure real-valued expressions in $x^{1/2}$ and to avoid the singularity at $x = 0$ in x^{-1}, we restrict consideration to $x > 0$.

Now the other equations in (10.2) provide recurrence relations involving r that give c_n in terms of c_{n-1}, $n = 1, 2, \ldots$:

$$c_1 = \frac{3r}{2r(r+1) + 3(r+1) - 1} c_0$$

$$c_2 = \frac{3(r+1)}{2(r+2)(r+1) + 3(r+2) - 1} c_1$$

$$\vdots$$

For each choice of r, we obtain a set of values for the constants c_n in the power series.

Solve for c's with $r = -1$.

Using $r = -1$ and setting $c_0 = 1$ for convenience, these equations give

$$c_1 = \frac{-3}{-1} c_0 = 3$$

$$c_2 = \frac{0}{-4} c_1 = 0.$$

Thus, the solution corresponding to $r = -1$ is

SOLUTION FOR $r = -1$

$$y_1 = x^{-1}(1 + 3x + \cdots), \qquad x > 0.$$

Solve for c's with $r = \frac{1}{2}$

Repeating these steps with $r = \frac{1}{2}$ and $c_0 = 1$ gives

$$c_1 = \frac{\frac{3}{2}}{\frac{10}{2}} c_0 = \frac{3}{10}$$

$$c_2 = \frac{-\frac{9}{2}}{\frac{29}{2}} c_1 = \frac{-9}{29}.$$

The solution corresponding to $r = 1/2$ is

SOLUTION FOR $r = \frac{1}{2}$

$$y_2 = x^{1/2}\left(1 + \frac{3x}{10} - \frac{9x^2}{29} + \cdots\right), \qquad x > 0.$$

Taking linear independence for granted (because of the different leading powers of x), a general solution of $2x^2 y'' + 3x(1 - x)y' - y = 0$ for $x > 0$ is

$$y_g = C_1 y_1 + C_2 y_2.$$

Had we kept the undetermined constant c_0 in either of the solutions for $r = -1, \frac{1}{2}$, it would have been absorbed into C_1 or C_2. ∎

Much of the value of the solution obtained in the preceding example is its picture of the behavior near the singular point $x = 0$. One solution ($y_2 = x^{1/2}$) is bounded. The other is not, growing like x^{-1}. Neither solution is differentiable at the origin.

This example illustrates the basic idea of the *method of Frobenius*.

Method of Frobenius   If $x = x_0$ is a regular singular point of

$$y'' + P(x)y' + Q(x)y = 0,$$

assume a solution of the form

$$y(x) = (x - x_0)^r \sum_{n=0}^{\infty} c_n (x - x_0)^n,$$

FROBENIUS SERIES

require $c_0 \neq 0$, and substitute into the differential equation.

Collect like powers of x. Apply the condition $c_0 \neq 0$ to the coefficient of the lowest power of x to obtain the *indicial equation*, a quadratic equation for r. Find the roots r_1, r_2 of the indicial equation, and determine one set of power series coefficients for each value of r from the recurrence relations giving c_n in terms of c_{n-1}, $n = 1, 2, \ldots$.

When r is other than a nonnegative integer, $(x - x_0)^r$ will either be singular at $x = x_0$ or complex-valued for $x < x_0$. The cases $x < x_0$ and $x > x_0$ can be considered separately and eventually consolidated using $|x - x_0|$ in place of $x - x_0$. For simplicity, we consider only $x > x_0$.

Unfortunately, several complications lie below the tip of this iceberg.

- If $r_1 = r_2$, there is no second independent Frobenius solution because the two sets of recurrence relations collapse into one.
- If $r_1 - r_2$ is an integer, then the series for one value of r may or may not be a multiple of the other series; the factor $x^{r_1 - r_2}$ might be absorbed into the power series part of one solution. A more detailed analysis is needed.

The following examples hint at some of these difficulties.

EXAMPLE 18   *Find series solutions about $x = 0$ of Bessel's equation of order one half,*

$$x^2 y'' + xy' + (x^2 - \frac{1}{4})y = 0.$$

Bessel's equation of order p is $x^2 y'' + xy' + (x^2 - p^2)y = 0$. It arises in the course of solving heat-flow problems in circular domains, among many other settings.

Assume Frobenius series solution.

We leave it to you to verify that $x = 0$ is a regular singular point. As with the previous example, we assume a solution of the form

$$y(x) = x^r \sum_{n=0}^{\infty} c_n x^n$$

$$= c_0 x^r + c_1 x^{r+1} + c_2 x^{r+2} + \cdots,$$

where sufficient powers of x have been factored from the sum to ensure $c_0 \neq 0$. Following the pattern of example 17, we differentiate, substitute in the differential equation, and collect like powers of x to obtain (through terms in x^{r+2})

Substitute.

$$c_0 r(r-1)x^r + c_1(r+1)r x^{r+1} + c_2(r+2)(r+1)x^{r+2} + \cdots$$

$$+ c_0 r x^r + c_1(r+1)x^{r+1} + \qquad c_2(r+2)x^{r+2} + \cdots$$

$$- \frac{c_0 x^r}{4} - \qquad \frac{c_1 x^{r+1}}{4} - \qquad \frac{c_2 x^{r+2}}{4} + \cdots$$

$$+ \qquad\qquad c_0 x^{r+2} + \cdots = 0$$

Equate like powers of x.

Equating to zero the coefficients of each power of x yields

$$x^r: \qquad c_0\left[r^2 - \frac{1}{4}\right] = 0$$

$$x^{r+1}: \qquad c_1\left[(r+1)^2 - \frac{1}{4}\right] = 0$$

$$x^{r+2}: \qquad c_2\left[(r+2)^2 - \frac{1}{4}\right] + c_0 = 0$$

$$\vdots$$

From the coefficient of x^r and the requirement that $c_0 \neq 0$, we have the indicial equation

Indicial equation from $c_0 \neq 0$.

$$r^2 - \frac{1}{4} = 0$$

and the roots

The indicial equation determines r.

$$r = \pm\frac{1}{2}.$$

These roots differ by 1.

The coefficients of x^{r+1} and x^{r+2}, as before, provide relations that define coefficients in the power series.

With $r = \frac{1}{2}$, we find from the coefficient of x^{r+1} that

Solve for c's with $r = \frac{1}{2}$.

$$c_1\left[\left(\frac{3}{2}\right)^2 - \frac{1}{4}\right] = 2c_1 = 0.$$

Hence, $c_1 = 0$ for $r = \frac{1}{2}$. From the coefficient of x^{r+2} with $r = \frac{1}{2}$, we find

$$c_2 = \frac{-c_0}{6}.$$

Setting $c_0 = 1$, we can take one solution of Bessel's equation of order one half to be

$$y_1 = x^{1/2}\left(1 - \frac{x^2}{6} + \cdots\right).$$

SOLUTION FOR $r = \frac{1}{2}$

With $r = -\frac{1}{2}$, we find a more surprising result from the coefficient of x^{r+1}:

Solve for c's with $r = -\frac{1}{2}$.

$$c_1\left[\left(\frac{1}{2}\right)^2 - \frac{1}{4}\right] = c_1 \cdot 0 = 0;$$

c_1 is arbitrary! With a bit of hesitation, perhaps, we choose $c_1 = 0$, the same value obtained when $r = \frac{1}{2}$. Then the coefficient of x^{r+2} gives

$$c_2 = \frac{-c_0}{2}.$$

Setting $c_0 = 1$, a second solution of Bessel's equation of order one half is

$$y_2 = x^{-1/2}\left(1 - \frac{x^2}{2} + \cdots\right),$$

SOLUTION FOR $r = -\frac{1}{2}$

and it appears to be linearly independent of y_1, even though the roots of the indicial equation differed by a positive integer. ∎

EXAMPLE 19 *Attempt to mimic the previous analysis to obtain solutions of Bessel's equation of order 2,*

$$x^2 y'' + x y' + (x^2 - 4)y = 0,$$

about $x = 0$.

Proceeding exactly as in example 18, we start with a Frobenius series solution, substitute in the differential equation, collect powers of x, and find the following relations from the coefficients of the lowest three powers of x:

Equate coefficients.

$$x^r : \qquad c_0[r^2 - 4] = 0$$

$$x^{r+1} : \qquad c_1[(r+1)^2 - 4] = 0$$

$$x^{r+2} : \qquad c_2[(r+2)^2 - 4] + c_0 = 0$$

$$\vdots$$

As usual, the first equation and $c_0 \neq 0$ produce the indicial equation

$$r^2 - 4 = 0$$

INDICIAL EQUATION

and the indicial roots

$$r = \pm 2.$$

The second relation forces $c_1 = 0$, and the third requires

$$c_2[2r + 4] + c_0 = 0.$$

When $r = 2$, we quickly find $c_2 = -c_0/8$. Setting $c_0 = 1$, we have one solution,

SOLUTION FOR $r = 2$

$$y_1 = x^2 \left(1 - \frac{x^2}{8} + \cdots \right).$$

But when $r = -2$, trouble arises. The relation $c_2[2r + 4] + c_0 = 0$ reduces to

FAILED RECURRENCE RELATION FOR $r = -2$

$$c_2 \cdot 0 + c_0 = 0.$$

The recurrence relations collapse at this point because there is no c_2 satisfying this equation as long as $c_0 \neq 0$. ∎

When separation of variables is applied to the heat equation in a circular domain, the resulting eigenvalue problem involves Bessel's equation of order zero. Consequently, we explore the solution of that equation in a bit more detail.

EXAMPLE 20 *Find a solution of Bessel's equation of order zero,*

$$x^2 y'' + xy' + x^2 y = 0,$$

subject to the initial conditions $y(0) = 1$, $y'(0) = 0$. Determine the behavior at $x = 0$ of a second linearly independent solution.

Since $x = 0$ is a regular singular point, we assume a solution in the form of a Frobenius series,

Assume Frobenius series solution.

$$y(x) = x^r \sum_{n=0}^{\infty} c_n x^n,$$

Substitute and collect coefficients.

where $c_0 \neq 0$. Substituting in the differential equation and collecting similar powers of x lead to

$$c_0[r(r-1) + r]x^r + c_1[(r+1)r + (r+1)]x^{r+1}$$
$$+ (c_2[(r+2)(r+1) + (r+2)] + c_0)x^{r+2} + \cdots$$
$$+ (c_n[(r+n)(r+n-1) + (r+n)] + c_{n-1})x^{r+n} + \cdots = 0.$$

The coefficient of x^r and $c_0 \neq 0$ yield the indicial equation,

INDICIAL EQUATION

$$r(r-1) + r = r^2 = 0.$$

The (repeated) indicial root is $r = 0$. Setting $r = 0$, the coefficient of x^{r+1} reduces to

Solve for c's.

$$c_1 = 0.$$

Setting to zero the coefficient of x^{r+n}, $n = 2, 3, \ldots$ and using $r = 0$ gives the general recurrence relation

$$c_n = \frac{-c_{n-2}}{n(n-1)+n} = \frac{-c_{n-2}}{n^2}, \qquad n = 2, 3, \ldots .$$

RECURRENCE RELATION

Since $c_1 = 0$, all coefficients with odd indices are zero.

Representing the remaining even indices as $n = 2m$, $m = 1, 2, \ldots$, the recurrence relation becomes

$$c_{2m} = \frac{-c_{2(m-1)}}{4m^2}, \qquad m = 1, 2, \ldots .$$

By writing $c_{2(m-1)}$ in terms of $c_{2(m-2)}$ and so on, we find

$$c_{2m} = \frac{(-1)^m c_0}{4^m (m!)^m}, \qquad m = 1, 2, \ldots .$$

Since $r = 0$, the solution $y(x) = x^r(c_0 + c_1 x + \cdots)$ can be evaluated at $x = 0$: $y(0) = c_0$. The initial condition $y(0) = 1$ forces $c_0 = 1$. We have found the solution commonly denoted $J_0(x)$, the **Bessel function of first kind of order zero**,

$$J_0(x) = \sum_{m=0}^{\infty} \frac{(-1)^m x^{2m}}{4^m (m!)^m}.$$

FIRST SOLUTION

Since $J_0(x) = 1 - x^2/4 + \cdots$ and $J_0' = -x/2 + \cdots$, it obviously satisfies the remaining initial condition, $J_0'(0) = 0$.

Since the indicial equation has only one root, we can not find a second linearly independent solution using a Frobenius series. Instead, we turn to reduction of order and assume the second solution has the form

SECOND SOLUTION VIA REDUCTION OF ORDER

$$y(x) = J_0(x)u(x),$$

where the function $u(x)$ is to be determined. Substituting into Bessel's equation and letting $v = u'$ in the usual way yield

$$x^2 J_0(x)v'(x) + \left(2x^2 J_0'(x) + x J_0(x)\right) v(x) = 0.$$

With caution because of the singularity at $x = 0$, we use separation of variables or an integrating factor to obtain the solution

$$v(x) = \frac{1}{x(J_0(x))^2}.$$

Hence,

$$u(x) = \int v(x)\, dx = \int \frac{dx}{x(J_0(x))^2}.$$

10.5   Regular Singular Points

Our only concern is the behavior of this second solution near $x = 0$. Consequently, we estimate the behavior of this integral rather than evaluate it exactly. Since $J_0(x) = 1 - x^2/4 + \cdots$, we have

$$\frac{1}{(J_0(x))^2} = \frac{1}{1 - x^2/2 + \cdots} \approx 1 + \frac{x^2}{2} + \cdots$$

for x small.

The last approximation used $(1 - z)^{-1} \approx 1 + z$, $|z| < 1$, a consequence of a Taylor expansion (or the binomial expansion).

Consequently, when x is small and positive, we find

$$u(x) = \int \frac{dx}{x(J_0(x))^2} \approx \int \frac{1}{x}\left(1 + \frac{x^2}{2} + \cdots\right) dx = \ln x + \cdots.$$

The terms other than $\ln x$ are all bounded near zero since they involve positive powers of x.

We conclude that a second linearly independent solution of Bessel's equation of order zero must behave near $x = 0$ as does $J_0(x) \ln x$. *The second solution of Bessel's equation of order zero has a logarithmic singularity at $x = 0$*, as does the second solution of a Cauchy-Euler equation with the repeated indicial root $r = 0$. ∎

A standard form for the second solution of Bessel's equation of order zero is the *Bessel function of second kind of order zero*. It is defined by

$$Y_0(x) = \frac{2}{\pi}\left(\ln\left(\frac{x}{2}\right) + \gamma\right) J_0(x) - \frac{2}{\pi} \sum_{m=1}^{\infty} \frac{(-1)^m x^{2m}}{4^m (m!)^2} S_m.$$

Here S_m is the partial sum of the harmonic series,

$$S_m = 1 + \frac{1}{2} + \frac{1}{3} + \cdots + \frac{1}{m},$$

and the Euler constant γ is defined by

$$\gamma = \lim_{n \to \infty} (S_n - \ln n).$$

Figure 10.2 shows graphs of J_0 and Y_0. Tables of values of these and other solutions of Bessel's equation appear in [1]. These functions are also well represented in such computer packages as *Macsyma* and *Mathematica*.

As the last example suggests, the problems that arise when indicial roots differ by zero or an integer can be overcome to produce a second linearly independent solution involving more than just a Frobenius series. Those issues are not pursued here.

When the *roots of the indicial equation do not differ by zero or a positive integer,* the method of Frobenius can be shown to *always yield two linearly independent solutions.*

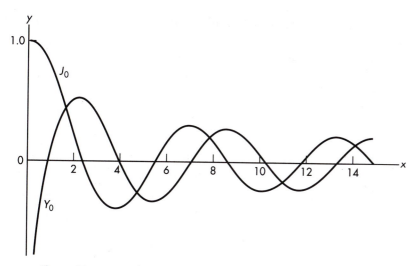

Figure 10.2: Graphs of the solutions J_0 and Y_0 of Bessel's equation of order zero.

10.5.1 Exercises

EXERCISE GUIDE

To gain experience . . .	Try exercises
Locating ordinary points, regular singular points	1–6, 9, 13–19
Finding indicial equations, indicial roots	1–6, 10–12, 22(a)–(b)
Identifying analytic functions	7–8
Using the method of Frobenius	10–11, 13–19, 20(b)
With the limitations of the method of Frobenius	19
With recurrence relations	20(a), 21
Finding radius of convergence	20(c)
With the method of Frobenius in a general setting	22

For each of the differential equations in exercise 1–6,

   i. Find all finite ordinary points,

   ii. Find all finite regular singular points,

   iii. Find the indicial equation and its roots at each regular singular point.

1. $x^2 y'' + 4xy' + (x^2 - 4)y = 0$

2. $(x - 4)^2 y'' + 4xy' + (x^2 - 4)y = 0$

3. $(x^2 - 4)y'' + 4xy' + (x^2 + 4)y = 0$

4. $y'' \sin 2x + xy' - y = 0$

5. $x^2 y'' + xy' + (x^2 - \frac{1}{16})y = 0$ (Bessel's equation of order $\frac{1}{4}$)

6. $4xy'' + 2y' + y = 0$

7. Verify the claim in the text that neither of the solutions

$$y_1 = x^{-1}, \qquad y_2 = x^{1/2}$$

of

$$y'' + \frac{3}{2x}y' - \frac{1}{2x^2}y = 0$$

are analytic at $x = 0$.

8. Example 17 used the method of Frobenius to find the solutions

$$y_1 = x^{-1}(1 - 3x + \cdots)$$

and

$$y_2 = x^{1/2}\left(1 + \frac{3x}{14} - \frac{9x^2}{31} + \cdots\right)$$

of $2x^2 y'' + 3x(1-x)y' - y = 0$ in a neighborhood of the regular singular point $x = 0$. Explain why neither of these solutions is
  (a) Analytic at $x = 0$,
  (b) A power series about $x = 0$.

9. Verify that $x = 0$ is a regular singular point of Bessel's equation of order p, $x^2 y'' + xy' + (x^2 - p^2)y = 0$. Does this equation have any other finite singular points?

10. Example 18 skips the details of substituting the Frobenius series into Bessel's equation of order one half, $x^2 y'' + xy' + (x^2 - \frac{1}{4})y = 0$. Supply them and verify that the resulting indicial equation and recurrence relations are correct.

11. Example 19 skips the details of substituting the Frobenius series into Bessel's equation of order two, $x^2 y'' + xy' + (x^2 - 4)y = 0$. Supply them and verify that the resulting indicial equation and recurrence relations are correct.

12. Show that the indicial equation for Bessel's equation of order p, $x^2 y'' + xy' + (x^2 - p^2) = 0$, is

$$r^2 - p^2 = 0.$$

What choices of p guarantee that the roots of the indicial equation will differ by *other* than zero or a positive integer?

In exercises 13–18, find the first three nonzero terms in series solutions of the given differential equations about a regular singular point.

13. $4xy'' + 2y' + y = 0$

14. $16x^2 y'' + 16xy' + (16x^2 - 1)y = 0$

15. $x^2 y'' + xy' + (x^2 - \frac{1}{16})y = 0$ (Bessel's equation of order $\frac{1}{4}$)

16. $2x^2 y'' + (1 - x)y' + y = 0$

17. $3(x - 2)y'' + y' - y = 0$

18. $6(x+1)^2 y'' + 5(x+1)y' + x(x+2)y = 0$

19. Try to apply the method of Frobenius to the equation $x^3 y'' + y = 0$ at $x = 0$. Is $x = 0$ a regular singular

point? Demonstrate how the method of Frobenius fails.

20. Building on the work of example 17,
  (a) Show that the terms in the Frobenius series about $x = 0$,

$$x^r \sum_{n=0}^{\infty} c_n x^n,$$

for the differential equation $2x^2 y'' + 3x(1-x)y' - y = 0$ are determined by the recurrence relation

$$c_{k+1} = \frac{3(r+k)}{2(r+k)(r+k+1) + 3(r+k+1) + 1} c_k.$$

  (b) Write the general terms in a pair of linearly independent solutions of this equation.
  (c) Determine the radius of convergence of the *power series* portion of these solutions.

21. Show that the general recurrence relation term for Bessel's equation of order p, $x^2 y'' + xy' + (x^2 - p^2) = 0$, is

$$c_n[(r+n)(r+n-1) + r + n - p^2] + c_{n-2} = 0,$$

$n = 2, 3, 4, \ldots$. Using the indicial equation $r^2 - p^2 = 0$, deduce that this relation reduces to

$$c_n[2nr + n^2] + c_{n-2} = 0.$$

What problem occurs whenever one of the roots of the indicial equation is one half a negative integer?

22. Suppose $x = 0$ is a regular singular point of the equation

$$y'' + P(x)y' + Q(x)y = 0.$$

Write the power series expansions for the analytic functions $xP(x)$ and $x^2 Q(x)$ as

$$xP(x) = \sum_{n=0}^{\infty} p_n x^n, \qquad x^2 Q(x) = \sum_{n=0}^{\infty} q_n x^n.$$

  (a) Show that the indicial equation for $y'' + P(x)y' + Q(x)y = 0$ is

$$r(r-1) + p_0 r + q_0 = 0;$$

i.e., it is the indicial equation for the Cauchy-Euler equation

$$x^2 y'' + p_0 x y' + q_0 y = 0$$

whose coefficients are the leading terms in the power series expansions for xP and $x^2 Q$.

(b) Use this result to recover the indicial equation for

$$2x^2 y'' + 3x(1-x)y' - y = 0,$$

the equation considered in example 17.

(c) State and derive the corresponding result for an arbitrary regular singular point $x = x_0$.

23. Find the radius of convergence of the series defining the Bessel function of first kind of order zero,

$$J_0(x) = \sum_{m=0}^{\infty} \frac{(-1)^m x^{2m}}{4^m (m!)^m}.$$

10.6 SOLUTION METHOD SUMMARY

The only general solution methods for *nonlinear* equations are numerical: Euler, Heun, Runge-Kutta, etc.

For *linear* equations, the options are as follows.

1. Homogeneous equation
   (a) Constant-coefficient equation
        i. Characteristic equation: $y = e^{rx}$
   (b) Variable-coefficient equation
        i. Cauchy-Euler equation: $y = x^r$
        ii. Reduction of order: $y_2 = u(x)y_1$
        iii. Power series: $y = c_0 + c_1(x - x_0) + \cdots$, x_0 an ordinary point
        iv. Frobenius: $y = (x - x_0)^r (c_0 + c_1(x - x_0) + \cdots)$, x_0 a regular singular point

(*continued*)

TABLE 10.2

A decision tree for choosing solution methods for second-order equations

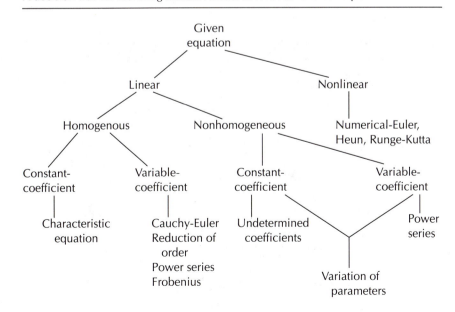

2. Nonhomogeneous equation
   (a) Constant-coefficient equation
      i. Undetermined coefficients: y_p similar to forcing terms of sine, cosine, exponential, polynomial
      ii. Variation of parameters: $y_p = u_1(x)y_1 + u_2(x)y_2$
   (b) Variable-coefficient equation
      i. Variation of parameters: $y_p = u_1(x)y_1 + u_2(x)y_2$
      ii. Power series: $y = c_0 + c_1(x - x_0) + \cdots$, x_0 an ordinary point, and forcing term expanded about x_0 as well.

Table 10.2 provides a decision tree for choosing among these possible solution methods. Note that all linear equations, regardless of order, can be written as systems so that the methods of chapter 8 can be applied.

10.7   CHAPTER EXERCISES

EXERCISE GUIDE

To gain experience . . .	Try exercises
Finding general solutions	1–10, 22–23
Solving initial-value problems	7–8
Locating ordinary points, regular singular points	11–21
Using reduction of order	24
Solving Cauchy–Euler equations	1–2, 5–6, 9
Using variation of parameters	2–6
Using series solution methods	7–8, 10, 22–23

In exercises 1–10, either solve the given initial-value problem or find a general solution of the differential equation. If you use a series method, find at least the first three nonzero terms in the series.

1. $4(x + 1)^2 y'' - 12(x + 1)y' + 16y = 0$

2. $x^2 y'' + 2xy' + 3y = \tan x$

3. $y'' + y = \sin t$

4. $y'' + y = \sin \pi t$

5. $t^2 u'' + 2tu' - 6u = 1$

6. $t^2 u'' + 2tu' - 6u = 1/t$

7. $xy'' + y = 0$, $y(1) = 2$, $y'(1) = 0$

8. $x^2 y'' + xy' + x^2 y = 0$, $y(1) = y'(1) = 1$

9. $(x - 1)^2 y'' - 3(x - 1)y' + 2y = 4(x - 1)^2$

10. $t^2 z'' - 3z' + 2z = (t - 1)^2$

For each of the differential equations in exercises 11–21, find all finite

i. Ordinary points,

ii. Regular singular points.

11. $x^2 y'' + xy' + x^2 y = 0$ (Bessel's equation of order zero)

12. $x^2 y'' + xy' + (x^2 - 1)y = 0$ (Bessel's equation of order one)

13. $x^2 y'' + xy' + (x^2 - p^2)y = 0$ (Bessel's equation of order p)

14. $x(x-1)y'' - y' - y = 0$ (A form of the hypergeometric equation)

15. $(1 - x^2)y'' - xy' + p^2 y = 0$ (Chebyshev's equation)

16. $xy'' + (1 - x)y' - y = 0$ (Another form of the hyper-geometric equation)

17. $xy'' + (1 - x)y' + ny = 0$, n a nonnegative integer (Laguerre's equation of order n)

18. $y'' + (3 - x^2)y = 0$ (A form of Hermite's equation)

19. $y'' + 4\pi^2 m(2E - kx^2)/h^2 = 0$, m, h, E, k all constants (A form of Schrödinger's wave equation; m is mass, h is Planck's constant, E is total energy, and $-k$ is a kind of spring constant.)

20. $(1 - x^2)y'' - 2xy' + n(n + 1)y = 0$, n a nonnegative integer (Legendre's equation)

21. $y'' - xy = 0$ (Airy's equation)

22. Find the first three nonzero terms in a pair of linearly independent series solutions about $x = 0$ of Hermite's equation,

$$y'' - 2xy' + 2py = 0,$$

where p is a constant.

23. Find the first three nonzero terms in a pair of linearly independent series solutions about $x = 0$ of Airy's equation,

$$y'' + xy = 0.$$

24. The purpose of this problem is to show via reduction of order that every constant-coefficient, linear, homogeneous, second-order equation

$$y'' + b_1 y' + b_0 y = 0$$

possessing the single repeated real characteristic root

r has the pair of linearly independent solutions e^{rx}, xe^{rx}.

(a) Show that if the characteristic equation of $y'' + b_1 y' + b_0 y = 0$ possesses a single repeated real root, then b_1 can be expressed in terms of b_0. (*Hint:* What is the value of the discriminant in this case?)

(b) Using e^{rx} as the first solution, construct a second solution of $y'' + b_1 y' + b_0 y = 0$ of the form $u(x)e^{rx}$ via reduction of order. Using part (a), show that $u(x) = x$ by deriving the equation $u'' = 0$.

(c) Verify that e^{rx} and xe^{rx} are indeed linearly independent.

25. The eigenvalue problem

$$xX''(x) + X'(x) + x\lambda X(x) = 0, \ X'(0) = X(L) = 0,$$

arises when separation of variables is applied to a model of heat flow in a thin circular disk of radius L; see exercise 12, page 397, of the chapter exercises for chapter 9.

(a) Show that solutions of the differential equation and the boundary condition $X'(0) = 0$ are $X(x) = J_0(\sqrt{\lambda}x)$, $\lambda > 0$. Recall that $J_0(x)$, the Bessel function of first kind of order zero, is a solution of Bessel's equation of order zero, $x^2 y''(x) + xy'(x) + x^2 y(x) = 0$.

(b) Argue that the eigenvalues are defined by $J_0(\sqrt{\lambda}L) = 0$. Use the graph of J_0 in figure 10.2, to write the first few eigenvalues in terms of L. How do those compare with the corresponding eigenvalues of $X''(x) + \lambda X(x) = 0$ subject to the same boundary conditions?

10.8 CHAPTER PROJECTS

1. **Cauchy-Euler equations for $x < 0$** The text treats the Cauchy-Euler equation

$$x^2 y'' + b_1 xy' + b_0 y = 0$$

for $x > 0$. This project asks you to extend that work to the case $x < 0$, thereby verifying the solution forms given in table 10.1, page 412.

Begin by assuming a solution of the form $y = (-x)^r$, $x < 0$. Using the chain rule *carefully*, compute dy/dx and d^2y/dx^2. Rewrite the coefficients x^2 and x in the differential equation in terms of $-x$. Substitute the assumed solution form and show that you obtain the same indicial equation as when $x > 0$. Write the appropriate forms of the solution when the indicial equation has distinct real roots, repeated real roots, and complex conjugate roots. Finally, use the definition of absolute value ($|x| = x$ for $x \geq 0$, $|x| = -x$ for $x < 0$) to obtain the entries in table 10.1. State your results in the form of a theorem.

2. **Change of variables in Cauchy-Euler equations** Complete steps (a–d) to show that the change of variables $x = e^t$ reduces the Cauchy-Euler equation

$$x^2 y''(x) + b_1 x y'(x) + b_0 y(x) = 0$$

to the constant-coefficient equation

$$\frac{d^2 Y}{dt^2} + (b_1 - 1)\frac{dY}{dt} + b_0 Y = 0,$$

which then may be solved by the characteristic equation method. What restrictions, if any, must be placed on x? On t?

(a) Define a new function $Y(t)$ by $Y(t) = y(e^t)$. Use the chain rule to show that

$$\frac{dY}{dt} = \frac{dy}{dx}\frac{dx}{dt} = xy'(x).$$

Derive a similar expression for $d^2 Y / dt^2$ beginning with

$$\frac{d^2 Y}{dt^2} = \frac{d(xy'(x))}{dt} = \frac{d(xy'(x))}{dx}\frac{dx}{dt}.$$

(b) Substitute the expressions obtained in part (a) into $x^2 y''(x) + b_1 x y'(x) + b_0 y(x) = 0$ to obtain the differential equation for Y.

(c) If the differential equation for Y has a solution $Y(t) = e^{rt}$, where r is a root of the characteristic equation

$$r^2 (b_1 - 1)r + b_0 = 0,$$

argue that x^r is a solution of $x^2 y''(x) + b_1 x y'(x) + b_0 y(x) = 0$.

(d) If r is a repeated characteristic root of the equation for Y, argue that the corresponding solution $t e^{rt}$ transforms into the solution $x^r \ln x$ of $x^2 y''(x) + b_1 x y'(x) + b_0 y(x) = 0$.

(e) Use this transformation technique to solve three Cauchy-Euler equations whose indicial equations have distinct real roots, repeated real roots, and complex conjugate roots.

3. **Proof of existence of analytic solutions** Following the steps outlined after the statement of theorem 10.1, page 435, prove that a second-order linear equation has a pair of linearly independent analytic solutions in a neighborhood of an ordinary point. In accordance with usual professional practice, give credit for any ideas you obtain from other sources.

11

THE LAPLACE TRANSFORM

One transformation leaves the way open for the introduction of others.
Niccolò Machiavelli

The Laplace transform is a useful tool for solving linear constant-coefficient differential equations, among other problems, especially when those equations involve square waves, ramps, and other unusual forcing functions. In addition, Laplace transforms succinctly describe the input-output behavior of systems modeled by such equations. For example, they provide a relation between the forcing frequency applied to a system and the frequencies that appear in the response, or output, of the system.

11.1 THE TRANSFORM IDEA

The next few pages introduce the idea of the Laplace transform. This device can transform constant-coefficient differential equations into algebraic equations, which can easily be solved for an alternative representation of the original unknown function.

> This transform is named for Pierre Simon de Laplace (1749-1827), a French mathematician and politician. (*Laplace* is pronounced L*aplaz* as in *plaza*, not *place*.) His contributions to mathematics are marred by his failure to give credit to others and by his greed as a politician. Those faults may be balanced by the value of his own work and by the assistance he gave to the young Cauchy, who established calculus much as we study it today.

The **Laplace transform** $\mathcal{L}\{f\}$ of a function $f(t)$ is defined by

$$\mathcal{L}\{f\} = \int_0^\infty f(t)e^{-st}\,dt, \tag{11.1}$$

for those values of s (if any) for which this improper integral exists.

The improper integral in (11.1) is defined by

$$\int_0^\infty f(t)e^{-st}\,dt = \lim_{b\to\infty} \int_0^b f(t)e^{-st}\,dt.$$

At the end of this section, we give conditions on f that guarantee this improper integral exists. These require that the integrand $f(t)e^{-st}$ decay to zero rapidly enough to ensure that the integral $\int_0^b f(t)e^{-st}\,dt$ approaches a finite limit for large b.

The improper integral $\int_0^\infty f(t)e^{-st}\,dt$ defining the Laplace transform is actually a function of the parameter s. Like the operations of differentiation and antidifferentiation, which transform one function into another, the Laplace transform takes one function, written here as a function of t, into another function, written here as a function of s. We usually write $\mathcal{L}\{f(t)\} = F(s)$, using upper case letters for Laplace transforms of functions denoted by lower case letters.

As another example of a transform, consider the *area transform*

$$\mathcal{A}\{f\} = \int_0^s f(t)\,dt.$$

It is defined for all functions f that are nonnegative and integrable for $t \geq 0$. The transformed function $F(s) = \mathcal{A}\{f(t)\}$ is just the area under the graph of f between $t = 0$ and $t = s$; $\mathcal{A}$ transforms a function into the area beneath its graph.

Although the Laplace transform lacks an obvious physical interpretation such as area, like the area transform and like integration, it is *linear*. More precisely, if $f(t)$, $g(t)$ are functions with Laplace transforms $F(s) = \mathcal{L}\{f\}$, $G(s) = \mathcal{L}\{g\}$, then

The Laplace transform is linear.

$$\mathcal{L}\{cf + dg\} = c\mathcal{L}\{f\} + d\mathcal{L}\{g\} = cF + dG$$

for any constants c, d.

Some examples of Laplace transforms follow. We show how to find transforms of common functions such as the exponential, how operations such as differentiation in time affect the transform, and how the transform of the solution of an initial-value problem can be obtained directly from the problem itself.

EXAMPLE 1   *Let a be a constant. Find $\mathcal{L}\{e^{at}\}$.*

We apply the definition (11.1) of Laplace transform with $f(t) = e^{at}$:

$$\mathcal{L}\{e^{at}\} = \int_0^\infty e^{at}e^{-st}\,dt$$

$$= \int_0^\infty e^{(a-s)t}\,dt$$

$$= \lim_{b \to \infty} \frac{e^{(a-s)t}}{a-s} \Big|_{t=0}^{t=b}$$

$$= \lim_{b \to \infty} \frac{e^{(a-s)b}}{a-s} - \frac{1}{a-s}$$

$$= \frac{1}{s-a},$$

providing $s > a$. (Why is this restriction on s needed? Where is it first required? See exercise 34.) Hence,

$$\boxed{\mathcal{L}\{e^{at}\} = \frac{1}{s-a}, \qquad s > a.}$$ ∎

EXAMPLE 2 *Find the Laplace transform of* $\sin at$, *a constant.*

Assuming $s > 0$, we apply the Laplace transform definition (11.1) and integrate by parts twice using $dv = e^{-st}\, dt$:

$$\mathcal{L}\{\sin at\} = \int_0^\infty (\sin at)e^{-st}\, dt$$

$$= -\frac{(\sin at)e^{-st}}{s} \Big|_0^\infty + \frac{a}{s} \int_0^\infty (\cos at)e^{-st}\, dt$$

$$= -\frac{a(\cos at)e^{-st}}{s^2} \Big|_0^\infty - \frac{a^2}{s^2} \int_0^\infty (\sin at)e^{-st}\, dt$$

$$= \frac{a}{s^2} - \frac{a^2}{s^2} \int_0^\infty (\sin at)e^{-st}\, dt.$$

We solve for the integral $\int_0^\infty (\sin at)e^{-st}\, dt$ defining the transform to obtain

$$\mathcal{L}\{\sin at\} = \int_0^\infty (\sin at)e^{-st}\, dt$$

$$= \left(1 + \frac{a^2}{s^2}\right)^{-1} \frac{a}{s^2}$$

$$= \frac{a}{a^2 + s^2}.$$

Hence, for $s > 0$,

$$\boxed{\mathcal{L}\{\sin at\} = \frac{a}{a^2 + s^2}.}$$ ∎

EXAMPLE 3 *Let f be a differentiable function. Find $\mathcal{L}\{f'\}$ in terms of the transform $F(s)$ of $f(t)$.*

We apply the definition (11.1) of the Laplace transform and integrate by parts using $u = e^{-st}$ and $dv = f'(t)\,dt$:

$$\mathcal{L}\{f'\} = \int_0^\infty f'(t)e^{-st}\,dt$$

$$= \lim_{b\to\infty} f(t)e^{-st}\Big|_{t=0}^{t=b} + s\int_0^\infty f(t)e^{-st}\,dt$$

$$= \lim_{b\to\infty} f(b)e^{-sb} - f(0) + s\mathcal{L}\{f\}.$$

The integral in the middle line is just the definition of $\mathcal{L}\{f\}$; the factor s could be brought outside the integral because it is constant with respect to the variable of integration t. Evaluating the limit for $s > 0$ and writing $F(s) = \mathcal{L}\{f\}$ yields an important property of Laplace transforms,

$$\boxed{\mathcal{L}\{f'\} = sF(s) - f(0).}$$

∎

The next example illustrates one way in which Laplace transforms are useful.

EXAMPLE 4   *Find the Laplace transform of the solution of the initial-value problem*

$$y' + ky = e^{-3t}, \qquad y(0) = 4.$$

Assume that $y(t)$ does indeed have a transform and let $Y(s) = \mathcal{L}\{y(t)\}$. Apply the transform operator $\mathcal{L}$ to the entire differential equation, invoke the linearity of the transform, and use the formulas developed in the preceding examples. On the left, we find

Transform left-hand side.

$$\mathcal{L}\{y' + ky\} = \mathcal{L}\{y'\} + \mathcal{L}\{ky\}$$

$$= \mathcal{L}\{y'\} + k\mathcal{L}\{y\}$$

$$= \mathcal{L}\{y'\} + kY(s)$$

$$= sY(s) - y(0) + kY(s).$$

The last expression came from the previous example, $\mathcal{L}\{f'\} = sF(s) - f(0)$, with y in place of f. Finally, with $y(0) = 4$, we have

$$\mathcal{L}\{y' + ky\} = (s+k)Y - 4.$$

On the right side of the transformed differential equation, we find $\mathcal{L}\{e^{-3t}\}$. But example 1, $\mathcal{L}\{e^{at}\} = 1/(s-a)$, with $a = -3$ yields

Transform right-hand side.

$$\mathcal{L}\{e^{-3t}\} = \frac{1}{s+3}.$$

From $\mathcal{L}\{y' + ky\} = \mathcal{L}\{e^{-3t}\}$, we have

$$(s + k)Y - 4 = \frac{1}{s + 3}.$$

We solve for Y to find that the transform of the solution of the initial-value problem $y' + ky = e^{-3t}$, $y(0) = 4$, is

Solve for $Y(s)$.

$$Y(s) = \frac{1}{(s + k)(s + 3)} + \frac{4}{s + k}. \qquad \blacksquare$$

What good is the transform of the solution of an initial-value problem? If we could somehow invert the Laplace transform process—that is, if we could find the *inverse Laplace transform*

$$y(t) = \mathcal{L}^{-1} \left\{ \frac{1}{(s + k)(s + 3)} + \frac{4}{s + k} \right\}$$

—then we would have the solution of the given initial-value problem. We will address the inverse transform question in the next section.

A table of transforms of simple functions will be useful in that study. By applying the definition of the Laplace transform, we can construct each of the entries in table 11.1; e.g., example 1 derived entry 3, and example 2 derived entry 4. Exercises 18–25 ask you to derive the balance of the entries in table 11.1.

TABLE 11.1
Basic Laplace Transform Pairs

$f(t)$	$F(s) = \mathcal{L}f(t)$
1. 1	$\dfrac{1}{s}$
2. t^n, $n = 1, 2, \ldots$	$\dfrac{n!}{s^{n+1}}$, $s > 0$
3. e^{at}	$\dfrac{1}{s - a}$, $s > a$
4. $\sin at$	$\dfrac{a}{s^2 + a^2}$, $s > 0$
5. $\cos at$	$\dfrac{s}{s^2 + a^2}$, $s > 0$
6. $\sinh at$	$\dfrac{a}{s^2 - a^2}$, $s > a$
7. $\cosh at$	$\dfrac{s}{s^2 - a^2}$, $s > a$

The derivative expression obtained in example 3 is the first in a list of useful transform properties. Table 11.2 starts such a list, and exercise 29 asks you to find the second entry. The list will grow in subsequent sections.

To hint at some of the utility of Laplace transforms for solving initial-value problems, we find the transform of a different kind of forcing function.

TABLE 11.2
The Beginning of a List of Useful Properties of
Laplace Transforms

Properties of Laplace Transform

$$F(s) = \mathcal{L}f(t)$$

---

1. Transform of a derivative

$$\mathcal{L}\{f'(t)\} = sF(s) - f(0)$$

$$\mathcal{L}\{f''(t)\} = s^2 F(s) - sf(0) - f'(0)$$

EXAMPLE 5   *Let $r(t)$ be the ramp function*

$$r(t) = \begin{cases} t, & 0 \leq t \leq 1 \\ 1, & 1 < t < \infty \end{cases}$$

shown in figure 11.1. *Find the Laplace transform of r.*

   We apply the Laplace transform definition (11.1) and explicitly evaluate
the integrals that result. Note the requirement that s be positive in the evaluation of the integral $\int_1^\infty e^{-st}\, dt$.

$$\mathcal{L}\{r\} = \int_0^\infty r(t)e^{-st}\, dt$$

$$= \int_0^1 te^{-st}\, dt + \int_1^\infty e^{-st}\, dt$$

$$= \frac{1 - e^{-s}}{s^2} - \frac{e^{-s}}{s} + \frac{e^{-s}}{s}$$

Hence, the Laplace transform of the ramp function is

$$\mathcal{L}\{r\} = \frac{1 - e^{-s}}{s^2}, \qquad s > 0. \qquad \blacksquare$$

11.1.1   Existence of the Laplace Transform

The Laplace transform is defined by an improper integral,

$$\mathcal{L}\{f\} = \int_0^\infty f(t)e^{-st}\, dt = \lim_{b \to \infty} \int_0^b f(t)e^{-st}\, dt.$$

The existence of the transform thus hinges on two points,

- The existence of the finite integral $\int_0^b f(t)e^{-st}\, dt$ for each $b > 0$,
- The existence of the limit as $b \to \infty$.

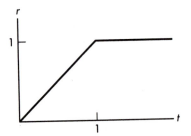

Figure 11.1:   The ramp function $r(t)$.

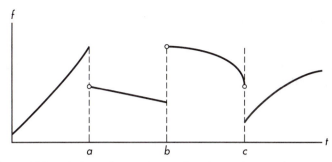

Figure 11.2: A piecewise-continuous function with jump discontinuities at $t = a, b, c$.

To accommodate the former, we require that f be **piecewise continuous**; – i.e., that f be continuous on every finite interval $0 \le t \le b$ except perhaps at a finite number of *jump discontinuities*, where the limits of f from the left and right both exist but differ in value.

Figure 11.2 illustrates a piecewise-continuous function. The integral over any finite interval of such a function certainly exists; it is just the sum of the integrals of the continuous pieces. The function $f(t) = 1/(1-t)$ is *not* piecewise continuous because neither one sided limit exists at $t = 1$.

The second condition, that the improper integral $\int_0^\infty f(t)e^{-st}\,dt$ exist, is assured if the integrand $f(t)e^{-st}$ decays to zero rapidly enough to ensure that the integral $\int_0^b f(t)e^{-st}\,dt$ approaches a finite limit for large b. To that end, we will require that f be of *exponential order*: f is of **exponential order** if for some constants $C, M > 0$ and for some $T \ge 0$,

$$|f(t)| \le Ce^{Mt}$$

for all $t \ge T$.

Figure 11.3 illustrates a function of exponential order. It is eventually bounded by e^{Mt}. If f is of exponential order, then the integrand $f(t)e^{-st}$ in the definition of the Laplace transform decays like $e^{-(s-M)t}$ providing $s > M$ and t is large enough. Hence, exponential order is a key to the existence of the Laplace transform.

Bounded functions are certainly of exponential order. On the other hand, e^{t^2} is not of exponential order because $e^{t^2} \le Ce^{Mt}$ requires $t^2 \le \ln C + Mt$. But that inequality fails because t^2 grows faster than t.

If f is of exponential order, we will show that the improper integral $\int_T^\infty |f(t)|\,dt$ exists, which guarantees that $\int_T^\infty f(t)\,dt$ converges. (The integral is said to *converge absolutely* in this case, analogous to the absolute convergence of series.)

With these ideas, we can state precise conditions for the existence of the Laplace transform.

Theorem 11.1 If f is piecewise continuous and of exponential order, then the Laplace transform of f defined by (11.1) exists for s sufficiently large.

PROOF Since f is of exponential order, $|f(t)| < e^{Mt}$ for some $M > 0$ and $t \ge T$. For large values of b, we have

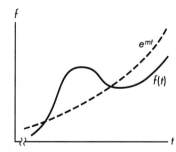

Figure 11.3: A function of exponential order.

$$\left| \int_0^b f(t)e^{-st}\, dt \right| \le \left| \int_0^T f(t)e^{-st}\, dt \right| + \left| \int_T^b f(t)e^{-st}\, dt \right|.$$

The first integral exists because its integrand is piecewise continuous, and its value is independent of b. For the second integral, the exponential bound on f gives

$$\left| \int_T^b f(t)e^{-st}\, dt \right| < \int_T^b \left| f(t)e^{-st} \right|\, dt$$

$$\le \int_T^b \left| e^{-(s-M)t} \right|\, dt$$

$$= \left. \frac{e^{-(s-M)t}}{M-s} \right|_T^b$$

$$= \frac{e^{-(s-M)T} - e^{-(s-M)b}}{s-M},$$

Requiring $s > M$, we find

$$\left| \int_T^\infty f(t)e^{-st}\, dt \right| = \lim_{b\to\infty} \left| \int_T^b f(t)e^{-st}\, dt \right|$$

$$\le \lim_{b\to\infty} \frac{e^{-(s-M)T} - e^{-(s-M)b}}{s-M}$$

$$= \frac{e^{-(s-M)T}}{s-M}.$$

Hence, if $s > M$, the improper integral defining the Laplace transform converges, and $\mathcal{L}\{f\}$ exists. ∎

11.1.2 Exercises

EXERCISE GUIDE

To gain experience . . .	Try exercises
Applying the definition of Laplace transform	1–5, 18–23, 26–29, 32–33, 35
With the linearity of transforms	1–5, 30–31, 36
Finding Laplace transforms of functions	1–17, 24–25, 34
Finding Laplace transforms of solutions of differential equations	14–17, 34
Using entries from table 11.1	1–13
Using entries from table 11.2	24–25

In exercises 1–5,

    i. Use the definition (11.1) to find the Laplace transform of each of the following functions. Give the values of s for which each transform is defined.

    ii. Use the linearity of the Laplace transform and the appropriate entries in table 11.1 to verify that the transform you obtained from the definition is correct. Indicate explicitly those steps which depend upon the linearity of the Laplace transform. Identify by number each entry from table 11.1 that you use.

1. $4 - 9e^{-4t}$

2. $3t^2$ (*Hint:* Integrate by parts.)

3. $7 + \pi$

4. $\cos \pi t$ (*Hint:* Integrate by parts twice as in example 2.)

5. $2t - 5$

    In exercises 6–13, use the linearity of the Laplace transform and the appropriate entries in table 11.1 to find the Laplace transform of each of the following functions. Indicate explicitly those steps which depend upon the linearity of the Laplace transform. Identify by number each entry from table 11.1 that you use.

6. 14

7. $2t^3$

8. $14 - 2t^3$

9. $t + 2e^{2t}$

10. $5t^3 + 3 \sin 2t$

11. $A \sin \omega t$, A, ω constants

12. $2 \sin 4t - 9 \cosh 6t$

13. $\sum_{n=1}^{4} nt^n$

    In exercises 14–17, assume that the Laplace transform of the solution of each initial value problem exists.

    i. Find the transform of the solution of the initial value problem working directly from the differential equation.

    ii. Solve each initial value problem by a method of your choice. Compute the transform of the solution you find and show that it is the same as the one you obtained directly from the differential equation in part (i).

14. $y' - 4y = \sin 2t$, $y(0) = 3$

15. $2y' + 7y = 5t + e^{-2t}$, $y(0) = -1$

16. $x' - 2x = t^2 - 3 \cos \pi t$, $x(0) = 3$

17. $y' - ky = \alpha e^{-t}$, $y(0) = y_i$, k, α constants

    In exercises 18–24, use the definition of Laplace transform to derive the indicated entry from table 11.1.

18. Entry 1; of what other entries is this a special case?

19. Entry 2 for $n = 1$

20. Entry 2 for $n = 2$

21. Entry 2 for n an arbitrary nonnegative integer (Use mathematical induction.)

22. Entry 5

23. Entry 6 (*Hint:* Use the definition of the hyperbolic sine function, entry 3 in table 11.1, and the linearity of the Laplace transform.)

24. Entry 7 (See the preceding hint.)

25. Find $\mathcal{L}\{\sinh at + \cosh at\}$ using entry 3.

    In exercises 26–27, use the indicated entry from table 11.2 to evaluate the Laplace transform of the given function. If the transform can be evaluated by a second method, do so to verify the accuracy of your answer. If a transform does not exist, explain why.

26. Use $\mathcal{L}\{f'(t)\} = sF(s) - f(0)$ with an appropriate choice of f to find the Laplace transform of
    (a) $4e^{4t}$
    (b) $3t^2$
    (c) $-a \cos at$

27. Use $\mathcal{L}\{f''(t)\} = s^2 F(s) - sf(0) - f'(0)$ with an appropriate choice of f to find the Laplace transform of
    (a) $-a^2 \sin at$
    (b) $a^2 e^{at}$
    (c) $6t$

    In exercises 28–29, use the definition of Laplace transform to derive the indicated entry in table 11.2. In each case, $F(s) = \mathcal{L}\{f(t)\}$, and all transforms are assumed to exist.

28. $\mathcal{L}\{f'(t)\} = sF(s) - f(0)$ (*Hint:* Integrate the definition of Laplace transform by parts using $dv = f'(t)\,dt$.)

29. $\mathcal{L}\{f''(t)\} = s^2 F(s) - sf(0) - f'(0)$ (*Hint:* Apply the result of exercise 26 to the function $g'(t)$, where $g(t) = f'(t)$.)

    In exercises 30–31, use the definition of Laplace transform to find the transform of the given function. Sketch the graph of each function before you attempt to find its transform.

30. $g(t) = \begin{cases} 4t, & 0 \le t \le 7 \\ 0, & 7 < t < \infty \end{cases}$

31. $g(t) = \begin{cases} 1, & 0 \le t \le 2\pi \\ \cos t, & 2\pi \le t \le 7\pi/2 \\ 0, & 7\pi/2 < t < \infty \end{cases}$

32. The text claims:

> The Laplace transform ... is linear ... like integration. If $f(t)$, $g(t)$ are functions with Laplace transforms $F(s) = \mathcal{L}\{f\}$, $G(s) = \mathcal{L}\{g\}$, then
>
> $$\mathcal{L}\{cf + dg\} = c\mathcal{L}\{f\} + d\mathcal{L}\{g\} = cF + dG$$
>
> for any constants c, d.

Verify this claim for $f(t) = r(t)$, the ramp function of example 5, and $g(t) = e^{at}$, a constant. Use the definition of Laplace transform to compute $\mathcal{L}\{cf + dg\}$. Then show that it is identical to $c\mathcal{L}\{f\} + d\mathcal{L}\{g\}$.

33. Verify that the area transform $\mathcal{A}(f) = \int_0^s f(t)\,dt$ is linear. That is, show that if $F(s) = \mathcal{A}(f)$, $G(s) = \mathcal{A}(g)$, then

$$\mathcal{A}(cf + dg) = c\mathcal{A}(f) + d\mathcal{A}(g) = cF + dG$$

for any constants c, d. Does the Laplace transform have this same property?

34. In example 1, the text computes $\mathcal{L}\{e^{at}\}$ from its definition,

$$\mathcal{L}\{e^{at}\} = \int_0^\infty e^{at} e^{-st}\,dt$$

$$= \int_0^\infty e^{(a-s)t}\,dt$$

$$= \lim_{b\to\infty} \frac{e^{(a-s)t}}{a-s}\Big|_{t=0}^{t=b}$$

$$= \lim_{b\to\infty} \frac{e^{(a-s)b}}{a-s} - \frac{1}{a-s}$$

$$= \frac{1}{s-a},$$

but requires $s > a$. Why is this restriction necessary? Where is it first required?

35. Example 3 shows that for any differentiable function f,

$$\mathcal{L}\{f'\} = sF(s) - f(0).$$

Mimic the calculations of that example for $f(t) = e^{at}$, a constant, to show that

$$\mathcal{L}\{ae^{at}\} = s\mathcal{L}\{e^{at}\} - 1.$$

What is the source of the 1 at the end of the last equation? Use the linearity of the Laplace transform to write $\mathcal{L}\{ae^{at}\}$ in terms of $\mathcal{L}\{e^{at}\}$ to confirm that your calculations are correct. (Use

$$\mathcal{L}\{cf + dg\} = c\mathcal{L}\{f\} + d\mathcal{L}\{g\} = cF + dG$$

with $c = a$, $d = 0$. Is it true that you can "bring constants across the transform sign" as you can with integrals?)

36. By working directly with the differential equation itself, we found in example 4 that the Laplace transform $Y(s)$ of the solution of the initial-value problem

$$y' + ky = e^{-3t}, \qquad y(0) = 4,$$

is

$$Y(s) = \frac{1}{(s+k)(s+3)} + \frac{4}{s+k}.$$

Find the solution $y(t)$ of this initial-value problem and verify that its Laplace transform is indeed the formula just given.

37. For $s > 0$ and a constant, example 2 finds $\mathcal{L}\{\sin at\}$ as follows:

$$\mathcal{L}\{\sin at\} = \int_0^\infty (\sin at)e^{-st}\,dt$$

$$= -\frac{(\sin at)e^{-st}}{s}\Big|_0^\infty + \frac{a}{s}\int_0^\infty (\cos at)e^{-st}\,dt$$

$$= -\frac{a(\cos at)e^{-st}}{s^2}\Big|_0^\infty - \frac{a^2}{s^2}\int_0^\infty (\sin at)e^{-st}\,dt$$

$$= \frac{a}{s^2} - \frac{a^2}{s^2}\int_0^\infty (\sin at)e^{-st}\,dt.$$

Solve for the integral $\int_0^\infty (\sin at)e^{-st}\,dt$ defining the transform to obtain

$$\mathcal{L}\{\sin at\} = \int_0^\infty (\sin at)e^{-st}\,dt$$

$$= \left(1 + \frac{a^2}{s^2}\right)^{-1}\frac{a}{s^2}$$

$$= \frac{a}{a^2 + s^2}.$$

Fill in the details of this calculation. Indicate clearly those step(s) that require(s) the condition $s > 0$.

38. For any differentiable function f, define the derivative transform $\mathcal{D}\{f(t)\} = f'(t)$. Verify that this transform is linear. That is, show that

$$\mathcal{D}\{cf + dg\} = c\mathcal{D}\{g\} + d\mathcal{D}\{g\}$$

for any differentiable functions f, g and any constants c, d.

11.2 INVERSE TRANSFORMS AND INITIAL-VALUE PROBLEMS

11.2.1 Inverse of the Laplace Transform

In example 4, of section 11.1, we worked directly with the differential equation to find the transform of the solution of the initial-value problem

$$y' + ky = e^{-3t}, \qquad y(0) = 4.$$

We assumed that the solution $y(t)$ did indeed have a transform, let $Y(s) = \mathcal{L}\{y(t)\}$, applied the transform operator $\mathcal{L}$ to the entire differential equation, and found that the transform of the solution of this initial-value problem is

$$Y(s) = \frac{1}{(s+k)(s+3)} + \frac{4}{s+k}.$$

We realized that if we could somehow invert the Laplace transform process — that is if we could find the **inverse Laplace transform**

$$y(t) = \mathcal{L}^{-1}\{Y(s)\} = \mathcal{L}^{-1}\left\{\frac{1}{(s+k)(s+3)} + \frac{4}{s+k}\right\},$$

— then we would have the solution of the initial-value problem.

The inverse transform $\mathcal{L}^{-1}\{Y\} = y$ exists only if there is exactly one function $y(t)$ associated with each transform $Y(s)$. To recall an analogy, the function $f(x) = x^2$ has no inverse because for each (positive) value of f, we can choose $x = +\sqrt{f}$ or $x = -\sqrt{f}$. The inverse relation is not single valued, so there is no inverse function. Although we will not prove it here, the Laplace transform is better behaved: Under suitable conditions in the appropriate setting, the inverse transform relation $Y(s) \to y(t)$ is single valued.

Although we have a straightforward definition for the Laplace transform itself in (11.1), we lack the mathematical tools to state a corresponding definition for the inverse transform. Instead, we must resort to a procedure much like antidifferentiation.

You are accustomed to inverting differentiation by evaluating an expression such as $\int g(t)\, dt$ to answer the question, What is a function whose derivative is g? The actual computations often require some deliberation and judgment, unlike differentiation, which usually entails no more than blind application of a formula.

Inverting Laplace transforms is much the same. We do have a formula (11.1) to compute the transform of a given function f. But evaluating an inverse transform $\mathcal{L}^{-1}\{F\}$ to answer the question, What is a function whose Laplace transform is F? requires careful thought and judgment. Just as a table of derivative formulas provides clues to antiderivatives, a list of simple Laplace transforms such as table 11.1 is a key to inverting transforms.

At the end of the last section, we found the transform of the solution of the initial-value problem $y' + ky = e^{-3t}$, $y(0) = 4$. We begin with an example which inverts that transform.

EXAMPLE 6   *Find the inverse transform*

$$y(t) = \mathcal{L}^{-1}\left\{\frac{1}{(s+k)(s+3)} + \frac{4}{s+k}\right\}.$$

Since $\mathcal{L}$ is linear, $\mathcal{L}^{-1}$ is linear as well. The inverse transform can be computed term by term, and constant factors can be brought outside the operator $\mathcal{L}^{-1}$, much like antiderivatives,

$$y(t) = \mathcal{L}^{-1}\left\{\frac{1}{(s+k)(s+3)}\right\} + \mathcal{L}^{-1}\left\{\frac{4}{s+k}\right\}$$

$$= \mathcal{L}^{-1}\left\{\frac{1}{(s+k)(s+3)}\right\} + 4\mathcal{L}^{-1}\left\{\frac{1}{s+k}\right\}.$$

To invert the second expression, we search table 11.1 for a transform that looks like $1/(s+k)$. Entry 3 with $-a = k$ is the obvious choice. Since

$$\mathcal{L}\{e^{-kt}\} = \frac{1}{s-(-k)} = \frac{1}{s+k}.$$

we have

$$\mathcal{L}^{-1}\left\{\frac{1}{s+k}\right\} = e^{-kt}.$$

The first term in the inverse is more of a problem, for we find nothing in the right-hand column of table 11.1 of the form

$$\frac{1}{(s+k)(s+3)}.$$

But we could invert the factors $1/(s+k)$ and $1/(s+3)$ if they stood alone. (We just found e^{kt} for the inverse of the first, and the second is the same with $k = 3$.)

An expression such as $1/(s+k)(s+3)$ can be separated into its component fractions using *partial fractions*. For some unknown constants A and B, write

Partial fraction decomposition; see section A.7.

$$\frac{1}{(s+k)(s+3)} = \frac{A}{s+k} + \frac{B}{s+3}$$

$$= \frac{A(s+3) + B(s+k)}{(s+k)(s+3)}.$$

Since the first and last numerator must be identical, we have

$$1 = A(s+3) + B(s+k).$$

466   11   The Laplace Transform

Equations for A and B result when we equate coefficients of like powers of s (See section A.7.):

$$s^0 : \quad 3A + kB = 1$$

$$s^1 : \quad A + B = 0,$$

Solving this pair of simultaneous equations, we find

$$A = \frac{1}{3 - k}, \qquad B = \frac{1}{k - 3}.$$

Using these values of A and B, the inverse transform is

$$\mathcal{L}^{-1} \left\{ \frac{1}{(s + k)(s + 3)} \right\} = A\mathcal{L}^{-1} \left\{ \frac{1}{s + k} \right\} + B\mathcal{L}^{-1} \left\{ \frac{1}{s + 3} \right\}$$

$$= Ae^{-kt} + Be^{-3t}$$

$$= \frac{e^{-kt} - e^{-3t}}{3 - k}. \tag{11.2}$$

Combining the inverse transforms of the two individual terms, we obtain the desired inverse transform,

$$y(t) = \mathcal{L}^{-1} \left\{ \frac{1}{(s + k)(s + 3)} \right\} + 4\mathcal{L}^{-1} \left\{ \frac{1}{s + k} \right\}$$

$$= \frac{e^{-kt}}{3 - k} + \frac{e^{-3t}}{k - 3} + 4e^{-kt}$$

$$= \frac{(14k - 13)e^{-kt} + e^{-3t}}{k - 3}, \qquad k \neq 3. \qquad \blacksquare$$

We can use Laplace transforms and their inverses to solve linear differential equations with constant coefficients. The steps are as follows:

LAPLACE TRANSFORM SOLUTION PROCESS

1. Transform the entire differential equation, using any initial values that are given. Solve for the transform of the solution.
2. Invert the transform to find the solution of the differential equation or initial-value problem.

Unfortunately, step 2 is easier said than done!

EXAMPLE 7 *Illustrate these steps by solving the initial-value problem*

$$y' + ky = e^{-3t}, \qquad y(0) = 4.$$

The first step was completed in example 4 of the previous section, where from $\mathcal{L}\{y' + ky\} = \mathcal{L}\{e^{-3t}\}$ we found

Transform differential equation and solve for $Y(s)$.

$$Y(s) = \frac{1}{(s + k)(s + 3)} + \frac{4}{s + k}.$$

Recall that the initial conditions were incorporated in the transform of the derivative term,

$$\mathcal{L}\{y'\} = sY(s) - y(0) = sY(s) - 4.$$

The second step was completed in the preceding example by inverting the transform to find

The second step was completed in the preceding example by inverting the transform to find

Invert $Y(s)$ to find the solution of the initial-value problem.

$$y(t) = \mathcal{L}^{-1}\{Y(s)\} = \mathcal{L}^{-1}\left\{\frac{1}{(s+k)(s+3)} + \frac{4}{s+k}\right\}$$

$$= \frac{(14k - 13)e^{-kt} + e^{-3t}}{k - 3}, \qquad k \neq 0. \qquad \blacksquare$$

EXAMPLE 8 *Solve the initial-value problem*

$$x''(t) + 4x(t) = 0, \qquad x(0) = 2, \qquad x'(0) = -1.$$

Transform differential equation.

Let $X(s) = \mathcal{L}\{x(t)\}$. Use entry 1 of table 11.2 and the initial conditions to transform the entire differential equation:

$$\mathcal{L}\{x''(t) + 4x(t)\} = \mathcal{L}\{x''(t)\} + 4\mathcal{L}\{x(t)\}$$

$$= s^2 X(s) - sx(0) - x'(0) + 4X(s)$$

$$= s^2 X - 2s + 1 + 4X$$

Solve for $X(s)$.

$$= (s^2 + 4)X - 2s + 1 = 0.$$

Solve for X:

$$X(s) = \frac{2s - 1}{s^2 + 4}.$$

Invert $X(s)$.

Now we seek the inverse transform,

$$x(t) = \mathcal{L}^{-1}\{X(s)\} = \mathcal{L}^{-1}\left\{\frac{2s - 1}{s^2 + 4}\right\}$$

$$= 2\mathcal{L}^{-1}\left\{\frac{s}{s^2 + 4}\right\} - \mathcal{L}^{-1}\left\{\frac{1}{s^2 + 4}\right\}.$$

With the aid of entries 4 and 5 in table 11.1, we find

$$\mathcal{L}^{-1}\left\{\frac{s}{s^2 + 4}\right\} = \cos 2t$$

$$\mathcal{L}^{-1}\left\{\frac{1}{s^2 + 4}\right\} = \frac{1}{2}\mathcal{L}^{-1}\left\{\frac{2}{s^2 + 4}\right\} = \frac{1}{2}\sin 2t.$$

Hence, we have the solution of the given initial-value problem,

SOLUTION OF INITIAL VALUE PROBLEM

$$x(t) = 2\cos 4t - \frac{1}{2}\sin 2t. \qquad \blacksquare$$

11.2.2 The Convolution Theorem

Some additional Laplace transform properties can ease the inversion in step 2 of the Laplace transform solution process. We develop two of those properties in the next few pages and illustrate their application to the solution of initial-value problems. (These and similar results are summarized in table 11.3.)

The motivation for the first property, *the convolution theorem*, is the problem encountered in the first example, finding the inverse transform of the product of two transforms.

> **Theorem 11.2 (Convolution)** *Let f and g be two functions of exponential order. Denote their Laplace transforms by*
>
> $$F(s) = \mathcal{L}\{f(t)\}, \qquad G(s) = \mathcal{L}\{g(t)\}.$$
>
> *Then the transform of the **convolution integral** of f and g, $\int_0^t f(r)g(t-r)\,dr$ is*
>
> $$\boxed{\mathcal{L}\{\int_0^t f(r)g(t-r)\,dr\} = F(s)G(s).}$$

PROOF The expression $F(s)G(s)$ is the product of two improper integrals:

$$F(s)G(s) = \left(\int_{r=0}^\infty e^{-sr}f(r)\,dr \right) \left(\int_{q=0}^\infty e^{-sq}g(q)\,dq \right).$$

(We have used r and q for the variables of integration rather than t to avoid confusion later.) Writing this product as an iterated double integral yields

$$F(s)G(s) = \int_{r=0}^\infty \int_{q=0}^\infty e^{-s(r+q)}g(q)f(r)\,dq\,dr.$$

TABLE 11.3
Useful Properties of Laplace Transform—First Revision

Properties of Laplace Transforms
$F(s) = \mathcal{L}\{f(t)\}, \ G(s) = \mathcal{L}\{g(t)\}$

1. Transform of a derivative

    $$\mathcal{L}\{f'(t)\} = sF(s) - f(0)$$

    $$\mathcal{L}\{f''(t)\} = s^2 F(s) - sf(0) - f'(0)$$

2. Convolution theorem

    $$\mathcal{L}\left\{ \int_0^t f(r)g(t-r)\,dr \right\} = F(s)G(s)$$

3. Shifted transform

    $$\mathcal{L}\{e^{at}f(t)\} = F(s-a)$$

Now change variables in the inner integral. Treating r as fixed, let $t = r + q$ to obtain

$$F(s)G(s) = \int_{r=0}^{\infty} \int_{t=r}^{\infty} e^{-st} g(t-r) f(r) \, dt \, dr.$$

This double integral covers the shaded triangle in the rt-plane, as illustrated in the left-hand graph in figure 11.4. But that same triangle can be covered equally well by a double integral with the order of integration reversed, as the right-hand graph in the same figure shows:

$$F(s)G(s) = \int_{t=0}^{\infty} \int_{r=0}^{t} e^{-st} g(t-r) f(r) \, dr \, dt$$

$$= \int_{t=0}^{\infty} e^{-sr} \left(\int_{r=0}^{t} g(t-r) f(r) \, dr \right) dt.$$

The inner integral is the convolution of f and g, and the outer integral is the definition of the Laplace transform. ∎

Careful justification of reversing the order of integration of the improper integrals depends on the *absolute convergence* of these integrals, which is guaranteed by the hypothesis that f and g are of exponential order.

EXAMPLE 9   *Use the convolution theorem 11.2 to find the inverse transform*

$$\mathcal{L}^{-1}\left\{ \frac{1}{(s+k)(s+3)} \right\},$$

which appears in example 6.

The given expression is the product of two transforms F and G:

$$F(s)G(s) = \frac{1}{(s+k)(s+3)}$$

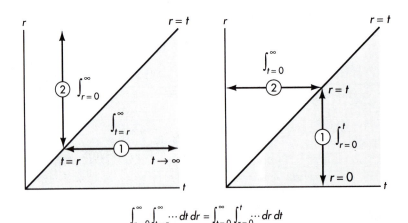

$$\int_{r=0}^{\infty} \int_{t=r}^{\infty} \cdots dt \, dr = \int_{t=0}^{\infty} \int_{r=0}^{t} \cdots dr \, dt$$

Figure 11.4:   The double integrals $\int_{r=0}^{\infty} \int_{t=0}^{\infty} \cdots dt \, dr$ and $\int_{t=0}^{\infty} \int_{r=0}^{t} \cdots dr \, dt$ cover the same region in the rt-plane. The inner integral is numbered 1 and the outer integral is numbered 2 in each graph.

with

$$F(s) = \frac{1}{s+k}, \qquad G(s) = \frac{1}{s+3}.$$

From entry 3 of table 11.1, the inverses of these individual transforms are

$$f(t) = \mathcal{L}^{-1}\{F(s)\} = e^{-kt}, \qquad g(t) = \mathcal{L}^{-1}\{G(s)\} = e^{-3t}.$$

From the convolution theorem, we have

$$\mathcal{L}^{-1}\{F(s)G(s)\} = \int_0^t f(r)g(t-r)\,dr,$$

or

$$\mathcal{L}^{-1}\left\{\frac{1}{(s+k)(s+3)}\right\} = \int_0^t e^{-kr}e^{-3(t-r)}\,dr$$

$$= e^{-3t}\int_0^t e^{(3-k)r}\,dr$$

$$= \frac{e^{-kt} - e^{-3t}}{3-k},$$

precisely as in equation (11.2) of example 6. ∎

11.2.3   The Shifted Transform

Another useful transform property is the *shifted transform theorem*:

> **Theorem 11.3 (Shifted Transform)**   Let $f(t)$ be of exponential order, and let $F(s)$ denote its Laplace transform. Then for any real number a, the transform of $e^{at}f(t)$ exists, and
>
> $$\boxed{\mathcal{L}\{e^{at}f(t)\} = F(s-a).}$$

PROOF   If f is of exponential order, the product $e^{at}f(t)$ is, too, with constant $M+a$. Hence, $\mathcal{L}\{e^{at}f(t)\}$ exists.

Evaluating the transform requires only a simple identification of the transform variable in the definition of Laplace transform:

$$\mathcal{L}\{e^{at}f(t)\} = \int_0^\infty e^{-st}e^{at}f(t)\,dt$$

$$= \int_0^\infty e^{-(s-a)t}f(t)\,dt = F(s-a).$$ ∎

EXAMPLE 10   *Use the shifted transform theorem to evaluate*

$$\mathcal{L}\{e^{-2t}\cos 6t\}.$$

To use the shifted transform theorem with $f(t) = \cos 6t$ and $a = -2$, note that $F(s) = \mathcal{L}\{\cos 6t\} = s/(s^2 + 36)$. Then

$$\mathcal{L}\{e^{-2t}\cos 6t\} = F(s + 2)$$

$$= \frac{s + 2}{(s + 2)^2 + 36} = \frac{s + 2}{s^2 + 4s + 40}. \tag{11.3}$$

For future reference, note that we could complete the square in the final denominator $s^2 + 4s + 40$ to recover $(s + 2)^2 + 36$, from which we could identify $a = -2$ and the denominator $s^2 + 36$ of $F(s)$. ∎

EXAMPLE 11 *Solve the initial-value problem*

$$x'' + 4x' + 40x = 0, \qquad x(0) = 3, \qquad x'(0) = 12.$$

Transform the differential equation.

Let $X(s) = \mathcal{L}\{x(t)\}$. Transform the differential equation,

$$\mathcal{L}\{x'' + 4x' + 40x\} = s^2 X - sx(0) - x'(0) + 4(sX - x(0)) + 40X$$

$$= (s^2 + 4s + 40)X - 3s - 24 = 0,$$

Solve for $X(s)$.

and solve for X,

$$X(s) = \frac{3s + 24}{s^2 + 4s + 40}.$$

Invert $X(s)$.

Now we seek the inverse transform

$$x(t) = \mathcal{L}^{-1}\{X(s)\} = \mathcal{L}^{-1}\left\{\frac{3s + 24}{s^2 + 4s + 40}\right\}$$

$$= 3\mathcal{L}^{-1}\left\{\frac{s}{s^2 + 4s + 40}\right\} + 24\mathcal{L}^{-1}\left\{\frac{1}{s^2 + 4s + 40}\right\}. \tag{11.4}$$

The denominator here is the same as that of (11.3), a transform we obtained with the shifted function theorem.

To work backward to find the form of $F(s - a)$, rewrite the denominator by completing the square,

$$s^2 + 4s + 40 = \left[s^2 + 4s + \left(\frac{4}{2}\right)^2\right] - \left(\frac{4}{2}\right)^2 + 40$$

$$= (s^2 + 4s + 4) + 36$$

$$= (s + 2)^2 + 36.$$

The second term in (11.4), for example, is now,

$$\mathcal{L}^{-1}\left\{\frac{1}{s^2 + 4s + 40}\right\} = \mathcal{L}^{-1}\left\{\frac{1}{(s + 2)^2 + 36}\right\},$$

an expression dangerously close to a shifted sine transform, entry 4 in table 11.1

with $a = 6$. A simple manipulation even puts the correct number in the numerator:

$$\mathcal{L}^{-1}\left\{\frac{1}{(s+2)^2 + 36}\right\} = \frac{1}{6}\mathcal{L}^{-1}\left\{\frac{6}{(s+2)^2 + 36}\right\}$$

$$= \frac{1}{6}F(s+2),$$

with

$$F(s) = \frac{6}{s^2 + 36}.$$

Hence, we can use the shifted transform theorem with

$$f(t) = \mathcal{L}^{-1}\{F(s)\} = \sin 6t$$

to write

$$\mathcal{L}^{-1}\left\{\frac{1}{s^2 + 4s + 40}\right\} = \frac{1}{6}\mathcal{L}^{-1}\left\{\frac{6}{(s+2)^2 + 36}\right\} = \frac{1}{6}\mathcal{L}^{-1}\{F(s+2)\}$$

$$= \frac{1}{6}e^{-2t}f(t) = \frac{1}{6}e^{-2t}\sin 6t. \tag{11.5}$$

The same logic can be applied to the first term in (11.4). We find

$$\mathcal{L}^{-1}\left\{\frac{s}{s^2 + 4s + 40}\right\} = \mathcal{L}^{-1}\left\{\frac{s}{(s+2)^2 + 36}\right\}.$$

But to use the shifted transform in entry 3 of table 11.3, *all* appearances of the argument s must be shifted, *including the one in the numerator.* Adding and subtracting 2 in the numerator yields

$$\mathcal{L}^{-1}\left\{\frac{s}{s^2 + 4s + 40}\right\} = \mathcal{L}^{-1}\left\{\frac{s}{(s+2)^2 + 36}\right\}$$

$$= \mathcal{L}^{-1}\left\{\frac{s+2-2}{(s+2)^2 + 36}\right\}$$

$$= \mathcal{L}^{-1}\left\{\frac{s+2}{(s+2)^2 + 36}\right\} - 2\mathcal{L}^{-1}\left\{\frac{1}{(s+2)^2 + 36}\right\}$$

$$= \mathcal{L}^{-1}\{F(s+2)\} - \frac{1}{3}e^{-2t}\sin 6t,$$

where

$$F(s) = \frac{s}{s^2 + 36}, \qquad f(t) = \mathcal{L}^{-1}\{F(s)\} = \cos 6t.$$

(The $\sin 6t$ term comes from (11.5).)
  Entry 3 of table 11.3 with $-a = 2$ yields

$$\mathcal{L}^{-1}\{F(s+2)\} = e^{-2t}f(t) = e^{-2t}\cos 6t.$$

Combining the expressions we have obtained for the two inverse trans-forms in (11.4) finally gives the solution

$$x(t) = \mathcal{L}^{-1}\{X(s)\}$$

$$= 3\mathcal{L}^{-1}\left\{\frac{s}{s^2 + 4s + 40}\right\} + 24\mathcal{L}^{-1}\left\{\frac{1}{s^2 + 4s + 40}\right\}$$

$$= 3e^{-2t}\cos 6t + 3e^{-2t}\sin 6t. \qquad\blacksquare$$

The examples in this section have decomposed quadratic denominators of transforms by partial fractions and by completing the square. Which technique is appropriate when?

■ Use partial fractions when the quadratic has real roots; the individual terms lead to exponentials in time. (Alternatively, apply the convolution theorem 11.2 to the product of the factors.)
■ Complete the square when the quadratic has complex roots; the corresponding shifted transform leads to the product of an exponential and a sine or cosine.

Exercises 26 and 27 ask you to prove these claims.

More generally, if the denominator of a transform contains factors you can identify, use partial fractions to decompose it into simpler expressions. The next example illustrates these ideas.

EXAMPLE 12   *Invert the transform*

$$Y(s) = \frac{s^3 - 33s^2 + 90s - 11}{(s^2 - 6s + 13)(s^2 + 4s - 12)}.$$

Since factors of the denominator are given, we begin with a partial-fractions decomposition:

$$Y(s) = \frac{s^3 - 33s^2 + 90s - 11}{(s^2 - 6s + 13)(s^2 + 4s - 12)}$$

$$= \frac{As + B}{s^2 - 6s + 13} + \frac{Cs + D}{s^2 + 4s - 12},$$

where A, B, C, and D are constants to be determined. Finding a common de-nominator for the sum and equating coefficients of powers of s in the resulting numerator and the original numerator lead to

$$s^0: \qquad -12B + 13D \qquad = -11$$

$$s^1: \quad -12A + 4B + 13C - 6D = 90$$

$$s^2: \qquad 4A + B - 6C + D \qquad = -33$$

$$s^3: \qquad A + C \qquad = 1.$$

We use the first and fourth equations to eliminate A and B from the middle two equations and then solve that pair of equations for C and D. We obtain

$$A = -3, \qquad B = 2, \qquad C = 4, \qquad D = 1.$$

Hence,

$$Y(s) = \frac{2 - 3s}{s^2 - 6s + 13} + \frac{4s + 1}{s^2 - 4s - 12},$$

and we must find two inverse transforms,

$$y_1(t) = \mathcal{L}^{-1} \left\{ \frac{2 - 3s}{s^2 - 6s + 13} \right\}, \qquad (11.6)$$

$$y_2(t) = \mathcal{L}^{-1} \left\{ \frac{4s + 1}{s^2 - 4s - 12} \right\}. \qquad (11.7)$$

The denominator $s^2 - 6s + 13$ of the first transform (11.6) has complex roots because the discriminant $6^2 - 4 \cdot 1 \cdot 13$ is negative. Consequently, we complete the square rather than factor:

$$s^2 - 6s + 13 = (s - 3)^2 + 4.$$

We have a transform $F(s - a)$ with $a = 3$, and we should prepare to use the shifted transform theorem, $\mathcal{L}\{e^{at} f(t)\} = F(s - a)$.

We write $F(s - 3)$ in terms of $s - 3$ in order to identify $F(s)$. We find

$$
\begin{aligned}
F(s - 3) &= \frac{2 - 3s}{(s - 3)^2 + 4} \\
&= \frac{-7 - 3(s - 3)}{(s - 3)^2 + 4} \\
&= -\frac{7}{2} \frac{2}{(s - 3)^2 + 4} - 3 \frac{(s - 3)}{(s - 3)^2 + 4}
\end{aligned}
$$

and

$$F(s) = -\frac{7}{2} \frac{2}{s^2 + 4} - 3 \frac{s}{s^2 + 4}.$$

Hence, the function f in $\mathcal{L}\{e^{at} f(t)\} = F(s - a)$ is

$$f(t) = \mathcal{L}^{-1}\{F(s)\} = -\frac{7}{2} \sin 2t - 3 \cos 2t.$$

Finally, the shifted transform theorem 11.3 yields

$$y_1(t) = e^{3t} \left(-\frac{7}{2} \sin 2t - 3 \cos 2t \right).$$

Since the denominator of the second transform (11.7) can be factored,

$$\frac{4s + 1}{s^2 + 4s - 12} = \frac{4s + 1}{(s - 2)(s + 6)},$$

we can either find a partial-fraction decomposition or use the convolution theorem 11.2. We choose partial fractions and seek constants A, B satisfying

$$\frac{4s + 1}{(s - 2)(s + 6)} = \frac{A}{s - 2} + \frac{B}{s + 6}.$$

Finding a common denominator and equating coefficients of powers of s lead to

$$s^0 : \quad 6A - 2B = 1$$

$$s^1 : \quad A + B = 4.$$

We have $A = \frac{9}{8}$, $B = \frac{23}{8}$, and

$$y_2(t) = \mathcal{L}^{-1} \left\{ \frac{4s + 1}{(s - 2)(s + 6)} \right\}$$

$$= \mathcal{L}^{-1} \left\{ \frac{9}{8} \frac{1}{s - 2} + \frac{23}{8} \frac{1}{s + 6} \right\}$$

$$= \frac{9}{8} e^{2t} + \frac{23}{8} e^{-6t}.$$

Combining the results for (11.6) and (11.7) yields the desired inverse transform,

$$\mathcal{L}^{-1}\{Y(s)\} = y_1(t) + y_2(t)$$

$$= e^{3t} \left(-\frac{7}{2} \sin 2t - 3 \cos 2t \right) + \frac{9}{8} e^{2t} + \frac{23}{8} e^{-6t}. \qquad \blacksquare$$

11.2.4 Summary

The following steps use Laplace transforms to solve a constant-coefficient differential equation:

1. Transform the entire differential equation, using any initial values that are given. Solve for the transform of the solution.
2. Invert the transform to find the solution of the differential equation or the initial-value problem.

Two useful tools for finding inverse transforms are:

CONVOLUTION THEOREM

$$\boxed{\mathcal{L}\left\{ \int_0^t f(r)g(t - r)\, dr \right\} = F(s)G(s),}$$

SHIFTED TRANSFORM

$$\boxed{\mathcal{L}\{e^{at} f(t)\} = F(s - a).}$$

EXERCISE GUIDE

To gain experience . . .	Try exercises
Using the convolution theorem	1–6, 25, 27
Using partial fractions	1–6, 27
With shifted transforms	8, 10–11, 13, 28
Finding inverse transforms	1–12
Finding Laplace transforms of functions	13, 23, 24(c)
Solving initial-value problems	14–24
With the fundamental ideas of inverse transforms	26

In exercises 1–6,

   i. Write the given transform as an appropriate product and use the convolution theorem to find its inverse.

   ii. Use partial fractions to write the given transform as an appropriate sum and find its inverse.

1. $\dfrac{4}{s(s+4)}$

2. $\dfrac{3}{(s-2)(s+3)}$

3. $\dfrac{1}{s^2 - 16}$

4. $\dfrac{s}{(s+1)(s^2+4)}$

5. $\dfrac{s}{(s-1)(s^2-4)}$

6. $\dfrac{1}{s^2+9}$ *Hint:* use $s^2 + 9 = (s+3i)(s-3i)$ and $e^{i3t} = \cos 3t + i \sin 3t$.

In exercises 7–12, find the inverse of the given transform.

7. $\dfrac{6}{s^2 - 5s + 4}$

8. $\dfrac{3s}{2s^2 - 3s + 5}$

9. $\dfrac{s-1}{3s^2 - 15s + 12}$

10. $\dfrac{2s+3}{s^2 - 2s + 2}$

11. $\dfrac{3-2s}{s^2 - 2s + 5}$

12. $\dfrac{2+s}{s^2 + 4s + 1}$

13. Use $\mathcal{L}\{e^{at} f(t)\} = F(s-a)$ with an appropriate choice of f to find the Laplace transform of each of the following.
   (a) $e^{3t} t$
   (b) $e^{-4t} \cos \pi t$
   (c) $e^{-3t/2}(t^2 + 4)$

In exercises 14–21,

   i. Working directly from the differential equation, find the transform of the solution of the given initial-value problem.

   ii. Invert that transform to find the solution of the initial-value problem.

   iii. Find the solution of the initial-value problem by another method, and verify that the solution obtained by Laplace transforms is correct.

14. $y' - 6y = 3$, $y(0) = 2$

15. $y' - 6y = \sin 3t$, $y(0) = 5$

16. $y' + 4y = \cos \pi t$, $y(0) = 0$

17. $2y' + 8y = 6e^{-3t}$, $y(0) = -2$

18. $3x''/16 + 1.41x' + 2x = 0$, $x(0) = 0.75$, $x'(0) = 0$
   (See example 21 of section 6.4.2.)

19. $x'' + 4x = \cos \pi t$, $x(0) = 1$, $x'(0) = 4$

20. $x'' + 4x = \sin 2t$, $x(0) = 1$, $x'(0) = 4$

21. $x'' - 4x' + 4x = 0$, $x(0) = 2$, $x'(0) = 2$

22. Example 7 used Laplace transforms to solve the initial-value problem

$$y' + ky = e^{-3t}, \quad y(0) = 4.$$

It found

$$y(t) = \frac{(14k - 13)e^{-kt} + e^{-3t}}{k - 3}.$$

(a) Verify by direct substitution that this solution is correct.

(b) Solve this initial-value problem by another method and show that you obtain the same solution.

(c) Obviously, this solution is valid only for $k \neq 3$. Other than avoiding dividing by zero in this formula, why must this particular value of k be singled out?

(d) Use Laplace transforms to solve this initial-value problem when $k = 3$.

23. Example 8 uses Laplace transforms to find the solution

$$x(t) = 2\cos 4t - \frac{1}{2}\sin 2t$$

of the initial-value problem

$$x''(t) + 4x(t) = 0, \qquad x(0) = 2, \qquad x'(0) = -1.$$

(a) Verify by direct substitution that this solution is correct.

(b) Solve this initial-value problem by another method and show that you obtain the same solution.

(c) Find the Laplace transform of the solution you found in part (b). Show that it agrees with the transform of the solution obtained in example 8 directly from the initial-value problem itself,

$$X(s) = \frac{3s + 24}{s^2 + 4s + 40}.$$

24. For any differentiable function f, define the derivative transform $\mathcal{D}\{f(t)\} = f'(t)$. Show that this transform does *not* have an inverse. That is, show that

$$\mathcal{D}\{f\} = \mathcal{D}\{g\}$$

for two differentiable functions f, g with $f \neq g$.

25. Example 11 obtains the solution

$$x(t) = 3e^{-2t}\cos 6t + 4e^{-2t}\sin 6t.$$

of the initial-value problem

$$x'' + 4x' + 40x = 0, \quad x(0) = 3, \quad x'(0) = 12.$$

(a) Verify by direct substitution that this solution is correct.

(b) Solve this initial-value problem by another method and show that you obtain the same solution.

(c) Find the Laplace transform of the solution you found in part (b). Show that it agrees with the transform of the solution obtained in example 11 directly from the initial-value problem itself,

$$X(s) = \frac{3s + 24}{s^2 + 4s + 40}.$$

26. Verify the following claim made in the text:

Complete the square when the quadratic has complex roots; the corresponding shifted transform leads to the product of an exponential and a sine or cosine.

(a) Consider

$$Y(s) = \frac{1}{s^2 + c_1 s + c_0}.$$

What conditions on c_1, c_0 guarantee that the quadratic denominator has complex roots? Denote those roots by $r_1, r_2 = \alpha \pm i\beta$.

(b) Show that completing the square leads to

$$s^2 + c_1 s + c_0 = (s - \alpha)^2 + \beta^2.$$

(c) Apply the shifted transform theorem to

$$Y(s) = \frac{1}{(s - \alpha)^2 + \beta^2}$$

to find $\mathcal{L}^{-1}\{Y(s)\}$.

27. Verify the following claim made in the text:

Use partial fractions when a quadratic denominator has real roots; the individual terms lead to exponentials in time. (Alternatively, apply the convolution theorem 11.2 to the product of the factors.)

(a) Consider

$$Y(s) = \frac{1}{s^2 + c_1 s + c_0}.$$

What conditions on c_1, c_0 guarantee that the quadratic denominator has real roots? Denote the real roots by r_1, r_2: $s^2 + c_1 s + c_0 = (s - r_1)(s - r_2)$.

(b) Use a partial-fraction expansion

$$\frac{1}{s^2 + c_1 s + c_0} = \frac{A}{s - r_1} + \frac{B}{s - r_2}$$

to find $\mathcal{L}^{-1}\{Y(s)\}$.

(c) Apply the convolution theorem to

$$Y(s) = \left(\frac{1}{s - r_1}\right)\left(\frac{1}{s - r_2}\right)$$

to find $\mathcal{L}^{-1}\{Y(s)\}$.

11.3 OTHER PROPERTIES OF LAPLACE TRANSFORMS

This section lists some additional properties of Laplace transforms, supplementing those in table 11.3, page 469. The indicated exercises guide your derivation of each of the formulas given.

11.3.1 Transform of a Periodic Function

Exercise 23 asks you to prove that if $f(t)$ is a *periodic function with period T*, then

$$\boxed{\mathcal{L}\{f\} = \frac{1}{1 - e^{-sT}} \int_0^T e^{-st} f(t)\, dt.}$$

TRANSFORM OF A PERIODIC FUNCTION

EXAMPLE 13 *Find the Laplace transform of the square wave of period T,*

$$f(t) = \begin{cases} a, & 0 \le t < \frac{T}{2}, \\ -a, & \frac{T}{2} \le t < T \end{cases},$$

shown in figure 11.5.

Using the definition of this function, we have

$$\int_0^T e^{-st} f(t)\, dt = \int_0^{T/2} a e^{-st}\, dt - \int_{T/2}^T a e^{-st}\, dt$$

$$= -\frac{a}{s} e^{-st}\Big|_{t=0}^{t=T/2} + \frac{a}{s} e^{-st}\Big|_{t=T/2}^{t=T}$$

$$= \frac{a}{s}\left(e^{-sT} - 2e^{-sT/2} + 1\right)$$

$$= \frac{a}{s}\left(e^{-sT/2} - 1\right)^2.$$

Since $1 - e^{-sT} = (1 - e^{-sT/2})(1 + e^{-sT/2})$, the transform of the square wave is

$$\mathcal{L}\{f(t)\} = \frac{a(1 - e^{-sT/2})}{s(1 + e^{-sT/2})}.$$

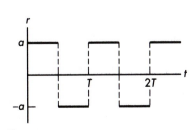

Figure 11.5: A square wave with amplitude a and period T.

■

EXAMPLE 14 *Find the Laplace transform of the solution of the initial value problem*

$$x'' + 4\pi^2 x = f(t), \qquad x(0) = 0, \qquad x'(0) = -2,$$

when f is a square wave of period 1,

$$f(t) = \begin{cases} a, & 0 \le t < \frac{1}{2}, \\ -a, & \frac{1}{2} \le t < 1 \end{cases} \tag{11.8}$$

TRANSFORM OF FORCING TERM

From the results of the preceding example, we find that the transform of the forcing term is

$$F(s) = \mathcal{L}\{f(t)\} = \frac{a(1 - e^{-s/2})}{s(1 + e^{-s/2})}.$$

Transform the differential equation.

Letting $X(s) = \mathcal{L}\{x(t)\}$ and transforming the differential equation in the usual way yield

$$s^2 \left[X(s) - sx(0) - x'(0)\right] + 4\pi^2 X = F(s).$$

Solve for $X(s)$.

Using the initial conditions and solving for X, we find the transform of the solution of this initial-value problem

$$X(s) = \frac{-2}{s^2 + 4\pi^2} + \frac{a(1 - e^{-s/2})}{s(1 + e^{-s/2})(s^2 + 4\pi^2)}. \qquad \blacksquare$$

EXAMPLE 15 *Invert the transform of the previous example and find the solution of the initial-value problem given there.*

Write that transform as $X(s) = X_1(s) + X_2(s)$, where

$$X_1(s) = \frac{-2}{s^2 + 4\pi^2}, \qquad X_2(s) = \frac{a(1 - e^{-s/2})}{s(1 + e^{-s/2})(s^2 + 4\pi^2)}.$$

The first term is easy to invert:

$$\mathcal{L}^{-1} X_1(s) = \mathcal{L}^{-1} \left\{ \frac{-2}{s^2 + 4\pi^2} \right\} = -\frac{2}{2\pi} \mathcal{L}^{-1} \left\{ \frac{2\pi}{s^2 + 4\pi^2} \right\} = -\frac{1}{\pi} \sin 2\pi t.$$

We recognize that the second term is the product of two terms whose inverse transforms we know,

$$X_2(s) = \left(\frac{a(1 - e^{-s/2})}{s(1 + e^{-s/2})} \right) \left(\frac{1}{2\pi} \frac{2\pi}{s^2 + 4\pi^2} \right).$$

an observation that suggests using the convolution theorem, entry 2 of table 11.3. Since the first factor is the transform of the square wave $f(t)$ defined in (11.8) and the second is the transform of $(\sin 2\pi t)/2\pi$, we can write $X_2(s)$ as the convolution integral

$$X_2(s) = \mathcal{L}\left\{\frac{1}{2\pi}\int_0^t f(r)\sin 2\pi(t-r)\,dr\right\}.$$

To evaluate this convolution integral, suppose $n \le t < n+1$. Then we can write the integral $\int_0^t \cdots dt$ as a sum of integrals over the common period of f and the sine function:

$$\int_0^t f(r)\sin 2\pi(t-r)\,dr = \int_0^1 f(r)\sin 2\pi(t-r)\,dr + \int_1^2 f(r)\sin 2\pi(t-r)\,dr + \cdots$$

$$+ \int_{n-1}^n f(r)\sin 2\pi(t-r)\,dr + \int_n^t f(r)\sin 2\pi(t-r)\,dr \qquad (11.9)$$

Because $f(r)$ and $\sin 2\pi(t-r)$ are both periodic in r with period 1, the integrals with integer limits are all equal. The sum collapses:

$$\int_0^t f(r)\sin 2\pi(t-r)\,dr$$

$$= n\int_0^1 f(r)\sin 2\pi(t-r)\,dr + \int_n^t f(r)\sin 2\pi(t-r)\,dr.$$

Now expand $\sin 2\pi(t-r)$ using the formula for the sine of a difference of angles and recall the definition (11.8) of $f(t)$:

$$\int_0^1 f(r)\sin 2\pi(t-r)\,dr = \sin 2\pi t\int_0^1 f(r)\cos 2\pi r\,dr - \cos 2\pi t\int_0^1 f(r)\sin 2\pi r\,dr$$

$$= a\sin 2\pi t\left(\int_0^{1/2}\cos 2\pi r\,dr - \int_{1/2}^1\cos 2\pi r\,dr\right)$$

$$- a\cos 2\pi t\left(\int_0^{1/2}\sin 2\pi r\,dr - \int_{1/2}^1\sin 2\pi r\,dr\right)$$

$$= \frac{a\sin 2\pi t}{2\pi}\left(\sin 2\pi r\big|_0^{1/2} - \sin 2\pi r\big|_{1/2}^1\right)$$

$$+ \frac{a\cos 2\pi t}{2\pi}\left(\cos 2\pi r\big|_0^{1/2} - \cos 2\pi r\big|_{1/2}^1\right)$$

$$= -\frac{2a\cos 2\pi t}{\pi}.$$

The last integral in (11.9) can be treated similarly. If $n \le t < n + \frac{1}{2}$, then

$$\int_n^t f(r)\sin 2\pi(t-r)\,dr = \sin 2\pi t\int_n^t f(r)\cos 2\pi r\,dr - \cos 2\pi t\int_n^t f(r)\sin 2\pi r\,dr$$

$$= a\sin 2\pi t\int_n^t\cos 2\pi r\,dr - a\cos 2\pi t\int_n^t\sin 2\pi r\,dr$$

$$= \frac{a\sin 2\pi t}{2\pi}\left(\sin 2\pi r\big|_n^t\right) + \frac{a\cos 2\pi t}{2\pi}\left(\cos 2\pi r\big|_n^t\right)$$

$$= \frac{a}{2\pi}\left(\sin^2 2\pi t + \cos^2 2\pi t - \cos 2\pi t\right)$$

$$= \frac{a(1 - \cos 2\pi t)}{2\pi}.$$

Likewise, if $n + \frac{1}{2} \le t < n + 1$, then

$$\int_n^t f(r) \sin 2\pi(t-r)\, dr = \sin 2\pi t \int_n^t f(r) \cos 2\pi r\, dr - \cos 2\pi t \int_n^t f(r) \sin 2\pi r\, dr$$

$$= a \sin 2\pi t \left(\int_n^{n+1/2} \cos 2\pi r\, dr - \int_{n+1/2}^t \cos 2\pi r\, dr \right)$$

$$- a \cos 2\pi t \left(\int_n^{n+1/2} \sin 2\pi r\, dr - \int_{n+1/2}^t \sin 2\pi r\, dr \right)$$

$$= \frac{a \sin 2\pi t}{2\pi} \left(\sin 2\pi r |_n^{n+1/2} - \sin 2\pi r |_{n+1/2}^t \right)$$

$$+ \frac{a \cos 2\pi t}{2\pi} \left(\cos 2\pi r |_n^{n+1/2} - \cos 2\pi r |_{n+1/2}^t \right)$$

$$= -\frac{a}{2\pi} \left(\sin^2 2\pi t + \cos^2 2\pi t + 3 \cos 2\pi t \right)$$

$$= -\frac{a(1 + 3 \cos 2\pi t)}{2\pi}.$$

Collecting the results of this analysis gives the solution of the initial-value problem:

$$x(t) = -\frac{\sin 2\pi t + 2na \cos 2\pi t}{\pi} + \frac{a(1 - \cos 2\pi t)}{2\pi}$$

for $n \le t < n + \frac{1}{2}$, and

$$x(t) = -\frac{\sin 2\pi t + 2na \cos 2\pi t}{\pi} + \frac{a(1 - \cos 2\pi t)}{2\pi}$$

for $n + \frac{1}{2} \le t < n + 1$. Note the presence of the term proportional to n; the amplitude of this solution is growing with time. Could this system be in resonance? ∎

11.3.2 Derivative of a Transform

Exercises 24 and 25 ask you to prove

DERIVATIVE OF A TRANSFORM

$$\boxed{\mathcal{L}\{(-1)^n f(t)\} = F^{(n)}(s), \qquad n = 1, 2, \ldots .}$$

When $n = 1$, we have

$$\boxed{\mathcal{L}\{-tf(t)\} = F'(s).}$$

EXAMPLE 16 *Use the derivative of a transform formula to find $\mathcal{L}\{te^{at}\}$.*

The expression te^{at} has the form $-tf(t)$ with $f(t) = -e^{at}$. With $f(t) = -e^{at}$, we find

$$F(s) = \mathcal{L}\{-e^{at}\} = -\mathcal{L}\{e^{at}\} = \frac{-1}{s-a}, \qquad s > a.$$

482   11   The Laplace Transform

To use $\mathcal{L}\{-tf(t)\} = F'(s)$, we compute

$$F'(s) = \frac{d}{ds}\left(\frac{-1}{s-a}\right) = \frac{1}{(s-a)^2}.$$

Hence, $\mathcal{L}\{te^{at}\} = 1/(s-a)^2$, $s > a$. ∎

Alternatively, we could use the shifted transform theorem, entry 3 of table 11.3,

$$\mathcal{L}\{e^{at}t\} = F(s-a),$$

with $f(t) = t$. From

$$F(s) = \mathcal{L}\{t\} = \frac{1}{s^2},$$

we find

$$\mathcal{L}\{e^{at}t\} = F(s-a) = \frac{1}{(s-a)^2}.$$

EXAMPLE 17 *Use the derivative of a transform formula to argue that the undamped system*

$$x'' + \omega^2 x = A\cos\omega t, \qquad x(0) = x_i, \qquad x'(0) = v_i,$$

is in resonance; i.e., show that the solution of this system contains at least one term whose amplitude grows with t.

The transform of the solution of this initial-value problem is

$$X(s) = \frac{sx_i + v_i}{s^2 + \omega^2} + \frac{As}{(s^2 + \omega^2)^2}.$$

The second term can be written as the derivative of a transform we know:

$$\frac{As}{(s^2 + \omega^2)^2} = -\frac{A}{2\omega}\frac{d}{ds}\left(\frac{\omega}{s^2 + \omega^2}\right)$$

$$= -\frac{A}{2\omega}\mathcal{L}\{-t\sin\omega t\}.$$

Hence, the solution contains a term whose amplitude grows with time; the system is in resonance. ∎

11.3.3   Transform of an Integral

Exercise 26 asks you to prove

$$\boxed{\mathcal{L}\left\{\int_0^t f(r)\,dr\right\} = \frac{F(s)}{s}.}$$

TRANSFORM OF AN INTEGRAL

EXAMPLE 18   *Use the transform of an integral formula to invert*

$$Y(s) = \frac{1}{s(s^2 + 4)}.$$

We recognize that we can write Y as a known transform divided by s:

$$Y(s) = \frac{F(s)}{s},$$

where

$$F(s) = \frac{1}{s^2 + 4} = \frac{1}{2}\mathcal{L}\{\sin 2t\}.$$

With $f(t) = (\sin 2t)/2$, the transform of an integral formula yields

$$Y(s) = \mathcal{L}\left\{\frac{1}{2}\int_0^t \sin 2t \, dt\right\}$$

$$= \mathcal{L}\left\{\frac{1}{4}(\cos 2t - 1)\right\}. \qquad \blacksquare$$

EXAMPLE 19   *Given that $\mathcal{L}\{t^k\} = k!/s^{k+1}$, $s > 0$, for some positive integer k, show that*

$$\mathcal{L}\{t^{k+1}\} = \frac{(k+1)!}{s^{k+2}}.$$

(This is the key step in a proof by mathematical induction of entry 2 of table 11.1.)
Apply the transform of an integral formula with $f(t) = t^k$. Since

$$\int_0^t t^k \, dt = \frac{t^{k+1}}{k+1}$$

and

$$F(s) = \mathcal{L}\{t^k\} = \frac{k!}{s^{k+1}},$$

we have

$$\mathcal{L}\left\{\frac{t^{k+1}}{k+1}\right\} = \mathcal{L}\left\{\int_0^t t^k \, dt\right\} = \frac{F(s)}{s} = \frac{k!}{s^{k+2}}.$$

Multiplying by $k + 1$ completes the argument. $\qquad \blacksquare$

11.3.4   Integral of a Transform

Exercise 27 asks you to prove that if $f(t)/t$ is defined at $t = 0$, then

$$\boxed{\mathcal{L}\left\{\frac{f(t)}{t}\right\} = \int_s^\infty F(r) \, dr.}$$

INTEGRAL OF A TRANSFORM

EXAMPLE 20 *Argue that the function*

$$g(t) = \frac{\sin at}{t}$$

can be defined at $t = 0$ and use the integral of a transform formula to find its Laplace transform.

Define $g(0)$ by its limiting value:

$$g(0) \equiv \lim_{t \to 0} \frac{\sin at}{t} = a.$$

Then g is continuous at $t = 0$ and, hence, integrable for $0 \le t < \infty$. Certainly, g is of exponential order.

To compute the transform whose existence we have just established, use the integral of a transform formula with $f(t) = \sin at$. Since $F(s) = \mathcal{L}\{\sin at\} = a/(s^2 + a^2)$, we have

$$\mathcal{L}\{g(t)\} = \mathcal{L}\left\{\frac{\sin at}{t}\right\} = \int_s^\infty \frac{a}{r^2 + a^2}\, dr$$

$$= \tan^{-1}\frac{r}{a}\Big|_{r=s}^{r=\infty} = \frac{\pi}{a} - \tan^{-1}\left(\frac{s}{a}\right),$$

the desired transform. ∎

EXAMPLE 21 *Find*

$$\mathcal{L}\left\{\frac{4\cos at}{t}\right\}.$$

The expression $4\cos at/t$ has the form $f(t)/t$ with $f(t) = 4\cos at$, and

$$F(s) = \mathcal{L}\{4\cos at\} = \frac{4s}{s^2 + a^2}, \qquad s > 0.$$

To use the integral of a transform formula above, we compute

$$\int_s^\infty F(s)\, ds = \int_s^\infty \frac{4s}{s^2 + a^2}\, ds$$

$$= 2\ln\left(s^2 + a^2\right)\Big|_s^\infty.$$

But when we evaluate this last expression, we find that it diverges to infinity as $s \to \infty$. The Laplace transform of $4\cos at/t$ does not exist!

To explain this difficulty, note that the integrand

$$\frac{4\cos at}{t}e^{-st}$$

appearing in the definition of this Laplace transform is not defined at the lower limit of integration $t = 0$. Indeed, near $t = 0$, this expression is behaving like $1/t$, and $1/t$ can not be integrated at $t = 0$. This Laplace transform does not exist because the integral (11.1) required for its definition does not exist even for finite values of the upper limit b. ∎

Table 11.4 provides an enlarged list of useful properties of the Laplace transform.

TABLE 11.4
Useful Properties of Laplace Transforms—Second Revision

Properties of Laplace Transforms

$$F(s) = \mathcal{L}\{f(t)\}, \ G(s) = \mathcal{L}\{g(t)\}$$

1. Transform of a derivative

$$\mathcal{L}\{f'(t)\} = sF(s) - f(0)$$

$$\mathcal{L}\{f''(t)\} = s^2 F(s) - sf(0) - f'(0)$$

2. Convolution theorem

$$\mathcal{L}\left\{\int_0^t f(r)g(t - r) \, dr\right\} = F(s)G(s)$$

3. Shifted transform
$$\mathcal{L}\{e^{at} f(t)\} = F(s - a)$$

4. Transform of a periodic function of period T

$$\mathcal{L}\{f\} = \frac{1}{1 - e^{-sT}} \int_0^T e^{-st} f(t) \, dt$$

5. Derivative of a transform

$$\mathcal{L}\{-tf(t)\} = F'(s)$$

$$\mathcal{L}\{(-t)^n f(t)\} = F^{(n)}(s), \ n = 1, 2, \ldots$$

6. Transform of an integral

$$\mathcal{L}\left\{\int_0^t f(r) \, dr\right\} = \frac{F(s)}{s}$$

7. Integral of a transform

$$\mathcal{L}\left\{\frac{f(t)}{t}\right\} = \int_s^\infty F(r) \, dr$$

EXERCISE GUIDE

To gain experience . . .	Try exercises
Finding Laplace transforms	1–5, 9–15, 21
Finding inverse transforms	6–8
With transforms of periodic functions	1, 9–10, 19, 23
With derivatives of transforms	2–3, 6, 11–12, 14, 22, 24–25
With transforms of integrals	4, 7, 26
With integrals of transforms	5, 8, 18, 27
Using the convolution theorem	22
Verifying existence of transforms	16–17
Solving initial-value problems	20–21
Deriving transform properties from the definition of Laplace transform	23–27
Analyzing and interpreting behavior of solutions	20

In exercises 1–5, use the indicated transform property to evaluate the Laplace transform of the given function. If the transform can be evaluated by a second method, do so to verify the accuracy of your answer. If a transform does not exist, explain why.

1. Use

$$\mathcal{L}\{f\} = \frac{1}{1 - e^{-sT}} \int_0^T e^{-st} f(t)\, dt,$$

where f has period T, to find the Laplace transform of

(a) $\sin \pi t$

(b) The square wave of period 4, which is defined for $0 \le t \le 4$ by

$$g(t) = \begin{cases} 1, & 0 \le t \le 2 \\ -1, & 2 < t \le 4. \end{cases}$$

2. Use $\mathcal{L}\{-tf(t)\} = F'(s)$ to find the Laplace transform of

(a) t^2 with $f(t) = -t$

(b) t^3 with $f(t) = -t^2$

(c) $t \cos 2t$

3. Use $\mathcal{L}\{(-t)^n f(t)\} = F^{(n)}(s)$ with an appropriate choice of f to find the Laplace transform of

(a) $t^2 e^{at}$

(b) $2t^3 \cos \pi t$

(c) $5t^2 \sinh at$

4. Use $\mathcal{L}\{\int_0^t f(r)\, dr\} = F(s)/s$ to find the Laplace transform of

(a) $(\sin \pi t)/\pi$ with $f(t) = \cos \pi t$

(b) e^{at}/a with $f(t) = e^{at}$

(c) $t^5/5$ with $f(t) = t^4$

5. Use $\mathcal{L}\{f(t)/t\} = \int_s^\infty F(s)\, ds$ with an appropriate choice of f to find the Laplace transform of

(a) $t^{-1} \sinh 2t$

(b) $t^{-1} \sin 2t$

(c) $(1 - \cos t)/t$

In exercises 6–8, use the indicated transform property (and other properties as needed) to find the inverse of the given transform.

6. Use $\mathcal{L}\{-tf(t)\} = F'(s)$ to invert the Laplace transforms

(a) $\dfrac{4}{(s - 1)^2}$

(b) $\dfrac{2s}{(s^2 + 16)^2}$

(c) $\dfrac{3s}{(1 + s^2)^3}$

7. Use $\mathcal{L}\{\int_0^t f(r)\, dr\} = F(s)/s$ to invert the Laplace transforms

(a) $\dfrac{1}{s(s - 1)}$

(b) $\dfrac{6(s-1)}{s^2(s+2)}$

8. Use $\mathcal{L}\{f(t)/t\} = \int_s^\infty F(s)\,ds$ with an appropriate choice of f to invert the Laplace transforms

(a) $\dfrac{2}{s-2}$

(b) $\dfrac{4s}{(s^2+a^2)^2}$

9. Sketch a graph of the nonsymmetric square wave of period T

$$f(t) = \begin{cases} a, & 0 \le t < b, \\ -a, & b \le t < T \end{cases}$$

and find its Laplace transform.

10. Sketch a graph of the sawtooth wave of period T

$$f(t) = -a + \dfrac{2at}{T}, \qquad 0 \le t < T$$

and find its Laplace transform.

In exercises 11–15, find the indicated transform; a and ω are constants.

11. $\mathcal{L}\{te^{at}\}$

12. $\mathcal{L}\{t^n e^{at}\}$

13. $\mathcal{L}\{e^{at}\sin\omega t\}$

14. $\mathcal{L}\{t\sin\omega t\}$

15. $\mathcal{L}\{\sin\omega t - \omega t\cos\omega t\}$

16. Example 20 begins with the claim

$$\lim_{t\to 0}\dfrac{\sin at}{t} = a.$$

Use l'Hôpital's rule or some other technique to evaluate this limit.

17. Verify the claim made in example 20 that the function

$$f(t) = \dfrac{\sin at}{t}, \qquad t > 0,$$

is of exponential order.

18. Supply the details of the evaluation of the integral

$$\int_s^\infty \dfrac{a}{r^2+a^2}\,dr$$

omitted from example 20.

19. Work through example 15, which solved the initial-value problem forced by a square wave. Show all of the details which were omitted from the text.

20. Example 15 solves the initial-value problem

$$x'' + 4\pi^2 x = f(t), \qquad x(0) = 0, \qquad x'(0) = -2,$$

where f is a square wave of period 1:

$$f(t) = \begin{cases} a, & 0 \le t < \frac{1}{2}, \\ -a, & \frac{1}{2} \le t < 1 \end{cases}.$$

It concludes that the solution is growing with time and asks whether this system could be in resonance. Is it? What is this system's natural frequency? What is the forcing frequency?

21. Use Laplace transforms to find the complete solution of the initial-value problem

$$x'' + \omega^2 x = A\cos\omega t, \qquad x(0) = x_i, \qquad x'(0) = v_i,$$

considered in example 17.

22. Example 17 inverts $As/(s^2+\omega^2)^2$ using the derivative of a transform formula:

$$\dfrac{As}{(s^2+\omega^2)^2} = -\dfrac{A}{2\omega}\dfrac{d}{ds}\left(\dfrac{\omega}{s^2+\omega^2}\right)$$

$$= -\dfrac{A}{2\omega}\mathcal{L}\{-t\sin\omega t\}.$$

Obtain the same result using the convolution theorem.

In exercises 23–27, use the definition of Laplace transform to derive the indicated transform property. In every case, $F(s) = \mathcal{L}\{f(t)\}$ and all transforms are assumed to exist.

23. If f has period T, then

$$\mathcal{L}\{f\} = \dfrac{1}{1-e^{-sT}}\int_0^T e^{-st}f(t)\,dt.$$

(*Hint:* Argue that the definition of the Laplace transform can be written

$$\mathcal{L}\{f\} = \sum_{n=1}^\infty \int_{(n-1)T}^{nT} f(t)e^{-st}\,dt.$$

The change of variables $r = (n-1)T + t$ and the periodicity of f then reduces the sum to one involving the integral $\int_0^T e^{-sr}f(r)\,dr$ in every term. When this integral is factored out of the sum, all that remains is a geometric series in e^{-sT}. Its sum is $1/(1-e^{-sT})$.)

24. $\mathcal{L}\{-tf(t)\} = F'(s)$ (*Hint:* Assuming the derivative

d/ds can be interchanged with the improper integral in the definition of Laplace transform, compute $dF(s)/ds$.)

25. $\mathcal{L}\{(-t)^n f(t)\} = F^{(n)}(s)$ for $n = 2, 3$ (*Hint:* For $n = 2$, apply the result of exercise 24 to the function $g(t) = tf(t)$. Use a similar idea for $n = 3, \ldots$.)

26. $\mathcal{L}\{\int_0^t f(r)\,dr\} = F(s)/s$ (*Hint:* Integrate the definition of Laplace transform by parts using $dv = e^{-st}\,dt$

and $u = \int_0^t f(r)\,dr$. The statement of the fundamental theorem of calculus in section A.2 will help you compute du/dt.)

27. $\mathcal{L}\{f(t)/t\} = \int_s^\infty F(r)\,dr$ (*Hint:* To evaluate the integral that appears on the right, assume that it can be interchanged with the improper integral in the definition of the Laplace transform.)

11.4 RAMPS AND JUMPS

Many models have forcing terms that increase steadily for a time and then level off at a fixed value. A good example is the heat output rate of a wood stove when it is first lit; its temperature-versus-time graph could have the ramp-like profile of figure 11.6.

Other forcing terms can be modeled by expressions that jump from one value to another. For example, in chapter 2 we modeled emigration caused by the Irish potato famine as continuing at a constant rate for all time,

$$P' - 0.015P = -E(t), \qquad P(0) = 8,$$

where

$$E(t) = 0.209 \text{ million people/year.}$$

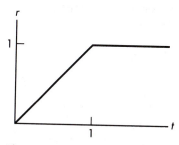

Figure 11.6: A ramp function.

But a better model might end emigration after 4 years, when the famine has subsided, by replacing the constant emigration rate with one that jumps to zero after 4 years have passed:

$$E(t) = \begin{cases} 0.209, & 0 \leq t < 4 \\ 0, & 4 \leq t < \infty. \end{cases}$$

The Laplace transform can solve initial-value problems having ramps, jumps, or similar functions in their forcing terms. This section develops the necessary transform tools.

> If the forcing term jumps, then the highest derivative in the differential equation must have a similar jump. For example, the emigration equation above has the form $P' + \cdots = -E(t)$. If $E(t)$ jumps suddenly from 0.209 to 0 at $t = 4$, then P' must exhibit a similar jump. Consequently, this first-order differential equation can not have a solution whose first derivative is continuous. We must relax our definition of *solution* to require that the *highest-order derivative be only piecewise continuous* and that the solution *satisfy the equation on the intervals where the highest-order derivative is continuous.*

11.4.1 The Unit Step Function

While transforms of ramps or jumps, such as the emigration function, can be computed directly from the definition of Laplace transform, we can also construct them conveniently using the **unit step function**

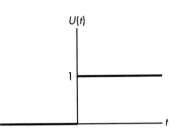

Figure 11.7: The unit step function $\mathcal{U}$ is "off" (value 0) for negative values of its argument and "on" (value 1) for positive values.

TRANSFORM OF A SHIFTED FUNCTION

$$\mathcal{U}(t) = \begin{cases} 0, & t < 0 \\ 1, & 0 \le t \end{cases}$$

shown in figure 11.7. The unit step function $\mathcal{U}$ is "off" for negative values of its argument and "on" for positive values.

The key transform result, which you are asked to derive in exercise 16, is

$$\boxed{\mathcal{L}\{f(t-a)\mathcal{U}(t-a)\} = e^{-as}F(s),}$$

where $F(s) = \mathcal{L}\{f(t)\}$ and $a > 0$. Before computing transforms involving the unit step function, we illustrate how it can be used to construct functions with switch points.

Figure 11.8 illustrates how the expression $f(t-a)\mathcal{U}(t-a)$ "turns on" at $t = a$. The upper curve shows $\mathcal{U}(t-a)$, an expression that switches from 0 to 1 when t reaches the value a. The middle curve in figure 11.8 shows the graph of an arbitrary function $f(t-a)$; it is the graph of $f(t)$ shifted from the origin to the right by a time units. The lower graph illustrates the product $f(t-a)\mathcal{U}(t-a)$ turning on at $t = a$. Because of the factor $\mathcal{U}$, the product is zero before $t = a$. After $t = a$, the product simply has the value $f(t-a)$.

Figure 11.9 illustrates two simple expressions using the unit step function. The function $\mathcal{U}(t-a)$ jumps from off to on at $t = a$,

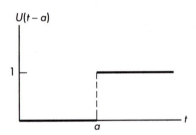

$$\mathcal{U}(t-a) = \begin{cases} 0, & 0 \le t < a \\ 1, & a \le t < \infty, \end{cases}$$

and $1 - \mathcal{U}(t-1)$ is its complement, jumping from on to off at $t = a$,

$$1 - \mathcal{U}(t-a) = \begin{cases} 1, & 0 \le t < a \\ 0, & a \le t < \infty. \end{cases}$$

A difference of unit step functions provides the off/on/off behavior of the **isolated pulse**

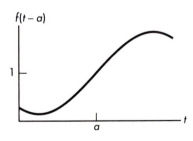

$$\mathcal{U}(t-a) - \mathcal{U}(t-b) = \begin{cases} 0, & 0 \le t < a \\ 1, & a \le t < b \\ 0, & b \le t < \infty \end{cases}$$

shown in figure 11.10. The term $\mathcal{U}(t-a)$ turns on at $t = a$. When $\mathcal{U}(t-b)$ turns on at a later time $t = b$, it subtracts 1 from $\mathcal{U}(t-a)$ to restore the net value of the expression to zero.

If b grows infinitely large, then the isolated pulse (figure 11.10) collapses to the off/on behavior of $\mathcal{U}(t-a)$, because $t-b$ is always negative and $\mathcal{U}(t-b) = 0$ for any finite value of t in the limit as b grows. If $a = 0$, then $\mathcal{U}(t-0) = \mathcal{U}(t) = 1$ for $t \ge 0$, and the isolated pulse has the on/off behavior of $1 - \mathcal{U}(t-b)$.

EXAMPLE 22   *Show that the ramp function*

$$r(t) = \begin{cases} t, & 0 \le t \le 1 \\ 1, & 1 < t < \infty \end{cases}$$

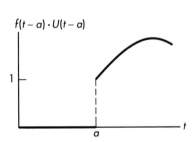

Figure 11.8: The "turn-on" property of the unit step function.

shown in figure 11.6 can be written using the unit step function $\mathcal{U}$.

We could think of this ramp function as composed of two parts, the ramp itself, $r(t) = t$, which appears only for $0 \leq t < 1$, and the flat section, $r(t) = 1$, which appears for $1 \leq t < \infty$. This decomposition is illustrated in figure 11.11.

Roughly, we could write

$$r(t) = t \times \text{(function that is on for } 0 \leq t < 1\text{)}$$
$$+ 1 \times \text{(function that is on for } 1 \leq t < \infty\text{)}.$$

The two functions that are "on" for fixed intervals are just isolated pulses. The function that is on for $0 \leq t < 1$ is

$$\mathcal{U}(t - 0) - \mathcal{U}(t - 1) = \begin{cases} 1, & 0 \leq t < 1 \\ 0, & 1 \leq t < \infty. \end{cases}$$

Its counterpart for $1 \leq t < \infty$ is

$$\mathcal{U}(t - 1) - \mathcal{U}(t - \infty) = \begin{cases} 0, & 0 \leq t < 1 \\ 1, & 1 \leq t < \infty. \end{cases}$$

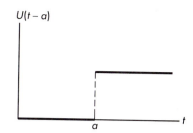

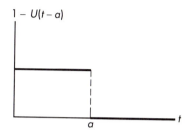

Figure 11.9: $\mathcal{U}(t - a)$ switches on at $t = a$; $1 - \mathcal{U}(t - a)$ switches off at $t = a$.

We have no business writing an expression such as $t - \infty$, because ∞ is a symbol used to denote unbounded limits, not a number with which we can perform arithmetic. Nonetheless, this abuse of notation is too suggestive to resist.

Using these expressions, the ramp function is

$$r(t) = t \times [\mathcal{U}(t - 0) - \mathcal{U}(t - 1)]$$
$$+ 1 \times [\mathcal{U}(t - 1) - \mathcal{U}(t - \infty)].$$

Since we are considering only $t \geq 0$, $\mathcal{U}(t - 0) = \mathcal{U}(t) = 1$. Likewise, since $t < \infty$, we have $\mathcal{U}(t - \infty) = 0$. The final form for the ramp function is

$$r(t) = t[1 - \mathcal{U}(t - 1)] + \mathcal{U}(t - 1).$$

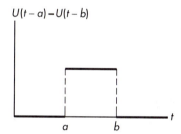

Figure 11.10: An isolated pulse.

The first factor turns on the linear function t for $0 \leq t \leq 1$. The second term turns on the constant 1, the flat part of the ramp function, when $t > 1$. This construction is also consistent with the on/off behavior of $1 - \mathcal{U}(t - 1)$ and $\mathcal{U}(t - 1)$ illustrated in figure 11.12 with $a = 1$. ∎

EXAMPLE 23   *Write the function*

$$g(t) = \begin{cases} 0, & 0 \leq t \leq 2 \\ \sin t, & 2 < t < \infty \end{cases}$$

using the unit step function.

Using the logic of example 22, we think of this function as composed of two sections, one for the interval $0 \leq t < 2$ and the other for $2 \leq t < \infty$. We write g as a sum of products, each product consisting of the desired value for the function multiplying a term that is unity when t is in the appropriate interval and zero otherwise:

$$g(t) = 0[\mathcal{U}(t - 0) - \mathcal{U}(t - 2)]$$
$$+ (\sin t)[\mathcal{U}(t - 2) - \mathcal{U}(t - \infty)].$$

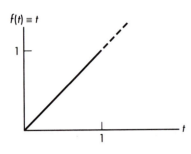

$f(t) \equiv t$

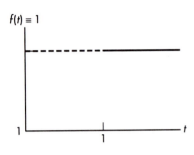

$f(t) \equiv 1$

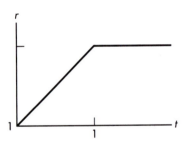

r

Figure 11.11: The ramp function, shown in the lower graph, is composed of a linear function that stops at $t = 1$ and a constant function that starts at $t = 1$.

The expression $\mathcal{U}(t - \infty)$ is poor notation, but it reduces to zero since t is finite. Hence, the final expression for g is

$$g(t) = (\sin t)\mathcal{U}(t - 2),$$

consistent with the off/on behavior of $\mathcal{U}(t - 2)$ illustrated in figure 11.9 with $a = 2$. ∎

EXAMPLE 24   *Find the Laplace transform of the function*

$$g(t) = \begin{cases} 0, & 0 \le t \le 2 \\ \sin t, & 2 < t < \infty \end{cases}$$

considered in the last example

From example 23, we have

$$g(t) = (\sin t)\mathcal{U}(t - 2).$$

To apply the shifted function transform

$$\mathcal{L}\{f(t - a)\mathcal{U}(t - a)\} = e^{-as}F(s),$$

we identify $a = 2$ and $f(t - a) = \sin t$. We must find a formula for $f(t)$, the function without its argument shifted, so that we can compute $F(s) = \mathcal{L}\{f(t)\}$.

A simple change of variables suffices. Let $r = t - 2$ in $f(t - 2) = \sin t$. Since $t = r + 2$, we have $f(r) = \sin(r + 2)$, or, if we replace the place holder r with t,

$$f(t) = \sin(t + 2).$$

Equivalently, we could argue that $f(t - 2) = \sin t$ means that $\sin t$ is the formula for f evaluated at t shifted back by 2 units. To obtain a formula for f at the present time, we must shift t forward by 2 units, replacing every occurrence of t by $t + 2$. Hence, $f(t) = \sin(t + 2)$.

Now we can compute the transform $F(s) = \mathcal{L}\{f(t)\}$:

$$F(s) = \mathcal{L}\{f(t)\} = \mathcal{L}\{\sin(t + 2)\}$$
$$= \mathcal{L}\{\sin t \cos 2 + \cos t \sin 2\}$$
$$= \frac{\cos 2}{s^2 + 1} + \frac{s \sin 2}{s^2 + 1}, \qquad s > 0.$$

With $F(s)$ in hand, we can use the shifted function transform to write

$$\mathcal{L}\{g(t)\} = \mathcal{L}\{f(t - a)\mathcal{U}(t - a)\}$$
$$= e^{-2s}\frac{\cos 2 + s \sin 2}{s^2 + 1}, \qquad s > 0. \quad ∎$$

EXAMPLE 25   *Find the Laplace transform of the ramp function*

$$r(t) = \begin{cases} t, & 0 \le t \le 1 \\ 1, & 1 < t < \infty \end{cases}$$

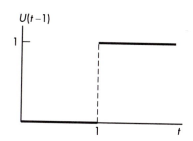

shown in figure 11.6.

Example 22 wrote this ramp function as

$$r(t) = t[1 - \mathcal{U}(t-1)] + \mathcal{U}(t-1).$$

From the linearity of the Laplace transform, we have

$$\mathcal{L}\{r(t)\} = \mathcal{L}\{t\} - \mathcal{L}\{t\mathcal{U}(t-1)\} + \mathcal{L}\{\mathcal{U}(t-1)\}.$$

We find in turn each of the transforms in this last expression.
Entry 2 of table 11.1 immediately yields

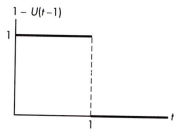

$$\mathcal{L}\{t\} = \frac{1}{s^2}.$$

The shifted function formula

$$\mathcal{L}\{f(t-a)\mathcal{U}(t-a)\} = e^{-as}F(s),$$

Figure 11.12: The off/on unit step function $\mathcal{U}(t-1)$ and its on/off complement $1 - \mathcal{U}(t-1)$ used to write the ramp function.

will provide the other two transforms involving the unit step function, $\mathcal{L}\{t\mathcal{U}(t-1)\}$ and $\mathcal{L}\{\mathcal{U}(t-1)\}$.
To treat $\mathcal{L}\{t\mathcal{U}(t-1)\}$, we identify $a = 1$ and

$$f(t-1) = t.$$

The change of variables $r = t - 1$ converts $f(t-1) = t$ to $f(r) = r + 1$, or, when we replace r with t,

$$f(t) = t + 1.$$

Entries 1 and 2 of table 11.1 immediately yield

$$F(s) = \mathcal{L}\{t + 1\} = \mathcal{L}\{t\} + \mathcal{L}\{1\}$$
$$= \frac{1}{s^2} + \frac{1}{s} = \frac{1+s}{s^2}.$$

Coupling this expression with $\mathcal{L}\{f(t-a)\mathcal{U}(t-a)\} = e^{-as}F(s)$, we find

$$\mathcal{L}\{t\mathcal{U}(t-1)\} = e^{-s}\frac{1+s}{s^2}.$$

The last transform we need is $\mathcal{L}\{\mathcal{U}(t-1)\}$. Now we use the shifted function transform with $a = 1$ and $f(t-1) = 1$. Obviously, $f(t) = 1$ and $F(s) = \mathcal{L}\{f(t)\} = 1/s$. Hence,

$$\mathcal{L}\{\mathcal{U}(t-1)\} = \frac{e^{-s}}{s}.$$

Combining these three expressions, which are transforms of t, $t\mathcal{U}(t-1)$, and $\mathcal{U}(t-1)$, we find the transform of the ramp function,

$$\mathcal{L}\{r(t)\} = \mathcal{L}\{t\} - \mathcal{L}\{t\,\mathcal{U}(t-1)\} + \mathcal{L}\{\mathcal{U}(t-1)\}$$

$$= \frac{1}{s^2} - e^{-s}\frac{1+s}{s^2} + \frac{e^{-s}}{s}$$

$$= \frac{1 - e^{-s}}{s^2}, \quad s > 0,$$

the same result as in example 5, which used the definition of Laplace transform. More elegantly, we might have noticed that $r(t) = [1 - \mathcal{U}(t-1)]\,t + \mathcal{U}(t-1)$ can be written

$$r(t) = t - (t-1)\mathcal{U}(t-1).$$

The shifted function result $\mathcal{L}\{f(t-a)\mathcal{U}(t-a)\} = e^{-as}F(s)$ can then be applied immediately with $f(t-1) = t - 1$ and $f(t) = t$. ∎

EXAMPLE 26 *Solve the initial-value problem*

$$y' + 2y = r(t), \qquad y(0) = 4,$$

where $r(t)$ is the ramp function shown in figure 11.6.

Transform the differential equation.

Let $Y(s) = \mathcal{L}\{y(t)\}$. Transforming the entire differential equation yields

$$(s+2)Y(s) - y(0) = \mathcal{L}\{r(t)\}.$$

Solve for $Y(s)$.

Using $y(0) = 4$ and the expression for $\mathcal{L}\{r(t)\}$ from example 25, we have

$$Y(s) = \frac{4}{s+2} + \frac{1}{s^2(s+2)} - e^{-s}\frac{1}{s^2(s+2)}. \tag{11.10}$$

Invert $Y(s)$.

The first term comes from table 11.1:

$$\mathcal{L}\{4e^{-2t}\} = \frac{4}{s+2}.$$

The second term in (11.10) can be evaluated using the convolution theorem or partial fractions. Using the latter, we find

$$\frac{1}{s^2(s+2)} = \frac{1}{4}\left(\frac{2}{s^2} - \frac{1}{s} + \frac{1}{s+2}\right).$$

Hence,

$$\frac{1}{s^2(s+2)} = \frac{1}{4}\mathcal{L}\{2t - 1 + e^{-2t}\}.$$

The third term in (11.10) contains an exponential in the transform variable s, a clue that we must use the shifted function theorem,

$$\mathcal{L}\{f(t-a)\mathcal{U}(t-a)\} = e^{-as}F(s)$$

Comparing $e^{-as}F(s)$ with the given transform,

$$e^{-s}\frac{1}{s^2(s+2)},$$

we identify $a = 1$ and

$$F(s) = \frac{1}{s^2(s+2)}.$$

But from our work in the preceding paragraph, we know

$$F(s) = \mathcal{L}\{f(t)\} = \frac{1}{4}\mathcal{L}\{2t - 1 + e^{-2t}\}.$$

If $f(t) = \frac{1}{4}(2t - 1 + e^{2t})$, then

$$f(t-1) = \frac{1}{4}\left(2(t-1) - 1 + e^{-2(t-1)}\right)$$

$$= \frac{1}{4}\left(2t - 3 + e^{-2(t-1)}\right).$$

We conclude that the third term in (11.10) must be the transform of $f(t-1)\mathcal{U}(t-1)$; that is, the transform of

$$\frac{1}{4}\left(2t - 3 + e^{-2(t-1)}\right)\mathcal{U}(t-1).$$

The solution of the given initial-value problem is the sum of the inverse transforms of the three terms in (11.10),

$$y(t) = 4e^{-2t} + \frac{1}{4}\left(2t - 1 + e^{-2t}\right) - \frac{1}{4}\left(2t - 3 + e^{-2(t-1)}\right)\mathcal{U}(t-1). \quad \blacksquare$$

11.4.2 Summary

The unit step function $\mathcal{U}$ has the value one when its argument is positive and the value zero when its argument is negative.

Functions such as ramps, that switch between several different formulas at various values of t, can be written as sums of isolated pulses. An isolated pulse of amplitude one is

$$\mathcal{U}(t - a) - \mathcal{U}(t - b).$$

It is unity when $a \leq t < b$ and zero otherwise.

Transforms involving the unit step function are computed using the shifted function transform

$$\boxed{\mathcal{L}\{f(t - a)\mathcal{U}(t - a)\} = e^{-as}F(s),}$$

TRANSFORM OF A SHIFTED FUNCTION

where $F(s) = \mathcal{L}\{f(t)\}$. An exponential in the s domain is a clue to the presence of a shift in the t domain.

A list of Laplace transform properties, including this latest formula, appears in table 11.5.

TABLE 11.5
A List of Properties of Laplace Transforms

Properties of Laplace Transforms

$$F(s) = \mathcal{L}\{f(t)\}, \ G(s) = \mathcal{L}\{g(t)\}$$

---

1. Transform of a derivative

$$\mathcal{L}\{f'(t)\} = sF(s) - f(0)$$

$$\mathcal{L}\{f''(t)\} = s^2 F(s) - sf(0) - f'(0)$$

2. Convolution theorem

$$\mathcal{L}\left\{ \int_0^t f(r)g(t-r)\,dr \right\} = F(s)G(s)$$

3. Shifted transform

$$\mathcal{L}\{e^{at} f(t)\} = F(s-a)$$

4. Transform of a periodic function of period T

$$\mathcal{L}\{f\} = \frac{1}{1 - e^{-sT}} \int_0^T e^{-st} f(t)\,dt$$

5. Derivative of a transform

$$\mathcal{L}\{-tf(t)\} = F'(s)$$

$$\mathcal{L}\{(-t)^n f(t)\} = F^{(n)}(s), \ n = 1, 2, \ldots$$

6. Transform of an integral

$$\mathcal{L}\left\{ \int_0^t f(r)\,dr \right\} = \frac{F(s)}{s}$$

7. Integral of a transform

$$\mathcal{L}\left\{ \frac{f(t)}{t} \right\} = \int_s^\infty F(r)\,dr$$

8. Transform of a shifted function using the unit step function $\mathcal{U}$

$$\mathcal{L}\{f(t-a)\mathcal{U}(t-a)\} = e^{-as} F(s), \ a > 0$$

EXERCISE GUIDE

To gain experience . . .	Try exercises
Finding transforms of shifted functions	1–11
Expressing functions in terms of the unit step function	4–11
With graphs involving the unit step function	8–11, 15
Solving initial-value problems	8–11
Finding transforms of solutions of initial-value problems	12–13
Using the convolution theorem	14
Deriving the shifted function transform from the definition of Laplace transform	16

In exercises 1–3, use $\mathcal{L}\{f(t-a)\mathcal{U}(t-a)\} = e^{-as}F(s)$ with an appropriate choice of f to find the Laplace transform of the given function.

1. $e^{4t}\mathcal{U}(t-3)$

2. $(2-3t)\mathcal{U}(t-2)$

3. the isolated pulse

$$g(t) = \begin{cases} 0, & 0 \le t \le a \\ 1, & a < t \le b \\ 0, & b < t \end{cases}$$

In exercises 4–7, . . .

   i. Rewrite the given function in terms of the unit step function $\mathcal{U}$.

   ii. Use that expression and $\mathcal{L}\{f(t-a)\mathcal{U}(t-a)\} = e^{-as}F(s)$ to find the Laplace transform of the given function.

4. $g(t) = \begin{cases} 4t, & 0 \le t \le 7 \\ 0, & 7 < t < \infty \end{cases}$

5. $g(t) = \begin{cases} mt, & 0 \le t \le b \\ 0, & b < t < \infty \end{cases}$

6. $g(t) = \begin{cases} 0, & 0 \le t \le 2 \\ t-2, & 2 \le t \le 4 \\ 6, & 4 < t < \infty \end{cases}$

7. $g(t) = \begin{cases} 1, & 0 \le t \le 2\pi \\ \cos t, & 2\pi \le t \le 7\pi/2 \\ 0, & 7\pi/2 < t < \infty \end{cases}$

In exercises 8–11,

   i. Graph the forcing term and write it in terms of the unit step function $\mathcal{U}$.

   ii. Solve the initial-value problem using Laplace transforms.

   iii. Give the interval(s) within which the highest derivative of the solution appearing in the equation is continuous.

   iv. Verify that your solution satisfies the differential equation within each interval of continuity.

8. $y' - 6y = g(t)$, $y(0) = 4$,

$$g(t) = \begin{cases} 0, & 0 \le t < 2 \\ 2, & 2 \le t < \infty. \end{cases}$$

9. $y' + 2y = g(t)$, $y(0) = 0$,

$$g(t) = \begin{cases} 2t - 1, & 0 \le t < 2 \\ 3, & 2 \le t < \infty. \end{cases}$$

10. $y'' + 4y = g(t)$, $y(0) = 0$, $y'(0) = -2$,

$$g(t) = \begin{cases} -2, & 0 \le t < 4 \\ 4, & 2 \le t < 8 \\ 0, & 8 \le t < \infty. \end{cases}$$

11. $P' - 0.015P = -E(t)$, $P(0) = 8$,

$$E(t) = \begin{cases} 0.209, & 0 \le t \le 4 \\ 0, & 4 < t < \infty \end{cases}$$

(This is the Irish potato famine emigration model with population P in millions and time t in years measured from 1847. In this form, emigration stops in 1851. See example 3 of section 2.2.1.)

In exercises 12–13, assume that the Laplace transform of the solution of each initial-value problem exists. Find the transform of the solution of the initial-value problem working directly from the differential equation.

12. $2y' - 5y = g(t)$, $y(0) = 3$, where $g(t)$ is the isolated pulse of exercise 3,

$$g(t) = \begin{cases} 0, & 0 \le t \le a \\ 1, & a < t \le b \\ 0, & b < t. \end{cases}$$

13. $y' - ky = r(t)$, $y(0) = y_i$, where k is a constant and $r(t)$ is the ramp function of example 25,

$$r(t) = \begin{cases} t, & 0 \le t \le 1 \\ 1, & 1 < t < \infty. \end{cases}$$

14. Use the convolution theorem to find the inverse transform of

$$\frac{1}{s^2(s+2)}.$$

Compare your answer with that in example 26 which used partial fractions.

15. On one graph, sketch both $\mathcal{U}(t - 2)$ and $-\mathcal{U}(t - 4)$. Illustrate graphically how they combine to form the isolated pulse

$$\mathcal{U}(t - 2) - \mathcal{U}(t - 4) = \begin{cases} 0, & 0 \le t < 2 \\ 1, & 2 \le t < 4 \\ 0, & 4 \le t < \infty. \end{cases}$$

16. Derive the shifted function transform

$$\mathcal{L}\{f(t - a)\mathcal{U}(t - a)\} = e^{-as}F(s),$$

where $F(s) = \mathcal{L}\{f(t)\}$ and $a > 0$. (*Hint:* Because of its definition, the unit step function $\mathcal{U}(t - a)$ changes to a nonzero value the lower limit of the integral in the definition (11.1) of the Laplace transform. Make a change of variable of integration that converts the lower limit back to zero and leaves the upper limit at infinity.)

11.5   THE UNIT IMPULSE FUNCTION

The upper end of a spring carrying a mass is jerked suddenly upward. A freely swinging pendulum is struck sharply with a hammer. A switch is closed in an electric circuit, causing an abrupt jump in voltage. A typhoon sweeps through Bangladesh, killing 125,000 people in the wink of an eye. This section develops the *unit impulse function*, a tool for modeling these kinds of instantaneous changes.

Whenever individuals are removed from a population, whether through the tragedy of a natural disaster or through a planned harvest in a fish farm, for example, the emigration model

$$P' = kP - E(t), \qquad P(0) = P_i,$$

is appropriate. Recall that $E(t)$ is the *rate* at which individuals are removed from the population; units of trout/minute might be appropriate for a fish farm.

If a finite number R of individuals is removed at time $t = a$, then the emigration function $E(t)$ could have the form

$$E(t) = Rd_h(t - a),$$

where $d_h(t - a)$ is an isolated pulse of width h and height $1/h$ centered at $t = a$. See figure 11.13.

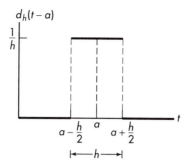

Figure 11.13: An isolated pulse of unit area centered at $t = a$

Because $E(t)$ is a *rate* of population removal, the number removed over all time is the integral

$$\int_{-\infty}^{\infty} E(t)\, dt.$$

But since the isolated pulse $d_h(t - a)$ has unit area, that number removed is

$$\int_{-\infty}^{\infty} E(t)\, dt = R \int_{-\infty}^{\infty} d_h(t)\, dt = R, \qquad (11.11)$$

the desired amount.

In a population model, where the action of removing individuals occurs over a brief but nonzero interval, the simple isolated pulse $d_h(t - a)$ might be adequate. But when the impulse is effectively instantaneous, such as a sudden jerk on the end of a spring, a different tool is needed. That tool is a limit of d_h as the width h of the pulse drops to zero while it retains unit area. Since such a limiting pulse of unit area can not have a finite value at $t = a$, we define it in terms of its action in integrals such as (11.11).

Let $\delta(t-a)$ represent this limiting pulse of zero width and unit area. Define $\delta(t - a)$ by

$$\int_{-\infty}^{\infty} \phi(t)\delta(t - a)\, dt = \lim_{h \to 0} \int_{-\infty}^{\infty} \phi(t)d_h(t - a)\, dt$$

$$= \lim_{h \to 0} \int_{a-h/2}^{a+h/2} \phi(t)d_h(t - a)\, dt, \qquad (11.12)$$

where ϕ is an arbitrary function, continuous for all t and $\phi(t) = 0$ for t sufficiently large. This limiting impulse is called the **Dirac delta function** or **unit impulse function**.

The functions ϕ are called *test functions* because they are used to test the action of the limiting impulse $\delta(t - a)$. They are said to have *compact support* since the length of the interval on which they are nonzero is finite.

The delta function is, in fact, not a function but a *distribution*, a concept defined in terms of its action on test functions rather than in terms of the input-output rules of a conventional function. Stakgold [18] provides a clear treatment of distributions.

To determine the action of $\delta(t - a)$ on an arbitrary ϕ, use the mean value theorem for integrals in the integral involving d_h in (11.12):

$$\int_{\infty}^{\infty} \phi(t)\delta(t - a)\, dt = \lim_{h \to 0} \phi(\tilde{t}) \int_{a-h/2}^{a+h/2} d_h(t - a)\, dt$$

$$= \lim_{h \to 0} \phi(\tilde{t}).$$

Here $a - h/2 \leq \tilde{t} \leq a + h/2$, and the last equality results because d_h has unit area. Since ϕ is continuous and $\tilde{t}$ approaches a as h shrinks to zero, we conclude that

$$\int_{\infty}^{\infty} \phi(t)\delta(t - a)\, dt = \phi(a).$$

Integrating a continuous function with $\delta(t - a)$ picks out the value of the function at $t = a$.

Now choose ϕ to be zero everywhere but in a small neighborhood of some point $t^* \neq a$, where ϕ is positive. Since the integral $\int \phi(t)d_h(t-a)\, dt$ in definition (11.12) is zero once h is small enough, we find

$$\int_{-\infty}^{\infty} \phi(t)\delta(t - a)\, dt = 0.$$

But in a neighborhood of t^*, $\phi(t)$ is positive. We conclude that the integral can be zero only if $\delta(t^* - a) = 0$. That is, $\delta(t - a) = 0, t \neq a$.

The two important properties of the delta function are as follows:

DELTA FUNCTION

$$\boxed{\begin{array}{l} \int_{-\infty}^{\infty} \phi(t)\delta(t - a)\, dt = \phi(a) \\[2mm] \delta(t - a) = 0, \qquad t \neq a. \end{array}}$$

Since $\delta(t - a) = 0$ for $t \neq a$, the function $\phi(t)$ can always be redefined away from $t = a$ without affecting the value of $\int \phi(t)\delta(t - a)\, dt$. Hence, the restriction in (11.12) that $\phi(t)$ be zero outside a finite interval can be dropped.

An immediate consequence of these properties is

$$\int_{-\infty}^{\infty} \delta(t - a)\, dt = 1.$$

The delta function *preserves the unit area property* of the finite pulses $d_h(t - a)$. The delta function somehow represents a pulse of zero width, infinite height, and unit area, requirements that can be met by no ordinary function. It is indeed a *unit impulse*.

EXAMPLE 27   *Use these properties to find the Laplace transform of $\delta(t - a)$, $a > 0$.*

Formally, the desired transform is defined by

$$\mathcal{L}\{\delta(t - a)\} = \int_0^\infty e^{-st} \delta(t - a) \, dt.$$

If $a > 0$, we can choose $\phi(t)$ in (11.12) so that $\phi(t) = e^{-st}$ near $t = a$ and $\phi(t)$ tapers quickly to zero elsewhere. Then the first property of the delta function immediately yields

$$\int_0^\infty e^{-st} \delta(t - a) \, dt = e^{-st}\Big|_{t=a} = e^{-as}.$$

Hence,

$$\boxed{\mathcal{L}\{\delta(t - a)\} = e^{-as}, \; a > 0.}$$ ■   **DELTA FUNCTION TRANSFORM**

We can also argue that $\mathcal{L}\{\delta(t)\} = 1$.

EXAMPLE 28   *Solve $y' = \delta(t - a)$, $y(0) = 0$, when $a > 0$.*

Since $\delta(t - a)$ is not defined at $t = a$, we must suspend our usual definition of *solution*. Proceeding formally, we let $Y(s) = \mathcal{L}\{y(t)\}$ and transform the differential equation to find

$$sY(s) = e^{-as}.$$

Since $\mathcal{L}^{-1}\{1/s\} = 1$, entry 8 of table 11.5 yields

$$y(t) = \mathcal{L}^{-1}\left\{ \frac{1}{s} e^{-as} \right\} = \mathcal{U}(t - a).$$

That is, in a sense we can not properly define here, *the derivative of the unit step function $\mathcal{U}(t - a)$ is the delta function $\delta(t - a)$.*

Since $\delta(t - a) = 0$ and $\mathcal{U}(t - a)$ is constant for $t \neq a$, the relation certainly seems reasonable. At $t = a$, $\mathcal{U}(t - a)$ is not even continuous, much less differentiable. But if we think of the derivative in some extended sense, the behavior of the derivative at a jump discontinuity is captured by the delta function's impulse of zero width and unit area.   ■

The delta function is the derivative of $\mathcal{U}(t - a)$ in the sense that the action of $\delta(t - a)$ on a test function $\phi(t)$ is the same as the action of the unit step function on the derivative $\phi'(t)$. Since $\phi(t)$ is zero for t sufficiently large, an integration by parts relation holds:

$$\int_{-\infty}^\infty \phi(t)\delta(t - a) \, dt = \int_{-\infty}^\infty \phi'(t)\mathcal{U}(t - a) \, dt = \int_a^\infty \phi'(t) \, dt = \phi(a).$$

EXAMPLE 29 *Solve the initial-value problem* $P' = kP - R\delta(t-a)$, $P(0) = P_i$, *a model of a population in which R individuals are removed suddenly at time* $t = a$.

Let $Q(s) = \mathcal{L}\{P(t)\}$. Formally transforming the equation yields

$$Q(s) = \frac{P_i}{s-k} - \frac{R}{s-k}e^{-as}.$$

The first term is inverted using entry 3 of table 11.1. The second is inverted using entry 8 of table 11.5. Hence, the solution is

$$P(t) = P_i e^{kt} - Re^{k(t-a)}\mathcal{U}(t-a).$$

The first term represents the familiar exponential population growth. The unit step function in the second term introduces a sudden drop in population at $t = a$, but the exponential factor leads immediately to renewed growth from the reduced population level. ∎

EXAMPLE 30 *At time* $t = a$, *the upper end of a vertical spring-mass system is jerked upward suddenly and returned to its original position. Write a model for this situation.*

The forced model (5.5–5.6) of section 5.1.3 applies with the position $h(t)$ of the upper end of the spring given by $h(t) = H\delta(t-a) - h(0)$ for some H. The model is

$$mx'' + kx = kH\delta(t-a), \qquad x(0) = x_i, \qquad x'(0) = v_i. \qquad ∎$$

To grasp the physical meaning of the equation $mx'' + \cdots = kH\delta(t-a)$, informally integrate both sides. Then the relation is $mx' + \cdots = kH$. The quantity on the left, mx', is the change in the momentum of the mass, and kH represents that change. Since mx'' is a force (mass times acceleration), $kH\delta(t-a)$ is an *impulsive force* of brief duration and sufficient magnitude to produce a change in momentum of kH.

Closing a switch at time $t = a$ in a series RLC circuit can cause a step function change in potential, $E(t) = V\mathcal{U}(t-a)$. The current model $Li'' + Ri' + i/C = E'(t)$ becomes

$$Li'' + Ri + \frac{1}{C}i = V\delta(t-a),$$

where the impulse $V\delta(t-a)$ is the consequence of a step increase V in potential at $t = a$. The time integral of $V\delta(t-a)$ is the magnitude V of the voltage change.

11.5.1 Exercises

EXERCISE GUIDE

To gain experience . . .	Try exercises
With transforms involving $\delta(t-a)$	1–3, 5–7, 9–13
Solving initial-value problems	2, 5–7, 6–12
Analyzing models involving impulses	4–6, 9–12
With $\delta(t-a)$ as the derivative of $\mathcal{U}(t-a)$	7–8, 11

1. Find $\mathcal{L}^{-1}\{1\}$.

2. Solve $y' + ky = 4\delta(t - 6)$, $y(0) = 2$.

3. Invert the transform $\dfrac{s}{s - 2}$.

4. Sketch a graph of the solution $P(t) = P_i e^{kt} - Re^{k(t-a)}\mathcal{U}(t - a)$ of the population model $P' = kP - R\delta(t - a)$, $P(0) = P_i$, considered in example 29. Explain the significance of the graph.

5. Consider the heat-loss model $T' + (Ak/cm)T = b\delta(t - a)$, $T(0) = 0$. Solve this initial-value problem. Use your solution to explain the significance of the parameter b. Of what physical situation might this initial-value problem be a model?

6. Solve the initial-value problem

$$mx'' + kx = kH\delta(t - a), \quad x(0) = x_i, \quad x'(0) = v_i,$$

developed in example 30. It models a spring-mass system in which the upper end of the spring is jerked suddenly. Consider $x_i = v_i = 0$. Use your solution to explain the significance of the parameter H.

7. Find the current in a series RLC circuit if the potential in the circuit is raised from $E = 0$ to $E = V$ at time $t = a$. For simplicity, impose zero initial conditions and neglect resistance.

8. Approximate the unit step function $\mathcal{U}(t - a)$ by

$$U_h(t - a) = \begin{cases} 0, & -\infty < t < a - \frac{h}{2} \\ (t - a + \frac{h}{2})/h, & a - \frac{h}{2} \leq t \leq a + \frac{h}{2} \\ 1, & a + \frac{h}{2} < t < \infty. \end{cases}$$

Sketch a graph of $U_h(t - a)$. Compute $dU_h(t - a)/dt$ and sketch its graph. For what values of t is this derivative not defined? What connections between $\delta(t - a)$ and $\mathcal{U}(t - a)$ does this approximation suggest?

9. Consider the linearized pendulum equation subject to an impulsive forcing term,

$$L\theta'' + g\theta = b\delta(t - a), \quad \theta(0) = \theta'(0) = 0.$$

(a) Solve this problem and explain the terms in the solution you obtain.

(b) Use the solution obtained in part (a) to explain the significance of the parameter b.

(c) If a pendulum is released from rest from the angle θ_i and left undisturbed, its amplitude is θ_i. Choose b so that the pendulum in this model also has amplitude θ_i for $t \geq a$.

10. An undamped spring-mass system subject to an impulsive disturbance at $t = a$ is modeled by $mx'' + kx = b\delta(t - a)$, $x(0) = x'(0) = 0$.

(a) Describe the motion of the mass for $t < a$. Solve the governing initial-value problem, confirm your description, and explain the terms that are nonzero for $t \geq a$.

(b) Choose the parameter b so that the mass achieves a prescribed amplitude A for $t \geq a$.

11. The potential in a series RLC circuit is $E(t) = V\mathcal{U}(t - a)$. Argue that the current in the circuit is governed by

$$Li'' + Ri' + \frac{1}{C}i = V\delta(t - a).$$

Solve this equation subject to zero initial conditions; for convenience, set $R = 0$. Determine the value of V that produces the same amplitude as the initial conditions $i(0) = I$, $i'(0) = 0$.

12. The *unit impulse response* of a system is the solution of the governing initial-value problem with zero initial conditions and the forcing function $\delta(t)$. Find the unit impulse response of the current in a series RLC circuit. How could the unit impulse response of such a circuit be realized physically?

13. Derive the formula for $\mathcal{L}\{\delta(t - a)\}$ from the definition (11.12) – that is, evaluate the integrals involving $e^{-st}d_h(t - a)$.

11.6 CHAPTER EXERCISES

EXERCISE GUIDE

To gain experience . . .	Try exercises
Using the definition of Laplace transform	1–2
Finding Laplace transforms	3–10
Finding inverse transforms	15–24
Expressing functions in terms of the unit step function	11–13
With graphs involving the unit step function	11–12, 14
Solving initial-value problems	11–13
Analyzing function behavior, given a Laplace transform	25–28
Solving systems of equations	29–36

In exercises 1–2, use the definition of Laplace transform to find the transform of the given function. Sketch the graph of each function before you attempt to find its transform.

1. $g(t) = \begin{cases} mt, & 0 \le t \le b \\ 0, & b < t < \infty \end{cases}$

2. $g(t) = \begin{cases} 0, & 0 \le t \le 2 \\ t - 2, & 2 \le t \le 4 \\ 6, & 4 < t < \infty \end{cases}$

In exercises 3–10, find the indicated transform; a and ω are constants.

3. $\mathcal{L}\{\cos 2t \sin 4t\}$

4. $\mathcal{L}\{\cos^2 3t\}$

5. $\mathcal{L}\{e^{-4t} \cos 2t \sin 4t\}$

6. $\mathcal{L}\{e^{-2t} \cos^2 3t\}$

7. $\mathcal{L}\{t(1 - t)\}$

8. $\mathcal{L}\{e^{at} \cos \omega t\}$

9. $\mathcal{L}\{t \cos \omega t\}$

10. $\mathcal{L}\{\sin \omega t + \omega t \cos \omega t\}$

In exercises 11–12,

  i. Graph the forcing term and write it in terms of the unit step function $\mathcal{U}$.

  ii. Solve the initial-value problem using Laplace transforms.

  iii. Give the interval(s) within which the high-

est derivative of the solution appearing in the equation is continuous.

  iv. Verify that your solution satisfies the differential equation within each interval of continuity.

11. $y' + 2y = g(t), y(0) = 5,$

$$g(t) = \begin{cases} -2, & 0 \le t < 4 \\ 4, & 4 \le t < 8 \\ 0, & 8 \le t < \infty \end{cases}$$

12. $y'' - 9y = g(t), y(0) = 4, y'(0) = 0,$

$$g(t) = \begin{cases} 0, & 0 \le t < 2 \\ 2, & 2 \le t < \infty \end{cases}$$

13. Assuming that it exists, find the Laplace transform of the solution of $x' + x = g(t), x(0) = -3$, where $g(t)$ is the isolated pulse and cosine wave of exercise 7, page 497.

$$g(t) = \begin{cases} 1, & 0 \le t \le 2\pi \\ \cos t, & 2\pi \le t \le 7\pi/2 \\ 0, & 7\pi/2 < t < \infty \end{cases}$$

Work directly from the differential equation.

14. Draw a graph like that of figure 11.8, illustrating the "turn-on" property of the unit step function product $f(t-a)\mathcal{U}(t-a)$ for the function appearing in example 24, page 492,

$$g(t) = \begin{cases} 0, & 0 \le t \le 2 \\ \sin t, & 2 < t < \infty. \end{cases}$$

In exercises 15–24, find the function of t whose transform is given.

15. $\dfrac{4}{s^2 + 5s + 6}$

16. $\dfrac{4e^{-2s}}{s^2 + 5s + 6}$

17. $\dfrac{2s}{s^2 - 4}$

18. $\dfrac{2s}{s^2 + 4}$

19. $\dfrac{2se^{-5s}}{s^2 + 4}$

20. $\dfrac{s - 1}{(s + 1)(s^2 - 4)}$

21. $\dfrac{4}{s^2 + 2s + 5}$

22. $\dfrac{3}{s^3 + 3s^2 + 2s}$

23. $\dfrac{2s + 6}{s^2 + 6s + 18}$

24. $\dfrac{3s}{s^2 - 3s + 10}$

In exercises 25–28, the transform of the response of a system is given. In each case,

    i. Determine whether the response contains oscillatory terms. If so give their frequencies.

    ii. Determine whether the response includes exponential decay or growth. If so give, the factor in the exponential (or the corresponding time constant).

    iii. Determine if the response contains any terms that switch on or off. If so, give the switching time(s).

Do *not* invert the transforms.

25. $\dfrac{s^2 + 1}{s^4 - 16}$

26. $\dfrac{s^2 e^{-2s}}{(s^2 + 4)(s^2 + 4s + 3)}$

27. $\dfrac{2 - 4e^{-3s}}{(s - 1)^3}$

28. $\dfrac{3}{(s - 3)(s^2 + 12s + 45)}$

Laplace transforms can be used to solve systems of constant-coefficient differential equations. The steps to follow are: Introduce variables for the transforms of the unknowns (e.g., $Y(s) = \mathcal{L}\{y(t)\}$, $Z(s) = \mathcal{L}\{z(t)\}$), transform the system, solve for $Y(s)$, $Z(s)$, and invert. Use this technique to solve exercises 29–36.

29. $y' = y + z$, $y(0) = 1$

    $z' = -y + z$, $z(0) = 2$

30. $y' = z$, $y(0) = 0$

    $z' = y$, $z(0) = 1$

31. $y' = -4y + 2z$, $y(0) = 3$

    $z' = -3y + 3z$, $z(0) = 4$

32. $y' = -4z$, $y(0) = 0$

    $z' = -4y$, $z(0) = 2$

33. $y' = y/2 + \sqrt{3}\, z/2 + 2t$, $y(0) = 6$

    $z' = \sqrt{3}\, y/2 - z/2 - 4t^2$, $z(0) = 4$

34. $y' = z + 4\cos t$, $y(0) = 1$

    $z' = y$, $z(0) = 0$

35. $y' = y + 3z + e^{2t}$, $y(0) = 1$

    $z' = 3y + z - 1$, $z(0) = 0$

36. $y' = y - z - 2$, $y(0) = 1$

    $z' = 2y + 4z - 2e^{-2t}$, $z(0) = 0$

Exercises 37–40 ask you to use Laplace transforms to determine the charge on the capacitor in the series RLC circuit of figure 7.11, section 7.3, when the voltage source E varies as described. Point out the terms in your solution that reflect dramatic changes in E.

The parameters of the circuit in figure 7.11 are $R = 200\ \Omega$, $L = 100\ \mu\mathrm{H}$, and $C = 100\ \mathrm{pF}$. Time is measured in microseconds and voltage in millivolts. Assume the voltage drop across the capacitor is initially zero and that no current is flowing at $t = 0$.

37. $E(t) = \begin{cases} 10, & 0 \le t < 4 \\ 0, & 4 \le t \end{cases}$

38. $E(t) = \begin{cases} t, & 0 \le t < 6 \\ 6, & 6 \le t \end{cases}$

39. $E(t) = \begin{cases} 0, & 0 \leq t < 2 \\ 5, & 2 \leq t \end{cases}$

40. E tapers linearly from 5 mV initially to 0 at 4 μs, where it remains thereafter.

41. Consider the series RLC circuit model with zero initial conditions: $Lq'' + Rq' + q/C = E(t)$, $q(0) = q'(0) = 0$; see section 5.2. Show that the solution of this initial-value problem can be written

$$q(t) = \int_0^t E(r)g(t - r)\, dr,$$

where

$$g(t) = \frac{e^{-Rt/2L}\sin \omega t}{\omega L}$$

is the inverse transform of the *transfer function*

$$G(s) = \frac{1}{Ls^2 + Rs + 1/C}.$$

Here $\omega = \sqrt{4L/C - R^2}/2L$. (*Hint:* Write the transform of the solution in terms of $G(s)$ and use the convolution theorem)

11.7 CHAPTER PROJECTS

1. **General inverse formula** Applying the Laplace transform to a differential equation usually leads to an expression of the form

$$Y(s) = \frac{N(s)}{D(s)},$$

where N and D are polynomials in s with real coefficients and the degree of N is no greater than the degree k of D. This project asks you to develop a general expression for the solution of such a differential equation by inverting $Y(s)$.

(a) Let $r_1, \ldots, r_k$ denote the roots of D, each repeated according to its multiplicity. Find the numerators A_i in the partial-fraction expansion

$$\frac{N(s)}{D(s)} = \frac{A_1}{s - r_1} + \cdots + \frac{A_k}{s - r_k}$$

in terms of $N(r_i), D'(r_i)$. (Multiply by $s - r_i$ and let $s \to r_i$.) Then write $y(t) = \mathcal{L}^{-1}\{Y(s)\}$ as a sum of exponentials using the A_i as coefficients. (How are these related to those of the characteristic polynomial?)

(b) Since complex roots must occur in conjugate pairs, argue that the corresponding partial-fraction coefficients A_i are also complex conjugate pairs. Further simplify the expression for $y(t)$ from part (a) by replacing complex exponentials with sines and cosines.

2. **Stability of a time-delayed logistic model** The familiar logistic equation $P' = (a - sP)P$ could be thought of as a simple population equation $P' = kP$ whose growth rate was modified to decrease with increasing population. That is, k is replaced by $a - sP$ to reflect decreasing birth rate as larger populations competed for the same resources.

In fact, there ought to be some delay in this feedback process. Beyond the fixed delay of gestation periods, a range of prior experiences should affect the present birth rate, with the most immediate and the most distant experiences having the least impact on today's birth rate. This argument suggests replacing k by

$$a - s \int_0^t P(r)Q(t-r)\,dr,$$

where Q might have the general shape shown in figure 11.14. The experience of large populations T years ago, near the peak in the Q curve, would tend to reduce today's birth rates.

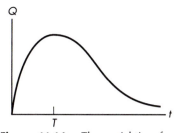

Figure 11.14: The weighting function Q in the modified logistic equation.

To normalize, assume $\int_0^\infty Q(r)\,dr = 1$. One choice of Q is

$$Q(r) = \frac{r^2}{T^2}e^{-r/T}.$$

In this new model, populations satisfy the *integrodifferential equation*

$$P'(t) = \left(a - s \int_0^t P(r)Q(t-r)\,dr\right)P(t).$$

(a) Derive this equation.
(b) Show that its equilibria are $P = 0, a/s$.
(c) Show that small perturbations $p(t)$ of either of these equilibria satisfy

$$p'(t) = -a \int_0^t p(r)Q(t-r)\,dr.$$

Use Laplace transforms to determine the behavior of p and, hence, the stability of these steady states.
(d) How does the behavior of this model differ from that of the simple logistic model?

3. **Transfer functions** Think of the forcing term in a differential equation as its *input*. Think of the forced response (the solution subject to *zero* initial conditions) as the *output*. Then the *transfer function* of the system modeled by the equation is the ratio of the Laplace transform of the output to the Laplace transform of the input. Analysis of the transfer function provides a picture of the response of the system.
(a) Show that the transfer function $G(s)$ of the generic second-order, linear, constant-coefficient equation

$$b_2 x'' + b_1 x' + b_0 x = f(t)$$

is

$$G(s) = \frac{X(s)}{F(s)} = \frac{1}{b_2 s^2 + b_1 s + b_0},$$

where $F(s) = \mathcal{L}\{f(t)\}, X(s) = \mathcal{L}\{x(t)\}$.
(b) Control engineers say, "In transform space, the output is the product of the transfer function and the input. Multiplication in transform space is convolution in the time domain." Explain these statements using the relation $X(s) = G(s)F(s)$.
(c) The *poles* or singularities of the transfer function $G(s)$ are the values of s at which its denominator is zero. What is the connection between the poles of the transfer function and the roots of the characteristic polynomial of the original differential equation?

Justify the following statements.

i. If the coefficients $b_2 > 0$ and $b_1, b_0 \geq 0$, then the poles of the transfer function lie in the left half of the complex plane; i.e., the poles all have nonnegative real part.

ii. If the poles of the transfer function are pure imaginary and $b_2 > 0$, $b_1, b_0 \geq 0$, then the system is undamped.

iii. If the poles of the transfer function are real and $b_2 > 0$, $b_1, b_0 \geq 0$, then the system is overdamped.

iv. If the poles of the transfer function are complex with negative real part and $b_2 > 0$, $b_1, b_0 \geq 0$, then the system is underdamped.

v. If a pole of the transfer function has a positive real part, then the response of the system can grow without bound.

(d) Find the transfer function for a *first-order* constant-coefficient system $b_1 x' + b_0 = f(t)$. Use it to justify the following statements about first-order systems:

i. The transfer function of a time-invariant (or constant-coefficient), first-order system can be written

$$G(s) = \frac{c}{1 + Ts}$$

where c is a constant and T is the time constant of the system. (Recall that the *time constant* of e^{-rt} is $1/r$.)

ii. At $t = T$ the *unit step response* of the system (the response to the forcing term $f(t) = \mathcal{U}(t)$ with zero initial condition) is 63.2% of its initial value.

iii. The initial slope of the unit step response is $1/T$.

APPENDIX

This appendix briefly reviews some of the ideas from calculus and algebra that are frequently used in studying differential equations.

A.1 DERIVATIVES

The derivative $f'(t)$ of the function $f(t)$ is defined by

$$f'(t) = \lim_{\Delta t \to 0} \frac{f(t + \Delta t) - f(t)}{\Delta t},$$

if the limit exists.

The derivative is often interpreted as a rate of change. If the position s of an object is given as a function of time t, then the derivative function ds/dt gives velocity as a function of time. When $ds/dt > 0$, then velocity is positive and position s is increasing. If $v = ds/dt$ defines the velocity function, then $dv/dt = d^2s/dt^2$ is acceleration. When $d^2s/dt^2 = d(ds/dt)/dt > 0$, then ds/dt is increasing; i.e., velocity v is increasing.

In a graphical setting, the value of the derivative $f'(x)$ of a function $f(x)$ is the slope of the line tangent to the graph of the function at x. The value of the second derivative $f''(x)$ gives the rate of change of the slope of the tangent line at x.

If $f'(x) > 0$, then the function is *increasing* at x; the slope of its graph is positive there. If $f''(x) = d(f'(x))/dx > 0$, then the rate of change of the slope is positive at x; the slope is increasing. If $f''(x) > 0$ over an interval, then the slope of the graph is increasing over that interval; the graph is *concave up* there.

Derivatives with respect to different variables are related by the *chain rule for derivatives*.

Theorem A.1 (Chain Rule for Derivatives) If s is a function of t and t is a function T and both are differentiable, then

$$\frac{ds}{dT} = \frac{ds}{dt}\frac{dt}{dT}.$$

For example, suppose s represents position measured in meters, t is time measured in seconds, and T is time measured in minutes. Then ds/dt is velocity in meters/second, and ds/dT is velocity in meters/minute.

The chain rule tells us how these two derivatives of the same function with respect to different independent variables are related. The functional relationship between t and T is

$$t = 60T.$$

Hence, $dt/dT = 60$ seconds/minute , and the chain rule yields

$$\frac{ds}{dT} \text{ meters/minute} = 60\frac{ds}{dt} \text{ meters/second.}$$

More frequently, you have used the chain rule to compute derivatives of composite functions, as in the following example.

EXAMPLE 1 *Find dy/dx if $y = e^{\cos x}$.*

To see clearly that this is a composite function, write it as $y = e^{(\cos x)}$. The "outside" function is the exponential, and the "inside" function is the cosine. If you were evaluating this function on a calculator, you would enter the value of x and then push the function keys *cos* and *exp* in that order, from inside out.

The chain rule can be stated informally as "derivative of the inside times derivative of the outside." To relate that phrase to the formal statement of the chain rule, let $u = \cos x$. Then we can write the given function in two parts as

$$y = e^u, \quad \text{and} \quad u = \cos x.$$

The chain rule yields

$$\frac{dy}{dx} = \frac{dy}{du}\frac{du}{dx}$$
$$= (e^u)(-\sin x) = -e^{\cos x}\sin x.$$

The term $e^u = e^{\cos x}$ is the derivative of the "outside" function, and $-\sin x$ is the derivative of the "inside" function. ∎

If necessary, review the section of your calculus text that discusses the chain rule as well as the derivative formulas for polynomials, trigonometric functions, the logarithm, and the exponential.

For functions of more than one variable, the rate of change with respect to one of those variables, independent of the others, is a partial derivative. For example, if t and y are independent variables, then we can define the partial derivatives

$$\frac{\partial f(t, y)}{\partial t} = \lim_{\Delta t \to 0} \frac{f(t + \Delta t, y) - f(t, y)}{\Delta t}$$

and

$$\frac{\partial f(t, y)}{\partial y} = \lim_{\Delta y \to 0} \frac{f(t + \Delta y, y) - f(t, y)}{\Delta y},$$

providing the limits in question exist.

Now suppose that y is not an independent variable but a function of t. Then the expression $z(t) = f(t, y(t))$ is a function of t alone. Another form of the chain rule provides a formula for the derivative $z'(t)$,

$$z'(t) = \frac{\partial f(t, y(t))}{\partial t} + \frac{\partial f(t, y(t))}{\partial y} \frac{dy(t)}{dt}.$$

A.2 ANTIDERIVATIVES

The antiderivative notation

$$\int f(x)\, dx = ?$$

asks the question,

> What is a function whose derivative with respect to x is f?

The answer always includes an additive constant (the famous "plus C") because adding a constant to a function does not change its derivative. More formally, if F is a function whose derivative with respect to x is f (i.e., if $dF/dx = f$), then the derivative of $F + C$ is also f:

$$\frac{d(F + C)}{dx} = \frac{dF}{dx} + \frac{dC}{dx} = f + 0 = f.$$

We can assign a specific value to C only if we have some added information, such as the value of the antiderivative at a specific point.

EXAMPLE 2 *Find a function whose derivative with respect to x is $5x^2$ and that assumes the value 6 at $x = 2$.*

To find the family of functions whose derivative with respect to x is $5x^2$, we evaluate

$$\int 5x^2\, dx = \frac{5x^3}{3} + C.$$

(In the notation of the paragraph preceding the example, $f(x) = 5x^2$ and $F(x) = 5x^3/3$.)

We must choose C so that $5x^3/3 + C$ has the value 6 when $x = 2$; i.e., so that $5 \cdot 8/3 + C = 6$. Hence, $C = -\frac{22}{3}$, and the desired function is

$$\frac{5x^3}{3} - \frac{22}{3}.$$

∎

A common error is writing an expression such as

$$\int t^2 \, dT = \frac{t^3}{3} + C.$$

This expression would be correct if the antiderivative on the left were $\int t^2 \, dt$. Since t and T are different variables, the expression $\int t^2 \, dT$ is a mixture of apples and oranges. It is asking for a function whose derivative with respect to T is t^2.

We can salvage this expression only when we know the functional relationship between t and T, as we did, for example, in the discussion in section A.1, where we had the relation $t = 60T$ between time t in seconds and time T in minutes.

In many cases, we cannot write a closed form expression for an antiderivative — e.g., $\int e^{x^2} \, dx$. Nonetheless, we often need notation for that one particular antiderivative which assumes a specified value at a given point. We make use of one part of the *fundamental theorem of calculus*:

> **Theorem A.2 (Fundamental Theorem of Calculus)**   If f is an integrable function and a is a constant, then
>
> $$\frac{d}{dx} \int_a^x f(s) \, ds = f(x).$$

In other words, $\int_a^x f(s) \, ds$ is an antiderivative of f. (The dummy variable s is used within the definite integral to avoid confusion with the independent variable x.)

If f is integrable, then $\int_a^x f(s) \, ds$ certainly defines a function of x. If f is positive-valued and $a \le x$, the value of the function for a given x is the area under the graph of f between a and x. In particular, the value of this function at $x = a$ is zero. For a given f, you could even evaluate the function $\int_a^x f(s) \, ds$ with a single keystroke on a calculator that includes a definite integral key.

EXAMPLE 3   *Find the function whose derivative with respect to x is $5x^2$ and that assumes the value 6 at $x = 2$.*

Using the fundamental theorem of calculus, we can immediately write one function whose derivative is $5x^2$ as

$$\int_a^x 5s^2 \, ds,$$

where a is a constant. That is,

$$\int 5x^2 \, dx = \int_a^x 5s^2 \, ds + C.$$

(Note that the $\int$ symbol on the left is an *antiderivative*, whereas the $\int_a^x$ on the right is a *definite integral*.)

We must determine C to give our particular antiderivative the value 6 at $x = 2$. That is, C must satisfy

$$\int_a^2 5s^2 \, ds + C = 6.$$

To avoid evaluating the integral, choose $a = 2$. Then we immediately have $C = 6$. The desired antiderivative is

$$F(x) = \int_2^x 5s^2 \, ds + 6.$$

Is this function the same as the one we obtained in example 2, which posed the same problem? Evaluating the definite integral defining $F(x)$ yields

$$F(x) = \int_2^x 5s^2 \, ds + 6$$

$$= \left. \frac{5s^3}{3} \right|_2^x + 6$$

$$= \frac{5x^3}{3} - \frac{40}{3} + 6$$

$$= \frac{5x^3}{3} - \frac{22}{3},$$

which agrees with example 2. ∎

EXAMPLE 4 *Find the function whose derivative is e^{x^2} and that has the value 2.6 at $x = -\pi$.*

We use the fundamental theorem. For ease of evaluation, choose $a = -\pi$. Add the constant 2.6 to give the antiderivative the proper value at $x = -\pi$. The desired function is

$$\int_{-\pi}^x e^{s^2} \, ds + 2.6. \qquad ∎$$

You may wish to review antiderivative formulas for polynomials, trigonometric functions, the exponential, and the logarithm. You should also refresh your understanding of such antidifferentiation methods as integration by parts and partial fractions. (See section A.7 for the latter.)

A.3 TAYLOR'S THEOREM

Complex functional behavior is often approximated by something simpler, such as a straight line. ("If I keep traveling at this rate, I'll be home by . . . ," and so on.) Taylor's theorem extends that approximation idea to polynomials of arbitrary order and it gives the error in such an approximation.

Theorem A.3 (Taylor's Theorem) Let $y(t)$ have n continuous derivatives on the interval $t_0 \leq t \leq t_1$. Then

$$y(t) = y(t_0) + y'(t_0)(t - t_0) + \frac{y''(t_0)(t - t_0)^2}{2}$$

$$+ \cdots + \frac{y^{(n-1)}(t_0)(t - t_0)^{n-1}}{(n-1)!} + R_n,$$

where the remainder term is

$$R_n = \frac{y^{(n)}(\xi)(t - t_0)^n}{n!} \qquad \text{(A.1)}$$

for some ξ, $t_0 \leq \xi \leq t$.

The remainder term R_n gives the error in approximating $y(t)$ by a polynomial in t of degree $n - 1$.

To formally validate the terms in the Taylor polynomial, evaluate both sides at $t = t_0$: $y(t_0) = y(t_0)$. Differentiating both sides and evaluating at $t = t_0$ again yields an identity, $y'(t_0) = y'(t_0)$, and so on.

EXAMPLE 5 *Find the first two nonzero terms in the Taylor polynomial for* $\cos 4t$ *about* $t_0 = 0$. *Illustrate the error in approximating this function by this Taylor polynomial.*

With $y(t) = \cos 4t$ and $t_0 = 0$, we find

$$y(t_0) = \cos 0 = 1$$

$$y'(t_0) = -4\sin 0 = 0$$

$$y''(t_0) = -16\cos 0 = -16.$$

In addition, for the remainder term we will need

$$y^{(3)}(t) = 64\sin 4t.$$

A total of three terms ($n = 3$) in the Taylor polynomial leaves us with two nonzero terms, as required. The Taylor polynomial is

$$\cos 4t = 1 - \frac{16}{2!}(t - 0)^2 + R_3 = 1 - 8t^2 + R_3,$$

where the remainder uses the third derivative of y,

$$R_3 = \frac{64\sin 4\xi}{3!}(t - 0)^3 = \frac{32\sin 4\xi}{3}t^3.$$

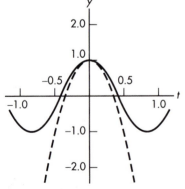

Figure A.1: The second-order Taylor polynomial approximation $1 - 8t^2$ (dotted curve) to $\cos 4t$ (solid curve).

Figure A.1 compares this quadratic Taylor polynomial with the original function. ∎

To approximate a function of two variables $f(t, y)$ in a neighborhood of (t_0, y_0), the formal analogue of the Taylor expansion is

$$f(t, y) = f(t_0, y_0) + f_t(t_0, y_0)(t - t_0) + f_y(t_0, y_0)(y - y_0)$$

$$+ f_{tt}(t_0, y_0)(t - t_0)^2 + 2f_{ty}(t_0, y_0)(t - t_0)(y - y_0)$$
$$+ f_{yy}(t_0, y_0)(y - y_0)^2 + \cdots.$$

Subscripts denote partial derivatives (e.g., $f_{ty} = \partial^2 f / \partial t \partial y$), which are all assumed to exist.

A.4 POWER SERIES

A **power series about** x_0 is an infinite sum of the form

$$\sum_{n=0}^{\infty} c_n(x - x_0)^n = c_0 + c_1(x - x_0) + c_2(x - x_0)^2 + \cdots.$$

Only nonnegative, integer powers of x may appear.

A power series **converges at** x if

$$\lim_{k \to \infty} \sum_{n=0}^{k} c_n(x - x_0)^n$$

exists; that is, the sequence of partial sums, each of which adds one more term to the sum, has a limit. A power series has **radius of convergence** R if it converges for each x in $|x - x_0| < R$ —i.e., for $x_0 - R < x < x_0 + R$. Within its radius of convergence, a power series may be *differentiated and integrated term by term.*

> The ability to interchange summation and differentiation or integration is a property of *uniformly convergent series.* Roughly, we can get a finite sum within a specified tolerance of the limiting value using a designated number of terms that is independent of x. Power series converge uniformly if they converge.

The **ratio test** provides the standard method for determining the radius of convergence of a power series.

Theorem A.4 (Ratio Test)   The radius of convergence of the power series

$$\sum_{n=0}^{\infty} c_n(x - x_0)^n$$

is

$$R = \lim_{n \to \infty} \left| \frac{c_n}{c_{n+1}} \right|$$

if this limit exists. The series diverges for $|x - x_0| > R$. If $c_n/c_{n+1} \to \infty$, then the power series converges for all x.

> Convergence at the end points $x = x_0 \pm R$ of the interval of convergence must be examined separately.

Suppose a power series has radius of convergence $R > 0$. Its limit is a function of x:

$$f(x) = \sum_{n=0}^{\infty} c_n(x - x_0)^n, \qquad |x - x_0| < R.$$

Since we can differentiate a convergent power series term by term, we easily recover the **Taylor coefficients,**

$$c_0 = f(x_0)$$
$$c_1 = f'(x_0)$$
$$c_2 = \frac{f''(x_0)}{2!}$$
$$\vdots$$
$$c_n = \frac{f(n)(x_0)}{n!}$$
$$\vdots$$

If $f(x)$ has derivatives of all orders for $|x - x_0| < R$ for some $R > 0$, then we can formally construct the **Taylor series,**

$$f(x) = \sum_{n=0}^{\infty} \frac{f^{(n)}(x_0)}{n!}(x - x_0)^n.$$

To determine if this series converges to f (as it often does), we ask whether the remainder term R_n in the Taylor polynomial of theorem A.3 decays to zero as n grows.

EXAMPLE 6   *Construct the Taylor series expansion of e^x about $x_0 = 0$ and determine its radius of convergence.*

Since e^x and all its derivatives are unity at $x_0 = 0$, the Taylor coefficients of this function are just

$$c_n = \frac{1}{n!}.$$

The formal Taylor series for e^x is

$$\sum_{n=0}^{\infty} \frac{x^n}{n!}.$$

The usual convention is $0! = 1$.

From the ratio test, its radius of convergence is

$$R = \lim_{n \to \infty} \frac{1/n!}{1/(n+1)!} = \lim_{n \to \infty} (n + 1).$$

Hence, this series converges for all x.

To determine if this series actually converges to e^x, examine the remainder term in the Taylor polynomial (A.1). It is

$$R_n = \frac{e^\xi}{n!}.$$

Suppose $|x| < S$ for any fixed $S > 0$. Then as $n \to \infty$,

$$|R_n| < \frac{e^S}{n!} \to 0.$$

On any bounded interval, the Taylor expansion converges to e^x. Hence, it converges to e^x for all x and we can write

$$e^x = \sum_{n=0}^{\infty} \frac{x^n}{n!}, \qquad |x| < \infty. \qquad \blacksquare$$

A.5   MATRICES AND DETERMINANTS

An $m \times n$ **matrix** is a rectangular array of numbers, called *elements* or *entries*, arranged in m rows and n columns,

$$\mathbf{A} = \begin{pmatrix} a_{11} & a_{12} & \cdots & a_{1n} \\ a_{21} & a_{22} & \cdots & a_{2n} \\ \vdots & \vdots & & \vdots \\ a_{m1} & a_{m2} & \cdots & a_{mn} \end{pmatrix}.$$

The arbitrary entry a_{ij} appears in the i-th row and the j-th column. *The first subscript denotes row; the second subscript denotes column.*

The **transpose** $\mathbf{A}^T$ is the $n \times m$ matrix whose j-th row is the j-th column of $\mathbf{A}$. For example, if

$$\mathbf{A} = \begin{pmatrix} 11 & 12 & 13 \\ 21 & 22 & 23 \end{pmatrix},$$

then

$$\mathbf{A}^T = \begin{pmatrix} 11 & 21 \\ 12 & 22 \\ 13 & 23 \end{pmatrix}.$$

A *square* matrix has an equal number of rows and columns. A matrix that consists of a single row or a single column is often called a *row vector* or a *column vector*. The transpose of a row vector is a column vector and vice versa.

Matrices of the same size are added or subtracted by adding or subtracting corresponding elements. Two matrices of the same size are equal if corresponding elements are equal.

Scalar multiplication is multiplication of a matrix by a single number. Each entry of the matrix is multiplied by that number. For example, using the matrix $\mathbf{A}$ defined previously,

$$-2\mathbf{A} = \begin{pmatrix} -22 & -24 & -26 \\ -42 & -44 & -46 \end{pmatrix}.$$

Matrix multiplication, multiplication of one matrix by another, is defined as follows. Suppose $\mathbf{A}$ is an $m \times n$ matrix and $\mathbf{B}$ is an $n \times p$ matrix. Then the $m \times p$ product matrix $\mathbf{C} = \mathbf{AB}$ is defined by

$$c_{ij} = \sum_{k=1}^{n} a_{ik} b_{kj}, \qquad i = 1, \ldots, m, \quad j = 1, \ldots, p.$$

Here a_{ik}, b_{kj}, and c_{ij} are entries in $\mathbf{A}$, $\mathbf{B}$, and $\mathbf{C}$, respectively. More loosely, matrix multiplication is *row times column*: The entry in the i-th row, j-th column of the product $\mathbf{AB}$ is the sum of the product of the corresponding entries from the i-th row of $\mathbf{A}$ and the j-th column of $\mathbf{B}$.

EXAMPLE 7 *Using the matrix $\mathbf{A}$ defined previously and*

$$\mathbf{B} = \begin{pmatrix} 1 & 2 & 3 \\ 3 & 4 & 5 \\ -1 & -2 & -3 \end{pmatrix},$$

find $\mathbf{AB}$. What can you say about $\mathbf{BA}$?

Since the number of columns of $\mathbf{A}$ matches the number of rows of $\mathbf{B}$ (three), the product $\mathbf{AB}$ is defined. Using row times column, we find

$$\mathbf{AB} = \begin{pmatrix} 11 & 12 & 13 \\ 21 & 22 & 23 \end{pmatrix} \begin{pmatrix} 1 & 2 & 3 \\ 3 & 4 & 5 \\ -1 & -2 & -3 \end{pmatrix}$$

$$= \begin{pmatrix} 11 \cdot 1 + 12 \cdot 3 + 13 \cdot -1 & 11 \cdot 2 + 12 \cdot 4 + 13 \cdot -2 & \cdots \\ 21 \cdot 1 + 22 \cdot 3 + 23 \cdot -1 & 21 \cdot 2 + 22 \cdot 4 + 23 \cdot -2 & \cdots \end{pmatrix}$$

$$= \begin{pmatrix} 34 & 44 & 54 \\ 64 & 84 & 104 \end{pmatrix}.$$

The size of the product is correct: a 2×3 matrix times a 3×3 matrix yields a 2×3 matrix.

The product $\mathbf{BA}$ is not defined; $\mathbf{B}$ has three columns, but $\mathbf{A}$ has only two rows. ∎

Matrix multiplication is *not commutative*: $\mathbf{AB} \neq \mathbf{BA}$. Indeed, in the previous example, $\mathbf{BA}$ is not even defined.

The **identity** matrix, denoted by $\mathbf{I}$, is a square matrix with 1s on the diagonal and 0s elsewhere; e.g., the 2×2 identity is

$$\mathbf{I} = \begin{pmatrix} 1 & 0 \\ 0 & 1 \end{pmatrix}.$$

Multiplication on the left or on the right by the identity matrix of the proper size leaves the matrix unchanged. You should verify that $\mathbf{I}_2 \mathbf{A} = \mathbf{A} \mathbf{I}_3 = \mathbf{A}$, where $\mathbf{A}$ is the 2×3 matrix defined previously and $\mathbf{I}_2$ and $\mathbf{I}_3$ are the 2×2 and 3×3 identity matrices.

If it exists, the **inverse** of a square matrix $\mathbf{A}$ is the matrix $\mathbf{A}^{-1}$ with the property that

$$\mathbf{A}^{-1}\mathbf{A} = \mathbf{A}\mathbf{A}^{-1} = \mathbf{I}.$$

A matrix that possesses an inverse is called **invertible** or **nonsingular**.

The **determinant** of a 2×2 matrix $\mathbf{A}$ is the number $\det \mathbf{A}$ calculated according to

$$\det \begin{pmatrix} a & b \\ c & d \end{pmatrix} = ad - bc.$$

Determinants of larger square matrices are computed recursively using the *minors* M_{ij} of $\mathbf{A}$; the **minor** M_{ij} of the $n \times n$ matrix $\mathbf{A}$ is the $n - 1 \times n - 1$ matrix obtained by deleting row i and column j from $\mathbf{A}$.

To compute the determinant of the $n \times n$ matrix $\mathbf{A}$ using its minors, chose a row (or column) k. Then

$$\det \mathbf{A} = \sum_{j=1}^{n} (-1)^{k+j} a_{kj} \det M_{kj} = \sum_{i=1}^{n} (-1)^{i+k} a_{ik} \det M_{ik}.$$

The first sum is the *expansion in minors along the k-th row*; the second is the expansion along the k-th column.

A proof is required to show that the value of the determinant is independent of the choice of row or column along which expansion occurs.

A matrix $\mathbf{A}$ is nonsingular if and only if $\det \mathbf{A} \neq 0$. If the entries in $\mathbf{A}^{-1}$ are denoted by c_{ij}, then

$$c_{ij} = \frac{(-1)^{i+j} M_{ji}}{\det \mathbf{A}}.$$

The determinant of a matrix is a powerful theoretical tool, and it is useful for hand calculations with tiny matrices. But the number of multiplications required to evaluate a determinant grows like $n!$, while matrices can be inverted with computational effort that grows only like n^3. To appreciate the difference, compare those two numbers when $n = 100$.

If the entries in a matrix are functions rather than numbers, all these ideas still apply. In addition, we can integrate or differentiate a matrix function by applying the appropriate operation to each of its elements.

A.6   CRAMER'S RULE

Cramer's rule uses determinants to solve systems of linear equations. To review determinants, read the concluding paragraphs of the previous section. We demonstrate Cramer's rule with a simple example.

EXAMPLE 8   *Solve*

$$2x + 3y = -4$$
$$6x + 4y = -2.$$

Cramer's rule gives the values of x and y as quotients. The denominator is the determinant of coefficients

$$D = \det \begin{pmatrix} 2 & 3 \\ 6 & 4 \end{pmatrix} = (2 \cdot 4) - (3 \cdot 6) = -10.$$

The numerator is the determinant formed by replacing the column corresponding to the desired unknown with the constants from the right-hand side of the system of equations. The numerator for the unknown x is

$$N_x = \det \begin{pmatrix} -4 & 3 \\ -2 & 4 \end{pmatrix} = (-4 \cdot 4) - (3 \cdot -2) = -10,$$

and Cramer's rule gives the value of x as

$$x = \frac{N_x}{D} = \frac{-10}{-10} = 1.$$

The numerator for the unknown y is

$$N_y = \det \begin{pmatrix} 2 & -4 \\ 6 & -2 \end{pmatrix} = (2 \cdot -2) - (-4 \cdot 6) = 20,$$

and Cramer's rule gives the value of y as

$$y = \frac{N_y}{D} = \frac{20}{-10} = -2.$$

Substituting $x = 1$, $y = -2$ into the original system confirms the accuracy of this answer. ∎

If the determinant of coefficients is *not* zero, then Cramer's rule provides the *unique* solution of the system of equations. In this case, *homogeneous* systems of linear equations, those with only zeros on the right-hand side, *can have only zero solutions*. The numerator for each unknown is zero because it is a determinant containing a column of zeros.

If the determinant of coefficients is zero, then Cramer's rule can not provide numerical values for the unknowns, for division by zero is either indeterminate or undefined. For homogeneous systems, Cramer's rule would give unknowns the indeterminate values $0/0$, so a homogeneous system whose determinant of coefficients is zero has *many* solutions. A *nonhomogeneous* system may have *many* solutions if the numerator for each unknown happens to be zero, or it may have *no* solutions if one of the numerators is not zero.

A.7 PARTIAL FRACTIONS

Partial fraction expansions reverse the common denominator process, separating a fraction with a complicated polynomial denominator into a sum of fractions with simpler denominators. Finding those simpler fractions can ease both integration and the inversion of Laplace transforms.

For example, we can easily add two fractions by finding a common denominator:

$$\frac{2}{s+1} + \frac{3}{s-2} = \frac{2(s-2)+3(s+1)}{(s+1)(s-2)} = \frac{5s-1}{s^2-s-2}.$$

A partial fraction expansion guides us in reversing that process, in going from the fraction on the right back to the simpler terms of which it is the sum.

The steps in the process are:

1. If necessary, divide the denominator into the numerator to obtain a **proper fraction**, a quotient of polynomials with the degree of the numerator *smaller* than the degree of the denominator.

2. Factor the denominator completely.

3. Write the original fraction as a sum of fractions whose denominators are the factors of the original denominator. Provide each simpler (proper) fraction with a numerator containing appropriate unknown coefficients.

> The numerator of a linear factor is a constant. The numerator of a quadratic factor with no real roots is linear, and so on.

4. Find a common denominator and sum the simpler fractions. To find the unknown coefficients, equate the numerator of the sum to the original numerator.

EXAMPLE 9 *Find a partial-fraction expansion for*

$$\frac{5s-1}{s^2-s-2}.$$

Since the degree of the numerator is 1 (from $5s$) and the degree of the denominator is 2 (from s^2), the given fraction is proper.

Factor the denominator:

$$s^2 - s - 2 = (s+1)(s-2).$$

Write the original fraction as a sum involving the factors of the denominator:

$$\frac{5s-1}{s^2-s-2} = \frac{A}{s+1} + \frac{B}{s-2}.$$

The coefficients A, B are to be determined. Note that each of the fractions on the right is proper. Because the highest power in the denominator is 1, the highest power in the numerator is 0.

Find a common denominator and sum:

$$\frac{5s-1}{s^2-s-2} = \frac{A}{s+1} + \frac{B}{s-2}$$

$$= \frac{A(s-2) + B(s+1)}{(s+1)(s-2)}$$

$$= \frac{(A+B)s + B - 2A}{s^2 - s - 2}.$$

The first and last fractions can be equal only if their numerators are equal:

$$(A+B)s + B - 2A = 5s - 1.$$

Equate like powers of s to find A, B:

$$s^0: \quad -2A + B = -1$$

$$s^1: \quad A + B = 5.$$

Solving this pair of equations by Cramer's rule or other methods yields $A = 2$, $B = 3$.

Hence, the desired partial-fraction expansion is

$$\frac{5s - 1}{s^2 - s - 2} = \frac{2}{s+1} + \frac{3}{s-2}. \qquad \blacksquare$$

If the denominator contains a repeated factor, e.g., $(s-2)^3$, then the fraction could have been formed by combining simple terms with the denominators $(s-2)^i$, $i = 1, 2, 3$, each with a constant numerator. This idea applies to repeated quadratic factors as well, but the numerators are linear. For example, we would construct the expansion

$$\frac{4 - 2s}{(s^2 + 1)(s-1)^2} = \frac{As + B}{s^2 + 1} + \frac{C}{s-1} + \frac{D}{(s-1)^2}.$$

Then A, B, C, D would be determined by finding a common denominator and equating like powers of s in the numerators.

APPENDIX EXERCISES

1. Use the fundamental theorem of calculus to find df/dx:
   (a) $f(x) = \exp\left(\int_0^x \cos \pi s \, ds\right)$
   (b) $f(x) = \exp\left(\int_{-1}^x (s+1)^3 \, ds\right)$

2. Evaluate the definite integral in each part of exercise 1, and recompute the derivative directly. Show that each answer agrees with that obtained using the fundamental theorem of calculus.

3. Use partial fractions to evaluate

$$\int_1^x \frac{dr}{ar - sr^2},$$

where a and s are constants, $0 < s < a$. Is there an upper limit on x?

4. Use the fundamental theorem of calculus

$$\frac{d}{dx} \int_a^x f(s) \, ds = f(x),$$

to write solutions of the following initial-value problems.
   (a) $y'(x) = e^{x^2}$, $y(0) = 2$
   (b) $y'(x) = \cos(\ln x)$, $y(1) = -3$

5. Find the first three nonzero terms in a Taylor expansion about $x = \pi/2$ of $\sin x$.

6. Use Taylor's theorem to estimate the error in the approximation $\sqrt{1-x} \approx 1 - x/2$, $0 \le x \le \frac{1}{2}$.

7. Find the determinant and the inverse of the matrix function

$$\mathbf{A}(x) = \begin{pmatrix} 1 & 1 \\ e^x & e^{-x} \end{pmatrix}, \, x \neq 0.$$

Confirm that $\mathbf{A}^{-1}\mathbf{A} = \mathbf{A}\mathbf{A}^{-1} = \mathbf{I}$.

8. Use Cramer's rule to solve

$$3x - 2y = 13$$

$$4x + 2y = 8.$$

9. Use Cramer's rule to determine conditions on the constant a that guarantee that the system

$$C_1 - aC_2 = 0$$

$$C_1 + C_2 = 0$$

has

(a) Exactly one solution (C_1, C_2),
(b) Infinitely many solutions.

10. Find a partial-fraction expansion for

$$\frac{4 - 2s}{(s^2 + 1)(s - 1)^2}.$$

11. Find a partial-fraction expansion for

$$\frac{s^3 - s - 2s^2 - 2}{s^4 - 1}.$$

BIBLIOGRAPHY

1. M. Abramowitz and I. A. Stegun (eds.), *Handbook of Mathematical Functions*, Dover Publishing Co., New York, 1965

2. G. L. Baker and J. P. Golub, *Chaotic Dynamics: an Introduction*, Cambridge University Press, Cambridge, 1990

3. Bureau of the Census, *Statistical Abstract of the United States*, 109th ed., Washington, D. C. 1989

4. E. A. Coddington and N. Levinson, *Theory of Ordinary Differential Equations*, McGraw-Hill, New York, 1955

5. Department of International Economic and Social Affairs, *Demographic Yearbook 1986*, United Nations, New York, 1988

6. J. Frauenthal, *Introduction to Population Modeling*, UMAP Monographs, Birkhauser, Boston, 1979

7. J. Gleick, *Chaos: Making a New Science*, Viking Penguin, Inc., New York, 1987

8. *Handbook of Chemistry and Physics*, various editions, The Chemical Rubber Publishing Co., Cleveland

9. W. Hurewicz, *Lectures on Ordinary Differential Equations*, The MIT Press, Cambridge, 1958

10. A. C. Lazer and P. J. McKenna, Large-Amplitude Periodic Oscillations in Suspension Bridges: Some New Connections with Nonlinear Analysis, *SIAM Review*, **32**(4), 537–578

11. C. C. Lin and L. A. Segel, *Mathematics Applied to Deterministic Problems in the Natural Sciences*, Macmillan Publishing Co., New York, 1974, cor-

rected reprinting by the Society for Industrial and Applied Mathematics, Philadelphia, 1988

12. J. D. Murray, *Mathematical Biology*, Springer-Verlag, New York, 1989

13. J. M. Ortega and W. G. Poole, Jr., *An Introduction to Numerical Methods for Differential Equations*, Pitman Publishing, Inc., Marshfield, Massachusetts, 1981

14. D. A. Sánchez, *Ordinary Differential Equations and Stability Theorey: An Introduction*, W. H. Freeman and Co., San Francisco, 1968

15. G. Schwartz and P. W. Bishop, *ed.*, *Moments of Discovery*, Basic Books, Inc., New York, 1958

16. G. F. Simmons, *Differential Equations with Applications and Historical Notes*, 2d ed., McGraw-Hill, Inc., New York, 1991

17. D. A. Smith, Human population growth: stability or extinction?, *Math. Magazine* **50** (1977), 186-197

18. I. Stakgold, *Green's Functions and Boundary Value Problems*, John Wiley & Sons, New York, 1979

19. B. van der Pol, On "Relaxation-Oscillations," *Phil. Mag.*, **2**(7) (1927), 978–992

20. H. F. Weinberger, *A First Course in Partial Differential Equations with Complex Variables and Transform Methods*, Blaisdell Publishing Co., Waltham, Masschusetts, 1965

ANSWERS TO SELECTED EXERCISES

CHAPTER 1

Chapter Exercises

1. Use $v(t_p) = 0$ to find $t_p = v_i/g$; t_p is the t-intercept in figure 1.3.

3. $v(0) = v_i \Rightarrow v_1(t) = -gt + v_i$; $v(t_p) = 0 \Rightarrow v_2(t) = g(t_p - t)$; $v_1(t) \equiv v_2(t)$ only if $t_p = v_i/g$. The conditions in this problem specify the slope $(v' = -g)$ and *two* points $((0, v_i)$ and $(t_p, 0))$ of the line in figure 1.3.

5. (a) Use $y' = v$; $y'' = -g \Rightarrow y(t) = -gt^2/2 + C_1 t + C_2$. You could determine C_1, C_2 from $y(0) = y_i$, $y'(0) = v_i$.

5. (b) If t_r is time to return, then $y(t_r) = y_i \Rightarrow t_r = 0, 2v_i/g$. Clearly, $t_r = 2v_i/g$; $y'(t_r) = -v_i$.
A parabola has constant second derivative; three items of data (e.g., curvature y'', y-intercept y_i, and initial slope v_i) define it.

5. (c) Using $y'(t_p) = 0$ yields $t_p = v_i/g$. Then maximum height is $y(t_p) = v_i^2/2g$; e.g., doubling the initial velocity increases the maximum height by 4.

5. (d) $y(t) = -gt^2/2 + v_i t + y_i$.

7. (a) $y = \sin t$.

(b) $w = 4t^{5/2} - 2t + 2$.

(c) $y = 4e^x - 4 + \pi$.

9. (a) Use a vertical coordinate system with up as positive; $F_g = -mg$. Since air resistance opposes velocity, $F_a = kv^2$ (acts upward against falling pellet).

9. (b) $F_t = F_g + F_a \Rightarrow mv' = -mg + kv^2$. Since shot is dropped, $v(0) = 0$.

9. (c) $v' = -g + kv^2/m > -g \Rightarrow$ slope of the v versus t curve for this model is always greater than that of model 3. Also, as v grows in magnitude, slope increases (decreases in magnitude) toward zero. See figure on page 527.

9. (d) $v' = 0 \Rightarrow -g + kv^2/m = 0 \Rightarrow v = \pm\sqrt{mg/k}$. Terminal velocity is $v_{\text{ter}} = -\sqrt{mg/k}$.

11. Draw a careful version of figure 1.3.

13. (a) $y = Ce^{-8t}$, $-\infty < t < \infty$.

13. (b) $P = Ce^{0.0102t}$, $-\infty < t < \infty$.

13. (c) $P = Ce^{kt}$, $-\infty < t < \infty$.

13. (d) $P = Ce^{0.015t} + 0.209/0.105$, $-\infty < t < \infty$.

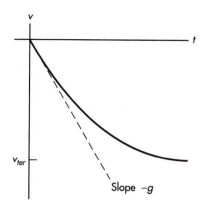

13. (e) $y = \pm(1 - Ce^{2t})^{-1/2}$, valid for t such that $1 - Ce^{2t} > 0$.

13. (f) $T = Ce^{-0.0002t} + 5$, $-\infty < t < \infty$.

13. (g) $T = e^{-Akt/cm} + T_{\text{out}}$, $-\infty < t < \infty$.

13. (h) $y = e^{2x}/2 + C$, $-\infty < x < \infty$.

13. (i) $y = \ln(1/\sqrt{C - 2x})$, $-\infty < x < C/2$.

13. (j) $u = y^3/3 - y + C$, $-\infty < y < \infty$.

15. Equations (a)–(c), and (e) have the trivial solution; the others do not.

17. See section 1.5.

19. (a) $y = y(0)e^{kt}$; clearly increasing if $y(0) > 0$ and $k > 0$. The exponential with a positive argument is increasing.

19. (b) The exponential with a negative argument is decreasing.

19. (c) If $k > 0$, then $y' = ky > 0$ whenever $y > 0$. If $y(0) > 0$, then y starts positive and remains positive as it increases.

19. (d) Reverse the preceding argument.

CHAPTER 2

Section 2.1

1. (a) Substitute in differential equation to verify solution; substitute $t = 1970$ to verify initial condition.

1. (b) $P(1988) = 246.37$ versus 246.11 in table 2.2

3. (a) (i) $P' = 0.0183P$, $P(1978) = 4,258$; (ii) $P(t) = 4,258e^{0.0183(t-1978)}$; (iii) $P(1985) = 4,840$ versus $4,837$ in table 2.1

5. $P' = -0.0012P$, $P(1985) = 16.64 \Rightarrow P(t) = 16.64e^{-0.0012(t-1985)}$; $P(t_{1/2}) = 8.32 \Rightarrow t_{1/2} - 1985 = (\ln 2)/0.0012 \Rightarrow t_{1/2} = 2563$. Using

the 1978 data with $k = -0.0007$, $t_{1/2} = 1978 + (\ln 2)/0.0007 = 2968$.

7. (c) $P(t_2) = P_i e^{kt_2} = 2P_i \Rightarrow t_2 = (\ln 2)/k$.

9. (a) $N' = -sN$ and $s, N > 0 \Rightarrow N' < 0$.

9. (b) $N = N_i e^{-st}$ is decaying.

9. (c) See 7(c) with $\frac{1}{2}$ replacing 2; $t_{1/2} = (\ln 2)/s$.

11. (a) $N' = iN$ and $i, N > 0 \Rightarrow N' > 0$.

11. (b) $N = N_i e^{it}$ is increasing.

11. (c) See 7(c); $t_2 = (\ln 2)/i$.

13. (a) $k \approx (5.30 - 3.93)/(3.93 \cdot 10) = 0.035$; a model is $P' = 0.035P$, $P(1790) = 3.93$.

13. (b) $P(t) = 3.93e^{0.035(t-1790)} \Rightarrow P(1970) = 2,140.17$, a little higher than the recorded 204.88 million! Addition of land area increased population in 1790–1800, a factor omitted from the model being fit to this data.

13. (c) $k \approx (31.51-23.26)/(23.26 \cdot 10) = 0.035$; a model is $P' = 0.035P$, $P(1850) = 23.26$. Then $P(t) = 23.26e^{0.035(t-1850)} \Rightarrow P(1970) = 1,551.12$, still much too high. Perhaps net birth rate is decreasing as population increases.

15. $P_1 = P_0 + kP_0 = (1 + k)P_0$, $P_2 = P_1 + kP_1 = (1 + k)P_1 = (1 + k)^2 P_0, \ldots, P_n = (1 + k)^n P_0$. Use $n = 10$, $k = 0.01$ or $k = 0.03$.

17. Use the argument of exercise 15 with $P_0 = P(1979)$, $P_1 = P(1980)$, etc.; e.g., the first line has slope $0.0282P_0$. See section 3.5.

19. $P(t+\Delta t) - P(t) = P(\tilde{t})$, $t < \tilde{t} < t + \Delta t \Rightarrow P' = kP$ because $\tilde{t} \to t$ as $\Delta t \to 0$.

Section 2.2

1. (a) Left side is $\ln|0.015P - .209|/0.015$, right is $t + c$.

1. (b) Multiply by 0.015 before exponentiating; use $C = e^{0.015c}$.

1. (c) Substitute $t = 1847$, $P = 8$.

3. $P' = 0 \Rightarrow 0.015P - 0.209 = 0$ or $P = 0.209/0.015 = 0.0031$ million people per year.

5. (a) Follow the derivation of the emigration model with $E = 12,000$ people per day $= 4.39$ million people per year; $k = k_b - k_d = -0.0012$ million per year (using the only available value, that for 1985). Find $P' = -0.0012P - 4.39$, $P(1989) = 49.92$ (again using a 1985 value).

5. (b) $P(t) = (49.92 + 4.39/0.0012)e^{-0.0012(t-1989)} - 4.39/0.0012 = 3,708.25e^{-0.0012(t-1989)} - 3,658.33$; $P(t) = 0 \Rightarrow t = 1989 - [\ln(3,658.33/$

3, 708.25)] /0.0012 ≈ 2,000.29. The country would have been deserted early in the year 2000.

5. (c) $P' = 0 \Rightarrow -0.0012P - 4.39 = 0$; no $P > 0$ satisfies this condition. The population was already declining because $k = -0.0012 < 0$.

7. Follow the derivation of the emigration model but use $I(t)$, the *immigration* rate, as a rate of population *increase* in place of *emigration* rate E as a rate of population loss; i.e., $+I\Delta t$ replaces $-E\Delta t$ in conservation expression. Find $P' = kP + I$, $P(t_0) = P_i$.

9. (a) $bA\Delta t$.

9. (b) $dA^2\Delta t$.

9. (c) $r(A - A_{\text{out}})\Delta t$.

9. (d) Net change equals number added less number lost $\Rightarrow A' = bA + r(A - A_{\text{out}}) - dA^2$, $A(0) = A_i$.

11. Even if interest rate I is 0, deposits increase at rate m. Reject (b), (d) because $m' = -D < 0$ when $I = 0$. Analogously, reject (a) (and (d)) because $m' = -Im < 0$ when $D = 0, m > 0$. Hence, accept model (c) as plausible.

13. $m(t) = (S + D/I)e^{It} - D/I$.

15. $P' = 0 \Rightarrow kP - E = 0 \Rightarrow P_{\text{crit}} = E/k$; $P_i > E/k \Rightarrow P'(0) > 0 \Rightarrow P(t) > E/k \Rightarrow P'(t) > 0$ for all t. Reverse inequalities for $P_i < E/k$. Alternatively, the maximum emigration rate satisfies $kP_i - E_{\max} = 0$, or $E_{\max} = kP_i$.

17. See exercise 15. In addition, E constant in $P' = kP - E \Rightarrow P'' = kP'$.

Since P'' is defined for all t and never zero unless $P' = 0$, there are no local maxima or minima. The case $P' = 0$ is the steady state solution $P \equiv E/k$. If $P_i < E/k$, the solution function can decrease to negative values.

19. See solution to exercise 19, section 2.1. If $E(t)$ is continuous, $E(\tilde{t}) \to E(t)$ as $\tilde{t} \to t$.

Section 2.3

3. Use separate terms in conservation law for rate of loss through walls and roof: $-A_w k_w(T - T_{\text{out}}) - A_r k_r(T - T_{\text{out}})$. Find $T' = -[(A_w k_w + A_r k_r)/cm] (T - T_{\text{out}})$, $T(0) = T_i$.

7. To estimate m, first estimate volume of wood, plaster, and other building materials, then use density values from a source such as [8]. Add an estimate of mass of interior furnishings.

13. Conservation of energy: $Q(t + \Delta t) - Q(t) = S\Delta t - Ak(T - T_{\text{out}})$. Divide by Δt, use $Q = cmT$, etc. to obtain $T' = -(Ak/cm)(T - T_{\text{out}}) + S/cm$.

Chapter Exercises

1. (a) Conservation of mass with zero rate of adding isotope: $y(t + \Delta t) - y(t) = 0 - ky\Delta t$. Divide by Δt, take limit as $\Delta t \to 0$ to find $y' = -ky$, $y(0) = y_i$.

1. (b) $y_i > 0 \Rightarrow y'(0) < 0 \Rightarrow y$ is decaying, as expected with radioactive decay.

1. (c) $y = y_i e^{-kt} \to 0$ as t grows, as expected.

1. (d) Use $y(t_{1/2}) = y_i/2$ in solution found in previous part to find $t_{1/2} = (\ln 2)/k$.

1. (e) y might have units of mass; k, units of inverse time.

3. (a) $v_R = Rdq/dt, v_C = q/C$.

3. (b) $v_R + v_C = 0$.

3. (c) Substitute 3(a) into 3(b); $q(0) = q_i$.

5. (a) $q' = -q/RC \Rightarrow q > 0$ decays; negative charge grows. In $q' = -q/RC + E/R$ growth or decay depends on the sign of $-q/RC + E/R$.

5. (b) $q' = -q/RC, q(0) = q_i \Rightarrow q(t) = q_i e^{-t/RC}$; all solutions decay to zero. With a constant imposed voltage E, $q' = -q/RC + E/R$, $q(0) = q_i \Rightarrow q(t) = (q_i - EC)e^{-t/RC} + EC$; all solutions decay to EC.

5. (c) Steady state is $q = EC$.

7. Nothing in the application of the conservation law hinges on the relative values of T and T_{out}; the derivation of the model is virtually identical to that in the text.

9. (a) Rate of addition of mass of chemical per unit volume due to reaction is kc; rate of loss of concentration due to dilution is d. Conservation of mass requires $c(t + \Delta t) - c(t) = kc(t)\Delta t - d\Delta t$. Usual arguments yield $c' - kc = -d$, $c(0) = c_i$. (Analogous to emigration model.)

9. (b) $c(t) = (c_i - d/k)e^{kt} + d/k$. Growth or decay of c over time depends on sign of $c_i - d/k$; i.e., on whether initial concentration is sufficiently large so that growth from reaction exceeds loss due to dilution.

9. (c) Initial-value problems (a) and (b) of exercise 8 have decaying solutions $c(t) = (c_i \mp d/k)e^{-kt} \pm d/k$; solution of (d), $c(t) = (c_i + d/k)e^{kt} - d/k$, *grows* if $c_i = 0$.

11. Reject (b), (d) because $c' \cdots = -i < 0$ means *decay* due to neutron beam; reject (c), (d) because $c' = kc > 0$ means *growth* in the absence of the neutron beam ($i = 0$). Hence, (a) is the only reasonable choice.

13. (a) Let c denote BOD concentration, k the decay constant, d the dumping rate. Conservation of BOD: $c(t + \Delta t) - c(t) = d\Delta t - kc(t)\Delta t$. Usual manipulations yield $c' = -kc + d$, $c(0) = c_i$.

13. (b) $c(t) = (c_i - d/k)e^{-kt} + d/k \to d/k$; BOD level in the lake approaches an equilibrium as rate of natural decay balances rate of addition of pollution.

13. (c) With $d = 0$, $c \to 0$; lake cleanses itself with time constant $1/k$.

15. Need $y' - gy = \cdots$ for growth, so reject (b), (d). Need $y' \cdots = -d$ for loss due to insecticide so reject (a), (b). Accept (c). (Analogous to emigration model.)

17. k is growth (positive) or decay (negative) constant; e.g., k in population models, $k = -Ak/cm$ in heat loss model; f is source (positive) or sink (negative); e.g., $f = -E$ in emigration model, $f = AkT_{out}/cm$ in heat loss model.

19. $F' = kF + A - R$, $F(0) = F_i$.

21. (a) heat loss $S\Delta t$, heat gain $-Ak(T - T_{out})$.

21. (b) heat loss through walls to cold water.

21. (c) heat loss through walls to cold outside, heat gain from stove.

CHAPTER 3

Section 3.1

1. Linear, nonhomogeneous; $f(x, y) = -9.8$.

3. Nonlinear, homogeneous; $f(x, y) = y^2 \cos \pi x$.

5. Linear, nonhomogeneous; $f(x, y) = 4e^x$.

7. Nonlinear, nonhomogeneous; $f(x, y) = -g - (k/m)y|y|$.

9. Linear, nonhomogeneous; $f(x, y) = -0.0002(y - 5)$.

11. Linear, nonhomogeneous; $f(x, y) = x^2 - 1$.

13. Linear, homogeneous; $f(x, y) = ky$.

15. Linear, nonhomogeneous; $f(x, y) = -(Ak/cm)(y - T_{out})$.

17. Linear, nonhomogeneous; $f(x, y) = -3y - 3x$.

19. Linear, homogeneous; $f(x, y) = -xy - 3y$.

21. Nonlinear, homogeneous; $f(x, y) = y^2 - y$.

23. Linear, nonhomogeneous; $f(x, y) = (\sin \pi x)/x^3$.

25. $dT/dt + (Ak/cm)T = (Ak/cm)T_{out}$. Replace $(Ak/cm)T_{out}$ with 0 to obtain homogeneous form.

27. $T_{out} = 0$, $k = 0$, $A = 0$, no values of c and m.

29. (a) $a_1 \to 1$, $a_o \to (Ak/cm)$, $r \to (Ak/cm)T_{out}$, $y \to T$, $x \to t$.

29. (b) Same as part (a) with $r = 0$.

29. (c) $T = 0 \Rightarrow 0 = (Ak/cm)T_{out} \Rightarrow$ nonhomogeneous (See Exercise 27).

29. (d) $T = 0 \Rightarrow 0 = 0 \Rightarrow$ homogeneous.

29. (e) Substitute $T = T_{out}$ into (3.3).

29. (f) Substitute $T = e^{-Akt/cm}$ into (3.4).

31. 0.

33. (a) $y' + y = 0$, linear, homogeneous; $y' + y = x$, linear, nonhomogeneous.

33. (b) Substitute $y = 0$, obtain $0 = r(x)$; have trivial solution only if $r(x) \equiv 0$.

35. (a) $P = Ce^{kt} = $ (constant)×(nontrivial solution of homogeneous).

35. (b) $P = P_i e^{k(t - t_0)}$.

37. (a) $v = -gt + C$; $v_p = -gt$, $v_h = 1$.

37. (b) $v = -gt + v_i$.

39. Use $dP_h/dt = kP_h$; $dP_p/dt = kP_p - E$.

41. (a) $(E/k)' = k(E/k) - E$.

41. (b) $P' = kP$; $(e^{kt})' = ke^{kt}$.

41. (c) $(E/k + Ce^{kt})' = k(E/k + Ce^{kt}) - E$.

43. (b) $dT_{out}/dt = -(Ak/cm)(T_{out} - T_{out}) \Rightarrow 0 = -(Ak/cm)(0)$.

43. (c) $T_h = Ce^{-Akt/cm}$.

43. (d) $(Ce^{-Akt/cm})' + (Ak/cm)(Ce^{-Akt/cm}) = 0 \Rightarrow 0 = 0$.

43. (e) $T_g = Ce^{-Akt/cm} + T_{out}$.

45. (a) Need linearity for general solution.

45. (b) $y = Ce^x + \sqrt{x}$.

45. (c) $y = Ch(x) + p(x)$.

45. (d) Insufficient information — no homogeneous solution given (Or use $y_h = e^x$.)

47. Use $y'(x) = C \exp\left(-\int_0^x (a_0(s)/a_1(s))\,ds\right)$ $(-a_0(x)/a_1(x))$.

49. $(\Rightarrow)$ $y' = f(x, y)$ homogeneous $\Rightarrow 0 = f(x, 0) = 0$. $(\Leftarrow)$ $f(x, 0) = 0$, $y \equiv 0 \Rightarrow y' = f(x, y) \Rightarrow$ homogeneous.

Section 3.2

1. $y = -1/(\sin x + C)$, $\sin x + C \neq 0$.

3. $P = P_i e^{k(t - t_0)}$ for all t.

5. $T = (T_i - T_{out})e^{-Akt/cm} + T_{out}$ for all t.

7. $P = (8 - 0.209/0.015)e^{0.015(t-1847)} + .209/0.015$ for all t.

9. $y = Ce^{2x} - \frac{1}{2}$ for all x.

11. $y = \tan(\sqrt{2}x + C)/\sqrt{2}$.

15. (a) $\mu = Ce^{x^2/2}$.

15. (b) $\mu = Ce^{\sin x}$.

15. (c) $\mu = x^2$.

15. (d) $\mu = Ce^{Akt/cm}$.

15. (e) $\mu = e^{\int g(x)dx}$.

17. (a) Substitute given expression for y.

(b) $3\cos x + 8 \neq 0, \infty \Rightarrow y \neq 0$.

(c) Find $y = (3\cos x + 1/8)^{-1/3}$ — undefined when $\cos x = -1/24$.

19. Use partial fractions for integral; $\dfrac{1}{aP - sP^2} = \dfrac{A}{P} + \dfrac{B}{a - sP}$, $A = 1/a$, $B = s/a$. See exercise 3 of the appendix.

21. (a) $P_g = Ce^{kt} + E/k$.

21. (b) $P_g = (P_i - E/k)e^{k(t-t_0)} + E/k$.

21. (c) $P_i > E/k \Rightarrow P' = kP - E > 0 \Rightarrow P$ increasing. $P_i < E/k \Rightarrow P' < 0$. Hence $P_{CR} = E/k$.

23. (a) Use $y' = C\exp\left(-\int_0^x (a_0/a_1)\,ds\right) (-a_0(x)/a_1(x))$.

23. (b) From (a), $\exp\left(-\int_0^x (a_0/a_1)\,ds\right)$ is a nontrivial solution to homogeneous equation.

23. (c) Choose $C = y_i$.

25. (a) $\dfrac{a_1}{f - a_0 y}dy = dx$.

25. (b) $y = Ce^{-a_0 x/a_1} + f/a_0$.

25. (c) Choose $f = ka_0(x)$.

Section 3.3

1. iii. $y_g = e^{x^2}\left(\int_0^x e^{-s^2}\,ds + C\right)$; **(v)** $C = 3$.

3. iii. $y_g = C - gt$; **(v)** $C = v_i$.

5. iii. $x_g = Ce^{-(\cos \pi t)/\pi} - 4$; **(v)** $C = 3e^{\cos(\pi^2/4)/\pi}$.

7. iii. $y_g = e^x(C + \int_1^x \sqrt{s}e^{-s}\,ds)$; **(v)** $C = 4/e$.

9. iii. $P_g = Ce^{kt} + (2\pi A/R)\sin 2\pi t - (kA/R)\cos 2\pi t$, where $R = 4\pi^2 + k^2$; **(v)** $C = P_i + kA/R$.

11. Use $y_h' = -Cg(x)\exp\left(-\int_a^x g(s)\,ds\right)$.

13. (b) Key: $y_h' + g(x)y_h = 0$.

15. (a) Write $y = \exp\left(-\int_a^x g(s)\,ds\right)\int_a^x f(s)\exp\left(\int_a^s g(r)dr\right)\,ds + C\exp\left(-\int_a^x g(s)\,ds\right)$. Compute

y' using the product rule and the fundamental theorem.

15. (b) First term is the particular solution and second term is the homogeneous solution.

15. (c) $-\int_a^x g(s)\,ds + \int_a^s g(r)dr = \int_x^a g(r)dr + \int_a^s g(r)dr = \int_x^s g(r)dr$.

17. (a) Idea: $y(0) = y_i \Rightarrow C = y_i$; $y_i = 0 \Rightarrow y_h \equiv 0$; $y_i \neq 0 \Rightarrow y_h \neq 0$.

17. (b) Idea: $g \geq \delta \Rightarrow$ exponential has increasingly negative argument.

19. $g(x)$ continuous $\Rightarrow$ integrable.

21. Use product rule.

23. $y_p = \pi, y_h = e^{-\sin x}$.

25. To ensure integrability, which is sufficient.

27. (a) $y_p = x^2, y_h = 1/x^2$.

27. (b) $y_g = x^2 + 5/x^2$.

29. Let $a = 1$.

31. $(\mu y)' = \mu y' + \mu' y = \mu y' + \mu g(x)y \Rightarrow \mu' = \mu g(x) \Rightarrow \ln|\mu| = \int_a^x g(s)\,ds$. Choose $a = 1 \Rightarrow \mu = \exp\left(\int_1^x g(s)\,ds\right)$.

33. $\mu = e^{b_0 x/b_1} \Rightarrow (e^{b_0 x/b_1} y)' = e^{b_0 x/b_1} f(x)/b_1 \Rightarrow y = b_1^{-1} e^{-b_0 x/b_1} \int_0^x e^{b_0 s/b_1} f(s)\,ds + Ce^{-b_0 x/b_1}$. $y(0) = y_i \Rightarrow C = y_i$.

Section 3.4

1. Unique solution for $-\infty < t < \infty$.

3. Nonlinear.

5. Unique solution for $-\infty < t < \infty$.

7. Unique solution for $-\infty < t < \infty$.

9. Unique solution for $-\infty < t < \infty$.

11. First term is particular solution and second term is homogeneous solution.

13. Derive y_h using separation of variables, $y_p = uy_h$ using variation of parameters.

15. $y(0) = \frac{1}{2} \Rightarrow$ undefined at $x = 2$.

17. Show $(u - v)' + g(x)(u - v) = 0$ using $u' + g(x)u = f$, etc.

19. $(x^{3/2})' = (3/2)x^{1/2} = (3/2)(x^{3/2})^{1/3}$, etc.

23. (a) $|y_g| = |C|\exp\left(-\int_0^x g(s)\,ds\right) \leq |C|\exp(-\delta x) \to 0$

23. (b) All nontrivial solutions have the form shown with an appropriate choice of C.

Section 3.5

1. (a) 0.03076.

1. (b) Error $= -0.09595$.

3. (a) 1.1875.

3. (b) Error $= .2185$.

5. $y = -1.0252$.

7. (a) $t = 1/k$.

7. (b) $t = \ln 2/k$; error $= (1 - \ln 2)/k$.

9. (a) $t = cm/2Ak$.

9. (b) $t = (cm/Ak)\ln 2$.

11. Exact solution is $P(t) = 115.40e^{0.0282(t-1978)}$.

n	t	P_n	Error
1	1988	147.9428	5.0521
2	1998	189.6627	13.1747
3	2008	243.1476	25.7699
4	2018	311.7152	44.8098
5	2028	399.6189	73.0545
6	2038	512.3114	114.3488
7	2048	656.7832	174.0294

13. (b) The efficient form saves one evaluation of $f(t, y)$ at each step; here, $f = 0.0282P$, saving one multiplication per step. If evaluating f requires 100 multiplications, then 100 multiplications are saved per step.

15. (a) (i) 161.8616; (iii) $P(t) = 115.40e^{0.0282(t-1978)} \Rightarrow$ 161.8718; (iv) 0.0102.

15. (b) (i) 10.0374; (iii) $T(t) = 18e^{-0.72t} + 10 \Rightarrow$ 10.0134; (iv) 0.0240.

15. (c) (i) 7.7269; (iii) $P(t) = -5.93e^{0.015(t-1847)} + 13.93 \Rightarrow 7.7269$; (iv) 0.0000.

17. $P' = aP - sP^2$, $P(0) = P_i$ with $a = 1$, $s = 0.5$, $P_i = 2.5, 1.1, 0.1$

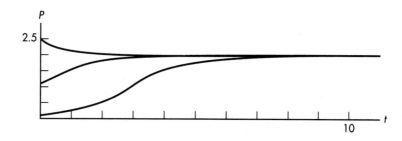

19. $y' = y^3 - y$, $y(0) = y_i$, $y_i = 1.05, 0.99, 0.3, -0.99,$
-1.05

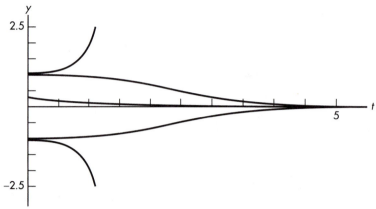

21. $y' = 1 - y^2$, $y(0) = 0$

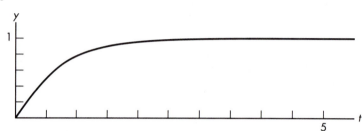

23. $y' = y^2$, $y(0) = 1$

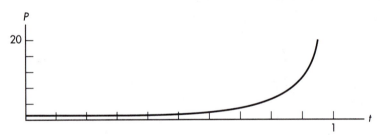

25. (b) $n = 6$: Error decreased by a factor of 4.

27. (a) Euler with $f = kP \Rightarrow P(t_{n+1}) = (1+k\Delta t)P_n = (1 + k\Delta t)^2 P_{n-1} = \cdots = (1 + k\Delta t)^{n+1} P_0.$

27. (b) Taylor expansion $\Rightarrow P(t_{n+1}) = P_0 e^{k(n+1)\Delta t} = P_0(1 + k(n+1)\Delta t + [k(n+1)\Delta t]^2/2 + \cdots$. Binomial expansion $\Rightarrow P_{n+1} = (1 + k\Delta t)^{n+1} P_0 = P_0(1 + k(n+1)\Delta t + n(n+1)(k\Delta t)^2 + \cdots$. Could also evaluate the limit of P_{n+1} as $n \to \infty$ with $n\Delta t$ fixed.

27. (c) At the end of each time step, the population is increased by the factor $1 + k$, rather than the "continuous compounding" of the actual model.

29. Since the logistic equation is nonlinear, can't obtain a simple recurrence relation as in exercise 27.

33. $P_{n+1} = P_n(1 + k\Delta t + (k\Delta t)^2/2)$ (Note resemblance to Taylor polynomial for $P_n e^{k\Delta t}$.)

Section 3.6

1. Exact solution is $y = 4 - 2e^{-x^2/2}$.

n	Δt	Error $\times 10^4$	Error/Δt
20	0.2	6.6519	3.3260×10^{-3}
40	0.1	5.3620	5.3620×10^{-3}
60	0.0667	4.1890	6.2832×10^{-3}
80	0.05	3.3951	6.7902×10^{-3}

2. Exact solution is $y = 4 - 2e^{-x^2/2}$.

n	Δt	Error $\times 10^5$	Error/Δt^2
20	0.2	63.088	1.5772×10^{-2}
40	0.1	9.6375	9.6875×10^{-3}
60	0.0667	3.8275	8.6110×10^{-3}
80	0.05	2.0475	8.1900×10^{-3}

3. Exact solution is $y = 4 - 2e^{-x^2/2}$.

n	Δt	Error $\times 10^8$	Error/Δt^4
20	0.2	1071.4	6.6961×10^{-3}
40	0.1	46.172	4.6172×10^{-3}
60	0.0667	1.5118	7.6537×10^{-4}
80	0.05	1.5118	2.4189×10^{-3}

5. See exercise 2, which considers $y' + xy = 4$, $y(0) = 2$.

7. Second-order Runge-Kutta requires two function evaluations per step, Euler, one. So using Euler with step size Δt takes about as much effort as second-order Runge-Kutta with step size $2\Delta t$. Halving Δt doubles the effort for either method, but it reduces global error in Runge-Kutta by four, offering the potential of equal accuracy with fewer steps and, hence, less effort: Compare $n = 40$ error in exercise 2 (Runge-Kutta) with $n = 80$ error in exercise 1 (Euler).

9. (a) Backwards Euler uses the slope at the *end* of the step, not the beginning as with the usual Euler method.

9. (b) Taylor expansion and $y(0) = y_0 \Rightarrow y(\Delta t) - y_1 = y(0) + y'(0)\Delta t + y''(\xi)\Delta t^2/2 - y_0 - f(t_0, y_0)\Delta t = y''(\xi)\Delta t^2/2$. (Use $y'(0) = f(t_0, y(0))$.)

9. (c) Exact solution is $y = e^{-t}$.

n	Δt	y_n	Error	Error/Δt
10	0.2	0.1615	0.0262	0.131
20	0.1	0.1486	0.0133	0.131
40	0.05	0.1420	0.0067	0.131

11. $k_1/\Delta t$ is slope at beginning of step; $k_2/\Delta t$ is slope at end of Euler step.

13. $[3(f_{yy}f^2 + 2f_{yt}f + f_{tt})/2 - d^2f/dt^2]\Delta t^3$

15. (a) Averaging slope at beginning of step with slope at end of step *as taken by this method*.

15. (b) Use $P_{n+1} = P_n + (0.0282P_n + 0.0282P_{n+1})\Delta t/2$; solve for P_{n+1}.

15. (c) Must use quadratic equation or otherwise solve *nonlinear* equation for P_{n+1}.

15. (d) Applying this method to a nonlinear differential equation adds an extra step, solving the nonlinear implicit Euler equation for P_{n+1}.

Section 3.7

1. $y = Ce^{6x}$.

3. $u = 6e^{-7(x+2)/2}$

5. $w = Ce^x$.

7. Nonhomogeneous.

9. $y = 4e^{3x}$.

11. Nonhomogeneous.

13. (a) Assume $y = e^{rx}$. Then $b_1y' + b_0y = b_1re^{rx} + b_0e^{rx} = e^{rx}(b_1r + b_0) = 0 \Rightarrow b_1r + b_0 = 0$.

13. (b) If $y = Ce^{-b_0x/b_1}$ then $y' = (-b_0/b_1)Ce^{-b_0x/b_1}$, so $b_1y' + b_0y = -b_0Ce^{-b_0x/b_1} + b_0Ce^{-b_0x/b_1} = 0$.

15. Use $\mu = e^{b_0x/b_1}$ to obtain $y = Ce^{-b_0x/b_1}$.

17. (a) Time to double is $(\ln 2)/k$. Time constant is $1/k$. See section 2.1, exercise 7(c).

17. (b) Time constant is cm/Ak; e.g., increasing area A decreases time constant, time required for house to cool.

19. $y = e^{rx} \Rightarrow r - 2 = x^3e^{-rx}$; r is not constant, contrary to assumption.

21. $y = e^{rx} \Rightarrow r + 4e^{-rx}\sin e^{rx} = 0$; r is not constant, contrary to assumption. Differential equation is homogeneous: $0' + 4\sin 0 = 0$.

Section 3.8

1. $4e^{\pi x}/(\pi - 4)$.

3. Tangent function forcing term.

5. $(4x + 1)/4$.

7. Nonconstant coefficients.

9. Homogeneous.

11. $-3(2x + 1)/4$.

13. Nonlinear.

15. Nonlinear.

17. $t - 3$.

19. (a) (1) $y_h = Ce^{4x}$; (5) $y_h = Ce^{4x}$; (11) $y_h = Ce^{2x}$; (17) $y_h = Ce^t$.

19. (b) (1) $4e^{\pi x}/(\pi - 4) + Ce^{4x}$; (3) $e^{4x}(4\int_0^x e^{-4s}\tan(\pi s)\,ds + C)$; (5) $Ce^{4x} + x + 1/4$; (11) $Ce^{2x} - 3x/2 - 3/4$; (17) $Ce^t + t - 3$.

19. (c) (1) $C = 1 - 4/(\pi - 4)$; (3) $C = 1$; (5) $C = 5/4$; (11) $C = 7/4$; (17) $C = 4$.

21. (a) $Ax^4 + Bx^3 + Cx^2 + Dx + E + Fe^{3x} + G\cos 3x + H\sin 3x$.

21. (b) $Ax^4 + Bx^3 + Cx^2 + Dx + E + Fe^{-3x} + G\cos 3x + H\sin 3x$.

21. (c) $Ax^4 + Bx^3 + Cx^2 + Dx + E + Fe^{-3x} + G\cos 3x + H\sin 3x$.

21. (d) $Ax^2 + Bx + C + D\sin 2x + E\cos 2x + F\sin(\pi x/4) + G\cos(\pi x/4)$.

21. (e) $Ax^3 + Bx^2 + Cx + D + Ee^{-3x}$.

21. (f) $Be^{kt} + C\sin kt + D\cos kt$.

23. $b_1 y' - b_0 y = 0 \Rightarrow$ characteristic equation is $b_1 r - b_0 = 0 \Rightarrow r = b_0/b_1 \Rightarrow y = Ce^{b_0/b_1}$ if $b_0 \neq 0$ and $y = C$ if $b_0 = 0$.

25. $4Axe^{4x} + Ae^{4x} = Ae^{4x} \Rightarrow e^{4x}(4Ax + A) = Ae^{4x} \Rightarrow 4A = 0, A = 1$. No value of A satisfies these two equations.

27. (a) (i) Guess $y_p = (Ax + B)e^x$; (ii) $(x/2 - 1/4)e^x$.

27. (b) (i) Guess $y_p = x(Ax + B)e^{-x}$; (ii) $x^2 e^{-x}/2$.

27. (c) (i) Guess $y_p = (A_1 x + A_0)\cos 2x + (B_1 x + B_0)\sin 2x + D_0 \cos x + E_0 \sin x$; (ii) $((10x - 4)\sin 2x + (5x + 3)\cos 2x)/25 - (\sin x - \cos x)/2$.

27. (d) (i) Guess $y_p = e^x(A\cos 2x + B\sin 2x)$; (ii) $e^x(\sin 2x + \cos 2x)/4$.

Section 3.9

1. (a) $f(x, y) = ky - y^3; \partial f/\partial y = k - 3y^2$ for all y.

1. (b) $f(x, y) = x^2 - 1 + \sin xy; \partial f/\partial y = x\cos xy$ for all y.

1. (c) $f(x, y) = ay - sy^2; \partial f/\partial y = a - 2sy$ for all y.

1. (d) $f(x, y) = ky - E(x); \partial f/\partial y = k$ for all y.

1. (e) $f(x, y) = -g - ky|y|/m$. If $y > 0, y|y| = y^2$ and $\partial f/\partial y = -2ky/m$. If $y < 0, y|y| = -y^2$ and $\partial f/\partial y = 2ky/m$. The derivative does not exist at $y = 0$.

3. $f(x, y) = y^\alpha \Rightarrow \partial f/\partial y = \alpha y^{\alpha-1}$. The derivative does not exist at $y = 0$ when $\alpha - 1 < 0$. Hence, uniqueness theorem applies when $\alpha \geq 1$.

5. $f(x, y) = -g(x)y \Rightarrow \partial f/\partial y = -g(x)$. Since the partial f_y is continuous for all x, y, the uniqueness theorem applies.

7. Multiplication by a negative factor would reverse the inequality.

9. $y = (1 - \alpha x)^{1/\alpha}; y$ is undefined at $x = 1/\alpha$.

11. $y_0 = y_i; y_1 = y_i + \int_0^x (-2y_i)ds = y_i(1 - 2x); y_2 = y_i(1 - 2x + (-2x)^2/2)$, etc. to give the Taylor series about $x = 0$ of the exact solution $y = y_i e^{-2x}$.

13. $y = 2 + \int_0^x s^2 ds = 2 + x^3/3; f(x, y) = x^2$, independent of y.

15. Successive approximation integrals begin at x_i, and intervals in x are centered around x_i.

Chapter Exercises

1. (a) $y' = y(x + y)$.

1. (b) No such equation.

1. (c) $y' + y = 0$.

1. (d) $y' + xy^2 = \cos x$.

1. (e) $y' + xy = 0$.

1. (f) $y'' + 3y' + y = 6x$.

3. $y = e^{(e^x - 1)} + e^x \int_0^x e^{-e^s} \cos s \, ds$.

5. $y = -e^{-\sin x}$.

7. $y = 3x^4$.

9. $y = x^4(3 - \int_1^x s^{-5}\cos(s^3) \, ds)$.

11. $z_g = Qx^2/2 + C$.

13. $y_g = \pi/(x + C)$.

15. $y = e^{-x^2/2} \int_0^x e^{s^2/2}\cos s \, ds$.

17. $y = 5e^{2x} - 3xe^{2x} - (2x^2 + 2x + 1)$.

19. $P = Ce^{kt} - E(\omega \sin \omega t - k\cos \omega t)/(\omega^2 + k^2)$.

21. $q = Ke^{-t/RC} + A(RC\omega \cos \omega t - \sin \omega t)$, where $A = E/(CR^2\omega^2 + 1/C)$, K arbitrary

23. $P = Ce^{at}$.

25. (a) $S(0) \neq 0 \Rightarrow S$ a nontrivial solution of homogeneous equation; $y_g = R(x) + CS(x)$.

25. (b) $C = 4$.

27. (a) $T = (T_i - T_{\text{out}})e^{-Akt/cm} + T_{\text{out}}$.

27. (b) From the solution formula given, $T(t) < T_{\text{out}}$ if $T_i < T_{\text{out}}$ and $T \to T_{\text{out}}$. Similarly, for $T_i > T_{\text{out}}$.

29. (a) $c(t) = \alpha e^{-kt} + d/k, \alpha$ an arbitrary constant.

29. (b) $c(t) = \alpha e^{-kt} - d/k, \alpha$ an arbitrary constant.

29. (c) $c(t) = \alpha e^{kt} - d/k, \alpha$ an arbitrary constant.

29. (d) $c(t) = \alpha e^{kt} + d/k, \alpha$ an arbitrary constant.

31. 1904.

33. (a) Q/k.

33. (b) $Ce^{-kt} + Q/k$.

33. (c) $C = y_i - Q/k$.

33. (d) mass $\to Q/k$.

35. (a) $(\sin \omega t - \omega CR\cos \omega t)/(CR^2\omega^2 + 1/C)$.

35. (b) $\alpha e^{-t/CR} + A(\sin \omega t - \omega CR\cos \omega t)/(CR^2\omega^2 + 1/C), \alpha$ an arbitrary constant.

35. (c) $\alpha = q_i + \omega/(R\omega^2 + 1/C^2R)$.

35. (d) Charge oscillates with frequency ω.

37. $y = \alpha - [12LmC\sin(\pi t/2)/\pi + akt^2 - (2\ell + L)mCt]/2mC, \alpha$ an arbitrary constant.

CHAPTER 4

Section 4.1

1. **(a)** All arrows slope downward with constant slope $-g$.

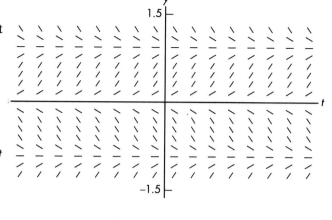

(b) Arrows would have shallower negative slope.

3. **(a–b)** See figure 4.9. Draw curves accordingly.

3. **(c)** All solution curves that start near a/s approach a/s so long term trend of population is to a/s, the population the environment can sustain.

5. $0 < u(0) < 1 \Rightarrow u'(0) < 0$ so solution decreases; $1 < u(0) \Rightarrow u'(0) > 0$ so solution increases; $u(0) = 1 \Rightarrow u(t) \equiv 1$ (a steady state).

7. **(b)** and **(c)** can be eliminated since $y < 1 \Rightarrow y' > 0$; **(a)** can be eliminated since $y'' = -2yy' \Rightarrow y'' < 0$; **(d)** is the only possible solution.

9. **(c)** and **(d)** can be eliminated since $y' > 0$ for all y; **(a)** can be eliminated since $y'' > 0$ for $y > 0$; **(b)** is the only possible solution.

11. Use the right-hand side of the differential equation to determine the sign of the first derivative. Compute the second derivative using the chain rule (e.g., $u' = u^2 - u \Rightarrow u'' = 2uu' - u'$) and find its sign.

11. (a)

11. (b)

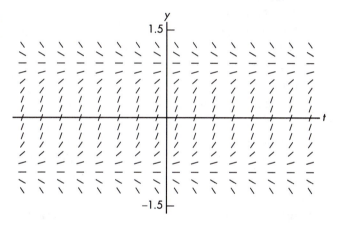

11. (c)

11. (d)

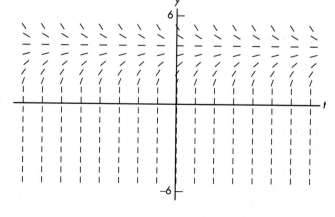

11. (e)

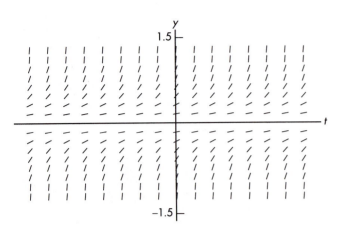

Section 4.2

1. Solution curve now increases to steady state $T = T_{\text{out}}$. Model is better described as *heat gain*.

3. Intuitively, adding insulation increases decay time. Hence, time constant cm/Ak should increase, as it does since added insulation decreases k.

5. (a) $v' = 0 \Rightarrow kv|v| = -gm \Rightarrow v_{ss} = -\sqrt{gm/k}$.

5. (b) Stable; falling rock reaches a terminal velocity v_{ss}.

7. Obtain B_s, B_c from undetermined coefficients.

9. $P(t) = 0 + p(t) \Rightarrow p' = kp$, so p grows with t; in this model, *any* population grows, no matter how small.

11. (a) $c_{ss} = i/k$.

11. (b) Solve the initial-value problem and show that $c(t) \to c_{ss}$ as $t \to \infty$.

11. (c) Rate of isotope creation by neutron beam balances natural decay of isotope.

11. (d) $c_{\text{for}} = i(1 - e^{-kt})/k$; $c_{\text{free}} = c_i e^{-kt}$.

13. (a) Substitute $0, a/s$ in logistic equation.

13. (b) $P_{ss} = 0$ is unstable; $P_{ss} = a/s$ is stable.

13. (c) Instability of $P \equiv 0$ shows any small (positive) population will grow. Stability of $P \equiv a/s$ exhibits *carrying capacity* of environment.

15. $g(x) \neq 0$ and f/g identically constant.

17. (a) $y_{ss} = h/b_o$.

17. (b) $b_1 \neq 0$, $b_0/b_1 > 0$ and $y(0) \neq h/b_0$.

17. (c) $b_1 \neq 0$ and $b_0/b_1 > 0$.

17. (d) (i) $b_1 = cm$, $b_0 = Ak$, $h = AkT_{\text{out}}$; (ii) $b_1 = 1$, $b_0 = -k$, $h = -E$; (iii) $b_1 = R$, $b_0 = 1/C$, $h = E$; (iv) $b_1 = 1$, $b_0 = k/m$, $h = -g$.

19. (a) $y_{ss} = 1$, stable, $p' = -p$, exact.

19. (b) $y_{1ss} = 1$, stable, $p' = -2p$, approximate; $y_{2ss} = -1$, unstable, $p' = 2p$, approximate.

19. (c) $y_{1ss} = 0$, unstable, $p' = p$, approximate; $y_{2ss} = -1$, unstable, $p' = 2p$, approximate; $y_{3ss} = 1$, stable, $p' = -2p$, approximate.

19. (d) $w_{1ss} = 2$, stable, $p' = -6p$, approximate; $w_{2ss} = -4$, unstable, $p' = 6p$, approximate.

19. (e) $P_{1ss} = 0$, unstable, $p' = ap$, approximate; $P_{2ss} = -\sqrt{a/s}$, stable, $p' = -2ap$, approximate; $P_{3ss} = \sqrt{a/s}$, stable, $p' = -2ap$, approximate (provided $a > 0$).

19. (f) No equilibria.

19. (g) $v_{1ss} = \sqrt{gm/k}$, unstable, $p' = 2\sqrt{gk/mp}$, approximate; $v_{2ss} = -\sqrt{gm/k}$, stable, $p' = -2\sqrt{gk/mp}$, approximate.

19. (h) $y_{ss} = 2$, unstable, $p' = p$, approximate.

21. *Flat response* is identically zero, the trivial solution; *full response* is the solution itself.

23. (a) Set $y' = 0$.

23. (b) 2; no.

23. (c) For example, $y' = y^2 - 4$.

23. (d) 0, 1, 2, 3 or 4.

Section 4.3

1. $dF/dt = -(Ak/cm)(F - F_{\text{out}})$; $F(0) = 9(T_i - 273)/5$.

3. $k(T_d - T_{\text{out}})$ is the rate of heat loss per unit area of wall or roof when the house temperature is T_d; hence, $Ak(T_d - T_{\text{out}})$ is the total rate of heat loss.

5. Integrating factor is $\mu = e^{Akt/cm}$, as for any equation of the form $T' + AkT/cm = \cdots$. See example 17 for the solution.

7. (a) $T_{ss} = H/Ak + T_m$; $T = T_m + H/Ak + (T_i - T_m - H/Ak)e^{-Akt/cm}$.

7. (b) Steady state is determined by the differential equation alone, independent of the initial conditions.

9. (a) $dT/dt = -(Ak/cm)(T - T_{\text{out}}) - S/cm$, where constant S is the rate the air conditioner absorbs heat; $T(0) = T_i$.

9. (b) $T_{ss} = T_m - S/Ak$; $S = Ak(T_m - T_d)$, where T_d is the desired temperature.

9. (c) The steady-state solution balances the rate of heat absorption S of the air conditioner with the rate of heat entry through the roof and walls.

11. (a) $\sin[\pi t/12+\phi]=\sin[\pi t/12]\cos\phi+\cos[\pi t/12]\sin\phi$, etc.

11. (c) The house

13. (a) $P_i=0\Rightarrow P(t)\equiv 0$; similarly for $P_i=a/s$.

13. (b) Yes; $P_i<0$.

13. (c) Emigration equation has exactly one equilibrium, which is unstable.

Chapter Exercises

1. $y_{\text{tran}}=3e^{-x^2}/2;\ y_{ss}=\frac{1}{2}$.

3. $y_{\text{free}}=0,\ y_{\text{for}}=xe^{-x^2}/2;\ y_{\text{tran}}=xe^{-x^2}/2$, no steady-state.

5. (a) $y_{\text{tran}}=(x+1)e^{-x}$, no (constant) steady-state; $y_{\text{free}}=e^{-x},\ y_{\text{for}}=xe^{-x}$.

5. (b) No transient, no steady-state; $y_{\text{free}}=0,\ y_{\text{for}}=(3x+1-e^{3x})/3$.

5. (c) $y_{\text{tran}}=2e^{(1-x^2)/2},\ y_{ss}=1;\ y_{\text{free}}=3e^{(1-x^2)/2},\ y_{\text{for}}=1-e^{(1-x^2)/2}$.

5. (d) $y_{\text{tran}}=(1-A/k)e^{-kx},\ y_{ss}=A/k;\ y_{\text{free}}=e^{-kx},\ y_{\text{for}}=(A/k)(1-e^{-kx})$.

7. (a) $y_{\text{free}}=3e^{-t^3/3},\ y_{\text{for}}=2-2e^{-t^3/3}$.

7. (b) $y_{\text{tran}}=e^{-t^3/3},\ y_{ss}=2$.

9. (Unstable) equilibrium solution $1,000/0.02$ is P_t; $P_i=P_t=1,000/0.02\Rightarrow P(t)\equiv P_t$.

11. (a) $k>0$.

11. (b) A sink

13. Could model a house with an air conditioner; $-S/cm$ is a heat sink, reducing T in the absence of other effects. The transient is independent of S, since the transient solves the homogeneous equation. Changing the sign of the S/cm term reduces the steady temperature below T_{out}, as desired with air conditioning.

15. (a) $a'\pm 0.000493a=\pm b$ has equilibrium a/b; $a'\pm 0.000493a=\mp b$ has equilibrium $-a/b$. Equilibrium is stable for $a'+0.00493a=\cdots$, unstable otherwise.

15. (b) Expect amount owed will increase even if $b=0$. Hence, choose (ii) since it has no positive equilibrium and solution grows for all $a>0$.

15. (c) All have steady states; only (i) and (iii) are stable.

17. (a) See 15(a).

17. (b) Expect same behavior as emigration model; insecticide acts like emigration to remove individuals from the population. Hence, choose (iv).

17. (c) All have steady states; (i) and (iii) are stable.

19. $y'=-ky,\ k>0,\ \Rightarrow y$ and y' have opposite sign; positive y decays toward zero, negative y grows toward zero. The uniqueness theorem prevents solutions with different initial values crossing. Since $y(0)=0\Rightarrow y\equiv 0,\ y(0)\neq 0\Rightarrow y(t)\neq 0,\ 0\leq t<\infty$.

CHAPTER 5

Section 5.1

1. (a) $2x''+122.5x=0,\ x(0)=-0.04,\ x'(0)=0$.

1. (b) $7x''+0.9x'+122.5x=0,\ x(0)=0,\ x'(0)=0.2$.

1. (c) $2x''+122.5x=9.335\sin(3\pi t/2),\ x(0)=-0.04,\ x'(0)=0$. (Use (5.5) with $k=122.5$ N/m, $h=3\sin(3\pi t/2)$ in $=0.0762\sin(3\pi t/2)$ m.)

3. $x=\sin(\sqrt{k/mt}),\cos(\sqrt{k/mt}),\ r=\sqrt{k/m}$.

5.

(a–b) Graph shows one period of sine function.

5. (c) Since displacement is a solution of this differential equation, both the function and the mass should have same period, $2\pi\sqrt{m/k}$.

5. (d) Mass oscillates faster with stiffer spring, faster with smaller mass. Replace springs of rapidly bouncing car with weaker springs or fill the trunk with rocks to increase period. (But *amplitude* of oscillations may change, leading to other problems.)

7. (a) $mx''(t)+px'(t)+kx(t)=k(h(t)-h(0)),\ x(0)=x_i,\ x'(0)=v_i$.

7. (b) $F_g=-mg,\ F_s=kz=k(h(t)-h(0)-x(t)+mg/k),\ F_d=-px'(t)$; Total force $F_t=F_g+F_s+F_d\Rightarrow mx''(t)+px'(t)+kx(t)=k(h(t)-h(0)),\ x(0)=x_i,\ x'(0)=v_i$.

9. $F_s=-kx$.

11. Use $h(t)=h_0+(1/4)\sin(3\pi t/2)$ ft.

13. (a) 0.5 inches above the fixed position.

13. (b) $2x''+8gx/3=(4g\sin\omega t)/3,\ x(0)=0$ inches, $x'(0)=0$.

Section 5.2

1. $L\to m,\ R\to p,\ C\to 1/k,\ E\to k(h(t)-h(0)),\ q\to x,\ q'\to x'$; for example, resistance is like damping.

3. (a) Substitute $i=\sin(t/\sqrt{LC})$ and $i=\cos(t/\sqrt{LC})$ into the equation.

3. (b) Substitute $i=C_1\sin(t/\sqrt{LC})+C_2\cos(t/\sqrt{LC})$ into the equation.

3. (c) $C_1 = v_L(0)\sqrt{C/L}$ and $C_2 = i_i$.

5. Kirchoff's voltage law $\Rightarrow V_R + V_C + E = 0$. Substituting the experimental fact yields $R(dq/dt) + q/C = 0$.

6. Following the solution of the previous problem, we add the imposed emf E to Kirchoff's voltage law to obtain $V_R + V_C + E = 0$. Substituting the experimental fact as before yields $R(dq/dt) + q/C = -E$ with initial condition $q(0) = q_i$.

7. See exercise 5 of chapter exercises in chapter 2.

8. Treat as series RC circuit as in figure 5.18 with potential E between kite and ground, resistance R in kite string, capacitance C of Leyden jar. Model is that of exercise 6. Steady state solution is $q = EC$. Measure charge in Leyden jar to estimate $E = q/C$.

Section 5.3

1. $L \to m, g \to k, \theta \to x, \theta' \to x', \theta'' \to x'', \theta_i \to x_i, \omega_i \to v_i$; for example, initial linear velocity v_i and initial angular velocity ω_i are analogous.

3. Substitute to verify solution claims.

5. (a) Accurate to within 0.001.

5. (b) 1.752 s.

7. Measuring angle ϕ in degrees, the displacement of the mass along the arc of its path is $s = L\pi\phi/180$. Then follow the derivation in the text to obtain $L\phi'' + (g180/\pi)\sin(\pi\phi/180) = 0$.

9. $L\theta'' + p\theta' + g\theta = 0, \theta(0) = \theta_i, \theta'(0) = \omega_i$.

Section 5.4

1. (a) $x_2 = x_1 - d_2 \Rightarrow F_{t1} = -k_1 x_1 + m_2 g$. Then $x_1 = x_{1u} - z_1 \Rightarrow F_{t1} = k_1 z_1 - m_1 g$. The only force on mass 1 is spring 1 plus gravity, since spring 2 is not extended.

1. (b) $x_1 = x_{1u}, x_2 = x_1 - d_2 \Rightarrow F_{t1} = -k_1 x_{1u} + m_2 g = -m_1 g = -m_1 g$. The only force on mass 1 is gravity, since neither spring is extended.

3. Divide the second equation by k_2. In the limit $k_2 \to \infty$ (an infinitely stiff spring), obtain $-x_1 + x_2 = -d_2$, as expected if the second spring can not stretch. Add the equations and use $x_1'' = x_2''$ to obtain $(m_1 + m_2)x_1'' + k_1 x_1 = 0$. The effective mass is $m_1 + m_2$, and the effective spring constant is k_1.

5. Spring constant is $k_1 k_2/(k_1 + k_2)$.

7. Use $h(t)$ is place of h in every spring force relation involving an extension of spring 1.

9. In the conservation-of-rabbits expression, add the loss term $-d_R R \Delta t$. Replace b_R with the *net* birth rate $b_R - d_R$.

Chapter Exercises

1. (a) If we chose up as the positive direction for this coordinate system and place its origin at the location of the end of the spring *before* the mass is suspended from it, then the extension z appearing in Hooke's law is just $-y$. Hence, the force exerted by the spring is $F_s = -ky$. The force of gravity on the mass remains $F_g = -mg$. Using Newton's law and $a = y''$, we then employ the usual logic: $F_t = F_s + F_g$, or $ma = -ky - mg$, or $my'' + ky = -mg$. The two initial conditions $y(0) = y_i, y'(0) = v_i$ complete this second-order model.

1. (b) Use $y(t) = x(t) - x_u$. Then substitute $y' = x'$, etc. into the differential equation for y to obtain the equation for x.

2. Same as forced spring mass system with forcing $h(t) = h_0 + b\sin(\pi St/L)$. (Car traveling at velocity S crosses S/L periods of the bumps in the road per unit time; i.e., the frequency of the forcing term is $\pi S/L$.)

3. Use Kirchhoff's current law to write $i_R + i_C + i_L = 0$. Differentiate once to obtain equation.

4. The term $g\sin\theta$ must represent the spring force for a displacement of θ. But $\sin\theta$ varies periodically with θ, while spring force does not. In particular, spring force does not decrease and then change sign with increasing displacement.

6. Similar to model in section 5.4.1 in text.

CHAPTER 6

Section 6.1

1. Linear, homogeneous, constant coefficients.

3. Linear, nonhomogeneous, constant coefficients.

5. Linear, homogeneous, constant coefficients.

7. Nonlinear, homogeneous.

9. Linear, nonhomogeneous, variable coefficients.

11. Both equations have constant coefficients.

13. Substitute given solution.

15. (a) y_{ss} satisfies only the nonhomogeneous equation. It is a particular solution. It is neither the *free* nor the *forced* response. Steady state satisfies $y(0) = y_{ss}, y'(0) = 0$.

15. (b) y_{tr} satisfies the homogeneous equation. It is not the *forced* response. It could be the *free* response if it satisfied the initial conditions.

15. (c) Use y_{tr} as homogeneous solution, y_{ss} as particular solution.

15. (d) $y_{\text{free}} = 0.02e^{-t}\sin 2t$, $y_{\text{for}} = -1.96 + e^{-t}$ $(1.96\cos 2t + .98\sin 2t)$.

15. (e) $\lim_{t\to\infty} y(t) = -1.96$ for all values of C_1, C_2.

17. The initial-value problem could represent the motion of a mass of 1 kg suspended vertically from a spring in a damping medium. The spring constant is 5 N/m. The damping coefficient is 2 N·s/m. The upper end of the spring oscillates with frequency $4/\pi$ Hz and amplitude $185/5 = 17$m(!). The mass is raised 9m above its rest position and released with downward velocity 47 m/s.

19. Saying applies to second derivatives as well. Roughly, superposition says, "a linear differential expression acting on a sum is the sum of the differential expressions."

Section 6.2

1. (a) $W = \omega \Rightarrow$ linearly independent for all x.

1. (b) $W = 3\sin x = 0$ at $x = 0, \pm\pi, \dots$. Hence, Wronskian guarantees linear independence only for $x \neq \pm n\pi$, n an integer. But we can apply the Wronskian test in any open interval between these points to evaluate $C_1 = C_2 = 0$ and conclude that these functions are linearly independent for all x.

1. (c) Linearly dependent since $W(\cos 2x, \sin(2x - \pi/2)) = 0$.

1. (d) $W = (s - r)e^{(r+s)x} \Rightarrow$ linearly independent for all x.

1. (e) $W = e^{2rx} \Rightarrow$ linearly independent for all x.

1. (f) Linearly dependent since $W(e^{rx}, e^{r(x-1)}) = 0$.

3. (a) (i) $W = -2 \Rightarrow$ linearly independent for all x; (iii) $C_1e^x + C_2e^{-x}$.

3. (b) $W = -2e \neq 0 \Rightarrow$ linearly independent; $e^{1-x} = e \cdot e^{-x}$, so this pair of functions is a constant multiple of pair in part (a). The general solution is $C_1e^x + C_2e^{1-x}$.

3. (c) $W = 0 \Rightarrow$ linearly dependent; $e^{1-x} = e \cdot e^{-x}$, so these functions are constant multiples of one another and, hence, linearly dependent.

5. $x_g = C_1x^2 + C_2x^{-2}$, $x \neq 0$; $x_g = 1 + C_1x^2 + C_2x^{-2}$, $x \neq 0$.

7. (a) Substitute, using fundamental theorem of calculus to evaluate $W'(x)$.

7. (b) Use separation of variables.

9. Show that the exponential expression for W is either identically zero or never zero.

11. Theorem 6.2 asserts that the Wronskian of smooth solutions is either never zero or always zero. Theorem 6.3 asserts linear dependence if $W = 0$ at one point. Theorem 6.1 and 6.2 together assert linear independence if $W \neq 0$ at one point.

Section 6.3

1. $y = -e^{2x} + 2e^{-x}$.

3. $y = \frac{1}{2}\left(e^{(\sqrt{2}-1)x} + e^{(-\sqrt{2}-1)x}\right)$.

5. $w = 2e^{x/2}$.

7. $y = (7x - 3)e^x$.

9. $u = (-4/11)e^{(x-1)/4} + (15/11)e^{2(1-x)/3}$.

11. $y'' - y = 0$.

13. $y'' - 2y' + y = 0$.

15. $y'' - 3y' + 2y = 0$.

17. $y'' - y' = 0$; choose second solution as a constant.

19. $W = (r_2 - r_1)e^{(r_1+r_2)x} \neq 0$ if and only if $r_1 \neq r_2$, whether r_i are real or complex.

21. Use $b_1^2 - 4b_2b_0 = 0$ to write $r = -b_1/2b_2$.

23. (a) $b_1^2 - 4b_0b_2 \neq 0 \Rightarrow$ linearly independent solutions are $\exp\left((-b_1 + \sqrt{b_1^2 - 4b_2b_0})x/2b_2\right)$, $\exp\left((-b_1 - \sqrt{b_1^2 - 4b_2b_0})x/2b_2\right)$.

23. (b) $b_1^2 - 4b_0b_2 = 0 \Rightarrow$ solutions are $\exp(-b_1x/2b_2)$, $x\exp(-b_1x/2b_2)$.

25. $u'' = 0 \Rightarrow u = C_1x + C_2$; choose $C_1 = 1, C_2 = 0$.

Section 6.4

1. (i) $r^2 - 9 = 0$; (ii) e^{3x}, e^{-3x}; (iv) $y = C_1e^{3x} + C_2e^{-3x}$.

3. (i) $r^2 + 9 = 0$; (ii) $\sin 3x, \cos 3x$; (iv) $w = C_1\sin 3x + C_2\cos 3x$.

5. (i) $r^2 - 4r + 40 = 0$; (ii) $e^{2x}\cos 6x, e^{2x}\sin 6x$; (iv) $y = e^{2x}(C_1\cos 6x + C_2\sin 6x)$.

7. (i) $r^2 + r - 6 = 0$; (ii) e^{-3t}, e^{2t}; (iv) $x = C_1e^{-3t} + C_2e^{2t}$.

9. Nonlinear.

11. Nonhomogeneous.

13. Nonlinear.

15. (i) $mr^2 + pr + k = 0$; (ii) $C_1 e^{-pt/2m}$, $C_2 te^{-pt/2m}$;
(iv) $x = C_1 e^{-pt/2m} + C_2 te^{-pt/2m}$.

17. (i) $u_g = e^{2t}(C_1 \cos 6t + C_2 \sin 6t)$; (ii) $u = e^{2t-\pi/3}(\frac{1}{2}\sin 6t - \cos 6t)$.

19. (i) $y_g = e^{-x}(C_1 \sin 2x + C_2 \cos 2x)$; (ii) $y = -2e^{\pi-x}\sin 2x$.

21. (i) $x_g = C_1 \sin 3t + C_2 \cos 3t$; (ii) $x = 3\sin 3t - 3\cos 3t$.

23. (i) $x_g = C_1 e^{3t} + C_2 e^{-3t}$; (ii) $x = 12e^{3t} + 6e^{-3t}$.

25. (a) (i) If $p^2 - 4mk > 0$, $x_g = C_1 \exp\left((-p + \sqrt{p^2 - 4mk})t/2m\right) + C_2 \exp\left((-p - \sqrt{p^2 - 4mk})t/2m\right)$. (ii) If $p^2 - 4mk = 0$, $x_g = C_1 e^{-pt/2m} + C_2 te^{-pt/2m}$. (iii) If $p^2 - 4mk < 0$, $x_g = e^{-pt/2m}(C_1 \cos \omega t + C_2 \sin \omega t)$, where $\omega = \sqrt{p^2 - 4mk}/2m$.

25. (b) e^{-rt}, $r > 0$ for all cases, converges to 0.

25. (c) It oscillates for $p^2 - 4mk < 0$. Frequency $\sqrt{4mk - p^2}/2m$ is different than that of the undamped model.

25. (d) (i) If $p^2 - 4mk > 0$, $C_1 = x_i(1 + (p - B)/2B) + mv_i/B$, $C_2 = -mv_i/B + x_i(-p + B)/2B$, where $B = \sqrt{p^2 - 4mk}$. (ii) If $p^2 - 4mk = 0$, $C_1 = x_i - (v_i + wx_i)/(w + 1)$, $C_2 = (v_i + wx_i)/(w + 1)$, where $w = p/2m$. (iii) If $p^2 - 4mk < 0$, $C_1 = 2m(v_i + px_i)/B$, $C_2 = x_i$, where $B = \sqrt{p^2 - 4mk}$.

27. (a) $C_1 e^{3.27it} + C_2 e^{-3.27it}$.

27. (b) $C_1 = -0.076i$, $C_2 = 0.076i$.

27. (c) Use values of C_1, C_2 above and $e^{\pm 3.27it} = \cos 3.27t \pm i \sin 3.27t$.

29. Use $C_1 = \frac{1}{2}$, $C_2 = 1/2i$ and Euler's formula to obtain $e^{\alpha x}\cos \beta x$ term; $C_1 = \frac{1}{2}$, $C_2 = -1/2i$ for sine term.

31. Homogeneous equation can have nonzero steady state if its characteristic polynomial has root $r = 0$; otherwise, zero is only steady state.

33. $b_1/b_2 > 0$.

Section 6.5

1. (i) $y_p = Ax + B$; (ii) $y_p = (2x - 14)/9$; (iii) $y_g = C_1 e^{3x} + C_2 e^{-3x} + y_p$.

3. (i) $y_p = Ae^{-2x}$; (ii) $y_p = -e^{-2x}/5$; (iii) $y_g = C_1 e^{3x} + C_2 e^{-3x} + y_p$.

5. (i) $y_p = A + Bxe^{-3x}$; (ii) $y_p = -4/9 + xe^{-3x}/6$; (iii) $y_g = C_1 e^{3x} + C_2 e^{-3x} + y_p$.

7. (i) $y_p = Axe^{3x} + Bxe^{-3x} + Cx + D$; (ii) $y_p = xe^{3x}/3 + xe^{-3x}/6 + 5x/9$; (iii) $y_g = C_1 e^{3x} + C_2 e^{-3x} + y_p$.

9. (i) $x_p = A\sin(t/2) + B\cos(t/2)$; (ii) $x_p = 3(\sin(t/2) - \cos(t/2))/4$; (iii) $x_g = e^{-t}(C_1 \cos(t/2) + C_2 \sin(t/2)) + x_p$.

11. (i) $w_p = x(Ax + B)\sin 3x + x(Cx + D)\cos 3x + E\sin 4x + F\cos 4x$; (ii) $w_p = (x\sin 3x)/36 - (x^2 \cos 3x)/12 - (2\cos 4x)/7$; (iii) $w_g = C_1 \cos 3x + C_2 \sin 3x + w_p$.

13. (i) $w_p = (Ax + B)e^{-x}$; (ii) $w_p = (5x + 1)e^{-x}/50$; (iii) $w_g = C_1 \cos 3x + C_2 \sin 3x + w_p$.

15. (i) $w_p = A + x(B\sin 3x + C\cos 3x)$; (ii) $w_p = \frac{2}{9} - (x\sin 3x)/6$; (iii) $w_g = C_1 \cos 3x + C_2 \sin 3x + w_p$.

17. (i) $z_p = Ate^{3t}$; (ii) $z_p = 6te^{3t}/5$; (iii) $z_g = C_1 e^{3t} + C_2 e^{t/2} + z_p$.

19. (i) $z_p = e^{t/2}(A\sin \pi t + B\cos \pi t) + Cte^{3t} + D$; (ii) $z_p = e^{t/2}(-(2/R)\sin \pi t + (5/\pi R)\cos \pi t) + 6te^{3t}/5 - 1/3$, where $R = 25 + 4\pi^2$; (iii) $z_g = C_1 e^{3t} + C_2 e^{t/2} + z_p$.

21. (i) $y_p = (Ax + B)e^{-2x}$; (ii) $y_p = xe^{-2x}/36$; (iii) $y_g = e^{-2x}(C_1 \cos 6x + C_2 \sin 6x)$.

23. (i) $x_p = At + B$; (ii) $x_p = t/2 + \frac{1}{12}$; (iii) $x_g = C_1 e^{2t} + C_2 e^{-3t} + x_p$.

25. (i) $y_p = (Ax + B)e^{-2x/3}$; (ii) $y_p = (-2x - 3)e^{-2x/3}/32$; (iii) $y_g = C_1 e^{(-2+4\sqrt{2}/3)t} + C_2 e^{(-2-4\sqrt{2}/3)t} + y_p$.

27. (i) $w_p = Axe^{2x} + Bxe^{-3x}$; (ii) $w_p = (xe^{2x} + xe^{-3x})/5$; (iii) $w_g = C_1 e^{2x} + C_2 e^{-3x} + w_p$.

29. Show that $-gt^2/2$ is a particular solution and that t and 1 are solutions of the homogeneous equation.

31. Show that $x^3 e^{-2x}$ is a particular solution and that e^{-2x}, xe^{-2x} are solutions of the homogeneous equation.

33. (a) $x_g = x_h + \dfrac{A(k - m\omega^2)}{(p\omega)^2 + (m\omega^2 - k)^2}\sin \omega t - \dfrac{Ap\omega}{(p\omega)^2 + (m\omega^2 - k)^2}\cos \omega t$, where: (i) $x_h = C_1 \exp\left((-p + \sqrt{p^2 - 4mk})t/2m\right) + C_2 \exp\left((-p - \sqrt{p^2 - 4mk})t/2m\right)$, if $p^2 > 4mk$; (ii) $x_h = (C_1 t + C_2)e^{-pt/2m}$, if $p^2 = 4mk$; (iii) $x_h = e^{-pt/2m}(C_1 \cos \lambda t + C_2 \sin \lambda t$, if $p^2 < 4mk$, where $\lambda = \sqrt{4mk - p^2}/2m$.

33. (b) No; yes, bounded for all time.

35. $y_{p1} = (3x/52 - 159/1352)\cos 2x - (15x/32 + 3/169)\sin 2x$, $y_{p2} = (25/78)\cos 3x + (5/78)\sin 3x$,

$$y_{p3} = (-x^4/20 - x^3/25 - 3x^2/125 - 6x/625)e^{-2x},$$
$$y_{p4} = (-x^3/4 - 9x^2/16 - 39x/32 - 153/128)e^{2x},$$
$$y_{p5} = x^4/6 - x^3/9 + 7x^2/18 - 13x/54 - \pi/6 + 55/324;$$
$$y_p = y_{p1} + y_{p2} + y_{p3} + y_{p4} + y_{p5}.$$

37. Direct substitution of y_{p1}, y_{p2} makes the left-hand side zero. Direct substitution shows y_{p3} does not solve the homogeneous equation; y_{p1} and y_{p2} are unsatisfactory candidates for particular solutions because they are, in fact, solutions of the homogeneous equation.

39. $y_g = (C_1 x + C_2)e^{-2x} + y_p.$

41. (a) $A \cosh px + B \sinh px.$

41. (b) $(A_n x^n + \cdots + A_1 x + A_0)\cosh px + (B_n x^n + \cdots + B_1 x + B_0)\sinh px.$

41. (c) $Ae^{qx}\cosh px + Be^{qx}\sinh px.$ All the above entries do not really add any new information, because $\cosh$ and $\sinh$ functions can be expressed in terms of exponentials.

Section 6.6

1. (a) Use direct substitution and check the initial conditions.

1. (b) $x(t) = (16/B)\sin(3\pi t/2) - (6\pi/B)\sin 4t,$
$v(t) = (24\pi/B)\cos(3\pi t/2) - (24\pi/B)\cos 4t,$
where $B = 64 - 9\pi^2.$

3. $y' = z, z' = f(t, y, z)$; Euler method : $y_{n+1} = y_n + z_n \Delta t$, $z_{n+1} = z_n + f(t_n, y_n, z_n)\Delta t$; $y_o = y(t_o)$, $z_o = z(t_o).$

5. $y' = z, z' = f(t, y, z)$; Runge-Kutta method : $k_1 = z_n \Delta t$, $l_1 = f(t_n, y_n, z_n)\Delta t$; $k_2 = (z_n + l_1/2)\Delta t$, $l_2 = f(t_n + \Delta t/2, y_n + k_1/2, z_n + l_1/2)\Delta t$; $k_3 = (z_n + l_2/2)\Delta t$, $l_3 = f(t_n + \Delta t/2, y_n + k_2/2, z_n + l_2/2)\Delta t$; $k_4 = (z_n + l_3)\Delta t$, $l_4 = f(t_n + \Delta t, y_n + k_3, z_n + l_3)\Delta t.$ $y_{n+1} = y_n + (k_1 + 2k_2 + 2k_3 + k_4)/6$, $z_{n+1} = z_n + (l_1 + 2l_2 + 2l_3 + l_4)/6.$

7. Exact solution of $x' = v, v' = -40x - 4v - \sin 6t,$ $x(0) = 0, v(0) = 3/74$, is $x(t) = (-3e^{-2x}/74)\sin 6t - (\sin 6t + 6\cos 6t)/148$; $x(2) = -0.030186.$ Using Heun with $\Delta t = 0.5$, $x(2) \approx x_4 = -0.473463$; $e_4 = 0.443277.$

9. Solving $mx'' + kx = 0$, $x(0) = 0.1, x'(0) = 0$, with $m = 0.2, k = 0.2 \cdot 9.8/0.08$; $x(t) = 0.1\cos 11.07t$; $x(1) = 0.007233.$ Using Euler with $\Delta t = 0.25$, $x(1) \approx x_4 = 1.368066$; $e_4 = -1.360833.$

11. See exercise 9. Using Runge-Kutta with $\Delta t = 0.25$, $x(1) \approx x_4 = -0.015825$; $e_4 = 0.023058.$

13. Exact solution is $x(t) = \cos \pi t + (\sin \pi t)/\pi$; $x(2) = 1.$

n	Δt	Error	Error/Δt
10	0.2	−2.078721	−10.393605
20	0.1	−1.356142	−13.561420
40	0.05	−0.599732	−11.994640

15. See exercise 13.

n	Δt	Error	Error/Δt^4
10	0.2	0.006312	3.945000
20	0.1	0.000289	2.890000
40	0.05	0.000014	2.240000

17. k_1 is slope of y at beginning of step; k_2 is slope of y at midpoint of Euler step; k_3 is slope at midpoint of Euler step using slopes k_2 and l_2; k_4 is slope after a full step using slopes k_3 and l_3. Similarly, interpret l_i as slope of z.

19. Yes. Rising fox population cuts off rising rabbit population.

Chapter Exercises

1. A particular solution is any function that satisfies the differential equation, whether it is constant or otherwise.

3. (a) $W = x^2 \Rightarrow$ linearly independent for $x \neq 0.$

3. (b) $W = -2 \Rightarrow$ linearly independent for all $x.$

3. (c) Linearly dependent since $W(\ln|x|, \ln(x^2)) = 0.$

5. (a) $W = x^3 \Rightarrow$ linearly independent for $x \leq -2.$

5. (b) $W = 2x^3 \Rightarrow$ linearly independent for $x \neq 0$; $x(1-x)(1+x) = x - x^3$, so this pair is a linear combination of pair in part (a).

5. (c) Same as 5(b); $y_g = C_1 x + C_2 x(4 + x^2).$

7. $y = C_1 e^{3x} + C_2 e^{-3x} - xe^{-3x}/6.$

9. $y = e^{2x}(C_1 \sin 6x + C_2 \cos 6x) - e^{-2x}/52 + e^{2x}/36.$

11. $z = (3 + \pi/36)\sin 3t + (\pi^2/12 - 3)\cos 3t + (t\sin 3t)/36 - (t^2 \cos 3t)/12.$

13. $w = C_1 e^{3t} + C_2 e^{t/2} - 1/3.$

15. (a) Need pure imaginary characteristic roots: $\alpha = 0$, $\beta > 0.$

15. (b) Need characteristic roots with negative real part: either $\alpha > 0, \alpha^2 - 4\beta < 0$ (complex roots) or $\alpha > 0, \alpha^2/4 \geq \beta > 0$ (negative real roots).

15. (c) Need real characteristic roots: $\alpha^2/4 - \beta \geq 0.$

15. (d) Need characteristic roots with positive real part: $\alpha < 0, \alpha^2 - 4\beta < 0$ (complex roots) or $\alpha < 0$, $\beta > 0, \alpha^2 - 4\beta > 0$ (positive real roots).

17. (a) Need pure imaginary characteristic roots: $a^2 - 4b < 0$.

17. (b) Forcing term has period π, so we need characteristic roots with negative real part; see 15(b).

17. (c) Need homogeneous solution with frequency $2\pi/4 = \pi/2$, so we require pure imaginary characteristic roots with imaginary part $\pi/2$: $a = 0, b = (\pi/2)^2$.

17. (d) Need undamped system in resonance with forcing frequency 2, so we require pure imaginary characteristic roots with imaginary part 2: $a = 0, b = 4$.

19. $y_g = C_1 + C_2 e^{2x} + C_3 e^{-2x}$

20. $y_g = C_1 e^x + C_2 e^{-x} + C_3 x e^{-x}$

21. $y_g = C_1 e^{3x} + C_2 \cos 2x + C_3 \sin 3x$

22. $y_g = C_1 e^{2x} + C_2 e^{-2x} + C_3 \cos 2x + C_4 \sin 2x$

23. $y_g = e^{4x} + C_1 + C_2 e^{2x} + C_3 e^{-2x}$

24. $y_g = 2\cos x + C_1 e^{2x} + C_2 e^{-2x} + C_3 \cos 2x + C_4 \sin 2x$

CHAPTER 7

Section 7.1

1. (a) $x(t) = 0.06 \sin(6.39t + \pi/2)$.

1. (b) $x(t) = 0.07 \sin(3.5t + \pi/2)$.

3. (a) No.

3. (b) No (initial position must be specified).

5. (a) $x_g = C_1 \cos \omega_n t + C_2 \sin \omega_n t$; so regardless of the values of arbitrary constants C_1, C_2 (determined by the initial conditions), the solution will be the sum of two simple periodic functions.

5. (b) $T = 2\pi/\omega_n = 2\pi/\sqrt{k/m}$.

5. (c) When $x_i = v_i = 0$, only the trivial solution exists.

7. (a) The system will oscillate slower, provided it is underdamped.

7. (b) $k \le 4$.

9. (a) Roots of the characteristic equation are real and distinct. Verify by showing $p^2 - 4mk > 0$.

9. (b) No.

9. (c) Yes; use analysis similar to part (b).

9. (d) For example, $x_i = -0.75$, $v_i = 0$, or $x_i = 0$, $v_i = -0.25$.

9. (e) For $x_i > 0$, impossible; for $x_i < 0$, any v_i would do; for $x_i = 0$, v_i must be negative.

11. $r < 0 \Rightarrow e^{rt}, te^{rt} \to 0$ (Use l'Hôpital's rule on the second exponential.)

13. $A \sin(\omega t + \phi) = A \sin \omega t \cos \phi + A \cos \omega t \sin \phi \Rightarrow C_1 = \sin \phi, C_2 = \cos \phi$. To obtain just $\cos \omega t$, choose $\phi = \pm \pi/2, \ldots$ to force $C_2 = 0$, etc.

15. Use $A \sin(\omega t + \phi) = A \sin \omega t \cos \phi + A \cos \omega t \sin \phi = C_1 \cos \omega t + C_2 \sin \omega t$ to find A in terms of C_1 and C_2.

17. $t = \phi T/2\pi$, where T is the period.

19. $3.76 = p/2m = (1.41)16/6$, $2.68 = \omega_n = \sqrt{4mk - p^2}(16)/6$. Use formula for the sine of a sum of angles to show equivalence of solutions; $A = 1.3$, $\phi = 35.3°$.

21. $A = \sqrt{C_1^2 + C_2^2}, \phi = \tan^{-1}(C_1/C_2)$.

Section 7.2

1. (a) A spring-mass system with mass of 2 gm (say), a damping coefficient of 3 dyne/cm-s and spring constant of 1 dyne/cm; it starts 1 cm above the equilibrium point with initial velocity of 2 cm/s downwards.

1. (b) A spring-mass system, similar to part (a), with the top of the spring oscillating with frequency ω and amplitude 5 cm.

3. (a) ω_{res} does not depend on x_i.

3. (b) ω_{res} does not depend on v_i.

5. $\omega_{\text{res}} = 3.5$ rad/s.

7. The form of the solution at resonance does not change.

11. $x(t) = -\dfrac{2F\omega}{16 - \omega^2} \sin 4t + \dfrac{8F}{16 - \omega^2} \sin \omega t$.

13. (a) $k > 2.65 \approx \frac{8}{3}$.

13. (b) $A = F/D$, where $D = \sqrt{(1.41\omega)^2 + (k - 3\omega^2/16)^2}$.

13. (c) Set $A' = 0$ to find the maximal ω_{for}. Then show that $A'' < 0$ at this maximal value of ω for $k > \frac{8}{3}$. The resonant frequency is $\omega_{\text{res}} = 8\sqrt{3k/4 - 4(1.41^2)}/3$. The system is overdamped for $k < \frac{8}{3}$.

17. An overdamped system does not oscillate.

Section 7.3

1. (i) Underdamped: $R^2 - 4L/C < 0$; (ii) critically damped: $R^2 - 4L/C = 0$; (iii) overdamped: $R^2 - 4L/C > 0$.

3. $C \to m, 1/R \to p, 1/L \to k, v \to x, v' \to x', v'' \to x''$.

5. $C \to L$, $1/R \to R$, $1/L \to 1/C$, $v \to q$, $v' \to q' = i$, $v'' \to i'$.

7. (i) underdamped: $1/R^2 - 4C/L < 0$; (ii) critically damped: $1/R^2 - 4C/L = 0$; (iii) overdamped: $1/R^2 - 4C/L > 0$.

9. $\phi = \tan^{-1} \dfrac{-1/R}{1/L - C\omega^2}$.

11. (a) $t = 4.6 \ \mu s$.

11. (b) $t = \ln(1/\delta) \ \mu s$.

11. (c) $t = (L \ln(1/\delta))/3$ s.

15. Yes.

17. $A = 7.6 \times 10^{-13}$ V; it agrees with the periodic response graph; R, L, C, ω and V affect the amplitude. Doubling V results in doubling the amplitude.

Section 7.4

1. They are both homogeneous.

3. $\theta(t) = \theta_i \cos(\sqrt{g/L}t) + \omega_i \sqrt{g/L} \sin(\sqrt{g/L}t)$. Frequency is the magnitude of the imaginary part of the solution of the characteristic equation, and it is independent of the initial conditions.

5. Characteristic equation $r^2 L e^{rt} + g \sin(e^{rt}) = 0 \Rightarrow r$ is not constant.

7. (a) (i) The natural frequency $\omega_n = \sqrt{g/L}$ is independent of initial position θ_i and initial speed ω_i. (ii) The amplitude of the response ($A = \sqrt{\theta_i^2 + \omega_i^2 L/g}$) depends on θ_i, ω_i, L and g.

7. (b) The pendulum here has initial speed ω_i.

7. (c) Numerical investigation is needed.

7. (d) Numerical investigation is needed.

9. Period $T = 2\pi\sqrt{L/g}$ increases on the moon. Amplitude $A = \sqrt{\theta_i^2 + \omega_i^2 L/g}$ also increases. For large oscillations, a nonlinear model must be used.

11. (a) $T = 2\pi\sqrt{L/g} = 2$ agrees with period values from graph.

11. (b) A pendulum with pendulum arm length of 0.993 units. θ_i must be *small*.

13. (a) Approximation is reasonably good.

13. (b) No difference in period.

13. (c) Yes.

15. $\theta_{ss} = -\pi$ is unstable; this is consistent with our intuition.

17. $\theta_{ss} = 0$, stable.

19. p could represent friction in the pivot or air resis-

tance. Since $F = ma = mL\theta''$, in the form $L\theta'' + p\theta' + \cdots$, p would involve m^{-1}.

21. (a) $p_g = C_1 \cos(\sqrt{g/Lt}) + C_2 \sin(\sqrt{g/Lt})$; $p_g = C_1 e^{\sqrt{g/Lt}} + C_2 e^{-\sqrt{g/Lt}}$.

21. (b) Solutions of the first equation are always bounded because sine and cosine are always bounded.

21. (c) The term $C_1 e^{\sqrt{g/Lt}}$ guarantees exponential growth unless $C_1 = 0$, which happens if initial conditions are such that $p'(0) = -\sqrt{g/L}p(0)$.

23. $x_{1ss} = 1$, neutrally stable; $x_{2ss} = -1$, unstable.

Section 7.5

1. Substitute $\omega = \theta'$.

3. (a) $(\theta_{ss}, \omega_{ss}) = ((0, \pm\pi, \pm 2\pi, ...), 0)$.

3. (b) $(\theta_{ss}, \omega_{ss}) = ((0, \pm\pi, \pm 2\pi, ...), 0)$.

3. (c) $(\theta_{ss}, \omega_{ss}) = (0, 0)$.

3. (d) $(x_{ss}, y_{ss}) = (0, 0)$.

3. (e) $(F_{ss}, R_{ss}) = (0, 0), (0, d_F/\alpha), (b_R/\beta, 0), (b_R/\beta, d_F/\alpha)$.

5. (a) $\theta = \theta_i \cos(\sqrt{g/Lt})$, $\omega = -\sqrt{g/L}\theta_i \sin(\sqrt{g/Lt})$.

5. (b) t increases clockwise.

5. (c) An ellipse with equation $\theta^2/\theta_i^2 + \omega^2/(g\theta_i^2/L) = 1$.

5. (d) Amplitude $A = \theta_i$.

5. (e) The trajectories are similar.

7. (a) $\theta = \theta_i \cos\sqrt{g/Lt} + \omega_i \sqrt{L/g} \sin\sqrt{g/Lt}$.

7. (b) t increases clockwise.

7. (c) An ellipse, $g\theta^2/L + \omega^2 = g\theta_i^2/L + \omega_i^2$.

7. (d) Amplitude $A = \sqrt{\theta_i^2 + \omega_i^2 L/g}$. (It always exists.)

7. (e) The trajectories are similar.

9. (a) It follows the same trajectory in the same direction (clockwise).

9. (b) The velocity is positive; the pendulum will be swinging to the right. The trajectory is the same with a clockwise direction.

9. (c) The velocity is negative; the pendulum will be swinging to the left. The trajectory is the same with a clockwise direction.

11. (a) $x' = v$, $mv' = -kx$.

11. (b) The periodic solutions are ellipses around the origin (stationary point (0,0)). If the initial values are *large*, the spring breaks.

13. Clockwise.

15. A ellipse around -2π in a clockwise direction.

17. A ellipse around 2π in a clockwise direction.

19. Clockwise.

21. Trajectories decay toward the origin until they are close enough to begin a spiral around a stable equilibrium point.

Chapter Exercises

1. $t = 0.46$ s.

3. The system is damped; thus, the solution decays with time—i.e. $x \to 0$.

5. $x'_i = 0.83$ m/s; yes.

7. $x'(0) = 0$.

9. Undamped system has amplitude growing linearly with time (until the spring breaks). Damped system has maximum, but finite, amplitude at resonant frequency.

11. (a) $x_p = (k - m\omega^2)(F/D^2) \sin \omega t - p\omega(F/D^2) \cos \omega t$. use the formula for the sine of a sum of angles to show $A = F/D$.

11. (b) Since $A = F/D$, A is maximum (at resonance) when D is minimum: $D_{\min} = \sqrt{kp^2/m - p^4/4m^2}$.

11. (c) As $p \to 0$, $\omega_r \to \omega_n = \sqrt{k/m}$; $(\omega_r < \omega_n)$.

11. (d) $\omega_r = 3.77$ rad/s.

13. $\omega_{\max} = \sqrt{1/CL - R^2/2L}$.

15. Use $\sin(\pm 2n\pi + p) = \sin p \approx p$, $\sin(\pm(2n-1)\pi + p) = -\sin p \approx -p$, $n = 1, 2, \ldots$.

17. Two trajectories of an autonomous system cannot cross unless they are identical.

CHAPTER 8

Section 8.1

1. (a) $y' = w$, $4w' = 16w - 2y$.

1. (b) $y' = w$, $4w' = 16w - 2y + e^{2t}$; $y(0) = 1$, $w(0) = -2$.

1. (c) $y' = w$, $4w' = w(16y^2 - 15)$.

1. (d) $x' = v$, $mv' = -pv - kx$; $x(0) = x_i$, $v(0) = v_i$.

1. (e) $x' = v$, $mv' = -kx + A \sin \omega t$.

1. (f) $q' = i$, $Ri' = -Ri - q/C + E(t)$.

1. (g) $\theta' = w$, $Lw' = -g\theta$; $\theta(0) = \theta_i$, $w(0) = w_i$.

1. (h) $\theta' = w$, $Lw' = -pw - g \sin \theta$.

3. $y' = z$, $z' = f(t, y, z)$.

5. (a) Linear, homogeneous, constant coefficients.

5. (b) Linear, nonhomogeneous, constant coefficients.

5. (c) Nonlinear, homogeneous.

5. (d) Linear, homogeneous, constant coefficients.

5. (e) Linear, nonhomogeneous, constant coefficients.

5. (f) Linear, nonhomogeneous, constant coefficients.

5. (g) Nonlinear, homogeneous.

5. (h) Nonlinear, homogeneous.

5. (i) Nonlinear, nonhomogeneous.

5. (j) Linear, nonhomogeneous, constant coefficients.

5. (k) Linear, homogeneous, constant coefficients.

7. Substitute each pair (y_1, z_1) and (y_2, z_2) into the system and show each satisfies the equations. $y'' - 16y = 0$ is equivalent to the given first-order system; therefore, y_1 and y_2 are two solutions of this equation.

9. ($\Rightarrow$) If the system is homogeneous, by definition $(0, 0)$ is a solution of it. Therefore, substituting $(0, 0)$ into it, we obtain $F = 0$, $G = 0$. ($\Leftarrow$) If $G = F = 0$, the system becomes $y' = ky + lz$ and $z' = py + qz$. Obviously $(0, 0)$ is a solution.

11. (i) Substitute each of (y_1, z_1), (y_2, z_2) and (y_3, z_3) into the system and show they satisfy both equations. **(ii)** Pairs $((a), (c))$ and $((b), (c))$ are linearly independent. **(iii)** $(y_{1g}, z_{1g}) = (C_1 y_1 + C_2 y_3, C_1 z_1 + C_2 z_3)$; $(y_{2g}, z_{2g}) = (C_3 y_2 + C_4 y_3, C_3 z_2 + C_4 z_3)$.

13. (a) Substitute $(C_1 y_1 + C_2 y_2, C_1 z_1 + C_2 z_2)$ into the system.

13. (b) Yes.

17. (a) $x_1 = \cos 2t$.

17. (b) $x' = y$, $y' = -4x$; $(x_1, y_1) = (\cos 2t, -2 \sin 2t)$.

17. (c) $x_g = C_1 \cos 2t + C_2 \sin 2t$.

17. (d) $(x_g, y_g) = (C_1 \cos 2t + C_2 \sin 2t, -2C_1 \sin 2t + 2C_2 \cos 2t)$.

17. (e) $W(t) = 2 \neq 0$.

17. (f) $x_p = 2t$.

17. (g) $x' = y$, $y' = -4x + 8t$; $(x_p, y_p) = (2t, 2)$.

17. (h) $(x_g, y_g) = (C_1 \cos 2t + C_2 \sin 2t + 2t, -2C_1 \sin 2t + 2C_2 \cos 2t + 2)$.

19. $C_1 y_1 + C_2 y_2 = 0$, $C_1 z_1 + C_2 z_2 = 0$. Cramer $\Rightarrow C_1 = 0/W(t) = 0$ if $W \neq 0$. Same for C_2.

21. (a) Using the initial conditions, $C_1 y_1 + C_2 y_2 = 0 \Rightarrow C_1 = 0$ and $C_1 z_1 + C_2 z_2 = 0 \Rightarrow C_2 = 0$. Therefore, the two solutions are linearly independent by definition.

21. (b) Use determinant : $y_1(0)z_2(0) - y_2(0)z_1(0) \neq 0$.

23. *Free response* is the solution of the corresponding homogeneous system : $y' = k_1 y + k_2 z$, $z' = \ell_1 y + \ell_2 z$; $y(0) = y_i$, $z(0) = z_i$. *Forced response* is the solution to the system with zero initial conditions : $y' = k_1 y + k_2 z + F$, $z' = \ell_1 y + \ell_2 z + G$; $y(0) = 0$, $z(0) = 0$.

25. (a) Introduce $y_1 = u$, $y_2 = u'$, $y_3 = u''$ and $y_4 = u'''$ to obtain the first-order system: $b_4 y_4' + b_3 y_4 + b_2 y_3 + b_1 y_2 + b_0 y_1 = h(t)$, $y_3' = y_4$, $y_2' = y_3$, $y_1' = y_2$.

25. (b) $y_4' = f(t, y_1, y_2, y_3, y_4)$, $y_3' = y_4$, $y_2' = y_3$, $y_1' = y_2$.

25. (c) Equation in part (b) includes equation in part (a) as a special case (f linear).

Section 8.2

1. (i) $(y_g, z_g) = C_1(e^{2t} \cos 4t, -e^{2t} \sin 4t) + C_2(e^{2t} \sin 4t, e^{2t} \cos 4t)$; (ii) $C_1 = 2$, $C_2 = -2$.

3. (i) $(y_g, z_g) = C_1(e^{2\sqrt{7}t}, (\sqrt{7}/2 - 1)e^{2\sqrt{7}t}) + C_2(e^{-2\sqrt{7}t}, (-\sqrt{7}/2 - 1)e^{-2\sqrt{7}t})$; (ii) $C_1 = 2 + 5\sqrt{7}/7$, $C_2 = 2 - 5\sqrt{7}/7$.

5. (i) $(y_g, z_g) = C_1(e^{-3t} \cos \omega t, (-2/5)e^{-3t} \cos \omega t + (1/5)e^{-3t} \sin \omega t) + C_2(e^{-3t} \sin \omega t, (\sqrt{6}/5)e^{-3t} \cos \omega t - (2/5)e^{-3t} \sin \omega t)$, where $\omega = \sqrt{6}/5$; (ii) $(y, z) = (0, 0)$.

7. (i) $(y_g, z_g) = C_1(e^t, \sqrt{3}e^t/3) + C_2(e^{-t}, -\sqrt{3}e^{-t}/3)$; (ii) $C_1 = 3 + 2\sqrt{3}$, $C_2 = 3 - 2\sqrt{3}$.

9. (i) $(x_g, v_g) = C_1(\cos \sqrt{k/mt}, -\sqrt{k/m} \sin \sqrt{k/mt}) + C_2(\sin \sqrt{k/mt}, \sqrt{k/m} \cos \sqrt{k/mt})$; (ii) $C_1 = x_i$, $C_2 = 0$.

11. (1) If $p^2 > 4mk$, (i) $(x_g, v_g) = C_1(e^{Qt}, Qe^{Qt}) + C_2(e^{Rt}, Re^{Rt})$, where $Q = (-p + \sqrt{p^2 - 4mk})/2m$ and $R = (-p - \sqrt{p^2 - 4mk})/2m$; (ii) $C_1 = (1/2 - p/2\sqrt{p^2 - 4mk})x_i$, $C_2 = (1/2 + p/2\sqrt{p^2 - 4mk})x_i$. (2) If $p^2 = 4mk$, (i) $(x_g, v_g) = C_1(e^{-pt/2m}, (-p/2m)e^{-pt/2m}) + C_2(te^{-pt/2m}, (-p/2m)te^{-pt/2m} + e^{-pt/2m})$; (ii) $C_1 = x_i$, $C_2 = (p/2m)x_i$. (3) If $p^2 < 4mk$, (i) $(x_g, v_g) = C_1(e^{-dt} \cos \omega t, -de^{-dt} \cos \omega t - \omega e^{-dt} \sin \omega t) + C_2(e^{-dt} \sin \omega t, -de^{-dt} \sin \omega t + \omega e^{-dt} \cos \omega t)$; (ii) $C_1 = x_i$, $C_2 = dx_i/\omega$, where $\omega = \sqrt{4mk - p^2}/2m$ and $d = p/2m$.

13. Show that $W(t) \neq 0$ for all t.

15. Show that $W(t) \neq 0$ for all t.

17. Follow similar steps, as in example 5 for $r = -4i$.

19. Determinant of coefficients $= (a - r)(d - r) - bc$. Cramer $\Rightarrow p, q = 0/((a - r)(d - r) - bc)$. Hence p, $q \neq 0$ only if $(a - r)(d - r) - bc = 0$.

21. (a) $W(t) = (p_1 q_2 - p_2 q_1)e^{(r_1 + r_2)t}$. For linearly independent solutions : $p_1 q_2 - p_2 q_1 \neq 0$. Use $p_i/q_i = b/(r_i - a)$, $r_1 \neq r_2$, to show we are sure the condition are satisfied.

21. (b) $\text{Re}(y, z) = (e^{\alpha t}(P_1 \cos \beta t - P_2 \sin \beta t), e^{\alpha t}(Q_1 \cos \beta t - Q_2 \sin \beta t)$; $\text{Im}(y, z) = (e^{\alpha t}(P_2 \cos \beta t + P_1 \sin \beta t), e^{\alpha t}(Q_2 \cos \beta t + Q_1 \sin \beta t)$. $W(t) = P_1 Q_2 - P_2 Q_1 \neq 0$ for linearly independent solutions. To show the condition is satisfied, use the same argument as in part (a).

23. $z' = -2z \Rightarrow z = Ce^{-2t} \Rightarrow y' - y = 3z = 3Ce^{-2t} \Rightarrow y = Ce^{-2t} + De^t$.

Section 8.3

1. (i) $(e^{2t}, 3e^{2t})$, $(e^{-3t}, e^{-3t}/2)$; (ii) $u_1 = (3t/2 + 19/4)e^{-2t}$, $u_2 = (6t + 2/3)e^{3t}$; (iii) $(y_p, z_p) = ((90t + 65)/12, (90t + 175)/12)$; (iv) $(y_g, z_g) = C_1(e^{2t}, 3e^{2t}) + C_2(e^{-3t}, (1/2)e^{-3t}) + (y_p, z_p)$; (v) $C_1 = -15/4$, $C_2 = 4/3$.

3. (i) $(e^{2t} \cos 4t, -e^{2t} \sin 4t)$, $(e^{2t} \sin 4t, e^{2t} \cos 4t)$; (ii) $u_1 = e^{-2t} \cos 4t$, $u_2 = e^{-2t} \sin 4t$; (iii) $(y_p, z_p) = (1, 0)$; (iv) $(y_g, z_g) = C_1(e^{2t} \cos 4t, -e^{2t} \sin 4t) + C_2(e^{2t} \sin 4t, e^{2t} \cos 4t) + (1, 0)$; (v) $C_1 = 1$, $C_2 = -2$.

5. (i) $(e^{4t}, \sqrt{3}e^{4t}/3)$, $(e^{-2t}, -\sqrt{3}e^{-2t}/3)$; (ii) $u_1 = (-24 + 3\sqrt{3} + 12\sqrt{3}t)e^{-4t}/32$, $u_2 = (6\sqrt{3}t + 4 - \sqrt{3})e^{6t}/8$; (iii) $(y_p, z_p) = ((36\sqrt{3}t - 9\sqrt{3} - 8)/32, -(60t + 24\sqrt{3} - 39)/32)$; (iv) $(y_g, z_g) = C_1(e^{4t}, \sqrt{3}e^{4t}/3) + C_2(e^{-2t}, -\sqrt{3}e^{-2t}) + (y_p, z_p)$; (v) $C_1 = 21/4 - 3\sqrt{3}/32$, $C_2 = 3\sqrt{3}/8 + 1$.

7. Let $\omega = \sqrt{6}/2$. (i) $(e^{-3t} \cos \omega t, (-2e^{-3t}/5) \cos \omega t + (\sqrt{6}e^{-3t}/5) \sin \omega t)$, $(e^{-3t} \sin \omega t, (-\sqrt{6}e^{-3t}/5) \cos \omega t - (2e^{-3t}/5) \sin \omega t)$; (ii) $u_1 = (-2\sqrt{6}e^{3t}/21) \sin \omega t - (3e^{3t}/7) \cos \omega t$, $u_2 = (3e^{3t}/7) \sin \omega t - (2\sqrt{6}e^{3t}/21) \cos \omega t$; (iii) $(y_p, z_p) = (3/7, -2/7)$; (iv) $(y_g, z_g) = C_1(e^{-3t} \cos \omega t, (-2e^{-3t}/5) \cos \omega t + (\sqrt{6}e^{-3t}/5) \sin \omega t) + C_2(e^{-3t} \sin \omega t, -(\sqrt{6}/5) \cos \omega t - (2e^{-3t}/5) \sin \omega t) + (1/7, 6/35)$; (v) $C_1 = -3/7$, $C_2 = 2\sqrt{6}/21$.

9. (i) $(\cos \sqrt{k/mt}, -\sqrt{k/m} \sin \sqrt{k/mt})$, $(\sin \sqrt{k/mt}, \sqrt{k/m} \cos \sqrt{k/mt})$; (ii) $u_1 = -F\sqrt{\dfrac{m}{k}}\left(\dfrac{\sin(\omega - \sqrt{k/m})t}{2(\omega - \sqrt{k/m})} - \dfrac{\sin(\omega + \sqrt{km})t}{2(\omega + \sqrt{k/m})}\right)$, $u_2 = F\sqrt{\dfrac{m}{k}}\left(-\dfrac{\cos(\omega + \sqrt{k/m})t}{2(\omega + \sqrt{k/m})} - \dfrac{\cos(\omega - \sqrt{k/m})t}{2(\omega - \sqrt{k/m})}\right)$; (iii) $(x_p, v_p) = (u_1 \cos \sqrt{k/mt} + u_2 \sin \sqrt{k/mt}, -\sqrt{k/m}u_1 \sin \sqrt{k/mt} + \sqrt{k/m}u_2 \cos \sqrt{k/mt})$; (iv) $(x_g, v_g) = C_1(\cos \sqrt{k/mt}, -\sqrt{k/m} \sin \sqrt{k/mt})$

$+ C_2(\sin \sqrt{k/m}t, \sqrt{k/m}\cos \sqrt{k/m}t) + (x_p, v_p);$
(v) $C_1 = 0, C_2 = \sqrt{m/k}(v_i - F\omega/(\omega^2 - k/m)).$

11. (i) $(\cos \sqrt{g/L}t, -\sqrt{g/L}\sin \sqrt{g/L}t),$
$(\sin \sqrt{g/L}t, \sqrt{g/L}\cos \sqrt{g/L}t);$ (ii) $u_1 =$
$-2\sqrt{\dfrac{L}{g}}\left(-\dfrac{\cos(\pi + \sqrt{g/L})t}{2(\pi + \sqrt{g/L})} - \dfrac{\cos(\sqrt{g/L} - \pi)t}{2(\sqrt{g/L} - \pi)}\right),$

$u_2 = 2\sqrt{\dfrac{L}{g}}\left(\dfrac{\sin(\pi + \sqrt{g/L})t}{2(\pi + \sqrt{g/L})} + \dfrac{\sin(\sqrt{g/L} - \pi)t}{2(\sqrt{g/L} - \pi)}\right);$

(iii) $(\theta_p, \omega_p) = (u_1 \cos \sqrt{g/L}t + u_2 \sin \sqrt{g/L}t,$
$-u_1\sqrt{g/L}\sin \sqrt{g/L}t + u_2 \cos \sqrt{g/L}t);$
(iv) $(\theta_g, \omega_g) = C_1(\cos \sqrt{g/L}t, -\sqrt{g/L}\sin \sqrt{g/L}t)+$
$C_2(\sin \sqrt{g/L}t, \sqrt{g/L}\cos \sqrt{g/L}t)+(\theta_p, \omega_p);$ (v) $C_1 =$
$\theta_i - 2/(g/L - \pi^2), C_2 = 0.$

15. (a) Substitute (y_g, z_g) into the system.

15. (b) $(y, z) = (\frac{85}{18} - 5t/3 + 2e^{-3t} + e^{2t}, \frac{65}{18} - 10t/3 + e^{-3t} + 3e^{2t}).$

17. $W(t) = 0$, and the only solution is the trivial solution.

Section 8.4

1. $r_1 = 3, \mathbf{p}_1 = (\ 0\ \ 1\ )^T; r_2 = 1, \mathbf{p}_2 = (\ 0\ \ 3\ )^T$
$\mathbf{Y}(t) = \begin{pmatrix} 0 & -e^{-t} \\ e^{3t} & 2e^t \end{pmatrix}$
$\mathbf{y}_g = \mathbf{Yc}; \mathbf{c} = (\ 4\ \ -4\ )^T$

3. $r_1 = 12, \mathbf{p}_1 = (\ 2\ \ 1\ )^T; r_2 = 9, \mathbf{p}_2 = (\ 1\ \ -1\ )^T$
$\mathbf{Y}(t) = \begin{pmatrix} 2e^{12t} & e^{9t} \\ e^{12t} & -e^{9t} \end{pmatrix}$
$\mathbf{y}_g = \mathbf{Yc}; \mathbf{c} = (\ 0\ \ 4\ )^T$

5. $r_i = 4 \pm 3i, \mathbf{p}_i = (\ 1\ \ \pm i\ )^T$
$\mathbf{Y}(t) = \begin{pmatrix} e^{4t}\cos 3t & e^{4t}\sin 3t \\ -e^{4t}\sin 3t & e^{4t}\cos 3t \end{pmatrix}$
$\mathbf{y}_g = \mathbf{Yc}; \mathbf{c} = (\ 4\ \ -4\ )^T$

7. multiply $\mathbf{Bp}_2$

9. $\det(\mathbf{B}-r\mathbf{I}) = (r-a)^2 \Rightarrow r = a.$ Substituting $r = a$ in $(\mathbf{B} - r\mathbf{I})\mathbf{p} = 0$ yields $0 = 0$; p, q are arbitrary. Choose $p = 1, q = 0$ for one eigenvector, $p = 0, q = 1$ for other.

11. Compute $\mathbf{Y}'$ term by term. Compute product $\mathbf{BY}$ and verify equality.

13. Find $r_1 = -3, \mathbf{p}_1 = (\ 2\ \ 1\ )^T, r_2 = 2, \mathbf{p}_2 = (\ 1\ \ 3\ )^T$; lead to same fundamental matrix as text. Eigenvectors – hence, columns of $\mathbf{Y}$ – can differ by a constant multiple; order of columns is arbitrary. Solve initial-value problem with $\mathbf{y} = \mathbf{Y}(\ 1\ \ -1\ )^T.$

15. $\mathbf{y}_p = \begin{pmatrix} -2e^{2t}/3 & -1/4 \\ e^{2t}/3 & t \end{pmatrix}$

17. Using $\mathbf{Y}(t)$ from exercise 1, $\mathbf{y}_p = \mathbf{Y}(t)\mathbf{c}+(-4e^{-2t}/3 - 6e^{2t} - 8e^t/3 + 16e^{-2t}/15\ )^T$

19. Using $\mathbf{Y}(t)$ from exercise 3, $\mathbf{y}_p = \mathbf{Y}(t)\mathbf{c} + (-2t - 277/18\ \ 11t + 95/36\ )^T$

21. $\mathbf{Yc} = c_1(\ y_1\ \ z_1\ )^T + c_2(\ y_2\ \ z_2\ )^T = (\ c_1y_1 + c_2y_2\ \ c_1z_1 + c_2z_2\ )^T$

23. Form scalar second-order equation, $b_2r^2 + b_1r + b_0 = 0.$ As system $\mathbf{y}' = \mathbf{By}$,

$$\mathbf{B} = \begin{pmatrix} 0 & 1 \\ -b_0/b_2 & -b_1/b_2 \end{pmatrix}.$$

$\det(\mathbf{B} - r\mathbf{I}) = -r(b_1/b_2 + r) - b_0/b_2 = 0 \Rightarrow b_2r^2 + b_1r + b_0 = 0.$

25. If r is an eigenvalue of $\mathbf{B}$, then the determinant of coefficients of the system of equations for the components of the eigenvector is $\det(\mathbf{B} - r\mathbf{I}) = 0.$

Chapter Exercises

1. (a) Consider the following systems: (1) $y' = f(t, y, z, w), z' = g(t, y, z, w), w' = h(t, y, z, w);$ (2) $y_i' = f_i(t, y_1, y_2, \cdots, y_n), i = 1, 2, \cdots, n.$ (i[ii]) A *solution* of a system of 3 $[n]$ first-order equations is a triple $[n$-tuple$]$ of sufficiently smooth functions $(y(t), z(t), w(t))$ $[(y_1(t), y_2(t), \cdots, y_n(t))]$ which reduce all 3 $[n]$ equations to identities when substituted for the corresponding dependent variables.

1. (b) (i[ii]) A system of 3 $[n]$ first-order equations is *homogeneous* if it has the trivial solution. Otherwise, it is *nonhomogeneous*.

1. (c) (i[ii]) A system of 3 $[n]$ first-order equations is *linear* if (f, g, h) $[(f_1, f_2, \cdots, f_n)]$ are linear in the dependent variables. Otherwise, it is *nonlinear*.

1. (d) (i[ii]) A linear system of 3 $[n]$ first-order equations has *constant*-coefficients if the coefficients in the expansions $f = k_1(t)y + k_2(t)z + k_3(t)w,$ $g = \ell_1(t)y + \ell_2(t)z + \ell_3(t)w, h = m_1(t)y + m_2(t)z + m_3w$ $[f_i = k_1^iy_1 + k_2^iy_2 + \cdots + k_n^iy_n, i = 1, 2, \cdots, n]$ are constants. Otherwise, it has *variable* coefficients.

1. (e) (i[ii]) Three $[n]$ solutions $(y_1, z_1, w_1), (y_2, z_2, w_2),$ (y_3, z_3, w_3) $[(y_1^i, y_2^i, \cdots, y_n^i), i = 1, 2, \cdots, n]$ of a first-order system of 3 $[n]$ equations are *linearly independent* on $t_0 \le t \le t_1$, if $C_1y_1 + C_2y_2 + C_3y_3 = 0, C_1z_1 + C_2z_2 + C_3z_3 = 0$ and $C_1w_1 + C_2w_2 + C_3w_3 = 0 \Rightarrow C_1 = C_2 = C_3 = 0$ $[C_1y_i^1 + C_2y_i^2 + \cdots + C_ny_i^n = 0, i = 1, 2, \cdots, n \Rightarrow$

$C_1 = C_2 = \cdots = C_n = 0$]. Otherwise, they are *linearly dependent*.

1. (f) (i[ii]) If $(y_1, z_1, w_1), (y_2, z_2, w_2), (y_3, z_3, w_3)$ $[(y_1i, y_2^i, \cdots, y_n^i), i = 1, 2, \cdots, n]$ are linearly independent solutions of a homogeneous system of 3 $[n]$ first-order equations, then the *general solution* of the homogeneous system is $(y_h, z_h, w_h) = C_1(y_1, z_1, w_1) + C_2(y_2, z_2, w_2) + C_3(y_3, z_3, w_3)$ $[(y_{1h}, y_{2h}, \cdots, y_{nh}) = \sum_{i=1}^{n} C_i(y_1^i, y_2^i, \cdots, y_n^i)]$. A given solution (y_p, z_p, w_p) $[(y_{1p}, y_{2p}, \cdots, y_{np})]$ of a nonhomogeneous linear system of 3 $[n]$ first-order equations is called a *particular solution*. A *general solution* of a nonhomogeneous system of 3 $[n]$ first-order equations is the sum of a particular solution and a general solution of the corresponding homogeneous system.

2. (i) $(y_g, z_g) = C_1(e^t, e^t) + C_2(e^{-t}, -e^{-t})$; **(ii)** $C_1 = \frac{1}{2}$, $C_2 = -\frac{1}{2}$.

3. (i) $(y_g, z_g) = C_1(e^{2t}, 3e^{2t}) + C_2(e^{-3t}, \frac{1}{2}e^{-3t})$; **(ii)** $C_1 = 1, C_2 = 2$.

4. (i) $(y_g, z_g) = C_1(-e^{-4t}, e^{4t}) + C_2(e^{-4t}, e^{-4t})$; **(ii)** $C_1 = C_2 = 1$

5. (i) $(y_g, z_g) = C_1(e^t \cos t, -e^t \sin t) + C_2(e^t \sin t, e^t \cos t)$; **(ii)** $C_1 = 1, C_2 = 2$

7. (i) $(e^t, \sqrt{3}e^t/3), (e^{-t}, -\sqrt{3}e^{-t}/3)$; **(ii)** $u_1 = (2\sqrt{3}t^2 + (4\sqrt{3} - 1)t + 4\sqrt{3} - 1)e^{-t}, u_2 = (2\sqrt{3}t^2 + (1 - 4\sqrt{3})t + 4\sqrt{3} - 1)e^t$; **(iii)** $(y_p, z_p) = 4\sqrt{3}t^2 + 8\sqrt{3} - 2, (8 - 2\sqrt{3}/3)t)$; **(iv)** $(y_g, z_g) = C_1(e^t, \sqrt{3}e^t/3) + C_2(e^{-t}, -\sqrt{3}e^{-t}/3) + (y_p, z_p)$; **(v)** $C_1 = 4 - 2\sqrt{3}$, $C_2 = 4 - 6\sqrt{3}$.

9. (i) $(e^{-2t}, -e^{-2t}), (e^{4t}, e^{4t})$; **(v)** $(y, z) = (5e^{4t} - e^{2t} + e^{-2t} + 3, 5e^{4t} - 3e^{2t} - e^{-2t} - 1)/8$

10. (i) $(e^{2t}, -e^{2t}), (-e^{3t}, 2e^{3t})$.

11. Consider the system $y' = k_1y + k_2z + k_3w + F, z' = \ell_1y + \ell_2z + \ell_3w + G, w' = m_1y + m_2z + m_3w + H$. Substitute $y_p = u_1y_1 + u_2y_2 + u_3y_3, z_p = u_1z_1 + u_2z_2 + u_3z_3$ and $w_p = u_1w_1 + u_2w_2 + u_3w_3$ into the system, to obtain $u_1'y_1 + u_2'y_2 + u_3'y_3 = F, u_1'z_1 + u_2'z_2 + u_3'z_3 = G, u_1'w_1 + u_2'w_2 + u_3'w_3 = H$. Solve for u_i'.

12. $\mathbf{Y} = \begin{pmatrix} e^{6t} & e^{4t} \\ -e^{6t} & e^{4t} \end{pmatrix}$

13. $\mathbf{y}_g = \begin{pmatrix} -e^{-t} & e^{-4t} \\ 2e^{-t} & -e^{-4t} \end{pmatrix} \mathbf{c} + \begin{pmatrix} -3t + 12 \\ 9t - 27 \end{pmatrix}$

14. $\mathbf{y} = \begin{pmatrix} -e^{-t} & -2e^{-4t} \\ 2e^{-t} & 5e^{-4t} \end{pmatrix}$

15. $\mathbf{y}_g = \begin{pmatrix} -e^{-t} & -2e^t \\ 2e^{-t} & 5e^t \end{pmatrix} \mathbf{c} + \begin{pmatrix} 2te^t + 5te^{-t} \\ -5te^t - 10te^{-t} \end{pmatrix}$

16. $\mathbf{y}_g = \begin{pmatrix} \cos 4t & \sin 4t \\ -\sin 4t & \cos 4t \end{pmatrix} \mathbf{c}$

17. $\mathbf{y}_g = \begin{pmatrix} -\cos 4t - \sin 4t & -\sin 4t + \cos 4t \\ 2\cos 4t & 2\sin 4t \end{pmatrix} \mathbf{c}$

18. $\mathbf{y}_g = \begin{pmatrix} -\cos t + \sin t & -\cos t - \sin t \\ 2\cos t & 2\sin t \end{pmatrix} \mathbf{c}$

19. $\mathbf{y}_g =$
$\begin{pmatrix} e^{-2t}(5\cos 3t - \sin 3t) & e^{-2t}(\sin 3t - 5\cos 3t) \\ 6e^{-2t}\cos 3t & 6e^{-2t}\sin 3t \end{pmatrix} \mathbf{c}$

20. $\mathbf{y}_g =$
$\begin{pmatrix} e^{-2t}(-2\cos 3t - \sin 3t) & e^{-2t}(-2\sin 3t + \cos 3t) \\ 3e^{-2t}\cos 3t & 3e^{-2t}\sin 3t \end{pmatrix} \mathbf{c}$

21. $\mathbf{y}_g =$
$\begin{pmatrix} e^{-t}(-\cos 2t - \sin 2t) & e^{-t}(\sin 2t + \cos 2t) \\ 2e^{-t}\cos 2t & 2e^{-t}\sin 2t \end{pmatrix} \mathbf{c}$

22. $\mathbf{y}_g = \begin{pmatrix} 2e^{3t} & -2e^t & e^{-4t} \\ e^{3t} & e^t & -3e^{-4t} \\ -2e^{3t} & 4e^t & 13e^{-4t} \end{pmatrix} \mathbf{c}$

23. $\mathbf{y}_g = \begin{pmatrix} 2e^{3t} & -2e^t & e^{-4t} \\ e^{3t} & e^t & -3e^{-4t} \\ -2e^{3t} & 4e^t & 13e^{-4t} \end{pmatrix} \mathbf{c} + \begin{pmatrix} -1/12 \\ 1/12 \\ 1/12 \end{pmatrix}$

24. $\mathbf{y}_g = \begin{pmatrix} 0 & -e^t & 2e^{2t} \\ e^{2t} & 2e^t & 0 \\ 0 & 0 & -e^{2t} \end{pmatrix} \mathbf{c}$

25. $\mathbf{y}_g = \begin{pmatrix} e^t & e^{3t} & e^{-4t} \\ e^t & 3e^{3t} & -4e^{-4t} \\ e^t & 9e^{3t} & 16e^{-4t} \end{pmatrix} \mathbf{c}$

26. $\mathbf{y}_g = \begin{pmatrix} e^{-t} & e^{-2t} & e^{6t} \\ -e^{-t} & -2e^{-2t} & 6e^{6t} \\ e^{-t} & 4e^{-2t} & 36e^{6t} \end{pmatrix} \mathbf{c}$

27. $\mathbf{y}_g = \begin{pmatrix} 2e^{3t} & e^{-2t} & 2e^t \\ e^{3t} & e^{-2t} & -e^t \\ -e^{3t} & -e^{-2t} & -e^t \end{pmatrix} \mathbf{c}$

CHAPTER 9

Section 9.1

1. Repeat the derivation in the text.

3. The derivation is the same with $D(x)$ replacing D. Do not factor $D(x)$ out of derivative expressions.

5. It is linear and nonhomogeneous. It has constant coefficients if A, V are independent of x.

7. Rate of flow in through the inner boundary is $-A(x)Dc'(x)$, out through outer boundary $-A(x+\Delta x)Dc'(x+\Delta x)$ with $A(x) = 2\pi dx$. Apply conservation law, etc. as in text.

8. (a) Rate of heat energy flow per unit area at x is $DT'(x)$ in the direction of $-T'$. Hence, rate of heat flow in at x is $-DAT'(x)$, rate of heat flow out at $x+\Delta x$ is $-DAT'(x+\Delta x)$.

8. (b) Net change in heat energy is zero.

8. (c) Divide expression of part (b) by Δx; let $\Delta x \to 0$.

8. (d) Divide out constants D, A. Specify temperature of ends of bar at $x = 0, L$.

8. (e) $T'' = 0 \Rightarrow T = C_1 x + C_2 \Rightarrow T = T_0(L-x)/L + T_1 x/L$.

9. Repeat the preceding arguments. At the insulated end of the bar, no heat flows. Hence, $-DAT'(L) = 0$.

11. Use the same technique as exercise 7. No heat flow at the origin $\Rightarrow T'(0) = 0$, as in exercise 9.

13. Use conservation of energy with addition of energy at the rate of $H(x)A\Delta x$: $-DAT'(x) + DAT'(x+\Delta x) = H(x)A\Delta x$. Obtain boundary conditions as usual.

Section 9.2

1. $y_g = C_1 e^{-x} + C_2 e^x$; $C_1 = -e^4/(1-e^4)$, $C_2 = 1/(1-e^4)$.

3. $y_g = C_1 e^{-x} + C_2 e^x - \dfrac{1}{1+\pi^2}\sin \pi x$; $C_1 = C_2 = 0$.

5. $y_g = C_1 e^{-x} + C_2 e^x - \dfrac{1}{1+\pi^2}\sin \pi x$;
$C_1 = \dfrac{e^2((e^2-4)(\pi^2+1)-\pi)}{(e^4+1)(\pi^2+1)}$,
$C_2 = \dfrac{e^2(\pi^2+\pi+4)+\pi^2}{(e^4+1)(\pi^2+1)}$.

7. $y_g = C_1 e^t + C_2 t e^t + t^2 e^t/2$; $C_1 = 0$, $C_2 = 1/e - \frac{1}{2}$.

9. $T = T_0$; insulation at one end of bar makes temperature constant throughout.

11. $T = C_1$, C_1 arbitrary. Body is perfectly insulated, but we need additional information to find its (constant) temperature.

13. $T = (x^4/12 - Lx^3/6 - L^4/12 - DAT_L)/DA$.

17. Redraw figure with $c \equiv c_L$, a horizontal line.

Section 9.3

1. $0.11 \text{ cm}^2/\text{s}$

3. A partial differential equation is *homogeneous* if it has the trivial solution. Otherwise, it is *nonhomogeneous*; $0_t = \kappa 0_{xx}$ proves homogeneous, etc.

5. $T_t = \kappa T_{xx}$, $T_x(0,t) = T_x(L,t) = 0$, $T(x,0) = Mx(L-x)$

7. Assume $A(x)$ is constant: $c\rho T_t = (k(x)T_x)_x$ with same boundary and initial conditions as in example.

9. $T_t = \kappa T_{yy} + H/c\rho$ with same boundary and initial conditions as in example.

11. Insulated sides: $T_x(0,y,t) = T_x(L,y,t) = 0$. Insulated top: $T_x(x,h,t) = 0 = 0$

13. If $H < 0$, then rod is absorbing heat; e.g., cooling fluid is being pumped through its core.

Section 9.4

1. i. $X'' + \lambda X = 0$, $X'(0) = X(L) = 0$; $\Theta' + \kappa \lambda \Theta = 0$.
ii. $X_n = \cos n\pi x/2L$, $\lambda_n = n^2\pi^2/4L^2$, n an odd integer; $\Theta_n = \exp(-\kappa\lambda_n^2 t/L^2)$; **iii.** $\sum_{n \text{ odd}} c_n X_n(x)\Theta_n(t)$

3. Temperature decay is dominated by $n = 1$ term: $e^{-\kappa\pi^2 t/L^2}$. Ratio of temperatures copper:wood is $e^{(\kappa_w - \kappa_c)\pi^2 t/L^2} = e^{-0.1081\pi^2 t/L^2}$.

5. $\lambda = 0 \Rightarrow X_g = C_1 + C_2 x$; $X(0) = 0 \Rightarrow C_1 = 0$; $X(L) = 0 \Rightarrow C_2 L = 0 \Rightarrow C_2 = 0$. Hence, $X(x) \equiv 0$.

7. $c_n = (-1)^{n-1}/n\pi$, $n = 1, 2, \ldots$. Series converges in the mean to the $2L$-periodic extension of $T_L x/L$, $-L \le x < L$.

9. $c_0 = L/2$, $c_n = 2L/n^2\pi^2$, n odd, $c_n = 0$, n even. For $-L \le x \le 0$, series converges in the mean to the even extension of $L - x$, $0 \le x \le L$. For other x, it converges to its $2L$-periodic extension.

11. $X_n(x) = \cos nx/2$, $\lambda = n^2/4$, n an odd integer, or $X_m(x) = \cos(2m-1)x/2$, $m = 1, 2, \ldots$;

$$c_m = \frac{8}{(2m-1)^2\pi^2}\left(\frac{(-1)^{m-1}(2m-1)\pi}{2} - 1\right).$$

Series converges to the even, 2π-periodic extension of x/π, $0 \le x \le \pi$.

13. (b) Use $T_n = X_n(x)\Theta_n(t)$, where $X_n'' + (n\pi/L)^2 X_n = 0$, $X(0) = X(L) = 0$, where $X_n(x) = \sin n\pi x/L$.

17. Eigenvalue problem is $X'' + \lambda X = 0$, $X'(0) = X'(4) = 0$. Eigenvalues and eigenfunctions are

$X_n(x) = \cos n\pi x/4$, $\lambda_n = n^2\pi^2/16$, $n = 0, 1, 2, \ldots$; see exercise 9 with $L = 4$. $T(x,t) = 4 + \sum_{n=1}^{\infty} c_n \exp(-\kappa n^2\pi^2 t/16) \cos n\pi x/4$, where $c_n = 8/n^2\pi^2$, n odd, $c_n = 0$, n even.

19. Eigenvalues and eigenfunctions are those of exercise 11. $T(x,t) = \sum_{m=1}^{\infty} c_m \exp(-\kappa(2m-1)^2 t/4) \cos(2m-1)x/2$, where c_m is given in exercise 11.

21. Solve $E''(x) = 0$, $E(0) = 0$, $E(L) = 8$ to obtain $E(x) = 8x/L$. Then solve the homogeneous initial-boundary value problem $u_t = \kappa u_{xx}$, $u(0,t) = u(L,t) = 0$, $u(x,0) = -2 - 8x/L$, where $T(x,t) = u(x,t) + E(x)$. Find $u(x,t) = \sum_{n=1}^{\infty} c_n \exp(-\kappa n^2\pi^2/L^2) \sin n\pi x/L$, where $c_n = -12/n\pi$, n odd, $c_n = 8/n\pi$, n even. (See exercise 7.)

23. $E_g = -Hx^2/2\kappa + C_1 + C_2 x$; $E(0) = 0 \Rightarrow C_1 = 0$; $E(L) = 0 \Rightarrow C_2 = HL/2\kappa$; $E(x) = Hx(L-x)/2\kappa$.

25. Substitute, using the differential equations defining E, T. Solution formulas are not required, and H need not be constant.

27. Use $\int \sin x \cos x \, dx = (\sin^2 x)/2$, etc.

29. $a_n = (1/2L)\int_{-L}^{L} f(x) \cos n\pi x/L \, dx = 0$, either by direct evaluation or because integrand is an odd function.

Chapter Exercises

1. (a) In figure 9.25, construct a control volume using two vertical lines at $x, x + \Delta x$. Rate of heat flow in at x per unit area is $-DT'(x)$; rate per unit area out at $x + \Delta x$ is $-DT'(x + \Delta x)$. Conservation of energy $\Rightarrow -DT'(x) + DT'(x + \Delta x) = 0$. Divide by Δx, let $\Delta x \to 0$ to obtain $T'' = 0$. Boundary conditions are temperatures at left and right walls.

1. (b) Repeat (a), but do not bring variable $D(x)$ outside of derivative.

3. $y_g = C_1 e^{-x} + C_2 e^x - 4$; $C_1 = (3e^2 - 4e^4)/(e^4 + 1)$, $C_2 = (4 + 3e^2)/(e^4 + 1)$.

5. $y \equiv 0$

7. $y = x + x^2$

9. Construct control volume 1 unit deep, Δx wide, Δy high. Put lower left corner at (x, y). Heat flows only through sides. Then derivation is same as for insulated bar with $A(x) = 1 \cdot \delta y$.

11. Temperature on left side of end cap is $T(L,t)$; temperature on right is T_L. Hence, as with house heat loss model of section 2.3, rate of heat energy flow through a unit are of this "wall" is $\delta(T(L,t) - T_L)$. Since this heat loss balances the rate of energy flow through the end of the cylinder, $-kT_x(L,t) = \delta(T(L,t) - T_L)$. Obtain $T(L,t) + aT_x(L,t) = b$, with $a = k/\delta$, $b = T_L/\delta$.

13. Eigenfunctions and eigenvalues of this problem are those of exercise 9, section 9.4; $\Theta_n(t) = e^{-\kappa\lambda_n t} \to 0$, $n = 1, 2, \ldots$. Hence, $T(x,t) = \sum_{n=0}^{\infty} c_n X_n(x), \Theta_n(t) \to c_0 X_0(x)\Theta_0(t) = c_0 = (1/L)\int_0^L f(x) \, dx$.

15. $X_m(x) = \sin(2m-1)\pi x/2L$, $\lambda = (2m-1)^2\pi^2/L^2$, $m = 1, 2, \ldots$; $2 = \sum_{m=1}^{\infty} c_m X_m(x) \Rightarrow c_m = 4/(2m-1)\pi$.

17. Problem could model temperature in a thin bar with insulated sides, left end at zero temperature, right end insulated. Using eigenfunctions, eigenvalues, and c_m from exercise 15, find $T(x,t) = \sum_{m=1}^{\infty} c_m X_m(x) e^{-\kappa\lambda t}$.

19. $E(y) = T_0(h - y)$; $T(y,t) = u(y,t) + E(y)$; u solves $u_t = \kappa u_{yy}$, $u(0,t) = u(h,t) = 0$, $u(y,0) = -T_0(h - y)$. Find $u(y,t) = \sum_{n=1}^{\infty} c_n e^{-\kappa n^2\pi^2 t/h^2} \sin n\pi y/h$, $c_n = -2T_0 h/n^2\pi^2$, n odd, $c_n = 0$, n even.

CHAPTER 10

Section 10.1

1. x.

3. e^{-4x}.

5. x^2.

7. $x \sin(\ln x)$.

9. $x = (\frac{38}{63})e^{3t} - (\frac{33}{56})e^{-4t} - (12t + 1)/72$.

11. $\mu = x \Rightarrow v' = Cx \Rightarrow u = x^2$.

13. $\mu = y_1^2 + \int P \, ds$.

Section 10.2

1. $W = -x^2 \Rightarrow x^2$, x are linearly independent for $x \neq 0$.

3. $W = -2x \Rightarrow x \sin(2\ln|x|)$, $x \cos(2\ln|x|)$ are linearly independent for $x \neq 0$.

5. $W = -x \Rightarrow x\sin(\ln x)$, $x\cos(\ln x)$ are linearly independent for $x \neq 0$.

7. $W = (5 + \sqrt{5})(t - 3/2)^{(1+\sqrt{5})/4}/8 - (5 - \sqrt{5})(t - 3/2)^{(1-\sqrt{5})/4}/8 \big|_{t=5/2} = \sqrt{5}/4 \neq 0 \Rightarrow (t - 3/2)^{(5+\sqrt{5})/8}$, $(t - 3/2)^{(5-\sqrt{5})/8}$ are linearly independent for $t \neq 3/2$.

9. $x^2 y'' - xy' - y = 0$.

11. $x^2 y'' - xy' + 5y = 0$.

13. $x \neq 0$.

15. $W(x^{r_1}, x^{r_2}) = (r_2 - r_1)x^{r_1+r_2-1} \neq 0 \Rightarrow$ linearly independent for $x \neq 0$, $r_1 \neq r_2$, r_i real or complex.

17. $x > 0 \Rightarrow x^{\alpha+i\beta} = (e^{\ln x})^{\alpha+i\beta} = e^{\alpha x}[\cos(\beta \ln x) + i\sin(\beta \ln x)]$. Find a similar expression for the other root. Form linear combinations to isolate the cosine and sine factors. Use the Wronskian to verify linear independence for $x \neq 0$.

19. (a) See example 5: $x^2 T'' + xT' = 0$, $T = x^r \Rightarrow r^2 = 0 \Rightarrow T_g = C_1 + C_2 \ln x$; $T'(0) = 0 \Rightarrow C_2 = 0$; $T(L) = T_L \Rightarrow C_1 = T_L \Rightarrow T(x) = T_L$.

19. (b) Circular disk has uniform equilibrium temperature determined by (uniform) temperature at boundary.

21. See exercise 19(a) or example 5: $T_g = C_1 + C_2 \ln x$. Boundary conditions require $C_1 + C_2 \ln r = T_s$, $C_1 + C_2 \ln R = T_{\text{out}}$. Solve these equations, for example by Cramer's rule, to obtain $T(x) = (T_s \ln R - T_{\text{out}} \ln r)/\ln(R/r) + [(T_{\text{out}} - T_s)/\ln(R/r)]\ln x$.

Section 10.3

1. $C_1 e^x + C_2 e^{-x} - 1$.

3. $C_1 e^{3x} + C_2 e^{-3x} - xe^{-3x}/6$.

5. $C_1 e^{3x} + C_2 e^{-3x} + xe^{3x}/12 - xe^{-3x}/12$.

7. $C_1 \sin 3x + C_2 \cos 3x + e^{3x}/9$.

9. $C_1 \sin t + C_2 \cos t + 4t$.

11. $C_1 t^2 + C_2 t^{-3} - (t + 1)/6t$.

13. Could not uniquely solve linear equations defining u_1', u_2'.

15. Compute indicated derivatives.

17. $y = C_1 x^2 + C_2 x^{-2} - (\ln x)/4$; $u_1' y_1 + u_2' y_2 = 0$, $u_1' y_1' + u_2' y_2' = R(x)$.

19. $u_1' = \dfrac{-y_2 r(x)}{a_2(x)W}$; $u_2' = \dfrac{y_1 r(x)}{a_2(x)W}$, where $W = y_1 y_2' - y_1' y_2$.

Section 10.4

1. $c_0' = 0$

5. (i) $x + 2$ is analytic at $x = -2$.

5. (ii) $y(x) = c_0 + c_1(x + 2) - c_0(x + 2)^3/6 - c_1(x + 2)^4/12$.

7. (i) Constant coefficients are analytic everywhere.

7. (ii) $y = -1 + 4(x - 1) - (x - 1)^2/2 + 2(x - 1)^3/3$.

9. (i) x^2 analytic at $x = 0$.

9. (ii) $y = c_0 - c_0 x + (1/2)c_0 x^2 - (1/6)c_0 x^3$.

11. $f''(x_0) = \sum_{n=2}^{\infty} c_n n(n - 1)(x - x_0)^{n-2}\big|_{x=x_0} = 2c_2$, etc.

13. (a) Let $\cos x = f(x)$, then $\sin x = f(x)$. Plug into $c_n = f^{(n)}(x_0)/n!$ for $x_0 = 0$.
(b) Let $y_1 = \cos x$ and $y_2 = \sin x$.
(c) $y_g = C_1 y_1 + C_2 y_2$.
(d) Compute $(1 - x^2/2 + \cdots)'' = -1$, etc. and substitute into differential equation.
(e) $y(0) = C_1$, $y'(0) = C_2$.

15. Using full series permits determination of radius of convergence.

19. $f(x) = a_0 + a_1(x - x_0) + a_2(x - x_0)^2 + \cdots \Rightarrow f'(x_0) = a_1 + 2a_2(x - x_0) + \cdots$.

Section 10.5

1. (i-ii) All real x, $x \neq 0$, are finite ordinary points; $x = 0$ is a finite regular singular point; (iii) $r^2 + 3r - 4 = 0 \Rightarrow r_1 = 1, r_2 = -4$.

3. (i-ii) $x = \pm 2$ are finite regular singular points; all other real x are ordinary points; (iii) For $x = \pm 2$, $r^2 + r = 0 \Rightarrow r_1 = 0, r_2 = -1$.

5. (i-ii) $x = 0$ is a regular singular point; all other real x are ordinary points; (iii) $r^2 - 1/16 = 0 \Rightarrow r_1 = 1/4$, $r_2 = -1/4$.

7. $y_1 = x^{-1}$ and $y_2' = x^{-1/2}$ do not exist at $x = 0$.

9. $P = 1/x$, $Q = 1 - p^2/x^2$ do not exist at $x = 0$, but $xP = 1$ and $x^2 Q = x^2 - p^2$ are analytic at $x = 0$.

13. $y_1 = 1 - x/2 + x^2/24 + \cdots$, $y_2 = x^{1/2}(1 + x/20 + x^2/42 + \cdots)$.

15. $y_1 = x^{1/4}(1 - x/5 + x^2/90 + \cdots)$, $y_2 = x^{-1/4}(1 - x/5 + x^2/90 + \cdots)$.

17. $y_1 = (x - 2)^{2/3}(1 + (x - 2)/5 + (x - 2)^2/80 + \cdots)$, $y_2 = 1 + (x - 2)/2 + (x - 2)^2/8 + \cdots$.

19. $x = 0$ is not a regular singular point since $x^2 Q = 1/x$ is not analytic at $x = 0$; $c_0, c_1, c_2, \ldots = 0$.

21. When $n = -2r$, recurrence relation becomes $c_n \cdot 0 + c_{n-2} = 0$, contradicting $c_{n-2} \neq 0$. A different solution form is required.

23. $4^{m+1}[(m+1)!]^{m+1}/4^m(m!)^m = 4(m+1)!(m+1)^m \rightarrow \infty \Rightarrow J_0$ series converges for all x.

Chapter Exercises

1. $(x+1)^2(C_2 \ln(x+1) + C_1)$.

3. $C_1 \sin t + C_2 \cos t - (t \cos t)/2$.

5. $C_1 t^2 + C_2 t^{-3} - \frac{1}{6}$.

7. $y = 2 + x - x^3/3 - x^4/12 + \cdots$.

9. $y = (4x^2 - 8x + 4)\ln(x-1) - 4x^2 + 4x + C_1(x-1)^2 + C_2(x-1)$.

11. $x = 0$ is a regular singular point; all other x are ordinary points.

13. $x = 0$ is a regular singular point; all other x are ordinary points.

15. $x = \pm 1$ are regular singular points; all other x are ordinary points.

17. $x = 0$ is a regular singular point; all other x are ordinary points.

19. All x are ordinary points.

21. All x are ordinary points.

23. $y = a_0 + a_1 x + a_0 x^3/6$.

25. (a) Using chain rule, substitute $X(x) = J_0(\sqrt{\lambda}x)$ into differential equation.

25. (b) The boundary condition $X(L) = J_0(\sqrt{\lambda}L) = 0$ forces $\sqrt{\lambda}L = 2.40, 5.52, 8.65, 11.79, \ldots$, the zeros of J_0. Hence, $\lambda = (2.40/L)^2, \ldots = 5.76/L^2, 30.47/L^2, 74.82/L^2, 139.00/L^2, \ldots$. Eigenvalues of $X'' + \lambda X = 0$ with the same boundary conditions are defined by $\cos \sqrt{\lambda}L = 0$; $\lambda = n^2\pi^2/L^2 = 9.87/L^2, 39.48/L^2, 88.83/L^2, 157.91/L^2, \ldots$.

CHAPTER 11

Section 11.1

1. $4/s - 9/(s+4)$.

3. $(\pi + 7)/s$.

5. $2/s^2 - 5/s$.

7. $12/s^4$.

9. $1/s^2 + 2/(s-2)$.

11. $A\omega/(s^2 + \omega^2)$.

13. $1/s^2 + 4/s^3 + 18/s^4 + 96/s^5$.

15. $Y(s) = \dfrac{-2s^3 - 3s^2 + 5s + 10}{s^2(s+2)(2s+7)}$.

17. $Y(s) = \dfrac{\alpha}{(s+1)(s-k)} + \dfrac{y_i}{s-k}$.

19. Integrate $\int_0^\infty te^{-st}\,dt$ by parts using $u = t$, $dv = e^{-st}\,dt$.

21. Use induction and integration by parts; see 19.

23. $\mathcal{L}\{\sinh t\} = (\mathcal{L}\{e^t\} - \mathcal{L}\{e^{-t}\})/2$.

27. (a) $-a^3/(s^2 + a^2)$.

27. (b) $a^2/(s-a)$.

27. (c) $6/s^2$.

29. Let $g(t) = f'(t)$. Then $\mathcal{L}\{g'(t)\} = sG(s) - g(0) = s\mathcal{L}\{f'(t)\} - f'(0) = s(sF(s) - f(0)) - f'(0) = s^2F(s) - sf(0) - f'(0)$.

31. $\dfrac{1}{s} - \dfrac{e^{-2\pi s}}{s} + \dfrac{se^{-2\pi s}}{s^2+1} + \dfrac{e^{-7\pi s/2}}{s^2+1}$.

33. $\mathcal{A}(cf + dg) = \int_0^s (cf(t) + dg(t))dt = c\int_0^s f(t)dt + d\int_0^s g(t)dt = c\mathcal{A}(f) + d\mathcal{A}(g)$; the Laplace operator is linear.

35. $\mathcal{L}\{ae^{at}\} = \int_0^\infty ae^{at}e^{-st}dt = \lim_{b\to\infty} e^{at}e^{-st}|_{t=0}^{t=b} + s\int_0^\infty e^{at}e^{-st}dt = \lim_{b\to\infty} e^{ab}e^{-sb} - 1 + s\mathcal{L}\{e^{at}\} = s\mathcal{L}\{e^{at}\} - 1$. Since a is constant, $\mathcal{L}\{ae^{at}\} = a\mathcal{L}\{e^{at}\}$ for all a.

Section 11.2

1. $1 - e^{-4t}$.

3. $e^{4t}/8 - e^{-4t}/8$.

5. $(e^{2t} + 2e^{-t} - 3e^{-2t})/6$.

7. $2e^{4t} - 2e^t$.

9. $e^{4t}/3$.

11. $e^t(\sin 2t - 4\cos 2t)/2$.

13. (a) $1/(s-3)^2$.

13. (b) $(s+4)/(s^2 + 8s + \pi^2 + 16)$.

13. (c) $4/(s+3/2) + 2/(s+3/2)^3$.

15. $(76/15)e^{6t} - (2\sin 3t + \cos 3t)/15$.

17. $(3e^t - 5)e^{-4t}$.

19. $\cos \pi t/(\pi^2 - 4) + 2\sin 2t + (\pi^2 - 5)\cos 2t/(\pi^2 - 4)$.

21. $(2 - 2t)e^{2t}$.

25. (b) Use the characteristic equation $r^2 + 4r + 40 = 0$; $r = -2 \pm 6i$.

27. (a) $c_1^2 - 4c_0 \geq 0$.

27. (b) $A = -1/(r_2 - r_1)$, $B = 1/(r_2 - r_1)$.

27. (c) $\dfrac{e^{r_2 t} - e^{r_1 t}}{r_2 - r_1}$.

Section 11.3

1. (a) $\dfrac{\pi}{s^2 + \pi^2}$.

1. (b) $\dfrac{1}{1 - e^{-4s}}\left(\dfrac{1 + e^{-4s}}{s} - \dfrac{2e^{-2s}}{s}\right) = \dfrac{1 - e^{-2s}}{s(1 + e^{-2s})}$.

3. (a) $2/(s-a)^3$.

3. (b) $12\left[\dfrac{1}{(s^2 + \pi^2)^2} - \dfrac{8s^2}{(s^2 + \pi^2)^3} + \dfrac{8s^4}{(s^2 + \pi^2)^4}\right]$.

3. (c) $10a\left(\dfrac{4s^2}{(s^2 - a^2)^3} - \dfrac{1}{(s^2 - a^2)^2}\right)$.

5. (a) $\dfrac{1}{2}\ln\left(\dfrac{s+2}{s-2}\right)$.

5. (b) $\pi/2 - \arctan(s/2)$.

5. (c) $(1/2)\ln(1 + 1/s^2)$.

7. (a) $e^t - 1$.

7. (b) $-9e^{-2t}/2 - 3t + 9/2$.

9. $\dfrac{a(e^{-sT} - 2e^{-bs} + 1)}{s(1 - e^{-sT})}$.

11. $1/(s-a)^2$.

13. $\omega/(s^2 - 2as + \omega^2 + a^2)$.

15. $2\omega/(s^2 + \omega^2) - 2\omega s^2/(s^2 + \omega^2)^2$.

17. $(\sin at)/t \le ae^{at}$.

21. $(v_i/\omega + At/2\omega)\sin\omega t + x_i \cos\omega t$.

23. $\mathcal{L}\{f\} = \sum_{n=1}^{\infty} e^{-(n-1)sT}\int_0^T e^{-sr} f(r)\,dr$.

25. $\mathcal{L}\{(-t)^2 f(t)\} = \mathcal{L}\{(-t)(-tf(t))\} = d\mathcal{L}\{-tf(t)\}/ds$ $= F''(s)$, etc.

27. $\int_s^\infty F(r)\,dr = \int_0^\infty \int_s^\infty f(t)e^{-rt}\,dr\,dt = \int_0^\infty (-f(t)$ $e^{-rt}/t)\mid_{r=s}^\infty dt = \int_0^\infty f(t)e^{-st}/t\,dt$

Section 11.4

1. $e^{(12-3s)}/(s-4)$

3. $g(t) = \mathcal{U}(t-a) - \mathcal{U}(t-b)$; $G(s) = e^{-as}/s - e^{-bs}/s$.

5. $g(t) = mt - m(t-b)\,\mathcal{U}(t-b) - mb\,\mathcal{U}(t-b)$; $G(s) = m/s^2 - e^{-bs}(m/s^2 + mb/s)$.

7. $g(t) = 1 - \mathcal{U}(t - 2\pi) + \cos t\,\mathcal{U}(t - 2\pi) - \cos t\,\mathcal{U}(t - 7\pi/2)$; $G(s) = \dfrac{1}{s} - \dfrac{e^{-2\pi s}}{s} + \dfrac{se^{-2\pi s}}{s^2 + 1} - \dfrac{e^{-7\pi s/2}}{s^2 + 1}$.

9. $g(t) = (2t - 1)(\mathcal{U}(t - 0) - \mathcal{U}(t - 2)) + 3\,\mathcal{U}(t - 2)$; $y = 1 + e^{-2t} + \mathcal{U}(t - 2)/2 - e^{-2(t-2)}/2$.

11. $E(t) = 0.209 - 0.209\,\mathcal{U}(t - 4)$; $y = 22e^{-0.015t} + 14(1 - e^{-0.015(t-4)})\,\mathcal{U}(t - 4) - 14$.

13. $Y(s) = \dfrac{1 - e^{-s}}{s^2(s-k)} + \dfrac{y_i}{s - k}$.

Section 11.5

1. $\delta(t)$

3. $\delta(t) + 2e^{2t}$

5. $T(t) = b\,\mathcal{U}(t-a)e^{-Ak(t-a)/cm}$; b represents the magnitude of the change in T_{out} due to an impulse of heat energy.

7. $i(t) = V\sqrt{C/L}\,\mathcal{U}(t - a)\sin[(t - a)/\sqrt{LC}]$

9. (a) $\theta(t) = (b/\sqrt{Lg})\mathcal{U}(t - a)\sin[\sqrt{g/L}\,(t - a)]$

9. (b) Since $b/\sqrt{Lg}$ is an angle, b is a change in velocity around the circumference of the circle described by the pendulum.

9. (c) $b = \sqrt{Lg}\theta_i$

11. $E'(t) = V\delta(t - a)$; see exercise 7 for the solution formula. Choose $V = \sqrt{L/C}I$ to achieve amplitude I.

13. Use $\int_{-\infty}^\infty e^{-st}d_h(t - a)\,dt = \int_{a-h/2}^{a+h/2} e^{-st}d_h(t - a)\,dt$, etc.

Chapter Exercises

1. $\dfrac{m}{s^2} - e^{-bs}\left(\dfrac{m}{s^2} + \dfrac{mb}{s}\right)$

3. $(4s^2 + 48)/(s^4 + 40s^2 + 144)$

5. $\dfrac{4s^2 + 32s + 112}{s^4 + 16s^3 + 136s^2 + 576s + 1040}$

7. $s^{-2} - 2s^{-3}$

9. $2s^2/(s^2 + \omega^2)^2 - 1/(s^2 + \omega^2)$

11. $y = 6e^{-2t} - 1 - 3(e^{-2(t-4)} - 1)\,\mathcal{U}(t - 4) + 2(1 - e^{-2(t-8)})\,\mathcal{U}(t - 8)$.

13. $X(s) = \dfrac{1}{s(s+1)} - \dfrac{e^{-2\pi s}}{s(s+1)} + \dfrac{se^{-2\pi s}}{(s+1)(s^2 + 1)} - \dfrac{e^{-7\pi s/2}}{(s+1)(s^2 + 1)} - \dfrac{3}{s+1}$.

15. $4e^{-2t} - 4e^{-3t}$

17. $e^{2t} + e^{-2t}$

19. $2\cos[2(t - 5)]\,\mathcal{U}(t - 5)$

21. $2e^{-t}\sin 2t$

23. $2e^{-3t}\cos 3t$

25. Factor $s^2 + 4 \Rightarrow$ sine and/or cosine of frequency 2; $(s - 2)(s + 2) \Rightarrow$ exponentials with factor ± 2; no on-off behavior.

27. No oscillations; $s - 1 \Rightarrow$ exponential with factor 1; $e^{-3s} \Rightarrow$ on-off behavior at $t = 3$.

29. $(e^t \cos t + 2e^t \sin t, -e^t \sin t + 2e^t \cos t)$

30. $(e^t - e^{-t}, e^t + e^{-t})/2$

31. $(e^{2t} + 2e^{-3t}, 3e^{2t} + e^{-3t})$

32. $(-e^{4t} + e^{-4t}, e^{4t} + e^{-4t})$

33. $(4 - 2\sqrt{3})(e^t, \sqrt{3}e^t/3) + (4 - 6\sqrt{3})(e^{-t}, -\sqrt{3}e^{-t}/3) + (4\sqrt{3}t^2 + 8\sqrt{3} - 2, (8 - 2\sqrt{3}/3)t)$

35. $(y, z) = (5e^{4t} - e^{2t} + e^{-2t} + 3, 5e^{4t} - 3e^{2t} - e^{-2t} - 1)/8$

37. $q(t) = 10[h(t - 6) - h(t)]/\omega L$, where $h(t) = e^{-at}(a \sin \omega t + \omega \cos \omega t)/(\omega^2 + a^2)$, $a = R/2L$, $\omega = \sqrt{4L/C - R^2}/2L \approx 9.9499$ rad/μs. The $t - 6$ term reflects the change in $E(t)$ at $t = 6$.

39. $q(t) = 5(\omega/(\omega^2 + a^2) - h(t - 2))/\omega L$, where $h(t)$, a, and ω are as in exercise 37. The $t - 2$ term reflects the change in $E(t)$ at $t = 2$.

41. Since $\mathcal{L}\{q(t)\} = G(s)\mathcal{L}\{E(t)\}$, the convolution theorem yields the desired result. To obtain $g(t)$, note that the roots of the denominator of $G(s)$ are $r = -R/2L \pm i\omega$ and use partial fractions (or complete the square).

APPENDIX

Appendix Exercises

1. (a) $\cos \pi x \exp \left(\int_0^x \cos \pi s \, ds \right)$

1. (b) $(x + 1)^3 \exp \left(\int_{-1}^x (s + 1)^3 \, ds \right)$

3. Let $\dfrac{1}{ar - sr^2} = \dfrac{A}{r} + \dfrac{B}{a - sr}$. Then $A = 1/a$, $B = s/a$. Using logarithms, integral is $\dfrac{1}{a} \ln \left(\dfrac{(a - s)x}{a - sx} \right)$.

5. $\sin x = 1 - (x - \pi/2)^3/3! + (x - \pi/2)^5/5! - \cdots$

7. $\det \mathbf{A} = e^{-x} - e^x$

$$\mathbf{A}^{-1} = (e^{-x} - e^x)^{-1} \begin{pmatrix} e^{-x} & -1 \\ -e^x & 1 \end{pmatrix}$$

9. (a) To have exactly one solution, the determinant of coefficients $1 + a$ must be nonzero. (That solution is $C_1 = C_2 = 0$.)

9. (b) $1 + a = 0$

11. $\dfrac{s}{s^2 + 1} + \dfrac{1}{s + 1} - \dfrac{1}{s - 1}$

INDEX

biological oxygen demand, 47
birth rate, 22, 34
blow up of solution, 131
BOD, 47
boundary condition, 354, 360, 369
    Dirichlet, 357
    first kind, 357
    Neumann, 357, 372
    second kind, 357, 372
    third kind, 397
boundary conditions, 354, 360
boundary-value problem, 354, 360
    numerical method, 398
    solution, 360
    uniqueness, 360, 399–400
British thermal unit, 39
Btu, 39
budworm population model, 168
building insulation, 42

C

C, coulomb, 181
calorie, 39
capacitance, 45, 181
capacitor:
    charge on, 135
    potential drop across, 181, 286
    voltage drop graph, 286
carrying capacity, 49, 143
casting, 38
Cauchy, 455
Cauchy sequence, 130
Cauchy-Euler equation, 406–7, 425, 451, 453
    summary, 411
    transformation to constant-coefficient equation, 453
Celsius temperature scale, 39
chain rule for derivatives, 510
chaos, 137
chaotic motion, 306
characteristic equation, 81, 107, 118–19, 121, 133, 178, 185, 188, 218–19, 232, 322, 337, 338, 363
    Cauchy-Euler equation, 453
    first-order, 107
    order n, 262
    second-order, 218-33
    systems, 318, 322
characteristic equation method summary:
    first-order equation, 107–8

matrix system, 342
    second-order equation, 232
    system, 322, 342
characteristic polynomial, 338–39
characteristic roots, 218–19, 322
characteristic value, 337 *see also* eigenvalue
characteristic vector, 337 *see also* eigenvector
charge, 180
Chebyshev's equation, 452
chemical reaction model, 46, 154
circular frequency, 229, 265
Coddington, E. A., 131, 393, 524
column vector, 517
compact support, 499
competition population model, 33–34
complete set of eigenvectors, 340
completing the square, 472, 474
complex conjugate, 219, 340
complex eigenvalue, 342
complex eigenvector, 342
complex number, 219
concentration gradient, 351
conjugate, complex, 219, 340
conservation law, 22, 40, 192
    energy, 40, 182, 366, 370
    heat energy, 40, 359, 366
    mass, 353
    population, 22, 32, 34, 192
        word equation, 32
constant-coefficient, 106, 199
constant-coefficient system, 309, 316, 337
continuous, piecewise, 461
continuous population function, 24
contraction mapping, 129
control volume, 353
convection, 356
convergent power series, 515
convolution, 507
    integral, 469
    theorem, 469, 476
corresponding homogeneous equation, 58
cosine series, 393, 427
coulomb, 181
coupled spring-mass model, 191
coupled system, 191
Cramer's rule, 212, 317, 330, 337, 344, 377, 519
critical damping, 305
critically damped, 231, 269
current, 180–81
cycles per second, 229
cyclic frequency, 229, 265

D

damped pendulum equation, 295–96
damped pendulum model, 189
damped spring-mass model, 175
damped spring-mass system, 173, 268
damped system, 175
damping coefficient, 174, 263
damping ratio, 305
death rate, 22, 34
decay, 19
decay equation, 48, 165
degree of polynomial, 116
delta function, 499
    as derivative of unit step function, 501
    Laplace transform, 501
    properties, 500
dependent variable, 6, 13, 52, 308
derivative transform, 465, 478
derivative, Laplace transform of, 482
determinant, 204, 519
dielectric, 181
differentiable population function, 24
differential equation, 3, 5–6, 13
    first-order, 51
    order n, 349
    ordinary, 369
    partial, 369
    second-order, 198
differential inequality, 126
diffusion, 351, 356
diffusion, time-dependent, 366
diffusion coefficient, 352
diffusion equation, 354, 409
    with source term, 357
diffusion model, 354
dimensionless variables, 166, 169
Dirac delta function, *see* delta function, 499
direction field, 142
discontinuity, jump, 461
discriminant, 219, 226, 268
dissipation, diffusion as, 375
distribution, 500
droid, 43

E

eigenfunction, 377, 378
eigenfunction expansion, 375, 379, 381, 383–84, 391
    theorem, 383

eigenvalue, 337, 339, 377, 378
    complex, 342
    summary, 339
eigenvalue problem, 377, 383
eigenvector, 337
    complete set, 340
    complex, 342
    summary, 339
electric potential, 181
elimination, 336
emf, 180
emigration model, 31–32, 88, 134, 136, 498
    direction field, 142
    solution graphs, 140
equation of order n, 349
equidimensional equation, *see* Cauchy-Euler
    equation, 407
equilibrium, 147, 208, 291, 298 *see also* steady state
    autonomous system, 350
    heat-loss model, 148
    logistic equation, 148
    pendulum equation, 291
    stable, 147
equilibrium boundary-value problem, 389
Euler constant, 448
Euler method, 81, 90, 92, 250–51, 257, 398
    error, 94, 101
    geometric interpretation, 93
    global error, 101
    implicit, 106
    local truncation error, 101
    modified *see* Heun method, 95, 253
    order, 101
Euler's formula, 228, 320, 342, 410
even function, 387, 393
existence of:
    analytic solution, 435
    Laplace transform, 460–61
    power series solution, 432, 435
existence theorem, 85, 128
    and uniqueness, 87, 130
    first-order linear equation, 85
    for systems, 350
experimental fact, 2, 13, 171
exponential order, 461, 469, 471, 488
extension of spring, 171–72, 174, 175

F

F, farad, 182
fact, 2

I

identity matrix, 518
Im, *see* imaginary part, 219
imaginary part, 219, 321
immigration population model, 36
implicit Euler method, 106
implicit method, 106
imposed emf, 46, 180, 186
independent variable, 6, 13, 52, 308
independent voltage source, 184
indicial equation, 407–9, 412, 441, 443–44, 453
inductance, 182
inductor, 181
    potential drop across, 182
initial condition, 4, 7, 14, 52, 172, 369
    for first-order system, 308
    homogeneous, 54
    nonhomogeneous, 53
initial value, 8, 14
initial velocity, 4
initial-boundary value problem, 369
initial-value problem, 7, 14, 52, 198, 308, 336
    for first-order equation, 52
    for first-order system, 308
    for second-order equation, 198
    solution, 8, 52, 198, 308
insect population model, 167
insecticide, 47
insulated boundary condition, 372
insulation:
    boundary condition, 372
    building, 41
    R value, 41
integral:
    of a Laplace transform, 484
    Laplace transform of, 483
integral equation, 127
integrating factor, 70, 76–77, 81, 85–86, 110, 126, 132–33, 135, 163
integrodifferential equation, 507
interpretation of a model, 1, 25, 32, 41, 133 *ff.*, 233 *ff.*, 506
    heat-loss model, 41
    population model, 25, 32
interval of existence, 127, 131
intrinsic growth rate, 24
inverse Laplace transform, 465
inverse matrix, 518
invertible matrix, 519
Ireland, potato famine, 32

irregular singular point, 440
isolated pulse, 490–91, 498
isotope, radioactive, 135

J

jump discontinuity, 392, 461

K

Kelvin temperature scale, 39
Kirchhoff's current law, 195
Kirchhoff's voltage law, 45, 182

L

l'Hôpital's rule, 271, 488
lag, 282
Laguerre's equation, 453
Laplace transform, 455
    definition, 456
    existence, 460–61
    existence of inverse, 465
    inverse, 465
    linearity, 456
    of derivative, 482
    of integral, 483
    of periodic function, 479
    of ramp function, 492
    of unit step function, 493
    properties, 460, 469, 486, 496
    solution of differential equation, 467, 476
    solution of systems, 505
    solution process summary, 467, 476
    table, 459, 496
Laplace, Pierre Simon, 455
law, 2, 13
law of conservation, *see* conservation law . . .
Lazer, A. C., 196, 280
lead, 282
Legendre's equation, 453
Levinson, N., 131, 393, 524
Leyden jar, 46, 181, 186
limiting state, periodic, 151

Lin, C. C., 27, 524
linear, 199
linear combination, 202, 310
linear differential equation, 61
linear equation, 54
    of order n, 349
linear first-order equation, 54, 71
linear independence, 211
    test, 215, 310
linear partial differential equation, 380
linear pendulum equation, 290, 292
linear second-order equation, 199
linear stability analysis, 149, 168–69, 292, 347, 507
linear system, 309
    differential equations, matrix formulation, 334
    order n, 349
linearity of Laplace transform, 456
linearized equation, 149
linearly dependent, 202
    solutions of first-order system, 310
linearly independent, 201, 334
    solutions of first-order system, 310
local existence theorem, 130
local truncation error, 100
    for the Euler method, 101
logistic equation, 34, 54, 67, 70, 137, 142, 148, 292
    direction field, 142
    equilibrium, 148
    solution graphs, 140
logistic map, 137
logistic population model, 34, 49, 67, 88
    analysis, 49
    time-delayed, 506

M

Malthus, Thomas, 27
matrix, 517
    addition, 517
    determinant, 519
    element, 517
    entry, 517
    equality, 517
    function, 334
    identity, 518
    inverse, 518
    invertible, 519
    minor, 519
    multiplication, 518

nonsingular, 519
    subtraction, 517
    transpose, 517
maximum interval of existence, 127, 131
maximum principle, 400
McKenna, P. J., 196, 280
mean square convergence, 384
method of Frobenius, 440
    summary, 443
method of undetermined coefficients, *see*
    undetermined coefficients, 112
mid-point method, 105–6
mid-point rule, 137
minimum principle, 399–400
minor of matrix, 519
model, 1, 13
    analyzing, 1, 25, 32, 41, 49, 138 *ff.*, 263 *ff.*, 360 *ff.*,
        506–8
    formulating, 1, 14, 24, 32, 38, 48, 66, 88, 170 *ff.*,
        351 *ff.*, 366 *ff.*, 506
    interpreting, 1, 25, 32, 41, 133 *ff.*, 233 *ff.*, 506
modified Euler method, *see* Heun method, 95, 253
moles, 29
molten metal, 38
molten plastic cooling model, 47
Murray, J. D., 27, 525

N

natural frequency, 265, 279, 294, 305
natural period, 265
net growth rate, 24, 32
neutrally stable, 347, 350
    steady state, 293, 298
neutron beam, 45, 47, 134–35
Newton's law, 2, 13, 171–72, 186, 189
node of circuit, 184, 195
nonhomogeneous, 53, 199
nonhomogeneous differential equation, 61
nonhomogeneous heat equation, 374
nonhomogeneous initial condition, 53
nonhomogeneous system, 309
nonlinear, 54, 199
nonlinear differential equation, 61, 187, 199, 262, 290,
    305, 309
nonlinear force-extension law, 196
nonlinear oscillations, 262
nonlinear pendulum equation, 290
nonlinear spring, 196, 305

nonlinear system, 309
nonsingular, 334
nonsingular matrix, 519
nontrivial solution, 219
numerical analysis, 94
numerical methods, 81, 133, 136, 249

O

odd extension, 382
odd function, 382, 393
ohm, 182
order:
   differential equation, 6, 13
   numerical method, 101, 104
   polynomial, 116
ordinary differential equation, 369
ordinary point, 433, 439
origin of coordinate system, 171
Ortega, J. M., 105, 525
orthogonal functions, 381, 383, 392
   with weight p, 398
orthogonality, 381, 383, 392
   sine, 381
   sine and cosine, 392
oscillations, nonlinear, 262
overdamped, 230–31, 268–69
overdetermined, 17

P

packed bed, 38
parallel circuit, 184
   RLC circuit equation, 184
parameter, 6, 13
partial differential equation, 366, 369
partial fraction expansion, 67, 466, 474, 520
   summary, 521
particular solution, 58, 60–61, 111, 203, 236, 311, 344
   methods, *see* integrating factor, variation of
      parameters, undetermined coefficients
pendulum, 186
   damped, 189
pendulum equation:
   damped, 295–96
   linear, 290
   nonlinear, 290

pendulum model, 186
   approximate, 186
period, 186, 229, 383
   doubling, 137
   linear pendulum equation, 290, 292
   nonlinear pendulum equation, 290
periodic boundary conditions, 391
periodic extension, 383, 387
periodic function, 383
   Laplace transform, 479
periodic limiting state, 151, 262
periodic solution, 208–9, 279, 286, 287
   as closed curve in phase plane, 296
perturbation, 149, 292
phase, 161, 267
phase angle, 164, 266, 287
phase angle-amplitude form:
   general solution, 267
   particular solution, 286
phase difference, 287
phase plane, 149–50, 296, 350
phase shift, 164
piecewise-continuous, 392, 461
   solution, 489
Pisa, 3
pole of transfer function, 507
Poole, W. G., 105
population curve, 29
population data:
   United States, 23
   world, 19
population equation:
   linear, 54
   nonlinear, 54
population model, 24, 32, 48, 66, 88
   amoeba, 37
   analysis, 25, 32
   competition, 33, 34
   direction field, 142
   emigration, 31–32
   immigration, 36
   insect, 47, 167
   interpretation, 25, 32
   limited growth, 34
   logistic, 33
   solution graphs, 140
   summary, 34, 48
   yeast, 36–37, 135, 154
potato famine model, 32, 89, 498
potential, 181
power series, 423, 515
   convergent, 515

thermal resistance, 41
time constant, 110, 146, 150, 505, 508
time-invariant, 508
time to peak in rock model, 4
time-dependent diffusion, 366–74
   summary, 373
trajectory, 296
transfer function, 506–7
transient, 146, 208, 271, 279, 287
   RC circuit, 150
transpose of matrix, 517
trial particular solution, 112, 236
trivial solution, 10, 53, 61, 309

U

undamped, 232
undamped spring-mass model, 173
undamped spring-mass system, 171, 173, 264
underdamped, 232, 268–69
undetermined coefficients, 81, 111–12, 116, 133, 223,
     236, 364
   for first-order equation, 112
   for order n, 262
   for second-order equation, 236
   for systems, 332–33, 347, 349
   summary, 116, 236
   table, 117, 236
unforced equation, 151
unforced response, 151
unheated house, 38
uniform convergence, 130
unique solution of linear equations, 520
uniqueness, 86, 125, 399–400
   for boundary-value problem, 399–400
   theorem for first-order initial-value problem, 125
   theorem for first-order linear initial-value
     problem, 86
unit impulse function, 498
unit impulse response, 503
unit step function, 489, 491, 508
   derivative of, 501
   Laplace transform of, 490, 493
unit step response, 508
unstable, 347, 350

unstable equilibrium, 147–48
unstable steady state, 298–99
unweighted length of spring, 171

V

v-intercept in rock model, 4
V, volt, 181
van der Pol equation, 262
van der Pol, B., 262
variable coefficient, 199
variable-coefficient system, 309
variation of constants, 73, 416
variation of parameters:
   for first-order equation, 73, 81, 133
   for second-order equation, 414, 451
     summary, 418–19, 421
   for systems, 328, 343
     summary, 330, 344
vector function, 334
velocity versus time in rock model, 3
Verhulst, P. F., 34
volt, 181
voltage drop:
   constant, 135
   exponential, 135
   sinusoidal, 135
voltage source, 46, 180, 186

W

wastewater model, 47
Weinberger, H. F., 393, 525
word problem, 14
Wronskian, 211, 213, 310, 330, 334, 338, 344, 417
   as solution of differential equation, 213

Y

yeast population model, 36-37, 135, 154